# Spon's Mechanical and Electrical Services Price Book

# Spon's Mechanical and Electrical Services Price Book

*Edited by*
DAVIS LANGDON & EVEREST
*Chartered Quantity Surveyors*

**1998**

*Twenty ninth edition*

**E & FN SPON**
An Imprint of Thomson Professional
London · Weinheim · New York · Tokyo · Melbourne · Madras

**Published by E & FN Spon, an imprint of Thomson Science & Professional
2–6 Boundary Row, London SE1 8HN**

Thomson Science & Professional, 2–6 Boundary Row, London SE1 8HN, UK

Thomson Science & Professional, Pappelallee 3, 69469 Weinheim, Germany

Thomson Science & Professional, 115 Fifth Avenue, New York, NY 10003, USA

Thomson Science & Professional, ITP-Japan, Kyowa Building, 3F, 2-2-1 Hirakawacho, Chiyoda-ku, Tokyo 102, Japan

Thomson Science & Professional, 102 Dodds Street, South Melbourne, Victoria 3205, Australia

Thomson Science & Professional, R. Seshadri, 32 Second Main Road, CIT East, Madras 600 035, India

First edition 1968
Twenty ninth edition 1998
© 1997 E & FN Spon

Printed in Great Britain by TJ International Ltd, Padstow, Cornwall

ISBN 0 419 23080 7
ISSN 0305-4543

Apart from any fair dealing for the purposes of research or private study, or criticism or review, as permitted under the UK Copyright Designs and Patents Act, 1988, this publication may not be reproduced, stored, or transmitted, in any form or by any means, without the prior permission in writing of the publishers, or in the case of reprographic reproduction only in accordance with the terms of the licences issued by the Copyright Licensing Agency in the UK, or in accordance with the terms of licences issued by the appropriate Reproduction Rights Organization outside the UK. Enquiries concerning reproduction outside the terms stated here should be sent to the publishers at the London address printed on this page.
   The publisher makes no representation, express or implied, with regard to the accuracy of the information contained in this book and cannot accept any legal responsibility or liability for any errors or omissions that may be made.

A catalogue record for this book is available from the British Library

# Preface

The twenty ninth edition of *Spon's Mechanical and Electrical Services Price Book* has a presentation and organisation of information as the last edition. All material and labour pricing data has been developed and stored by electronic means to allow future updating and distribution. The order of the book reflects the order of estimating, first outline costs, followed by more detailed costs, then unit rate items. The major sections of the book, Rates of Wages and Working Rules and Materials Costs/Measured Work Prices, are sub-divided into Mechanical Installations and Electrical Installations.

Before referring to prices or other information in the book, readers are advised to study the `Directions' which precede each section of the Materials Costs/Measured Work Prices. As before, no allowance has been made in any of the sections for Value Added Tax.

Approximate Estimating sections give a broad scope of outline costs for various building types and all-in rates for certain elements and selected specialist activities; these are followed by elemental analysis and then by a quantified analysis of a specific building type.

The prime purpose of the Materials Costs/Measured Work Prices part is to provide average prices for mechanical and electrical and other engineering services enabling a Bill of Quantities to be priced, giving a reasonably accurate indication of the likely project cost. Supplementary information is included which will enable readers to make adjustments to suit their own requirements. It cannot be emphasised too strongly that it is not intended that prices should be used in the preparation of an actual tender without adjustment for the circumstances of the particular project such as productivity, locality, project size and current market conditions. Adjustments should be made to standard rates for time, location, local conditions, site constraints and any other factor likely to affect the costs of a specific scheme. Readers are referred to the build up of the gang rates, where allowances are included for supervision, labour related insurances etc., and to the pages which introduce each section, where the percentage allowances for overheads and profit are defined. However, it should be noted that other than these items, no other allowances have been included for specific preliminaries. In addition Materials Costs/Measured Work Prices for the electrical services have been printed on different coloured paper to distinguish them from the mechanical services section.

Readers are reminded of the service available in *Spon's Price Book Update*, where details of significant changes to the published information are given. The printed update is circulated free of charge every 3 months, until the publication of next year's *Price Book*, to those readers who have requested it. In order to receive this the coloured card bound in with this volume should be completed and returned.

Readers are advised that the measured rates section of this book with "drill down" associated rate build-ups, are now available on a Windows™ cost estimating package named **reSPONse**. This new software was launched in February 1996 and not only allows users to display and examine rates at different price levels, but also allows them to alter standard library into specific rates, once items have been transferred across into project estimates, etc. Facilities exist for users to prepare their own estimates and tenders as well as developing their own rate libraries alongside any selected *Spons* libraries. For further information refer to advertisements elsewhere in this book.

*Preface*

As with previous editions the Editors invite the views of readers, critical or otherwise, which might usefully be considered when preparing future editions of this work.

While every effort is made to ensure the accuracy of the information given in this publication, neither the Editors nor Publishers in any way accept liability for loss of any kind resulting from the use made by any person of such information.

In conclusion, the Editors record their appreciation of the indispensable assistance received from many individuals and organisations in compiling this book.

DAVIS LANGDON & EVEREST
*Chartered Quantity Surveyors*
*Princes House*
*39 Kingsway*
*London WC2B 6TP*

*Telephone:* 0171 497 9000
*Facsimile:* 0171 497 8858

*e-Mail:* London@DLE.CO.UK

# Contents

| | |
|---|---|
| Preface | page v |
| Acknowledgements | page xi |

## PART ONE: APPROXIMATE ESTIMATING

| | |
|---|---|
| Directions | 3 |
| Cost Indices | 4 |
| Outline Costs | 6 |
| Elemental and All-in-Rates | 9 |
| Elemental Costs | 34 |

## PART TWO: RATES OF WAGES AND WORKING RULES

Mechanical Installations

| | |
|---|---|
| Rates of Wages | 53 |
| Working Rules (HVCA) | 60 |
| Working Rules (JIB PMES) | 72 |

Electrical Installations

| | |
|---|---|
| Rates of Wages | 83 |
| Working Rules | 89 |

## PART THREE: MATERIAL COSTS/MEASURED WORK PRICES

Mechanical Installations

| | |
|---|---|
| Directions | 101 |
| S41 : Fuel Oil Storage/Distribution : Storage Tanks and Vessels | 105 |
| S60 : Fire Hose Reels : Hose Reels | 106 |
| S61 : Dry Risers : Inlet Box, Outlet Boxes and Valves | 107 |
| S63 : Sprinklers : Sprinkler Heads | 108 |
| S65 : Fire Hydrants : Extinguishers | 110 |
| S65 : Fire Hydrants : Hydrants and Hydrant Boxes | 111 |
| T10 : Gas/Oil Fired Boilers : Boiler Plant and Ancillaries | 112 |
| T11 : Coal Fired Boilers : Boiler Plant and Ancillaries | 116 |
| T10-T11 : Gas/Oil and Coal Fired Boilers : Chimneys and Flues | 117 |
| T13 : Packaged Steam Generators | 127 |
| T30-32 : Medium and Low Temperature Hot Water Heating : Perimeter Heating | 128 |
| T30-32 : Medium and Low Temperature Hot Water Heating : Fan Convectors | 130 |
| T30-32 : Medium and Low Temperature Hot Water Heating : Radiant Panels | 132 |
| T30-32 : Medium and Low Temperature Hot Water Heating : Radiant Strip Heaters | 134 |
| T42 : Local Heating Units : Unit Heaters | 135 |
| U70 : Air Curtains : Air Heaters | 136 |
| Y10 : Pipelines - Plastics - Soil and Waste | 138 |
| Y10 : Pipelines - Plastics - Water and Gas | 149 |

## Contents

Mechanical Installations *(continued)*

| | |
|---|---:|
| Y10 : Pipelines - Cast and Ductile Iron | page 161 |
| Y10 : Pipelines - Black Steel - Screwed | 163 |
| Y10 : Pipelines - Black Steel - Welded | 174 |
| Y10 : Pipelines - Galvanised Steel | 181 |
| Y10 : Pipelines - Malleable Grooved Jointing System | 192 |
| Y10 : Pipelines - Carbon Steel | 194 |
| Y10 : Pipelines - Stainless Steel | 197 |
| Y10 : Pipelines - Copper | 200 |
| Y10 : Pipelines - Pipe Fixings | 217 |
| Y11 : Pipeline Ancillaries : Regulating Valves | 225 |
| Y11 : Pipeline Ancillaries : Radiator Valves | 227 |
| Y11 : Pipeline Ancillaries : Ball Float Valves | 229 |
| Y11 : Pipeline Ancillaries : Check Valves | 231 |
| Y11 : Pipeline Ancillaries : Commissioning Valves | 233 |
| Y11 : Pipeline Ancillaries : Control Valves | 234 |
| Y11 : Pipeline Ancillaries : Lubricated Plug Valves | 236 |
| Y11 : Pipeline Ancillaries : Gate Valves | 237 |
| Y11 : Pipeline Ancillaries : Globe Valves | 240 |
| Y11 : Pipeline Ancillaries : Safety and Relief Valves | 241 |
| Y11 : Pipeline Ancillaries : Drain Cocks and Vent Cocks | 242 |
| Y11 : Pipeline Ancillaries : Automatic Air Vents | 246 |
| Y11 : Pipeline Ancillaries : Steam Traps and Strainers | 247 |
| Y11 : Pipeline Ancillaries : Sight Glasses | 248 |
| Y11 : Pipeline Ancillaries : Expansion Joints | 249 |
| Y11 : Pipeline Ancillaries : Gauges | 251 |
| Y20 : Pumps : Pumps, Circulators and Accelerators | 255 |
| Y20 : Pumps : Pressurised Cold Water Supply Set | 260 |
| Y21 : Water Tanks/Cisterns : Sectional Glass Fibre Tanks | 261 |
| Y21 : Water Tanks/Cisterns : Moulded Fibreglass Cisterns | 262 |
| Y21 : Water Tanks/Cisterns : Moulded Plastic Cisterns | 263 |
| Y22 : Heat Exchangers : Rotary Air to Air Wheel | 264 |
| Y23 : Storage Cylinders/Calorifiers : Copper Direct Cylinders | 265 |
| Y23 : Storage Cylinders/Calorifiers : Steel Indirect Cylinders | 266 |
| Y23 : Storage Cylinders/Calorifiers : Copper Indirect Cylinders | 267 |
| Y23 : Storage Cylinders/Calorifiers : Steel Non Storage Calorifier | 269 |
| Y23 : Storage Cylinders/Calorifiers : Steel Storage Calorifier | 271 |
| Y23 : Storage Cylinders/Calorifiers : Copper Storage Calorifier | 272 |
| Y30 : Air Ductline : Mild Steel Ductwork | 273 |
| Y30 : Air Ductline Ancillaries : Dampers | 299 |
| Y30 : Air Ductline Ancillaries : Access Openings, Doors and Covers | 303 |
| Y40 : Air Handling Units : Sound Attenuators | 304 |
| Y40 : Air Handling Units : Air Distribution Equipment | 306 |
| Y41 : Fans : Axial Flow Fans | 308 |
| Y41 : Fans : Toilet Extract Fans | 310 |
| Y41 : Fans : Kitchen Extract Fans | 311 |
| Y41 : Fans : Multi Vent Units | 312 |
| Y42 : Air Filtration : Filters | 313 |
| Y45 : Silencers/Acoustic Treatment : Acoustic Louvres | 315 |
| Y46 : Grilles/Diffusers/Louvres : Grilles | 317 |
| Y46 : Grilles/Diffusers/Louvres : Diffusers | 322 |
| Y46 : Grilles/Diffusers/Louvres : External Louvres | 326 |
| Y50 : Thermal Insulation : Preformed Rigid Sections and Slabs | 327 |
| Y50 : Thermal Insulation : Valve Boxes | 333 |
| Y50 : Thermal Insulation : Insulation of Ductwork | 336 |
| Y52 : Vibration Insulation Mounting | 337 |
| Y53 : Control Components - Mechanical | 338 |

**Contents**

Electrical Installations

| | | |
|---|---|---|
| Directions | page | 341 |
| T32 : Low Temperature Hot Water Heating (Small Scale) : Control Components | | 344 |
| V10 : Electricity Generation Plant : Generating Plant | | 345 |
| V11 : HV Supply/Installation/Public Utility : Transformers | | 346 |
| V32 : Uninterrupted Power Supply : U.P.S. Systems/Inverters/Batteries/Chargers | | 349 |
| W10 : Telecommunications : Telecommunication Wiring | | 351 |
| W20 : Radio/TV/CCTV : Coaxial Cables | | 352 |
| W23 : Clocks : Clock Systems | | 354 |
| W30 : Data Transmissions : Guidelines for Network Design | | 356 |
| W30 : Data Transmission : Data Communications | | 362 |
| W50 : Fire Detection and Alarm : Fire Detection Systems | | 378 |
| W52 : Lightning Protection : General | | 380 |
| Y24 : Trace Heating : Electric Trace Heating | | 382 |
| Y60 : Conduit and Cable Trunking : Metal and Plastics Conduit | | 383 |
| Y61 : HV/LV Cables and Wiring : Cable | | 404 |
| Y61 : HV/LV Cables and Wiring : Flexible Cords | | 421 |
| Y62 : Busbar Trunking : Busbars | | 424 |
| Y63 : Support Components - Cables : Cable Tray | | 428 |
| Y63 : Support Components - Cables : Cable Rack | | 434 |
| Y70 : HV Switchgear : Circuit Breakers | | 436 |
| Y71 : LV Switchgear and Distribution Boards : Switches | | 437 |
| Y71 : LV Switchgear and Distribution Boards : Circuit Breakers | | 439 |
| Y71 : LV Switchgear and Distribution Boards : Distribution Boards | | 443 |
| Y71 : LV Switchgear and Distribution Boards : Consumer Units | | 445 |
| Y71 : LV Switchgear and Distribution Boards : Voltage Stabilisers/Line Conditioners | | 449 |
| Y71 : LV Switchgear and Distribution Boards : Fusible Boards | | 451 |
| Y72 : Contactors and Starters : Contactors/Push Buttons | | 452 |
| Y73 : Luminaires and Lamps : Luminaires | | 453 |
| Y74 : Accessories for Electrical Services : Switches | | 455 |
| Y74 : Accessories for Electrical Services : Socket Outlets | | 458 |
| Y74 : Accessories for Electrical Services : Connector Units | | 460 |
| Y74 : Accessories for Electrical Services : Lampholders, Outlets and Control Units | | 461 |
| Y74 : Accessories for Electrical Services : Cable Tiles | | 463 |
| Y89 : Sundry Common Electrical Items : Junction Boxes | | 464 |

### PART FOUR: DAYWORK

| | |
|---|---|
| Heating and Ventilating Industry | 467 |
| Electrical Industry | 472 |
| Building Industry Plant Hire Costs | 476 |

### PART FIVE: FEES FOR PROFESSIONAL SERVICES

| | |
|---|---|
| Quantity Surveyors Fees | 489 |
| Tables and Memoranda | 495 |
| Index | 520 |

# SPON'S PRICE BOOKS 1998

## SPON'S ARCHITECTS' AND BUILDERS' PRICE BOOK 1998

**123rd edition**
*Davis Langdon & Everest*

The only price book geared to market conditions that affect building tender prices. Spon's A & B provides comprehensive coverage of construction costs from small scale alterations and repairs to the largest residential or commercial developments.

944 pages
October 1997
Hardback
0-419-23060-2
£72.50

## SPON'S MECHANICAL AND ELECTRICAL SERVICES PRICE BOOK 1998

**29th edition**
*Davis Langdon & Everest*

"An essential reference for everybody concerned with the calculation of costs of mechanical and electrical works" - Cost Engineer

Outline costs for a wide variety of engineering services are followed by more detailed elemental and quantified analysis.

560 pages
October 1997
Hardback
0-419-23080-7
£75.00

## SPON'S LANDSCAPE AND EXTERNAL WORKS PRICE BOOK 1998 — NEW LAYOUT

**17th edition**
*Derek Lovejoy Partnership and Davis Langdon & Everest*

Now completely revised and expanded. Every rate has been reworked and recalculated. Now includes for the first time a realistic labour rate as well as plant and material costs.

"Surely there can be no office without this publication" - **Landscape Design**

288 pages
October 1997
Hardback
0-419-23070-X
£62.50

## SPON'S CIVIL ENGINEERING AND HIGHWAY WORKS PRICE BOOK 1998

**12th edition**
*Davis Langdon & Everest*

"Unquestionably, this will be required reading by all estimators involved in civil engineering works. Quantity surveyors will also find it essential for their shelves." - Civil Engineering Surveyor

640 pages
October 1997
Hardback
0-419-23050-5
£95.00

FREE UPDATES every 3 months during books' currency

E & FN SPON an imprint of Thomson Professional

## ReSPONsE —

*SPON's price data now available in an electronic estimating system*
Call **0171-522-9966** for details

# Acknowledgements

The editors wish to record their appreciation of the assistance given by many individuals and organisations in the compilation of this edition.

Manufacturers, Distributors and Sub-Contractors who have contributed this year include:-

ABB-Wylex Sales Ltd
Wylex Works
Wythenshaw
Manchester M22 4RA
**Cabling Systems,**
**Switchgear & Distribution**
Tel: 0161 998 5454
Fax: 0161 945 1587

A E L
Acoustics & Environmetrics Ltd
1 Berkeley Court
Manor Park
Runcorn
Cheshire WA7 1TQ
**Energy Recovery Systems**
Tel: 01928 579068
Fax: 01928 579523

Allaway Acoustics Ltd
Old Police Station
1 Queens Road
Hertford
Herts SG14 1EN
**Air Distribution Equipment**
Tel: 01992 550825
Fax: 01992 554982

Angus Fire Armour Ltd
Thame Park Road
Thame
Oxon OX9 3RT
**Fire Fighting & Detection Equipment**
Tel: 01844 214545
Fax: 01844 213511

APV-Vent Axia Industrial Division
Fleming Way
Crawley
West Sussex RH10 2NN
**Fans**
Tel: 01293 526062
Fax: 01293 560257

S G Baldwin
Barholm Road
Tallington
Stamford
Lincolnshire PEG 4RL
**Concrete Cable Protection**
Tel: 01778 345455
Fax: 01778 345949

BICC Cables Ltd
Energy Cables Division
P O Box 1
Warrington Road
Prescot
Merseyside L34 5SZ
**Cables**
Tel: 0151 430 8080
Fax: 0151 762676

BICC Pyrotenax Ltd
P O Box 20
Prescot
Merseyside L34 5GB
**MI Cables**
Tel: 0151 430 4000
Fax: 0151 430 4004

Biddle Air Curtains
St Mary's Road
Nuneaton
Warwickshire CV11 5AU
**Air Distribution Equipment**
Tel: 01203 384233
Fax: 01203 373621

Bill Switchgear Ltd
Aston Lane
Perry Barr
Birmingham B20 3BT
**Switchgear & CCT Protection**
Tel: 0121 356 6001
Fax: 0121 356 1962

## *Acknowledgements*

Braithwaite Eng Ltd
Neptune Works
Uskway
Newport
Gwent
NP9 2UY
**Liquid Storage Systems**
Tel: 01633 262141
Fax: 01633 250631

S Brannan & Sons Ltd
Leconfield Industrial Estate
Cleator Moor
Cumbria
CA25 5QE
**Thermometers**
Tel: 01946 810413
Fax: 01946 813694

Brights of London Ltd
Westgate Business Park
Westgate Carr Road
Pickering
N Yorks YO18 8LX
**Clock Systems**
Tel: 0181 7868466
Fax: 0181 7868477

British Fittings Ltd
Stanton House
PO Box 297
Holyhead Road
Birmingham B21 0AH
**Pipes, Fittings, Fixings & Sundries**
Tel: 0121 553 6666
Fax: 0121 525 0952

British Steel Plc
Tubes & Pipes
Weldon Road
Corby Works
PO Box 101
Northants NN17 5UA
**Pipes, Fittings, Fixings & Sundries**
Tel: 01536 402121
Fax: 01536 404111

Broadcrown Ltd
Alliance Works
Mill Street
Stone
Staffordshire ST15 8BA
**Standy Engine Power**
Tel: 01785 817513
Fax: 01785 812474

BSS UK Ltd (Wirral Branch)
Rossmore Road East
Rossmore Road Industrial Estate
Ellesmere Port
South Wirral L65 3DD
**Pipeline Fittings**
Tel: 0151 355 8281
Fax: 0151 375 2001

Brush Transformers Ltd
P O Box 20
Loughborough
Leics LE11 1HN
**Switchgear & Distribution**
Tel: 01509 612822
Fax: 01509 612825

Bryan Donkin Co Ltd
Derby Road
Chesterfield
S40 2EB
**Gas Valves**
Tel: 012460273153
Fax: 012460235273

Caradon MK Electric Ltd
The Arnold Centre
Paycocke Road
Basildon
Essex SS14 3EA
**Cabling Systems, Switchgear & Distribution, Wiring, Fittings & Accessories**
Tel: 01268 563000
Fax: 01268 563563

Caradon Stelrad Ideal Boilers
PO Box 103
National Avenue
Kingston-upon-Hall
North Humberside
HU5 4JN
**Boilers/Heating Products**
Tel: 01482 492251
Fax: 01482 448858

CMP Products Ltd
Glasshouse Street
St Peters
Newcastle Upon Tyne NE6 1BE
**Cabling Systems, Wiring, Fittings & Accessories**
Tel: 0191 265 7411
Fax: 0191 265 0581

# Acknowledgements

Coley Instruments
Burnside Industrial Estate
Kilsyth
Glasgow
G65 9JX
**Thermometers**
Tel: 01236 826040
Fax: 01236 824090

Conex Sanbra Ltd
Whitehall Road
Tipton
West Midlands DY4 7JU
**Copper Pipes, Fittings, Fixings & Sundries**
Tel: 0121 557 2831
Fax: 0121 520 8778

Constant Power Services Ltd
Unit 8A, The Cam Centre
Wilbury Way
Hitchin
Herts SG4 0TW
**UPS Systems**
Tel: 01462 422955
Fax: 01462 422754

Crabtree Electrical Industries Ltd
Lincoln Works
Walsall WS1 2DN
**Switchgear & Distribution,**
**Wiring, Fittings & Accessories**
Tel: 01922 721202
Fax: 01922 721321

Crane Fluid Systems
National Sales Centre
Nacton Road
Ipswich
Suffolk
1PE 9QH
**Valves, Fittings, Actuators**
Tel: 01473 270222
Fax: 01473 270393

Cumbria Heating Components
Block Two
Brymau Three Industrial Estate
River Lane
Saltney
Chester CH4 8RH
**Domestic & Industrial Heating Equipment**
Tel: 01244 671877
Fax: 01244 671305

Cutler Hammer Limited
Mill Street
Ottery St Mary
Devon
EX11 1AG
**Cable Management Systems**
Tel: 01404 812131
Fax: 01404 815471

Davis Group Limited
Rookwood Way
Havenhill
Suffolk CB9 8PB
**Cable Management System**
Tel: 01440 704411
Fax: 01440 702822

Delta Crompton Cables Ltd
Millmarsh Lane
Brimsdown
Enfield
Middlesex EN3 7QD
**Rubber and Plastic Cables**
Tel: 0181 804 2468
Fax: 0181 970 0062

Dimplex Heating Limited
Marketing Dept
Millbrook
Southampton S09 2DP
**Electrical Heating**
Tel: 01703 777117
Fax: 01703 771096

Donald Brown(Brownall)
Stretford Road
Manchester
M16 9AR
**Lab.Tap**
Tel: 0161 872 6941
Fax: 0161 848 7681

Dorman-Smith Switchgear Ltd
Blackpool Road
Preston PR2 2DQ
**Switchgear & Distribution**
Tel: 01772 728271
Fax: 01772 726276

Drainage Systems Ltd
Cray Avenue
St Mary Cray
Orpington
Kent BR5 3RH
**C.I. Pipes**
Tel: 01689 890900
Fax : 01689 822372

## Acknowledgements

Dunham Bush
Brittannia House
Newton Road
Hoylake
Merseyside L47 3DG
**Air Distribution, Heating**
Tel: 0151 632 3393
Fax: 0151 632 4453

Durapipe Glynwed Plastics Ltd
Walsall Road
Norton Canes
Cannock
Staffs WS11 3NS
**Thermoplastic Piping**
Tel: 01543 279909
Fax: 01543 279450

Engineering Appliances Ltd
Unit 11
Sunbury Cross Ind Est
Brooklands Close
Sunbury On Thames
TW16 7DX
**Expansion Joints**
Tel: 01932 788889
Fax: 01932 761263

Euro-Diesel (UK) Ltd
Strato House
Somerfield Court
Somerfield Road
Circencester
Gloustershire GL7 1TW
**UPS Systems**
Tel: 01285 640879
Fax: 01285 652 509

Farnell Electrical Services
Edinburgh Way
Harlow
Essex
CM20 2DF
**Cables**
Tel: 01279 441144
Fax: 01279 441687

Friatec House
Old Parkbury Lane
Colney Street
Saint Albarns
Herts
AL2 2ED
**PVCC Pipes and Fittings**
Tel: 01923 857878
Fax : 01923 853434

Furse and Company Limited
Wilford Road
Nottingham NG2 1EB
**Lightning Protection**
Tel: 0115 863471
Fax: 0115 9860071

GEC-Alsthom Installation Equipment
East Lancashire Road
Liverpool L10 5HB
**Switchgear & Distribution**
Tel: 0151 525 8371
Fax: 0151 523 7007

Gent Ltd
Unit 11
Pavilion Business Park
Royds Hall Road
Lower Wortley
Leeds
LS12 6AJ
**Fire Detection Systems**
Tel: 0116 246 2400
Fax: 0116 246 2017

Hattersley, Newman, Hender Ltd
Burscough Road
Ormskirk
Lancashire L39 2XG
**Valves**
Tel: 01695 577199
Fax: 01695 578775

Hawker Siddeley Switchgear Ltd
P O Box 19
Falcon Works
Loughborough
Leicestershire LE11 1HL
**Switchgear & Distribution**
Tel: 01509 611311
Fax: 01509 610404

Heap and Partners (Head Office)
Brittannia House
Newton Road
Hoylake
Wirral
Merseyside L47 3DG
**Fluid Control Engineer**
Tel: 0151 632 3393
Fax: 0151 632 4453

## Acknowledgements

Hepworth Building Products
Hazle Head
Stocksbridge
Sheffield
S30 5HG
**Clay and Plastic Drainage**
Tel: 01226 763561
Fax: 01226 764827

Horne Engineering Ltd
Rankine Street
Johnstone
Renfrewshire
Scotland PA5 8BD
**Valves**
Tel: 01505 321455
Fax: 01505 336287

Horseley Bridge Tanks
27/29 Thornleigh Trading Estate
Blowers Green
Dudley
West Midlands DY2 8UB
**Storage Tanks**
Tel: 01384 459119
Fax: 01384 459117

HRP Ltd
Rougham Industrial Estate
Rougham
Bury St. Edmunds
Suffolk IP30 9ND
**Air Conditioning**
Tel: 01359 271131
Fax: 01359 271132

HVCA Publications
Old Mansion House
Eamont Bridge
Penrith
Cumbria
CA10 2BX
**Labour**
Tel: 01768 864771
Fax :01768 867138

IMI Rycroft Ltd
Duncombe Road
Bradford
BD8 9TB
**Calorifiers**
Tel: 01274 490911
Fax: 01274 482565

IMI Yorkshire Fittings Ltd
PO Box 166
Leeds LS1 1RD
**Copper Pipes, Fittings, Fixings
& Sundries**
Tel: 0113 2706945
Fax: 0113 2705644

Jasun Filtration
Unit 2B/2c
Yeo Road
Bridgewater
Somerset
TA6 5NA
**Air Filtration**
Tel: 01278 452277
Fax: 01278 450873

JIB- For the Electrical Industry
Kingswood House
47/51 Sidcup Hill
Kent
DA14 6HP
**Labour**
Tel: 0181 302 0031
Fax: 01081 309 1103

JIB-Plumb & Mech Eng Industry
The Joint Industry Board For Plumbing
Brook House
Brook Street
St.Neots
Huntingdon
PE19 2HW
**Labour**
Tel: 01480 476925
Fax: 01480 403081

John Millar UK Limited
Brittania House
Newton Road
Hoylake
Wirral
Merseyside L47 3DG
**Flexible Joints**
Tel: 0151 632 3393
Fax: 0151 632 4453

Kiddie Thorn Fire Protection
Century Buildings
1 Summers Road
Brunswick Business Park
Liverpool L3 4BL
**Fire Protection**
Tel: 0151 207 1243
Fax: 0151 709 6062

## Acknowledgements

Kiloheat Ltd
Enterprise Way
Edenbridge
Kent TN8 6HF
**Fans**
Tel: 01732 866000
Fax: 01295 257477

Knowsley SK
Centre-Point
Trafford Park Road
Trafford Park
Manchester M17 1AE
**Fire Fighting Equipment**
Tel: 0161 872 7511
Fax: 0161 848 8508

Kopex International Ltd
189 Bath Road
Slough
Berks SL1 4AR
**Wiring, Fittings & Accessories**
Tel: 01753 534931
Fax: 01753 693521

Lamatherm Products Ltd
Forge Factory Estate
Maesteg
Mid Glamorgan CF34 0AU
**Sound Attenuators**
Tel: 01656 730833
Fax: 01656 730115

Lancashire Fittings Limited
The Science Village
Claro Road
Harrogate
North Yorkshire HG1 4AF
**Stainless Steel Fittings**
Tel: 01423 522355
Fax: 01423 506111

Lennox Industries Ltd
PO Box 174
Westgate Interchange
Northampton
NN5 5AG
**Heating Units**
Tel: 01604 591159
Fax: 01604 587536

Marley Extrusions Ltd
Dickley Lane
Lenham
Maidstone
Kent ME17 2DE
**UPVC, Pipes, Fittings, Fixings & Sundries**
Tel: 01622 858888
Fax: 01622 858725

Marshall Tufflex Ltd
Ponswood Industrial Estate
Hastings
East Sussex TN34 1YJ
**Cabling Management Systems**
Tel: 01424 427691
Fax: 01424 720670

MC Integ Limited
Integ House
Rougham Industrial Estate
Barry St Edmonds
Suffolk IP30 9ND
**Tanks and Vessels**
Tel: 01359 270610
Fax: 01359 270458

MEM Ltd
Whitegate
Broadway, Chadderton
Oldham OL9 9QG
**Wiring, Fittings & Accessories**
Tel: 0161 652 1111
Fax: 0161 626 1709

Menvier Ltd
Southam Road
Banbury
Oxon OX16 7RX
**Fire Detection Systems**
Tel: 01295 256363
Fax: 01295 270102

Mita (UK) Ltd
Bodelwyddan Business Park
Bodelwyddan
Clwyd
North Wales
LL18 5SX
**Cabling Management Systems**
Tel: 01745 826224
Fax: 01745 833161

Mucklow Bros Ltd
Narrow Lane
Halesowen
West Midlands B62 9PA
**Flanges, Pipes, Fittings, Fixings & Sundries**
Tel: 0121 423 1212
Fax: 0121 423 2020

Myson RCMOld Wolverton Road
Milton Keynes
Buckinghamshire MK12 5PT
**Skirting Radiator**
Tel: 01908 321155
Fax: 01908 317387

## Acknowledgements

National Vulcan Ltd
St Mary's Parsonage
Manchester M6 9AP
**Valves**
Tel: 0161 834 8124
Fax: 0161 839 3227

Netco Ltd
4 Watt House
Pensnett Estate
Kingswinford
West Midlands DY6 8XZ
**Trace Heating**
Tel: 01384 400750
Fax: 01384 400314

Ottermill Ltd
Mill Street
Ottery St Mary
Devon EX11 1AG
**Cabling Systems, Switchgear & Distribution**
Tel: 01404 812131
Fax: 01404 815471

Owens-Corning Building Products (UK) Ltd
PO Box10
Stafford Road
St Helens
Merseyside LA10 3NS
**Thermal & Accoustics Insulation**
Tel: 01744 24022
Fax: 01744612007

Ozonair Limited
Quarrywood Industrial Estate
Aylesford
Maidstone
Kent ME20 7NB
**Air Distribution Equipment**
Tel: 01622 717861
Fax: 01622 719291

Pegler Ltd
St Catherines Avenue
Doncaster
South Yorkshire DN4 8DF
**Radiators, Heaters & Controls, Valves**
Tel: 01302 329777
Fax: 01302 730515

Philmac PTY Ltd
Diplocks Way
Hailsham
East Sussex
BN27 3JF
**Pipes**
Tel: 01323 847323
Fax: 01323 844775

Pipeline Centre
Textilose Road
Trafford Park
Manchester M17 1WA
**Pipes & Fittings**
Tel: 0161 872 8431
Fax: 0161 872 7208

Pirelli Cables Ltd
Special Cables Division
PO Box 30
Chickenhall Lane
Eastleigh
Hants SO5 5XA
**Cables**
Tel: 01703 644544
Fax: 01703 649649

Pirelli Cables Ltd
PO Box 6
Leigh Road
Eastleigh
Hampshire
S050 9YE
**Power Cables**
Tel: 01703 295555
Fax: 01703 295444

Plumb Centre Ltd
Boroughbridge Road
Ripon
N. Yorks
HG4 1SL
**General**
Tel: 01765 690690
Fax: 01765 690100

Plumb Centre (Saltney, Chester)
Units 1 & 2
Marley Way
Off High Street
Saltney
Chester CH4 8SX
**Merchants, Plumbing, Electrical**
Tel: 01244 674352
Fax: 01244 680937

## Acknowledgements

Plumb Centre (Wrexham)
Rivulet Road
Wrexham
Clwyd
North Wales
LL13 8DT
**Fuel & Storage / Distribution**
Tel: 01978 291722
Fax: 01978 291310

Polypipe
Broomhouse Lane
Edlington
Doncaster DN12 1ES
**UPVC Pipes & Fittings**
Tel: 01709 770000
Fax: 01709 770001

E Poppleton & Son Limited
Conway Road
Colwyn Bay
Clwyd Ll28 4AS
**Ductwork**
Tel: 01492 546061
Fax: 01492 544076

Potterton Myson Limited
Eastern Avenue
Team Valley Trading Estate
Gateshead
Newcastle upon Tyne NE11 0PG
**Heating Products**
Tel: 0191 491 7500
Fax: 0191 491 756

Pullen Pumps Ltd
Unit 2, Mercury Park
Mercury Way (off Barton Dock Road)
Urmston
Manchester M41 7LR
**Pumps, Circulators**
Tel: 0161 248 4801
Fax: 0161 866 8901

Richard Vanspall Associates Ltd
2nd Floor
Drayton Bridge House
Tavistock Road
West Drayton
Middlesex UB7 7BB
**Cooling Water Specialists**
Tel: 01895 440546
Fax: 01895 431944

Salamandre Plc
Hunts Rise
South Marston Park
South Marston
Swindon
Wiltshire SN3 4RE
**Cabling Trunking Systems**
Tel: 01793 828000
Fax: 01793 828597

Sauter Automation
86 Talbot Road
Old Trafford
Manchester M16 0PG
**Butterfly Control Valves**
Tel: 0161 8724791
Fax: 0161 8480855

Selkirk Manufacturing Ltd
Bassett House
High Street
Banstead
Surrey SM7 2LZ
**Chimney Systems, Exhausts**
Tel: 01737 353388
Fax: 01737 362501

Sheffield Insulation
Off Westinghouse Road
Trafford Park
Manchester M17 1PY
**Insulation General**
Tel: 0161 876 4776
Fax: 0161 876 4775

South Wales Transformers
PO Box 20
Loughborough
Leicestershire
LE11 1HN
**Switchgear & Distribution**
Tel: 01509 612822
Fax: 01509 612825

Space Air Conditioning
Willway Court
1 Deacon Field
Guildford
Surrey
GU2 5YT
**Air Conditioning**
Tel: 0143 504883
Fax: 01483 574835

Spirax-Sarco Ltd
Charlton House
Cheltenham
Gloustershire
GL53 8ER
**Traps and Valves**
Tel: 01242 521361
Fax: 01242 573342

## Acknowledgements

Sunvic Controls Ltd
Bellshill Road
Glasgow
G71 6NP
**Heat Controls**
Tel: 01698 812944
Fax: 01698 813637

Swifts of Scarborough Limited
Cayton Low Road
Eastfield
Scarborough
North Yorkshire
YO11 3BY
**Cable Trays**
Tel: 01723 583131
Fax: 01723 584625

Tecra Limited
Bumpers Lane
Sealand Industrial Estate
Chester CH1 4LT
**Copper Pipe**
Tel: 01244 377539
Fax: 01244 378644

Teddington Bellows Ltd
Teilo Works
Portardulais
Swansea
SA4 1RP
**Expansion Joints**
Tel: 01792 882591
Fax: 01792 885199

Telemechanique
University of Warwick Science Park
Sir William Lyons Road
Coventry CV4 7EZ
**Busbars**
Tel: 01203 416255
Fax: 01203 417517

Thorn Lighting Ltd
Elstree Way
Borehamwood
Herts WD6 1HZ
**Lighting**
Tel: 0181 905 1313
Fax: 0181 905 1287

Vent-Axia Limited
Unit 2
Caledonia Way
Stretford Motorway Estate
Barton Dock Road
Manchester M32 0ZH
**Ventilation/Cooling**
Tel: 0161 865 8421
Fax: 0161 865 0098

Victaulic Systems Ltd
PO Box 13
46-48 Wilbury Way
Hitchin
Herts SG4 0UD
**Pipe Jointing Systems**
Tel: 01462 422622
Fax: 01462 422072

Vokes Ltd
Henley Park
Guildford
Surrey GU3 2AF
**Air Filters**
Tel: 01483 699711
Fax: 01483 236419

Walsall Conduits Ltd
Dial Lane
West Bromwich
West Midlands B70 0EB
**Cabling Management Systems**
Tel: 0121 557 1171
Fax: 0121 557 5631

Waterloo Air Management
Quarry Wood Industrial Estate
Aylesford
Maidstone
Kent
ME20 7NB
**Air Ductwork**
Tel: 01622 717861
Fax: 01622 710648

Waterloo Air Distribution/Control
14 Parsons Road
Manor Trading Estate
South Benfleet
Essex SS7 4PT
**Air Diffusers/Controls**
Tel: 01268 794121
Fax: 01268 751841

Wellman Robey Ltd
Newfield Road
Oldbury
West Midlands
B69 3ET
**Boilers**
Tel: 0121 5523311
Fax: 0121 5524571

## Acknowledgements

Weidmuller (Klippon Products) Ltd
Power Station Road
Sheerness
Kent ME12 3AB
**Wiring Accessories**
Tel: 01795 580999
Fax: 01795 580115

Wibe Ltd
Unit 8D, Castle Vale Ind. Estate
Maybrook Road
Minworth
Sutton Coldfield
West Midlands
B76 1AL
**Cabling Support Systems**
Tel: 0121 313 1010
Fax: 0121 313 1020

Woods of Colchester
Tufnell Way
Colchester
Essex CO4 5AR
**Air Distribution**
Tel: 01206 441222
Fax: 01206 574434

PART ONE

# Approximate Estimating

Directions, *page* 3
Cost Indices, *page* 4
Outline Costs, *page* 6
Elemental and All-in-Rates, *page* 9
Elemental Costs, *page* 34

# ENERGY MANAGEMENT AND OPERATING COSTS IN BUILDINGS

## Keith J. Moss

Managing the consumption and conservation of energy in buildings must now become the concern of both building managers and occupants. The provision of lighting, hot water supply, communications, cooking, space heating and cooling accounts for 45 per cent of UK energy consumption.

**Energy Management and Operating Costs in Buildings** introduces the reader to the principles of managing and conserving energy consumption in buildings people use for work or leisure. Energy consumption is considered for the provision of space heating, hot water supply ventilation and air conditioning. The author introduces the use of standard performance indicators and energy consumption yardsticks, and discusses the use and application of Degree Days. Following an introduction to the preparation of the energy audit, monitoring and targeting techniques are investigated and analysed.

Readers are not expected to have prior knowledge of the design of building services. Each chapter of the book is set out with:

- nomenclature:
- introduction;
- worked examples and case studies;
- data, text and illustrations appropriate to each topic.

This is a key text for undergraduates on building services engineering, energy management, environmental engineering and related construction courses. It will also be an invaluable work for professional energy managers and building services engineers.

Preface. Acknowledgements. Introduction. The economics of space heating plants. Estimating energy consumption - space heating. Intermittent space heating. Estimating the annual cost for the provision of a hot water supply. Energy consumption for cooling loads. Performance indicators. Energy conservation strategies. Cost benefit analysis. Energy audits. Monitoring and targeting. Appendices. Bibliography. Index

**Keith Moss** is a consultant engineer and a visiting lecturer at the University of Bath and City of Bath College.

246x189mm, 200 pages
25 line illustrations
Paperback ISBN 0-419-21770-3
£24.99 June 1997

## New From

### E & FN SPON
An Imprint of Thomson Professional
2-6 Boundary Row
London SE1 8HN

## *Approximate Estimating*

## DIRECTIONS

Prices shown are average prices on a fluctuating basis for typical buildings during the second quarter, of 1997. Unless otherwise shown, they exclude external services and professional fees.

The information in this section has been arranged to follow more closely the order in which estimates may be developed:

a) Cost Indices and Regional Variations - gives indices and variations to be applied to estimates in general

b) Outline Costs - gives a range of data (based on a rate per square metre) for all-in engineering costs associated with a wide variety of building functions.

For certain specified building types these initial all-in costs are further subdivided to give a broad division between the main work elements.

c) Elemental and All-in-Rates - given for a number of items and complete component parts i.e. boiler plant, ductwork and mains and switchgear together with some other items not covered in parts of this or other sections of the book i.e. lifts, escalators, kitchen equipment, medical gases.

d) Elemental Costs - are shown for four building types - a speculative office block, a hotel, a data processing centre and a district general hospital. In each case, a full analysis of engineering services costs is given to show the division between all elements and their relative costs to the total building area.

A further analysis is given to indicate the additional stage of cost information required when, for a complex building such as a hospital, prices are also required for each functional area or department.

Prices should be applied to the total floor area of all storeys of the building under consideration. The area should be measured between the external walls without deduction for internal walls and staircases/lift shafts.

Although prices are reviewed in the light of recent tenders it has only been possible to provide a range of prices for each building type. This should serve to emphasise that these can only be average prices for typical requirements and that such prices can vary widely depending upon a number of features. Rates per square metre should not therefore be used indiscriminately and each case must be assessed on its merits.

The prices do not include for incidental builder's work nor for profit and attendance by a main contractor where the work is executed as a sub-contract: they do however include for profit and overheads for the services contractor and for 2.5% cash discount for the main contractor. Capital contributions to statutory authorities and public undertakings and the cost of work carried out by them have been excluded.

## Approximate Estimating

### COST INDICES

The following tables reflect the major changes in cost to contractors but do not necessarily reflect changes in tender levels. In addition to changes in labour and materials costs tenders are affected by other factors such as the degree of competition in the particular industry and area where the work is to be carried out, the availability of labour and the general economic situation. This has meant in recent years that, when there has been an abundance of work, tender levels have often increased at a greater rate than can be accounted for by increases in basic labour and material costs and, conversely, when there is a shortage of work this has often resulted in keener tenders. Allowances for these factors are impossible to assess on a general basis and can only be based on experience and knowledge of the particular circumstances. In compiling the tables the cost of labour has been calculated on the basis of a notional gang as set out elsewhere in the book. The proportion of labour to materials has been assumed as follows:
Mechanical Services - 30:70 / Electrical Services - 50:50 (1976 = 100)

### Mechanical Services

| Year | First Quarter | Second Quarter | Third Quarter | Fourth Quarter |
|---|---|---|---|---|
| 1979 | 138 | 141 | 150 | 156 |
| 1980 | 165 | 171 | 172 | 174 |
| 1981 | 181 | 185 | 187 | 190 |
| 1982 | 197 | 201 | 201 | 201 |
| 1983 | 204 | 205 | 208 | 211 |
| 1984 | 215 | 221 | 223 | 226 |
| 1985 | 232 | 236 | 236 | 236 |
| 1986 | 238 | 240 | 239 | 243 |
| 1987 | 244 | 249 | 251 | 254 |
| 1988 | 257 | 266 | 268 | 272 |
| 1989 | 276 | 284 | 284 | 286 |
| 1990 | 288 | 298 | 299 | 301 |
| 1991 | 312 | 316 | 317 | 319 |
| 1992 | 323 | 325 | 326 | 329 |
| 1993 | 333 | 334 | 336 | 338 |
| 1994 | 341 | 342 | 345 | 350 |
| 1995 | 357 | 358 | 360 | 365 |
| 1996 | 364 | 361 | P 357 | P 360 |
| 1997 | F 362 | F 364 | F 365 | F 370 |
| 1998 | F 374 | F 376 | F 378 | F 384 |

### Electrical Services

| Year | First Quarter | Second Quarter | Third Quarter | Fourth Quarter |
|---|---|---|---|---|
| 1979 | 143 | 145 | 147 | 154 |
| 1980 | 169 | 169 | 177 | 192 |
| 1981 | 195 | 197 | 199 | 200 |
| 1982 | 202 | 211 | 211 | 212 |
| 1983 | 214 | 225 | 225 | 226 |
| 1984 | 228 | 229 | 236 | 237 |
| 1985 | 240 | 240 | 247 | 250 |
| 1986 | 251 | 249 | 249 | 257 |
| 1987 | 268 | 269 | 271 | 274 |
| 1988 | 286 | 289 | 290 | 293 |
| 1989 | 306 | 306 | 306 | 308 |
| 1990 | 323 | 324 | 326 | 330 |
| 1991 | 350 | 350 | 350 | 350 |
| 1992 | 367 | 368 | 368 | 370 |
| 1993 | 373 | 377 | 378 | 378 |
| 1994 | 380 | 381 | 383 | 389 |
| 1995 | 397 | 397 | 398 | 399 |
| 1996 | 406 | 404 | 401 | P 401 |
| 1997 | F 411 | F 411 | F 412 | F 414 |
| 1998 | F 426 | F 427 | F 428 | F 430 |

(P = Provisional)
(F = Forecast)

**Approximate Estimating**

## COST INDICES

Regional Variations

The table of regional variation factors have been derived from data rebased so that Outer London = 1.00. The figures have been extracted from an analysis of contract prices for schemes tendered between mid 1982 and the end of 1995. As such it is important to realise that the factors represent the average state of affairs over the thirteen year period, rather than an exact indication of the regional differences at this precise point in time.

The factors include an adjustment to the general building factors to reflect the lower level of regional variations that exist for engineering sub-contract work, as compared to main contract work.

| Region | Factor | Region | Factor |
|---|---|---|---|
| Outer London | 1.00 | Inner London | 1.06 |
| Kent, Surrey & Sussex | 0.99 | Beds, Essex & Herts | 0.97 |
| Berks, Bucks, Herts & Oxon | 0.95 | South West | 0.92 |
| East Midlands | 0.88 | West Midlands | 0.89 |
| East Anglia | 0.91 | Yorkshire & Humberside | 0.90 |
| North West | 0.93 | Northern | 0.91 |
| Scotland | 0.92 | Wales | 0.88 |
| Northern Ireland | 0.77 | Channel Islands | 1.09 |

## OUTLINE COSTS

## SQUARE METRE RATES FOR BASIC SERVICES BY BUILDING TYPE
*The undernoted examples indicate the range of rates within which normal engineering services would be contained for each of the more common building types.*

|  | Square metre £ |
|---|---|
| **Industrial Buildings (Cl/SfB 2)** | |
| Factories | |
|    for letting | 33 to 75 |
|    owner occupation | 50 to 102 |
| Warehouses | |
|    high bay for owner occupation | 50 to 107 |
| **Administrative, public, commercial and office buildings; general (Cl/SfB 3)** | |
| Civic offices | |
|    fully air conditioned | 294 to 421 |
| Offices for letting | |
|    non air conditioned | 139 to 210 |
|    fully air conditioned | 228 to 384 |
| Offices for owner occupation | |
|    medium rise, air conditioned | 237 to 462 |
|    high rise, air conditioned | 237 to 484 |
| Fitted out offices including air conditioning | 305 to 460 |
| Fitted out department store ditto | 240 to 372 |
| **Health and welfare facilities (Cl/SfB 4)** | |
| District general hospitals | 318 to 451 |
| Private hospitals | 336 to 529 |
| **Refreshment, entertainment, recreation buildings (Cl/SfB 5)** | |
| Arts and drama centres | 239 to 326 |
| Theatres, large | 310 to 421 |
| **Educational, scientific and information buildings (Cl/SfB 7)** | |
| Secondary/middle schools | 132 to 195 |
| Universities | |
|    arts buildings | 172 to 217 |
|    science buildings | |
|       physics | 172 to 273 |
|       biology | 242 to 286 |
|       chemistry | 273 to 320 |
| Museums | |
|    national | 584 to 730 |
| **Residential buildings (Cl/SfB 8)** | |
| Local authority schemes | |
|    two storey houses | 57 to 79 |
|    medium rise flats | 78 to 102 |

# DAVIS LANGDON & EVEREST
**Authors of Spon's Price Books**

**Davis Langdon & Everest** is an independent practice of Chartered Quantity Surveyors with over 500 staff in 19 UK offices.

We believe that cost awareness and control are responsibilities that must be accepted by all members of the team, but the ability to act on that responsibility is dependent upon the quality of the cost advice given. Our approach to cost consultancy is therefore:

* to be positive and creative in our advice, rather than simply reactive;

* to concentrate on value for money and value engineering rather than on superficial cost-cutting;

* to give advice that is matched to the Client's own criteria, rather than to impose standard or traditional solutions;

* to see cost as one component of a successful design solution, which needs to be balanced with many others, and to work as an integrated member of a design team in achieving that balance;

* to pay attention to the life-long costs of owning and operating a facility, rather than simply the initial capital cost.

We work in small teams, under the active leadership of a Partner, to provide continuity of involvement throughout the stages of a project. At the same time, we can draw on the resources of a large firm, and upon departments able to give specialist advice on engineering services, life cycle costing, project analyses, cost data and cost research.

**Davis Langdon & Everest** - a first point of call for anyone contemplating a construction project. **Consult the Source.**

**TRUST DLE TO PROVIDE THE DEFINITIVE PRICE DATA**

# DAVIS LANGDON & EVEREST
CHARTERED QUANTITY SURVEYORS
CONSTRUCTION COST CONSULTANTS

**LONDON**
Princes House
39 Kingsway
London
WC2B 6TP
Tel : (0171) 497 9000
Fax : (0171) 497 8858

**BRISTOL**
St Lawrence House
29/31 Broad Street
Bristol BS1 2HF
Tel : (0117) 927 7832
Fax : (0117) 925 1350

**CAMBRIDGE**
36 Storey's Way
Cambridge CB3 ODT
Tel : (01223) 351258
Fax : (01223) 321002

**CARDIFF**
3 Raleigh Walk
Brigantine Place
Atlantic Wharf
Cardiff CF1 5LN
Tel : (01222) 471306
Fax : (01222) 471465

**CHESTER**
Ford Lane Farm
Lower Lane
Aldford
Chester CH3 6HP
Tel : (01244) 620222
Fax : (01244) 620303

**EDINBURGH**
74 Great King Street
Edinburgh
EH3 6QU
Tel : (0131) 557 5306
Fax : (0131) 557 5704

**GATESHEAD**
11 Regent Terrace
Gateshead
Tyne and Wear
NE8 1LU
Tel : (0191) 477 3844
Fax : (0191) 490 1742

**GLASGOW**
Cumbrae House
15 Carlton Court
Glasgow G5 9JP
Tel : (0141) 429 6677
Fax : (0141) 429 2255

**IPSWICH**
17 St Helens Street
Ipswich IP4 1HE
Tel : (01473) 253405
Fax : (01473) 231215

**LEEDS**
Duncan House
14 Duncan Street
Leeds
LS1 6DL
Tel : (0113) 243 2481
Fax : (0113) 242 4601

**LIVERPOOL**
Cunard Building
Water Street
Liverpool
L3 1JR
Tel : (0151) 236 1992
Fax : (0151) 227 5401

**MANCHESTER**
Boulton House
Chorlton Street
Manchester M1 3HY
Tel : (0161) 228 2011
Fax : (0161) 228 6317

**MILTON KEYNES**
6 Bassett Court
Newport Pagnell
Buckinghamshire
MK16 OJN
Tel : (01908) 613777
Fax : (01908) 210642

**NEWPORT**
34 Godfrey Road
Newport
Gwent NP9 4PE
Tel : (01633) 259712
Fax : (01633) 215694

**NORWICH**
63 Thorpe Road
Norwich NR1 1UD
Tel : (01603) 628194
Fax : (01603) 615928

**OXFORD**
Avalon House
Marcham Road
Abingdon
Oxford OX14 1TZ
Tel : (01235) 555025
Fax : (01235) 554909

**PLYMOUTH**
3 Russell Court
St Andrews Street
Plymouth
Devon PL6 2AX
Tel : (01752) 668372
Fax : (01752) 221219

**PORTSMOUTH**
Kings House
4 Kings Road
Portsmouth PO5 3BQ
Tel : (01705) 815218
Fax : (01705) 827156

**SOUTHAMPTON**
Clifford House
New Road
Southampton
SO14 OAB
Tel : (01703) 333438
Fax : (01703) 226099

**DAVIS LANGDON CONSULTANCY**
Princes House
39 Kingsway
London WC2B 6TP
Tel : (0171) 379 3322
Fax : (0171) 379 3030

A MEMBER OF
## DAVIS LANGDON & SEAH INTERNATIONAL

## Approximate Estimating

### OUTLINE COSTS

**ELEMENTAL RATES FOR BASIC SERVICES BY BUILDING TYPE**

There follows typical examples of how rates per square metre for the basic engineering services of a project may be allocated to individual elements.

|  | Factories Speculative £ | Prestige £ |
|---|---|---|
| Cold water services | 2.66 | 2.66 |
| Hot water services | 1.97 | 1.97 |
| Heating and ventilation | 20.36 | 45.24 |
| Fire protection | 1.97 | 1.97 |
| Mains and switchgear | 9.49 | 9.72 |
| Power | 4.98 | 6.26 |
| Lighting | 8.22 | 9.37 |
| Luminaires | 10.00 | 12.15 |
| Fire protection | 2.55 | 4.40 |
| Communications | 0.00 | 1.28 |
| Total approximate cost of above services | 62.20 | 95.02 |

|  | Speculative £ | Offices Owner occupied £ | Prestige £ |
|---|---|---|---|
| Cold water services | 6.83 | 7.76 | 8.56 |
| Hot water services | 5.21 | 6.13 | 6.83 |
| Heating and ventilation | 60.64 | 0.00 | 0.00 |
| Air conditioning | 0.00 | 199.51 | 229.94 |
| Fire protection | 5.31 | 5.31 | 5.31 |
| Mains and switchgear | 13.89 | 17.24 | 20.72 |
| Power | 27.77 | 31.59 | 33.21 |
| Lighting | 8.44 | 9.95 | 11.34 |
| Luminaires | 22.56 | 27.66 | 36.80 |
| Emergency lighting | 4.51 | 6.36 | 7.99 |
| Telephone wireways | 0.81 | 1.05 | 1.15 |
| Clock | 0.00 | 0.00 | 0.46 |
| Public address and television | 0.00 | 0.00 | 1.73 |
| Fire alarms | 6.49 | 6.49 | 7.76 |
| Security | 0.00 | 3.81 | 6.24 |
| Lightning protection | 0.81 | 0.81 | 0.81 |
| External lighting | 0.23 | 1.05 | 1.73 |
| Total approximate cost of above services | 163.50 | 324.72 | 380.58 |

## Approximate Estimating

### OUTLINE COSTS

**ELEMENTAL RATES FOR BASIC SERVICES BY BUILDING TYPE** *continued*

|  | Non-teaching hospitals £ | Secondary schools £ | University arts & administration buildings £ | Local authority houses £ | Local authority flats £ |
|---|---|---|---|---|---|
| Cold water services | 28.00 | 12.72 | 20.72 | 7.76 | 10.30 |
| Hot water services | 20.83 | 11.93 | 13.77 | 7.06 | 9.02 |
| Heating | 61.10 | 67.23 | 63.77 | 29.28 | 40.03 |
| Ventilation | 55.08 | 12.38 | 0.00 | 0.00 | 2.78 |
| Fire protection | 2.78 | 2.20 | 2.78 | 0.00 | 0.00 |
| Ancillary services | 38.53 | 4.51 | 0.00 | 2.31 | 2.31 |
| Mains and switchgear | 22.22 | 8.10 | 12.15 | 2.55 | 3.93 |
| Power | 28.58 | 12.15 | 16.90 | 6.83 | 7.99 |
| Lighting | 19.44 | 11.34 | 12.85 | 7.76 | 9.02 |
| Luminaires | 26.50 | 15.15 | 18.86 | 3.24 | 4.06 |
| Fire protection | 11.34 | 7.86 | 7.64 | 0.00 | 0.81 |
| Communications | 23.14 | 5.21 | 4.86 | 1.38 | 1.73 |
| Ancillary services | 2.78 | 0.00 | 8.00 | 0.00 | 0.00 |
| Total approximate cost of above services | 340.32 | 170.78 | 182.30 | 68.17 | 91.98 |

### Notes

The above prices do not include any public utility charges that may be made.

Prices for local authority flats and houses are based upon tenders for large schemes and on dwellings accommodating 4 persons.

Cold and hot water services exclude the cost of sanitary fittings.

Heating in the case of non-teaching hospitals excludes the heat source. The price for heating to Local Authority flats assumes individual gas fired boilers serving panel radiators.

Fire protection includes extinguishers, hose reels, wet and dry risers and foam inlets. Where required sprinklers would cost an additional £11.71 to £20.38 per m$^2$ of area protected.

Ancillary services includes, where applicable, natural gas, compressed air, treated water, refrigeration, laboratory services, special piped services, steam, and condense and medical vacuum installations. The prices do not allow for such items as kitchen and laundry equipment, hoists and balers.

In the case of non-teaching hospitals, mains and switchgear includes stand-by facilities.

Fire protection includes alarms and detectors.

Communications, includes where applicable, burglar and security alarms, clocks, telephone conduit, television and radio aerials and outlets, public address and call systems.

Ancillary services include common services trunking and cable trays.

## Approximate Estimating

### ELEMENTAL AND ALL-IN-RATES

### ELEMENTAL RATES FOR ALTERNATIVE MECHANICAL SERVICES SOLUTIONS FOR OFFICES

*The undernoted examples, when considered within the context of office accommodation, indicate the range of rates for alternative design solutions for each of the basic mechanical services elements.*

|  | Square metre £ |
|---|---|
| **Sanitary and disposal installation** | |
| Normal services for low rise building (up to 1,000 m²) | 3 to 11 |
| Normal services for low rise building (over 1,000 m²) | 8 to 11 |
| Normal services for medium rise building (up to 1,000 m²) | 8 to 13 |
| Normal services for medium rise building (over 1,000 m²) | 11 to 17 |
| Normal services for high rise building (over 1,000 m²) | 11 to 17 |
| **Hot and cold water installation** | |
| Services to speculative offices (up to 1,000 m²) | 4 to 13 |
| Services to speculative offices (medium to high rise) | 7 to 17 |
| Services to owner occupied offices (up to 1,000 m²) | 7 to 15 |
| Services to owner occupied offices (medium to high rise) | 11 to 19 |
| **Heating installation** | |
| LPHW radiator system to speculative offices (up to 1,000 m²) | 40 to 53 |
| LPHW radiator system to speculative offices (over 1,000 m²) | 44 to 62 |
| LPHW radiator system to owner occupied offices (up to 1,000 m²) | 44 to 62 |
| LPHW radiator system to owner occupied offices (over 1,000 m²) | 49 to 66 |
| LPHW convector system to speculative offices (up to 1,000 m²) | 36 to 56 |
| LPHW convector system to speculative offices (over 1,000 m²) | 49 to 66 |
| LPHW convector system to owner occupied (up to 1,000 m²) | 50 to 62 |
| LPHW convector system to owner occupied (over 1,000 m²) | 54 to 73 |
| Ventilated heating system | 54 to 71 |
| **Ventilation and air conditioning** | |
| Ventilation to internal WC's only | 2 to 10 |
| Warm air heating and ventilation to offices | 98 to 128 |
| Comfort cooling, fan coil, speculative offices | 133 to 181 |
| Comfort cooling, fan coil, owner occupied offices | 141 to 198 |
| Comfort cooling, variable air volume, speculative offices | 150 to 227 |
| Comfort cooling, variable air volume, owner occupied offices | 158 to 249 |
| Full air conditioning, fan coil, speculative offices | 150 to 212 |
| Full air conditioning, fan coil, owner occupied offices | 157 to 239 |
| Full air conditioning, variable air volume, speculative offices | 172 to 248 |
| Full air conditioning, variable air volume, owner occupied offices | 174 to 265 |
| **Special services** | |
| Firefighting installation - hose reels and dry risers | 3 to 11 |
| Single level sprinkler installation | 11 to 19 |
| Double level sprinkler installation | 21 to 26 |
| **External services** | |
| Incoming mains services (gas, water) | 2 to 10 |

## Approximate Estimating

## ELEMENTAL AND ALL-IN-RATES

## ELEMENTAL RATES FOR AIR CONDITIONING FOR OFFICES

Approximate costs of air conditioning installations for two specimen office blocks with floor areas of 6,000 m² and 15,000 m² in four and eight storeys respectively. The types of installation selected are the variable air volume and the induction system.
The costs allow for all plant and equipment, distribution ductwork, pipework for heating, chilled and cooling water, automatic controls, fire protection systems and all associated electrical work.

|  | Floor area | |
|---|---|---|
|  | 6,000 m² | 15,000 m² |
|  | £ | £ |
| Variable air volume system, per m² | 225.00 | 181.00 |
| Induction system, per m² | 199.00 | 152.00 |

Air conditioning installations vary considerably according to the type of plant selected, the ancillary services chosen, the methods of heating and cooling, the sophistication of automatic controls, the requirements for fire protection, the type of fuel available for heating and many other considerations. No two buildings will have precisely the same requirements.
The following information has been compiled to indicate the average cost of a number of different design solutions. Brief specification notes are provided to enable the user to make his own adjustments to the cost of individual elements to take into account his own design criteria. It has been assumed that the building is double glazed, has better than average insulation and a window to wall ratio not exceeding 50%.

## Approximate Estimating

### ELEMENTAL AND ALL-IN-RATES

**ELEMENTAL RATES FOR AIR CONDITIONING FOR OFFICES** *continued*

**VARIABLE AIR VOLUME SYSTEM**

| Elements | Office block of 6,000 m² Cost of element £ | Office block of 6,000 m² Cost of element per m² of floor area £ | Office block of 15,000 m² Cost of element £ | Office block of 15,000 m² Cost of element per m² of floor area £ | Spec'n Ref. |
|---|---|---|---|---|---|
| Boilers | | | | | |
|   Plant and instruments | 33,660.00 | 5.61 | 58,640.00 | 3.91 | A |
|   Flue | 11,350.00 | 1.89 | 17,590.00 | 1.17 | B |
| Water treatment | 11,350.00 | 1.89 | 22,350.00 | 1.49 | C |
| Gas installation | 5,200.00 | 0.87 | 5,230.00 | 0.35 | D |
| Space heating | | | | | |
|   Distribution pipework | 32,780.00 | 5.46 | 65,930.00 | 4.40 | E |
|   Convectors and/or radiators | 6,400.00 | 1.07 | 15,690.00 | 1.05 | F |
| Heating to batteries | 64,540.00 | 10.76 | 116,490.00 | 7.77 | H |
| Chilled water to batteries | 48,810.00 | 8.14 | 89,390.00 | 5.96 | J |
| Condenser cooling water | | | | | |
|   Distribution pipework | 22,950.00 | 3.83 | 39,620.00 | 2.64 | K |
| Cooling plant | | | | | |
|   Chillers | 116,270.00 | 19.38 | 196,210.00 | 13.08 | L |
|   Cooling towers | 13,000.00 | 2.17 | 19,810.00 | 1.32 | M |
| Automatic controls | 101,560.00 | 16.93 | 159,920.00 | 10.66 | N |
| Ductwork | | | | | |
|   Supply | 493,660.00 | 82.28 | 1,124,170.00 | 74.94 | P |
|   Extract | 104,290.00 | 17.38 | 235,830.00 | 15.72 | P |
| Air conditioning plant | | | | | |
|   Heating batteries | 33,660.00 | 5.61 | 59,120.00 | 3.94 | Q |
|   Humidifiers & cooling batteries | 59,910.00 | 9.99 | 103,650.00 | 6.91 | Q |
|   Fans and filters | 91,920.00 | 15.32 | 157,060.00 | 10.47 | R |
|   Sound attenuation | 44,630.00 | 7.44 | 109,360.00 | 7.29 | S |
| Fire protection | 26,060.00 | 4.34 | 47,390.00 | 3.16 | T |
| Electrical work in connection | 30,240.00 | 5.04 | 57,370.00 | 3.82 | V |
| Totals | 1,352,240.00 | 225.40 | 2,700,820.00 | 180.05 | |
| | say | 225.00 | say | 180.00 | |

## Approximate Estimating

### ELEMENTAL AND ALL-IN-RATES

**ELEMENTAL RATES FOR AIR CONDITIONING FOR OFFICES** *continued*

**INDUCTION SYSTEM**

| Elements | Office block of 6,000 m² | | Office block of 15,000 m² | | Spec'n Ref. |
|---|---|---|---|---|---|
| | Cost of element £ | Cost of element per m² of floor area £ | Cost of element £ | Cost of element per m² of floor area £ | |
| Boilers | | | | | |
|   Plant and instruments | 33,660.00 | 5.61 | 58,640.00 | 3.91 | A |
|   Flue | 11,350.00 | 1.89 | 17,590.00 | 1.17 | B |
| Water treatment | 11,350.00 | 1.89 | 22,350.00 | 1.49 | C |
| Gas installation | 5,200.00 | 0.87 | 5,230.00 | 0.35 | D |
| Space heating | | | | | |
|   Distribution pipework | 32,780.00 | 5.46 | 65,930.00 | 4.40 | E |
|   Convectors and/or radiators | 6,400.00 | 1.07 | 15,690.00 | 1.05 | F |
|   Induction units | 120,900.00 | 20.15 | 323,640.00 | 21.58 | G |
| Heating to batteries | 133,320.00 | 22.22 | 221,570.00 | 14.77 | H |
| Chilled water | | | | | |
|   Distribution pipework | 171,360.00 | 28.56 | 317,140.00 | 21.14 | J |
| Condenser cooling water | | | | | |
|   Distribution pipework | 22,890.00 | 3.82 | 38,040.00 | 2.54 | K |
| Cooling plant | | | | | |
|   Chillers | 90,280.00 | 15.05 | 157,220.00 | 10.48 | L |
|   Cooling towers | 11,790.00 | 1.97 | 17,750.00 | 1.18 | M |
| Automatic controls | 112,080.00 | 18.68 | 200,970.00 | 13.40 | N |
| Ductwork | | | | | |
|   Supply | 144,800.00 | 24.13 | 261,190.00 | 17.41 | P |
|   Extract | 72,840.00 | 12.14 | 155,800.00 | 10.39 | P |
| Air conditioning plant | | | | | |
|   Heating batteries | 25,300.00 | 4.22 | 46,120.00 | 3.07 | Q |
|   Humidifiers & cooling batteries | 43,810.00 | 7.30 | 67,360.00 | 4.49 | Q |
|   Fans and filters | 62,640.00 | 10.44 | 114,590.00 | 7.64 | R |
|   Sound attenuation | 33,660.00 | 5.61 | 87,010.00 | 5.80 | S |
| Fire protection | 23,010.00 | 3.84 | 41,050.00 | 2.74 | T |
| Electrical work in connection | 29,920.00 | 4.99 | 50,720.00 | 3.38 | V |
| Totals | 1,199,340.00 | 199.91 | 2,285,600.00 | 152.38 | |
| | say | 200.00 | say | 152.00 | |

## ELEMENTAL AND ALL-IN-RATES

## ELEMENTAL RATES FOR AIR CONDITIONING FOR OFFICES *continued*

**Brief Specification Notes**

*Ref.*

**Boilers**
A    Plant and instruments: Three gas-fired boilers each of approximately 250 and 580 kW capacity for the two buildings respectively; together with burners, pumps, direct-mounted instruments, feed and expansion tanks. Normal standby facilities are included.
B    Flue: Mild steel insulated in boiler house, internal lining to vertical builders' stack.
C    **Water treatment**: Chemical dosage equipment.
D    **Gas installation**: Pipework internal to building, meter, solenoid valves.
      **Space heating**
E    Distribution pipework: pipework from boilers to terminal equipment, all valves, fittings and supports, insulation.
F    Convector and/or radiators: Panel radiators, or natural convectors in circulation areas and staircases.
G    **Induction units**: High velocity units suitable for four-pipe system utilizing ducted fresh air.
H    **Heating to air heater batteries**: Distribution pipework to batteries, valves, fittings and supports, insulation.
J    **Chilled water to batteries and induction units**: Distribution pipework, valves, fittings and supports, insulation.
K    **Condenser cooling water:** Distribution pipework, valves, fittings and supports, insulation.
      **Cooling plant**
L    Chillers: Centrifugal chiller units of approximately 190 and 470 tons total capacity for the two buildings respectively, including mountings and supports, insulation and pumps. Normal standby facilities are included.
M    Cooling towers: Forced or induced draught fans, roof-mounted cooling towers with supports.
N    **Automatic controls**: Pneumatic controls including motorised valves, all thermostats, control panels, actuators, interconnecting wiring and tubing.
      **Ductwork**
P    Supply and extract: Galvanised mild steel ductwork, fittings and supports, terminal units (for V.A.V system), dampers, grilles and diffusers, insulation.
      **Air conditioning plant**
Q    Heating and cooling batteries: Humidifiers, batteries and casing and connections.
R    Fans and filters: Centrifugal and axial flow fans with casings and connections, and automatic roll type filters.
S    Sound attenuation: Silencers and duct lining (short lengths only).
T    **Fire protection**: Heat detectors, smoke detectors, gas detectors, control panel, interconnecting wiring (excluding other fire protection services not directly associated with air conditioning installation).
V    **Electrical work in connection**: Electrical supplies to control panels and mechanical plant, mechanical services distribution board.

## Approximate Estimating
## ELEMENTAL AND ALL-IN-RATES

### AIR CONDITIONING DESIGN LOADS

Recommended air conditioning loads for various applications:
- Computer rooms ........................................ 10 m² of floor area per ton
- Restaurant ............................................. 16 m² of floor area per ton
- Banks (main area) ..................................... 22 m² of floor area per ton
- Large Office Buildings (exterior zone) ................ 25 m² of floor area per ton
- Supermarkets .......................................... 30 m² of floor area per ton
- Large Office Block (interior zone) .................... 32 m² of floor area per ton
- Small Office Block (interior zone) .................... 35 m² of floor area per ton

### ALL IN RATES FOR PRICING MECHANICAL APPROXIMATE QUANTITIES

### COLD WATER

Light gauge copper tube to B.S. 2871 part 1 table X with joints as described including allowance for waste, fittings and brackets assuming average runs.

*Cost per metre*
£

Capillary Joints
- 15mm ............................................................18.03
- 22mm ............................................................16.65
- 28mm ............................................................20.36
- 35mm ............................................................23.46
- 42mm ............................................................33.94
- 54mm ............................................................35.79

Compression Joints
- 15mm ............................................................17.02
- 22mm ............................................................16.05
- 28mm ............................................................20.36
- 35mm ............................................................26.11
- 42mm ............................................................37.33
- 54mm ............................................................41.34

Bronze Welded Joints
- 76mm ...........................................................124.63

### HOT WATER

Light gauge copper tube to B.S. 2871 part 1 table X with joints as described including allowance for waste, fittings and brackets assuming average runs.

*Cost per metre*
£

Capillary Joints
- 15mm ............................................................18.03
- 22mm ............................................................13.40
- 28mm ............................................................17.89
- 35mm ............................................................20.23
- 42mm ............................................................29.63
- 54mm ............................................................33.94

Compression Joints
- 15mm ............................................................16.79
- 22mm ............................................................12.53
- 28mm ............................................................16.79
- 35mm ............................................................24.24
- 42mm ............................................................35.72
- 54mm ............................................................39.61

Bronze Welded Joints
- 76mm ...........................................................113.65

*Approximate Estimating*

## ELEMENTAL AND ALL-IN-RATES

## ALL IN RATES FOR PRICING MECHANICAL APPROXIMATE QUANTITIES *continued*

### DISTRIBUTION PIPEWORK HEATING OUTSIDE PLANT ROOM

Mild steel tube to B.S. 1387 with joints in the running length allowance for waste, fittings and brackets assuming average runs.

|  | Cost per metre Screwed £ | Welded £ |
|---|---:|---:|
| **Black Medium Weight** | | |
| 15mm | 37.83 | 28.56 |
| 20mm | 16.53 | 19.56 |
| 25mm | 18.38 | 21.34 |
| 32mm | 18.08 | 21.34 |
| 40mm | 18.94 | 21.84 |
| 50mm | 42.82 | 51.02 |
| 65mm | 50.66 | 61.02 |
| 80mm | 62.50 | 75.15 |
| 100mm | 51.96 | 65.89 |
| 125mm | 79.33 | 90.81 |
| 150mm | 106.55 | 144.92 |
| **Black Heavy Weight** | | |
| 15mm | 39.74 | 40.28 |
| 20mm | 17.94 | 20.43 |
| 25mm | 19.74 | 23.82 |
| 32mm | 19.62 | 24.01 |
| 40mm | 20.45 | 24.64 |
| 50mm | 46.03 | 54.41 |
| 65mm | 51.27 | 69.10 |
| 80mm | 65.58 | 80.69 |
| 100mm | 70.21 | 64.22 |
| 125mm | 71.87 | 93.16 |
| 150mm | 113.34 | 121.55 |

## Approximate Estimating

### ELEMENTAL AND ALL-IN-RATES

### ALL IN RATES FOR PRICING MECHANICAL APPROXIMATE QUANTITIES *continued*

#### TANKS, CALORIFIERS, PUMPS AND VALVES

These have not been detailed in this section, for sizes and types see 'Prices for Measured Work' section.

#### HEAT SOURCE

#### BOILERS

Low pressure hot water boiler plant of welded construction comprising 2 No. boilers with instruments and boiler mountings and burners for gas combustion including delivery and commissioning, 2 No. heating circulating pumps, boiler control panel and thermostatic controls for heating circuit, automatic controls, pipework, fittings and valves with all necessary insulation, feed and expansion tank, cold feed, vent pipes and flue.

|  | Total cost of plant £ |
|---|---|
| Total boiler capacity 275 kW | 50,560.00 - 67,040.00 |
| Total boiler capacity 1100 kW | 69,240.00 - 92,320.00 |
| Total boiler capacity 2250 kW | 94,510.00 - 119,790.00 |

Low pressure hot water boiler plant of welded construction comprising 2 No. boilers with instruments and boiler mounting and burners for gas combustion including delivery and commissioning, 2 No. heating circulating pumps, 2 No. hot water service circulation pumps, 2 No. calorifiers, boiler control panel and thermostatic controls for heating and hot water service circuits, automatic controls, all interconnecting pipework, fittings and valves, with all necessary insulation, feed and expansion tank, cold feed and vent pipes and flue.

|  | Total cost of plant £ |
|---|---|
| Total boiler capacity 275 kW | 56,050.00 - 75,820.00 |
| Total boiler capacity 1100 kW | 79,130.00 - 103,300.00 |
| Total boiler capacity 2250 kW | 101,110.00 - 127,480.00 |

## Approximate Estimating

### ELEMENTAL AND ALL-IN-RATES

## ALL IN RATES FOR PRICING MECHANICAL APPROXIMATE QUANTITIES *continued*

**BOILERS** *continued*

Low pressure hot water boiler plant of welded construction comprising 2 No. boilers with instruments and boiler mountings and burners for 35 second oil including delivery and commissioning, 2 No. heating circulating pumps, 2 No. oil storage tanks, boiler control panel and thermostatic controls for heating and hot water service circuits, automatic controls, all interconnecting pipework, fittings and valves with all necessary insulation, feed and expansion tank, cold feed and vent pipes and flue.

|  | Total cost of plant £ |
|---|---|
| Total boiler capacity 275 kW | 53,850.00 - 72,540.00 |
| Total boiler capacity 1100 kW | 72,540.00 - 103,300.00 |
| Total boiler capacity 2250 kW | 101,110.00 - 131,870.00 |

Low pressure hot water boiler plant of welded construction comprising 2 No. boilers with instruments and boiler mountings and burners for 35 second oil including delivery and commissioning, 2 No. heating circulation pumps, 2 No. hot water service circulating pumps, 2 No. calorifiers, 2 No. oil storage tanks, boiler control panel and thermostatic controls for heating and hot water service circuits, automatic controls, all interconnecting pipework, fittings and valves to feed and expansion tank, cold feed and vent pipes and flue.

|  | Total cost of plant £ |
|---|---|
| Total boiler capacity 275 kW | 61,550.00 - 85,720.00 |
| Total boiler capacity 1100 kW | 84,620.00 - 112,090.00 |
| Total boiler capacity 2250 kW | 108,800.00 - 140,680.00 |

## Approximate Estimating

## ELEMENTAL AND ALL-IN-RATES

## ALL IN RATES FOR PRICING MECHANICAL APPROXIMATE QUANTITIES *continued*

### SPACE HEATING AND AIR TREATMENT

### DOMESTIC CENTRAL HEATING

Solid fuel central heating installation comprising either open fire room heater or boiler, fuel storage, pump, small or microbore distribution pipework, steel radiators, valves, expansion tank, room thermostat, and all insulation:

|  | £ |
|---|---|
| Heating for 3 rooms comprising 2 radiators plus hot water service | 1,760.00 - 2,030.00 |
| Heating for 4 rooms comprising 3 radiators plus hot water service | 2,000.00 - 2,410.00 |
| Heating for 5 rooms comprising 4 radiators plus hot water service | 2,410.00 - 2,720.00 |
| Heating for 6 rooms comprising 5 radiators plus hot water service | 2,600.00 - 2,840.00 |
| Heating for 7 rooms comprising 6 radiators plus hot water service | 2,840.00 - 3,130.00 |
| Heating for 7 rooms comprising 7 radiators plus hot water service | 3,060.00 - 3,510.00 |

Gas fired central heating installation comprising boiler, pump, small or microbore distribution pipework, steel radiators, valves, expansion tank, room thermostat and all insulation:

|  | £ |
|---|---|
| Heating for 3 rooms comprising 3 radiators plus hot water service | 1,980.00 - 2,630.00 |
| Heating for 4 rooms comprising 4 radiators plus hot water service | 2,590.00 - 2,870.00 |
| Heating for 5 rooms comprising 5 radiators plus hot water service | 2,870.00 - 3,050.00 |
| Heating for 6 rooms comprising 6 radiators plus hot water service | 3,050.00 - 3,230.00 |
| Heating for 7 rooms comprising 7 radiators plus hot water service | 3,230.00 - 3,930.00 |

Oil fired central heating installation comprising oil storage tank and supports, pump, room thermostats, small or microbore distribution pipework, steel radiators, valves and insulation together with hot water cylinder for hot water supply:

|  | £ |
|---|---|
| Heating for 7 rooms comprising 7 radiators and boiler plus hot water service | 3,480.00 - 4,160.00 |
| Heating for 10 rooms comprising 10 radiators and boiler plus hot water service | 3,810.00 - 4,380.00 |

*Approximate Estimating*

## ELEMENTAL AND ALL-IN-RATES

## ALL IN RATES FOR PRICING MECHANICAL APPROXIMATE QUANTITIES *continued*

### RADIATORS

| Pressed steel panel type radiator prime finish including brackets and fixing, taking down once for painting by others (cost per m² of heating surface) | | 305mm high £ | 432mm high £ | 584mm high £ | 740mm high £ |
|---|---|---|---|---|---|
| Single panel | m² | 67.37 | 48.47 | 49.34 | 48.79 |
| Double panel | m² | 50.00 | 37.58 | 37.03 | 67.92 |

For floor or wall mounted individual convector and unit heaters see 'Materials Costs/Measured Work Prices' section.

### CHILLED AND COOLING WATER

Chilled water installation comprising two refrigeration compressors, cooling towers, pumps, pipework valves and fittings, controls, thermostatic controls, insulation, anti-vibration mountings and starters

Cost per ton
of cooling
capacity
£
1,220.00 - 2,430.00

### DUCTWORK

To calculate the weight of ductwork multiply the duct length (measured on the centre line overall fittings) by their respective girths and then apply the appropriate sheeting weights to the superficial areas without making any allowance for the additional sheeting in joints, seams, welts, waste, etc. The rates below allow for ductwork and for all other labour and material in fabrication fittings, supports and jointing to equipment, stop and capped ends, elbows, bends, diminishing and transition pieces, regular and reducing couplings, branch diffuser and 'snap on' grille connections, ties, 'Ys', crossover spigots, etc., turning vanes, regulating dampers, access doors and openings, handholes, test holes and covers, blanking plates, flanges, stiffeners, tie rods and all supports and brackets fixed to concrete or brickwork.

Per tonne
£

Rectangular low velocity galvanised mild steel ductwork in accordance with table 5 of the HVCA DW 142 (metric).     8,790.00

Rectangular high velocity galvanised mild steel ductwork in accordance with tables 7 and 8 of the HVCA DW 142 (metric)     10,330.00

### PACKAGE AIR HANDLING UNITS, FANS AND ROOF EXTRACT UNITS

These have not been detailed in this section, for sizes and types see 'Material Costs/Measured Work Prices' section.

## Approximate Estimating

### ELEMENTAL AND ALL-IN-RATES

## ALL IN RATES FOR PRICING MECHANICAL APPROXIMATE QUANTITIES *continued*

### PROTECTIVE INSTALLATIONS

### SPRINKLER INSTALLATION

Sprinkler installation including sprinkler head and all associated pipework, valve sets, booster pumps and water storage:

|  | £ |
|---|---|
| Price per sprinkler head | 149.00 |

Recommended maximum area coverage per sprinkler head:
Extra light hazard, 21 m² of floor area
Ordinary hazard, 12 m² of floor area
Extra high hazard, 9 m² of floor area

### HOSE REELS, WET AND DRY RISERS

100mm dry riser main including dry riser breeches horizontal inlet with 2 No. 64mm male instantaneous coupling and landing valve. Complete with padlock and leather strap.

|  | £ |
|---|---|
| Price per landing | 1,087.00 |

Hose reel pedestal mounted type with 37 metres of hose including approximately 15 metres of pipework:

| Price per hose reel | 627.00 |
|---|---|

### THERMAL INSULATION

Foil-faced pre-formed mineral fibre glass fibre rigid sections, 25mm thick, fixed with adhesive and aluminium bands at 450mm intervals all joints and ends vapour sealed, including all pipe fittings, flanges and valves.

| Nominal size | Cost per metre £ |
|---|---|
| 15mm | 8.35 |
| 20mm | 8.90 |
| 25mm | 9.45 |
| 32mm | 10.11 |
| 40mm | 10.33 |
| 50mm | 14.95 |
| 65mm | 18.46 |
| 80mm | 23.19 |
| 100mm | 25.94 |
| 125mm | 30.22 |
| 150mm | 35.93 |

*Approximate Estimating*

## ELEMENTAL AND ALL-IN-RATES

## ALL IN RATES FOR PRICING MECHANICAL APPROXIMATE QUANTITIES *continued*

**THERMAL INSULATION** *continued*

Canvas-faced pre-formed mineral fibre glass fibre rigid sections, fixed with aluminium bands at 450mm intervals with 0.5mm thick aluminium sheet secured with rivets or self tapping screws, including all pipe fittings, flanges and valves.

| Nominal size | Cost per metre £ |
| --- | --- |
| 15mm | 22.30 |
| 20mm | 23.40 |
| 25mm | 24.72 |
| 32mm | 25.94 |
| 40mm | 26.82 |
| 50mm | 32.53 |
| 65mm | 36.93 |
| 80mm | 45.82 |
| 100mm | 50.55 |
| 125mm | 56.60 |
| 150mm | 64.73 |

## Approximate Estimating

## ELEMENTAL AND ALL-IN-RATES

## ALL IN RATES FOR PRICING ELECTRICAL APPROXIMATE QUANTITIES

## MAINS AND SWITCHGEAR

### Indoor substation equipment.

The cost of an indoor substation is governed by various factors e.g. the duty it is required to perform, local electricity authority requirements, earthing requirements dependent on the soil resistivity of the ground in the vicinity of the substation and the actual location of the site etc.

An accurate cost of a proposed substation can be estimated only when full design details and all relevant information is available. However for budgetary purposes cost figures given below in respect of typical equipment utilized in a modern substation may serve as a helpful guide.

### Cubicle pattern 11kV switchboards.

The cost of cubicle pattern switchgear will depend on the type of circuit breakers employed and the level of protection and metering required. As an approximate guide, the cost per section might be of the order of the following:

|  | Cost per Section £ |
|---|---|
| Utilizing oil circuit breakers | 13,080.00 |
| Utilizing vacuum circuit breakers | 14,300.00 |

### Cubicle pattern LV switchboards.

The cost of cubicle pattern LV switchboards will depend on quality, short- circuit rating, level of metering and protection required, and the number and ratings of incoming and outgoing circuit breakers and fuse switches.

In the absence of such information an approximate cost can be determined on the basis of installed capacity. For general applications this will range between £19.78 and £25.94 per kVA.

### 2MVA - 2 Transformer sub-station

The total approximate cost of a sub-station can therefore be determined as follows:

|  | £ |
|---|---|
| 5 Panel HV switchboard using oil circuit breakers | 66,340.00 |
| 2 x 1000 kVA ON AN type transformers at £13,700.00each | 27,400.00 |
| LV switchboard - 2000 kVA at say £24.72 per kVA | 49,440.00 |
| HV and LV cabling, say | 3,740.00 |
| Earthing, say | 3,190.00 |
| Total approximate cost of 2 MVA sub-station | £150,110.00 |

## ELEMENTAL AND ALL-IN-RATES

## ALL IN RATES FOR PRICING ELECTRICAL APPROXIMATE QUANTITIES *continued*

### MAINS AND SWITCHGEAR *continued*

#### Rising main busbar trunking

Four-pole mild steel sheet rising main busbar trunking with copper busbars, complete with tap-off points, fire barriers at 3.5 m intervals, one thrust block, one expansion joint assembly and one cable entry chamber with glands for incoming cables and including tap-off cable clamps, interconnections, internal wiring, shields, phase buttons, labels and external earth tape; cost of installations based on a 60 m length of rising main.

|   | Per Metre £ |
|---|---|
| 200 amp rising main | 210.00 |
| 300 amp rising main | 230.00 |
| 400 amp rising main | 248.00 |
| 600 amp rising main | 316.00 |

#### Overhead plug-in busbar trunking

Four-pole overhead plug-in busbar trunking with copper busbars enclosed within extended aluminium casing. Cost of installation based on network system of 300 metres running length including various fittings, e.g. feed units, bends, tees, etc.

|   | Per Metre £ | PC each £ |
|---|---|---|
| 100 amp plug-in overhead busbar system fixed and connected | 50.91 |  |
| 20 amp T.P. & N. plug-in unit |  | 17.65 |
| 30 amp T.P. & N. plug-in unit |  | 45.01 |
| 60 amp T.P. & N. plug-in unit |  | 73.65 |

|   | Per Metre £ | PC each £ |
|---|---|---|
| 300 amp plug-in overhead busbar system fixed and connected | 83.98 |  |
| 30 amp T.P. & N. plug-in unit |  | 114.32 |
| 60 amp T.P. & N. plug-in unit |  | 137.36 |
| 100 amp T.P. & N. plug-in unit |  | 150.04 |

**Note** P.C. values of plug-in units are inclusive of HRC fuse links.

### STANDBY GENERATING SETS

#### Diesel

For ratings up to 1000 kVA see 'Prices For Measured Work' section.

For ratings over 1000 kVA, approximate installed cost .......................... £125.00 per kVA

#### Gas Turbine

Approximate installed cost
50 - 1000 kVA .......................... £249.00 per kVA
Over 1000 kVA .......................... £199.00 per kVA

## Approximate Estimating
## ELEMENTAL AND ALL-IN-RATES

## ALL IN RATES FOR PRICING ELECTRICAL APPROXIMATE QUANTITIES continued

### POWER
Approximate prices for wiring of power points complete, including socket outlets with plugs but excluding consumer control units.

|  | Per Point £ |
|---|---|
| **13 amp Socket outlets** | |
| Wired in PVC insulated and PVC sheathed cable in flats and houses on a ring main circuit protected, where buried, by heavy gauge PVC conduit | 50.08 |
| As above but in commercial property | 61.49 |
| Wired in PVC insulated cable in screwed welded conduit on a ring main circuit in flats and houses | 71.00 |
| As above but in commercial property | 82.84 |
| As above but in industrial property | 95.20 |
| Wired in M.I.C.C. cable on a ring main circuit in flats and houses | 72.06 |
| As above but in commercial property | 84.42 |
| As above but in industrial property (PVC sheathed cable) | 107.56 |

|  | Each £ |
|---|---|
| **Cooker and control units** | |
| 45 amp Circuit including control unit, wired in PVC insulated and PVC sheathed cable, protected where buried by heavy gauge PVC conduit | 97.84 |
| As above but wired in PVC insulated cable in screwed welded conduit | 142.85 |
| As above but wired in M.I.C.C. cable | 160.50 |

### Low voltage power circuits
Three phase four wire circuit feeding an individual load, wired with PVC insulated cable drawn in heavy gauge black enamelled screwed welded conduit including standard associated fittings and flexible PVC sheathed metallic conduit, not exceeding one metre long, fixed to brickwork or concrete with distance saddles in surface work, *per 10 metre run.*

| Cable size $mm^2$ | Conduit size mm | £ |
|---|---|---|
| 1.5 | 20 | 147.29 |
| 2.5 | 20 | 150.57 |
| 4 | 20 | 160.50 |
| 6 | 25 | 196.74 |
| 10 | 25 | 223.15 |
| 16 | 32 | 252.84 |

Three phase four wire circuit feeding an individual load item, wired with four core M.I.C.C./PVC cable including terminations, with glands and cable loop for vibration-free connection, fixed with PVC covered clips, *per 10 metre run.*

| Cable size $mm^2$ | £ |
|---|---|
| 1.5 | 139.58 |
| 2.5 | 172.54 |
| 4 | 195.68 |
| 6 | 223.15 |
| 10 | 268.17 |
| 16 | 353.86 |

*Approximate Estimating*

## ELEMENTAL AND ALL-IN-RATES

## ALL IN RATES FOR PRICING ELECTRICAL APPROXIMATE QUANTITIES *continued*

**POWER** *continued*

£

**Power and communications common underfloor ducting**
225 x 25 mm three compartment underfloor steel
cable duct including fittings, laid in screed,
excluding the builder's work ............................................. per metre   31.91

**Power and communications common flush floor trunking**
395 x 60 mm three compartment flush floor trunking,
including associated fittings, fixed to floor ................................ per metre   62.66

**Flush floor outlet boxes**
Three compartment flush floor outlet box complete
with twin socket outlet, twin telephone outlet
plate, blank plate for data section, power circuit
wiring only.
Fixed to underfloor ducting ............................................. per box   112.11

Ditto but fixed to flush floor trunking .................................... per box   137.36

Ditto but fixed to cavity floor panel and complete
with flexible metallic conduit, not exceeding two
metres long ......................................................... per box   149.51


**Wiring and connections only, to mechanical services.**
The cost of electrical connections to mechanical services equipment will vary depending on the type of building and complexity of the equipment but and allowance of £6.59 per m² of total floor area should be a useful guide to allow for power and control wiring in the conduit system including local isolators and remote push button stations.

## Approximate Estimating

## ELEMENTAL AND ALL-IN-RATES

## ALL IN RATES FOR PRICING ELECTRICAL APPROXIMATE QUANTITIES *continued*

## LIGHTING

Approximate prices for wiring of lighting points complete, including accessories but excluding lighting fittings.

|  | Per Point £ |
|---|---|
| **Final circuits** | |
| Wired in PVC insulated and PVC sheathed cable in flats and houses installed in cavities and roof space protected, where buried, by heavy gauge PVC conduit | 45.12 |
| As above but in commercial property | 54.62 |
| Wired in PVC insulated cable in screwed welded conduit in flats and houses | 94.04 |
| As above but in commercial property | 116.12 |
| As above but in industrial property | 132.39 |
| Wired in M.I.C.C. cable in flats and houses | 79.45 |
| As above but in commercial property | 95.20 |
| As above but in industrial property (PVC sheathed cable) | 109.89 |

*Approximate Estimating*

## ELEMENTAL AND ALL-IN-RATES

**ALL IN RATES FOR PRICING ELECTRICAL APPROXIMATE QUANTITIES** *continued*

### EMERGENCY LIGHTING

Self-contained, non maintained 3 hour duration fitting wired with 1.5 mm² single core PVC insulated cables drawn in 20 mm heavy gauge black enamelled screwed welded conduit.

|  |  | £ |
|---|---|---|
| tungsten | per fitting | 184.69 |
| fluorescent | per fitting | 214.28 |

### ELECTRIC HEATING
**Electric underfloor heating by heating cable**
The approximate cost of solid embedded underfloor heating including distribution fuseboard, time switches and thermostats is £23.08 per m² of floor area based on 145 watts per m².

**Storage heaters**
Storage heater with built in thermostat, temperature controller and safety cut-out.

| Capacity kW | | | Supply and installation cost of storage heating including wiring. Unit £ |
|---|---|---|---|
| 1.7 | P.C. | £149.51 | 324.16 |
| 2.55 | P.C. | £191.14 | 375.83 |
| 3.4 | P.C. | £230.87 | 425.28 |

**Note**
Cost of wiring is based on 10 metre run of circuit including a 25 amp double pole switch with metal box, share of distribution fuse board and 4 mm² PVC cable drawn into 20 mm conduit.
For wiring with PVC insulated and PVC sheathed cable the above costs may be reduced by £65.94 in each case.

### CLOCK SYSTEMS
**Clock circuits**
D.C. clock circuit wired with 2 x 1.5 mm² single core PVC insulated cable drawn in 20 mm heavy gauge black enamelled screwed welded conduit including standard associated fittings, clock connector and share of the battery unit fixed to brickwork or concrete with spacer bar saddles in surface work.
Cost based on an average circuit run of 30 metres per clock point (excluding provisions for clocks).

|  |  | £ |
|---|---|---|
|  | per clock point | 371.50 |
| Crystal controlled master clock | P.C. | 619.80 |
| 300 mm Wall mounted slave clock | P.C. | 69.21 |

A.C. mains clock circuit wired as above but cost based on an average circuit run of 4 m per clock point.

|  |  | £ |
|---|---|---|
|  | per clock point | 52.72 |
| 300 mm Wall mounted mains clock | P.C. | 56.11 |

## Approximate Estimating

### ELEMENTAL AND ALL-IN-RATES

## ALL IN RATES FOR PRICING ELECTRICAL APPROXIMATE QUANTITIES *continued*

### FIRE ALARMS

**Fire alarm circuits**
Fire alarm system wired in single core PVC insulated cables drawn into black enamelled conduit.

|  | *Average cost per point* £ |
|---|---|
| Bell | 137.36 |
| Break glass contact | 125.31 |
| Heat detector | 120.98 |
| Smoke detector | 180.26 |

For costs of zone control/indicator panels, battery chargers and batteries, see 'Prices for Measured Work' section.

### EXTERNAL LIGHTING

**Estate road lighting**
Post type road lighting lantern 80 watt MBF/U complete with 4.5 m high column with hinged lockable door, control gear and cut-out including all internal wiring, interconnections and earthing fed by 2.5 mm$^2$ two core PVC SWA PVC underground cable.
Approximate installed price *per metre road length* (based on 300 metres run) including time switch but excluding all trench and builder's work
    Columns erected on same side of road at 30 m intervals ............................. £23.03
    Columns erected on both sides of road at 60 m intervals
    in staggered formation ...................................................... £28.53

**Bollard lighting**
Bollard lighting fitting 50 watt MBF/U including controlgear,
all internal wiring, interconnections, earthing and 10 metres
of 2.5 mm$^2$ two core PVC SWA PVC underground cable
    Approximate installed price excluding trench work and
    builder's work ............................................................ £264.89 *each*

**Outdoor flood lighting**
Wall mounted outdoor flood light fitting complete with tungsten
halogen lamp, mounting bracket, wire guard and all internal
wiring; fixed to brickwork or concrete and connected
    Installed price 500 watt ..................................................... £96.68 *each*
    Installed price 1000 watt .................................................... £117.60 *each*

Pedestal mounted outdoor floor light fitting complete with
1000 watt MBF/U lamp, mounting bracket, control gear,
contained in weatherproof steel box, all internal wiring,
interconnections and earthing; fixed to brickwork or
concrete and connected
    Approximate installed price excluding builder's work .......................... £789.07 *each*

## Approximate Estimating

## ELEMENTAL AND ALL-IN-RATES

## ALL IN RATES FOR PRICING SPECIALIST APPROXIMATE QUANTITIES

### MEDICAL GASES

**Tubing and Fittings**
Light gauge copper tube to B.S.2871, Part 1, Table X degreased and sealed with silver brazed end feed fittings including allowance for waste, fittings and brackets, assuming average runs

| Nominal size | Cost per metre £ |
|---|---|
| 12mm | 17.65 |
| 15mm | 19.76 |
| 22mm | 19.76 |
| 28mm | 24.20 |
| 35mm | 37.40 |
| 42mm | 42.90 |
| 54mm | 51.67 |
| 67mm | 60.44 |
| 76mm | 73.65 |
| 108mm | 104.50 |

| | Cost each £ |
|---|---|
| Flush mounted terminal unit, with 12 mm copper tail brazed to terminal inlet and distribution pipe | 75.86 |

Lockable isolating valve, with lockable head mounted in flush mounted box

| | Cost each £ |
|---|---|
| 15mm | 25.25 |
| 22mm | 31.91 |
| 28mm | 40.68 |
| 35mm | 50.61 |
| 42mm | 63.71 |
| 54mm | 75.86 |

**Equipment**

| | Total Plant Cost £ |
|---|---|
| Cylinder manifold, 2 x 5 cylinders, comprising two header assemblies, bottom frame, inlet pipe and control panel, regulator, non-return and isolating valves, relief valves, spanners and spanner rack, and one spare cylinder rack with ten spare cylinders | 3,770.00 |
| 2 cylinder emergency supply manifold with header and header support regulator, non-return and isolating valves, two relief valves | 407.00 |

## Approximate Estimating

## ELEMENTAL AND ALL-IN-RATES

**ALL IN RATES FOR PRICING SPECIALIST APPROXIMATE QUANTITIES** *continued*

**MEDICAL GASES** *continued*

**Equipment** *continued*

|  | Total Plant Cost £ |
|---|---|
| Packaged compressed air unit, one duty and one stand-by compressor, two stage reciprocating piston type, duty of 1587 litres/min of oil free air at 7 bar from each compressor, complete with horizontal air receiver, separator, silencers, air dryers, air cooled coolers and cylinders, controls and control panel, all necessary valves, gauges, alarms, motors and flexible connections | 33,360.00 |
| Packaged medical vacuum plant, one duty and one standby direct driven vacuum pumps having a duty of 1200 litres/min of free air displacement at 500 mm Hg below atmosphere, total design flow of 1480 litres/min of free air, with horizontal vacuum reservoir bacterial filters, silencers, controls and control panel and all valves, gauges, alarms, motors and meters, flexible connections | 18,600.00 |
| Liquid oxygen supply plant, with vacuum insulated evaporator (VIE), liquid capacity of 3080 litres, standby emergency oxygen twin manifold, air vaporiser, control and control panel, safety valves, isolating and non-return valves, flexible connections | 13,290.00 |

**Approximate Estimating**

## ELEMENTAL AND ALL-IN-RATES

**ALL IN RATES FOR PRICING SPECIALIST APPROXIMATE QUANTITIES** *continued*

**CATERING EQUIPMENT**

Commercial Catering Establishment, 215m², 175 Nr covers/places, including cold storage preparation, cooking, display/servery, waste disposal, cash till point, dishwashing/food disposal and cleaning facilities and local electrical/mechanical connections only. Excludes dining furniture/fitments.

**ELEMENTAL COST PLAN**

|  | *Total Cost* £ | *Cost/m²* £ | *Cost/Cover* £ |
|---|---|---|---|
| Food Storage (Cold rooms/ storage racks, etc.) | 15,250.00 | 70.93 | 87.14 |
| Food Preparation/Cooking | 24,580.00 | 114.33 | 140.46 |
| Display Counters/Serving | 19,670.00 | 91.49 | 112.40 |
| Washing/Clearing/Disposal | 16,780.00 | 78.05 | 95.89 |
| Cash Tills/Counters | 5,150.00 | 23.95 | 29.43 |
| Total | 81,430.00 | 378.75 | 465.32 |

## Approximate Estimating

## ELEMENTAL AND ALL-IN-RATES

## ALL IN RATES FOR PRICING SPECIALIST APPROXIMATE QUANTITIES *continued*

### LIFT INSTALLATIONS

The cost of lift installations will vary depending upon a variety of circumstances. The following prices assume a floor to floor height of 3.5 metres and standard finishes to cars, doors and gates.

### Passenger lifts

Electrically operated A.C. driven 8 or 10 person lift serving 6 levels with directional collective controls and a speed of 1 metre per second ............................ £63,180.00 to £76,050.00

Add to above for
- Bottom motor room ................................ £5,250.00 to £7,590.00
- Extra levels served ................................ £2,420.00 to £3,740.00
- Increase speed of travel from 1 to 1.60 metres per second ........................................ £2,420.00 to £5,050.00
- Increase speed of travel from 1 to 2.50 metres per second ........................................ £32,970.00 to £37,910.00
- Enhanced finish to car ............................. £3,740.00 to £10,110.00

Electrically operated A.C. driven 21 person lift serving 6 levels with directional collective controls and a speed of 1.60 metres per second ....................... £101,381.00 to £126,720.00

Add to above for
- Bottom motor room ................................ £6,266.00 to £12,530.00
- Extra levels served ................................ £3,740.00 to £5,060.00
- Increased speed of travel from 1.60 to 2.50 metres per second ................................. £37,921.00 to £44,180.00
- Enhanced finish to car ............................. £6,266.00 to £12,530.00

For any floor bypassed add 45% of the cost of an an extra level served

### Goods lifts

Electrically operated two speed general purpose goods lift serving 5 levels, to take 500 kg load, inter manually operated shutters and automatic push button control and a speed of 0.25 metres per second ................................................... £44,398.00 to £63,420.00

Add to above for
- Extra levels served ................................ £2,420.00 to £3,740.00
- Increased capacity up to 2000 kg .................. £12,531.00 to £19,120.00
- Increase speed of travel from 0.25 to 0.50 metres per second ................................. £6,266.00 to £12,530.00

Oil hydraulic operated goods lift 5 levels, to take 500 kg load with manually operated shutter and automatic push button control and a speed of 0.25 metres per second ........................................ £50,558.00 to £69,790.00

Add to above for
- Increased capacity up to 1000 kg .................. £3,740.00 to £6,270.00
- Increased capacity up to 1500 kg .................. £6,266.00 to £12,530.00
- Increased capacity up to 2000 kg .................. £12,531.00 to £19,010.00

*Approximate Estimating*

### ELEMENTAL AND ALL-IN-RATES

**ALL IN RATES FOR PRICING SPECIALIST APPROXIMATE QUANTITIES** *continued*

**ESCALATOR INSTALLATIONS**

30°Pitch escalator with a rise of 3 to 5 metres and enamelled sheet steel or glass balustrades

| | | |
|---|---|---|
| 600mm step width | £50,560.00 to | £63,420.00 |
| 800mm step width | £54,510.00 to | £67,260.00 |
| 1000mm step width | £57,040.00 to | £69,790.00 |

Add to above for

| | | |
|---|---|---|
| Stainless steel balustrades | £5,060.00 to | £7,590.00 |
| Glass balustrades with balustrade lighting | £3,740.00 to | £6,380.00 |

# Keep your figures up to date, free of charge

This section, and most of the other information in this Price Book, is brought up to date every three months, until the next annual edition, in the *Price Book Update*.

The *Update* is available free to all Price Book purchasers.

To ensure you receive your copy, simply complete the reply card from the centre of the book and return it to us.

## Approximate Estimating

### ELEMENTAL COSTS

OFFICE BUILDING

Speculative ten storey city office block with a gross floor area of 6,585 m², upon which this cost analysis is based. Fully air conditioned with heating and cooling, and 3 ten person passenger lifts.

### Cost Summary

| El. Ref. | Element | Total Cost £ | Cost / m² £ |
|---|---|---|---|
| 5D | Water Installations | | |
| | Cold water services | 57,785.00 | 8.78 |
| | Hot water services | 37,630.00 | 5.71 |
| | Chilled water | 205,263.00 | 31.17 |
| 5E | Heating and Heat Source | 272,048.00 | 41.31 |
| 5G | Ventilating Systems | | |
| | Air conditioning | 635,964.00 | 96.58 |
| | Automatic controls | 135,374.00 | 20.56 |
| 5H | Electrical Installation | | |
| | Electrical source and power supplies | 112,552.00 | 17.09 |
| | To mechanical installation | 107,304.00 | 16.30 |
| | General lighting | 85,404.00 | 12.97 |
| | Lighting fittings | 155,813.00 | 23.66 |
| 5J | Lift and Conveyor Installation | | |
| | Lifts | 245,899.00 | 37.34 |
| 5K | Protective Installation | | |
| | Sprinkler installation | 143,505.00 | 21.79 |
| | Fire-fighting installation | 28,006.00 | 4.25 |
| | Lighting protection | 6,662.00 | 1.01 |
| 5L | Communication Installation | | |
| | Telephones | 2,521.00 | 0.38 |
| | Fire alarms and security | 41,554.00 | 6.31 |
| | Summary total | 2,273,284.00 | 345.21 |

## Approximate Estimating
### ELEMENTAL COSTS

HOTEL BUILDING

The Hotel has a gross floor area of 10,153 m², upon which the cost analysis is based, and associated car parking area of 4,149 m².

The construction is on seven levels, having three half area floors containing service yard and goods entrance, reception areas and public areas, with the remaining Hotel on four full floors.

The private and associated public car parking is constructed on the half area levels.

The Hotel provides a total of 150 bedrooms, comprising 116 double/twin bedded rooms, 24 single/syndicate rooms, 6 interconnecting rooms and 4 suites. All of the bedrooms have en-suite bathrooms and are fitted to a high standard but are not air conditioned.

The public areas of the Hotel include a restaurant with 150 covers, a lounge bar and cocktail bar, all fully air conditioned. The major facilities also include a Banqueting Hall and Foyer to seat 300 and two large Conference / Promotion rooms to cater for medium sized meetings. The main Hall and one of the meeting rooms is capable of being subdivided into two smaller units if required. Two smaller Conference rooms are provided as part of a suite which allows the option of incorporating a number of related syndicate rooms. The Banqueting and Conference facilities, like the public areas are fully air conditioned.

Five lifts are provided, two main passenger lifts serving all floors, two passenger lifts serving lower and car park floors and a service/goods lift.

### Cost Summary

| El. Ref. | Element | Total Cost £ | Cost / m² £ |
|---|---|---|---|
| 5D | Water Installations | | |
|    | Hot and cold water services | 381,108.00 | 37.54 |
| 5E | Heating and Heat Source | 384,416.00 | 37.86 |
|    | Carried forward | 765,524.00 | 75.40 |

**Approximate Estimating**

**ELEMENTAL COSTS**

HOTEL BUILDING (Cont'd)

**Cost Summary**

| El. Ref. | Element | Total Cost £ | Cost / m² £ |
|---|---|---|---|
| | Brought forward | 765,524.00 | 75.40 |
| 5G | Ventilating Systems | | |
| | Car park ventilation | 15,468.00 | 1.52 |
| | Bathroom ventilation | 76,339.00 | 7.52 |
| | Air conditioning | 639,496.00 | 62.99 |
| 5H | Electrical Installation | | |
| | Electrical source and power supplies | 288,604.00 | 28.43 |
| | To mechanical installation | 40,960.00 | 4.03 |
| | General lighting | 439,312.00 | 43.27 |
| | Emergency lighting | 52,936.00 | 5.21 |
| | Car park lighting | 117,287.00 | 11.55 |
| | External lighting | 33,526.00 | 3.30 |
| 5J | Lift and Conveyor Installation | | |
| | Lifts | 287,900.00 | 28.36 |
| 5K | Protective Installation | | |
| | Fire-fighting installation | 45,502.00 | 4.48 |
| | Lighting protection | 11,693.00 | 1.15 |
| 5L | Communication Installation | | |
| | Telephones | 5,992.00 | 0.59 |
| | Fire alarms | 182,006.00 | 17.93 |
| 5M | Special Services | | |
| | Whirlpool, sauna and steam room | 81,159.00 | 7.99 |
| | Kitchen equipment | 236,806.00 | 23.32 |
| | Summary total | 3,320,510.00 | 327.04 |

*Approximate Estimating*

## ELEMENTAL COSTS

DATA PROCESSING CENTRE

Commercial computer centre comprising three distinct areas - energy centre, main computer hall / air handling corridors, and a two storey office complex with atrium, over a total floor area of 5,300 m², upon which the cost analysis is based. The plant room for the Office Complex being located with the main computer hall / A.H.C. areas. The external area of the building site including access roads, car parks and footpaths is 13,525 m².

Piping systems have been designed and installed to permit the installation of additional plant items (eg. PCU's, Dri-Coolers, Chillers), to cope with the increased computer load without the existing plant having to be shut down. The infra structure has been ultimately designed for 750 watts/m² although the internal process cooling plant has been sized to handle 400 watts/m² (with an allowance of 25% spare capacity).

All of the electrical systems are designed to avoid system failure. A single circuit or component failure will not effect more equipment than the redundancy facilities allow, and will not therefore impair the operation of the Centre.

As computer loads increase additional plant items (eg. additional generator set and UPS set) can be added into the system without the existing system having to be shut-down.

Common services, trunking, trays and ladder racks, also have inbuilt allowances of 25% spare capacity.

### Cost Summary

| El. Ref. | Element | Total Cost £ | Cost / m² £ |
|---|---|---|---|
| 5D | Water Installations | | |
| | Mains supply | 9,142.00 | 1.72 |
| | Cold water services | 28,941.00 | 5.46 |
| | Hot water services | 8,327.00 | 1.57 |
| | Steam and condensate | 10,254.00 | 1.93 |
| | Carried forward | 56,664.00 | 10.68 |

## Approximate Estimating

## ELEMENTAL COSTS

DATA PROCESSING CENTRE (Cont'd)

### Cost Summary

| El. Ref. | Element | Total Cost £ | Cost / m² £ |
|---|---|---|---|
| | Brought forward | 56,664.00 | 10.68 |
| 5E | Heat Source | | |
| | Condenser water installation | 305,092.00 | 57.56 |
| | Chilled water installation | 418,361.00 | 78.94 |
| | Pressurisation installation | 19,446.00 | 3.67 |
| | L.P. hot water installation | 60,634.00 | 11.44 |
| | Water treatment system | 7,051.00 | 1.33 |
| 5F | Space Heating and Air Treatment | | |
| | Water and/or steam | 83,961.00 | 15.84 |
| | Heating with cooling air (treated locally) | 358,755.00 | 67.69 |
| 5G | Ventilating Systems | | |
| | Mechanical extract system | 62,489.00 | 11.79 |
| | Halon extract system | 45,459.00 | 8.58 |
| 5H | Electrical Installation | | |
| | Electrical source and power supplies | 1,838,061.00 | 346.80 |
| | Electric power supplies | 661,043.00 | 124.73 |
| | Electric lighting and fittings | 255,451.00 | 48.20 |
| 5J | Lift and Conveyor Installation | 38,248.00 | 7.22 |
| 5K | Protective Installation | | |
| | Halon extinguishing system | 64,135.00 | 12.10 |
| | Lighting protection | 18,040.00 | 3.40 |
| 5L | Communication Installation | | |
| | Warning installation (security and fire) | 473,291.00 | 89.30 |
| | Visual and audio installation | 22,895.00 | 4.32 |
| | Telephone installation | 9,407.00 | 1.77 |
| 5M | Special Services | | |
| | Energy management system | 484,582.00 | 91.43 |
| | Summary total | 5,283,065.00 | 996.79 |

*Approximate Estimating*

***ELEMENTAL COSTS***

DISTRICT GENERAL HOSPITAL

The Hospital is based on the `Nucleus' design concept and comprises four `cruciform' templates on two /three levels arranged on both sides of a communication corridor (Street).

Adjoining the templates is a Service and Energy Centre consisting of Linen Distribution, Kitchen, Switch Rooms and Generators, Boiler Hall and Incinerator / Waste Handling Facility.

The total floor area of the Hospital is 15,520 m², upon which the cost analysis is based.

**Cost Summary**

| El. Ref. | Element | Total Cost £ | Cost / m² £ |
|---|---|---|---|
| 5B | Services Equipment | | |
| | (Mechanical) | | |
| | Cold room installation | 123,185.00 | 7.94 |
| | Services equipment | 217,331.00 | 14.00 |
| | (Electrical) | | |
| | Equipment - all areas | 29,025.00 | 1.87 |
| 5B | Water Installation | | |
| | (Mechanical) | | |
| | Boosted mains cold water | 106,885.00 | 6.89 |
| | Potable water storage | 29,468.00 | 1.90 |
| | Non potable water storage | 22,617.00 | 1.46 |
| | Utility water storage | 19,696.00 | 1.27 |
| | Boilerhouse wash down | 3,225.00 | 0.21 |
| | Incinerator wash down | 3,684.00 | 0.24 |
| | Incinerator bin wash | 3,073.00 | 0.20 |
| | Emergency shower | 367.00 | 0.02 |
| | Kitchen potable cold water | 16,864.00 | 1.09 |
| | Potable tank cold water | 189,981.00 | 12.24 |
| | Kitchen D.H.W.S. (cold feed) | 4,201.00 | 0.27 |
| | Kitchen D.H.W.S | 55,605.00 | 3.58 |
| | Non potable cold storage | 33,834.00 | 2.18 |
| | D.H.W.S | 417,066.00 | 26.87 |
| | Carried forward | 1,276,107.00 | 82.23 |

## Approximate Estimating
### ELEMENTAL COSTS

DISTRICT GENERAL HOSPITAL (Cont'd)

### Cost Summary

| El. Ref. | Element | Total Cost £ | Cost / m² £ |
|---|---|---|---|
| | Brought forward | 1,276,107.00 | 82.23 |
| 5B | Water Installation (cont'd) | | |
| | Hydrotherapy cold water | 5,146.00 | 0.33 |
| | Patient hoist | 1,226.00 | 0.08 |
| | Cascade fountains | 12,375.00 | 0.80 |
| | Incoming water | 3,094.00 | 0.20 |
| 5E | Heat Source | | |
| | (Mechanical) | | |
| | Soft water | 33,471.00 | 2.16 |
| | Boiler feed water | 41,287.00 | 2.66 |
| | INT blowdown (cold feed) | 8,776.00 | 0.57 |
| | INT blowdown | 30,444.00 | 1.96 |
| | CONT blowdown | 6,764.00 | 0.44 |
| | Boiler sight glass/aux. drains | 2,415.00 | 0.16 |
| | Incinerator boosted cold water | 1,672.00 | 0.11 |
| | Heavy fuel oil | 159,076.00 | 10.25 |
| | High pressure steam | 255,970.00 | 16.49 |
| | Flash steam | 2,836.00 | 0.18 |
| | HP steam (temp) | 56,600.00 | 3.65 |
| | Existing steam | 120.00 | 0.01 |
| | HP condensate | 8,506.00 | 0.55 |
| | Pumped condensate | 15,954.00 | 1.03 |
| | HP condensate (temp) | 15,047.00 | 0.97 |
| | Pumped condensate (temp) | 45,993.00 | 2.96 |
| | Existing condensate | 466.00 | 0.03 |
| | MPHW cold feed | 3,566.00 | 0.23 |
| | MPHW heating | 126,794.00 | 8.17 |
| | LPHW to tank farm | 8,481.00 | 0.55 |
| | LPHW common | 55,680.00 | 3.59 |
| | LPHW 12 hour | 30,421.00 | 1.96 |
| | LPHW 24 hour | 29,035.00 | 1.87 |
| | LPHW CT | 8,315.00 | 0.54 |
| | LPHW kitchen | 6,267.00 | 0.40 |
| | LPHW hydrotherapy pool | 5,096.00 | 0.33 |
| | Chilled water installation | 191,340.00 | 12.33 |
| | Heat recovery | 16,352.00 | 1.05 |
| | Incinerator installation | 168,484.00 | 10.86 |
| | Connections to equipment | 6,187.00 | 0.40 |
| | Carried forward | 2,639,363.00 | 170.10 |

*Approximate Estimating*

**ELEMENTAL COSTS**

DISTRICT GENERAL HOSPITAL (Cont'd)

**Cost Summary**

| El. Ref. | Element | | Total Cost £ | Cost / m² £ |
|---|---|---|---:|---:|
| | Brought forward | | 2,639,363.00 | 170.10 |
| 5F | Space Heating | | | |
| | (Mechanical) | | | |
| | | Low pressure steam | 3,409.00 | 0.22 |
| | | Humidifier steam | 20,251.00 | 1.30 |
| | | Humidifier condensate | 10,715.00 | 0.69 |
| | | MPHW heating | 148,059.00 | 9.54 |
| | | LPHW to tank farm | 20,271.00 | 1.31 |
| | | LPHW 12 hour | 146,746.00 | 9.46 |
| | | LPHW 24 hour | 118,413.00 | 7.63 |
| | | LPHW CT | 65,819.00 | 4.24 |
| | | LPHW kitchen | 26,689.00 | 1.72 |
| | | LPHW hydrotherapy pool | 2,495.00 | 0.16 |
| | | Chilled water (hydrotherapy pool) | 5,474.00 | 0.35 |
| | | Low velocity supply vent | 370,437.00 | 23.87 |
| | | Low velocity supply (kitchen) | 49,844.00 | 3.21 |
| | | Hydrotherapy pool supply | 19,191.00 | 1.24 |
| | | Plant actuation | 12,375.00 | 0.80 |
| | | Acoustic work | 9,281.00 | 0.60 |
| | | Fire dampers | 99,002.00 | 6.38 |
| 5G | Ventilating Systems | | | |
| | (Mechanical) | | | |
| | | Clean extract | 119,089.00 | 7.67 |
| | | Dirty extract | 123,477.00 | 7.96 |
| | | Fume cupboard extract | 7,301.00 | 0.47 |
| | | Fume canopy extract | 6,728.00 | 0.43 |
| | | Clean extract (kitchen) | 16,089.00 | 1.04 |
| | | Dirty special/extract (kitchen) | 9,540.00 | 0.61 |
| | | Special extract | 3,211.00 | 0.21 |
| | | Hydrotherapy pool extract | 9,581.00 | 0.62 |
| | | Waste handling dirty extract | 1,066.00 | 0.07 |
| | Carried forward | | 4,063,916.00 | 261.90 |

## Approximate Estimating

### ELEMENTAL COSTS

DISTRICT GENERAL HOSPITAL (Cont'd)

**Cost Summary**

| El. Ref. | Element | Total Cost £ | Cost / m² £ |
|---|---|---:|---:|
| | Brought forward | 4,063,916.00 | 261.90 |
| 5H | **Electrical Installation** | | |
| | (Electrical) | | |
| | High voltage installation | 113,562.00 | 7.32 |
| | Standby generator | 275,788.00 | 17.77 |
| | Sub-main distribution | 334,507.00 | 21.55 |
| | General power | 139,445.00 | 8.98 |
| | Lighting installation | 279,870.00 | 18.03 |
| | Luminaires | 228,691.00 | 14.74 |
| | Cable trunking/tray | 67,493.00 | 4.35 |
| | Earthing | 38,856.00 | 2.50 |
| | Canopy lighting | 12,375.00 | 0.80 |
| | Emergency lighting plant room and roof | 12,375.00 | 0.80 |
| | Emergency lighting hydrotherapy pool | 9,281.00 | 0.60 |
| | Additional lighting to artwork | 12,375.00 | 0.80 |
| | Additional fire barriers for electrical work | 12,375.00 | 0.80 |
| | Additional connections | 9,281.00 | 0.60 |
| | Cable supports future use | 2,475.00 | 0.16 |
| | Services to external signs | 12,375.00 | 0.80 |
| | Additional switchgear | 12,375.00 | 0.80 |
| 5I | **Gas Installation** | | |
| | (Mechanical) | | |
| | Natural Gas Incoming (Attendance) (See 6C) | | |
| | Interruptible natural gas | 24,627.00 | 1.59 |
| | Uninterruptable natural gas | 18,697.00 | 1.20 |
| 5K | **Protective Installations** | | |
| | (Mechanical) | | |
| | Fire hose reels | 58,928.00 | 3.80 |
| | Charged riser | 29,581.00 | 1.91 |
| | Hydrant mains (see also 6C) | 4,984.00 | 0.32 |
| | Smoke input ventilation | 25,768.00 | 1.66 |
| | Smoke extract ventilation | 33,309.00 | 2.15 |
| | (Electrical) | | |
| | Lightning protection | 16,977.00 | 1.09 |
| | Carried forward | 5,850,286.00 | 377.02 |

*Approximate Estimating*

*ELEMENTAL COSTS*

DISTRICT GENERAL HOSPITAL (Cont'd)

**Cost Summary**

| El. Ref. | Element | Total Cost £ | Cost / m² £ |
|---|---|---|---|
| | Brought forward | 5,850,286.00 | 377.02 |
| 5L | Communications Installations | | |
| | *(Electrical)* | | |
| | Telephone installation | 38,215.00 | 2.46 |
| | Intercommunication system | 25,159.00 | 1.62 |
| | Radio distribution system | 31,005.00 | 2.00 |
| | Television system | 9,227.00 | 0.59 |
| | Nurse call system | 143,519.00 | 9.25 |
| | Fire alarm system | 307,011.00 | 19.78 |
| | Intruder alarm system | 3,303.00 | 0.21 |
| | Panic and lift alarm system | 2,790.00 | 0.18 |
| | Call bell system | 682.00 | 0.04 |
| | Computer system | 2,870.00 | 0.18 |
| | Cable trunking/tray | 83,257.00 | 5.36 |
| | Additional fire alarm needs | 24,751.00 | 1.59 |
| 5M | Special Installations | | |
| | *(Mechanical)* | | |
| | Proportional chemical dosing | 1,421.00 | 0.09 |
| | Shot chemical dosing | 4,406.00 | 0.28 |
| | Diesel fuel oil | 65,649.00 | 4.23 |
| | Hydrotherapy pool | 54,962.00 | 3.54 |
| | Medical gas alarms | 10,443.00 | 0.67 |
| | Oxygen installation | 48,577.00 | 3.13 |
| | Compressed air installation | 52,181.00 | 3.36 |
| | Vacuum installation | 52,497.00 | 3.38 |
| | Condenser installation | 2,878.00 | 0.19 |
| | Bin cleaning equipment | 8,663.00 | 0.56 |
| | Lifting beams | 12,375.00 | 0.80 |
| 6C | External Services | | |
| | *(Mechanical)* | | |
| | Incoming water main | 161.00 | 0.01 |
| | Existing hydrant main connection | 1,024.00 | 0.07 |
| | Fire hydrant main | 11,186.00 | 0.72 |
| | Incoming natural gas | 160.00 | 0.01 |
| | *(Electrical)* | | |
| | Main electric supply | 219.00 | 0.01 |
| | External lighting and luminaires | 39,351.00 | 2.54 |
| | Courtyard lighting | 10,864.00 | 0.70 |
| | Summary total | 6,899,092.00 | 444.57 |

## ELEMENTAL COSTS

DISTRICT GENERAL HOSPITAL (Example 2)

(The details which follow show how costs per square metre based upon the functional or departmental area of the building can fluctuate within each element, reflecting the varying demands for different services within different departments.)

For this example, an alternative Hospital (to that used for example 1) has been used. This Hospital, again designed around the `Nucleus' concept, comprises 5 Nr. two storey templates to either side of the communications corridor together with a two storey Service Centre to one end of the corridor. Auxiliary works - generators, fuel tanks and minor control rooms are sited outside the main buildings.

The total floor area is 16,970 m², upon which this cost analysis is based but it should be noted that this is also the first phase of a planned larger development. In addition, the Service Centre also caters for an adjacent existing Hospital and as a consequence major areas (Kitchens etc) are oversized in relation to the template (department) areas.

**Approximate Estimating**

## ELEMENTAL COSTS

DISTRICT GENERAL HOSPITAL (cont'd)

| Element | Adult Acute 168 beds Area m² 3,094 Cost £ | Rate m² | Children 46 beds Area m² 1,031 Cost £ | Rate m² | Operating Theatres 4 theatres Area m² 855 Cost £ | Rate m² |
|---|---|---|---|---|---|---|
| 5A Sanitary Appliances | 93,283 | 30.15 | 33,101 | 32.11 | 30,358 | 35.51 |
| 5B Services Equipment | 19,407 | 6.27 | 5,819 | 5.64 | 5,763 | 6.74 |
| 5C Disposal Installation | 46,538 | 15.04 | 15,898 | 15.42 | 14,685 | 17.18 |
| 5D Water Installation | 60,201 | 19.46 | 25,590 | 24.82 | 30,530 | 35.71 |
| 5E Heat Source | | | | | | |
| 5F Space Heating/Air Treatment | 111,814 | 36.14 | 37,449 | 36.32 | 116,670 | 136.46 |
| 5G Ventilating System | 100,156 | 32.37 | 24,401 | 23.67 | 56,361 | 65.92 |
| 5H Electrical Installation | 209,398 | 67.68 | 60,227 | 58.42 | 88,524 | 103.54 |
| 5I Gas Installation | | | | | | |
| 5J Lift/Conveyor Installation | | | | | | |
| 5K Protective Installations | 58,224 | 18.82 | 19,401 | 18.82 | 16,090 | 18.82 |
| 5L Communication Installation | 130,476 | 42.17 | 36,569 | 35.47 | 13,833 | 16.18 |
| 5M Special Installations | 95,354 | 30.82 | 43,727 | 42.41 | 102,144 | 119.47 |
| 5N Builders Work to Services | 112,974 | 36.51 | 36,306 | 35.21 | 55,209 | 64.57 |
| 5O Builders Profit and Attendance on Services | 58,893 | 19.03 | 18,923 | 18.35 | 28,834 | 33.72 |
| Totals | 1,096,718 | 354.46 | 357,411 | 346.66 | 559,001 | 653.82 |

| Element | Adult Day Care 15 beds Area m² 482 Cost £ | Rate m² | Accident and Emergency 46 beds Area m² 787 Cost £ | Rate m² | Fracture Clinic 27 sessions Area m² 222 Cost £ | Rate m² |
|---|---|---|---|---|---|---|
| 5A Sanitary Appliances | 21,809 | 45.25 | 25,317 | 32.17 | 10,850 | 48.87 |
| 5B Services Equipment | 3,040 | 6.31 | 5,447 | 6.92 | | |
| 5C Disposal Installation | 10,468 | 21.72 | 12,000 | 15.25 | 6,327 | 28.50 |
| 5D Water Installation | 26,485 | 54.95 | 11,020 | 14.00 | 10,298 | 46.39 |
| 5E Heat Source | | | | | | |
| 5F Space Heating/Air Treatment | 16,566 | 34.37 | 23,375 | 29.70 | 30,665 | 138.13 |
| 5G Ventilating System | 14,937 | 30.99 | 15,666 | 19.91 | 13,072 | 58.88 |
| 5H Electrical Installation | 30,903 | 64.11 | 38,382 | 48.77 | 28,226 | 127.14 |
| 5I Gas Installation | | | | | | |
| 5J Lift/Conveyor Installation | | | | | | |
| 5K Protective Installations | 9,070 | 18.82 | 14,810 | 18.82 | 4,178 | 18.82 |
| 5L Communication Installation | 8,629 | 17.90 | 24,554 | 31.20 | 12,357 | 55.66 |
| 5M Special Installations | 10,991 | 22.80 | 35,650 | 45.30 | 8,790 | 39.59 |
| 5N Builders Work to Services | 18,249 | 37.86 | 24,994 | 31.76 | 14,558 | 65.58 |
| 5O Builders Profit and Attendance on Services | 9,515 | 19.74 | 13,018 | 16.54 | 7,620 | 34.32 |
| Totals | 180,662 | 374.82 | 244,233 | 310.34 | 146,941 | 661.88 |

## Approximate Estimating

### ELEMENTAL COSTS

DISTRICT GENERAL HOSPITAL (cont'd)

| Element | Intensive Therapy Unit 8 beds Area m² 381 | | Outpatients Department 67.5 beds Area m² 649 | | Pharmacy (excl. manf.) 300 beds Area m² 278 | |
|---|---|---|---|---|---|---|
| | Cost £ | Rate m² | Cost £ | Rate m² | Cost £ | Rate m² |
| 5A Sanitary Appliances | 7,270 | 19.08 | 15,475 | 23.84 | 1,992 | 7.17 |
| 5B Services Equipment | 2,723 | 7.15 | 31,491 | 48.52 | | |
| 5C Disposal Installation | 5,916 | 15.53 | 8,567 | 13.20 | 2,872 | 10.33 |
| 5D Water Installation | 11,431 | 30.00 | 25,950 | 39.98 | 4,051 | 14.57 |
| 5E Heat Source | | | | | | |
| 5F Space Heating/Air Treatment | 66,309 | 174.04 | 55,014 | 84.77 | 17,177 | 61.79 |
| 5G Ventilating System | 38,069 | 99.92 | 17,678 | 27.24 | 11,361 | 40.87 |
| 5H Electrical Installation | 54,336 | 142.61 | 53,422 | 82.31 | 15,694 | 56.45 |
| 5I Gas Installation | | | | | 1,955 | 7.03 |
| 5J Lift/Conveyor Installation | | | | | | |
| 5K Protective Installations | 7,171 | 18.82 | 12,214 | 18.82 | 5,230 | 18.81 |
| 5L Communication Installation | 11,553 | 30.32 | 20,780 | 32.02 | 4,933 | 17.74 |
| 5M Special Installations | 21,094 | 55.36 | 27,397 | 42.21 | 1,921 | 6.91 |
| 5N Builders Work to Services | 26,298 | 69.02 | 31,754 | 48.93 | 8,285 | 29.80 |
| 5O Builders Profit and Attendance on Services | 13,765 | 36.13 | 16,620 | 25.61 | 4,335 | 15.59 |
| Totals | 265,935 | 697.98 | 316,362 | 487.45 | 79,806 | 287.06 |

| Element | X-Ray 4 - R/D rooms Area m² 565 | | Administration 9 points Area m² 597 | | Main Entrance 9 points Area m² 306 | |
|---|---|---|---|---|---|---|
| | Cost £ | Rate m² | Cost £ | Rate m² | Cost £ | Rate m² |
| 5A Sanitary Appliances | 8,922 | 15.79 | 2,974 | 4.98 | 5,159 | 16.86 |
| 5B Services Equipment | 5,447 | 9.64 | | | 1,523 | 4.98 |
| 5C Disposal Installation | 7,961 | 14.09 | 2,971 | 4.98 | 3,107 | 10.15 |
| 5D Water Installation | 20,769 | 36.76 | 6,923 | 11.60 | 8,649 | 28.26 |
| 5E Heat Source | | | | | | |
| 5F Space Heating/Air Treatment | 36,143 | 63.97 | 31,421 | 52.63 | 1,547 | 5.06 |
| 5G Ventilating System | 15,883 | 28.11 | 12,858 | 21.54 | 479 | 1.57 |
| 5H Electrical Installation | 41,155 | 72.84 | 39,396 | 65.99 | 8,031 | 26.25 |
| 5I Gas Installation | | | | | | |
| 5J Lift/Conveyor Installation | | | | | | |
| 5K Protective Installations | 10,634 | 18.82 | 11,234 | 18.82 | 5,758 | 18.82 |
| 5L Communication Installation | 17,385 | 30.77 | 3,186 | 5.34 | 1,481 | 4.84 |
| 5M Special Installations | 15,212 | 26.92 | 1,544 | 2.59 | | |
| 5N Builders Work to Services | 21,636 | 38.29 | 14,324 | 23.99 | 4,851 | 15.85 |
| 5O Builders Profit and Attendance on Services | 11,324 | 20.04 | 7,497 | 12.56 | 2,539 | 8.30 |
| Totals | 212,471 | 376.04 | 134,328 | 225.02 | 43,124 | 140.94 |

## Approximate Estimating

### ELEMENTAL COSTS

DISTRICT GENERAL HOSPITAL (cont'd)

| Element | Medical Records 16 points Area m² 204 Cost £ | Rate m² | Staff Changing 436 places Area m² 156 Cost £ | Rate m² | Med./Nursing Management 5 offices Area m² 64 Cost £ | Rate m² |
|---|---|---|---|---|---|---|
| 5A Sanitary Appliances |  |  | 13,702 | 87.83 |  |  |
| 5B Services Equipment |  |  |  |  |  |  |
| 5C Disposal Installation | 574 | 2.81 | 6,385 | 40.93 | 180 | 2.81 |
| 5D Water Installation |  |  | 17,281 | 110.78 |  |  |
| 5E Heat Source |  |  |  |  |  |  |
| 5F Space Heating/Air Treatment | 12,982 | 63.64 | 13,930 | 89.29 | 12,226 | 191.03 |
| 5G Ventilating System | 3,808 | 18.67 | 5,101 | 32.70 | 5,019 | 78.42 |
| 5H Electrical Installation | 15,202 | 74.52 | 8,705 | 55.80 | 26,006 | 406.34 |
| 5I Gas Installation |  |  |  |  |  |  |
| 5J Lift/Conveyor Installation |  |  |  |  |  |  |
| 5K Protective Installations | 3,840 | 18.82 | 2,935 | 18.81 | 1,203 | 18.80 |
| 5L Communication Installation | 3,392 | 16.63 | 2,497 | 16.01 | 4,189 | 65.45 |
| 5M Special Installations |  |  |  |  | 1,031 | 16.11 |
| 5N Builders Work to Services | 5,184 | 25.41 | 8,440 | 54.10 | 5,697 | 89.02 |
| 5O Builders Profit and Attendance on Services | 2,713 | 13.30 | 4,419 | 28.33 | 3,008 | 47.00 |
| Totals | 47,695 | 233.80 | 83,395 | 534.58 | 58,559 | 914.98 |

| Element | Education 2 rooms Area m² 33 Cost £ | Rate m² | Rehab. (part) (excl. gym pool & workshop) Area m² 537 Cost £ | Rate m² | On call Suite 3 rooms Area m² 29 Cost £ | Rate m² |
|---|---|---|---|---|---|---|
| 5A Sanitary Appliances |  |  | 11,956 | 22.26 |  |  |
| 5B Services Equipment |  |  | 5,447 | 10.14 |  |  |
| 5C Disposal Installation | 92 | 2.79 | 9,081 | 16.91 | 80 | 2.76 |
| 5D Water Installation |  |  | 24,306 | 45.26 |  |  |
| 5E Heat Source |  |  |  |  |  |  |
| 5F Space Heating/Air Treatment | 1,807 | 54.76 | 38,473 | 71.64 | 6,028 | 207.86 |
| 5G Ventilating System | 796 | 24.12 | 25,446 | 47.39 | 3,460 | 119.31 |
| 5H Electrical Installation | 1,371 | 41.55 | 36,405 | 67.79 | 9,658 | 333.03 |
| 5I Gas Installation |  |  | 652 | 1.21 |  |  |
| 5J Lift/Conveyor Installation |  |  |  |  |  |  |
| 5K Protective Installations | 620 | 18.79 | 10,105 | 18.82 | 544 | 18.76 |
| 5L Communication Installation | 1,143 | 34.64 | 13,418 | 24.99 | 2,055 | 70.86 |
| 5M Special Installations | 1,031 | 31.24 | 2,573 | 4.79 | 3,227 | 111.28 |
| 5N Builders Work to Services | 858 | 26.00 | 21,374 | 39.80 | 2,879 | 99.28 |
| 5O Builders Profit and Attendance on Services | 450 | 13.64 | 11,187 | 20.83 | 1,508 | 52.00 |
| Totals | 8,168 | 247.53 | 210,423 | 391.83 | 29,439 | 1015.14 |

## Approximate Estimating
## ELEMENTAL COSTS

DISTRICT GENERAL HOSPITAL (cont'd)

| Element | Dental (Excl. W'shop Recovery) 16 sessions Area m² 23 | | E.N.T. (Part) Area m² 93 | | Laboratory Services (Outstation) Area m² 69 | |
|---|---|---|---|---|---|---|
| | Cost £ | Rate m² | Cost £ | Rate m² | Cost £ | Rate m² |
| 5A Sanitary Appliances | | | | | | |
| 5B Services Equipment | | | | | | |
| 5C Disposal Installation | 168 | 7.30 | 260 | 2.80 | 499 | 7.23 |
| 5D Water Installation | | | | | | |
| 5E Heat Source | | | | | | |
| 5F Space Heating/Air Treatment | 1,547 | 67.26 | 11,609 | 124.83 | 6,869 | 99.55 |
| 5G Ventilating System | 479 | 20.83 | 3,593 | 38.63 | 4,542 | 65.83 |
| 5H Electrical Installation | 1,607 | 69.87 | 12,046 | 129.53 | 6,276 | 90.96 |
| 5I Gas Installation | 1,829 | 79.52 | | | | |
| 5J Lift/Conveyor Installation | | | | | | |
| 5K Protective Installations | 433 | 18.83 | 1,751 | 18.83 | 1,299 | 18.83 |
| 5L Communication Installation | | | 1,975 | 21.24 | 985 | 14.28 |
| 5M Special Installations | 2,057 | 89.43 | 1,031 | 11.09 | 15,335 | 222.25 |
| 5N Builders Work to Services | 973 | 42.30 | 3,864 | 41.55 | 4,192 | 60.75 |
| 5O Builders Profit and Attendance on Services | 508 | 22.09 | 2,023 | 21.75 | 2,196 | 31.83 |
| Totals | 9,601 | 417.43 | 38,152 | 410.25 | 42,193 | 611.51 |

| Element | Mortuary/P.M. 18 places 2 P.M. rooms Area m² 198 | | Dining 600 meals Area m² 460 | | Kitchen 1200 meals Area m² 726 | |
|---|---|---|---|---|---|---|
| | Cost £ | Rate m² | Cost £ | Rate m² | Cost £ | Rate m² |
| 5A Sanitary Appliances | 28,819 | 145.55 | | | 17,434 | 24.01 |
| 5B Services Equipment | 1,114 | 5.63 | 31,453 | 68.38 | 232,634 | 320.43 |
| 5C Disposal Installation | 13,147 | 66.40 | 1,293 | 2.81 | 5,873 | 8.09 |
| 5D Water Installation | 17,269 | 87.22 | 14,375 | 31.25 | 131,035 | 180.49 |
| 5E Heat Source | | | | | | |
| 5F Space Heating/Air Treatment | 68,962 | 348.29 | 51,673 | 112.33 | 84,617 | 116.55 |
| 5G Ventilating System | 18,983 | 95.87 | 14,976 | 32.56 | 24,695 | 34.02 |
| 5H Electrical Installation | 12,912 | 65.21 | 19,756 | 42.95 | 60,884 | 83.86 |
| 5I Gas Installation | | | 4,124 | 8.97 | 11,241 | 15.48 |
| 5J Lift/Conveyor Installation | | | | | | |
| 5K Protective Installations | 3,726 | 18.82 | 8,657 | 18.82 | 13,663 | 18.82 |
| 5L Communication Installation | 3,329 | 16.81 | 3,745 | 8.14 | 11,652 | 16.05 |
| 5M Special Installations | 59,345 | 299.72 | | | 41,823 | 57.61 |
| 5N Builders Work to Services | 26,329 | 132.97 | 18,837 | 40.95 | 77,443 | 106.67 |
| 5O Builders Profit and Attendance on Services | 13,770 | 69.55 | 9,828 | 21.37 | 40,476 | 55.75 |
| Totals | 267,705 | 1352.04 | 178,717 | 388.53 | 753,470 | 1037.83 |

**Approximate Estimating**

**ELEMENTAL COSTS**

DISTRICT GENERAL HOSPITAL (cont'd)

| Element | H.D.S.U. 4 theatres 300 beds Area m² 569 | | Stores Area m² 749 | | Linen Store at source 300 beds Area m² 96 | |
|---|---|---|---|---|---|---|
| | Cost £ | Rate m² | Cost £ | Rate m² | Cost £ | Rate m² |
| 5A  Sanitary Appliances | 2,062 | 3.62 | 454 | 0.61 | | |
| 5B  Services Equipment | 263,121 | 462.43 | 2,313 | 3.09 | | |
| 5C  Disposal Installation | 2,074 | 3.64 | 2,302 | 3.07 | 269 | 2.80 |
| 5D  Water Installation | 142,338 | 250.15 | 1,841 | 2.46 | | |
| 5E  Heat Source | | | | | | |
| 5F  Space Heating/Air Treatment | 62,897 | 110.54 | 36,743 | 49.06 | 10,497 | 109.34 |
| 5G  Ventilating System | 18,422 | 32.38 | 9,887 | 13.20 | 2,823 | 29.41 |
| 5H  Electrical Installation | 13,188 | 23.18 | 33,395 | 44.59 | 5,566 | 57.98 |
| 5I  Gas Installation | | | 3,373 | 4.50 | | |
| 5J  Lift/Conveyor Installation | | | | | | |
| 5K  Protective Installations | 10,709 | 18.82 | 14,096 | 18.82 | 1,807 | 18.82 |
| 5L  Communication Installation | 3,331 | 5.85 | 7,489 | 10.00 | 1,248 | 13.00 |
| 5M  Special Installations | 3,427 | 6.02 | 21,892 | 29.23 | | |
| 5N  Builders Work to Services | 63,564 | 111.71 | 18,830 | 25.14 | 2,956 | 30.79 |
| 5O  Builders Profit and Attendance on Services | 33,225 | 58.39 | 9,855 | 13.16 | 1,547 | 16.11 |
| Totals | 618,358 | 1086.73 | 162,470 | 216.93 | 26,713 | 278.25 |

| Element | Works Dept. 300 beds Area m² 425 | | Telephone Exchange 400 extensions Area m² 96 | | Auxiliary Buildings Area m² 399 | |
|---|---|---|---|---|---|---|
| | Cost £ | Rate m² | Cost £ | Rate m² | Cost £ | Rate m² |
| 5A  Sanitary Appliances | 4,647 | 10.93 | | | | |
| 5B  Services Equipment | 15,221 | 35.81 | | | | |
| 5C  Disposal Installation | 5,090 | 11.98 | 269 | 2.80 | 753 | 1.89 |
| 5D  Water Installation | 13,192 | 31.04 | | | | |
| 5E  Heat Source | | | | | | |
| 5F  Space Heating/Air Treatment | 52,491 | 123.51 | 15,747 | 164.03 | | |
| 5G  Ventilating System | 14,123 | 33.23 | 4,236 | 44.13 | | |
| 5H  Electrical Installation | 44,527 | 104.77 | 5,566 | 57.98 | 35,037 | 87.81 |
| 5I  Gas Installation | | | | | | |
| 5J  Lift/Conveyor Installation | | | | | | |
| 5K  Protective Installations | 7,997 | 18.82 | 1,807 | 18.82 | 7,508 | 18.82 |
| 5L  Communication Installation | 4,576 | 10.77 | 416 | 4.33 | 7,416 | 18.59 |
| 5M  Special Installations | | | | | | |
| 5N  Builders Work to Services | 20,184 | 47.49 | 3,608 | 37.58 | 8,097 | 20.29 |
| 5O  Builders Profit and Attendance on Services | 10,565 | 24.86 | 1,889 | 19.68 | 4,238 | 10.62 |
| Totals | 192,613 | 453.21 | 33,538 | 349.35 | 63,049 | 158.02 |

## Approximate Estimating
### ELEMENTAL COSTS

DISTRICT GENERAL HOSPITAL (cont'd)

| Element | Communications Area m² 2,432 | | Stores Area m² 365 | | TOTAL Area m² 16,970 | |
|---|---|---|---|---|---|---|
| | Cost £ | Rate m² | Cost £ | Rate m² | Cost £ | Rate m² |
| 5A Sanitary Appliances | | | | | 335,587 | 19.78 |
| 5B Services Equipment | | | | | 631,960 | 37.24 |
| 5C Disposal Installation | 21,785 | 8.96 | 5,097 | 13.96 | 212,582 | 12.53 |
| 5D Water Installation | | | 130,097 | 356.43 | 733,630 | 43.23 |
| 5E Heat Source | | | 439,022 | 1202.80 | 439,022 | 25.87 |
| 5F Space Heating/Air Treatment | 68,906 | 28.33 | 226,277 | 619.94 | 1,328,431 | 78.28 |
| 5G Ventilating System | 29,770 | 12.24 | 39,452 | 108.09 | 550,532 | 32.44 |
| 5H Electrical Installation | 56,821 | 23.36 | 189,286 | 518.59 | 1,261,911 | 74.36 |
| 5I Gas Installation | | | 8,824 | 24.18 | 31,997 | 1.89 |
| 5J Lift/Conveyor Installation | 179,074 | 73.63 | | | 179,074 | 10.55 |
| 5K Protective Installations | 45,765 | 18.82 | 6,869 | 18.82 | 319,348 | 18.82 |
| 5L Communication Installation | 4,374 | 1.80 | 4,449 | 12.19 | 367,418 | 21.65 |
| 5M Special Installations | 106,922 | 43.96 | | | 623,517 | 36.74 |
| 5N Builders Work to Services | 68,733 | 28.26 | 73,714 | 201.96 | 805,194 | 47.45 |
| 5O Builders Profit and Attendance on Services | 35,975 | 14.79 | 39,202 | 107.40 | 421,465 | 24.84 |
| Totals | 618,125 | 254.15 | 1,162,289 | 3184.36 | 8,241,668 | 485.67 |

PART TWO

# Rates of Wages and Working Rules

*Mechanical Installations, page 53*
*Electrical Installations, page 83*

# Site Management of Building Services Contractors

**Jim Wild, Building Services Management Consultant, Berkshire, UK**

There is an increasingly wide range of building engineering services. When installed, these must sustain and protect a specified internal environment to the satisfaction of the client, designers, insurers and the relevant authorities.

Managing building services contractors can prove to be a minefield. The most successful jobs will always be those where building site managers have first built teams focused on tackling issues that might cause adversarial attitudes later on and jeopardize the project.

The author shows how a simple common management approach can improve site managers' competency in overseeing building services contractors, sub traders and specialists, and maximize the effectiveness of time spent on building services. By providing an account of building services from the site management rather than the corporate viewpoint, this book breaks new ground.

**Site Management of Building Services Contractors:**
- provides step by step guidance from pre award to post handover;
- emphasises system based nature of building services contracts;
- covers risk management throughout the process.

**Contents:**
Overview of building services. Basic planning strategy. Project plans for quality, safety and the environment. Planning and programming. Schedules. Supervision and inspection. Assessing construction progress. Commissioning and its management. The management of defects. Handover. Getting help. Summary. Index

January 1997: 246x189mm
384 pp: 32 line illustrations
Hardback: 0 419 20450 4: £35.00

# E & FN SPON

An Imprint of
Thomson Professional

# Mechanical Installations
## RATES OF WAGES AND WORKING RULES

**RATES OF WAGES**

HEATING, VENTILATING, AIR CONDITIONING, PIPING AND DOMESTIC ENGINEERING INDUSTRY

**Extracts from National Agreement made between:**

| | |
|---|---|
| Heating and Ventilating and<br>Contractor's Association<br>ESCA House,<br>34 Palace Court,<br>Bayswater,<br>London W2 4JG<br>Telephone: 0171-229 2488 | Manufacturing Science Finance<br>Union (MSF)<br>Park House,<br>64-66 Wandsworth Common<br>North Side<br>London SW18 2SH<br>Telephone: 0181-871 2100 |

WAGE RATES, ALLOWANCES AND OTHER PROVISIONS

**Hourly rates of wages**
All districts of the United Kingdom

| Main grades | % of Fitters rate | From 2 September 1996*<br>p/hr | From 11 November 1996<br>p/hr |
|---|---|---|---|
| Pipework Foreman | 134 | 729 | 747 |
| Ductwork Erector Foreman | 129 | 703 | 721 |
| Chargehand | 124 | 677 | 694 |
| Advanced Fitter | 109 | 598 | 613 |
| Fitter | 100 | 546 | 560 |
| Improver | 95 | 520 | 533 |
| Assistant | 90 | 490 | 502 |
| Mate over 18 | 80 | 438 | 449 |
| Junior Mate 17-18 | 51 | 281 | 288 |
| Junior Mate up to 17 | 37 | 203 | 208 |

*Note*: Ductwork Erection Operatives are entitled to the same rates and allowances as the parallel Fitter grades shown (Note 1)

| Craft Apprentice | From 2 September 1996*<br>p/hr | From 11 November 1996<br>p/hr |
|---|---|---|
| Year 1 - if taken on as an apprentice | 234 | 240 |
| Year 2 | 281 | 288 |
| Year 3 | 359 | 368 |
| Year 4 | 438 | 449 |

* 2 September 1996 agreement promulgated as JCC letter 62.

## RATES OF WAGES

Junior Ductwork Erectors (Probationary)
Trainee Rates of Pay

| Age at entry | From 2 September 1996* p/hr | From 11 November 1996 p/hr |
|---|---|---|
| 17 | 234 | 240 |
| 18 | 281 | 288 |
| 19 | 359 | 368 |
| 20 | 438 | 449 |

Junior Ductwork Erectors (Year of Training)

| | From 2 September 1996* | | | From 11 November 1996 | | |
|---|---|---|---|---|---|---|
| | 1 yr p/h | 2 yr p/hr | 3 yr p/hr | 1 yr p/hr | 2 yr p/hr | 3 yr p/hr |
| 17 | 281 | 359 | 438 | 288 | 368 | 449 |
| 18 | 359 | 438 | 467 | 368 | 449 | 479 |
| 19 | 438 | 442 | 489 | 449 | 453 | 501 |
| 20 | 438 | 464 | 490 | 449 | 476 | 502 |

* 2 September 1996 agreement promulagated as JCC letter 62.

| Welding supplements | From 2 September 1996 p/hr | From 11 November 1996 p/hr |
|---|---|---|
| gas/arc | 52 | 53 |
| gas or arc | 26 | 27 |

Payable to Fitters and Advanced Fitters qualified in accordance with Clause 8f and becomes part of the normal hourly rate

**Daily travelling allowance**
C: Craftsmen including Improvers
M&A: Assistants, Mates, Junior Mates and Craft Apprentices
Direct distance from centre to job in miles

| | | From 2 September 1996 | | From 1 September 1997 | |
|---|---|---|---|---|---|
| Over | Not exceeding | C p/hr | M&A p/hr | C p/hr | M&A p/hr |
| 0 | 2 | - | - | - | - |
| 2 | 5 | - | - | - | - |
| 5 | 10 | - | - | - | - |
| 10 | 15 | 179 | 152 | 310 | 267 |
| 15 | 20 | 302 | 260 | 310 | 267 |
| 20 | 25 | 427 | 366 | 557 | 481 |
| 25 | 30 | 543 | 469 | 557 | 481 |
| 30 | 35 | 629 | 556 | 734 | 632 |
| 35 | 40 | 716 | 617 | 734 | 632 |
| 40 | 45 | 805 | 692 | 915 | 784 |
| 45 | 50 | 893 | 765 | 915 | 784 |

## RATES OF WAGES

*From 30 September 1996*

|  | £ a | £ b | £ c | £ d | £ e | £ f | £ g | £ h | £ i |
|---|---|---|---|---|---|---|---|---|---|
| Weekly Holiday Credit | 31.73 | 30.77 | 28.75 | 27.54 | 25.19 | 22.12 | 18.46 | 13.27 | 8.85 |
| Weekly Welfare Premium | 3.85 | 3.85 | 3.85 | 3.85 | 3.85 | 3.85 | 3.85 | 3.85 | 3.85 |
| Total | 35.58 | 34.62 | 32.60 | 31.39 | 29.04 | 25.97 | 22.31 | 17.12 | 12.70 |

|  | *From* 2 September 1996 | *From* 11 November 1996 |
|---|---|---|
| **Daily abnormal conditions money** p per day | 250 | 256 |
| **Exposed work at heights money** Over 125 ft, p per day | 100 | 102 |
| **Swings, cradles and ladders money** p per hour | 26 | 27 |
| **Lodging allowance** £ per day | 19.60 | 20.50 |

Weekly Sickness and Accident Benefit - Payable in accordance with the Rates of the WELPLAN Welfare and Holiday Scheme Supplement to the National Agreement.

|  | *From 16 October 1995* | | | *From 30 September 1996* | |
|---|---|---|---|---|---|
|  | Weeks 1-28 | Weeks 29-52 |  | Weeks 1-28 | Weeks 29-52 |
| Category a | 86.17 | 58.66 | Category a | 153.30 | 76.65 |
| Category b | 78.94 | 43.26 | Category b | 142.17 | 71.05 |
| Category c | 59.99 | 27.79 | Category c | 127.33 | 63.70 |
| Category d | 40.74 | 9.38 | Category d | 116.20 | 58.10 |
| Category e | 39.20 | - | Category e | 105.14 | 52.57 |
|  |  |  | Category f | 84.00 | 42.00 |
|  |  |  | Category g | 61.25 | 30.59 |
|  |  |  | Category h | 23.03 | - |
|  |  |  | Category i | 4.20 | - |

|  | *From 16 October 1995* |
|---|---|
| Death Benefit for Dependants | 20,500 |
| Accidental Dismemberent | 13,800 |
| **Permanent Total Disability Benefit** | |
| - Upto and including age 54 | 14,000 |
| - Between ages 55-59 | 8,400 |
| - Between ages 60-64 | 4,200 |
| **Index Benefits** | |
| - Loss of four fingers or thumb | 2,500 |
| - Loss of index finger | 1,500 |
| - Loss of any other finger | 250 |
| - Loss of big toe | 500 |
| - Loss of any other toe | 135 |

# RATES OF WAGES

**Notes**

1. The grades of Ductwork Erection Operatives as defined in Clause 6g of the National Agreement are:
   Ductwork Chargehand Erector
   Ductwork Advanced Erector
   Ductwork Erector

2. From October 1995, the grades of Operatives covered by the range of credit values and entitled to the different rates of Sickness and Accident Benefit, Weekly Holiday Credit and Welfare Contribution are as follows:

| a | b | c | d | e |
|---|---|---|---|---|
| Foreman Chargehand* | Advanced Fitter qualified gas/arc Advanced Fitter qualified gas or arc Advanced Fitter* Fitter qualified gas/arc | Fitter qualified gas or arc Fitter* Improver 4th year Apprentice | Assistant Mate 3rd Year Apprentice | 1st and 2nd Year Apprentices and Junior Mates aged 17/18 and 16/17 |

* Ductwork Erection Operatives are entitled to the same credit values and rates of sickness and accident benefit as the parallel Fitter grade shown.

3. From 30 September 1996, the grades of Operatives covered by the range of credit values and entitled to the different rates of Sickness and Accident Benefit, Weekly Holiday Credit and Welfare Contribution are as follows:

| a | b | c | d | e |
|---|---|---|---|---|
| Foreman (Pipefitter) | Foreman (Ductwork) Chargehand | Advanced fitter qualified gas/arc Advanced fitter qualified gas or arc | Advanced fitter Fitter qualified gas/arc | Fitter qualified gas or arc Fitter* |

| f | g | h | i |
|---|---|---|---|
| Improver Assistant 4th year Apprentice Senior Modern Apprentice | Mate over 18 3rd year Apprentice | 2nd year Apprentice Intermedite Modern Apprentice Junior Mate aged 17/18 and 16/17 | 1st year Apprentice Junior Modern Apprentice |

4. Payment of death benefit is subject to Inland Revenue requirements which currently provide that it may not exceed four times annual earnings of the deceased, subject to a minimum of £5,000.

*Rates of Wages and Working Rules - Mechanical Installations*      57

## RATES OF WAGES

**Authorised rates of wages agreed by the Joint Industry Board for the Plumbing Mechanical Engineering Services Industry in England and Wales**

The Joint Industry Board for Plumbing Mechanical
Engineering Services in England and Wales,
Brook House, Brook Street,
St Neots, Huntingdon, Cambs. PE19 2HW      Telephone: 01480 476925

### WAGE RATES, ALLOWANCES AND OTHER PROVISIONS

EFFECTIVE FROM 1 MAY 1997

All districts of the United Kingdom

|  | Hourly rate £ |
|---|---|
| Technical plumber and gas service technician | 6.78 |
| Advanced plumber and gas service engineer | 6.07 |
| Trained plumber and gas service fitter | 5.42 |
| Apprentices |  |
| 1st year of training | 2.02 |
| 2nd year of training | 2.58 |
| 3rd year of training | 3.19 |
| 3rd year of training with NVQ level 2* | 3.89 |
| 4th year of training | 3.92 |
| 4th year of training with NVQ level 2* | 4.47 |
| 4th year of training with NVQ level 3* | 4.93 |

* Where Apprentices have achieved NVQ's, the appropriate rate is payable from the date of attainment except that it shall not be any earlier than the commencement of the promulgation year of training in which it applies.

| Adult Trainees |  |
|---|---|
| 1st 6 months of employment | 4.26 |
| 2nd 6 months of employment | 4.54 |
| 3rd 6 months of employment | 4.74 |

### Overtime

Overtime premium rates shall be paid after 39 hours are worked. The working week shall be 37.5 hours.

### Daily travelling allowance plus return fares

All daily travel allowances are to be paid at the daily rate as follows (as at 5 May 1997):

|  | Miles | | | | | | | | |
|---|---|---|---|---|---|---|---|---|---|
| Over | 5 | 10 | 15 | 20 | 25 | 30 | 35 | 40 | 45 |
| Not exceeding | 10 | 15 | 20 | 25 | 30 | 35 | 40 | 45 | 50 |
|  | £ | £ | £ | £ | £ | £ | £ | £ | £ |
| Technical plumber, etc | 3.21 | 4.03 | 4.82 | 5.64 | 6.42 | 7.22 | 7.52 | 7.75 | 7.98 |
| Advanced plumber, etc | 2.81 | 3.52 | 4.22 | 4.92 | 5.65 | 6.34 | 6.65 | 6.85 | 7.05 |
| Trained plumber, etc | 2.55 | 3.21 | 3.83 | 4.49 | 5.12 | 5.76 | 6.06 | 6.24 | 6.42 |
| Apprentice |  |  |  |  |  |  |  |  |  |
| 1st year | 0.77 | 0.95 | 1.16 | 1.35 | 1.53 | 1.72 | 1.82 | 1.87 | 1.92 |
| 2nd year | 1.15 | 1.43 | 1.71 | 1.99 | 2.27 | 2.56 | 2.71 | 2.79 | 2.87 |
| 3rd year | 1.40 | 1.74 | 2.09 | 2.45 | 2.79 | 3.16 | 3.36 | 3.46 | 3.56 |
| 4th year | 1.76 | 2.23 | 2.66 | 3.13 | 3.57 | 4.02 | 4.27 | 4.40 | 4.53 |

## Rates of Wages and Working Rules - Mechanical Installations

### RATES OF WAGES

For all distances over 50 miles operatives to be paid lodging allowance in accordance with the rules. The travel allowances set out are to be paid when public transport is used.

| | |
|---|---|
| Mileage allowance (see new working rule 8.4.3(c)) | 0.26 per mile |
| **Abnormal conditions** | £2.00 per day |
| **Lodging allowance (Note 1)** | £18.00 per night |
| **Subsistence Allowance (London Only) (Note 1)** | £4.18 per night |

Note 1 Important - Taxation Treatment

Please note that by way of concession from the Inland Revenue the Lodging Allowance of £18.00 is payable without the deduction of income tax. The subsistence allowance is subject to Schedule E Income Tax through PAYE System.

| | |
|---|---|
| **Responsibility money** | £0.33 per hour |
| **Plumbers welding supplement** | |
| Possession of Gas or Arc Certificate | £0.25 per hour |
| Possession of Gas and Arc Certificate | £0.44 per hour |
| **Tool allowance** | £2.24 per week |

### Sickness with Pay and Accident Benefits (Rule 9)
Effective from 3 April 1995

| | Weeks 1-28 Daily £ | Weeks 1-28 Weekly £ | Weeks 29-52 Daily £ | Weeks 29-52 Weekly £ |
|---|---|---|---|---|
| Technical Plumber and Gas Service Technician | 10.20 | 71.40 | 7.15 | 50.05 |
| Advanced Plumber and Gas Service Engineer | 8.70 | 60.90 | 5.65 | 39.55 |
| Trained Plumber and Gas Service Fitter | 7.50 | 52.50 | 4.40 | 30.80 |
| Adult Trainee | 7.50 | 52.50 | 4.40 | 30.80 |
| Apprentice in last year of training | 7.50 | 52.50 | 4.40 | 30.80 |
| Apprentice 2nd to 3rd year of training | 6.00 | 42.00 | 3.00 | 21.00 |
| 1st year Apprentice | 1.00 | 7.00 | N/A | N/A |
| Ancillary employee | 6.50 | 45.50 | 3.60 | 25.20 |

**Accident Permanent Total Disability/Dismemberment Benefit (Rule 9)** £3.50 per benefit

Death Benefit (as at 1 May 1995)

| | £ |
|---|---|
| Technical Plumber and Gas Service Technician | 24,687 |
| Advanced Plumber and Gas Service Engineer | 22,074 |
| Trained Plumber and Gas Service Fitter | 19,734 |
| Apprentice 1st year of Training | 6,162 |
| Apprentice 2nd year of Training | 9,477 |
| Apprentice 3rd year of Training | 11,700 |
| Apprentice 3rd year of Training with NVQ Level 2 | 14,313 |
| Apprentice 4th year of Training | 14,430 |
| Apprentice 4th year of Training with NVQ Level 2 | 16,419 |
| Apprentice 4th year of Training with NVQ level 3 | 18,096 |
| Adult Trainee first 6 months of training | 15,678 |
| Adult Trainee second 6 months of training | 16,692 |
| Adult Trainee third 6 months of training | 17,433 |
| Ancillary Employees | (Twice the current basic annual pay) |

*Rates of Wages and Working Rules - Mechanical Installations*

## RATES OF WAGES

### Weekly Holiday Credit/Benefit Stamp as from 3 April 1995

|  | Gross Value | Holiday with Pay |
|---|---|---|
| Technical Plumber and Gas Service Technician | 21.42 | 19.28 |
| Advance Plumber and Gas Service Engineer | 19.13 | 16.93 |
| Trained Plumber and Gas Service Fitter | 17.60 | 15.37 |
| Adult Trainee | 17.60 | 15.37 |
| Apprentice in last year of training | 17.60 | 15.37 |
| Apprentice 2nd and 3rd year | 7.85 | 6.63 |
| Apprentice 1st year | 4.79 | 4.63 |
| Working Principal | 11.48 | 9.97 |
| Ancillary Employee | 14.13 | 12.70 |

### Additional Holiday Pay

With effect from 3 April 1995 it has been agreed that Operatives and Apprentices aged over 18 who are in current membership of the PMES Section of the AEEU at the time on Annual Holiday is taken will be entitled to receive an Additional Holiday Payment, based on the number of stamps on each part of their Benefits Stamp card for the year 1995-96 commencing with the Winter Holiday 1996.

| | |
|---|---|
| Operatives | £1.65 per stamp |
| Apprentices over 18 | £1.30 per stamp |

# Rates of Wages and Working Rules - Mechanical Installations

## WORKING RULES

### Extracts from National Working Rules (HVCA)

#### HOURS OF WORK (CLAUSE 3)

(a) The normal working week shall consist of 38 hours to be worked in five days from Monday to Friday inclusive. The length of each normal working day shall be determined by the Employer but shall not be less than six hours or more than eight hours unless otherwise agreed between the Employer and the Operative concerned.

(b) The Employer and the Operative concerned may agree to extend the working hours to more than 38 hours per week for particular jobs, provided that overtime shall be paid in accordance with Clause 9. (Attention is also drawn to the provisions for containing overtime - see Clause 9a).

#### MEAL AND TEA BREAKS (CLAUSE 4)

(a) The meal break which is not included in normal working hours shall, under normal circumstances, be unpaid and shall be one hour except when such a break would make it impossible for the normal working day to be worked, in which case the break may be reduced to not less than half an hour.

(b) An Operative directed to start work before his normal starting time or to continue work after his normal finishing time shall be entitled to a quarter of an hour meal interval with pay at the appropriate overtime rate for each two hours of working (or part thereof exceeding one hour) in excess of the normal working day, which on Saturdays and Sundays shall mean eight hours. Where an Operative is entitled to a morning and/or evening meal interval under this clause, the meal interval shall replace the morning and/or afternoon tea break referred to in 4(c).

(c) A tea break shall, subject to 4(b), be allowed in the morning and in the afternoon without loss of pay, provided that Operatives co-operate with the Employer in minimising the interruption to production. To this end the duration of the tea break shall be limited to the time necessary to drink tea and the tea shall be drunk at the Operatives workplace wherever possible.

#### GUARANTEED WEEK (CLAUSE 5)

(a) Subject to the provisions of this clause an Operative who has been continuously employed by the same Employer for not less than two weeks is guaranteed wages equivalent to his inclusive hourly normal time earnings for 38 hours in any normal working week; provided that during working hours he is capable of, available for and willing to perform satisfactorily the work associated with his usual occupation, or reasonable alternative work if his usual work is not available.

(b) In the case of a week in which holidays recognised by agreement, custom or practice occur, the guaranteed week shall be reduced for each day of holiday by the normal working day as determined in Clause 3(a).

(c) In the event of a dislocation of production as a result of industrial action the guarantee shall be automatically suspended. In the event of such dislocation being caused by Operatives working under other Agreements and the Operatives covered by this Agreement not being parties to the dislocation, the Employers shall, in accordance with Clause 5(d), endeavour to provide other work or if not able to do so will provide for the return of the Operatives to the shop or office from which they were sent. The Operatives will receive instructions as soon as is practicable as to proceedings to other work or return to shop.

(d) The basis upon which the Employer shall endeavour to provide alternative work as required in Clause 5c shall be as follows:
   (i) where possible the Employer shall try to organise work on each job so as to provide a normal day's work for five days, Monday to Friday.
   (ii) where this is not possible on any particular job, the Employer shall endeavour to arrange to transfer Operatives to other sites to make up working hours to a normal day's work for five days, Monday to Friday.
   (iii) where an Employer finds it impossible to provide a normal day's work for five days, Monday to Friday, he should rearrange the working hours in agreement with the Operatives concerned so that normal time earnings for 38 hours in the normal working week can be earned but in less than five days
   (iv) where it is not possible to provide Operatives with a minimum of 38 hours during the week, rather than resort to dismissals a reduced working week may be agreed.

## WORKING RULES

GUARANTEED WEEK (CLAUSE 5) (CONTD)

(e)　In the event of dislocation of production as a result of civil commotion, the guarantee shall be automatically suspended at the termination of the pay week after the dislocation first occurs and the Operative may be required by the Employer to register as an unemployed person.

GRADES AND SKILLS OF OPERATIVES (CLAUSE 6)

(a)　Operatives covered by this Agreement shall be graded in accordance with the definitions in Clauses 6(f) and 6(g).

(f)　The definitions of the grades (other than ductwork grades in Clause 6(g)) shall be:

*Junior Mate*
A Junior Mate is a Mate under 18 years of age.

*Mate*
A Mate must be at least 18 years of age.

*Assistant*
An assistant must have the following qualifications:
  (i)　Have worked in the trade at least three years
  (ii)　Be at least 20 years of age
  (iii)　Be capable of performing semi-skilled tasks including support work for craftsmen without constant direct supervision
  (iv)　Have had at least six months' continuous employment prior to appointment as an assistant with the Employer so appointing him.

With due consideration of the requirements of firms specialising in repetitive or prefabricated work the ratio of Assistants to craftsmen shall not normally exceed one to two employed in each company as a whole (not a branch office or a job by job basis).

*Improver*
An Improver must have either Qualification A or Qualification B.

　Qualification A:
  (i)　Have worked in the trade for at least five years
  (ii)　Have been employed with his current employer for not less than one year
  (iii)　Be at least 24 years of age
  (iv)　Conform with such other requirements that may be laid down from time to time by the Heating Ventilating and Domestic Engineers' National Joint Industrial Council (NJIC) and have received the prior approval of the NJIC. Applications for up-grading shall be made on the form approved by the NJIC, who shall keep a register of Operatives up-graded under this Clause.

　Qualification B:
  (i)　Be at least 24 years of age
  (ii)　Have satisfactorily completed the Department of Employment Adult Training Course in Heating and Ventilating Fitting Craft Practice as approved by the NJIC.

*Craft Apprentice*
A Craft Apprentice shall be undertaking an approved course of apprenticeship by duly executed Agreements of Service in the form prescribed by the NJIC.

*Fitter*
A Fitter must have one of the following qualifications:
  (i)　Have successfully completed an apprenticeship approved by the NJIC, and have passed the practical examination of an appropriate City and Guilds of London Institute basic craft course which has been recognised by the NJIC and approved by the Parties
　　or
  (ii)　Have successfully completed an improvership of one year under Qualification A or 18 months under Qualification B in the reference to Improvers or
  (iii)　Be already employed as a craftsman in the industry on 24 February 1969.

A Fitter who is qualified in accordance with Clause 8(f) shall receive the appropriate welding supplement.

## WORKING RULES

*Advanced Fitter*
An Advanced Fitter must have one of the qualifications of a Fitter and in addition:
- (i) Must have had at least five years' service in the industry as a craftsman or since successful completion of an approved basic craft course as defined in the Fitter grade
- (ii) Must have technical knowledge and skill beyond that of a Fitter including <u>either</u> successful completion of an appropriate City and Guilds of London Institute advanced craft course which has been recognised by the NJIC and approved by the Parties <u>or</u> competence, both practical and theoretical, in at least one of the following:
  - Commissioning and testing of simple systems
  - Layout and installation of plant and associated pipework
  - fault diagnosis
  - installation of large bore pipework over six inches
  - installation of high pressure steam and hot water pipework
- (iii) Must have general competence and organising ability beyond that of a Fitter so that he is able without detailed supervision to set out jobs from drawings and specifications, requisition sundry materials, work in an efficient and economical manner and liaise effectively with other trades.

An Advanced Fitter who is qualified in accordance with Clause 8(f) shall receive the appropriate welding supplement.

*Chargehand*
A craftsman who is designated by the Employer as a Chargehand to carry out Chargehand duties in the course of working with the tools of the trade, shall receive remuneration dependent on the character of the charge as provided below:
- (i) Where the Chargehand duties require the Chargehand to take sole responsibility for smaller contracts with an average labour force including three other craftsmen and/or of a gang of that size on larger contracts, the Chargehand shall receive the appropriate rate for the Chargehand grade and his rate shall include for any welding skill.
- (ii) Where the charge is of a lesser character than that detailed in (i) above, the Chargehand shall receive a remuneration which shall be agreed between the Chargehand and the Employer.

A Chargehand may be appointed on a temporary or short term basis to cover peak periods of working or to allow a craftsman to gain experience in that role and shall be paid in accordance with (i) or (ii) above, as appropriate.

*Foreman*
A craftsman who satisfies the qualifications of an Advanced Fitter may be designated by the employer as a Foreman and shall receive the rate for the Foreman grade, provided he is competent to perform all the duties listed below (or the vast majority of them as appropriate to and in accordance with the requirements of the Employer):
- (i) Assign tasks to the Chargehands and other Operatives under his direct control
- (ii) Redeploy Chargehands or Operatives under his direct control, in order to achieve the optimum productivity including on-site batch production and fabrication
- (iii) Decide methods to be used for individual operations and instruct Operatives accordingly
- (iv) Ensure variation work does not proceed without authority from the office
- (v) Maintain site contract control procedure
- (vi) Requisition and progress supply of necessary equipment and materials to Operatives when required
- (vii) Ensure that Operatives take all reasonable steps to safeguard, maintain and generally take care of Employer's tools and materials
- (viii) Maintain day to day liaison and programme of work with main contractor and other sub-contractors
- (ix) Inspect and review progress of work of sub-contractors
- (x) Monitor progress of main contractor, in order that agreed programme is met
- (xi) Measure and record progress of work
- (xii) Inspect the work of Operatives for quality, progress and satisfactory completion
- (xiii) Check weekly progress against programme and identify deviations therefrom
- (xiv) Verify bookings on time and job cards and despatch them promptly to the office
- (xv) Notify office of impending delays likely to affect progress or give rise to a claim
- (xvi) Establish reasons for delays to work and notify office
- (xvii) Provide information for cost variation investigations when necessary

*Rates of Wages and Working Rules - Mechanical Installations*

## WORKING RULES

- (xviii) Forecast labour requirements
- (xix) Ensure company instructions and standards of discipline, workmanship and safety are maintained on site
- (xx) Ensure that the conditions of the National Agreement and any other conditions of employment are complied with
- (xxi) Supervise training of Craft Apprentices assigned to his control
- (xxii) Take overall charge of all his Employer's labour on site and act where necessary as the Employer's site agent
- (xxiii) Evolve and/or agree order of work within overall programme and control its progress
- (xxiv) Decide or agree locations of site office, site stores, site workshop and other work stations and adjust same to suit site progress and changing conditions
- (xxv) Ensure compliance of all work, whether executed by own Operatives or sub-contractors with drawings and specifications
- (xxvi) Organise, supervise and record such tests (e.g. hydraulic) and/or inspections as are required during progress of contract
- (xxvii) Requisition or otherwise procure such attendances and facilities as are required of the main contractors and/or of other sub-contractors
- (xxviii) Attend site meetings (if so required by the Employer)
- (xxix) Ensure that safe methods of work are adopted by Operatives under his direct control
- (xxx) Ensure clearance of rubbish as specified
- (xxxi) Arrange and supervise testing on completion, including compliance with specifications, snagging and operational handing over as directed and final site clearance
- (xxxii) Such other duties as are reasonably required by the Employer

(g) The definitions of the ductwork grades shall be:

*Junior Ductwork Trainee*

A Junior Ductwork Trainee shall undertake the approved in-company scheme of training as set out in Clause 23.

*Ductwork Erector*

A Ductwork Erector shall be at least 20 years of age; or shall have trained in accordance with the training agreement set out in Clause 23; or shall have undergone training in the ductwork industry for not less than four years; shall be able, without constant direct supervision, to carry out the installation of ductwork and have the skill to perform the following duties (or such others as may from time to time be agreed upon between the Association and the Union):

- (i) Set out duct runs and plant equipment from drawings to datum lines
- (ii) Make simple supports on site and fix supports of all types
- (iii) Fix items of plant within the duct runs, together with grilles, diffusers, terminal units etc., and make final connections to ductwork
- (iv) Position and assemble sections of air handling units, complete with fans and associated equipment
- (v) Operate all machines and equipment used in ductwork erection other than those used for manual welding (Note - It is agreed that the spot welding of steel sheet shall be included in the list of the Ductwork Erector's skills)
- (vi) Carry out on-site minor modifications and repairs
- (vii) Measure and specify any simple make up pieces of ductwork
- (viii) Blank off and prepare ductwork runs for pressure testing of high-velocity high-pressure ductwork
- (ix) Fit thermostats, temperature probes and control equipment
- (x) Sling and hoist loads of various shapes and sizes, provided adequate training has been given.
- (xi) Rig portable and temporary working platforms to conform with safety regulations
- (xii) Assemble and erect light steel structures and associated sheet metal cladding, including insulation
- (xiii) Have knowledge (by instructions and training) of and comply with all site safety regulations
- (xiv) Be responsible for grades junior to the Ductwork Erector who directly assist him in carrying out any of the above operations

## WORKING RULES

(xv) Undertake any training in relation to his own grading and co-operate in the training of the other Operatives junior to him
(xvi) Carry out any other operations incidental to the foregoing

*Ductwork Advanced Erector*
A Ductwork Advanced Erector shall be at least 22 years of age and shall have had at least two years' service as a Ductwork Erector with his Employer at the time of upgrading to a Ductwork Advanced Erector. He shall have the qualifications of a Ductwork Erector as set out above and in addition the technical knowledge and skill including competence, both theoretical and practical, in the following duties (or others as may from time to time be agreed between the Association and the Union):

(i) Appreciation of a contract situation and of his Employer's obligations thereunder
(ii) General understanding of the operation of ducted air systems and equipment
(iii) Installation of the ductwork system in sequence with other services
(iv) Measuring and specifying any modifications of ductwork as required
(v) Assigning tasks to Operatives under his control and deploying labour on site as necessary
(vi) Pressure testing of high-velocity high-pressure ductwork systems to the requirements of any relevant HVCA specification
(vii) Installation fault diagnosis in the ductwork system
(viii) Recording, measuring and documenting variations from contract.

He shall have general competence and organising ability beyond that of a Ductwork Erector as defined herein, so that he is able without detailed supervision to set out jobs from drawings and specifications; to requisition sundry materials; to work in an efficient and economical manner; and to liaise effectively with other trades.

A Ductwork Advanced Erector may be designated a 'Temporary Chargehand', for example, to cater for peak periods of work volume, high labour force and unusual circumstances. Where such appointment is made it shall be for a minimum period of two weeks and the grade rate shall apply during the period which he is designated as Temporary Chargehand, and at the end of the period he shall revert to his previous grade.

*Ductwork Chargehand Erector*
A craftsman who has the qualifications of a Ductwork Advanced Erector may be designated by the Employer as a Ductwork Chargehand Erector to carry out Chargehand duties in the course of working with the tools of the trade, and shall receive the rate for the grade, if he is competent to perform all or most of the duties listed below in accordance with the requirements of the Employer:

(i) Assign tasks to the Ductwork Advanced Erectors and other Operatives under his control
(ii) Deploy Operatives under his control so as to achieve the optimum productivity including off-site assembly and minor modifications
(iii) Decide method to be used for individual operations and instruct Operatives accordingly
(iv) Requisition and progress supply of necessary equipment and materials to Operatives as required
(v) Ensure that Operatives take all reasonable steps to safeguard, maintain and generally take care of the Employer's tools and materials
(vi) Inspect the work of Operatives for quality, progress and satisfactory completion
(vii) Ensure that the Employer's instructions and standards of discipline, workmanship and safety are maintained on site
(viii) Ensure that variation work does not proceed without authority from the Employer
(ix) Verify bookings on time and job cards and despatch them promptly to the Employer
(x) Notify the Employer of any delay likely to affect progress or give rise to additional costs
(xi) Maintain such records of installation progress and variations as may be required by the Employer
(xii) Forecast labour requirements on the job in hand
(xiii) Supervise craft training of Operatives assigned to his control
(xiv) Organise, supervise and record such tests (for example, leakage testing) and/or inspections as are required during the progress of the contract
(xv) Requisition or otherwise procure such attendances, services or facilities as are required from the main contractor and/or other sub-contractors
(xvi) Ensure that safe methods of work are adopted by Operatives under his control
(xvii) Ensure clearance of rubbish and scrap for which the Employer is responsible
(xviii) Ensure compliance with specifications, carry out any necessary remedial work and hand over as required by the contract and clear site of Employer's equipment
(xix) Carry out any other operations incidental to the above.

## WORKING RULES

It is accepted that a Ductwork Chargehand Erector does not necessarily carry this grading with him to a new employer. It is understood that the Ductwork Chargehand Erector's rate of pay shall be dependent on the character of the charge, for example, where his duties require him to take sole responsibility for a small contract or a section or sections of a large contract, and according to the number of Operatives.

(h) Erection, Alteration and Dismantling of Scaffolding and Mobile Towers
Operatives shall, where properly trained or supervised, undertake the erection, alteration and dismantling of mobile towers and easy-fix scaffolding as part of their normal work. The Employer shall ensure that such supervision is undertaken by Operatives who are properly instructed in the necessary working and safety procedures.

BALANCE OF GANGS (CLAUSE 7)

The balance of gangs as between Craftsmen, Assistants, Mates and Craft Apprentices shall be on the basis that:
- (i) Assistants may do semi-skilled tasks including support work for craftsmen without constant direct supervision and shall perform the manual work of fetching and carrying, receiving and checking materials as required by the Employer.
- (ii) Support work for skilled men may be done by the skilled men themselves or by Assistants, Mates or Craft Apprentices in order to secure the maximum utilisation of labour and the optimum economic production; thus one Mate can be used to do the support work for two or more Craftsmen or conversely two or more Mates may work with one Craftsman.
- (iii) Mates shall not be confined to the manual work of fetching and carrying. They shall within their capacity, carry out semi-skilled tasks; one object being to improve productivity and the other being to permit those who wish to do so to qualify for consideration for regrading as Assistants.
- (iv) In order to provide Craft Apprentices with appropriate practical experience and to permit them to make the fullest possible contribution to production, they shall be permitted to work with the tools with the maximum of supervision necessary but always on work which is under the control of a recognised Craftsman. In the case of welding, this shall mean that Craft Apprentices shall not weld until they have completed the appropriate City and Guilds welding course. Craft Apprentices shall not be employed solely and continuously on heavy labouring work.

WAGES AND ALLOWANCES (CLAUSE 8)

(f) *Welding Supplement*
A Fitter or an Advanced Fitter who holds one or both current certificates of Competency issued by the Heating, Ventilating and Domestic Engineer's National Joint Industrial Council, in oxy-acetylene welding and/or metal arc welding and who is competent in such welding to the standard(s) required by such certificate(s) shall receive an hourly welding supplement for one welding skill or for both welding skills as appropriate, which shall be the following percentage of the Fitter's hourly rate:
One welding skill - 5%
Two welding skills - 10%

(g) *Merit Money*
Payment of merit money to an Operative may be made at the option of the Employer for mobility, loyalty, long service etc., and for special skill over and above that detailed in definition of the Operative's grade.

(h) *Payment of Wages*
- (i) Unless otherwise agreed between the Operative and the Employer, the pay week shall end at midnight on Friday and wages shall be paid on the following Thursday.
- (ii) The Employer at his discretion may pay each Operative to the nearest £1 upwards each week carrying the credit forward, deducting it from the next wage payment which is again paid to the nearest £1 upwards.
- (iii) Where any packets cannot be calculated on time sheets, the Employer shall make an assessed payment for the days worked. The pay packet shall be corrected the following week.

## Rates of Wages and Working Rules - Mechanical Installations

### WORKING RULES

(i)  Junior Mate
Junior Mates shall be paid the same rates as Craft Apprentices. They shall be subject to the conditions of employment set out in this Agreement.

(j)  Abnormal Conditions
Operatives engaged on exceptionally dirty work, or work under abnormal conditions, of such a character as to be equally onerous, shall receive an allowance extra per day or part of a day. The determination of the conditions to which this allowance shall apply shall be agreed between the Employer and the Operative concerned in each case. The allowance shall be agreed from time to time by the Association and the Union and shall be enumerated in an Appendix to this Agreement. It shall be in addition to any part payment for exposed work at heights made under Clause 8(k).

(k)  Exposed Work at Heights
An Operative working in exposed conditions at heights over 125 feet on an unclad building having no other form of protection from weather conditions shall be paid an allowance extra per day or part of a day. The allowance shall be agreed from time to time by the Association and the Union and shall be enumerated in an Appendix to this Agreement.

(l)  Swings, Cradles, Ladders
An Operative working in swings or cradles shall be paid an allowance extra per hour for the time actually worked in those conditions. An Operative working on ladders shall be paid an allowance extra per hour for the time actually worked at a height of 20 feet and an additional allowance per hour for each additional 10 feet. The height shall be measured from the nearest fixed flooring or fixed scaffolding to the actual work. The allowances shall be agreed from time to time by the Association and the Union and shall be enumerated in an Appendix to this Agreement.

(m)  Target Incentive Schemes
Where it has been agreed between the Employer and the majority of his workforce, that target incentive schemes shall be operated in connection with works on which they are or are to be employed, such schemes shall be operated in accordance with general principles established by the Association and the Union for their operation, which are set out in Appendix B.

OVERTIME (CLAUSE 9)

(a) It is accepted by the parties that overtime must be contained. To this end, except in cases of urgency or emergency, actual working hours should not exceed:
 (i) An average of 45 per week in the case of travelling men who should only work on Saturdays and/or Sundays in cases of urgency or emergency
 (ii) An average of 55 per week in the case of lodging men whose work on Saturdays and/or Sundays should be reasonably contained
(b) It is also accepted by the Parties that the reference period for calculating compliance with the provisions of the European Directive on Working Time (93/104/EEC) concerning certain aspects of working time and UK legislation deriving therefrom shall be a period of 12 months.
(d) The difference between the normal hourly rate and the overtime rate shall be known as the 'premium' payment.
(e) For the purpose of calculating overtime, and regardless of the length of the normal working day as determined under Clause 3, time worked on Monday to Friday inclusive shall be paid for as follows:
 (i) First eight hours worked after normal starting time - normal hourly rate.
 (ii) Thereafter, until four hours after the normal finishing time - time-and-a-half.
 (iii) Thereafter, until normal starting time next morning - double time.
 (iv) If time is lost through the fault of the Operative the time lost shall be added to the normal starting time and the resultant time shall be used for the purposes of calculating overtime payable at time-and-a-half.
 (v) An Operative directed to start work before the normal starting time shall be paid the appropriate overtime rates for all hours worked before the normal starting time, but if through the action of the Operative the normal working day is not worked, ordinary hourly rates shall be paid for all hours worked.

## WORKING RULES

(vi) The calculation of overtime for any day shall not be affected by any hours of absence arising from,
-certified sickness
-absence with the concurrence of the Employer
-absence for which the Operative can produce evidence to the satisfaction of the Employer that his absence was due to causes beyond his control.

(vii) An Operative called back to work at any time between the period commencing two hours after the normal finishing time and until two hours before normal starting time shall be paid such overtime rates as would apply had work been continuous from normal finishing time and shall be paid a minimum of two hours at the appropriate rate.

(f) Time worked on Saturday and Sunday shall be paid for as follows:
    (i) Saturday - first five hours, time and a half; after the first five hours, double time but if time is lost through the fault of the Operative the double time rate shall not apply until time lost has been made up.
    (ii) Sunday - double time for all hours worked until starting time on Monday Morning.

PAYMENT FOR HOLIDAYS WORKED (CLAUSE 10)

(a) This clause applies to all recognised holidays as defined in Clause 18, and in Scotland - three days of the Winter Holiday period as defined in Clause 10(c).
(b) An Operative who works on any of the days in Clause 10(a) shall be paid a minimum of two hours at the appropriate rate. In addition an Operative shall be granted a day's holiday with pay for each holiday day worked as provided in Clause 18(c).
(c) Time worked on such days shall be paid as follows:

*In England and Wales*
New Year's Day, Good Friday, Easter Monday, May Bank Holiday, Spring Bank Holiday, Late Summer Bank Holiday, Christmas Day, Boxing Day:
    - Double time for all hours worked.

*In Scotland*
Three consecutive days of the Winter Holiday period including New Year's Day and the one or two holiday days which immediately follow it (if any), Spring Bank Holiday, Friday before Spring Bank Holiday, May Bank Holiday, Autumn Holiday (one day): Boxing Day.
    - Double time for all hours worked.
Christmas Day and the one day of recognised holiday to be agreed locally.
The normal working day as determined in Clause 3(a), time and a half, thereafter double time.
Friday before Autumn Holiday.
The normal working day as determined in Clause 3(a), normal hourly rates; thereafter overtime rates in accordance with Clause 9.

*In Northern Ireland*
The eight days of recognised holidays agreed by the Northern Ireland Branches of the Association and Union. Double time for all hours worked.

(d) The general conditions of the Agreement shall apply to men called back to work on these holidays.

NIGHT SHIFTS AND NIGHT WORK (CLAUSE 11)

(a) For an Operative who works for at least five consecutive nights:
    (i) The basic rate, called the night shift rate, shall be one and a third times the normal rate.
    (ii) Overtime rates and conditions shall be as for normal working days (provided in clause 9) but the basic rate shall be the night shift.
(b) An Operative who works for less than five nights and does not work during the day shall be paid at overtime rates as if the normal day had already been worked.

**Rates of Wages and Working Rules - Mechanical Installations**

## WORKING RULES

CONTINUOUS SHIFT WORK (CLAUSE 12)

Where jobs have to be continuously operated the work shall be carried out in two or three shifts of eight hours each according to requirements. The Operatives concerned shall be paid time and a third in cases where a six day shift is worked and time and a half in cases where a seven day shift is worked, overtime and night shift rates being compounded in these rates. Arrangements shall be made to change the shifts worked by each Operative.

PROVISION OF TOOLS (CLAUSE 13)

(a) The Operative shall provide a rule and spirit level. Other tools shall be provided by the Employer but the Operative shall take all reasonable steps to safeguard, maintain and generally take care of the Employer's tools.

(b) The Operative shall co-operate in the implementation of reasonable procedures properly designed to prevent loss of or damage to tools.

ALLOWANCES TO OPERATIVES WHO TRAVEL DAILY (CLAUSE 15)

(a) Except where his centre is the job, an Operative who is required by the Employer to travel daily up to 50 miles to the job shall be paid fares and travelling time as stated in (i) and (ii) below:

   (i) Return daily travelling fares from his centre to the job. Where cheap daily or period fares or other cheap travel arrangements by public transport are available the Employer may pay fares on that basis. Where, however, a change in such travel arrangements results from a change in the working arrangements the Employer must pay the Operative for any additional cost. The Employer at his option may provide suitable conveyance for the Operative to and from the job in which case fares shall not be paid.

   (ii) Allowances for travelling time, provided that the normal hours are worked on the job. The allowances for travelling time shall be agreed from time to time by the Association and the Union and shall be enumerated in an Appendix to this Agreement. When a reasonably direct journey is not possible, a claim for special consideration may be made by the Operative and in case of dispute the matter shall be referred to the Chief Officials of the parties, whose decision shall be final.

(b) Except where his centre is the job, payment to the Operative of allowances for travelling time and fares for journeys beyond fifty miles daily from his centre to the job will be for agreement between the Employer and the Operative concerned.

ALLOWANCES TO OPERATIVES WHO LODGE (CLAUSE 16)

(a) Where an Operative is sent to a job to which it is impracticable to travel daily and where the Operative lodges away from his place of residence he shall (except if he is engaged at the job or if his centre is the job) be paid the items in (i) to (v) below where appropriate:

   (i) A nightly lodging allowance including the night of the day of return and when on week-end leaves in accordance with Clause 17(a). The nightly lodging allowance shall be agreed from time to time by the Association and the Union and shall be enumerated in an Appendix to this Agreement. The lodging allowance shall not be paid when an Operative is absent from work without the concurrence of the Employer, nor when suitable lodging is arranged by the Employer at no expense to the Operative, nor during the annual holidays defined in Clause 20 including the week of Winter Holiday. The Operative shall provide the Employer with a statement signed by himself to the effect that he is in lodgings for the period of payment of lodging allowance under this clause. Without such evidence, the Employer shall deduct tax on lodging allowance paid.

   (ii) Any V.A.T. charged on the cost of lodgings, subject to the provision by the Operative of a valid tax invoice on which the Employer can claim input credit.

   (iii) When suitable lodgings are not available within two miles from the job, daily return fares from lodgings to job. The Employer at his option may provide suitable conveyance for the Operative between the lodgings and the job, in which case the fares shall not be paid.

# Heating and Water Services Design in Buildings

**K. Moss**, Visiting Lecturer to City of Bath College and the University of Bath, UK

This book addresses practical design procedures and solutions in heating and water services in buildings. The reader is encouraged to participate in the worked solutions to the numerous examples and case studies given. This book has been written following the authors twenty seven years experience in the industry as a teacher and consultant. Moss has worked with college students, university undergraduates and open learning candidates of all ages.

- **does not include sections of mathematical theory which do not have a practical relevance**

### Contents
Heat requirements of heated buildings in temperate climates. Low-temperature hot water heating systems. Pump and system. High temperature hot water systems. Steam systems. Plant connections and controls. The application of probability and demand units in design. Hot and cold water supply systems utilising the static head. Hot and cold water supply systems using booster pumps. Loose ends.

*E & F N Spon*
*246x189: 264pp :98 line illus: May 1996*
*Paperback: 0-419-20110-6: £19.99*

# Building Services Engineering Spreadsheets

**D.V. Chadderton**, Chartered building services engineer, Victoria, Australia, formerly at the Southampton Institute of Higher Education, Southampton, UK

Building Services Engineering Spreadsheets is a versatile, user friendly tool for design calculations. Spreadsheet application software is readily understandable since each formula is readable in the location where it is used. Each step in the development of these engineering solutions is fully explained.

- fills the gap between manual calculation methods using a calculator and specifically engineered software costing thousands of pounds

**Contents**

Contents. Preface. Acknowledgements. Introduction. Units and constants. Symbols. Chapter 1. Computer and spreadsheet use. Chapter 2. Thermal transmittance. Chapter 3. Heat gain. Chapter 4. Combustion of a fuel. Chapter 5. Building heat loss. Chapter 6. Fan and system selection. Chapter 7. Air duct network. Chapter 8. Water pipe sizing. Chapter 9. Lighting. Chapter 10. Electrical cable sizing. References. Index. Answers.

*E & F N Spon*
*most major IBM compatible spreadsheet applications*
*246x189: approx. 300pp: 63 line illus, 3 halftone illus: September 1997*
*Paperback: 0-419-22620-6: c. £29.95*

## WORKING RULES

- (iv) Time spent in travelling to and from the centre at the commencement and completion of the job at the normal hourly rate but when an excessive number of hours of travelling is necessarily incurred, a claim for special consideration may be made by the Operative to the Employer or by the Employer to the Operative and in case of dispute the matter shall be referred to the Chief Officials of the parties, whose decision shall be final.
- (v) Fares between his centre and the job at the commencement and the completion of the job. Return fares shall be used when available.
- (vi) Week-end leaves in accordance with Clause 17(a).

(b) An Operative whose employment is terminated in accordance with Clause 2a during the course of a job, shall be entitled to travelling time and a single fare for the journey from the job to his centre. This condition shall not apply to an Operative who is discharged for misconduct or who leaves the job without the concurrence of his Employer.

WEEK-END LEAVES (CLAUSE 17)

(a) An Operative who is in receipt of lodging allowance in accordance with Clause 16 shall be allowed a week-end leave every two weeks. Such Operative shall be entitled to return to his respective centre for the recognised holidays prescribed in Clause 18 and to facilitate this, the nearest normal week-end leave shall, where necessary, be deferred or brought forward to coincide with the holiday.

(b) Unless the Employer and the Operative agree otherwise the week-end leave shall be from normal finishing time on Friday to normal starting time on Monday.

(c) An Operative shall not normally be required to start his return journey before 6.00a.m. on the appropriate day of return to the job but shall, where the return journey makes it impossible to commence work at the normal starting time, agree with his Employer the working arrangements for the day.

(d) Weekend return fares shall be paid for weekend leaves. If an Operative does not elect to return to his centre a single fare from the job to his centre shall be paid.

(e) The following travelling time arrangements shall apply to an operative on weekend leave for journeys to and from his centre:
- (i) Where the job is up to 150 miles from his centre, he shall travel in his own time from the job to his centre, but travelling time from his centre to the job shall be paid at the normal hourly rate.
- (ii) Where the job is 150 miles or more from his centre, he shall be paid four hours at the normal hourly rate from the job to his centre, and travelling time from his centre to the job shall be paid at the normal hourly rate.

If an Operative elects to stay at the job travelling time shall not be paid.

(f) When a reasonably direct journey is not possible or when an excessive number of hours travelling is necessarily incurred on jobs more than 150 miles from an Operative's centre, a claim for special consideration in respect of travelling time may be made by the Operative to the Employer or by the Employer to the Operative and in case of dispute the matter shall be referred to the Chief Officials of the parties, whose decision shall be final.

(g) An Operative on week-end leaves (including holidays, provided under Clause 18), shall be paid the nightly lodging allowance, provided that the leave is within this Agreement or is agreed with the Employer.

RECOGNISED HOLIDAYS (CLAUSE 18)

(a) The following days have been designated as recognised holidays and shall be paid in accordance with WELPLAN the HVACR Welfare and Holiday Scheme. The pay shall consist of the appropriate holiday credits standing to the credit of the Operative

If any of these days comes within the annual holidays as provided in Clause 20, mutual arrangements shall be made to substitute some other day for the day or days included.

*In England and Wales*
New Year's Day; Good Friday; Easter Monday; May Bank Holiday; Spring Bank Holiday; Late Summer Bank Holiday; Christmas Day; Boxing Day.

## WORKING RULES

*In Scotland*
New Years Day; Spring Holiday; May Holiday; Autumn Holiday (two days); Christmas Day; Boxing Day; plus one other day to be agreed locally.

*In Northern Ireland*
Note: The days when recognised holidays are to be taken in Northern Ireland are subject to discussion between the Northern Ireland Branches of the Association and the Union.

(b) Any Operative who has insufficient credits in his WELPLAN account to pay for the three days of recognised holiday included in the winter holiday period, because he entered the industry after the commencement of the appropriate stamping period, shall be entitled to three days pay at the normal hourly rate for eight hours. The Employer shall be responsible for paying the difference between this sum and the value of any holiday credits that may have been accrued in the appropriate stamping period.

(c) Operatives who work on a recognised holiday as set out in Clause 18(a) shall be paid in accordance with Clause 10(c) and shall be entitled to a day's holiday in lieu, at a mutually agreed time, payment for which shall be the sum of the appropriate holiday credits for that day of recognised holiday.

(d) The general conditions of the Agreement shall apply to Operatives called back to work on these holidays.

### WELPLAN THE HVACR WELFARE AND HOLIDAY SCHEME (CLAUSE 19)

(a) The Employer shall notify WELPLAN of all Operatives to be included in the Scheme. The rules of the Scheme which are incorporated into and form part of this Agreement are set out in a separate Supplement. Operatives are entitled to a weekly credit subject to the rules of the scheme to be purchased by the Employer by means of a four weekly return to WELPLAN. The credit shall cover:
  (i) A weekly credit in respect of annual and recognised holidays (the value of the credit shall be agreed from time to time between the Association and the Union and shall be enumerated in an Appendix to this Agreement).
  (ii) A weekly premium in respect of welfare benefits (the value of the premium shall be determined from time to time by the Association and shall be enumerated in an Appendix to this Agreement).

(b) Variation or Amendment: Clauses 19, 20, 21 and 22 of this Agreement may be varied or amended by agreement of the Parties but any variation or amendment shall, subject to the rules of the WELPLAN, only become operative at the beginning of a new accounting period. Notice of any proposed variation must be given in writing to each of the other Parties at least six months prior to the commencement of any accounting period.

(c) Termination: Either of the Parties to this Agreement may terminate Clauses 19, 20, 21 or 22 at the end of any accounting period by giving notice in writing to the other Party at least 12 months before the end of the accounting period. In the event of termination of the `Annual and Recognised Holidays Provision' the Parties agree to provide the holiday facilities and holiday payments until such time as the rights acquired by the Operatives in respect of holiday credits under this section have been met.

### ANNUAL HOLIDAYS (CLAUSE 20)

All Operatives shall have annual holidays with pay in accordance with WELPLAN. The pay shall consist of the appropriate annual holiday credits standing to the credit of the Operative.

Annual holidays shall consist of:

England and Wales
(i) Four days of Spring Holiday
(ii) Two weeks of Summer Holiday
(iii) Seven days of Winter Holiday

Scotland
(i) Four days of SpringHoliday
(ii) Eleven days of Summer Holiday
(iii) Six days of Winter Holiday

## WORKING RULES

Northern Ireland
- (i) Three days of Spring Holiday
- (ii) Two weeks of Summer Holiday
- (iii) Two weeks of Autum Holiday
- (iv) Seven days of Winter Holiday

ANNUAL AND RECOGNISED HOLIDAY CREDITS - PAYMENT (CLAUSE 21)

The sum standing to the credit of each Operative, being the sum of the weekly credits less any administrative charge approved by the Parties to the Agreement, shall be paid to the Operative on taking his annual and/or recognised holidays by the Employer in accordance with the WELPLAN Scheme.

WELFARE BENEFITS - ENTITLEMENT AND PAYMENT (CLAUSE 22)

(a) All Operatives shall be entitled to sickness and accident benefit and other welfare benefits in accordance with the WELPLAN (HVACR Welfare and Holiday Scheme).

# Keep your figures up to date, free of charge

This section, and most of the other information in this Price Book, is brought up to date every three months, until the next annual edition, in the *Price Book Update*.

The *Update* is available free to all Price Book purchasers.

To ensure you receive your copy, simply complete the reply card from the centre of the book and return it to us.

## Rates of Wages and Working Rules - Mechanical Installations

### WORKING RULE

**Extracts from National Working Rules (JIB PMES)**

WORKING HOURS (CLAUSE 1)

(1.1) Working Week

The normal working week shall be 37½ hours consisting of five working days, Monday to Friday inclusive, each day to consist of 7½ working hours. The normal starting time shall be 8.00 am but this may be varied nationally by the National JIB or locally by a Regional JIB, according to the circumstances.

(1.2) Utilisation of Working Hours

The hours set out in Rule 1.1 are working hours which shall be fully utilised and shall not be subject to unauthorised 'breaks'. Time permitted for tea breaks shall not be exceeded. Bad time-keeping and/or unauthorised absence from the place of work during working hours shall be the subject of disciplinary action.

Meetings of Operatives shall not be held during working hours except by arrangement with the Job, Shop or Site Representative and with the Employer or the Employer's Representative.

(1.3) Meal Intervals

Meal intervals in each such working day shall be as follows:

(1.3.1) a midday break of not less than half an hour, to be taken at a time fixed by the Employer, and shall be unpaid.

(1.3.2) two paid tea-breaks, not exceeding ten minutes each in duration, which shall be taken at times determined by the Employer.

GUARANTEED WEEK (CLAUSE 2)

(2.1) Guarantee

An Operative who has been continually employed by the same Employer for not less than one full week is guaranteed wages equivalent to his inclusive normal graded earnings for 37½ hours in any normal working week, provided that during working hours he is capable of, available for, and willing to perform satisfactorily the work associated with his usual occupation, or reasonable alternative work if his usual work is not available, as determined by the Employer and his Representative.

(2.2) Holidays

In the case of a week in which recognised holidays occur, the guaranteed week shall be reduced by 7½ hours for each day of holiday.

(2.3) Industrial Action

In the event of an interruption of work as a result of industrial action by the Operatives in the employ of an Employer who is a party to the Agreement, the guarantee shall be automatically suspended in respect of Operatives affected on the site, job or shop where the industrial action is taking place.

In the event of such interruption being caused by Operatives working under other Agreements and the Operatives covered by this Agreement not being parties to the interruption, the Employers will try to provide other work or if not able to do so will provide for the return of the Operatives to the shop or office from which they were sent. The Operatives will receive instructions as soon as is practicable as to proceeding to other work or return to the shop. If other work is not available the Employer may suspend Operatives for a temporary period and the provisions of Rule 2.1 shall be suspended until normal working is restored or alternative work becomes available.

*Rates of Wages and Working Rules - Mechanical Installations*

**WORKING RULES**

(2.4) Temporary Lay-Off

Where an Operative is not provided with work for a complete payweek, although remaining available for work, he shall be paid in accordance with the Guarantee set out in Rule 2.1. Thereafter, following consultation with a written confirmation of the agreement from the Representatives of the AEEU, the Employer may require the Operative to register as an unemployed person and Rule 2.1 will not apply for the duration of the stoppage of work. Where an Operative, who has been subject to temporary layoff, is restarted, his employment will be deemed to have been continuous and no break in service will apply.

OVERTIME (CLAUSE 3)

(3.1) Hours

Overtime is strongly discouraged by the Board and systematic overtime should only be introduced to meet specific circumstances and be limited to the period to which the circumstances apply, i.e. breakdown, urgent maintenance and repairs or country jobs.

(3.2) Payment

(3.2.1) 39 hours shall be worked at normal rates in any one week (Monday to Friday) before any overtime is calculated. Overtime shall be paid at time-and-a-half. Overtime after midnight shall be paid at double time. After midnight a 7½ hour rest period, starting at the time at which the work ceased and paid at normal rates of pay, shall follow night work. All hours worked between 1.00 pm on Saturday until normal starting time on Monday shall be paid at double time. To qualify for the rest period the night work must be within the twenty-four hour period in which the normal shift occurs.

Exceptions: For the purpose of overtime payment, an Operative shall be deemed to have worked normal hours on days when, although no payment is made by the Employer, the Operative:-

a) has lost time through certified sickness;
b) was absent with the Employer's permission;

or when the Operative

c) was on statutory holiday;
d) was on a rest period for the day following continuous working all the previous night.

(3.2.2) Any Operative who has not worked five days from Monday to Friday taking into account the exceptions detailed under 3.2.1 is precluded from working the following Saturday or Sunday unless specifically required to do so by his Employer due to a shortage of labour. In such a case overtime rates as in 3.2.1 shall be payable.

(3.2.3) Payday: Wages shall be paid on Thursday of the week following the week in which hours were worked unless a statutory holiday intervenes or if strikes or other circumstances beyond the Employer's control cause delay in payment, when payday may be Friday.

Payment of wages by credit transfer should be encouraged to avoid delays.

(3.3) Call Out

When an Operative is called to return to work after his normal finishing time and before his next normal starting time, he shall be guaranteed the equivalent of 4 hours payment at the Operative's JIB Graded Rate of Wages in accordance with the overtime provisions contained in Rule 3.2.1, provided the Operative has completed 39 hours work in the relevant week.

Payment for call out will be for all hours inclusive home to home.

## WORKING RULES

NIGHT SHIFT WORK (CLAUSE 4)

(4.1)   Hours

Night shift is where Operatives, other than day shift Operatives, work throughout the night for not less than three consecutive nights.

A full night shift pattern shall consist of 37½ working hours worked on 4 or 5 nights by mutual agreement with breaks for meals each night to be mutually arranged. The Employer shall agree the working hours (including breaks) on each contract.

(4.2)   Payment

Night shift shall be paid at the rate of time-and-a-third for all hours worked up to 39 hours in any one week (Monday to Friday inclusive). Overtime shall be at the rate of double basic time.

HOLIDAYS (CLAUSE 5)

(5.1)   Annual Holiday and Stamp Credit Card Scheme

Operative Plumbers and Apprentice Plumbers shall be entitled to 21 working days paid annual holiday in each year to be taken in conformity with Rules 5.1 and 5.1.4. Employers of Operative Plumbers and Apprentice Plumbers **must in all** circumstances purchase, from the Board on behalf of their Employees, weekly credit stamps in respect of annual holidays and covering for the sickness and other benefits referred to in Rule 9. The values of credit stamps for each contribution period (from the beginning of April to the end of March in the succeeding year) shall be determined from time to time by the Board (see Appendix B). The credit stamps shall be affixed weekly to a card supplied by the Board and the stamped cards shall be regarded as the property of the Employee but retained in safe keeping by the Employer during the currency of employment. Upon termination of that employment, the cards, stamped up to the date of termination, shall be handed to the Operative who shall provide acknowledgement of receipt of the stamped cards. To qualify for a holiday stamp in any one week an Operative must work a minimum of four full days. Statutory holidays or authorised absence including Jury Service, count as worked days.

Excepting Apprentices (see below), at the time of the Operative taking his holiday the Employer shall pay to the Operative the holiday credit due, calculable by reference to the "Holiday Credit Calculator" and the number of stamps affixed to the stamped card. in the case of Apprentices, the Employer shall pay to the Apprentice the normal wages due to the Apprentice according to the wage for year of training scale for each day of annual holiday. The Board will, upon receipt of the stamped cards from the Employer, reimburse the Employer the amount of holiday credit which has been properly paid on the total of stamps affixed thereon, in accordance with the terms of this agreement, provided the Board is satisfied that the claim has been made in accordance with the object and rules of the Scheme.

The object of the Scheme is to secure a rest period with pay and it is not a means by which an Employee can achieve double pay by not taking all or part of his prescribed holiday. According to the obligations of membership of the Board, participation in the Scheme is obligatory upon all Employers. HOLIDAYS MUST BE TAKEN.

(5.1.1)   Employees Covered by the Scheme

The Scheme shall apply to all Operatives whose rates of wages and working conditions are determined by the Board and to Apprentices in accordance with the terms of their Board Apprentice Training Agreement and failure to comply will be a breach of the Agreement.

Any Employee in the Plumbing Mechanical Engineering Services Industry may be admitted to the Scheme at the Director's discretion.

## *Rates of Wages and Working Rules - Mechanical Installations*

**WORKING RULES**

(5.1.2) Annual Holiday Periods

Annual holidays cannot be carried forward from one Holiday Period to the next and no Employees shall in any circumstances be eligible to receive in any Holiday Period payment of Holiday Credits except those made on his behalf during the relevant Contribution Period. "Contribution Period" means the period of one year from the commencement of the week in which 1st April falls.

"Holiday Period" means the period 1st April to 31st March immediately following the Contribution Period.

(5.1.3) Extent of Annual Holidays

All Employees covered by this Scheme shall be entitled to Annual Holidays consisting of 21 working days for which payment shall be made as provided in Rule 5.1. Ten days of holiday are to be taken in the period 1st May to 30th September, the balance of the holiday to be taken at a time to be agreed with the Employer.

Where possible, the holidays shall commence at normal finishing time on Friday. For the purpose of this Scheme no Saturday or Sunday shall be considered as a working day.

(5.1.4) Notice

Operatives shall give the Employer at least six weeks notice of the dates on which a holiday is to be taken. Where the dates are mutually agreed a shorter period of notice is permissible.

(5.2) Statutory and Other Holidays

(5.2.1) Holiday Dates and Qualification for Payment

Seven-and-a-half hours pay at the appropriate JIB graded rate of wages shall be paid for 8 days of holiday per annum additional to the annual holidays in accordance with Rule 5.1. In general, the following shall constitute such paid holidays:

New Years Day
Good Friday
Easter Monday
May Day
Spring Bank Holiday
Late Summer Bank Holiday
Christmas Day
Boxing Day

In areas where any of these days are not normally observed as holidays in the Industry, traditional local holidays may be substituted by the appropriate Regional Board.

In order to qualify payment, Operatives must work full-time for the normal day on the working days preceding and following the holiday, except where the Operative is absent with the Employer's permission or has lost time through certified sickness. In no case shall the Operative lose pay in excess of a single day of the holiday.

For the purpose of this Rule Operatives shall be deemed to have worked on one or both of the qualifying days if they comply with the following conditions:

i) were absent through certified sickness;
ii) were on a rest period for the day following continuous working all the previous night;
iii) were absent with the Employer's permission;
iv) except where Rule 6.4 applies;
v) where the preceding working day was a statutory holiday when the Operatives would have to be available for work only on the day following the holiday;

## Rates of Wages and Working Rules - Mechanical Installations

## WORKING RULES

vi) were absent on Jury Service.

(5.2.2) Payment for Working During Statutory and/or Public Holidays

When Operatives are required to work on a paid holiday within the scope of this Agreement, they shall receive wages at the following rates for all hours worked.

Christmas Day: Double time and a day or shift off in lieu for which they shall be paid wages at bare time rates for the hours constituting a normal working day. The alternative day hereunder shall be mutually agreed between the Employer and the Operative concerned.

In respect of all other days of holidays:
Time-and-a-half plus a day or shift off in lieu for which they shall be paid wages at bare time rates for the hours constituting a normal working day. The alternative day hereunder shall be mutually agreed between the Employer and the Operative concerned. In the case of night shift workers required to work on a holiday, the overtime rates mentioned above shall be calculated upon the night shift rate. Time off in lieu of statutory holidays shall be paid at bare time day rates.

(5.2.3) Coincidence of Annual and Other Holidays

Where a holiday under Rule 5.2 coincides with part of an Operative's annual holiday, the annual holiday shall be paid at the rate laid down in Rule 8.1 and time off in lieu of the day or days of statutory or other holiday shall be granted at a later date mutually agreed between the Employer and the Operative concerned, and shall be paid for at ordinary time rates. Payment in lieu of the holiday shall not be permitted.

## WAGES AND ALLOWANCES (CLAUSE 8)

(8.1) Graded Rates of Wages

The national standard rates of wages for each grade of Operative and Apprentices, shall be those made by the Board and currently applicable.

Employers are not permitted to pay, nor Operatives to receive, any rate other than the national standard rate. Additional payments and deductions shall be as permitted by these Working Rules.

The current national standard graded rates of wages shall be separately published. (See Appendix A, Section 1).

(8.2) Responsibility Money

Advanced Plumbers designated by the Employer to be in charge of other Operatives shall be paid responsibility money at an hourly rate for the period of such responsibility. (See Appendix A, Section 1).

(8.3) Incentive Bonus Schemes

Bona fide incentive bonus schemes only may be operated, subject to the approval of the Board. (See Appendix C).

(8.4) Daily Travelling Allowances

(8.4.1) Journeys of less than 50 miles

Except where his Centre is the Job, an Operative who is required by his Employer to travel daily up to 50 miles to the Job, shall be paid fares and travelling time as stated below.

*Rates of Wages and Working Rules - Mechanical Installations*

## WORKING RULES

Fares

Return travelling fares (cheapest available) from his Centre to the Job. The Employer at his option may provide suitable conveyance for the Operative to and from the Job in which case fares shall not be paid.

Travelling allowance

Allowances for travelling time, provided that the normal hours are worked on the Job, shall be agreed from time to time by the Board and shall be promulgated and separately published by the Board. When a reasonably direct journey is not possible, a claim for special consideration may be made by the Operative and in case of dispute the matter shall be referred to the Board whose decision shall be final. (See Appendix A, Section 1).

(8.4.2) Employers' Own Transport

    a) It is the Employer's responsibility, if he or the main contractor provides suitable free transport to a site, to get the Operatives to and from the job on time. The Employer's liability for late arrival of the provided transport shall be agreed between the parties.

    b) Mileage Allowance - Use of Private Vehicles on Company Business

    The use of Operatives' private vehicles on employer business will be subject to compliance with the following:-

        i) that prior agreement regarding use of such vehicles be reached between employer and operative
        ii) it is the responsibility of the operative to provide insurance for the vehicle and to ascertain that the insurance is adequate for such use
        iii) the rate of payment for use of vehicle for distance travelled will be as promulgated and that this payment be made in lieu of fares
        iv) use of private vehicle is not a condition of employment

(8.4.3) Definition of Centre

The Centre for determining distances under this Rule must be agreed between the Employer and the Operative and must be either:

    a) the Job on which the Operative is for the time being employed, if the Operative is engaged on the understanding that his Centre will be the Job,
    or
    b) a convenient Centre near the Operative's place of residence. Unless otherwise agreed such Centre must be the nearest convenient public transport boarding point to the Operative's place of residence.

(8.4.4) Change of Centre

When an Operative whose Centre is the Job, as in (a) above, is an Operative in Regular Employment as defined below, his Centre may, by agreement between Employer and Operative, be transferred to one located in accordance with his place of residence as in (b) above.

An Operative after 28 days continuous employment or who, after his first engagement, is transferred to another Job or who is re-engaged by the same firm or any of its subsidiaries within 28 days of ceasing work for any reason with the said firm, shall be regarded as an Operative in Regular Employment for the purpose of this Rule.

Any change of residence after an Operative has been engaged and a Centre established in accordance with Rule 8.4.3 which substantially varies the journey time/distance shall not alter the Centre without prior agreement.

## Rates of Wages and Working Rules - Mechanical Installations
## WORKING RULES

(8.4.5) Measurement of Travelling Distances

All distances referred to in this Rule are to be calculated in a straight line, point to point. When a reasonably direct journey is not possible, a claim for special consideration may be made by the Operative to the Employer.

(8.4.6) Allowances to Operatives who Lodge

(a) When an Operative is sent to a Job to which it is impracticable to travel daily and where the Operative lodges away from his place of residence, he shall (except if he is engaged at the Job or if his Centre is the Job) be paid the items (i) to (v) below, where appropriate.

i) A nightly lodging allowance including the night of the day of return and when on week-end leaves in accordance with Rule 8.4.7.(i). The nightly lodging allowance shall enumerated in a separate appendix to this Agreement. (See Appendix A, Section 1). The lodging allowance shall not be paid when an Operative is absent from work without the concurrence of the Employer nor when suitable lodging is arranged by the Employer at no expense to the Operative, nor during the annual holidays defined in Rule 5 including the week of Winter Holiday. The Operative shall provide the Employer with a statement signed by himself to the effect that he is in lodgings for the period of payment of lodging allowance under this Rule. Without such evidence the Employer shall deduct tax on lodging allowance paid.

ii) When suitable lodgings are not available within two miles from the Job, daily return fares from lodgings to Job. The Employer at his option may provide suitable conveyance for the Operative between lodgings and the Job, in which case fares shall not be paid.

iii) Travelling time for the time spent in travelling from the Centre at the commencement and completion of the Job at the normal time rates but when an excessive number of hours of travelling is necessarily incurred, a claim for special consideration may be made. In case of dispute, the matter shall be referred to the Board.

iv) Fares between his Centre and the Job at the commencement and the completion of the job. Return fares shall be used when available.

v) Week-end leaves in accordance with Rule 8.4.7.

(b) An Operative whose employment is terminated by proper notice on either side during the course of a Job, shall be entitled to travelling time back to his Centre and a single fare for the journey from the Job to his Centre. This condition shall not apply to an Operative who is discharged for misconduct or who leaves the Job without the concurrence of his Employer.

(8.4.7) Week-End Leaves

i) An Operative who is in receipt of lodging allowance in accordance with Rule 8.4.6 shall be allowed a week-end leave every two weeks. Such Operative shall be entitled to return to his respective Centre for the recognised holidays prescribed in Rule 5 and to facilitate this, the nearest normal week-end leave shall, where necessary, be deferred or brought forward to coincide with the holiday.

ii) Unless the Employer and the Operative agree otherwise, the week-end leave shall be from normal finishing time on Friday to normal starting time on Monday.

iii) An Operative shall not be required to start his return journey before 6.00 am on the appropriate day of return to the Job.

*Rates of Wages and Working Rules - Mechanical Installations*

**WORKING RULES**

iv) Week-end return fares shall be paid for week-end leaves. If an Operative does not elect to return to his Centre, a single fare from the Job to his Centre shall be paid.

v) An Operative on a week-end leave whose work is up to 150 miles from his Centre shall travel home in his own time, but travelling time from the Centre to the Job shall be paid at the normal time rate. An Operative whose work is 150 miles or more from his Centre shall be paid travelling time for four hours at normal rate from the Job to the Centre; travelling time for the journey back to the Job to be paid at normal time rate. If an Operative elects to stay at the Job, travelling time shall not be paid.

vi) When a reasonably direct journey is not possible or when an excessive number of hours travelling is necessarily incurred on jobs more than 150 miles from an Operative's Centre, a claim for special consideration in respect of travelling time may be made. In case of dispute, the matter shall be referred to the Board.

vii) An Operative on week-end leave (including statutory holidays provided under Rule 5) shall be paid the nightly lodging allowance provided that the leave is within this agreement or is agreed with the Employer.

(8.5) Tool Allowance

Tool allowance being in respect of the provision, maintenance and upkeep of tools provided by the Operatives, is not deemed to be wage-payment. Where employment starts after Monday, or is terminated in accordance with Rule 6.3 otherwise than on a Friday, the amount of tool allowance paid shall be the appropriate proportion of the weekly allowance for each day worked. Tool allowance (see Appendix A, Section 1) is payable to all Operatives who provide and maintain the following set of tools:

| | |
|---|---|
| Allen Keys - 1 set | Hand Drill/Brace |
| Adjustable Spanner - up to 6" | Junior Hacksaw/Saw |
| Adjustable Spanner - up to 12" | Mole Wrench |
| Basin Key | Padsaw/Compass Saw |
| Bending Spring - 15mm | Pipewrench - |
| Bending Spring - 22mm | Stillsons - |
| Blow Torch and Nozzle - complete | up to type 14" |
| (similar to Primus Type B) | Pilers - Insulated - |
| Bolster Chisel - 2½" | General Type |
| Bossing Mallet | Pocket Knife - |
| Bossing Stick | (Stanley type) |
| Bradawl | Putty Knife* |
| Chisel - Brick - up to 20" | Rasp |
| Chisel - Wood | Rule - 3m Tape |
| Dresser for Lead | Screwdriver - Large |
| Flooring Chisel | Screwdriver - Small |
| Footprints - 9" | Shave Hook |
| Gas or Adjustable Piliers | Snips |
| Glass Cutter* | Spirit Level - 600mm |
| Hacking Knife* | Tank Cutter |
| Hacksaw Frame | Tool Bag |
| Hammer - Large | Trowel |
| Hammer - Small | Tube Cutter |
| | Wiping Cloth - (one) |

* Only if glazing is normally done by plumbers in the district.

(8.6) Storage Accommodation for Tools

Where reasonably practical a lock-fast and weather-proof place shall be provided on all jobs where tools can be left at the owner's risk. The Board provides for financial assistance to be given to Operatives in replacing lost or stolen tools, subject to the conditions of the scheme. (See Appendix A, Section 8).

*Rates of Wages and Working Rules - Mechanical Installations*

**WORKING RULES**

(8.7) Abnormal Conditions

Operatives required to work in exceptionally dirty conditions or under other abnormal conditions as listed in the Schedule attached to these Rules, (see Appendix D) shall be paid a daily supplement, as shown in Appendix A, Section 1. This supplement is a fixed daily amount and does not vary with the hours worked or with different combinations of abnormal conditions.

(8.8) Recognition of Certificates of Competency in Welding

Graded Operatives who hold one or more current JIB Certificates of Competency in Welding or such other qualifications as the Board may require, shall be paid a differential rate as determined by the Board, in addition to the graded rate of wages. (See Appendix E).

For the purposes of this Rule the holder of JIB Certificate of Competency in Oxy-acetylene Welding of Mild Steel Pipework and/or in Bronze Welding of Copper Sheet and Tube shall be classified as a certified "Gas Welder". The holder of JIB Certificate of Competency in Metal Arc Welding of Mild Steel Pipework shall be classified as a certified "Arc Welder".

GRADING DEFINITIONS (CLAUSE 14)

(14.1) Trained Plumber, Mechanical Pipe Fitter and Gas Service Fitter

i) Must have obtained a National Vocational Qualification (NVQ) - MES/Plumbing Level 2 or City & Guilds of London Institute (CGLI) 603/1 Craft Certificate or such other qualifications as are acceptable to the Board.

ii) Entry by way of a recognised and registered form of training (usually a 4 year term as a JIB registered apprentice) or other accepted method of entry into the Industry.

iii) Must be at least 20 years of age.

iv) Must be able to carry out all such installation and maintenance work to the recognised standard and level of productivity expected from an Operative working under minimum supervision.

(14.2) Advanced Plumber, Mechanical Pipe Fitter and Gas Service Engineer

i) Must have obtained a National Vocational Qualification (NVQ) - MES/Plumbing Level 3 or City & Guilds of London Institute (CGLI) 603/2 Advanced Craft Certificate or such other qualifications as are acceptable to the Board.

ii) After achieving **NVQ-MES/Plumbing Level 3** must have **at least one year's experience** working as a Trained Plumber, Mechanical Pipe Fitter or Gas Service Fitter and must be **at least 21 years old**.

iii) After obtaining the CGLI **603/2 Advanced Craft Certificate** or other such qualifications as are acceptable to the Board must have **at least two years experience** working as a Trained Plumber, Mechanical Pipe Fitter or Gas Service Fitter and must be **at least 22 years old**.

iv) Entry by way of a recognised and registered form of training (usually a 4 year term as a JIB registered apprentice) or other accepted method of entry into the industry.

v) Must possess particular and productive skills and be able to work without supervision in the most efficient and economical manner and must be able to set out jobs from working drawings and specifications and requisition the necessary installation materials and/or have technical and supervisory knowledge and skill beyond that expected of a Trained Plumber, Mechanical Pipe Fitter or Gas Service Fitter.

## WORKING RULES

(14.3) Technical Plumber, Mechanical Pipe Fitter and Gas Service Technician

i) Entry by way of a recognised and registered form of training or other accepted method of entry into the Industry.

ii) Must have obtained such academic qualifications or such other qualifications as are acceptable to the Board.

iii) Must have superior technical skill, ability and experience beyond that expected of an Advanced Plumber, Mechanical Pipe Fitter or Gas Service Engineer, and be able to lay out and prepare contract work in accordance with the Building Regulations and Water Bye-laws, take off quantities and measure work, assess labour requirements and control and supervise all manner of plumbing or other relevant installations in the most economic and effective way and achieve a high level of productivity.

AND EITHER

iv) Must be at least 27 years of age.

v) Must have had at least five years experience as an Advanced Plumber, Mechanical Pipe Fitter or Gas Service Engineer, with a minimum of three years in a supervisory capacity in charge of plumbing or other relevant installations of such a complexity and size as to require wide technical experience and organised ability.

OR

vi) May not have reached 27 years of age as in (v) or have the full experience as required in (v) but is otherwise fully qualified in accordance with (i), (ii) and (iii) and his present Employer wishes to have him graded as a Technical Plumber, Mechanical Pipe Fitter or Gas Service Technician in which event he may be granted this grade by the Board.

# The Technology of Building Defects

**J. Hinks**, Reader in Facilities Management, Department of Building Engineering and Surveying, Heriot-Watt University, Edinburgh, UK
**G. Cook**, Senior Lecturer, Department of Construction Management and Engineering, University of Reading, UK

The Technology of Building Defects has been developed to provide a unique stand alone review of building defects. It gives the reader a comprehensive understanding of how and why building defects occur. Defects are considered as part of the whole building rather than in isolation. General education objectives are set out which offer the reader the opportunity of self-assessment and build up an understanding of a range of technical topics concerned with building defects. This is a one stop resource which dispenses with the need to consult a mass of different information sources.

*E & F N Spon*
*246x189: c. 304pp: 95 line illus, 46 halftone illus: October 1997*
*Paperback: 0-419-19780-X: £24.99*

# Electrical Installations
## RATES OF WAGES AND WORKING RULES

**RATES OF WAGES**

ELECTRICAL CONTRACTING INDUSTRY

Extracts from National Working Rules determined by:

The Joint Industry Board for the Electrical Contracting Industry
Kingswood House
47/51 Sidcup Hill, Sidcup, Kent, DA14 6HP
Telephone 0181-302 0031

### Operatives in possession of XVth Edition Grade Cards

From and including 4th January 1997 the JIB rates of wages will be as set out below;

| Electrician, Instrument Mechanic, Instrument Pipefitter (or equivalent specialist grade) | Approved Electrician, Approved Instrument Mechanic, Approved Instrument Pipefitter (or equivalent specialist grade) | Technician (or equivalent specialist grade) | Labourer |
|---|---|---|---|
| **National Standard Rates** | | | |
| £ 6.13 | £ 6.64 | £ 7.69 | £ 4.78 |
| **London Rates** | | | |
| 6.55 | 7.06 | 8.11 | 5.20 |

### On Shore Agreement

Operatives engaged upon On Shore Work in connection with oil and gas exploration from the seabed, as laid down in section 5.3 of the JIB Handbook.

**Electrician**, Instrument Mechanic, Instrument Pipefitter (or equivalent specialist grade) ....... £8.65

**Approved Electrician**, Approved Instrument Mechanic, Approved Instrument Pipefitter, (or equivalent specialist grade) ........................................................ £9.16

**Technician**, (or equivalent specialist grade) ........................................ £10.21

**Labourer** ................................................................................ £7.30

## RATES OF WAGES

### 1983 Joint Industry Board Apprentice Training Scheme
Apprentice rates effective from and including 7 October 1996

| | |
|---|---|
| Junior Apprentice | Training allowance of £60.75 per week |
| Senior Apprentice (stage 1) | £2.71 per hour |
| Senior Apprentice (stage 2) | £4.00 per hour |

Senior Apprentices qualifying for London Weighting shall receive the amount determined, from time to time, by the Joint Industry Board. This is currently 23p per hour. Junior Apprentices will not receive London Weighting.

Apprentices who obtain a Pass with Distinction in both relevant components of the City & Guilds 236 Part I Examination (or an equivalent examination approved by the Joint Industry Board) shall be paid an additional amount of 13p per hour as a Senior Apprentice (Stage 1) until becoming a Senior Apprentice (Stage 2).

Apprentices who obtain a pass with Distinction in both relevant components of the City & Guilds 236 Part II Examination (or an equivalent examination approved by the Joint Industry Board) and an overall pass shall be paid an additional amount of 17p per hour until termination of the Training Contract.

### 1989 Joint Industry Board Adult Craft Training Scheme

| | *Adult Trainee (under 21)* | *Adult Trainee (over 21)* | *Senior Graded Electrical Trainee* |
|---|---|---|---|
| **From and including 4th January 1997** | £ | £ | £ |
| National Standard Rate | 3.59 | 4.78 | 5.52 |
| London Weighting | 4.01 | 5.20 | 5.94 |

*Rates of Wages and Working Rules - Electrical Installations*

## RATES OF WAGES

**Travel time and travel allowances from 2 January 1992**

Operatives who are required to start and finish at the normal starting and finishing time on jobs which are up to and including 35 miles from the shop - in a straight line - shall receive both travel allowances and payment for travelling time as follows:

| Distance Point to Point for journeys of | Total daily trav. allowance | Total daily travelling time | | | |
|---|---|---|---|---|---|
| | | Tech | Apvd Elec | Elec | Lab |
| **(a) National Standard Rate:** | | | | | |
| **From 2 January 1992** | | | | | |
| Return journey of | | | | | |
| Up to 1 mile each way | 83p | Nil | Nil | Nil | Nil |
| Up to 2 miles each way | 95p | 143p | 123p | 113p | 89p |
| Up to 3 miles each way | 109p | 238p | 205p | 189p | 148p |
| Up to 4 miles each way | 121p | 285p | 246p | 227p | 177p |
| Up to 5 miles each way | 171p | 379p | 327p | 302p | 237p |
| Between 5 and 10 miles each way | 241p | 569p | 491p | 454p | 354p |
| Between 10 and 15 miles each way | 343p | 759p | 656p | 605p | 472p |
| Between 15 and 20 miles each way | 482p | 854p | 738p | 680p | 531p |
| Between 20 and 25 miles each way | 552p | 948p | 819p | 756p | 591p |
| Between 25 and 35 miles each way | 622p | 1043p | 901p | 832p | 649p |
| **(b) London area:** | | | | | |
| **From 2 January 1992** | | | | | |
| Return journey of | | | | | |
| Up to 1 mile each way | 83p | Nil | Nil | Nil | Nil |
| Up to 2 miles each way | 95p | 150p | 131p | 122p | 97p |
| Up to 3 miles each way | 109p | 251p | 218p | 203p | 161p |
| Up to 4 miles each way | 121p | 301p | 262p | 243p | 193p |
| Up to 5 miles each way | 171p | 401p | 349p | 324p | 258p |
| Between 5 and 10 miles each way | 241p | 602p | 524p | 486p | 387p |
| Between 10 and 15 miles each way | 343p | 802p | 699p | 648p | 515p |
| Between 15 and 20 miles each way | 482p | 903p | 786p | 729p | 580p |
| Between 20 and 25 miles each way | 552p | 1002p | 873p | 810p | 645p |
| Between 25 and 35 miles each way | 622p | 1103p | 960p | 891p | 708p |

*Rates of Wages and Working Rules - Electrical Installations*

**RATES OF WAGES**

**Travel time and travel allowance from 12 August 1995**

Apprentices who are required to start and finish at the normal starting and finishing time on jobs which are up to and including 35 miles from the shop - in a straight line - shall receive travel allowances and payment for travelling time as follows:

| Distance Point to Point for journeys of | Total daily travel allow- ance | Total daily travelling time | |
|---|---|---|---|
| | | Senior Apprent. (Stage 1) | Senior Apprent. (Stage 2) |
| **(a) National Standard Rate:** **From 12 August 1995** | | | |
| Up to 1 mile each way | 83p | Nil | Nil |
| Up to 2 miles each way | 95p | 63p | 93p |
| Up to 3 miles each way | 109p | 105p | 155p |
| Up to 4 miles each way | 121p | 126p | 186p |
| Up to 5 miles each way | 171p | 168p | 248p |
| Between 5 and 10 miles each way | 241p | 252p | 372p |
| Between 10 and 15 miles each way | 343p | 336p | 496p |
| Between 15 and 20 miles each way | 482p | 378p | 558p |
| Between 20 and 25 miles each way | 552p | 420p | 620p |
| Between 25 and 35 miles each way | 622p | 462p | 682p |
| **(b) London area:** **From 12 August 1995** | | | |
| Up to 1 mile each way | 83p | Nil | Nil |
| Up to 2 miles each way | 95p | 68p | 98p |
| Up to 3 miles each way | 109p | 114p | 164p |
| Up to 4 miles each way | 121p | 137p | 197p |
| Up to 5 miles each way | 171p | 182p | 262p |
| Between 5 and 10 miles each way | 241p | 273p | 393p |
| Between 10 and 15 miles each way | 343p | 364p | 524p |
| Between 15 and 20 miles each way | 482p | 410p | 590p |
| Between 20 and 25 miles each way | 552p | 455p | 655p |
| Between 25 and 35 miles each way | 622p | 501p | 721p |

Travelling costs for Junior Apprentices remain as laid down in the 1983 JIB Training Scheme.

*Rates of Wages and Working Rules - Electrical Installations*

## RATES OF WAGES

**Travel time and travel allowance from 4 January 1992**

Adult trainees who are required to start and finish at the normal starting and finishing time on jobs which are up to and including 35 miles from the shop - in a straight line - shall receive both travel allowances and payment for travelling time as follows:

| Distance Point to Point for journeys of | Total daily travel allowance | Total daily travelling time *(1) | *(2) | *(3) |
|---|---|---|---|---|
| **(a) National Standard Rate:** From 4 January 1992 | | | | |
| Up to 1 mile each way | 83p | Nil | Nil | Nil |
| Up to 2 miles each way | 95p | 67p | 89p | 103p |
| Up to 3 miles each way | 109p | 111p | 148p | 171p |
| Up to 4 miles each way | 121p | 133p | 177p | 204p |
| Up to 5 miles each way | 171p | 177p | 237p | 272p |
| Between 5 and 10 miles each way | 241p | 266p | 354p | 408p |
| Between 10 and 15 miles each way | 343p | 354p | 472p | 544p |
| Between 15 and 20 miles each way | 482p | 399p | 531p | 612p |
| Between 20 and 25 miles each way | 552p | 443p | 591p | 680p |
| Between 25 and 35 miles each way | 622p | 487p | 649p | 748p |
| **(b) London area:** From 4 January 1992 | | | | |
| Up to 1 mile each way | 83p | Nil | Nil | Nil |
| Up to 2 miles each way | 95p | 75p | 97p | 110p |
| Up to 3 miles each way | 109p | 124p | 161p | 184p |
| Up to 4 miles each way | 121p | 149p | 193p | 220p |
| Up to 5 miles each way | 171p | 199p | 258p | 294p |
| Between 5 and 10 miles each way | 241p | 298p | 387p | 441p |
| Between 10 and 15 miles each way | 343p | 397p | 515p | 588p |
| Between 15 and 20 miles each way | 482p | 447p | 580p | 661p |
| Between 20 and 25 miles each way | 552p | 497p | 645p | 734p |
| Between 25 and 35 miles each way | 622p | 546p | 708p | 808p |

Note: *(1) Adult Trainee (under 21)
  *(2) Adult Trainee (over 21)
  *(3) Senior Graded Electrical Trainee

## RATES OF WAGES

**Country Allowance**
£18.63 from and including 4th January 1997

**Lodging Allowances**
£19.15 from and including 4th January 1997

**Lodgings weekend retention fee, maximum reimbursement**
£19.15 from and including 4th January 1997

**Annual Holiday Lodging Allowance Retention**
A maximum of £4.14 per night (£28.98 per week) from and including 4th January 1997

**Responsibility money**
Approved Electricians in charge of work, who undertake supervision of other operatives, shall be paid "Responsibility Money" of not less than 2.5p and not more than 50p per hour.

From and including 4 January 1992 responsibility payments shall be enhanced by overtime and shift premiums where appropriate.

**Combined JIB Benefits Stamp Value** (from week commencing 30 September 1996)

|  | *Weekly stamp value* |
|---|---|
| Technician | £29.57 |
| Approved Electrician | £26.39 |
| Electrician | £24.85 |
| Senior Graded Electrical Trainee | £23.00 |
| Labourer & Adult Trainee | £20.76 |
| Adult Trainee(Under 21) | £17.15 |

### JIB Welfare Benefits

Sick Pay, Death Benefit, Accidental Death Benefit

(a) The following will apply with effect from Monday 1 October 1990

    (i) Sick Pay in addition to Statutory Sick Pay

| | |
|---|---|
| 1st three days | No payment |
| 4th to the 14th day inclusive | £6.00 per day |
| 3rd to the 28th week inclusive | £24.50 per week |
| 29th to the 52nd week inclusive | £21.00 per week |

    Benefit ceases after 52 weeks.

    (ii) Death Benefit
    £15,000.00 for death from any cause.

    (iii) Accidental Death Benefits with effect from 4 January 1992
    £12,500.00 for adult operatives (£6,250.00 for apprentices) for death from an accident either at work, or travelling to or from work, making a total of £27,500.00 (£21,250 for apprentices) for death due to an accident at work.

## WORKING RULES

## INTRODUCTION

The JIB National working Rules are made under rule 80 of the Rules of the Joint Industry Board (Section 1) as the National Joint Industrial Council for the Electrical Contracting Industry.

The principal objects of the Joint Industry Board are to regulate the relations between employers and employees engaged in the industry and to provide all kinds of benefits for persons concerned with the Industry in such ways as the Joint Industry Board may think fit, for the purpose of stimulating and furthering the improvement and progress of the industry for the mutual advantage of the employers and employees engaged therein, and in particular, for the purpose aforesaid, and in the public interest, to regulate and control employment and productive capacity within the Industry and the levels of skill and proficiency, wages, and welfare benefits of persons concerned in the Industry.

"The Industry" means the Electrical Contracting Industry in all its aspects in England, Wales, Northern Ireland, the Isle of Man and Channel Islands and such other places as may from time to time be determined by the Joint Industry Board, including the design, manufacture, sale, distribution, installation, erection, maintenance, repair and renewal of all kinds of electrical installations, equipment and appliances and ancillary plant activities.

### 1: General

These JIB National Working Rules and Industrial Determinations supersede previous Rules and Agreements made between the constituent parties of the National Joint Industrial Council for the Electrical Contracting Industry and shall govern and control the conditions for electrical instrumentation and control engineering, data and communications transmission work, its installation, maintenance and its dismantling and other ancillary activities covered by the Joint Industry Board for the Electrical Contracting Industry ("JIB") and shall come into effect in respect of work performed on and after Monday, 2 March 1970. These Rules apply nationally and in such a manner as may be determined from time to time by the JIB National Board.

### 2: Grading

Graded operatives shall comply in all respects with the Grading Definitions set out in Section 4 in carrying out the work of the Industry, erect their own mobile scaffolds and use such power operated and other tools, plant, etc, as may be provided by their Employer and the standard JIB Graded Rates of Wages shall be paid. Grading shall only be valid by the possession of a Grade Card issued by the Joint Industry Board.

Nothing in these rules shall prevent the maximum flexibility in the employment of skilled operatives.

### 3: Working Hours

The working week shall be 37.5 hours per week worked on five days, Monday to Friday inclusive.

The normal day shall be not more than eight hours worked during any consecutive nine hours between 7.30 am and 6.30 pm. Where shifts are required which fall outside these limits, payments and conditions of work shall be determined by the Joint Industry Board.

Meal breaks, including washing time, shall be unpaid of one hour duration or lesser period at the Employers discretion and shall not be exceeded. The Employer shall declare the working days and hours (including breaks) on each job.

### 4: Utilisation of Working Hours

There shall be full utilisation of working hours which shall not be subject to unauthorized "breaks". Time permitted for tea breaks shall not be exceeded.

Bad timekeeping and/or unauthorized absence from the place of work, during working hours, shall be construed as Industrial Misconduct.

## Rates of Wages and Working Rules - Electrical Installations

## WORKING RULES

## INTRODUCTION (contd)

### 4: Utilisation of Working Hours (contd)

Meeting of operatives shall not be held during working hours except by arrangement with the Job/Shop Representative and with the prior permission of the Employer or the Employers site management or the Employers representative.

### 5: Tools

The Employer shall provide all power-operated and expendable tools as required; operatives shall act with the greatest possible responsibility in respect of the use, maintenance and safe-keeping of tools and equipment of their Employers.

The operative shall have a kit of hand tools appropriate for carrying out efficiently the work for which he is employed; the kit shall include a lockable tool box.

The Employer shall provide, where practicable, suitable and lockable facilities for storing operatives tool-kits.

### 6: Wages (Graded Operatives)

(a) The National Standard JIB Graded Rates of Wages (hereinafter called "the JIB Rates of Wages") shall be those from time to time determined by the Joint Industry Board pursuant to Rule 80 of the Rules of the Joint Industry Board.

The JIB Rates of Wages appropriate to operatives shall be such rates for their grades as the Joint Industry Board may from time to time determine to be appropriate for their grade in the place where they are working and they shall be paid no more and no less wages.

This rule does not permit the introduction of any scheme of payment by result of production bonuses, except as may be determined from time to time by the JIB National Board.

(b) London Weighting

      (i) Definition of the London Zone
      That area lying within and including the M25 London Orbital Motorway.
      (ii) Application
      London Weighting as determined from time to time by the Joint Industry Board shall apply to all operatives and (at separate rates) to apprentices working on jobs in the London Zone as defined above in Rule 6 (b) (i) and for avoidance of doubt it is intended that such London Weighting shall be deemed to be part of the standard JIB Rates of Wages.

Such London Weighting shall also apply to any operative or apprentice who has been working from a London based shop in the London Zone for not less than 12 weeks and who is sent by his employer to a job out of the London Zone for a period of not more than 12 weeks or for the duration of one particular contract, whichever is the longer.
London Weighting shall apply to all paid hours including overtime and shift premium payments (National Working Rules 9 and 10), statutory holiday payments (National Working Rule 13) and travelling time payments (National Working Rule 12(b)), but not to incentive payments under JIB authorized schemes.

      (iii) The amount of London Weighting shall be
          from and including 4 January 1997:
          Graded Operative . . . . . . . . . . . . . . . . . . . . . 42p per hour

## *Rates of Wages and Working Rules - Electrical Installations*

## WORKING RULES

(c) Responsibility Money

Approved Electricians in charge of work, who undertake the supervision of other operatives, shall be paid "responsibility money" as determined from time to time by the JIB National Board, currently not less than 2.5p and not more than 50p per hour.

From and including **4th January, 1992** responsibility payments shall be enhanced by overtime and shift premiums where appropriate.

### 7: Payment of Wages

Wages shall normally be paid by Credit Transfer. Alternatively, another method of payment may be adopted by mutual arrangement between Employer and Operative.

Wages shall be calculated for weekly periods and paid to the Operative within 5 normal working days of week termination, unless alternative arrangements are agreed.

Each operative shall receive an itemised written pay statement in accordance with the Employment Protection (Consolidation) Act 1978.

### 9: Overtime

(a) Hours

Overtime is deprecated by the Joint Industry Board; systematic overtime in particular is to be avoided.

A Regional Joint Industry Board may, from time to time, declare permissible hours of overtime in the Region which shall not be exceeded without permission of the Regional Joint Industry Board.

Overtime will not be restricted in the case of Breakdowns or Urgent Maintenance and Repairs.

(b) Payment
    (i) The number of hours to be worked at normal rates in any one week (Monday to Friday) before any overtime premium is calculated shall be:-

        From and including 4 January 1992    38 hrs

    Premium time shall be paid at time-and-a-half. All hours worked between 1 p.m. on Saturday and normal starting time on Monday shall be paid at double time. Overtime premium payments shall be calculated on the appropriate standard rate of pay.

    Exceptions: For the purpose of premium payment, an Operative shall be deemed to have worked normal hours on days where, although no payment is made by the Employer, the Operative:
    (a)    has lost time through certified sickness.
    (b)    was on a rest period for the day following continuous working the previous night.
    (c)    was absent with the Employers permission.
    (ii)   Any Operative who has not worked five days (as determined in Rule 3) from Monday to Friday, taking into account the exceptions detailed above, is precluded from working the following Saturday or Sunday.

(c) Call out

Notwithstanding the previous Clause, when an operative is called upon to return to work after his normal finishing time and before his next normal starting time he shall be paid at time-and-a-half for all the hours involved (home-to-home), subject to the guaranteed minimum payment. Between 1 pm on Saturday, and the normal starting time on Monday the appropriate premium rate shall be double time.

The guaranteed minimum payment for a single call out under this clause shall be the equivalent of 4 hours at the Operatives JIB Rates of Wages.

## WORKING RULES

**INTRODUCTION (contd)**

**10: Shiftwork**

Operatives may be required to undertake shiftworking arrangements in order to meet the requirements of the job or client. Operatives may not be so required without reasonable notice.

(a) Permanent Night Shift
    (i) Night shift is where operatives (other than as overtime after the end of a day shift) work throughout the night for not less than three consecutive nights.
A full night shift shall consist of 37.5 hours worked on five nights, Monday night to Friday night inclusive, with unpaid breaks for meals each night, to be mutually arranged. The employer shall declare working hours including breaks on each contract.
    (ii) Payment
Night shifts shall be paid at the rate of time and one-third for all hours worked up to 37.5 in any one week, Monday to Friday.

(b) Double day Shift (Rotating)
    (i) The shift week will be from Monday to Friday. Each shift shall be of 7.5 hours worked with an unpaid half hour meal break. The distribution of the hours will be subject to local requirements. Shifts will normally be on an early and late basis.
    (ii) Payment
Rotating double-day shift working will be paid at the rate of time plus 20% for normal hours in the early shift and time plus 30% for normal hours worked in the late shift.

(c) Three Shift Working (Rotating)
    (i) The shift week will be from Monday to Friday. Each shift shall be of 7.5 hours duration with an unpaid half hour meal break. The distribution of the hours will be subject to local requirements. Shifts will normally be on an early, late and night shift basis.
    (ii) Payment
Rotating Three-Shift work will be paid at the rate of time plus 20%, time plus 30% and time plus 33 1/3% for the early, late and night shifts respectively.

(d) Three Shift Working (Seven day continuous)
    (i) Occasional
Where continuous shift work is occasionally required to cover both weekdays and weekends, weekend working shall attract the appropriate premiums contained in Rule 9(b) above. Weekday working shall attract the premiums contained in Rule 10(c) above. Generally speaking, "occasional" shiftwork shall be defined as a shiftwork requirement for a period of four weeks or less to meet some short term or emergency exigency.
    (ii) Rostered
Where continuous three shift working is required to cover a regular seven day working pattern the following conditions shall be observed:
        (i) Unless the requirements for continuous three shift working is specified in the operatives contract of employment or terms of engagement, four weeks prior notice shall be given before the introduction of a rostered three shift working system.

(d) Three Shift Working (Seven day continuous)(contd)

        (ii) Prior to the introduction of a rostered three shift working system the employer will discuss and agree with his employees representatives the most suitable pattern of hours to achieve the required cover.
        (iii) Subject to the above, rostered three shift working shall not be restricted.
        (iv) The normal shift week shall be from Monday to Sunday and will comprise a maximum 37.5 hours in any one week for which the employees shall be paid at time plus thirty per cent.
        (v) All hours rostered, or unrostered, in excess of 37.5 hours in any week, Monday to Sunday, shall fall within the terms of Rule 10(e) below.

*Rates of Wages and Working Rules - Electrical Installations*

## WORKING RULES

(e) Overtime on Shifts
The number of hours to be worked at the appropriate shift rates before overtime premium is calculated shall be:-

    From and including 4 January 1992    38 hrs

Premium payments shall be calculated on the appropriate standard rate of pay and not on the shift rate.

(f) Other Shift Arrangements
Detailed arrangements for any other shift system of those operating on sites covered by the JIB/NJC Treaty Arrangement will be as approved by the JIB.

### 12: Travelling Time and Travel Allowances, Country and Lodging Allowances

(a) Wages and Allowances
Operatives who are required to book on and off at the Employers Shop shall be entitled to time from booking on until booking off with overtime if the time so booked exceeds the normal working day. They shall also be entitled to a travel allowance as determined from time to time.

(b) Travelling Time and Travel Allowance from 4 January 1992
Operatives who are required to start and finish at the normal starting and finishing time on jobs which are up to and including 35 miles from the shop - in a straight line - shall receive both Travel Allowance and Payment for Travelling Time as determined.

N.B. Site Transport
Where an employer provides transport from shop to site and from site to shop or between sites, those operatives so transported shall not be entitled to Travel Allowance.

(c) Country Allowance
Operatives sent from the Employers Shop who are required to start and finish at the normal starting and finishing time on jobs over 35 miles from the Shop and who elect to travel daily to the job instead of taking lodgings will be paid Country Allowance in lieu of travelling time and Travel Allowance. In the case of London jobs this is over and above the JIB Rates of Wages for the London Zone.

(d) Lodging Allowance
    (i) Operatives sent from the Employers Shop who are required to start and finish at the normal starting and finishing time on jobs over 35 miles from the Shop, who elect to lodge away from home and provide proof of lodging will be paid Lodging Allowance.
    (ii) Travelling time and travel allowance between the lodgings and the job shall not normally be paid. Where it is proved to the Employers satisfaction that suitable lodging and accommodation is not available near the job, travelling time and travel allowances for any distance of more than 10 miles each way will be paid in accordance with the scale contained in paragraph 12(b) above on the excess distance.
    (iii) On being sent to the job the Operative shall receive his actual fare and travelling time at ordinary rates from the Employers Shop and when he returns to the Employers Shop except that when, of his own free will, he leaves the job within one calendar month from the date of his arrival and in cases where he is dismissed by the Employer for proved bad timekeeping, improper work or similar misconduct, no return travelling time or fares shall be paid.
    (iv) The payment of Lodging Allowance shall not be made when suitable board and lodging is arranged by the Employer at no cost to the Operative.
    (v) The payment of Lodging Allowance shall not be made during absence from employment unless a Medical Certificate is produced for the whole of the period claimed. When an operative is sent home by the firm at their cost the payment of Lodging Allowance shall cease.

## WORKING RULES

## INTRODUCTION (contd)

## 12: Travelling Time and Travel Allowances, Country and Lodging Allowances (contd)

(d) Lodging Allowance (contd)
- (vi) No payment for the retention of lodgings during Annual Paid Holiday shall be made by the Employer except in cases where the Operative is required to pay a retention fee during Annual Paid Holiday when reimbursement shall be of the amount actually paid to a maximum of £4.14 per day (£28.98 per week) from and including 4 January 1997 upon production of proof of payment to the Employers satisfaction.
- (vii) Where an Operative is away from his lodgings at a weekend under Rule 12(e) but has to pay a retention fee for his lodgings, reimbursement shall be the amount actually paid, to a maximum of £19.15 from and including 4 January 1997, upon production of proof of payment to the Employers satisfaction.

(e) Period Return Fares for Operatives who lodge:
- (i) On jobs up to and including 100 miles from the Employers Shop, return railway fares from the Job to the Employers Shop, without travelling time, shall be paid for every two weeks.
- (ii) On jobs over 100 miles and up to and including 250 miles from the Employers Shop, return railway fares from the Job to the Employers Shop, with 4 hours travelling time at ordinary rate time, shall be paid every four weeks.
- (iii) On jobs over 250 miles from the Employers Shop, return railway fares from the Job to the Employers Shop, with 7.5 hours travelling time at ordinary rates, shall be paid every four weeks.
- (iv) In cases under sub-clauses (ii) and (iii) above, where the Employer, through necessity or expediency, requires his Operatives to work during the specified weekend leave period, he shall arrange that they shall have another period in substitution but this provision shall not apply under sub-clause (i) above.

N.B. All distances shall be calculated in a straight line (point to point).

When Annual Holidays with pay are taken the period returns may be moved forward or backward from the date upon which they become due, to enable the period returns to coincide with the date of the Annual Paid Holiday.

Special consideration shall be given to Operatives where it is necessary for them to return home on compassionate grounds, e.g. domestic illness.

(f) Locally Engaged Labour:

Where an Employer does not have a Shop within 25 miles of the Job, he can engage labour domiciled within a 25 miles radius of that Job. Operatives shall receive the JIB Rates of Wages applicable to the Zone of the Job and travelling time and Travel Allowance in accordance with Clause (b), but with the exception of "home" being substituted for "shop" in Clause (b).

Locally engaged labour, domiciled within a 25 miles radius of the Job, can be transferred to other Jobs within that radius without affecting their entitlements under this Rule. Operatives transferred to a Job outside that radius and within the Zone Rate will be entitled to Country Allowance in accordance with the Rules.

## 13: Statutory Holidays

(a) Qualification

Seven and a half hours pay at the appropriate JIB Rates of Wages shall be paid for a maximum of eight Statutory Holidays per annum. In general, the following shall constitute such paid holidays:

New Years Day; Good Friday; Easter Monday; May Day; Spring Time Bank Holiday; Late Summer Bank Holiday; Christmas Day; Boxing Day.

In areas where any of these days are not normally observed as holidays in the Electrical Contracting Industry, traditional local holidays may be substituted by mutual agreement and subject to the determination of the appropriate Regional Joint Industry Board.

When Christmas Day and/or Boxing Day or New Years Day falls on a Saturday or Sunday, the following provisions apply:

### Rates of Wages and Working Rules - Electrical Installations

**WORKING RULES**

*Christmas Day*
When Christmas Day falls on a Saturday or a Sunday, the Tuesday next following shall be deemed to be a paid holiday.
*Boxing Day or New Years Day*
When Boxing Day or New Years Day falls on a Saturday or Sunday, the Monday next following shall be deemed to be a paid holiday.
In order to qualify for payment, operatives must work full time for the normal day on the working days preceding and following the holiday.

For the purpose of this Rule, an operative shall be deemed to have worked on one or both of the qualifying days when the Operative
    (i)    has lost time through certified sickness.
    (ii)    was on a rest period for the day following continuous working all the previous night.
    (iii)    was absent with the Employers permission.

(b) Payment for working Statutory Holidays
When operatives are required to work on a Paid Holiday within the scope of this Agreement, they shall receive wages at the following rates for all hours worked:
    CHRISTMAS DAY - Double time and a day or shift off in lieu for which they shall be paid wages at bare time rates for the hours constituting a normal working day. The alternative day hereunder shall be mutually agreed between the Employer and the Operatives concerned.
    In respect of all other days: either
    (a)    Time-and-a-half plus a day or shift off in lieu for which they shall be paid wages at bare time rates for the hours constituting a normal working day. The alternative day hereunder shall be mutually agreed between the Employer and the Operatives concerned; or
    (b)    at the discretion of the Employer 2.5 times the bare time rate in which event no alternative day is to be given.
In the case of night shift workers required to work on a Statutory Holiday, the premiums mentioned above shall be calculated upon the night shift rate of time-and-a-third. Time off in lieu of Statutory Holidays shall be paid at bare time day rates.

### 14: Annual Holidays

Operatives shall be entitled to payment for Annual Holidays as determined from time to time under the JIB Annual Holiday with Pay Scheme, depending upon their continuity of service in the Industry.

The JIB Annual Holiday with Pay Scheme is carried out on behalf of the Joint Industry Board by the Electrical Contracting Industry Benefits Agency (ECIBA) who operate a Benefits Credit collection system. Details of the scheme are shown in Section 9.

### GRADING DEFINITIONS

#### 1.1 Technician

**Qualification and Training**

Must have been a Registered Apprentice and have had practical training in electrical installation work and must have obtained the City and Guilds of London Institute Electrical Installation Work Part III Course Certificate (or approved equivalent), and either:

    (a)    Must be at least 27 years of age.
                Must have had at least five years' experience as a Approved Electrician with "responsibility money", including a minimum of three years in a supervisory capacity in charge of electrical engineering installations of such a complexity and dimension as to require wide technical experience and organisational ability.

## WORKING RULES

## GRADING DEFINITIONS (contd)

### 1.1 Technician (contd)

**Qualification and Training (contd)**

or      (b)   Must have exceptional technical skill, ability and experience beyond that expected of an Approved Electrician, so that his value to the Employer would be as if he were qualified as a Technician under (a) above and, with the support of his present Employer, may be granted this grade by the Joint Industry Board

**Duties**

Technicians must have knowledge of the most economical and effective layout of electrical installations together with the ability to achieve a high level of productivity in the work which they control. They must also be able to apply a thorough working knowledge of the National Working Rules for the Electrical Contracting Industry, of the current IEE Regulations for Electrical Installations, of the Electricity at Work Regulations 1989, the Electricity Supply Regulations, Installations (ie Regulations 22-29 inclusive and 31), of any Regulations dealing with Consumers Installations which may be issued, relevant British Standards and Codes of Practice, and of the Construction Industry Safety Regulations.

### 1.2 Approved Electrician

**Qualification and Training**

Must satisfy the following hree conditions :-

Must have been a Registered Apprentice or undergone some equivalent method of training and have had practical training in electrical installation work.
    and
Must have obtained the NVQ Level 3 - Installing and Commissioning Electrical Systems and Equipment, together with the necessary practical experience
    or
Must have obtained at least the City and Guilds 2360 Electrical Installation Theory Part 2 Course Certificate (or approved equivalent)
    and
have obtained Achievement Measurement 2 or must be able, with the application for Grading and any other relevant supporting evidence (i.e. the City and Guilds Electricians Certificate ) which may be required, to satisfy the Grading Committee of his experience and suitability
    and
must have had two years experience working as an Electrician subsequent to the satisfactory completion of training and immediately prior to the application for this grade, or be 22 years of age, whichever is the sooner.

**Duties**

Approved Electricians must possess particular practical, productive and electrical engineering skills with adequate technical supervisory knowledge so as to be able to work on their own proficiently and carry out electrical installation work without detailed supervision in the most efficient and economical manner; be able to set out jobs from drawings and specifications and requisition the necessary installation materials. They must also have a thorough working knowledge of the National Working Rules for the Electrical Contracting Industry, of the current IEE Regulations for Electrical Installations, of the Electricity Supply Regulations, 1988, issued by the Electricity Commissioners so far as they deal with Consumers' Installations (i.e. Regulations 22-29 inclusive and 31), of any Regulations dealing with Consumers' Installations which may be issued, relevant British Standards and Codes of Practice, and of the Construction Industry Safety Regulations.

*Rates of Wages and Working Rules - Electrical Installations*

## WORKING RULES

### 1.3. Electrician

**Qualification and Training**

Must have been a registered Apprentice or undergone some equivalent method of training and have had adequate practical training in electrical installation work
   and
Must have obtained the NVQ Level 3 - Installing and Commissioning Electrical Systems and Equipment, together with the necessary practical experience
   or
Must have completed the City and Guilds 2360 Electrical Installation Work Theory Part 2 Course (or approved equivalent) and have obtained Achievement Measurement 2 or must be able, with the application for grading and any other relevant supporting evidence which may be required, to satisfy the Grading Committee of his experience and suitability.
   and
Must be at least 21 years of age (which requirement may be waived if the applicant has obtained a pass in the City and Guilds 2360 Electrical Installation Theory Part 2 Course or approved equivalent).

**Duties**

Must be able to carry out electrical installation work efficiently in accordance with the National Working Rules for the Electrical Contracting Industry, the current IEE Regulations for Electrical Installations, and the Construction Industry Safety Regulations.

### 1.4. Labourer

Labourers may be employed to assist in the installation of cables in accordance with Section 5.1 - Cable Agreement: and to do other unskilled work under supervision provided that they should not be used to re-introduce pair working. Nothing in these rules should be taken to imply that labourers must be employed where there is not sufficient unskilled work to justify their employment, nor to prevent skilled men from doing a complete electrical installation job including the unskilled elements in these circumstances. On any Site at any time there shall be employed in total no more than one Labourer to four skilled JIB Graded Operatives. This particular requirement may be reviewed in the light of the particular circumstances in respect of a particular site upon application, by either Party to the appropriate Regional Joint Industry Board.

# Marketing for Architects and Engineers

*A new approach*
*B. Richardson, Director, Northern Architecture Centre Ltd,*
*Newcastle-upon-Tyne, UK*
**Foreword by Francis Duffy**

Professional services marketing is a relatively new form of marketing that has been recognised only since the late 1980s. Most of the attempts to write about marketing for professional services have been a regurgitation of the traditional marketing approach that has evolved since the 1960s and have concentrated on minor differences and adjustments. In many ways, what is needed is a fresh approach which takes into account the complex political, social, economic, legislative and cultural backdrop and provides a way for design professionals, such as architects and engineers, to look to the future. This book does just that.

- offers a way for architects and engineers to take charge of their own future

- provides a comprehensive, forward looking marketing discipline for design professionals

- introduces the new concepts of synthesis marketing and strategic mapping

**Contents:**

Introduction. Markets and marketing. Scenario planning. Synthesis marketing. Strategic mapping. A synthesis marketing programme. Architecture centres. Bibliography and references. Index.

**For further information and to order please contact**
The Marketing Dept., E & FN Spon, 2-6 Boundary Row, London SE1 8HN
Tel: 0171 865 0066  Fax: 0171 522 9621

**E & F N Spon**
**234x156: 152pp :25 line illus: November 1996**
**Paperback:0-419-20290-0: £17.50**

# PART THREE
# Material Costs/ Measured Work Prices

Mechanical Installations, *page 101*
Electrical Installations, *page 341*

# DAVIS LANGDON & EVEREST
## Authors of Spon's Price Books

**Davis Langdon & Everest** is an independent practice of Chartered Quantity Surveyors with over 500 staff in 19 UK offices.

We believe that cost awareness and control are responsibilities that must be accepted by all members of the team, but the ability to act on that responsibility is dependent upon the quality of the cost advice given. Our approach to cost consultancy is therefore:

*   to be positive and creative in our advice, rather than simply reactive;

*   to concentrate on value for money and value engineering rather than on superficial cost-cutting;

*   to give advice that is matched to the Client's own criteria, rather than to impose standard or traditional solutions;

*   to see cost as one component of a successful design solution, which needs to be balanced with many others, and to work as an integrated member of a design team in achieving that balance;

*   to pay attention to the life-long costs of owning and operating a facility, rather than simply the initial capital cost.

We work in small teams, under the active leadership of a Partner, to provide continuity of involvement throughout the stages of a project. At the same time, we can draw on the resources of a large firm, and upon departments able to give specialist advice on engineering services, life cycle costing, project analyses, cost data and cost research.

**Davis Langdon & Everest** - a first point of call for anyone contemplating a construction project. **Consult the Source.**

**TRUST DLE TO PROVIDE THE DEFINITIVE PRICE DATA**

# Mechanical Installations
## MATERIAL COSTS/MEASURED WORK PRICES

### DIRECTIONS

The following explanations are given for each of the column headings and letter codes. It should be noted that not only are full material costs per item declared but also the published list price, the latest published price increase (update).

| | |
|---|---|
| Unit | Prices for each unit are given as singular (1 metre, 1 nr). |
| Net price | Manufacturer's latest issued material/component price list, plus nominal allowance for fixings, plus any percentage uplift advised by manufacturer, plus where appropriate waste less percentage discount. The net price also reflects the applicable quantity for the minimum purchase of each batch of material and against which discounts etc. apply. |
| Material cost | Net price plus percentage allowance for overheads and profit. |
| Labour constant | Gang norm (in manhours) for each operation. |
| Labour cost | Labour constant multiplied by the appropriate all-in manhour cost. (See also relevant Rates of Wages Section) |
| Measured work price | Material cost plus Labour cost. |

### MATERIAL COSTS
The Material Costs given includes for delivery to sites in the London area at April/May 1997 with an allowance for waste, overheads and profit, and represents the price paid by contractors after the deduction of all trade discounts but excludes any charges in respect of VAT.

### MEASURED WORK PRICES
These prices are intended to apply to new work in the London area and include allowances for all charges, preliminary items and profit. The prices are for reasonable quantities of work and the user should make suitable adjustments if the quantities are especially small or especially large. Adjustments may also be required for locality (eg outside London) and for the market conditions (eg volume of work on hand or on offer) at the time of use.

## Material Costs/Measured Work Prices - Mechanical Installations
### DIRECTIONS

MECHANICAL INSTALLATIONS
The labour rate on which these prices have been based is £9.00 per man hour which is the London Standard Rate at May 1997 plus allowances for all other emoluments and expenses. To this rate has been added 25% to cover site and head office overheads and preliminary items together with 2½% for profit, resulting in an inclusive rate of **£11.48** per man hour. The rate of £9.00 per man hour has been calculated on a working year of 1,755.60 hours; a detailed build-up of the rate is given at the end of these Directions.

PLUMBING INSTALLATIONS
The labour rate on which these prices have been based is £9.00 per man hour which is the rate at May 1997 plus allowances for all other emoluments and expenses. To this rate has been added 25% to cover site and head office overheads and preliminary items together with 2½% for profit, resulting in an inclusive rate of **£11.48** per man hour.

DUCTWORK INSTALLATIONS
The labour rate on which these prices have been based is £9.52 per man hour which is the rate at November 1996. To this rate has been added 55% to cover shop, site and head office overheads and preliminary items together with 2½% for profit, resulting in an inclusive rate of **£14.99** per man hour.

In calculating the 'Measured Work Prices' the following assumptions have been made:
- (a) That the work is carried out as a sub-contract under the Standard Form of Building Contract and that such facilities as are usual would be afforded by the main contractor.
- (b) That, unless otherwise stated, the work is being carried out in open areas at a height which would not require more than simple scaffolding.
- (c) That the building in which the work is being carried out is no more than six storeys high.

Where these assumptions are not valid, as for example where work is carried out in ducts and similar confined spaces or in multi-storey structures when additional time is needed to get to and from upper floors, then an appropriate adjustment must be made to the prices. Such adjustment will normally be to the labour element only.

No allowance has been made in the prices for any cash discount to the main contractor.

## Material Costs/Measured Work Prices - Mechanical Installations

## DIRECTIONS

## LABOUR RATE - PLUMBING MECHANICAL ENGINEERING

The following detail shows how the labour rate of £9.00 per man hour has been calculated.

### The annual cost of notional eleven man gang.

|  | TECHNICAL PLUMBER/ TECHNICIAN | ADVANCED PLUMBER/ ENGINEER GAS/ARC | ADVANCED PLUMBER/ ENGINEER GAS OR ARC | ADVANCED PLUMBER/ ENGINEER | TRAINED PLUMBER/ FITTER | APPRENTICE 4TH YEAR NVQ LEVEL 3 | APPRENTICE 3RD YEAR NVQ LEVEL 2 | SUB-TOTALS |
|---|---|---|---|---|---|---|---|---|
|  | 1 NR | 1 NR | 2 NR | 3 NR | 2 NR | 1 NR | 1 NR |  |
| Hourly Rate from 05/05/1997 | 7.22 | 6.51 | 6.32 | 6.07 | 5.42 | 4.93 | 3.89 |  |
| Working hours/annum | 1,755.60 | 1,755.60 | 3,511.20 | 5,266.80 | 3,511.20 | 1,755.60 | 1,755.60 |  |
| x Hourly rate = £/annum | 12,675.43 | 11,428.96 | 22,190.78 | 31,969.48 | 19,030.70 | 8,655.11 | 6,829.28 | 112,779.74 |
| Incentive schemes @ 5% | 633.77 | 571.45 | 1,109.54 | 1,598.47 | 951.54 | 432.76 | 341.46 | 5,638.99 |
| Daily travel rate/day @ 5.10 | 4.03 | 3.52 | 3.52 | 3.52 | 3.21 | 2.23 | 1.74 |  |
| Days/annum | 231 | 231 | 462 | 693 | 462 | 231 | 231 |  |
| £/annum | 930.93 | 813.12 | 1,626.24 | 2,439.36 | 1,483.02 | 515.13 | 401.94 | 8,209.74 |
| Daily travel fare | 3.43 | 3.43 | 3.43 | 3.43 | 3.43 | 3.43 | 3.43 |  |
| Days/annum | 226 | 226 | 452 | 678 | 452 | 226 | 182 |  |
| £/annum | 775.18 | 775.18 | 1,550.36 | 2,325.54 | 1,550.36 | 775.18 | 624.26 | 8,376.06 |
| Weekly holiday/welfare (nr of weeks) | 52 | 52 | 104 | 156 | 104 | 52 | 52 |  |
| Stamp value | 21.42 | 19.13 | 19.13 | 19.13 | 17.60 | 17.60 | 7.85 |  |
| £/annum | 1,113.84 | 994.76 | 1,989.52 | 2,984.28 | 1,830.40 | 915.20 | 408.20 | 10,236.20 |
| National insurance contributions |  |  |  |  |  |  |  |  |
| Weekly gross pay EACH | 297.91 | 268.07 | 260.74 | 251.10 | 224.53 | 200.90 | 158.42 |  |
| % Contributions | 0.100 | 0.100 | 0.100 | 0.100 | 0.100 | 0.070 | 0.070 |  |
| £ Contributions EACH | 29.79 | 26.86 | 26.07 | 25.11 | 22.45 | 14.06 | 11.09 |  |
| x Nr of men = £ contributions/week | 29.79 | 26.86 | 52.14 | 75.33 | 44.90 | 14.06 | 11.09 |  |
| £ Contributions/annum | 1,423.96 | 1,283.91 | 2,492.29 | 3,600.77 | 2,146.22 | 672.07 | 530.10 | 12,149.32 |
| Tool and clothing allowance | 116.48 | 116.48 | 232.96 | 349.44 | 232.96 | 116.48 | 116.48 | 1,281.28 |

| | |
|---|---|
| SUB-TOTAL | 158,671.33 |
| TRAINING (INCLUDING ANY TRADE REGISTRATIONS) - SAY 1% | 1,586.71 |
| SEVERANCE PAY AND SUNDRY COSTS - SAY 1.5% | 2,403.87 |
| EMPLOYER'S LIABILITY AND THIRD PARTY INSURANCE - SAY 2% | 3,253.24 |
| ANNUAL COST OF NOTIONAL 11 MAN GANG (10.5 MEN ACTUALLY WORKING) | 165,915.15 |
| ANNUAL COST PER PRODUCTIVE MAN (ie÷10.5) | 15,801.44 |
| ALL IN MAN HOUR (BASED ON 1755.6 HOURS) | 9.00 |

Notes:
(1) The following assumptions have been made in the above calculations:-
    (a) The working week of 37.5 hours is made up of 7.5 hours Monday to Friday.
    (b) The actual hours worked are five days of 8.5 hours each.
    (c) Five days in the year are lost through sickness or similar reasons.
    (d) A working year of 1755.6 hours.
(2) National insurance contributions are those effective from April 1997.
(3) Weekly Holiday Credit/Welfare Stamp values are those effective from 3 April 1995.

## Material Costs/Measured Work Prices - Mechanical Installations

### DIRECTIONS

### LABOUR RATE - HEATING AND VENTILATION

The following detail shows how the labour rate of £9.52 per man hour has been calculated.

**The annual cost of notional seven man gang.**

|  | CHARGEHAND ERECTOR  1 NR | ADVANCED ERECTOR  1 NR | ERECTOR  4 NR | TRAINEE  1 NR | SUB-TOTALS |
|---|---|---|---|---|---|
| **Hourly Rate from 11/11/1996** | 6.94 | 6.13 | 5.60 | 5.02 | |
| Working hours/annum | 1,755.60 | 1,755.60 | 7,022.40 | 1,755.60 | |
| x Hourly rate = £/annum | 12,183.86 | 10,761.83 | 39,325.44 | 8,813.11 | 71,084.24 |
| **Incentive schemes @ 5%** | 609.19 | 538.09 | 1,966.27 | 440.66 | 3,554.21 |
| Daily travel rate/day | 3.45 | 3.45 | 3.34 | 2.92 | |
| Days/annum | 231 | 231 | 924 | 231 | |
| £/annum | 796.95 | 796.95 | 3,086.16 | 674.52 | 5,354.58 |
| Daily travel fare | 3.10 | 3.10 | 3.10 | 2.67 | |
| Days/annum | 226 | 226 | 904 | 226 | |
| £/annum | 700.60 | 700.60 | 2,802.40 | 603.42 | 4,807.02 |
| Weekly holiday/welfare (nr of weeks) | 52 | 52 | 208 | 52 | |
| Credit Value | 34.62 | 32.60 | 29.04 | 25.97 | |
| £/annum | 1,800.24 | 1,695.20 | 6,040.32 | 1,350.44 | 10,886.20 |
| National insurance contributions | | | | | |
| Weekly gross pay EACH | 284.31 | 253.07 | 232.10 | 207.70 | |
| % Contributions | 0.100 | 0.100 | 0.100 | 0.070 | |
| £ Contributions EACH | 28.43 | 25.31 | 23.21 | 14.54 | |
| x Nr of men = £ contributions/week | 28.43 | 25.31 | 92.84 | 14.54 | |
| £ Contributions/annum | 1,358.95 | 1,209.82 | 4,437.75 | 695.01 | 7,701.53 |
| **Tool and clothing allowance** | 75.00 | 75.00 | 300.00 | 75.00 | 525.00 |
| | | | | SUB-TOTAL | 103,912.78 |
| | | TRAINING (INCLUDING ANY TRADE REGISTRATIONS) - SAY 1% | | | 1,039.13 |
| | | SEVERANCE PAY AND SUNDRY COSTS - SAY 1.5% | | | 1,574.28 |
| | | EMPLOYER'S LIABILITY AND THIRD PARTY INSURANCE - SAY 2% | | | 2,130.52 |
| | | ANNUAL COST OF NOTIONAL 7 MAN GANG (6.5 MEN ACTUALLY WORKING) | | | 108,656.71 |
| | | ANNUAL COST PER PRODUCTIVE MAN (ie÷6.5) | | | 16,716.42 |
| | | ALL IN MAN HOUR (BASED ON 1755.6 HOURS) | | | 9.52 |

Notes:
(1) The following assumptions have been made in the above calculations:-
  (a) The working week of 38 hours is made up of 8 hours Monday and 7.5 hours Tuesday to Friday.
  (b) The actual hours worked are five days of 8.5 hours each.
  (c) Five days in the year are lost through sickness or similar reasons.
  (d) A working year of 1755.60 hours.
(2) National insurance contributions are those effective from April 1997.
(3) Weekly Holiday Credit/Welfare Stamp values are those effective from 30 September 1996.

## S:PIPED SUPPLY SYSTEMS

| Item | Net Price £ | Material £ | Labour hours | Labour £ | Unit | Total Rate £ |
|---|---|---|---|---|---|---|
| **S41 : FUEL OIL STORAGE/DISTRIBUTION : STORAGE TANKS AND VESSELS** | | | | | | |
| Fuel storage tanks; mild steel; with all necessary screwed bosses; oil resistant joint rings; includes placing in position | | | | | | |
| Rectangular | | | | | | |
| 1360 litres (300 gallon) capacity; 2mm plate | 108.00 | 118.80 | 10.00 | 114.75 | nr | 233.55 |
| 2730 litres (600 gallon) capacity; 2.5mm plate | 159.30 | 175.23 | 22.00 | 252.45 | nr | 427.68 |
| 4550 litres (1000 gallon) capacity; 3mm plate | 394.00 | 433.40 | 45.00 | 516.38 | nr | 949.78 |
| Fuel storage tanks; plastic; with all necessary screwed bosses; oil resistant joint rings; includes placing in position | | | | | | |
| Cylindrical; horizontal | | | | | | |
| 1250 litres (285 gallon) capacity | 171.56 | 188.72 | 2.00 | 22.95 | nr | 211.67 |
| 1350 litres (300 gallon) capacity | 177.09 | 194.80 | 2.00 | 22.95 | nr | 217.75 |
| 2500 litres (550 gallon) capacity | 294.58 | 324.04 | 2.00 | 22.95 | nr | 346.99 |
| Cylindrical; vertical | | | | | | |
| 1300 litres (300 gallon) capacity | 144.96 | 159.46 | 2.00 | 22.95 | nr | 182.41 |
| 2600 litres (570 gallon) capacity | 211.46 | 232.61 | 2.00 | 22.95 | nr | 255.56 |
| 3600 litres (800 gallon) capacity | 322.29 | 354.52 | 2.00 | 22.95 | nr | 377.47 |
| 5000 litres (1100 gallon) capacity | 444.20 | 488.62 | 2.00 | 22.95 | nr | 511.57 |
| 10000 litres (2200 gallon) capacity | 898.60 | 988.46 | 2.00 | 22.95 | nr | 1011.41 |
| Bunded tanks | | | | | | |
| 1150 litres (250 gallon) capacity | 444.20 | 488.62 | 2.00 | 22.95 | nr | 511.57 |
| 1350 litres (300 gallon) capacity | 488.53 | 537.38 | 2.00 | 22.95 | nr | 560.33 |
| 2500 litres (550 gallon) capacity | 571.65 | 628.82 | 2.00 | 22.95 | nr | 651.77 |
| 5000 litres (1100 gallon) capacity | 942.93 | 1037.22 | 2.00 | 22.95 | nr | 1060.17 |

*SPON's price data now available in an electronic estimating system call us on 0171-522-9966 for details*

## S:PIPED SUPPLY SYSTEMS

| Item | Net Price £ | Material £ | Labour hours | Labour £ | Unit | Total Rate £ |
|---|---|---|---|---|---|---|
| **S60 : FIRE HOSE REELS : HOSE REELS** | | | | | | |
| Hose reels; automatic; connection to 25mm screwed joint; reel with 30.5 metres, 19mm rubber hose; suitable for working pressure up to 7 bar | | | | | | |
| Reels | | | | | | |
|   Non-swing pattern | 159.25 | 175.18 | 2.50 | 28.69 | nr | **203.87** |
|   Recessed non-swing pattern | 327.49 | 360.24 | 2.50 | 28.69 | nr | **388.93** |
|   Swinging pattern | 207.65 | 228.41 | 2.50 | 28.69 | nr | **257.10** |
|   Recessed swinging pattern | 329.33 | 362.26 | 2.50 | 28.69 | nr | **390.95** |
| Hose reels; manual; connection to 25mm screwed joint; reel with 30.5 metres, 19mm rubber hose; suitable for working pressure up to 7 bar | | | | | | |
| Reels | | | | | | |
|   Non-swing pattern | 142.16 | 156.38 | 2.50 | 28.69 | nr | **185.07** |
|   Recessed non-swing pattern | 306.73 | 337.40 | 2.50 | 28.69 | nr | **366.09** |
|   Swinging pattern | 252.86 | 278.15 | 2.50 | 28.69 | nr | **306.84** |
|   Recessed swinging pattern | 299.81 | 329.79 | 2.50 | 28.69 | nr | **358.48** |

## S:PIPED SUPPLY SYSTEMS

| Item | Net Price £ | Material £ | Labour hours | Labour £ | Unit | Total Rate £ |
|---|---|---|---|---|---|---|
| **S61 : DRY RISERS : INLET BOX, OUTLET BOXES AND VALVES** | | | | | | |
| **Dry rising main; (Note: for tubing and flanged connections see previous sections.)** | | | | | | |
| Bronze/gunmetal inlet breeching for pumping in with 64mm dia. instantaneuos male coupling; with cap, chain and 25mm drain valve | | | | | | |
|   Single inlet with black pressure valve, screwed or flanged to steel | 127.65 | 140.41 | 1.15 | 13.20 | nr | **153.61** |
|   Double inlet with back pressure valve, screwed or flanged to steel | 155.65 | 171.22 | 1.28 | 14.69 | nr | **185.91** |
|   Quadruple inlet with back pressure valve, flanged to steel | 459.30 | 505.23 | 2.40 | 27.58 | nr | **532.81** |
| Steel dry riser inlet box with hinged wire glazed door suitably lettered (fixing by others) | | | | | | |
|   610 x 460 x 325mm; double inlet | 130.30 | 143.33 | 0.50 | 5.74 | nr | **149.07** |
|   610 x 610 x 356mm; quadruple inlet | 130.30 | 143.33 | 0.50 | 5.74 | nr | **149.07** |
| Bronze/gunmetal gate type outlet valve with 64mm dia. instantaneous female coupling cap and chain; wheel head secured by padlock and leather strap | | | | | | |
|   Flanged BS table D inlet (bolted connection to counter flanges measured separately) | 125.20 | 137.72 | 0.80 | 9.18 | nr | **146.90** |
| Bronze/gunmetal landing type outlet - valve, with 64mm dia. instantaneous female coupling; cap and chain; wheelhead secured by padlock and leatherstrap, bolted connections to counter flanges measured separately | | | | | | |
|   Horizontal, flanged BS table D inlet | 106.35 | 116.98 | 0.80 | 9.18 | nr | **126.16** |
|   Oblique, flanged BS table D inlet | 101.50 | 111.65 | 0.80 | 9.18 | nr | **120.83** |
| Air valve, screwed joint to steel | | | | | | |
|   25mm dia. | 11.40 | 12.54 | 0.80 | 9.18 | nr | **21.72** |

## Material Costs / Measured Work Prices - Mechanical Installations

### S:PIPED SUPPLY SYSTEMS

| Item | Net Price £ | Material £ | Labour hours | Labour £ | Unit | Total Rate £ |
|---|---|---|---|---|---|---|
| **S63 : SPRINKLERS : SPRINKLER HEADS** | | | | | | |
| **Sprinkler heads and valves** | | | | | | |
| Sprinkler heads; brass body; frangible glass bulb; manufactured to standard operating temperature of 57-141 degrees Celsius; standard response; RTI=100 | | | | | | |
| conventional pattern; 15mm dia. | 2.91 | 3.20 | 0.15 | 1.72 | nr | 4.92 |
| sidewall pattern; 15mm dia. | 3.20 | 3.52 | 0.15 | 1.72 | nr | 5.24 |
| conventional pattern; 15mm dia.; satin chrome plated | 3.24 | 3.56 | 0.15 | 1.72 | nr | 5.28 |
| sidewall pattern; 15mm dia.; satin chrome plated | 3.55 | 3.90 | 0.15 | 1.72 | nr | 5.62 |
| Sprinkler heads; brass body; frangible glass bulb; manufactured to standard operating temperature of 57-141 degrees Celsius; quick response; RTI<50 | | | | | | |
| conventional pattern; 15mm dia. | 5.72 | 6.29 | 0.15 | 1.72 | nr | 8.01 |
| sidewall pattern; 15mm dia. | 6.26 | 6.89 | 0.15 | 1.72 | nr | 8.61 |
| conventional pattern; 15mm dia.; satin chrome plated | 6.05 | 6.66 | 0.15 | 1.72 | nr | 8.38 |
| sidewall pattern; 15mm dia.; satin chrome plated | 6.61 | 7.27 | 0.15 | 1.72 | nr | 8.99 |
| Wet system alarm valves; including internal non-return valve; working pressure up to 12.5 bar; BS4504 PN16 flanged ends; bolted connections | | | | | | |
| 100mm dia. | 297.25 | 326.98 | 2.18 | 25.05 | nr | 352.03 |
| 150mm dia. | 333.13 | 366.44 | 2.69 | 30.85 | nr | 397.29 |
| Alternate system wet/dry alarm station; including butterfly valve, wet alarm valve, dry pipe differential pressure valve and pressure gauges; working pressure up to 10.5 bar; BS4505 PN16 flanged ends; bolted connections | | | | | | |
| 100mm dia. | 2047.95 | 2252.74 | 4.00 | 45.90 | nr | 2298.64 |
| 150mm dia. | 2405.68 | 2646.25 | 5.00 | 57.38 | nr | 2703.63 |
| Alternate system wet/dry alarm station; including electrically supervised butterfly valve, water supply accelerator set, wet alarm valve, dry pipe differential pressure valve and pressure gauges; working pressure up to 10.5 bar; BS4505 PN16 flanged ends; bolted connections | | | | | | |
| 100mm dia. | 2047.95 | 2518.84 | 6.00 | 68.85 | nr | 2587.69 |
| 150mm dia. | 2405.68 | 2912.34 | 7.00 | 80.33 | nr | 2992.67 |

## S:PIPED SUPPLY SYSTEMS

| Item | Net Price £ | Material £ | Labour hours | Labour £ | Unit | Total Rate £ |
|---|---|---|---|---|---|---|
| Water operated motor alarm and gong; stainless steel and aluminum body and gong; screwed connections | | | | | | |
| Connection to sprinkler system and drain pipework | 152.72 | 167.99 | 1.50 | 17.24 | nr | **185.23** |

# Keep your figures up to date, free of charge

This section, and most of the other information in this Price Book, is brought up to date every three months, until the next annual edition, in the *Price Book Update*.

The *Update* is available free to all Price Book purchasers.

To ensure you receive your copy, simply complete the reply card from the centre of the book and return it to us.

## S:PIPED SUPPLY SYSTEMS

| Item | Net Price £ | Material £ | Labour hours | Labour £ | Unit | Total Rate £ |
|---|---|---|---|---|---|---|
| **S65 : FIRE HYDRANTS : EXTINGUISHERS** | | | | | | |
| Fire extinguishers; hand held; BS 5423; placed in position | | | | | | |
| Water type; cartridge operated; for Class A fires | | | | | | |
| Water type, 9 litres capacity; 55gm $CO_2$ cartridge; Class A fires (fire rating 13A) | 34.72 | 38.19 | 0.50 | 5.74 | nr | **43.93** |
| Foam type, 9 litres capacity; 75 gm $CO_2$ cartridge; Class A & B fires (fire rating 13A:183B) | 50.44 | 55.48 | 0.50 | 5.74 | nr | **61.22** |
| Dry powder type; cartridge operated; for Class A, B & C fires and electrical equipment fires | | | | | | |
| Dry powder type, 1kg capacity; 12gm $CO_2$ cartridge; Class A, B & C fires (fire rating 5A:34B) | 18.75 | 20.63 | 0.50 | 5.74 | nr | **26.37** |
| Dry powder type, 2kg capacity; 28gm $CO_2$ cartridge; Class A, B & C fires (fire rating 13A:55B) | 31.44 | 34.58 | 0.50 | 5.74 | nr | **40.32** |
| Dry powder type, 4kg capacity; 90gm $CO_2$ cartridge; Class A, B & C fires (fire rating 21A:183B) | 42.73 | 47.00 | 0.50 | 5.74 | nr | **52.74** |
| Dry powder type, 9kg capacity; 190gm $CO_2$ cartridge; Class A, B & C fires (fire rating 43A:233B) | 53.41 | 58.75 | 0.50 | 5.74 | nr | **64.49** |
| Carbon dioxide type; for Class B fires and electrical equipment fires | | | | | | |
| $CO_2$ type with hose and horn, 2kg capacity, Class B fires (fire rating 34B) | 46.71 | 51.38 | 0.50 | 5.74 | nr | **57.12** |
| $CO_2$ type with hose and horn, 5kg capacity, Class B fires (fire rating 55B) | 82.06 | 90.27 | 0.50 | 5.74 | nr | **96.01** |
| Glass fibre blanket, in GRP container | | | | | | |
| 1100 x 1100mm | 11.55 | 12.71 | 0.50 | 5.74 | nr | **18.45** |
| 1200 x 1200mm | 13.44 | 14.78 | 0.50 | 5.74 | nr | **20.52** |
| 1800 x 1200mm | 17.59 | 19.35 | 0.50 | 5.74 | nr | **25.09** |

## S:PIPED SUPPLY SYSTEMS

| Item | Net Price £ | Material £ | Labour hours | Labour £ | Unit | Total Rate £ |
|---|---|---|---|---|---|---|
| **S65 : FIRE HYDRANTS : HYDRANTS AND HYDRANT BOXES** | | | | | | |
| **Fire hydrants; bolted connections** | | | | | | |
| 100mm diameter pillar hydrant | | | | | | |
|    cast iron with sluice valve | 970.00 | 1067.00 | 1.00 | 11.47 | nr | 1078.47 |
|    steel with sluice valve | 1520.00 | 1672.00 | 1.00 | 11.47 | nr | 1683.47 |
| Underground hydrants, complete with frost plug to BS 750 | | | | | | |
|    sluice valve pattern type 1 | 298.10 | 327.91 | 4.50 | 51.69 | nr | 379.60 |
|    screw down pattern type 2 | 145.05 | 159.56 | 4.50 | 51.69 | nr | 211.25 |
| Stand pipe for underground hydrant screwed base; light alloy | | | | | | |
|    Single outlet | 101.60 | 111.76 | 0.25 | 2.87 | nr | 114.63 |
|    Double outlet | 137.30 | 151.03 | 0.25 | 2.87 | nr | 153.90 |
| 64mm diameter bronze/gunmetal outlet valves | | | | | | |
|    Oblique flanged landing valve | 101.50 | 111.65 | 0.80 | 9.18 | nr | 120.83 |
|    Oblique screwed landing valve | 98.25 | 108.08 | 0.80 | 9.18 | nr | 117.26 |
| Cast iron surface box; fixing by others | | | | | | |
|    400 x 200 x 100mm | 55.65 | 61.22 | 0.50 | 5.74 | nr | 66.96 |
|    500 x 200 x 150mm | 79.55 | 87.50 | 0.50 | 5.74 | nr | 93.24 |
|    Frost Plug | 17.50 | 19.25 | 0.25 | 2.87 | nr | 22.12 |

## Material Costs / Measured Work Prices - Mechanical Installations

### T:MECHANICAL HEATING/COOLING/REFRIGERATION SYSTEMS

| Item | Net Price £ | Material £ | Labour hours | Labour £ | Unit | Total Rate £ |
|---|---|---|---|---|---|---|
| **T10 : GAS/OIL FIRED BOILERS : BOILER PLANT AND ANCILLARIES** | | | | | | |
| Domestic water boilers; stove enamelled casing; electric controls; placing in position; assembling and connecting; (electrical work elsewhere) | | | | | | |
| Gas fired; floor standing; connected to conventional flue | | | | | | |
| 30,000 to 40,000 Btu/Hr | 322.00 | 354.20 | 8.00 | 91.80 | nr | 446.00 |
| 40,000 to 50,000 Btu/Hr | 338.80 | 372.68 | 8.00 | 91.80 | nr | 464.48 |
| 50,000 to 60,000 Btu/Hr | 361.20 | 397.32 | 8.00 | 91.80 | nr | 489.12 |
| 60,000 to 70,000 Btu/Hr | 361.20 | 397.32 | 9.01 | 103.38 | nr | 500.70 |
| 70,000 to 80,000 Btu/Hr | 464.80 | 511.28 | 10.00 | 114.75 | nr | 626.03 |
| 80,000 to 100,000 Btu/Hr | 602.70 | 662.97 | 10.00 | 114.75 | nr | 777.72 |
| 100,000 to 125,000 Btu/Hr | 714.70 | 786.17 | 10.00 | 114.75 | nr | 900.92 |
| 125,000 to 140,000 Btu/Hr | 744.10 | 818.51 | 10.00 | 114.75 | nr | 933.26 |
| Gas fired; wall hung; connected to conventional flue | | | | | | |
| 20,000 to 30,000 Btu/Hr | 282.80 | 311.08 | 8.00 | 91.80 | nr | 402.88 |
| 30,000 to 40,000 Btu/Hr | 322.00 | 354.20 | 8.00 | 91.80 | nr | 446.00 |
| 40,000 to 50,000 Btu/Hr | 336.70 | 370.37 | 8.00 | 91.80 | nr | 462.17 |
| 45,000 to 60,000 Btu/Hr | 426.30 | 468.93 | 8.00 | 91.80 | nr | 560.73 |
| Gas fired; floor standing; connected to balanced flue | | | | | | |
| 30,000 to 40,000 Btu/Hr | 406.70 | 447.37 | 8.00 | 91.80 | nr | 539.17 |
| 40,000 to 50,000 Btu/Hr | 422.80 | 465.08 | 10.00 | 114.75 | nr | 579.83 |
| 50,000 to 60,000 Btu/Hr | 452.90 | 498.19 | 10.99 | 126.10 | nr | 624.29 |
| 60,000 to 70,000 Btu/Hr | 542.50 | 596.75 | 12.05 | 138.25 | nr | 735.00 |
| 70,000 to 80,000 Btu/Hr | 625.10 | 687.61 | 12.05 | 138.25 | nr | 825.86 |
| 80,000 to 100,000 Btu/Hr | 806.40 | 887.04 | 14.08 | 161.62 | nr | 1048.66 |
| 100,000 to 125,000 Btu/Hr | 1492.40 | 1641.64 | 15.87 | 182.14 | nr | 1823.78 |
| Gas fired; wall hung; connected to balanced flue | | | | | | |
| 20,000 to 30,000 Btu/Hr | 282.80 | 311.08 | 8.00 | 91.80 | nr | 402.88 |
| 30,000 to 40,000 Btu/Hr | 318.50 | 350.35 | 8.00 | 91.80 | nr | 442.15 |
| 40,000 to 50,000 Btu/Hr | 364.70 | 401.17 | 8.00 | 91.80 | nr | 492.97 |
| 50,000 to 60,000 Btu/Hr | 456.40 | 502.04 | 8.00 | 91.80 | nr | 593.84 |
| 60,000 to 75,000 Btu/Hr | 501.20 | 551.32 | 8.00 | 91.80 | nr | 643.12 |
| Gas fired; wall hung; connected to fan flue (incl. flue kit) | | | | | | |
| 20,000 to 30,000 Btu/Hr | 338.10 | 371.91 | 8.00 | 91.80 | nr | 463.71 |
| 30,000 to 40,000 Btu/Hr | 381.50 | 419.65 | 8.00 | 91.80 | nr | 511.45 |
| 40,000 to 50,000 Btu/Hr | 410.90 | 451.99 | 10.00 | 114.75 | nr | 566.74 |
| 50,000 to 60,000 Btu/Hr | 443.80 | 488.18 | 10.99 | 126.10 | nr | 614.28 |
| 60,000 to 70,000 Btu/Hr | 509.60 | 560.56 | 12.05 | 138.25 | nr | 698.81 |
| 60,000 to 80,000 Btu/Hr | 562.80 | 619.08 | 12.05 | 138.25 | nr | 757.33 |
| 80,000 to 100,000 Btu/Hr | 730.10 | 803.11 | 14.08 | 161.62 | nr | 964.73 |
| 100,000 to 120,000 Btu/Hr | 860.30 | 946.33 | 15.87 | 182.14 | nr | 1128.47 |

Material Costs / Measured Work Prices - Mechanical Installations

## T:MECHANICAL HEATING/COOLING/REFRIGERATION SYSTEMS

| Item | Net Price £ | Material £ | Labour hours | Labour £ | Unit | Total Rate £ |
|---|---|---|---|---|---|---|
| Oil fired; floor standing; connected to conventional flue | | | | | | |
| 40,000 to 50,000 Btu/Hr | 630.70 | 693.77 | 10.00 | 114.75 | nr | 808.52 |
| 50,000 to 65,000 Btu/Hr | 656.60 | 722.26 | 12.05 | 138.25 | nr | 860.51 |
| 70,000 to 85,000 Btu/Hr | 750.40 | 825.44 | 14.08 | 161.62 | nr | 987.06 |
| 88,000 to 110,000 Btu/Hr | 823.20 | 905.52 | 14.93 | 171.27 | nr | 1076.79 |
| 120,000 to 170,000 Btu/Hr | 930.30 | 1023.33 | 20.00 | 229.50 | nr | 1252.83 |
| Industrial cast iron sectional boilers; including controls, enamelled jacket and insulation; assembled on site and commissioned by supplier - costs included in material prices; (electrical work elsewhere) | | | | | | |
| Gas fired; on/off type | | | | | | |
| 16-26 kW; 3 sections; 125mm dia. flue | 1109.68 | 1220.65 | 8.00 | 91.80 | nr | 1312.45 |
| 26-33 kW; 4 sections; 125mm dia. flue | 1171.28 | 1288.41 | 8.00 | 91.80 | nr | 1380.21 |
| 33-40 kW; 5 sections; 125mm dia. flue | 1235.52 | 1359.07 | 8.00 | 91.80 | nr | 1450.87 |
| 35-50 kW; 3 sections; 153mm dia. flue | 1516.24 | 1667.86 | 8.00 | 91.80 | nr | 1759.66 |
| 50-65 kW; 4 sections; 153mm dia. flue | 1904.32 | 2094.75 | 8.00 | 91.80 | nr | 2186.55 |
| 65-80 kW; 5 sections; 153mm dia. flue | 2089.12 | 2298.03 | 8.00 | 91.80 | nr | 2389.83 |
| 80-100 kW; 6 sections; 180mm dia. flue | 2298.56 | 2528.42 | 8.00 | 91.80 | nr | 2620.22 |
| 100-120 kW; 7 sections; 180mm dia. flue | 2691.04 | 2960.14 | 8.00 | 91.80 | nr | 3051.94 |
| 105-140 kW; 5 sections; 180mm dia. flue | 3222.56 | 3544.82 | 8.00 | 91.80 | nr | 3636.62 |
| 140-180 kW; 6 sections; 180mm dia. flue | 3556.96 | 3912.66 | 8.00 | 91.80 | nr | 4004.46 |
| 180-230 kW; 7 sections; 200mm dia. flue | 4154.48 | 4569.93 | 8.00 | 91.80 | nr | 4661.73 |
| 230-280 kW; 8 sections; 200mm dia. flue | 4638.48 | 5102.33 | 8.00 | 91.80 | nr | 5194.13 |
| 280-330 kW; 9 sections; 200mm dia. flue | 5491.20 | 6040.32 | 8.00 | 91.80 | nr | 6132.12 |
| Gas fired; high/low type | | | | | | |
| 105-140 kW; 5 sections; 180mm dia. flue | 3652.88 | 4018.17 | 8.00 | 91.80 | nr | 4109.97 |
| 140-180 kW; 6 sections; 180mm dia. flue | 3987.28 | 4386.01 | 8.00 | 91.80 | nr | 4477.81 |
| 180-230 kW; 7 sections; 200mm dia. flue | 4416.72 | 4858.39 | 8.00 | 91.80 | nr | 4950.19 |
| 230-280 kW; 8 sections; 200mm dia. flue | 4900.72 | 5390.79 | 8.00 | 91.80 | nr | 5482.59 |
| 280-330 kW; 9 sections; 200mm dia. flue | 5880.16 | 6468.18 | 8.00 | 91.80 | nr | 6559.98 |
| 300-390 kW; 8 sections; 250mm dia. flue | 6667.76 | 7334.54 | 12.00 | 137.70 | nr | 7472.24 |
| 390-450 kW; 9 sections; 250mm dia. flue | 7239.76 | 7963.74 | 12.00 | 137.70 | nr | 8101.44 |
| 450-540 kW; 10 sections; 250mm dia. flue | 7744.88 | 8519.37 | 12.00 | 137.70 | nr | 8657.07 |
| 540-600 kW; 11 sections; 300mm dia. flue | 8399.60 | 9239.56 | 12.00 | 137.70 | nr | 9377.26 |
| 600-670 kW; 12 sections; 300mm dia. flue | 10043.44 | 11047.78 | 12.00 | 137.70 | nr | 11185.48 |
| 670-720 kW; 13 sections; 300mm dia. flue | 10709.60 | 11780.56 | 12.00 | 137.70 | nr | 11918.26 |
| 720-780 kW; 14 sections; 300mm dia. flue | 11914.32 | 13105.75 | 12.00 | 137.70 | nr | 13243.45 |
| 754-812 kW; 14 sections; 400mm dia. flue | 12423.84 | 13666.22 | 12.00 | 137.70 | nr | 13803.92 |
| 812-870 kW; 15 sections; 400mm dia. flue | 12963.28 | 14259.61 | 12.00 | 137.70 | nr | 14397.31 |
| 870-928 kW; 16 sections; 400mm dia. flue | 13662.00 | 15028.20 | 12.00 | 137.70 | nr | 15165.90 |
| 928-986 kW; 17 sections; 400mm dia. flue | 14790.16 | 16269.18 | 12.00 | 137.70 | nr | 16406.88 |
| 986-1,044 kW; 18 sections; 400mm dia. flue | 15452.80 | 16998.08 | 12.00 | 137.70 | nr | 17135.78 |
| 1,044-1,102 kW; 19 sections; 400mm dia. flue | 16016.88 | 17618.57 | 12.00 | 137.70 | nr | 17756.27 |
| 1,102-1,160 kW; 20 sections; 400mm dia. flue | 16617.04 | 18278.74 | 12.00 | 137.70 | nr | 18416.44 |
| 1,160-1,218 kW; 21 sections; exceeding 400mm dia. flue | 17343.04 | 19077.34 | 12.00 | 137.70 | nr | 19215.04 |
| 1,218-1,276 kW; 22 sections; exceeding 400mm dia. flue | 18142.08 | 19956.29 | 12.00 | 137.70 | nr | 20093.99 |
| 1,276-1,334 kW; 23 sections; exceeding 400mm dia. flue | 19171.68 | 21088.85 | 12.00 | 137.70 | nr | 21226.55 |
| 1,334-1,392 kW; 24 sections; exceeding 400mm dia. flue | 20027.04 | 22029.74 | 12.00 | 137.70 | nr | 22167.44 |

## T:MECHANICAL HEATING/COOLING/REFRIGERATION SYSTEMS

| Item | Net Price £ | Material £ | Labour hours | Labour £ | Unit | Total Rate £ |
|---|---|---|---|---|---|---|
| **T10 : GAS/OIL FIRED BOILERS : BOILER PLANT AND ANCILLARIES (contd)** | | | | | | |
| Industrial cast iron sectional boilers; including controls, enamelled jacket and insulation; assembled on site and commissioned by supplier - costs included in material prices; electrical work elsewhere (contd) | | | | | | |
| Gas fired; high/low type (contd) | | | | | | |
| 1,392-1,450 kW; 25 sections; exceeding 400mm dia. flue | 20596.40 | 22656.04 | 12.00 | 137.70 | nr | 22793.74 |
| Oil fired; on/off type | | | | | | |
| 16-26 kW; 3 sections; 125mm dia. flue | 970.64 | 1067.70 | 8.00 | 91.80 | nr | 1159.50 |
| 26-33 kW; 4 sections; 125mm dia. flue | 1032.24 | 1135.46 | 8.00 | 91.80 | nr | 1227.26 |
| 33-40 kW; 5 sections; 125mm dia. flue | 1096.48 | 1206.13 | 8.00 | 91.80 | nr | 1297.93 |
| 35-50 kW; 3 sections; 153mm dia. flue | 1377.20 | 1514.92 | 8.00 | 91.80 | nr | 1606.72 |
| 50-65 kW; 4 sections; 153mm dia. flue | 1565.52 | 1722.07 | 8.00 | 91.80 | nr | 1813.87 |
| 65-80 kW; 5 sections; 153mm dia. flue | 1750.32 | 1925.35 | 8.00 | 91.80 | nr | 2017.15 |
| 80-100 kW; 6 sections; 180mm dia. flue | 1959.76 | 2155.74 | 8.00 | 91.80 | nr | 2247.54 |
| 100-120 kW; 7 sections; 180mm dia. flue | 2062.72 | 2268.99 | 8.00 | 91.80 | nr | 2360.79 |
| 105-140 kW; 5 sections; 180mm dia. flue | 2560.80 | 2816.88 | 8.00 | 91.80 | nr | 2908.68 |
| 140-180 kW; 6 sections; 180mm dia. flue | 2895.20 | 3184.72 | 8.00 | 91.80 | nr | 3276.52 |
| 180-230 kW; 7 sections; 200mm dia. flue | 3264.80 | 3591.28 | 8.00 | 91.80 | nr | 3683.08 |
| 230-280 kW; 8 sections; 200mm dia. flue | 4092.88 | 4502.17 | 8.00 | 91.80 | nr | 4593.97 |
| 280-330 kW; 9 sections; 200mm dia. flue | 4499.44 | 4949.38 | 8.00 | 91.80 | nr | 5041.18 |
| Oil fired; high/low type | | | | | | |
| 105-140 kW; 5 sections; 180mm dia. flue | 2641.76 | 2905.94 | 8.00 | 91.80 | nr | 2997.74 |
| 140-180 kW; 6 sections; 180mm dia. flue | 2976.16 | 3273.78 | 8.00 | 91.80 | nr | 3365.58 |
| 180-230 kW; 7 sections; 200mm dia. flue | 3763.76 | 4140.14 | 8.00 | 91.80 | nr | 4231.94 |
| 230-280 kW; 8 sections; 200mm dia. flue | 4247.76 | 4672.54 | 8.00 | 91.80 | nr | 4764.34 |
| 280-330 kW; 9 sections; 200mm dia. flue | 4654.32 | 5119.75 | 8.00 | 91.80 | nr | 5211.55 |
| 300-390 kW; 8 sections; 250mm dia. flue | 5406.72 | 5947.39 | 12.00 | 137.70 | nr | 6085.09 |
| 390-450 kW; 9 sections; 250mm dia. flue | 5978.72 | 6576.59 | 12.00 | 137.70 | nr | 6714.29 |
| 450-540 kW; 10 sections; 250mm dia. flue | 6755.76 | 7431.34 | 12.00 | 137.70 | nr | 7569.04 |
| 540-600 kW; 11 sections; 300mm dia. flue | 7410.48 | 8151.53 | 12.00 | 137.70 | nr | 8289.23 |
| 600-670 kW; 12 sections; 300mm dia. flue | 8253.52 | 9078.87 | 12.00 | 137.70 | nr | 9216.57 |
| 670-720 kW; 13 sections; 300mm dia. flue | 8919.68 | 9811.65 | 12.00 | 137.70 | nr | 9949.35 |
| 720-780 kW; 14 sections; 300mm dia. flue | 9372.00 | 10309.20 | 12.00 | 137.70 | nr | 10446.90 |
| 754-812 kW; 14 sections; 400mm dia. flue | 9881.52 | 10869.67 | 12.00 | 137.70 | nr | 11007.37 |
| 812-870 kW; 15 sections; 400mm dia. flue | 10671.76 | 11738.94 | 12.00 | 137.70 | nr | 11876.64 |
| 870-928 kW; 16 sections; 400mm dia. flue | 11370.48 | 12507.53 | 12.00 | 137.70 | nr | 12645.23 |
| 928-986 kW; 17 sections; 400mm dia. flue | 11865.92 | 13052.51 | 12.00 | 137.70 | nr | 13190.21 |
| 986-1,044 kW; 18 sections; 400mm dia. flue | 12528.56 | 13781.42 | 12.00 | 137.70 | nr | 13919.12 |
| 1,044-1,102 kW; 19 sections; 400mm dia. flue | 13092.64 | 14401.90 | 12.00 | 137.70 | nr | 14539.60 |
| 1,102-1,160 kW; 20 sections; 400 mm dia. flue | 13692.80 | 15062.08 | 12.00 | 137.70 | nr | 15199.78 |
| 1,160-1,218 kW; 21 sections; exceeding 400mm dia. flue | 14418.80 | 15860.68 | 12.00 | 137.70 | nr | 15998.38 |
| 1,218-1,276 kW; 22 sections; exceeding 400mm dia. flue | 15103.44 | 16613.78 | 12.00 | 137.70 | nr | 16751.48 |
| 1,276-1,334 kW; 23 sections; exceeding 400mm dia. flue | 15619.12 | 17181.03 | 12.00 | 137.70 | nr | 17318.73 |
| 1,334-1,392 kW; 24 sections; exceeding 400mm dia. flue | 16474.48 | 18121.93 | 12.00 | 137.70 | nr | 18259.63 |

## T:MECHANICAL HEATING/COOLING/REFRIGERATION SYSTEMS

| Item | Net Price £ | Material £ | Labour hours | Labour £ | Unit | Total Rate £ |
|---|---|---|---|---|---|---|
| 1,392-1,450 kW; 25 sections; exceeding 400mm dia. flue | 17043.84 | 18748.22 | 12.00 | 137.70 | nr | 18885.92 |
| **Packaged water boilers; boiler mountings controls; enamelled casing; burner; insulation; all connections and commissioning** | | | | | | |
| Gas fired | | | | | | |
| 150 kW rating; on/off type | 4129.48 | 4542.43 | 20.00 | 229.50 | nr | 4771.93 |
| 350 kW rating; high/low type | 6700.36 | 7370.40 | 32.26 | 370.16 | nr | 7740.56 |
| 600 kW rating; high/low type | 8463.82 | 9310.20 | 40.00 | 459.00 | nr | 9769.20 |
| 1500 kW rating; high/low type | 15145.09 | 16659.60 | 50.00 | 573.75 | nr | 17233.35 |
| 3000 kW rating; high/low type | 26810.46 | 29491.51 | 50.00 | 573.75 | nr | 30065.26 |
| Oil fired | | | | | | |
| 150 kW rating; high/low type | 3644.42 | 4008.86 | 20.00 | 229.50 | nr | 4238.36 |
| 350 kW rating; high/low type | 5135.73 | 5649.30 | 32.26 | 370.16 | nr | 6019.46 |
| 600 kW rating; high/low type | 6915.27 | 7606.80 | 40.00 | 459.00 | nr | 8065.80 |
| 1500 kW rating; high/low type | 12686.69 | 13955.36 | 50.00 | 573.75 | nr | 14529.11 |
| 3000 kW rating; high/low type | 24207.45 | 26628.19 | 50.00 | 573.75 | nr | 27201.94 |

## Material Costs / Measured Work Prices - Mechanical Installations

### T: MECHANICAL HEATING/COOLING/REFRIGERATION SYSTEMS

| Item | Net Price £ | Material £ | Labour hours | Labour £ | Unit | Total Rate £ |
|---|---|---|---|---|---|---|
| **T11 : COAL FIRED BOILERS : BOILER PLANT AND ANCILLARIES** | | | | | | |
| Domestic water boilers; stove enamelled casing; electric controls; placing in position; assembling and connecting; (electrical work elsewhere) | | | | | | |
| Solid fuel fired cast iron water boiler; floor standing stove enamelled casing; thermostat; draught stabilizer; electric controls; conventional flue | | | | | | |
| 45,000 Btu/Hr | 955.50 | 1051.05 | 8.00 | 91.80 | nr | **1142.85** |
| 60,000 Btu/Hr | 1111.60 | 1222.76 | 10.00 | 114.75 | nr | **1337.51** |
| 80,000 Btu/Hr | 1311.10 | 1442.21 | 12.05 | 138.25 | nr | **1580.46** |
| 100,000 Btu/Hr | 1629.60 | 1792.56 | 14.08 | 161.62 | nr | **1954.18** |
| 125,000 Btu/Hr | 1748.60 | 1923.46 | 16.13 | 185.08 | nr | **2108.54** |
| Fire place mounted natural gas fire and back boiler; cast iron water boiler; electric control box; fire output 3kW with wood surround | | | | | | |
| 10.50kW Rating (45,000 Btu/Hr) | 830.20 | 913.22 | 8.00 | 91.80 | nr | **1005.02** |
| 10.50kW Rating (57,000 Btu/Hr) | 861.00 | 947.10 | 8.00 | 91.80 | nr | **1038.90** |

## T:MECHANICAL HEATING/COOLING/REFRIGERATION SYSTEMS

| Item | Net Price £ | Material £ | Labour hours | Labour £ | Unit | Total Rate £ |
|---|---|---|---|---|---|---|
| **T10 - T11 : GAS/OIL & COAL FIRED BOILERS: CHIMNEYS AND FLUES** | | | | | | |
| Chimneys; applicable to domestic, medium sized industrial and commercial oil and gas appliances; stainless steel, twin wall, insulated; for use internally or externally | | | | | | |
| Straight length; 120mm long; including one locking band | | | | | | |
| 127mm dia. | 29.39 | 32.33 | 0.46 | 5.28 | nr | 37.61 |
| 152mm dia. | 32.90 | 36.19 | 0.49 | 5.63 | nr | 41.82 |
| 175mm dia. | 38.14 | 41.97 | 0.53 | 6.10 | nr | 48.07 |
| 203mm dia. | 43.34 | 47.68 | 0.58 | 6.67 | nr | 54.35 |
| 254mm dia. | 51.41 | 56.56 | 0.64 | 7.36 | nr | 63.92 |
| 304mm dia. | 64.44 | 70.88 | 0.70 | 8.08 | nr | 78.96 |
| 355mm dia. | 92.38 | 101.62 | 0.78 | 8.90 | nr | 110.52 |
| Straight length; 300mm long; including one locking band | | | | | | |
| 127mm dia. | 45.10 | 49.61 | 0.50 | 5.74 | nr | 55.35 |
| 152mm dia. | 50.95 | 56.06 | 0.50 | 5.74 | nr | 61.80 |
| 178mm dia. | 57.69 | 63.47 | 0.54 | 6.20 | nr | 69.67 |
| 203mm dia. | 66.14 | 72.75 | 0.66 | 7.60 | nr | 80.35 |
| 254mm dia. | 73.64 | 81.00 | 0.76 | 8.76 | nr | 89.76 |
| 304mm dia. | 88.42 | 97.27 | 0.86 | 9.89 | nr | 107.16 |
| 355mm dia. | 97.34 | 107.07 | 0.97 | 11.14 | nr | 118.21 |
| 400mm dia. | 87.37 | 114.85 | 1.08 | 12.41 | nr | 127.26 |
| 450mm dia. | 100.59 | 131.31 | 1.08 | 12.41 | nr | 143.72 |
| 500mm dia. | 107.49 | 140.89 | 1.08 | 12.41 | nr | 153.30 |
| 550mm dia. | 160.71 | 201.21 | 1.08 | 12.41 | nr | 213.62 |
| 600mm dia. | 131.93 | 171.46 | 1.08 | 12.41 | nr | 183.87 |
| Straight length; 500mm long; including one locking band | | | | | | |
| 127mm dia. | 52.92 | 58.21 | 0.54 | 6.20 | nr | 64.41 |
| 152mm dia. | 58.97 | 64.88 | 0.54 | 6.20 | nr | 71.08 |
| 178mm dia. | 66.38 | 73.03 | 0.65 | 7.46 | nr | 80.49 |
| 203mm dia. | 77.52 | 85.27 | 0.65 | 7.46 | nr | 92.73 |
| 254mm dia. | 90.37 | 99.42 | 0.86 | 9.89 | nr | 109.31 |
| 304mm dia. | 108.35 | 119.19 | 0.98 | 11.25 | nr | 130.44 |
| 355mm dia. | 121.98 | 134.18 | 1.09 | 12.51 | nr | 146.69 |
| 400mm dia. | 116.38 | 146.76 | 1.20 | 13.78 | nr | 160.54 |
| 450mm dia. | 135.24 | 169.42 | 1.20 | 13.78 | nr | 183.20 |
| 500mm dia. | 145.40 | 182.59 | 1.20 | 13.78 | nr | 196.37 |
| 550mm dia. | 160.71 | 201.21 | 1.20 | 13.78 | nr | 214.99 |
| 600mm dia. | 165.62 | 208.52 | 1.20 | 13.78 | nr | 222.30 |

## T: MECHANICAL HEATING/COOLING/REFRIGERATION SYSTEMS

| Item | Net Price £ | Material £ | Labour hours | Labour £ | Unit | Total Rate £ |
|---|---|---|---|---|---|---|
| **T10 - T11 : GAS/OIL & COAL FIRED BOILERS: CHIMNEYS AND FLUES (contd)** | | | | | | |
| Chimneys; applicable to domestic, medium sized industrial and commercial oil and gas appliances; stainless steel, twin wall, insulated; for use internally or externally (contd) | | | | | | |
| Straight length; 1000mm long; including one locking band | | | | | | |
| 127mm dia. | 94.24 | 103.67 | 0.64 | 7.35 | nr | 111.02 |
| 152mm dia. | 104.99 | 115.50 | 0.71 | 8.15 | nr | 123.65 |
| 178mm dia. | 118.11 | 129.92 | 0.80 | 9.18 | nr | 139.10 |
| 203mm dia. | 139.04 | 152.96 | 0.87 | 10.00 | nr | 162.96 |
| 254mm dia. | 158.28 | 174.11 | 0.87 | 9.99 | nr | 184.10 |
| 304mm dia. | 182.77 | 201.05 | 1.13 | 12.97 | nr | 214.02 |
| 355mm dia. | 210.03 | 231.03 | 1.26 | 14.47 | nr | 245.50 |
| 400mm dia. | 204.94 | 244.18 | 1.39 | 15.96 | nr | 260.14 |
| 450mm dia. | 218.88 | 261.43 | 1.39 | 15.96 | nr | 277.39 |
| 500mm dia. | 237.35 | 283.73 | 1.39 | 15.96 | nr | 299.69 |
| 550mm dia. | 261.44 | 310.23 | 1.39 | 15.96 | nr | 326.19 |
| 600mm dia. | 274.10 | 327.84 | 1.39 | 15.96 | nr | 343.80 |
| Adjustable length; boiler removal; internal use only; including one locking band | | | | | | |
| 127mm dia. | 47.09 | 51.80 | 0.50 | 5.74 | nr | 57.54 |
| 152mm dia. | 53.22 | 58.55 | 0.54 | 6.20 | nr | 64.75 |
| 178mm dia. | 60.29 | 66.33 | 0.59 | 6.77 | nr | 73.10 |
| 203mm dia. | 69.15 | 76.06 | 0.66 | 7.57 | nr | 83.63 |
| 254mm dia. | 77.01 | 84.71 | 0.76 | 8.73 | nr | 93.44 |
| 304mm dia. | 92.43 | 101.67 | 0.86 | 9.88 | nr | 111.55 |
| 355mm dia. | 101.55 | 111.70 | 1.03 | 11.83 | nr | 123.53 |
| 400mm dia. | 205.46 | 244.75 | 0.93 | 10.62 | nr | 255.37 |
| 450mm dia. | 219.44 | 262.04 | 0.93 | 10.62 | nr | 272.66 |
| 500mm dia. | 239.24 | 285.81 | 0.93 | 10.62 | nr | 296.43 |
| 550mm dia. | 261.21 | 311.76 | 0.93 | 10.62 | nr | 322.38 |
| 600mm dia. | 272.81 | 326.43 | 0.93 | 10.62 | nr | 337.05 |
| Inspection length; 500mm long; including one locking band | | | | | | |
| 127mm dia. | 115.86 | 127.45 | 0.54 | 6.20 | nr | 133.65 |
| 152mm dia. | 125.33 | 137.86 | 0.54 | 6.20 | nr | 144.06 |
| 178mm dia. | 131.99 | 145.20 | 0.65 | 7.46 | nr | 152.66 |
| 203mm dia. | 139.77 | 153.76 | 0.65 | 7.46 | nr | 161.22 |
| 254mm dia. | 152.83 | 168.11 | 0.86 | 9.89 | nr | 178.00 |
| 304mm dia. | 168.65 | 185.52 | 0.98 | 11.25 | nr | 196.77 |
| 355mm dia. | 191.94 | 211.13 | 1.09 | 12.51 | nr | 223.64 |
| 400mm dia. | 263.41 | 308.50 | 1.20 | 13.78 | nr | 322.28 |
| 450mm dia. | 269.58 | 317.20 | 1.20 | 13.78 | nr | 330.98 |
| 500mm dia. | 295.80 | 348.03 | 1.20 | 13.78 | nr | 361.81 |
| 550mm dia. | 308.80 | 364.11 | 1.20 | 13.78 | nr | 377.89 |
| 600mm dia. | 312.99 | 370.62 | 1.20 | 13.78 | nr | 384.40 |

## T:MECHANICAL HEATING/COOLING/REFRIGERATION SYSTEMS

| Item | Net Price £ | Material £ | Labour hours | Labour £ | Unit | Total Rate £ |
|---|---|---|---|---|---|---|
| Adapters | | | | | | |
| 127mm dia. | 9.29 | 10.22 | 0.46 | 5.28 | nr | 15.50 |
| 152mm dia. | 10.16 | 11.18 | 0.49 | 5.63 | nr | 16.81 |
| 178mm dia. | 11.84 | 13.02 | 0.53 | 6.10 | nr | 19.12 |
| 203mm dia. | 13.48 | 14.83 | 0.58 | 6.67 | nr | 21.50 |
| 254mm dia. | 13.62 | 14.98 | 0.64 | 7.36 | nr | 22.34 |
| 304mm dia. | 16.98 | 18.68 | 0.70 | 8.08 | nr | 26.76 |
| 355mm dia. | 20.62 | 22.68 | 0.78 | 8.90 | nr | 31.58 |
| 400mm dia. | 25.21 | 27.73 | 0.90 | 10.33 | nr | 38.06 |
| 450mm dia. | 26.92 | 29.61 | 0.90 | 10.33 | nr | 39.94 |
| 500mm dia. | 28.59 | 31.45 | 0.90 | 10.33 | nr | 41.78 |
| 550mm dia. | 33.24 | 36.56 | 0.90 | 10.33 | nr | 46.89 |
| 600mm dia. | 39.95 | 43.95 | 0.90 | 10.33 | nr | 54.28 |
| **Chimney fittings** | | | | | | |
| 90 degree insulated tee; including two locking bands | | | | | | |
| 127mm dia.; including locking plug | 103.17 | 119.19 | 1.94 | 22.28 | nr | 141.47 |
| 152mm dia.; including locking plug | 119.50 | 137.46 | 2.14 | 24.57 | nr | 162.03 |
| 178mm dia.; including locking plug | 130.80 | 150.18 | 2.42 | 27.78 | nr | 177.96 |
| 203mm dia.; including locking plug | 153.81 | 175.69 | 2.65 | 30.44 | nr | 206.13 |
| 254mm dia.; including locking plug | 156.07 | 178.56 | 2.98 | 34.15 | nr | 212.71 |
| 304mm dia.; including locking plug | 193.97 | 222.47 | 3.39 | 38.90 | nr | 261.37 |
| 355mm dia.; including locking plug | 245.18 | 284.06 | 3.88 | 44.48 | nr | 328.54 |
| 400mm dia. | 317.83 | 387.10 | 4.33 | 49.69 | nr | 436.79 |
| 450mm dia. | 329.33 | 403.58 | 4.81 | 55.19 | nr | 458.77 |
| 500mm dia. | 373.93 | 456.62 | 5.29 | 60.70 | nr | 517.32 |
| 550mm dia. | 401.80 | 490.84 | 5.75 | 65.98 | nr | 556.82 |
| 600mm dia. | 416.48 | 510.80 | 6.25 | 71.72 | nr | 582.52 |
| 135 degree insulated tee; including two locking bands | | | | | | |
| 127mm dia.; including locking plug | 135.07 | 154.26 | 1.94 | 22.28 | nr | 176.54 |
| 152mm dia.; including locking plug | 146.16 | 166.78 | 2.14 | 24.57 | nr | 191.35 |
| 178mm dia.; including locking plug | 160.00 | 182.29 | 2.42 | 27.78 | nr | 210.07 |
| 203mm dia.; including locking plug | 204.18 | 231.10 | 2.65 | 30.44 | nr | 261.54 |
| 254mm dia.; including locking plug | 233.46 | 263.69 | 2.98 | 34.15 | nr | 297.84 |
| 304mm dia.; including locking plug | 279.51 | 316.57 | 3.39 | 38.90 | nr | 355.47 |
| 355mm dia.; including locking plug | 342.95 | 391.61 | 3.88 | 44.48 | nr | 436.09 |
| 400mm dia. | 451.08 | 533.68 | 4.33 | 49.68 | nr | 583.36 |
| 450mm dia. | 487.70 | 577.79 | 4.81 | 55.17 | nr | 632.96 |
| 500mm dia. | 570.62 | 672.98 | 5.29 | 60.71 | nr | 733.69 |
| 550mm dia. | 583.38 | 690.58 | 5.75 | 65.95 | nr | 756.53 |
| 600mm dia. | 614.74 | 728.88 | 6.25 | 71.72 | nr | 800.60 |
| Wall sleeve; for 135 degree tee through wall | | | | | | |
| 127mm dia. | 15.63 | 17.19 | 1.94 | 22.28 | nr | 39.47 |
| 152mm dia. | 20.96 | 23.06 | 2.14 | 24.57 | nr | 47.63 |
| 178mm dia. | 21.96 | 24.16 | 2.42 | 27.78 | nr | 51.94 |
| 203mm dia. | 24.67 | 27.14 | 2.65 | 30.44 | nr | 57.58 |
| 254mm dia. | 27.55 | 30.30 | 2.98 | 34.15 | nr | 64.45 |
| 304mm dia. | 32.10 | 35.31 | 3.39 | 38.90 | nr | 74.21 |
| 355mm dia. | 35.65 | 39.22 | 3.88 | 44.48 | nr | 83.70 |

## T:MECHANICAL HEATING/COOLING/REFRIGERATION SYSTEMS

| Item | Net Price £ | Material £ | Labour hours | Labour £ | Unit | Total Rate £ |
|---|---|---|---|---|---|---|
| **T10 - T11 : GAS/OIL & COAL FIRED BOILERS: CHIMNEYS AND FLUES (contd)** | | | | | | |
| **Chimney fittings (contd)** | | | | | | |
| 15 degree insulated elbow; including two locking bands | | | | | | |
| 127mm dia. | 65.13 | 83.02 | 1.61 | 18.48 | nr | 101.50 |
| 152mm dia. | 72.93 | 92.21 | 1.81 | 20.79 | nr | 113.00 |
| 178mm dia. | 78.19 | 98.59 | 2.06 | 23.66 | nr | 122.25 |
| 203mm dia. | 83.16 | 104.48 | 2.34 | 26.87 | nr | 131.35 |
| 254mm dia. | 85.74 | 108.09 | 2.79 | 31.96 | nr | 140.05 |
| 304mm dia. | 107.74 | 136.73 | 3.61 | 41.43 | nr | 178.16 |
| 355mm dia. | 140.47 | 183.25 | 5.13 | 58.85 | nr | 242.10 |
| 30 degree insulated elbow; including two locking bands | | | | | | |
| 127mm dia. | 65.13 | 83.02 | 1.44 | 16.53 | nr | 99.55 |
| 152mm dia. | 72.93 | 92.21 | 1.59 | 18.27 | nr | 110.48 |
| 178mm dia. | 78.19 | 98.59 | 1.85 | 21.25 | nr | 119.84 |
| 203mm dia. | 83.16 | 104.48 | 2.12 | 24.36 | nr | 128.84 |
| 254mm dia. | 85.74 | 108.09 | 2.40 | 27.58 | nr | 135.67 |
| 304mm dia. | 107.74 | 136.73 | 2.70 | 30.93 | nr | 167.66 |
| 355mm dia. | 140.47 | 183.25 | 3.08 | 35.31 | nr | 218.56 |
| 400mm dia. | 173.65 | 228.50 | 3.46 | 39.71 | nr | 268.21 |
| 450mm dia. | 180.24 | 239.58 | 3.83 | 43.97 | nr | 283.55 |
| 500mm dia. | 192.55 | 257.10 | 4.22 | 48.42 | nr | 305.52 |
| 550mm dia. | 210.44 | 280.35 | 4.61 | 52.88 | nr | 333.23 |
| 600mm dia. | 213.67 | 287.70 | 4.98 | 57.09 | nr | 344.79 |
| 45 degree insulated elbow; including two locking bands | | | | | | |
| 127mm dia. | 65.13 | 83.02 | 1.44 | 16.53 | nr | 99.55 |
| 152mm dia. | 72.93 | 92.21 | 1.44 | 16.53 | nr | 108.74 |
| 178mm dia. | 78.19 | 98.59 | 1.44 | 16.53 | nr | 115.12 |
| 203mm dia. | 83.16 | 104.48 | 1.44 | 16.53 | nr | 121.01 |
| 254mm dia. | 85.74 | 108.09 | 1.44 | 16.53 | nr | 124.62 |
| 304mm dia. | 107.74 | 136.73 | 1.44 | 16.53 | nr | 153.26 |
| 355mm dia. | 140.47 | 183.25 | 1.44 | 16.53 | nr | 199.78 |
| 400mm dia. | 173.65 | 228.50 | 1.44 | 16.53 | nr | 245.03 |
| 450mm dia. | 180.24 | 239.58 | 1.44 | 16.53 | nr | 256.11 |
| 500mm dia. | 192.55 | 257.10 | 1.44 | 16.53 | nr | 273.63 |
| 550mm dia. | 210.44 | 280.35 | 1.44 | 16.53 | nr | 296.88 |
| 600mm dia. | 213.67 | 287.70 | 1.44 | 16.53 | nr | 304.23 |
| **Chimney supports** | | | | | | |
| Wall support, galvanised; including plate and brackets | | | | | | |
| 127mm dia. | 30.90 | 33.99 | 2.21 | 25.39 | nr | 59.38 |
| 152mm dia. | 34.42 | 37.86 | 2.37 | 27.19 | nr | 65.05 |
| 178mm dia. | 37.94 | 41.73 | 2.48 | 28.47 | nr | 70.20 |
| 203mm dia. | 38.20 | 42.02 | 2.72 | 31.27 | nr | 73.29 |
| 254mm dia. | 43.30 | 47.63 | 3.03 | 34.77 | nr | 82.40 |
| 304mm dia. | 51.06 | 56.17 | 3.46 | 39.71 | nr | 95.88 |
| 355mm dia. | 66.57 | 73.23 | 4.10 | 47.03 | nr | 120.26 |

*Material Costs / Measured Work Prices - Mechanical Installations*

## T:MECHANICAL HEATING/COOLING/REFRIGERATION SYSTEMS

| Item | Net Price £ | Material £ | Labour hours | Labour £ | Unit | Total Rate £ |
|---|---|---|---|---|---|---|
| 400mm dia.; including 300mm support length and collar | 212.16 | 233.38 | 4.76 | 54.64 | nr | **288.02** |
| 450mm dia.; including 300mm support length and collar | 221.59 | 243.75 | 5.41 | 62.03 | nr | **305.78** |
| 500mm dia.; including 300mm support length and collar | 238.50 | 262.35 | 6.06 | 69.55 | nr | **331.90** |
| 550mm dia.; including 300mm support length and collar | 260.89 | 286.98 | 6.71 | 77.01 | nr | **363.99** |
| 600mm dia.; including 300mm support length and collar | 270.49 | 297.54 | 7.35 | 84.37 | nr | **381.91** |
| Ceiling/floor support | | | | | | |
| 127mm dia. | 15.80 | 17.38 | 1.90 | 21.82 | nr | **39.20** |
| 152mm dia. | 17.58 | 19.34 | 2.18 | 25.05 | nr | **44.39** |
| 178mm dia. | 15.80 | 17.38 | 1.90 | 21.82 | nr | **39.20** |
| 203mm dia. | 31.69 | 34.86 | 2.89 | 33.16 | nr | **68.02** |
| 254mm dia. | 36.14 | 39.75 | 3.32 | 38.12 | nr | **77.87** |
| 304mm dia. | 41.30 | 45.43 | 3.75 | 42.98 | nr | **88.41** |
| 355mm dia. | 49.23 | 54.15 | 4.37 | 50.11 | nr | **104.26** |
| 400mm dia. | 236.77 | 260.45 | 4.95 | 56.81 | nr | **317.26** |
| 450mm dia. | 250.43 | 275.47 | 5.56 | 63.75 | nr | **339.22** |
| 500mm dia. | 264.35 | 290.79 | 6.13 | 70.40 | nr | **361.19** |
| 550mm dia. | 289.39 | 318.33 | 6.76 | 77.53 | nr | **395.86** |
| 600mm dia. | 294.33 | 323.76 | 7.35 | 84.37 | nr | **408.13** |
| Ceiling/floor firestop spacer | | | | | | |
| 127mm dia. | 3.08 | 3.39 | 0.50 | 5.74 | nr | **9.13** |
| 152mm dia. | 3.43 | 3.77 | 0.53 | 6.10 | nr | **9.87** |
| 178mm dia. | 3.88 | 4.27 | 0.55 | 6.34 | nr | **10.61** |
| 203mm dia. | 4.59 | 5.05 | 0.59 | 6.77 | nr | **11.82** |
| 254mm dia. | 4.73 | 5.20 | 0.64 | 7.36 | nr | **12.56** |
| 304mm dia. | 5.70 | 6.27 | 0.69 | 7.91 | nr | **14.18** |
| 355mm dia. | 10.37 | 11.41 | 0.74 | 8.50 | nr | **19.91** |
| Wall band; internal or external use | | | | | | |
| 127mm dia. | 18.61 | 20.47 | 0.98 | 11.25 | nr | **31.72** |
| 152mm dia. | 19.44 | 21.38 | 1.02 | 11.71 | nr | **33.09** |
| 178mm dia. | 20.20 | 22.22 | 1.06 | 12.17 | nr | **34.39** |
| 203mm dia. | 21.34 | 23.47 | 1.12 | 12.86 | nr | **36.33** |
| 254mm dia. | 22.21 | 24.43 | 1.24 | 14.24 | nr | **38.67** |
| 304mm dia. | 23.99 | 26.39 | 1.38 | 15.83 | nr | **42.22** |
| 355mm dia. | 25.53 | 28.08 | 1.57 | 18.04 | nr | **46.12** |
| 400mm dia. | 27.65 | 30.41 | 1.76 | 20.20 | nr | **50.61** |
| 450mm dia. | 32.12 | 35.33 | 2.43 | 27.85 | nr | **63.18** |
| 500mm dia. | 38.60 | 42.46 | 2.14 | 24.57 | nr | **67.03** |
| 550mm dia. | 40.44 | 44.48 | 2.33 | 26.75 | nr | **71.23** |
| 600mm dia. | 42.76 | 47.04 | 2.53 | 28.98 | nr | **76.02** |

## T:MECHANICAL HEATING/COOLING/REFRIGERATION SYSTEMS

| Item | Net Price £ | Material £ | Labour hours | Labour £ | Unit | Total Rate £ |
|---|---|---|---|---|---|---|
| **T10 - T11 : GAS/OIL & COAL FIRED BOILERS: CHIMNEYS AND FLUES (contd)** | | | | | | |
| **Chimney flashings and terminals** | | | | | | |
| Insulated top stub; including one locking band | | | | | | |
| 127mm dia. | 42.33 | 46.56 | 1.44 | 16.53 | nr | 63.09 |
| 152mm dia. | 47.81 | 52.60 | 4.59 | 52.64 | nr | 105.24 |
| 178mm dia. | 51.47 | 56.63 | 1.85 | 21.25 | nr | 77.88 |
| 203mm dia. | 54.76 | 60.24 | 2.12 | 24.36 | nr | 84.60 |
| 254mm dia. | 58.38 | 64.23 | 2.40 | 27.52 | nr | 91.75 |
| 304mm dia. | 79.79 | 87.77 | 2.69 | 30.85 | nr | 118.62 |
| 355mm dia. | 105.50 | 116.05 | 3.07 | 35.20 | nr | 151.25 |
| 400mm dia. | 96.20 | 124.56 | 3.46 | 39.71 | nr | 164.27 |
| 450mm dia. | 100.46 | 131.16 | 3.83 | 43.97 | nr | 175.13 |
| 500mm dia. | 117.34 | 151.72 | 4.22 | 48.42 | nr | 200.14 |
| 550mm dia. | 122.11 | 158.75 | 4.61 | 52.88 | nr | 211.63 |
| 600mm dia. | 125.43 | 164.31 | 4.98 | 57.09 | nr | 221.40 |
| Rain cap; including one locking band | | | | | | |
| 127mm dia. | 24.52 | 26.97 | 1.44 | 16.53 | nr | 43.50 |
| 152mm dia. | 24.80 | 27.28 | 1.44 | 16.53 | nr | 43.81 |
| 178mm dia. | 28.21 | 31.04 | 1.59 | 18.27 | nr | 49.31 |
| 203mm dia. | 33.74 | 37.13 | 1.85 | 21.25 | nr | 58.38 |
| 254mm dia. | 44.36 | 48.80 | 2.40 | 27.58 | nr | 76.38 |
| 304mm dia. | 59.80 | 65.79 | 2.70 | 30.93 | nr | 96.72 |
| 355mm dia. | 80.12 | 88.13 | 3.08 | 35.31 | nr | 123.44 |
| 400mm dia. | 62.89 | 87.92 | 3.28 | 37.62 | nr | 125.54 |
| 450mm dia. | 68.16 | 95.63 | 3.83 | 43.97 | nr | 139.60 |
| 500mm dia. | 73.41 | 103.40 | 4.22 | 48.42 | nr | 151.82 |
| 550mm dia. | 78.67 | 110.97 | 4.61 | 52.88 | nr | 163.85 |
| 600mm dia. | 83.89 | 118.61 | 4.98 | 57.09 | nr | 175.70 |
| Round top; including one locking band | | | | | | |
| 127mm dia | 46.86 | 51.56 | 1.44 | 16.53 | nr | 68.09 |
| 152mm dia | 51.11 | 56.22 | 1.59 | 18.24 | nr | 74.46 |
| 178mm dia | 58.27 | 64.10 | 1.85 | 21.25 | nr | 85.35 |
| 203mm dia | 68.58 | 75.44 | 2.12 | 24.36 | nr | 99.80 |
| 254mm dia | 81.06 | 89.18 | 2.40 | 27.58 | nr | 116.76 |
| 304mm dia | 106.48 | 117.13 | 2.70 | 30.93 | nr | 148.06 |
| 355mm dia | 142.04 | 156.24 | 3.08 | 35.31 | nr | 191.55 |
| Coping cap; including one locking band | | | | | | |
| 127mm dia. | 26.25 | 28.88 | 1.44 | 16.53 | nr | 45.41 |
| 152mm dia. | 27.50 | 30.25 | 1.59 | 18.27 | nr | 48.52 |
| 178mm dia. | 30.27 | 33.30 | 1.85 | 21.25 | nr | 54.55 |
| 203mm dia. | 36.30 | 39.93 | 2.12 | 24.36 | nr | 64.29 |
| 254mm dia. | 44.36 | 48.80 | 2.40 | 27.52 | nr | 76.32 |
| 304mm dia. | 59.80 | 65.79 | 2.69 | 30.85 | nr | 96.64 |
| 355mm dia. | 80.12 | 88.13 | 3.08 | 35.31 | nr | 123.44 |

*Material Costs / Measured Work Prices - Mechanical Installations*

## T:MECHANICAL HEATING/COOLING/REFRIGERATION SYSTEMS

| Item | Net Price £ | Material £ | Labour hours | Labour £ | Unit | Total Rate £ |
|---|---|---|---|---|---|---|
| Storm collar | | | | | | |
| 127mm dia. | 4.99 | 5.49 | 0.50 | 5.74 | nr | 11.23 |
| 152mm dia. | 5.33 | 5.86 | 0.53 | 6.08 | nr | 11.94 |
| 178mm dia. | 5.91 | 6.50 | 0.55 | 6.31 | nr | 12.81 |
| 203mm dia. | 6.19 | 6.81 | 0.64 | 7.34 | nr | 14.15 |
| 254mm dia. | 7.74 | 8.51 | 0.64 | 7.34 | nr | 15.85 |
| 304mm dia. | 8.08 | 8.89 | 0.69 | 7.92 | nr | 16.81 |
| 355mm dia. | 8.63 | 9.49 | 0.74 | 8.49 | nr | 17.98 |
| 400mm dia. | 21.78 | 23.96 | 0.79 | 9.07 | nr | 33.03 |
| 450mm dia. | 23.96 | 26.36 | 0.84 | 9.64 | nr | 36.00 |
| 500mm dia. | 26.14 | 28.75 | 0.89 | 10.21 | nr | 38.96 |
| 550mm dia. | 28.32 | 31.15 | 0.94 | 10.79 | nr | 41.94 |
| 600mm dia. | 30.49 | 33.54 | 0.99 | 11.36 | nr | 44.90 |
| Flat flashing; including storm collar and sealant | | | | | | |
| 127mm dia. | 22.97 | 25.27 | 1.44 | 16.53 | nr | 41.80 |
| 152mm dia. | 23.54 | 25.89 | 1.59 | 18.27 | nr | 44.16 |
| 178mm dia. | 24.39 | 26.83 | 1.85 | 21.25 | nr | 48.08 |
| 203mm dia. | 27.00 | 29.70 | 2.12 | 24.36 | nr | 54.06 |
| 254mm dia. | 37.60 | 41.36 | 2.40 | 27.58 | nr | 68.94 |
| 304mm dia. | 46.11 | 50.72 | 2.70 | 30.93 | nr | 81.65 |
| 355mm dia. | 70.66 | 77.73 | 3.08 | 35.31 | nr | 113.04 |
| 400mm dia. | 84.08 | 92.49 | 3.46 | 39.71 | nr | 132.20 |
| 450mm dia. | 97.56 | 107.32 | 3.83 | 43.97 | nr | 151.29 |
| 500mm dia. | 105.56 | 116.12 | 4.22 | 48.42 | nr | 164.54 |
| 550mm dia. | 111.54 | 122.69 | 4.61 | 52.88 | nr | 175.57 |
| 600mm dia. | 114.50 | 125.95 | 4.98 | 57.09 | nr | 183.04 |
| 5 - 30 degree rigid adjustable flashing; including storm collar and sealant | | | | | | |
| 127mm dia. | 22.97 | 25.27 | 1.44 | 16.53 | nr | 41.80 |
| 152mm dia. | 23.54 | 25.89 | 1.59 | 18.27 | nr | 44.16 |
| 178mm dia. | 24.39 | 26.83 | 1.85 | 21.21 | nr | 48.04 |
| 203mm dia. | 27.00 | 29.70 | 2.12 | 24.36 | nr | 54.06 |
| 254mm dia. | 37.60 | 41.36 | 2.40 | 27.58 | nr | 68.94 |
| 304mm dia. | 46.11 | 50.72 | 2.70 | 30.93 | nr | 81.65 |
| 355mm dia. | 70.66 | 77.73 | 3.07 | 35.20 | nr | 112.93 |
| 400mm dia. | 108.31 | 119.14 | 3.46 | 39.71 | nr | 158.85 |
| 450mm dia. | 127.53 | 140.28 | 3.83 | 43.97 | nr | 184.25 |
| 500mm dia. | 137.50 | 151.25 | 4.22 | 48.42 | nr | 199.67 |
| 550mm dia. | 161.88 | 178.07 | 4.59 | 52.64 | nr | 230.71 |
| 600mm dia. | 185.11 | 203.62 | 4.98 | 57.09 | nr | 260.71 |

## T:MECHANICAL HEATING/COOLING/REFRIGERATION SYSTEMS

| Item | Net Price £ | Material £ | Labour hours | Labour £ | Unit | Total Rate £ |
|---|---|---|---|---|---|---|
| **T10 - T11 : GAS/OIL & COAL FIRED BOILERS: CHIMNEYS AND FLUES (contd)** | | | | | | |
| Domestic and small commercial; twin walled gas vent system suitable for gas fired appliances; domestic gas boilers; small commercial boilers with internal or external flues | | | | | | |
| 152mm long | | | | | | |
| 100mm dia. | 4.37 | 4.81 | 0.50 | 5.74 | nr | **10.55** |
| 125mm dia. | 5.38 | 5.92 | 0.50 | 5.74 | nr | **11.66** |
| 150mm dia. | 5.83 | 6.41 | 0.50 | 5.74 | nr | **12.15** |
| 305mm long | | | | | | |
| 100mm dia. | 6.63 | 7.29 | 0.50 | 5.74 | nr | **13.03** |
| 125mm dia. | 7.78 | 8.56 | 0.50 | 5.74 | nr | **14.30** |
| 150mm dia. | 9.23 | 10.15 | 0.50 | 5.74 | nr | **15.89** |
| 457mm long | | | | | | |
| 100mm dia. | 7.34 | 8.07 | 0.54 | 6.20 | nr | **14.27** |
| 125mm dia. | 8.25 | 9.07 | 0.54 | 6.20 | nr | **15.27** |
| 150mm dia. | 10.21 | 11.23 | 0.54 | 6.20 | nr | **17.43** |
| 914mm long | | | | | | |
| 100mm dia. | 13.12 | 14.43 | 0.64 | 7.35 | nr | **21.78** |
| 125mm dia. | 15.27 | 16.80 | 0.64 | 7.35 | nr | **24.15** |
| 150mm dia. | 17.51 | 19.26 | 0.64 | 7.35 | nr | **26.61** |
| 1524mm long | | | | | | |
| 100mm dia. | 18.94 | 20.83 | 0.80 | 9.18 | nr | **30.01** |
| 125mm dia. | 23.31 | 25.64 | 0.83 | 9.56 | nr | **35.20** |
| 150mm dia. | 25.01 | 27.51 | 0.83 | 9.56 | nr | **37.07** |
| Adjustable length 305mm long | | | | | | |
| 100mm dia. | 8.39 | 9.23 | 0.56 | 6.38 | nr | **15.61** |
| 125mm dia. | 9.43 | 10.37 | 0.56 | 6.38 | nr | **16.75** |
| 150mm dia. | 25.01 | 27.51 | 0.56 | 6.38 | nr | **33.89** |
| Adjustable length 457mm long | | | | | | |
| 100mm dia. | 11.31 | 12.44 | 0.56 | 6.38 | nr | **18.82** |
| 125mm dia. | 13.72 | 15.09 | 0.56 | 6.38 | nr | **21.47** |
| 150mm dia. | 15.26 | 16.79 | 0.56 | 6.38 | nr | **23.17** |
| Adjustable elbow 0 - 90 deg | | | | | | |
| 100mm dia. | 9.58 | 10.54 | 0.45 | 5.17 | nr | **15.71** |
| 125mm dia. | 11.32 | 12.45 | 0.45 | 5.16 | nr | **17.61** |
| 150mm dia. | 14.18 | 15.60 | 0.45 | 5.16 | nr | **20.76** |

## T:MECHANICAL HEATING/COOLING/REFRIGERATION SYSTEMS

| Item | Net Price £ | Material £ | Labour hours | Labour £ | Unit | Total Rate £ |
|---|---|---|---|---|---|---|
| Draughthood connector | | | | | | |
| 100mm dia. | 2.95 | 3.25 | 0.45 | 5.16 | nr | **8.41** |
| 125mm dia. | 3.33 | 3.66 | 0.45 | 5.16 | nr | **8.82** |
| 150mm dia. | 3.62 | 3.98 | 0.45 | 5.16 | nr | **9.14** |
| Adaptor | | | | | | |
| 100mm dia. | 7.16 | 7.88 | 0.45 | 5.16 | nr | **13.04** |
| 125mm dia. | 7.33 | 8.06 | 0.45 | 5.16 | nr | **13.22** |
| 150mm dia. | 7.48 | 8.23 | 0.45 | 5.16 | nr | **13.39** |
| Support plate | | | | | | |
| 100mm dia. | 5.18 | 5.70 | 0.45 | 5.16 | nr | **10.86** |
| 125mm dia. | 5.50 | 6.05 | 0.45 | 5.16 | nr | **11.21** |
| 150mm dia. | 5.89 | 6.48 | 0.45 | 5.16 | nr | **11.64** |
| Wall band | | | | | | |
| 100mm dia. | 4.69 | 5.16 | 0.45 | 5.16 | nr | **10.32** |
| 125mm dia. | 4.99 | 5.49 | 0.45 | 5.16 | nr | **10.65** |
| 150mm dia. | 6.33 | 6.96 | 0.45 | 5.16 | nr | **12.12** |
| Firestop | | | | | | |
| 100mm dia. | 2.04 | 2.24 | 0.45 | 5.16 | nr | **7.40** |
| 125mm dia. | 2.04 | 2.24 | 0.45 | 5.16 | nr | **7.40** |
| 150mm dia. | 2.33 | 2.56 | 0.45 | 5.16 | nr | **7.72** |
| Flat flashing | | | | | | |
| 100mm dia. | 14.68 | 16.15 | 0.45 | 5.16 | nr | **21.31** |
| 125mm dia. | 15.98 | 17.58 | 0.45 | 5.16 | nr | **22.74** |
| 150mm dia. | 17.60 | 19.36 | 0.45 | 5.16 | nr | **24.52** |
| Adjustable flashing 5-30 deg. | | | | | | |
| 100mm dia. | 14.68 | 16.15 | 0.45 | 5.16 | nr | **21.31** |
| 125mm dia. | 15.98 | 17.58 | 0.45 | 5.16 | nr | **22.74** |
| 150mm dia. | 17.60 | 19.36 | 0.45 | 5.16 | nr | **24.52** |
| Adjustable flashing 30-45 deg. | | | | | | |
| 100mm dia. | 14.68 | 16.15 | 0.45 | 5.16 | nr | **21.31** |
| 125mm dia. | 15.98 | 17.58 | 0.45 | 5.16 | nr | **22.74** |
| 150mm dia. | 17.60 | 19.36 | 0.45 | 5.16 | nr | **24.52** |
| Storm collar | | | | | | |
| 100mm dia. | 2.87 | 3.16 | 0.45 | 5.16 | nr | **8.32** |
| 125mm dia. | 2.94 | 3.23 | 0.45 | 5.16 | nr | **8.39** |
| 150mm dia. | 3.01 | 3.31 | 0.45 | 5.16 | nr | **8.47** |
| Gas vent terminal | | | | | | |
| 100mm dia. | 10.93 | 12.02 | 0.45 | 5.16 | nr | **17.18** |
| 125mm dia. | 12.01 | 13.21 | 0.45 | 5.16 | nr | **18.37** |
| 150mm dia. | 15.40 | 16.94 | 0.45 | 5.16 | nr | **22.10** |

*Material Costs / Measured Work Prices - Mechanical Installations*

## T:MECHANICAL HEATING/COOLING/REFRIGERATION SYSTEMS

| Item | Net Price £ | Material £ | Labour hours | Labour £ | Unit | Total Rate £ |
|---|---|---|---|---|---|---|
| **T10 - T11 : GAS/OIL & COAL FIRED BOILERS: CHIMNEYS AND FLUES (contd)** | | | | | | |
| **Domestic and small commercial; twin walled gas vent system suitable for gas fired appliances; domestic gas boilers; small commercial boilers with internal or external flues (contd)** | | | | | | |
| Twin wall galvanised steel flue box, 125mm dia.; (fitted where no chimney exists) for gas fire | | | | | | |
| Free standing | 56.76 | 62.44 | 2.07 | 23.76 | nr | **86.20** |
| Recess | 56.76 | 62.44 | 2.07 | 23.76 | nr | **86.20** |
| Back boiler | 41.82 | 46.00 | 2.40 | 27.58 | nr | **73.58** |

## T:MECHANICAL HEATING/COOLING/REFRIGERATION SYSTEMS

| Item | Net Price £ | Material £ | Labour hours | Labour £ | Unit | Total Rate £ |
|---|---|---|---|---|---|---|
| **T13 : PACKAGED STEAM GENERATORS** | | | | | | |
| Packaged steam boilers; boiler mountings centrifugal water feed pump; insulation; and sheet steel wrap around casing; plastic coated | | | | | | |
| Gas fired | | | | | | |
| 293 kW rating | 15513.65 | 17065.01 | 83.33 | 956.25 | nr | **18021.26** |
| 1465 kW rating | 24629.23 | 27092.15 | 142.86 | 1639.29 | nr | **28731.44** |
| 2930 kW rating | 35839.67 | 39423.64 | 200.00 | 2295.00 | nr | **41718.64** |
| Oil fired | | | | | | |
| 293 kW rating | 14094.65 | 15504.11 | 83.33 | 956.25 | nr | **16460.36** |
| 1465 kW rating | 23178.09 | 25495.90 | 142.86 | 1639.29 | nr | **27135.19** |
| 2930 kW rating | 33506.80 | 36857.48 | 200.00 | 2295.00 | nr | **39152.48** |

## T:MECHANICAL HEATING/COOLING/REFRIGERATION SYSTEMS

| Item | Net Price £ | Material £ | Labour hours | Labour £ | Unit | Total Rate £ |
|---|---|---|---|---|---|---|
| **T30-T32 : MEDIUM AND LOW TEMPERATURE HOT WATER HEATING : PERIMETER HEATING** | | | | | | |
| Perimeter heating metal casing standard finish top outlet; punched louvre grill; including backplates | | | | | | |
| Standard unit | | | | | | |
| 60 x 200mm | 19.73 | 21.70 | 2.00 | 22.95 | m | 44.65 |
| 60 x 300mm | 21.45 | 23.59 | 2.00 | 22.95 | m | 46.54 |
| 60 x 450mm | 26.59 | 29.25 | 2.00 | 22.95 | m | 52.20 |
| 60 x 525mm | 28.31 | 31.14 | 2.00 | 22.95 | m | 54.09 |
| 60 x 600mm | 30.88 | 33.97 | 2.00 | 22.95 | m | 56.92 |
| 90 x 250mm | 21.45 | 23.59 | 2.00 | 22.95 | m | 46.54 |
| 90 x 300mm | 22.30 | 24.53 | 2.00 | 22.95 | m | 47.48 |
| 90 x 375mm | 24.88 | 27.37 | 2.00 | 22.95 | m | 50.32 |
| 90 x 450mm | 27.46 | 30.21 | 2.00 | 22.95 | m | 53.16 |
| 90 x 525mm | 30.03 | 33.03 | 2.00 | 22.95 | m | 55.98 |
| 90 x 600mm | 31.75 | 34.92 | 2.00 | 22.95 | m | 57.87 |
| Perimeter heating metal casing standard finish sloping outlet; punched louvre grill; including backplates | | | | | | |
| Sloping outlet | | | | | | |
| 60 x 200mm | 19.73 | 21.70 | 2.00 | 22.95 | m | 44.65 |
| 60 x 300mm | 21.45 | 23.59 | 2.00 | 22.95 | m | 46.54 |
| 60 x 450mm | 26.59 | 29.25 | 2.00 | 22.95 | m | 52.20 |
| 60 x 525mm | 28.31 | 31.14 | 2.00 | 22.95 | m | 54.09 |
| 60 x 600mm | 30.88 | 33.97 | 2.00 | 22.95 | m | 56.92 |
| 90 x 250mm | 21.45 | 23.59 | 2.00 | 22.95 | m | 46.54 |
| 90 x 300mm | 22.30 | 24.53 | 2.00 | 22.95 | m | 47.48 |
| 90 x 375mm | 24.88 | 27.37 | 2.00 | 22.95 | m | 50.32 |
| 90 x 450mm | 27.46 | 30.21 | 2.00 | 22.95 | m | 53.16 |
| 90 x 525mm | 30.03 | 33.03 | 2.00 | 22.95 | m | 55.98 |
| 90 x 600mm | 31.75 | 34.92 | 2.00 | 22.95 | m | 57.87 |
| Perimeter heating metal casing standard finish flat front outlet; punched louvre grill; including backplates | | | | | | |
| 60 x 200mm | 19.73 | 21.70 | 2.00 | 22.95 | m | 44.65 |
| 60 x 300mm | 21.45 | 23.59 | 2.00 | 22.95 | m | 46.54 |
| 60 x 450mm | 26.59 | 29.25 | 2.00 | 22.95 | m | 52.20 |
| 60 x 525mm | 28.31 | 31.14 | 2.00 | 22.95 | m | 54.09 |
| 60 x 600mm | 30.88 | 33.97 | 2.00 | 22.95 | m | 56.92 |
| 90 x 250mm | 21.45 | 23.59 | 2.00 | 22.95 | m | 46.54 |
| 90 x 300mm | 22.30 | 24.53 | 2.00 | 22.95 | m | 47.48 |
| 90 x 375mm | 24.88 | 27.37 | 2.00 | 22.95 | m | 50.32 |
| 90 x 450mm | 27.46 | 30.21 | 2.00 | 22.95 | m | 53.16 |
| 90 x 525mm | 30.03 | 33.03 | 2.00 | 22.95 | m | 55.98 |
| 90 x 600mm | 31.75 | 34.92 | 2.00 | 22.95 | m | 57.87 |

## T:MECHANICAL HEATING/COOLING/REFRIGERATION SYSTEMS

| Item | Net Price £ | Material £ | Labour hours | Labour £ | Unit | Total Rate £ |
|---|---|---|---|---|---|---|
| **Perimeter heating metal casing standard finish top outlet; punched louvre grill; including backplates** | | | | | | |
| Standard unit | | | | | | |
|   60 x 200mm | 24.67 | 27.14 | 2.00 | 22.95 | m | **50.09** |
|   60 x 300mm | 26.81 | 29.49 | 2.00 | 22.95 | m | **52.44** |
|   60 x 450mm | 33.24 | 36.56 | 2.00 | 22.95 | m | **59.51** |
|   60 x 525mm | 35.39 | 38.93 | 2.00 | 22.95 | m | **61.88** |
|   60 x 600mm | 38.61 | 42.47 | 2.00 | 22.95 | m | **65.42** |
|   90 x 250mm | 26.81 | 29.49 | 2.00 | 22.95 | m | **52.44** |
|   90 x 300mm | 27.88 | 30.67 | 2.00 | 22.95 | m | **53.62** |
|   90 x 375mm | 31.11 | 34.22 | 2.00 | 22.95 | m | **57.17** |
|   90 x 450mm | 34.32 | 37.75 | 2.00 | 22.95 | m | **60.70** |
|   90 x 525mm | 37.54 | 41.29 | 2.00 | 22.95 | m | **64.24** |
|   90 x 600mm | 39.68 | 43.65 | 2.00 | 22.95 | m | **66.60** |
| Extra over for dampers | | | | | | |
|   Damper | 8.58 | 9.44 | 0.25 | 2.87 | m | **12.31** |
| Extra over for fittings | | | | | | |
|   60mm End caps | 6.86 | 7.55 | 0.25 | 2.87 | m | **10.42** |
|   90mm End caps | 11.15 | 12.27 | 0.25 | 2.87 | m | **15.14** |
|   60mm Corners | 14.59 | 16.05 | 0.25 | 2.87 | m | **18.92** |
|   90mm Corners | 21.45 | 23.59 | 0.25 | 2.87 | m | **26.46** |

## T:MECHANICAL HEATING/COOLING/REFRIGERATION SYSTEMS

| Item | Net Price £ | Material £ | Labour hours | Labour £ | Unit | Total Rate £ |
|---|---|---|---|---|---|---|
| **T30-T32 : MEDIUM AND LOW TEMPERATURE HOT WATER HEATING : FAN CONVECTORS** | | | | | | |
| Quality sheet steel cased units; extruded aluminium grilles for LPHW centrifugal fans; filter choice of 3 speeds; single phase thermostatic controls; 3/42" BSP connections; all 230mm deep complete with access locks | | | | | | |
| Free standing flat top 695mm high medium speed rating | | | | | | |
| E.A.T. at 18 degree | | | | | | |
| 695mm length 1 row 1.94 kW 75 l/sec 39C | 469.04 | 515.94 | 2.38 | 27.32 | nr | 543.26 |
| 695mm length 2 row 2.64 kW 75 l/sec 47C | 469.04 | 515.94 | 2.38 | 27.32 | nr | 543.26 |
| 895mm length 1 row 4.02 kW 150 l/sec 40C | 528.53 | 581.38 | 2.38 | 27.32 | nr | 608.70 |
| 895mm length 2 row 5.62 kW 150 l/sec 49C | 528.53 | 581.38 | 2.38 | 27.32 | nr | 608.70 |
| 1195mm length 1 row 6.58 kW 250 l/sec 40C | 601.74 | 661.91 | 2.38 | 27.32 | nr | 689.23 |
| 1195mm length 2 row 9.27 kW 250 l/sec 48C | 601.74 | 661.91 | 2.38 | 27.32 | nr | 689.23 |
| 1495mm length 1 row 9.04 kW 340 l/sec 40C | 671.53 | 738.68 | 2.38 | 27.32 | nr | 766.00 |
| 1495mm length 2 row 12.73 kW 340 l/sec 49C | 671.53 | 738.68 | 2.38 | 27.32 | nr | 766.00 |
| Free standing flat top 695mm high medium speed rating | | | | | | |
| E.A.T. at 18 degree | | | | | | |
| 695mm length 1 row 1.94 kW 75 l/sec 39C | 486.20 | 534.82 | 2.38 | 27.32 | nr | 562.14 |
| 695mm length 2 row 2.64 kW 75 l/sec 47C | 486.20 | 534.82 | 2.38 | 27.32 | nr | 562.14 |
| 895mm length 1 row 4.02 kW 150 l/sec 40C | 545.69 | 600.26 | 2.38 | 27.32 | nr | 627.58 |
| 895mm length 2 row 5.62 kW 150 l/sec 49C | 545.69 | 600.26 | 2.38 | 27.32 | nr | 627.58 |
| 1195mm length 1 row 6.58 kW 250 l/sec 40C | 618.90 | 680.79 | 2.38 | 27.32 | nr | 708.11 |
| 1195mm length 2 row 9.27 kW 250 l/sec 48C | 618.90 | 680.79 | 2.38 | 27.32 | nr | 708.11 |
| 1495mm length 1 row 9.04 kW 340 l/sec 40C | 688.69 | 757.56 | 2.38 | 27.32 | nr | 784.88 |
| 1495mm length 2 row 12.73 kW 340 l/sec 49C | 688.69 | 757.56 | 2.38 | 27.32 | nr | 784.88 |
| Free standing flat top 695mm high medium speed rating c/w plinth | | | | | | |
| 695mm length 1 row 1.94 kW 75 l/sec 39C | 492.49 | 541.74 | 2.38 | 27.32 | nr | 569.06 |
| 695mm length 2 row 2.64 kW 75 l/sec 47C | 492.49 | 541.74 | 2.38 | 27.32 | nr | 569.06 |
| 895mm length 1 row 4.02 kW 150 l/sec 40C | 554.95 | 610.45 | 2.38 | 27.32 | nr | 637.77 |
| 895mm length 2 row 5.62 kW 150 l/sec 49C | 554.95 | 610.45 | 2.38 | 27.32 | nr | 637.77 |
| 1195mm length 1 row 6.58 kW 250 l/sec 40C | 631.83 | 695.01 | 2.38 | 27.32 | nr | 722.33 |
| 1195mm length 2 row 9.27 kW 250 l/sec 48C | 631.83 | 695.01 | 2.38 | 27.32 | nr | 722.33 |
| 1495mm length 1 row 9.04 kW 340 l/sec 40C | 705.10 | 775.61 | 2.38 | 27.32 | nr | 802.93 |
| 1495mm length 2 row 12.73 kW 340 l/sec 49C | 705.10 | 775.61 | 2.38 | 27.32 | nr | 802.93 |
| Free standing sloping top 695mm high medium speed rating c/w plinth | | | | | | |
| 695mm length 1 row 1.94 kW 75 l/sec 39C | 509.65 | 560.62 | 2.38 | 27.32 | nr | 587.94 |
| 695mm length 2 row 2.64 kW 75 l/sec 47C | 509.65 | 560.62 | 2.38 | 27.32 | nr | 587.94 |
| 895mm length 1 row 4.02 kW 150 l/sec 40C | 572.11 | 629.32 | 2.38 | 27.32 | nr | 656.64 |
| 895mm length 2 row 5.62 kW 150 l/sec 49C | 572.11 | 629.32 | 2.38 | 27.32 | nr | 656.64 |
| 1195mm length 1 row 6.58 kW 250 l/sec 40C | 648.99 | 713.89 | 2.38 | 27.32 | nr | 741.21 |
| 1195mm length 2 row 9.27 kW 250 l/sec 48C | 648.99 | 713.89 | 2.38 | 27.32 | nr | 741.21 |
| 1495mm length 1 row 9.04 kW 340 l/sec 40C | 722.26 | 794.49 | 2.38 | 27.32 | nr | 821.81 |
| 1495mm length 2 row 12.73 kW 340 l/sec 49C | 722.26 | 794.49 | 2.38 | 27.32 | nr | 821.81 |

*Material Costs / Measured Work Prices - Mechanical Installations*

## T:MECHANICAL HEATING/COOLING/REFRIGERATION SYSTEMS

| Item | Net Price £ | Material £ | Labour hours | Labour £ | Unit | Total Rate £ |
|---|---|---|---|---|---|---|
| Wall mounted reversed air floor high level sloping discharge | | | | | | |
| 695mm length 1 row 1.94 kW 75 l/sec 39C | 514.80 | 566.28 | 2.38 | 27.32 | nr | **593.60** |
| 695mm length 2 row 2.64 kW 75 l/sec 47C | 514.80 | 566.28 | 2.38 | 27.32 | nr | **593.60** |
| 895mm length 1 row 4.02 kW 150 l/sec 40C | 529.67 | 582.64 | 2.38 | 27.32 | nr | **609.96** |
| 895mm length 2 row 5.62 kW 150 l/sec 49C | 529.67 | 582.64 | 2.38 | 27.32 | nr | **609.96** |
| 1195mm length 1 row 6.58 kW 250 l/sec 40C | 644.75 | 709.23 | 2.38 | 27.32 | nr | **736.55** |
| 1195mm length 2 row 9.27 kW 250 l/sec 48C | 644.75 | 709.23 | 2.38 | 27.32 | nr | **736.55** |
| 1495mm length 1 row 9.04 kW 340 l/sec 40C | 698.98 | 768.88 | 2.38 | 27.32 | nr | **796.20** |
| 1495mm length 2 row 12.73 kW 340 l/sec 49C | 698.98 | 768.88 | 2.38 | 27.32 | nr | **796.20** |
| Ceiling mounted sloping inlet/outlet 665mm width | | | | | | |
| 895mm length 1 row 4.02 kW 150 l/sec 40C | 582.30 | 640.53 | 4.00 | 45.90 | nr | **686.43** |
| 895mm length 2 row 5.62 kW 150 l/sec 49C | 582.30 | 640.53 | 4.00 | 45.90 | nr | **686.43** |
| 1195mm length 1 row 6.58 kW 250 l/sec 40C | 654.37 | 719.81 | 4.00 | 45.90 | nr | **765.71** |
| 1195mm length 2 row 9.27 kW 250 l/sec 48C | 654.37 | 719.81 | 4.00 | 45.90 | nr | **765.71** |
| 1495mm length 1 row 9.04 kW 340 l/sec 40C | 718.43 | 790.27 | 4.00 | 45.90 | nr | **836.17** |
| 1495mm length 2 row 12.73 kW 340 l/sec 49C | 718.43 | 790.27 | 4.00 | 45.90 | nr | **836.17** |
| Free standing unit extended height 1700/1900/2100mm | | | | | | |
| 895mm length 1 row 4.02 kW 150 l/sec 40C | 673.82 | 741.20 | 3.00 | 34.46 | nr | **775.66** |
| 895mm length 2 row 5.62 kW 150 l/sec 49C | 673.82 | 741.20 | 3.00 | 34.46 | nr | **775.66** |
| 1195mm length 1 row 6.58 kW 250 l/sec 40C | 787.07 | 865.78 | 3.00 | 34.46 | nr | **900.24** |
| 1195mm length 2 row 9.27 kW 250 l/sec 48C | 787.07 | 865.78 | 3.00 | 34.46 | nr | **900.24** |
| 1495mm length 1 row 9.04 kW 340 l/sec 40C | 866.01 | 952.61 | 3.00 | 34.46 | nr | **987.07** |
| 1495mm length 2 row 12.73 kW 340 l/sec 49C | 866.01 | 952.61 | 3.00 | 34.46 | nr | **987.07** |

## T:MECHANICAL HEATING/COOLING/REFRIGERATION SYSTEMS

| Item | Net Price £ | Material £ | Labour hours | Labour £ | Unit | Total Rate £ |
|---|---|---|---|---|---|---|
| **T30-T32 : MEDIUM AND LOW TEMPERATURES HOT WATER HEATING : RADIANT PANELS** | | | | | | |
| Pressed steel panel type radiators; fixed with and including brackets; taking down once for decoration; refixing | | | | | | |
| 300mm high; single panel | | | | | | |
| 500mm length | 17.40 | 19.14 | 1.40 | 16.07 | nr | 35.21 |
| 1000mm length | 33.09 | 36.40 | 1.40 | 16.07 | nr | 52.47 |
| 1500mm length | 48.27 | 53.10 | 1.40 | 16.07 | nr | 69.17 |
| 2000mm length | 63.09 | 69.40 | 2.00 | 22.95 | nr | 92.35 |
| 2500mm length | 77.62 | 85.38 | 3.00 | 34.46 | nr | 119.84 |
| 3000mm length | 92.01 | 101.21 | 3.21 | 36.78 | nr | 137.99 |
| 300mm high; double panel; convector | | | | | | |
| 500mm length | 33.63 | 36.99 | 1.60 | 18.36 | nr | 55.35 |
| 1000mm length | 64.99 | 71.49 | 1.60 | 18.36 | nr | 89.85 |
| 1500mm length | 96.34 | 105.97 | 1.60 | 18.36 | nr | 124.33 |
| 2000mm length | 126.55 | 139.21 | 2.20 | 25.27 | nr | 164.48 |
| 2500mm length | 156.57 | 172.23 | 3.21 | 36.78 | nr | 209.01 |
| 3000mm length | 186.09 | 204.70 | 3.40 | 39.03 | nr | 243.73 |
| 450mm high; single panel | | | | | | |
| 500mm length | 15.55 | 17.11 | 1.50 | 17.23 | nr | 34.34 |
| 1000mm length | 30.57 | 33.63 | 1.50 | 17.23 | nr | 50.86 |
| 1600mm length | 47.04 | 51.74 | 2.40 | 27.58 | nr | 79.32 |
| 2000mm length | 58.19 | 64.01 | 3.00 | 34.46 | nr | 98.47 |
| 2400mm length | 76.62 | 84.28 | 4.00 | 45.90 | nr | 130.18 |
| 3000mm length | 94.33 | 103.76 | 4.41 | 50.55 | nr | 154.31 |
| 450mm high; double panel; convector | | | | | | |
| 500mm length | 31.30 | 34.43 | 1.70 | 19.52 | nr | 53.95 |
| 1000mm length | 62.49 | 68.74 | 1.70 | 19.52 | nr | 88.26 |
| 1600mm length | 95.75 | 105.33 | 2.60 | 29.81 | nr | 135.14 |
| 2000mm length | 136.30 | 149.93 | 3.19 | 36.66 | nr | 186.59 |
| 2400mm length | 161.96 | 178.16 | 3.80 | 43.63 | nr | 221.79 |
| 3000mm length | 200.24 | 220.26 | 4.61 | 52.88 | nr | 273.14 |
| 600mm high; single panel | | | | | | |
| 500mm length | 20.04 | 22.04 | 1.70 | 19.52 | nr | 41.56 |
| 1000mm length | 38.78 | 42.66 | 2.20 | 25.27 | nr | 67.93 |
| 1600mm length | 60.16 | 66.18 | 3.60 | 41.28 | nr | 107.46 |
| 2000mm length | 83.12 | 91.43 | 4.61 | 52.88 | nr | 144.31 |
| 2400mm length | 98.50 | 108.35 | 5.21 | 59.77 | nr | 168.12 |
| 3000mm length | 121.16 | 133.28 | 6.99 | 80.24 | nr | 213.52 |
| 600mm high; double panel; convector | | | | | | |
| 500mm length | 39.61 | 43.57 | 1.90 | 21.82 | nr | 65.39 |
| 1000mm length | 79.14 | 87.05 | 1.90 | 21.82 | nr | 108.87 |
| 1600mm length | 141.75 | 155.93 | 3.80 | 43.63 | nr | 199.56 |
| 2000mm length | 174.93 | 192.42 | 4.81 | 55.17 | nr | 247.59 |
| 2400mm length | 208.01 | 228.81 | 5.41 | 62.03 | nr | 290.84 |
| 3000mm length | 257.07 | 282.78 | 7.25 | 83.15 | nr | 365.93 |

*Material Costs / Measured Work Prices - Mechanical Installations*

## T:MECHANICAL HEATING/COOLING/REFRIGERATION SYSTEMS

| Item | Net Price £ | Material £ | Labour hours | Labour £ | Unit | Total Rate £ |
|---|---|---|---|---|---|---|
| 700mm high; single panel | | | | | | |
| 500mm length | 23.95 | 26.34 | 1.80 | 20.68 | nr | **47.02** |
| 1000mm length | 45.70 | 50.27 | 3.00 | 34.46 | nr | **84.73** |
| 1600mm length | 77.12 | 84.83 | 4.81 | 55.17 | nr | **140.00** |
| 2000mm length | 95.38 | 104.92 | 5.99 | 68.71 | nr | **173.63** |
| 2400mm length | 112.97 | 124.27 | 6.02 | 69.13 | nr | **193.40** |
| 3000mm length | 138.99 | 152.89 | 7.25 | 83.15 | nr | **236.04** |
| 700mm high; double panel; convector | | | | | | |
| 500mm length | 46.25 | 50.88 | 2.00 | 22.95 | nr | **73.83** |
| 1000mm length | 89.81 | 98.79 | 3.51 | 40.26 | nr | **139.05** |
| 1600mm length | 164.23 | 180.65 | 5.00 | 57.38 | nr | **238.03** |
| 2000mm length | 200.37 | 220.41 | 5.41 | 62.03 | nr | **282.44** |
| 2400mm length | 238.22 | 262.04 | 5.81 | 66.72 | nr | **328.76** |
| 3000mm length | 294.38 | 323.82 | 6.41 | 73.56 | nr | **397.38** |

Material Costs / Measured Work Prices - Mechanical Installations

## T:MECHANICAL HEATING/COOLING/REFRIGERATION SYSTEMS

| Item | Net Price £ | Material £ | Labour hours | Labour £ | Unit | Total Rate £ |
|---|---|---|---|---|---|---|
| **T30-T32 : MEDIUM AND LOW TEMPERATURE HOT WATER HEATING : RADIANT STRIP HEATERS** | | | | | | |
| Black 1.25" steel tube, aluminium radiant plates including insulation, sliding brackets, cover plates, end closures; weld or screwed BSP ends | | | | | | |
| One tube | | | | | | |
| 1500mm long | 69.33 | 76.26 | 3.00 | 34.46 | nr | **110.72** |
| 3000mm long | 97.24 | 106.96 | 3.00 | 34.46 | nr | **141.42** |
| 4500mm long | 116.69 | 128.36 | 3.00 | 34.46 | nr | **162.82** |
| 6000mm long | 148.72 | 163.59 | 3.00 | 34.46 | nr | **198.05** |
| Two tube | | | | | | |
| 1500mm long | 128.13 | 140.94 | 4.00 | 45.90 | nr | **186.84** |
| 3000mm long | 179.61 | 197.57 | 4.00 | 45.90 | nr | **243.47** |
| 4500mm long | 215.07 | 236.58 | 4.00 | 45.90 | nr | **282.48** |
| 6000mm long | 273.42 | 300.76 | 4.00 | 45.90 | nr | **346.66** |

*Material Costs / Measured Work Prices - Mechanical Installations*

## T:MECHANICAL HEATING/COOLING/REFRIGERATION SYSTEMS

| Item | Net Price £ | Material £ | Labour hours | Labour £ | Unit | Total Rate £ |
|---|---|---|---|---|---|---|
| **T42 : LOCAL HEATING UNITS : UNIT HEATERS** | | | | | | |
| Unit heater; horizontal or vertical discharge; recirculating type for industrial and commercial user for heights up to 3m, normal speed; EAT 15C; fixed to existing suspension rods; complete with enclosures; includes connections or hot water services; (electrical work included elsewhere) | | | | | | |
| Low pressure hot water | | | | | | |
| 7.5 kW 265 l/sec | 263.34 | 289.67 | 6.02 | 69.13 | nr | 358.80 |
| 15.4 kW 575 l/sec | 318.78 | 350.66 | 6.99 | 80.24 | nr | 430.90 |
| 26.9 kW 1040 l/sec | 431.97 | 475.17 | 8.47 | 97.25 | nr | 572.42 |
| 48.0 kW 1620 l/sec | 569.41 | 626.35 | 9.01 | 103.38 | nr | 729.73 |
| Steam, 2 Bar | | | | | | |
| 9.2 kW 265l/sec | 376.53 | 414.18 | 6.02 | 69.13 | nr | 483.31 |
| 18.8 kW 575l/sec | 407.71 | 448.48 | 6.02 | 69.13 | nr | 517.61 |
| 34.8 kW 1040l/sec | 468.93 | 515.82 | 6.02 | 69.13 | nr | 584.95 |
| 51.6 kW 1625l/sec | 635.25 | 698.77 | 6.02 | 69.13 | nr | 767.90 |

## U:VENTILATION/AIR CONDITIONING SYSTEMS

| Item | Net Price £ | Material £ | Labour hours | Labour £ | Unit | Total Rate £ |
|---|---|---|---|---|---|---|
| **U70 : AIR CURTAINS : AIR HEATERS** | | | | | | |
| The selection of an air curtain requires consideration of the particular conditions involved; such as, climatic conditions, wind influence, construction and position; consultation with a specialist manufacturer is therefore, advisable. | | | | | | |
| Industrial grade air curtains; recessed or exposed units with rigid sheet steel casing; aluminium grilles; high quality motor/centrifugal fan assembly | | | | | | |
| Ambient temperature 240V single phase supply; mounting height 2.20m | | | | | | |
|   1000 x 555 x 312mm | 1652.70 | 1817.97 | 12.05 | 138.25 | nr | 1956.22 |
|   1500 x 555 x 312mm | 2145.15 | 2359.66 | 12.05 | 138.25 | nr | 2497.91 |
|   2000 x 555 x 312mm | 2614.50 | 2875.95 | 12.05 | 138.25 | nr | 3014.20 |
| Ambient temperature 240V single phase supply; mounting height 2.50m | | | | | | |
|   1000 x 555 x 312mm | 1927.80 | 2120.58 | 16.13 | 185.08 | nr | 2305.66 |
|   1500 x 555 x 312mm | 2486.40 | 2735.04 | 16.13 | 185.08 | nr | 2920.12 |
|   2000 x 555 x 312mm | 3119.55 | 3431.51 | 16.13 | 185.08 | nr | 3616.59 |
| Ambient temperature 240V single phase supply; mounting height 3.00m | | | | | | |
|   1000 x 686 x 392mm | 2545.20 | 2799.72 | 17.24 | 197.85 | nr | 2997.57 |
|   1500 x 686 x 392mm | 3476.55 | 3824.20 | 17.24 | 197.85 | nr | 4022.05 |
|   2000 x 686 x 392mm | 4317.60 | 4749.36 | 17.24 | 197.85 | nr | 4947.21 |
| Water heated 240V single phase supply; mounting height 2.20m | | | | | | |
|   1000 x 555 x 312mm; 3.80 - 15.60kW output | 1739.85 | 1913.84 | 12.05 | 138.25 | nr | 2052.09 |
|   1500 x 555 x 312mm; 5.80 - 24.00kW output | 2258.55 | 2484.41 | 12.05 | 138.25 | nr | 2622.66 |
|   2000 x 555 x 312mm; 7.80 - 31.00kW output | 2752.05 | 3027.26 | 12.05 | 138.25 | nr | 3165.51 |
| Water heated 240V single phase supply; mounting height 2.50m | | | | | | |
|   1000 x 555 x 312mm; 5.80 - 20.70kW output | 2029.65 | 2232.61 | 16.13 | 185.08 | nr | 2417.69 |
|   1500 x 555 x 312mm; 7.10 - 28.70kW output | 2617.65 | 2879.41 | 16.13 | 185.08 | nr | 3064.49 |
|   2000 x 555 x 312mm; 11.50 - 40.90kW output | 3283.35 | 3611.68 | 16.13 | 185.08 | nr | 3796.76 |
| Water heated 240V single phase supply; mounting height 3.00m | | | | | | |
|   1000 x 686 x 392mm; 9.90 - 34.20kW output | 2679.60 | 2947.56 | 17.24 | 197.85 | nr | 3145.41 |
|   1500 x 686 x 392mm; 15.10 - 52.10kW output | 3659.25 | 4025.18 | 17.24 | 197.85 | nr | 4223.03 |
|   2000 x 686 x 392mm; 20.20 - 67.20kW output | 4544.40 | 4998.84 | 17.24 | 197.85 | nr | 5196.69 |

*Material Costs / Measured Work Prices - Mechanical Installations*

## U:VENTILATION/AIR CONDITIONING SYSTEMS

| Item | Net Price £ | Material £ | Labour hours | Labour £ | Unit | Total Rate £ |
|---|---|---|---|---|---|---|
| Electrically heated 415V three phase supply; mounting height 2.20m | | | | | | |
| 1000 x 555 x 312mm; 4.00 - 8.00kW output | 2219.70 | 2441.67 | 12.05 | 138.25 | nr | 2579.92 |
| 1500 x 555 x 312mm; 5.40 - 10.70kW output | 2815.05 | 3096.55 | 12.05 | 138.25 | nr | 3234.80 |
| 2000 x 555 x 312mm; 8.00 - 16.10kW output | 3353.70 | 3689.07 | 12.05 | 163.91 | nr | 3852.98 |
| Electrically heated 415V three phase supply; mounting height 2.50m | | | | | | |
| 1000 x 555 x 312mm; 5.40 - 10.70kW output | 2661.75 | 2927.93 | 16.13 | 219.42 | nr | 3147.35 |
| 1500 x 555 x 312mm; 7.10 - 14.30kW output | 3299.10 | 3629.01 | 16.13 | 185.08 | nr | 3814.09 |
| 2000 x 555 x 312mm; 10.70 - 21.40kW output | 4028.85 | 4431.73 | 16.13 | 185.08 | nr | 4616.81 |
| **Commercial grade air curtains; recessed or exposed units with rigid sheet steel casing; aluminium grilles; high quality motor/centrifugal fan assembly** | | | | | | |
| Ambient temperature 415V three phase supply; including wiring between multiple units: horizontally or vertically mounted; opening maximum 6.00m | | | | | | |
| 1106 x 516 x 689mm; 1.2A supply | 2190.30 | 2409.33 | 17.24 | 197.85 | nr | 2607.18 |
| 1661 x 516 x 689mm; 1.8A supply | 3152.10 | 3467.31 | 17.24 | 197.85 | nr | 3665.16 |
| Water heated 415V three phase supply; including wiring between multiple units; horizontally or vertically mounted; opening maximum 6.00m | | | | | | |
| 1106 x 516 x 689mm; 1.2A supply; 34.80kW output | 2355.15 | 2590.66 | 17.24 | 197.85 | nr | 2788.51 |
| 1661 x 516 x 689mm; 1.8A supply; 50.70kW output | 3389.40 | 3728.34 | 17.24 | 197.85 | nr | 3926.19 |
| Water heated 415V three phase supply; including wiring between multiple units; vertically mounted in single bank for openings maximum 6.00m wide or opposing twin banks for openings maximum 10.00m wide | | | | | | |
| 1106 x 689mm; 1.2A supply; 34.80kW output | 2355.15 | 2590.66 | 17.24 | 197.85 | nr | 2788.51 |
| 1661 x 689mm; 1.8A supply; 50.70kW output | 3389.40 | 3728.34 | 17.24 | 197.85 | nr | 3926.19 |
| Remote mounted electronic controller unit; 415V three phase supply; excluding wiring to units | | | | | | |
| five speed; 7A | 741.30 | 815.43 | 15.00 | 172.13 | nr | 987.56 |

Material Costs / Measured Work Prices - Mechanical Installations

## Y:MECHANICAL AND ELECTRICAL SERVICES

| Item | Net Price £ | Material £ | Labour hours | Labour £ | Unit | Total Rate £ |
|---|---|---|---|---|---|---|
| **Y10 : PIPELINES : PLASTICS - SOIL & WASTE** | | | | | | |
| PVC-U (Unplasticized Polyvinyl Chloride) soil, waste and ventilating pipe; solvent welded joints; fixed with PVC or galvanised metal clips to backgrounds; BS 4514 | | | | | | |
| Pipe | | | | | | |
| 82mm dia. | 3.06 | 3.08 | 0.31 | 3.56 | m | 6.64 |
| 110mm dia. | 3.33 | 2.34 | 0.35 | 4.03 | m | 6.37 |
| 160mm dia. | 7.66 | 5.68 | 0.45 | 5.16 | m | 10.84 |
| Extra Over PVC-U pipe; PVC-U fittings, solvent welded joints | | | | | | |
| Pipe coupler; double socket | | | | | | |
| 82mm | 3.10 | 3.41 | 0.25 | 2.87 | nr | 6.28 |
| 110mm | 3.73 | 4.10 | 0.25 | 2.87 | nr | 6.97 |
| 160mm | 6.91 | 7.60 | 0.25 | 2.87 | nr | 10.47 |
| Reducer; double socket | | | | | | |
| 110 to 82mm | 2.64 | 2.90 | 0.25 | 2.87 | nr | 5.77 |
| Short access pipe; single socket | | | | | | |
| 110mm | 5.62 | 6.18 | 0.25 | 2.87 | nr | 9.05 |
| Short access pipe; double socket | | | | | | |
| 110mm | 5.62 | 6.18 | 0.25 | 2.87 | nr | 9.05 |
| 160mm | 15.47 | 17.02 | 0.25 | 2.87 | nr | 19.89 |
| Bend; double socket; 92.5 degree | | | | | | |
| 82mm | 3.61 | 3.97 | 0.35 | 4.02 | nr | 7.99 |
| 110mm | 4.27 | 4.70 | 0.38 | 4.36 | nr | 9.06 |
| 160mm | 9.57 | 10.53 | 0.58 | 6.66 | nr | 17.19 |
| Bend; double socket; 135 degree | | | | | | |
| 82mm | 3.61 | 3.97 | 0.35 | 4.02 | nr | 7.99 |
| 110mm | 4.27 | 4.70 | 0.38 | 4.36 | nr | 9.06 |
| 160mm | 9.57 | 10.53 | 0.58 | 6.67 | nr | 17.20 |
| Single branch; triple socket; 92.5 degree | | | | | | |
| 82mm | 5.04 | 5.54 | 0.41 | 4.70 | nr | 10.24 |
| 110mm | 5.48 | 6.03 | 0.50 | 5.74 | nr | 11.77 |
| 160mm | 18.16 | 19.98 | 0.58 | 6.66 | nr | 26.64 |
| Single branch; triple socket; 135 degree | | | | | | |
| 110mm | 5.60 | 6.16 | 0.50 | 5.74 | nr | 11.90 |
| 160mm | 18.93 | 20.82 | 0.58 | 6.66 | nr | 27.48 |

## Y:MECHANICAL AND ELECTRICAL SERVICES

| Item | Net Price £ | Material £ | Labour hours | Labour £ | Unit | Total Rate £ |
|---|---|---|---|---|---|---|
| Unequal single branch; triple socket; 92.5 degree | | | | | | |
| 160 x 160 x 110mm | 13.23 | 14.55 | 0.58 | 6.66 | nr | 21.21 |
| Unequal single branch; triple socket; 135 degree | | | | | | |
| 160 x 160 x 110mm | 13.77 | 15.15 | 0.58 | 6.66 | nr | 21.81 |
| Single boss pipe; single socket | | | | | | |
| 110 x 110 x 32mm | 1.74 | 1.91 | 0.25 | 2.87 | nr | 4.78 |
| 110 x 110 x 40mm | 1.74 | 1.91 | 0.25 | 2.87 | nr | 4.78 |
| 110 x 110 x 50mm | 1.74 | 1.91 | 0.25 | 2.87 | nr | 4.78 |
| Single boss pipe; double socket | | | | | | |
| 110 x 110 x 32mm | 1.74 | 1.91 | 0.25 | 2.87 | nr | 4.78 |
| 110 x 110 x 40mm | 1.74 | 1.91 | 0.25 | 2.87 | nr | 4.78 |
| 110 x 110 x 50mm | 1.74 | 1.91 | 0.25 | 2.87 | nr | 4.78 |
| Vent terminal | | | | | | |
| 82mm | 0.95 | 1.04 | 0.10 | 1.15 | nr | 2.19 |
| 110mm | 1.00 | 1.10 | 0.10 | 1.15 | nr | 2.25 |
| 160mm | 2.42 | 2.66 | 0.10 | 1.15 | nr | 3.81 |
| Weathering slate; angled; 610 x 610mm | | | | | | |
| 82mm | 13.77 | 15.15 | 1.00 | 11.47 | nr | 26.62 |
| 110mm | 13.77 | 15.15 | 1.00 | 11.47 | nr | 26.62 |
| Weathering slate; flat; 400 x 400mm | | | | | | |
| 82mm | 8.68 | 9.55 | 1.00 | 11.47 | nr | 21.02 |
| 110mm | 8.68 | 9.55 | 1.00 | 11.47 | nr | 21.02 |
| Air admittance valve | | | | | | |
| 82mm | 13.99 | 15.39 | 0.17 | 1.91 | nr | 17.30 |
| 110mm | 13.99 | 15.39 | 0.17 | 1.91 | nr | 17.30 |
| Boss adaptor; rubber; push fit | | | | | | |
| 32mm | 0.63 | 0.69 | 0.25 | 2.87 | nr | 3.56 |
| 40 mm | 0.63 | 0.69 | 0.25 | 2.87 | nr | 3.56 |
| 50 mm | 0.63 | 0.71 | 0.25 | 2.87 | nr | 3.58 |
| PVC-U (Unplasticized Polyvinyl Chloride) soil, waste and ventilating pipe; ring seal joints; fixed with PVC or galvanised metal clips to backgrounds; BS 4514 | | | | | | |
| Pipe | | | | | | |
| 82mm dia. | 3.06 | 3.08 | 0.31 | 3.56 | m | 6.64 |
| 110mm dia. | 3.33 | 2.34 | 0.35 | 4.03 | m | 6.37 |
| 160mm dia. | 7.66 | 5.68 | 0.45 | 5.16 | m | 10.84 |

## Material Costs / Measured Work Prices - Mechanical Installations

### Y:MECHANICAL AND ELECTRICAL SERVICES

| Item | Net Price £ | Material £ | Labour hours | Labour £ | Unit | Total Rate £ |
|---|---|---|---|---|---|---|
| **Y10 : PIPELINES : PLASTICS - SOIL & WASTE (contd)** | | | | | | |
| **Extra Over PVC-U pipe; PVC-U fittings; ring seal joints** | | | | | | |
| Pipe coupler; single socket | | | | | | |
| 82mm | 1.96 | 2.16 | 0.25 | 2.87 | nr | 5.03 |
| 110mm | 2.11 | 2.32 | 0.25 | 2.87 | nr | 5.19 |
| 160mm | 4.41 | 4.85 | 0.25 | 2.87 | nr | 7.72 |
| Pipe coupler; double socket | | | | | | |
| 82mm | 3.10 | 3.41 | 0.25 | 2.87 | nr | 6.28 |
| 110mm | 3.73 | 4.10 | 0.25 | 2.87 | nr | 6.97 |
| 160mm | 6.91 | 7.60 | 0.25 | 2.87 | nr | 10.47 |
| Reducer; single socket | | | | | | |
| 110 to 82mm | 3.54 | 3.89 | 0.25 | 2.87 | nr | 6.76 |
| Access pipe; single socket | | | | | | |
| 82mm | 5.32 | 5.85 | 0.25 | 2.87 | nr | 8.72 |
| 110mm | 7.64 | 8.40 | 0.25 | 2.87 | nr | 11.27 |
| 160mm | 16.53 | 18.18 | 0.25 | 2.87 | nr | 21.05 |
| Bend; single socket; 92.5 degree | | | | | | |
| 82mm | 4.22 | 4.64 | 0.35 | 4.02 | nr | 8.66 |
| 110mm | 5.04 | 5.54 | 0.38 | 4.36 | nr | 9.90 |
| 160mm | 10.21 | 11.23 | 0.58 | 6.66 | nr | 17.89 |
| Bend; single socket; 135 degree | | | | | | |
| 82mm | 4.22 | 4.64 | 0.35 | 4.02 | nr | 8.66 |
| 110mm | 5.04 | 5.54 | 0.38 | 4.36 | nr | 9.90 |
| 160mm | 10.21 | 11.23 | 0.58 | 6.66 | nr | 17.89 |
| Bend; double socket; 92.5 degree | | | | | | |
| 110mm | 5.60 | 6.16 | 0.38 | 4.36 | nr | 10.52 |
| Bend; double socket; 135 degree | | | | | | |
| 160mm | 12.39 | 13.63 | 0.58 | 6.66 | nr | 20.29 |
| Single branch; double socket, 2 boss; 92.5 degree | | | | | | |
| 82mm | 5.99 | 6.59 | 0.41 | 4.70 | nr | 11.29 |
| 110mm | 7.39 | 8.13 | 0.50 | 5.74 | nr | 13.87 |
| 160mm | 18.11 | 19.92 | 0.58 | 6.66 | nr | 26.58 |
| Single branch; double socket; 135 degree | | | | | | |
| 82mm | 5.99 | 6.59 | 0.41 | 4.70 | nr | 11.29 |
| 110mm | 7.51 | 8.26 | 0.50 | 5.74 | nr | 14.00 |

*Material Costs / Measured Work Prices - Mechanical Installations*

## Y:MECHANICAL AND ELECTRICAL SERVICES

| Item | Net Price £ | Material £ | Labour hours | Labour £ | Unit | Total Rate £ |
|---|---|---|---|---|---|---|
| Unequal single branch; double socket; 92.5 degree | | | | | | |
| 160 x 160 x 110mm | 14.50 | 15.95 | 0.58 | 6.66 | nr | 22.61 |
| Unequal single branch; double socket; 135 degree | | | | | | |
| 160 x 160 x 110mm | 14.50 | 15.95 | 0.58 | 6.66 | nr | 22.61 |
| Vent terminal | | | | | | |
| 82mm | 0.95 | 1.04 | 0.10 | 1.15 | nr | 2.19 |
| 110mm | 1.00 | 1.10 | 0.10 | 1.15 | nr | 2.25 |
| 160mm | 2.42 | 2.66 | 0.10 | 1.15 | nr | 3.81 |
| Weathering slate; angled; 610 x 610mm | | | | | | |
| 82mm | 13.77 | 15.15 | 1.00 | 11.47 | nr | 26.62 |
| 110mm | 13.77 | 15.15 | 1.00 | 11.47 | nr | 26.62 |
| Weathering slate; flat; 400 x 400mm | | | | | | |
| 82mm | 8.68 | 9.55 | 1.00 | 11.47 | nr | 21.02 |
| 110mm | 8.68 | 9.55 | 1.00 | 11.47 | nr | 21.02 |
| Air admittance valve | | | | | | |
| 82mm | 13.99 | 15.39 | 0.17 | 1.91 | nr | 17.30 |
| 110mm | 13.99 | 15.39 | 0.17 | 1.91 | nr | 17.30 |
| Boss adaptor; rubber; push fit | | | | | | |
| 32mm | 0.63 | 0.69 | 0.25 | 2.87 | nr | 3.56 |
| 40 mm | 0.63 | 0.69 | 0.25 | 2.87 | nr | 3.56 |
| 50 mm | 0.63 | 0.71 | 0.25 | 2.87 | nr | 3.58 |
| **ABS (Acrylonitrile Butadiene Styrene) waste pipe; solvent welded joints; fixed with PVC clips to backgrounds; BS 5255** | | | | | | |
| Pipe | | | | | | |
| 32mm dia. | 0.79 | 0.62 | 0.20 | 2.29 | m | 2.91 |
| 40mm dia. | 0.97 | 0.70 | 0.20 | 2.29 | m | 2.99 |
| 50mm dia. | 1.17 | 0.88 | 0.23 | 2.64 | m | 3.52 |
| **Extra Over ABS waste pipe; ABS fittings; solvent welded joints** | | | | | | |
| Screwed access plug | | | | | | |
| 32mm dia. | 0.45 | 0.49 | 0.20 | 2.29 | nr | 2.78 |
| 40mm dia. | 0.45 | 0.49 | 0.20 | 2.29 | nr | 2.78 |
| 50mm dia. | 0.68 | 0.75 | 0.30 | 3.44 | nr | 4.19 |
| Socket plug | | | | | | |
| 32mm dia. | 0.45 | 0.49 | 0.20 | 2.29 | nr | 2.78 |
| 40mm dia. | 0.45 | 0.49 | 0.20 | 2.29 | nr | 2.78 |
| 50mm dia. | 0.68 | 0.75 | 0.30 | 3.44 | nr | 4.19 |

## Material Costs / Measured Work Prices - Mechanical Installations

### Y: MECHANICAL AND ELECTRICAL SERVICES

| Item | Net Price £ | Material £ | Labour hours | Labour £ | Unit | Total Rate £ |
|---|---|---|---|---|---|---|
| **Y10 : PIPELINES : PLASTICS - SOIL & WASTE (contd)** | | | | | | |
| Extra Over ABS waste pipe; ABS fittings; solvent welded joints (contd) | | | | | | |
| Straight coupling | | | | | | |
| 32mm dia. | 0.42 | 0.46 | 0.25 | 2.87 | nr | **3.33** |
| 40mm dia. | 0.45 | 0.49 | 0.25 | 2.87 | nr | **3.36** |
| 50mm dia. | 0.68 | 0.75 | 0.25 | 2.87 | nr | **3.62** |
| Expansion coupling | | | | | | |
| 32mm dia. | 0.45 | 0.49 | 0.25 | 2.87 | nr | **3.36** |
| 40mm dia. | 0.45 | 0.49 | 0.25 | 2.87 | nr | **3.36** |
| 50mm dia. | 0.68 | 0.75 | 0.25 | 2.87 | nr | **3.62** |
| Threaded coupling | | | | | | |
| 32mm dia. | 0.45 | 0.49 | 0.25 | 2.87 | nr | **3.36** |
| 40mm dia. | 0.45 | 0.49 | 0.25 | 2.87 | nr | **3.36** |
| Reducer | | | | | | |
| 40 x 32mm dia. | 0.42 | 0.46 | 0.25 | 2.87 | nr | **3.33** |
| 50 x 32mm dia. | 0.67 | 0.74 | 0.25 | 2.87 | nr | **3.61** |
| 50 x 40mm dia. | 0.67 | 0.74 | 0.25 | 2.87 | nr | **3.61** |
| Swept bend; 92.5 degree | | | | | | |
| 32mm dia. | 0.45 | 0.49 | 0.25 | 2.87 | nr | **3.36** |
| 40mm dia. | 0.45 | 0.49 | 0.25 | 2.87 | nr | **3.36** |
| 50mm dia. | 0.68 | 0.75 | 0.30 | 3.44 | nr | **4.19** |
| Knuckle bend; 90 degree | | | | | | |
| 32mm dia. | 0.45 | 0.49 | 0.25 | 2.87 | nr | **3.36** |
| 40mm dia. | 0.45 | 0.49 | 0.25 | 2.87 | nr | **3.36** |
| 50mm dia. | 0.68 | 0.75 | 0.30 | 3.44 | nr | **4.19** |
| Obtuse bend; 45 degree | | | | | | |
| 40mm dia. | 0.45 | 0.49 | 0.25 | 2.87 | nr | **3.36** |
| 50mm dia. | 0.68 | 0.75 | 0.30 | 3.44 | nr | **4.19** |
| Spigot bend; 45 degree | | | | | | |
| 32mm dia. | 0.45 | 0.49 | 0.25 | 2.87 | nr | **3.36** |
| 40mm dia. | 0.45 | 0.49 | 0.25 | 2.87 | nr | **3.36** |
| 50mm dia. | 0.68 | 0.75 | 0.30 | 3.44 | nr | **4.19** |
| Junction; 45 degree | | | | | | |
| 32mm dia. | 0.45 | 0.49 | 0.25 | 2.87 | nr | **3.36** |
| 40mm dia. | 0.45 | 0.49 | 0.25 | 2.87 | nr | **3.36** |
| 50mm dia. | 0.68 | 0.75 | 0.30 | 3.44 | nr | **4.19** |

## Y:MECHANICAL AND ELECTRICAL SERVICES

| Item | Net Price £ | Material £ | Labour hours | Labour £ | Unit | Total Rate £ |
|---|---|---|---|---|---|---|
| Swept tee; 92.5 degree | | | | | | |
| 32mm dia. | 0.45 | 0.49 | 0.30 | 3.44 | nr | 3.93 |
| 40mm dia. | 0.45 | 0.49 | 0.30 | 3.44 | nr | 3.93 |
| 50mm dia. | 0.68 | 0.75 | 0.30 | 3.44 | nr | 4.19 |
| Cross tee; 92.5 degree | | | | | | |
| 40mm dia. | 0.45 | 0.49 | 0.50 | 5.74 | nr | 6.23 |
| 50mm dia. | 0.68 | 0.75 | 0.50 | 5.74 | nr | 6.49 |
| Adaptor; screwed iron; male | | | | | | |
| 32mm dia. | 0.45 | 0.49 | 0.25 | 2.87 | nr | 3.36 |
| 40mm dia. | 0.45 | 0.49 | 0.25 | 2.87 | nr | 3.36 |
| 50mm dia. | 0.68 | 0.75 | 0.30 | 3.44 | nr | 4.19 |
| Tank connector | | | | | | |
| 32mm dia. | 0.45 | 0.49 | 0.33 | 3.79 | nr | 4.28 |
| 40mm dia. | 0.45 | 0.49 | 0.33 | 3.79 | nr | 4.28 |
| **Polypropylene waste pipe; push fit joints; fixed with PVC clips to backgrounds; BS 5254** | | | | | | |
| Pipe | | | | | | |
| 32mm dia. | 0.41 | 0.47 | 0.17 | 2.01 | m | 2.48 |
| 40mm dia. | 0.51 | 0.53 | 0.17 | 2.01 | m | 2.54 |
| 50mm dia. | 0.85 | 0.75 | 0.25 | 2.87 | m | 3.62 |
| **Extra Over polypropylene waste pipe; polypropylene fittings; push fit joints** | | | | | | |
| Screwed access plug | | | | | | |
| 32mm dia. | 0.40 | 0.44 | 0.17 | 2.01 | nr | 2.45 |
| 40mm dia. | 0.40 | 0.44 | 0.17 | 2.01 | nr | 2.45 |
| Socket plug | | | | | | |
| 32mm dia. | 0.40 | 0.44 | 0.18 | 2.07 | nr | 2.51 |
| 40mm dia. | 0.40 | *0.44 | 0.18 | 2.07 | nr | 2.51 |
| 50mm dia. | 0.73 | 0.80 | 0.25 | 2.87 | nr | 3.67 |
| Straight coupling | | | | | | |
| 32mm dia. | 0.40 | 0.44 | 0.13 | 1.49 | nr | 1.93 |
| 40mm dia. | 0.40 | 0.44 | 0.13 | 1.49 | nr | 1.93 |
| 50mm dia. | 0.73 | 0.80 | 0.15 | 1.72 | nr | 2.52 |
| Threaded coupling | | | | | | |
| 32mm dia. | 0.40 | 0.44 | 0.13 | 1.49 | nr | 1.93 |
| 40mm dia. | 0.40 | 0.44 | 0.13 | 1.49 | nr | 1.93 |

## Y:MECHANICAL AND ELECTRICAL SERVICES

| Item | Net Price £ | Material £ | Labour hours | Labour £ | Unit | Total Rate £ |
|---|---|---|---|---|---|---|
| **Y10 : PIPELINES : PLASTICS - SOIL & WASTE (contd)** | | | | | | |
| **Extra Over polypropylene waste pipe; polypropylene fittings; push fit joints (contd)** | | | | | | |
| Reducer | | | | | | |
| 40 x 32mm dia. | 0.40 | 0.44 | 0.13 | 1.49 | nr | 1.93 |
| 50 x 32mm dia. | 0.73 | 0.80 | 0.13 | 1.49 | nr | 2.29 |
| 50 x 40mm dia. | 0.73 | 0.80 | 0.15 | 1.72 | nr | 2.52 |
| Swept bend; 92 degree | | | | | | |
| 32mm dia. | 0.40 | 0.44 | 0.13 | 1.49 | nr | 1.93 |
| 40mm dia. | 0.40 | 0.44 | 0.13 | 1.49 | nr | 1.93 |
| 50mm dia. | 0.73 | 0.80 | 0.15 | 1.72 | nr | 2.52 |
| Knuckle bend; 90 degree | | | | | | |
| 32mm dia. | 0.40 | 0.44 | 0.13 | 1.49 | nr | 1.93 |
| 40mm dia. | 0.40 | 0.44 | 0.13 | 1.49 | nr | 1.93 |
| 50mm dia. | 0.73 | 0.80 | 0.15 | 1.72 | nr | 2.52 |
| Obtuse bend; 45 degree | | | | | | |
| 32mm dia. | 0.40 | 0.44 | 0.13 | 1.49 | nr | 1.93 |
| 40mm dia. | 0.40 | 0.44 | 0.13 | 1.49 | nr | 1.93 |
| 50mm dia. | 0.73 | 0.80 | 0.15 | 1.72 | nr | 2.52 |
| Swept tee; 92 degree | | | | | | |
| 32mm dia. | 0.40 | 0.44 | 0.16 | 1.84 | nr | 2.28 |
| 40mm dia. | 0.40 | 0.44 | 0.16 | 1.84 | nr | 2.28 |
| 50mm dia. | 0.73 | 0.80 | 0.18 | 2.07 | nr | 2.87 |
| Tank connector | | | | | | |
| 32mm dia. | 0.40 | 0.44 | 0.13 | 1.49 | nr | 1.93 |
| 40mm dia. | 0.40 | 0.44 | 0.13 | 1.49 | nr | 1.93 |
| Vent terminal | | | | | | |
| 50mm dia. | 0.73 | 0.80 | 0.10 | 1.15 | nr | 1.95 |
| Universal waste pipe coupler | | | | | | |
| 32mm dia. | 0.46 | 0.51 | 0.13 | 1.49 | nr | 2.00 |
| 40mm dia. | 0.48 | 0.53 | 0.13 | 1.49 | nr | 2.02 |
| 50mm dia. | 0.62 | 0.68 | 0.15 | 1.72 | nr | 2.40 |
| **Polypropylene overflow pipe; push fit joints; fixed with PVC clips to backgrounds** | | | | | | |
| Pipe | | | | | | |
| 22mm dia. | 0.24 | 0.34 | 0.20 | 2.29 | m | 2.63 |

## Y:MECHANICAL AND ELECTRICAL SERVICES

| Item | Net Price £ | Material £ | Labour hours | Labour £ | Unit | Total Rate £ |
|---|---|---|---|---|---|---|
| **Extra Over polypropylene overflow pipe; polypropylene fittings; push fit joints** | | | | | | |
| Straight connector | | | | | | |
| 22mm dia. | 0.29 | 0.32 | 0.20 | 2.29 | nr | **2.61** |
| Bend; 90 degree | | | | | | |
| 22mm dia. | 0.29 | 0.32 | 0.20 | 2.29 | nr | **2.61** |
| Bend: 45 degree | | | | | | |
| 22mm dia. | 0.29 | 0.32 | 0.20 | 2.29 | nr | **2.61** |
| Tee; 90 degree | | | | | | |
| 22mm dia. | 0.29 | 0.32 | 0.20 | 2.29 | nr | **2.61** |
| Straight adaptor | | | | | | |
| 22mm dia. | 0.30 | 0.33 | 0.20 | 2.29 | nr | **2.62** |
| Bent adaptor | | | | | | |
| 22mm dia. | 0.30 | 0.33 | 0.20 | 2.29 | nr | **2.62** |
| Straight tank connector | | | | | | |
| 22mm dia. | 0.29 | 0.32 | 0.20 | 2.29 | nr | **2.61** |
| Bent tank connector | | | | | | |
| 22mm dia. | 0.29 | 0.32 | 0.20 | 2.29 | nr | **2.61** |
| **Water Byelaw 30 kit; complete with screened breather and warning pipes, dip tube bend and rubber grommet** | | | | | | |
| England and Wales | 2.99 | 3.29 | 0.50 | 5.74 | nr | **9.03** |
| Scotland | 4.82 | 5.30 | 0.50 | 5.74 | nr | **11.04** |
| **PVC-ABS overflow pipe; solvent welded joints; fixed with PVC clips to backgrounds** | | | | | | |
| Pipe | | | | | | |
| 22mm dia. | 0.24 | 0.34 | 0.20 | 2.29 | m | **2.63** |
| **Extra Over PVC-ABS overflow pipe; PVC-ABS fittings; solvent welded joints** | | | | | | |
| Straight connector | | | | | | |
| 22mm dia. | 0.28 | 0.31 | 0.20 | 2.29 | nr | **2.60** |
| Bend; 90 degree | | | | | | |
| 22mm dia. | 0.28 | 0.31 | 0.20 | 2.29 | nr | **2.60** |

## Y:MECHANICAL AND ELECTRICAL SERVICES

| Item | Net Price £ | Material £ | Labour hours | Labour £ | Unit | Total Rate £ |
|---|---|---|---|---|---|---|
| **Y10 : PIPELINES : PLASTICS - SOIL & WASTE (contd)** | | | | | | |
| **Extra Over PVC-ABS overflow pipe; PVC-ABS fittings; solvent welded joints (contd)** | | | | | | |
| Bend; 45 degree | | | | | | |
| 22mm dia. | 0.28 | 0.31 | 0.20 | 2.29 | nr | **2.60** |
| Tee; 90 degree | | | | | | |
| 22mm dia. | 0.28 | 0.31 | 0.20 | 2.29 | nr | **2.60** |
| Straight adaptor | | | | | | |
| 22mm dia. | 0.28 | 0.31 | 0.20 | 2.29 | nr | **2.60** |
| Bent adaptor | | | | | | |
| 22mm dia. | 0.28 | 0.31 | 0.20 | 2.29 | nr | **2.60** |
| Straight tank connector | | | | | | |
| 22mm dia. | 0.29 | 0.32 | 0.20 | 2.29 | nr | **2.61** |
| Bent tank connector | | | | | | |
| 22mm dia. | 0.29 | 0.32 | 0.20 | 2.29 | nr | **2.61** |
| **Polypropylene traps; fixed and connected to pipework; BS 3943** | | | | | | |
| Tubular trap; 38mm seal with cleaning eye | | | | | | |
| 32mm dia. | 1.13 | 1.24 | 0.20 | 2.29 | nr | **3.53** |
| 40mm dia. | 1.25 | 1.38 | 0.20 | 2.29 | nr | **3.67** |
| Tubular trap; 75mm seal with cleaning eye | | | | | | |
| 32mm dia. | 1.18 | 1.30 | 0.20 | 2.29 | nr | **3.59** |
| 40mm dia. | 1.30 | 1.43 | 0.20 | 2.29 | nr | **3.72** |
| Bottle trap; 38mm seal | | | | | | |
| 32mm dia. | 0.98 | 1.08 | 0.20 | 2.29 | nr | **3.37** |
| 40mm dia. | 1.13 | 1.24 | 0.20 | 2.29 | nr | **3.53** |
| Bottle trap; 75mm seal | | | | | | |
| 32mm dia. | 1.08 | 1.19 | 0.20 | 2.29 | nr | **3.48** |
| 40mm dia. | 1.25 | 1.38 | 0.25 | 2.87 | nr | **4.25** |
| Resealing bottle trap; 75mm seal | | | | | | |
| 32mm dia. | 1.46 | 1.61 | 0.20 | 2.29 | nr | **3.90** |
| 40mm dia. | 1.70 | 1.87 | 0.25 | 2.87 | nr | **4.74** |

*Material Costs / Measured Work Prices - Mechanical Installations*

## Y:MECHANICAL AND ELECTRICAL SERVICES

| Item | Net Price £ | Material £ | Labour hours | Labour £ | Unit | Total Rate £ |
|---|---|---|---|---|---|---|
| Bath trap; 20mm seal with cleaning eye | | | | | | |
| 32mm dia. | 0.98 | 1.08 | 0.20 | 2.29 | nr | 3.37 |
| 40mm dia. | 1.01 | 1.11 | 0.25 | 2.87 | nr | 3.98 |
| Bath trap, low level; 38mm seal | | | | | | |
| 40mm dia. | 1.39 | 1.53 | 0.25 | 2.87 | nr | 4.40 |
| Bath trap, low level; 38mm seal complete with overflow hose and rose | | | | | | |
| 40mm dia. | 2.93 | 3.22 | 0.25 | 2.87 | nr | 6.09 |
| Bath trap, low level; 75mm seal | | | | | | |
| 40mm dia. | 1.66 | 1.83 | 0.25 | 2.87 | nr | 4.70 |
| Bath trap, low level; 75mm seal complete with overflow hose and rose | | | | | | |
| 40mm dia. | 3.22 | 3.54 | 0.20 | 2.29 | nr | 5.83 |
| Shower trap; 19mm seal | | | | | | |
| 40mm dia.; 70mm dia. plastic grid | 2.42 | 2.66 | 0.25 | 2.87 | nr | 5.53 |
| 40mm dia.; 70mm dia. chrome grid | 3.00 | 3.30 | 0.25 | 2.87 | nr | 6.17 |
| Shower trap; 50mm seal with adjustable bend | | | | | | |
| 40mm dia.; 70mm dia. plastic grid | 2.90 | 3.19 | 0.25 | 2.87 | nr | 6.06 |
| 40mm dia.; 70mm dia. chrome grid | 3.48 | 3.83 | 0.25 | 2.87 | nr | 6.70 |
| Tubular swivel trap 'S'; 38mm seal | | | | | | |
| 32mm dia. | 1.32 | 1.45 | 0.20 | 2.29 | nr | 3.74 |
| 40mm dia. | 1.39 | 1.53 | 0.25 | 2.87 | nr | 4.40 |
| Tubular swivel trap 'S'; 75mm seal | | | | | | |
| 32mm dia. | 1.32 | 1.45 | 0.20 | 2.29 | nr | 3.74 |
| 40mm dia. | 1.46 | 1.61 | 0.25 | 2.87 | nr | 4.48 |
| Washing machine trap; 75mm seal | | | | | | |
| 40mm dia. | 2.09 | 2.30 | 0.25 | 2.87 | nr | 5.17 |
| Washing machine half trap; 75mm seal | | | | | | |
| 40mm dia. | 2.16 | 2.38 | 0.25 | 2.87 | nr | 5.25 |
| **Polypropylene anti-syphon traps; fixed and connected to pipework; BS 3943** | | | | | | |
| Bottle trap; 75mm seal | | | | | | |
| 32mm dia. | 1.58 | 1.74 | 0.20 | 2.29 | nr | 4.03 |
| 40mm dia. | 1.75 | 1.93 | 0.25 | 2.87 | nr | 4.80 |

## Y:MECHANICAL AND ELECTRICAL SERVICES

| Item | Net Price £ | Material £ | Labour hours | Labour £ | Unit | Total Rate £ |
|---|---|---|---|---|---|---|
| **Y10 : PIPELINES : PLASTICS - SOIL & WASTE (contd)** | | | | | | |
| **Polypropylene anti-syphon traps; fixed and connected to pipework; BS 3943 (contd)** | | | | | | |
| Bath trap, low level; 38mm seal | | | | | | |
| 40mm dia. | 1.90 | 2.09 | 0.25 | 2.87 | nr | **4.96** |
| Bath trap, low level; 75mm seal | | | | | | |
| 40mm dia. | 2.16 | 2.38 | 0.25 | 2.87 | nr | **5.25** |
| Washing machine trap; 75mm seal | | | | | | |
| 40mm dia. | 2.59 | 2.85 | 0.25 | 2.87 | nr | **5.72** |

# Keep your figures up to date, free of charge

This section, and most of the other information in this Price Book, is brought up to date every three months, until the next annual edition, in the *Price Book Update*.

The *Update* is available free to all Price Book purchasers.

To ensure you receive your copy, simply complete the reply card from the centre of the book and return it to us.

## Y:MECHANICAL AND ELECTRICAL SERVICES

| Item | Net Price £ | Material £ | Labour hours | Labour £ | Unit | Total Rate £ |
|---|---|---|---|---|---|---|
| **Y10 : PIPELINES : PLASTICS - WATER & GAS** | | | | | | |
| **MDPE (Medium Density Polyethylene) pipes for water distribution; laid underground; electrofusion joints in the running length; BS 6572** | | | | | | |
| Blue MDPE coiled service pipe | | | | | | |
| 20mm dia. | 0.43 | 0.49 | 0.37 | 4.25 | m | 4.74 |
| 25mm dia. | 0.55 | 0.62 | 0.41 | 4.70 | m | 5.32 |
| 32mm dia. | 0.91 | 1.03 | 0.47 | 5.39 | m | 6.42 |
| 50mm dia. | 2.20 | 2.49 | 0.53 | 6.08 | m | 8.57 |
| 63mm dia. | 3.37 | 3.82 | 0.60 | 6.89 | m | 10.71 |
| Blue MDPE mains service pipe | | | | | | |
| 90mm dia. | 5.80 | 6.57 | 0.90 | 10.33 | m | 16.90 |
| 125mm dia. | 11.18 | 12.67 | 1.20 | 13.78 | m | 26.45 |
| 180mm dia. | 22.58 | 25.58 | 1.50 | 17.23 | m | 42.81 |
| 250mm dia. | 43.26 | 49.01 | 1.75 | 20.10 | m | 69.11 |
| **Extra over blue MDPE water pipe; MDPE fittings, electrofusion joints** | | | | | | |
| Cap | | | | | | |
| 63mm dia. | 9.50 | 10.45 | 0.32 | 3.67 | nr | 14.12 |
| 90mm dia. | 16.84 | 18.52 | 0.37 | 4.25 | nr | 22.77 |
| 125mm dia. | 37.04 | 40.74 | 0.46 | 5.28 | nr | 46.02 |
| 180mm dia. | 70.85 | 77.94 | 0.60 | 6.88 | nr | 84.82 |
| Straight connector | | | | | | |
| 63mm dia. | 6.50 | 7.15 | 0.58 | 6.66 | nr | 13.81 |
| 90mm dia. | 9.57 | 10.53 | 0.67 | 7.69 | nr | 18.22 |
| 125mm dia. | 17.44 | 19.18 | 0.83 | 9.52 | nr | 28.70 |
| 180mm dia. | 31.20 | 34.32 | 1.25 | 14.34 | nr | 48.66 |
| 250mm dia. | 65.04 | 71.54 | 1.50 | 17.21 | nr | 88.75 |
| Reducing connector | | | | | | |
| 90 x 63mm dia. | 13.12 | 14.43 | 0.67 | 7.69 | nr | 22.12 |
| 125 x 90mm dia. | 26.55 | 29.20 | 0.83 | 9.52 | nr | 38.72 |
| 180 x 125mm dia. | 48.95 | 53.84 | 1.25 | 14.34 | nr | 68.18 |
| Bend; 45 degree | | | | | | |
| 63mm dia. | 14.34 | 15.77 | 0.58 | 6.66 | nr | 22.43 |
| 90mm dia. | 31.27 | 34.40 | 0.67 | 7.69 | nr | 42.09 |
| 125mm dia. | 38.53 | 42.38 | 0.83 | 9.52 | nr | 51.90 |
| 180mm dia. | 81.95 | 90.14 | 1.25 | 14.34 | nr | 104.48 |
| Bend; 90 degree | | | | | | |
| 63mm dia. | 14.11 | 15.52 | 0.58 | 6.66 | nr | 22.18 |
| 90mm dia. | 20.56 | 22.62 | 0.67 | 7.69 | nr | 30.31 |
| 125mm dia. | 43.55 | 47.91 | 0.83 | 9.52 | nr | 57.43 |
| 180mm dia. | 94.78 | 104.26 | 1.25 | 14.34 | nr | 118.60 |

## Y:MECHANICAL AND ELECTRICAL SERVICES

| Item | Net Price £ | Material £ | Labour hours | Labour £ | Unit | Total Rate £ |
|---|---|---|---|---|---|---|
| **Y10 : PIPELINES : PLASTICS - WATER & GAS (contd)** | | | | | | |
| **Extra over blue MDPE water pipe; MDPE fittings, electrofusion joints (contd)** | | | | | | |
| Equal tee | | | | | | |
| 63mm dia. | 16.16 | 17.78 | 0.75 | 8.61 | nr | 26.39 |
| 90mm dia. | 28.70 | 31.57 | 0.87 | 9.98 | nr | 41.55 |
| 125mm dia. | 53.60 | 58.96 | 1.08 | 12.39 | nr | 71.35 |
| 180mm dia. | 88.80 | 97.68 | 1.63 | 18.70 | nr | 116.38 |
| Unequal tee | | | | | | |
| 63 x 32mm dia. | 15.68 | 17.25 | 0.75 | 8.61 | nr | 25.86 |
| 90 x 32mm dia. | 15.68 | 17.25 | 0.87 | 9.98 | nr | 27.23 |
| 90 x 63mm dia. | 37.60 | 41.36 | 0.87 | 9.98 | nr | 51.34 |
| 125 x 63mm dia. | 37.60 | 41.36 | 1.08 | 12.39 | nr | 53.75 |
| 180 x 63mm dia. | 37.60 | 41.36 | 1.63 | 18.70 | nr | 60.06 |
| **Extra over blue MDPE water pipe; plastic fittings, compression joints** | | | | | | |
| Straight connector | | | | | | |
| 20mm dia. | 2.31 | 2.54 | 0.38 | 4.36 | nr | 6.90 |
| 25mm dia. | 2.58 | 2.84 | 0.45 | 5.16 | nr | 8.00 |
| 32mm dia. | 5.23 | 5.75 | 0.50 | 5.74 | nr | 11.49 |
| 50mm dia. | 13.31 | 14.64 | 0.68 | 7.81 | nr | 22.45 |
| 63mm dia. | 20.03 | 22.03 | 0.85 | 9.76 | nr | 31.79 |
| Reducing connector | | | | | | |
| 25mm dia. | 4.64 | 5.10 | 0.38 | 4.36 | nr | 9.46 |
| 32mm dia. | 7.50 | 8.25 | 0.45 | 5.16 | nr | 13.41 |
| 50mm dia. | 20.78 | 22.86 | 0.50 | 5.74 | nr | 28.60 |
| 63mm dia. | 28.99 | 31.89 | 0.62 | 7.12 | nr | 39.01 |
| Straight connector; polyethylene to MI | | | | | | |
| 20mm dia. | 2.10 | 2.31 | 0.31 | 3.56 | nr | 5.87 |
| 25mm dia. | 2.66 | 2.93 | 0.35 | 4.02 | nr | 6.95 |
| 32mm dia. | 3.76 | 4.14 | 0.40 | 4.59 | nr | 8.73 |
| 50mm dia. | 9.52 | 10.47 | 0.55 | 6.31 | nr | 16.78 |
| 63mm dia. | 13.42 | 14.76 | 0.65 | 7.46 | nr | 22.22 |
| Straight connector; polyethylene to FI | | | | | | |
| 20mm dia. | 2.83 | 3.11 | 0.31 | 3.56 | nr | 6.67 |
| 25mm dia. | 3.06 | 3.37 | 0.35 | 4.02 | nr | 7.39 |
| 32mm dia. | 3.55 | 3.90 | 0.40 | 4.59 | nr | 8.49 |
| 50mm dia. | 11.31 | 12.44 | 0.55 | 6.31 | nr | 18.75 |
| 63mm dia. | 15.87 | 17.46 | 0.75 | 8.61 | nr | 26.07 |

## Y:MECHANICAL AND ELECTRICAL SERVICES

| Item | Net Price £ | Material £ | Labour hours | Labour £ | Unit | Total Rate £ |
|---|---|---|---|---|---|---|
| Elbow | | | | | | |
| 20mm dia. | 3.10 | 3.41 | 0.38 | 4.36 | nr | 7.77 |
| 25mm dia. | 4.57 | 5.03 | 0.45 | 5.16 | nr | 10.19 |
| 32mm dia. | 6.32 | 6.95 | 0.50 | 5.74 | nr | 12.69 |
| 50mm dia. | 14.88 | 16.37 | 0.68 | 7.81 | nr | 24.18 |
| 63mm dia. | 20.21 | 22.23 | 0.80 | 9.18 | nr | 31.41 |
| Elbow; polyethylene to MI | | | | | | |
| 25mm dia. | 3.83 | 4.21 | 0.35 | 4.02 | nr | 8.23 |
| Elbow; polyethylene to FI | | | | | | |
| 20mm dia. | 2.81 | 3.09 | 0.31 | 3.56 | nr | 6.65 |
| 25mm dia. | 3.83 | 4.21 | 0.35 | 4.02 | nr | 8.23 |
| 32mm dia. | 5.55 | 6.11 | 0.42 | 4.82 | nr | 10.93 |
| 50mm dia. | 13.09 | 14.40 | 0.50 | 5.74 | nr | 20.14 |
| 63mm dia. | 17.17 | 18.89 | 0.55 | 6.31 | nr | 25.20 |
| Tank coupling | | | | | | |
| 25mm dia. | 4.92 | 5.41 | 0.42 | 4.82 | nr | 10.23 |
| Equal tee | | | | | | |
| 20mm dia. | 4.17 | 4.59 | 0.53 | 6.08 | nr | 10.67 |
| 25mm dia. | 6.52 | 7.17 | 0.55 | 6.31 | nr | 13.48 |
| 32mm dia. | 7.73 | 8.50 | 0.64 | 7.35 | nr | 15.85 |
| 50mm dia. | 18.30 | 20.13 | 0.75 | 8.61 | nr | 28.74 |
| 63mm dia. | 28.34 | 31.17 | 0.87 | 9.99 | nr | 41.16 |
| Equal tee; FI branch | | | | | | |
| 20mm dia. | 3.92 | 4.31 | 0.45 | 5.16 | nr | 9.47 |
| 25mm dia. | 6.25 | 6.88 | 0.50 | 5.74 | nr | 12.62 |
| 32mm dia. | 7.57 | 8.33 | 0.60 | 6.89 | nr | 15.22 |
| 50mm dia. | 17.32 | 19.05 | 0.68 | 7.81 | nr | 26.86 |
| 63mm dia. | 20.09 | 22.10 | 0.81 | 9.30 | nr | 31.40 |
| Equal tee; MI branch | | | | | | |
| 25mm dia. | 6.13 | 6.74 | 0.50 | 5.74 | nr | 12.48 |
| MDPE (Medium Density Polyethylene) pipes for gas distribution; laid underground; electrofusion joints in the running length; BS 6572 | | | | | | |
| Yellow MDPE coiled service pipe | | | | | | |
| 20mm dia. | 0.35 | 0.40 | 0.37 | 4.25 | m | 4.65 |
| 25mm dia. | 0.45 | 0.51 | 0.41 | 4.70 | m | 5.21 |
| 32mm dia. | 0.75 | 0.85 | 0.47 | 5.39 | m | 6.24 |
| 63mm dia. | 2.74 | 3.10 | 0.60 | 6.88 | m | 9.98 |
| 90mm dia. | 10.08 | 11.42 | 0.90 | 10.33 | m | 21.75 |

## Y:MECHANICAL AND ELECTRICAL SERVICES

| Item | Net Price £ | Material £ | Labour hours | Labour £ | Unit | Total Rate £ |
|---|---|---|---|---|---|---|
| Y10 : PIPELINES : PLASTICS - WATER & GAS (contd) | | | | | | |
| MDPE (Medium Density Polyethylene) pipes for gas distribution; laid underground; electrofusion joints in the running length; BS 6572 (contd) | | | | | | |
| Yellow MDPE mains service pipe | | | | | | |
| 63mm dia. | 3.20 | 3.63 | 0.60 | 6.88 | m | 10.51 |
| 90mm dia. | 10.06 | 11.40 | 0.90 | 10.33 | m | 21.73 |
| 125mm dia. | 12.25 | 13.88 | 1.20 | 13.77 | m | 27.65 |
| 180mm dia. | 18.59 | 21.06 | 1.50 | 17.21 | m | 38.27 |
| 250mm dia. | 35.89 | 40.66 | 1.75 | 20.08 | m | 60.74 |
| Extra over yellow MDPE gas pipe; MDPE fittings, electrofusion joints | | | | | | |
| Cap | | | | | | |
| 63mm dia. | 9.41 | 10.35 | 0.32 | 3.67 | nr | 14.02 |
| 90mm dia. | 16.67 | 18.34 | 0.37 | 4.25 | nr | 22.59 |
| 125mm dia. | 36.67 | 40.34 | 0.46 | 5.28 | nr | 45.62 |
| 180mm dia. | 70.14 | 77.15 | 0.60 | 6.88 | nr | 84.03 |
| Straight connector | | | | | | |
| 32mm dia. | 3.41 | 3.75 | 0.47 | 5.39 | nr | 9.14 |
| 63mm dia. | 6.43 | 7.07 | 0.58 | 6.66 | nr | 13.73 |
| 90mm dia. | 9.47 | 10.42 | 0.67 | 7.69 | nr | 18.11 |
| 125mm dia. | 17.26 | 18.99 | 0.83 | 9.52 | nr | 28.51 |
| 180mm dia. | 30.90 | 33.99 | 1.25 | 14.34 | nr | 48.33 |
| 250mm dia. | 64.40 | 70.84 | 1.50 | 17.21 | nr | 88.05 |
| Reducing connector | | | | | | |
| 90 x 63mm dia. | 12.99 | 14.29 | 0.67 | 7.69 | nr | 21.98 |
| 125 x 90mm dia. | 26.29 | 28.92 | 0.83 | 9.52 | nr | 38.44 |
| 180 x 125mm dia. | 48.46 | 53.31 | 1.25 | 14.34 | nr | 67.65 |
| Bend; 45 degree | | | | | | |
| 63mm dia. | 14.20 | 15.62 | 0.58 | 6.66 | nr | 22.28 |
| 90mm dia. | 21.18 | 23.30 | 0.67 | 7.69 | nr | 30.99 |
| 125mm dia. | 35.57 | 39.13 | 0.83 | 9.52 | nr | 48.65 |
| 180mm dia. | 75.52 | 83.07 | 1.25 | 14.34 | nr | 97.41 |
| Bend; 90 degree | | | | | | |
| 63mm dia. | 12.74 | 14.01 | 0.58 | 6.66 | nr | 20.67 |
| 90mm dia. | 18.93 | 20.82 | 0.67 | 7.69 | nr | 28.51 |
| 125mm dia. | 39.92 | 43.91 | 0.83 | 9.52 | nr | 53.43 |
| 180mm dia. | 86.42 | 95.06 | 1.25 | 14.34 | nr | 109.40 |
| Equal tee | | | | | | |
| 63mm dia. | 16.00 | 17.60 | 0.75 | 8.61 | nr | 26.21 |
| 90mm dia. | 28.45 | 31.30 | 0.87 | 9.98 | nr | 41.28 |
| 125mm dia. | 53.07 | 58.38 | 1.08 | 12.39 | nr | 70.77 |
| 180mm dia. | 87.92 | 96.71 | 1.63 | 18.70 | nr | 115.41 |

## Y:MECHANICAL AND ELECTRICAL SERVICES

| Item | Net Price £ | Material £ | Labour hours | Labour £ | Unit | Total Rate £ |
|---|---|---|---|---|---|---|
| Unequal tee | | | | | | |
| 63 x 32mm dia. | 13.02 | 14.32 | 0.75 | 8.61 | nr | 22.93 |
| 90 x 32mm dia. | 13.02 | 14.32 | 0.87 | 9.98 | nr | 24.30 |
| 90 x 63mm dia. | 30.90 | 33.99 | 0.87 | 9.98 | nr | 43.97 |
| 125 x 63mm dia. | 30.90 | 33.99 | 1.08 | 12.39 | nr | 46.38 |
| 180 x 63mm dia. | 30.90 | 33.99 | 1.63 | 18.70 | nr | 52.69 |
| **Extra over yellow MDPE gas pipe; malleable iron fittings, compression joints** | | | | | | |
| Straight connector | | | | | | |
| 20mm dia. | 6.14 | 6.75 | 0.38 | 4.36 | nr | 11.11 |
| 25mm dia. | 6.70 | 7.37 | 0.45 | 5.16 | nr | 12.53 |
| 32mm dia. | 7.50 | 8.25 | 0.50 | 5.74 | nr | 13.99 |
| 63mm dia. | 15.07 | 16.58 | 0.85 | 9.76 | nr | 26.34 |
| Straight connector; polyethylene to MI | | | | | | |
| 20mm dia. | 5.21 | 5.73 | 0.31 | 3.56 | nr | 9.29 |
| 25mm dia. | 5.67 | 6.24 | 0.35 | 4.02 | nr | 10.26 |
| 32mm dia. | 6.36 | 7.00 | 0.40 | 4.59 | nr | 11.59 |
| 63mm dia. | 10.65 | 11.71 | 0.65 | 7.46 | nr | 19.17 |
| Straight connector; polyethylene to FI | | | | | | |
| 20mm dia. | 5.02 | 5.52 | 0.31 | 3.56 | nr | 9.08 |
| 25mm dia. | 5.48 | 6.03 | 0.35 | 4.02 | nr | 10.05 |
| 32mm dia. | 6.14 | 6.75 | 0.40 | 4.59 | nr | 11.34 |
| 63mm dia. | 10.27 | 11.30 | 0.75 | 8.61 | nr | 19.91 |
| Elbow | | | | | | |
| 20mm dia. | 7.98 | 8.78 | 0.38 | 4.36 | nr | 13.14 |
| 25mm dia. | 8.70 | 9.57 | 0.45 | 5.16 | nr | 14.73 |
| 32mm dia. | 9.76 | 10.74 | 0.50 | 5.74 | nr | 16.48 |
| 63mm dia. | 19.60 | 21.56 | 0.80 | 9.18 | nr | 30.74 |
| Equal tee | | | | | | |
| 20mm dia. | 9.29 | 10.22 | 0.53 | 6.08 | nr | 16.30 |
| 25mm dia. | 10.80 | 11.88 | 0.55 | 6.31 | nr | 18.19 |
| 32mm dia. | 13.63 | 14.99 | 0.64 | 7.35 | nr | 22.34 |
| **ABS (Acrylonitrile Butadiene Styrene) pipes; solvent welded joints in the running length** | | | | | | |
| Class C (9 bar pressure) | | | | | | |
| 1" dia. | 2.69 | 3.05 | 0.30 | 3.44 | m | 6.49 |
| 1 1/4" dia. | 4.48 | 5.08 | 0.33 | 3.79 | m | 8.87 |
| 1 1/2" dia. | 5.70 | 6.46 | 0.36 | 4.13 | m | 10.59 |
| 2" dia. | 7.72 | 8.75 | 0.39 | 4.48 | m | 13.23 |
| 2 1/2" dia. | 9.46 | 9.46 | 0.39 | 4.48 | m | 13.94 |
| 3" dia. | 15.93 | 18.05 | 0.45 | 5.16 | m | 23.21 |
| 4" dia. | 26.10 | 29.57 | 0.52 | 5.97 | m | 35.54 |
| 6" dia. | 25.90 | 29.34 | 0.75 | 8.61 | m | 37.95 |
| 8" dia. | 46.59 | 52.79 | 0.95 | 10.91 | m | 63.70 |

## Y:MECHANICAL AND ELECTRICAL SERVICES

| Item | Net Price £ | Material £ | Labour hours | Labour £ | Unit | Total Rate £ |
|---|---|---|---|---|---|---|
| **Y10 : PIPELINES : PLASTICS - WATER & GAS (contd)** | | | | | | |
| **ABS (Acrylonitrile Butadiene Styrene) pipes; solvent welded joints in the running length (contd)** | | | | | | |
| Class E (15 bar pressure) | | | | | | |
| 1/2" dia. | 2.02 | 2.29 | 0.24 | 2.75 | m | **5.04** |
| 3/4" dia. | 3.12 | 3.54 | 0.27 | 3.10 | m | **6.64** |
| 1" dia. | 4.16 | 4.71 | 0.30 | 3.44 | m | **8.15** |
| 1 1/4" dia. | 6.19 | 7.01 | 0.33 | 3.79 | m | **10.80** |
| 1 1/2" dia. | 8.15 | 9.23 | 0.36 | 4.13 | m | **13.36** |
| 2" dia. | 10.23 | 11.59 | 0.39 | 4.48 | m | **16.07** |
| 3" dia. | 20.59 | 23.33 | 0.49 | 5.63 | m | **28.96** |
| 4" dia. | 33.09 | 37.49 | 0.57 | 6.54 | m | **44.03** |
| **Extra Over ABS pipe; ABS fittings; solvent welded joints** | | | | | | |
| Cap | | | | | | |
| 1/2" dia. | 0.57 | 0.63 | 0.17 | 1.95 | nr | **2.58** |
| 3/4" dia. | 0.66 | 0.73 | 0.20 | 2.29 | nr | **3.02** |
| 1" dia. | 0.75 | 0.82 | 0.23 | 2.64 | nr | **3.46** |
| 1 1/4" dia. | 1.25 | 1.38 | 0.26 | 2.98 | nr | **4.36** |
| 1 1/2" dia. | 1.98 | 2.18 | 0.29 | 3.33 | nr | **5.51** |
| 2" dia. | 2.48 | 2.73 | 0.32 | 3.67 | nr | **6.40** |
| 3" dia. | 7.44 | 8.18 | 0.37 | 4.25 | nr | **12.43** |
| 4" dia. | 11.40 | 12.54 | 0.46 | 5.28 | nr | **17.82** |
| Elbow 90 degree | | | | | | |
| 1/2" dia. | 0.82 | 0.90 | 0.30 | 3.44 | nr | **4.34** |
| 3/4" dia. | 0.97 | 1.07 | 0.35 | 4.02 | nr | **5.09** |
| 1" dia. | 1.33 | 1.46 | 0.41 | 4.70 | nr | **6.16** |
| 1 1/4" dia. | 2.26 | 2.49 | 0.47 | 5.39 | nr | **7.88** |
| 1 1/2" dia. | 2.94 | 3.23 | 0.53 | 6.08 | nr | **9.31** |
| 2" dia. | 4.40 | 4.84 | 0.58 | 6.66 | nr | **11.50** |
| 3" dia. | 12.80 | 14.08 | 0.67 | 7.69 | nr | **21.77** |
| 4" dia. | 19.12 | 21.03 | 0.83 | 9.53 | nr | **30.56** |
| 6" dia. | 76.80 | 84.48 | 1.25 | 14.34 | nr | **98.82** |
| 8" dia. | 123.20 | 135.52 | 1.50 | 17.23 | nr | **152.75** |
| Elbow 45 degree | | | | | | |
| 1/2" dia. | 1.51 | 1.66 | 0.30 | 3.44 | nr | **5.10** |
| 3/4" dia. | 1.56 | 1.72 | 0.35 | 4.02 | nr | **5.74** |
| 1" dia. | 1.98 | 2.18 | 0.41 | 4.70 | nr | **6.88** |
| 1 1/4" dia. | 2.85 | 3.13 | 0.47 | 5.39 | nr | **8.52** |
| 1 1/2" dia. | 3.54 | 3.89 | 0.53 | 6.08 | nr | **9.97** |
| 2" dia. | 4.92 | 5.41 | 0.58 | 6.66 | nr | **12.07** |
| 3" dia. | 11.60 | 12.76 | 0.67 | 7.69 | nr | **20.45** |
| 4" dia. | 24.00 | 26.40 | 0.83 | 9.53 | nr | **35.93** |
| 6" dia. | 49.60 | 54.56 | 1.25 | 14.34 | nr | **68.90** |
| 8" dia. | 112.16 | 123.38 | 1.50 | 17.23 | nr | **140.61** |

## Y:MECHANICAL AND ELECTRICAL SERVICES

| Item | Net Price £ | Material £ | Labour hours | Labour £ | Unit | Total Rate £ |
|---|---|---|---|---|---|---|
| Reducing bush | | | | | | |
| 3/4" x 1/2" dia. | 0.60 | 0.66 | 0.44 | 5.05 | nr | 5.71 |
| 1" x 1/2" dia. | 0.78 | 0.86 | 0.47 | 5.39 | nr | 6.25 |
| 1" x 3/4" dia. | 0.78 | 0.86 | 0.47 | 5.39 | nr | 6.25 |
| 1 1/4" x 1" dia. | 1.06 | 1.17 | 0.50 | 5.74 | nr | 6.91 |
| 1 1/2" x 3/4" dia. | 1.32 | 1.45 | 0.53 | 6.08 | nr | 7.53 |
| 1 1/2" x 1" dia. | 1.33 | 1.46 | 0.53 | 6.08 | nr | 7.54 |
| 1 1/2" x 1 1/4" dia. | 1.33 | 1.46 | 0.53 | 6.08 | nr | 7.54 |
| 2" x 1" dia. | 1.74 | 1.91 | 0.58 | 6.66 | nr | 8.57 |
| 2" x 1 1/4" dia. | 1.76 | 1.94 | 0.58 | 6.66 | nr | 8.60 |
| 2" x 1 1/2" dia. | 1.74 | 1.91 | 0.58 | 6.66 | nr | 8.57 |
| 3" x 1 1/2" dia. | 4.92 | 5.41 | 0.67 | 7.69 | nr | 13.10 |
| 3" x 2" dia. | 4.92 | 5.41 | 0.67 | 7.69 | nr | 13.10 |
| 4" x 3" dia. | 6.80 | 7.48 | 0.83 | 9.53 | nr | 17.01 |
| 6" x 4" dia. | 17.60 | 19.36 | 1.25 | 14.34 | nr | 33.70 |
| Union | | | | | | |
| 1/2" dia. | 3.18 | 3.50 | 0.35 | 4.02 | nr | 7.52 |
| 3/4" dia. | 3.46 | 3.81 | 0.40 | 4.59 | nr | 8.40 |
| 1" dia. | 4.60 | 5.06 | 0.45 | 5.16 | nr | 10.22 |
| 1 1/4" dia. | 5.64 | 6.20 | 0.52 | 5.97 | nr | 12.17 |
| 1 1/2" dia. | 7.76 | 8.54 | 0.59 | 6.77 | nr | 15.31 |
| 2" dia. | 10.20 | 11.22 | 0.64 | 7.35 | nr | 18.57 |
| Sockets | | | | | | |
| 1/2" dia. | 0.60 | 0.66 | 0.35 | 4.02 | nr | 4.68 |
| 3/4" dia. | 0.66 | 0.73 | 0.40 | 4.59 | nr | 5.32 |
| 1" dia. | 0.78 | 0.86 | 0.45 | 5.16 | nr | 6.02 |
| 1 1/4" dia. | 1.33 | 1.46 | 0.52 | 5.97 | nr | 7.43 |
| 1 1/2" dia. | 1.62 | 1.78 | 0.59 | 6.77 | nr | 8.55 |
| 2" dia. | 2.26 | 2.49 | 0.64 | 7.35 | nr | 9.84 |
| 3" dia. | 9.00 | 9.90 | 0.73 | 8.38 | nr | 18.28 |
| 4" dia. | 12.88 | 14.17 | 0.73 | 8.38 | nr | 22.55 |
| 6" dia. | 32.20 | 35.42 | 1.30 | 14.92 | nr | 50.34 |
| 8" dia. | 67.28 | 74.01 | 1.60 | 18.36 | nr | 92.37 |
| Barrel nipple | | | | | | |
| 1/2" dia. | 1.20 | 1.32 | 0.35 | 4.02 | nr | 5.34 |
| 3/4" dia. | 1.40 | 1.54 | 0.40 | 4.59 | nr | 6.13 |
| 1" dia. | 1.84 | 2.02 | 0.45 | 5.16 | nr | 7.18 |
| 1 1/4" dia. | 2.40 | 2.64 | 0.52 | 5.97 | nr | 8.61 |
| 1 1/2" dia. | 2.60 | 2.86 | 0.59 | 6.77 | nr | 9.63 |
| 2" dia. | 3.32 | 3.65 | 0.64 | 7.35 | nr | 11.00 |
| 3" dia. | 9.44 | 10.38 | 0.73 | 8.38 | nr | 18.76 |

## Y:MECHANICAL AND ELECTRICAL SERVICES

| Item | Net Price £ | Material £ | Labour hours | Labour £ | Unit | Total Rate £ |
|---|---|---|---|---|---|---|
| **Y10 : PIPELINES : PLASTICS - WATER & GAS (contd)** | | | | | | |
| **Extra Over ABS pipe; ABS fittings; solvent welded joints (contd)** | | | | | | |
| Tee 90 degree | | | | | | |
| 1/2" dia. | 0.92 | 1.01 | 0.42 | 4.82 | nr | 5.83 |
| 3/4" dia. | 1.25 | 1.38 | 0.49 | 5.62 | nr | 7.00 |
| 1" dia. | 1.74 | 1.91 | 0.57 | 6.54 | nr | 8.45 |
| 1 1/4" dia. | 2.53 | 2.78 | 0.66 | 7.57 | nr | 10.35 |
| 1 1/2" dia. | 3.68 | 4.05 | 0.74 | 8.49 | nr | 12.54 |
| 2" dia. | 5.64 | 6.20 | 0.81 | 9.29 | nr | 15.49 |
| 3" dia. | 16.48 | 18.13 | 0.94 | 10.79 | nr | 28.92 |
| 4" dia. | 24.20 | 26.62 | 1.16 | 13.31 | nr | 39.93 |
| 6" dia. | 84.80 | 93.28 | 1.75 | 20.08 | nr | 113.36 |
| 8" dia. | 138.00 | 151.80 | 2.10 | 24.10 | nr | 175.90 |
| **Extra Over ABS pipe; ABS flanges; solvent welded joints** | | | | | | |
| Full face flange | | | | | | |
| 1/2" dia. | 3.36 | 3.70 | 0.10 | 1.15 | nr | 4.85 |
| 3/4" dia. | 3.44 | 3.78 | 0.13 | 1.49 | nr | 5.27 |
| 1" dia. | 3.60 | 3.96 | 0.16 | 1.84 | nr | 5.80 |
| 1 1/4" dia. | 4.32 | 4.75 | 0.19 | 2.18 | nr | 6.93 |
| 1 1/2" dia. | 5.04 | 5.54 | 0.22 | 2.52 | nr | 8.06 |
| 2" dia. | 6.60 | 7.26 | 0.30 | 3.44 | nr | 10.70 |
| 3" dia. | 12.64 | 13.90 | 0.38 | 4.36 | nr | 18.26 |
| 4" dia. | 16.56 | 18.22 | 0.42 | 4.82 | nr | 23.04 |
| **PVC-U (Unplasticized Polyvinyl Chloride) pipes; solvent welded joints in the running length** | | | | | | |
| Class C (9 bar pressure) | | | | | | |
| 2" dia. | 6.16 | 7.11 | 0.42 | 4.82 | m | 11.93 |
| 3" dia. | 11.81 | 13.64 | 0.49 | 5.63 | m | 19.27 |
| 4" dia. | 20.95 | 24.20 | 0.52 | 5.97 | m | 30.17 |
| 6" dia. | 45.34 | 52.37 | 1.82 | 20.86 | m | 73.23 |
| Class D (12 bar pressure) | | | | | | |
| 1 1/4" dia. | 3.59 | 4.15 | 0.42 | 4.82 | m | 8.97 |
| 1 1/2" dia. | 4.93 | 5.69 | 0.44 | 5.05 | m | 10.74 |
| 2" dia. | 7.65 | 8.84 | 0.47 | 5.39 | m | 14.23 |
| 3" dia. | 16.38 | 18.92 | 0.50 | 5.74 | m | 24.66 |
| 4" dia. | 27.45 | 31.70 | 0.55 | 6.31 | m | 38.01 |
| 6" dia. | 50.92 | 58.81 | 0.60 | 6.89 | m | 65.70 |

*Material Costs / Measured Work Prices - Mechanical Installations*

## Y:MECHANICAL AND ELECTRICAL SERVICES

| Item | Net Price £ | Material £ | Labour hours | Labour £ | Unit | Total Rate £ |
|---|---|---|---|---|---|---|
| **Class E (15 bar pressure)** | | | | | | |
| 1/2" dia. | 1.75 | 2.02 | 0.39 | 4.48 | m | 6.50 |
| 3/4" dia. | 2.52 | 2.91 | 0.41 | 4.70 | m | 7.61 |
| 1" dia. | 2.92 | 3.37 | 0.42 | 4.82 | m | 8.19 |
| 1 1/4" dia. | 4.30 | 4.97 | 0.42 | 4.82 | m | 9.79 |
| 1 1/2" dia. | 5.58 | 6.44 | 0.44 | 5.05 | m | 11.49 |
| 2" dia. | 8.72 | 10.07 | 0.47 | 5.39 | m | 15.46 |
| 3" dia. | 18.86 | 21.78 | 0.49 | 5.63 | m | 27.41 |
| 4" dia. | 31.00 | 35.80 | 0.52 | 5.97 | m | 41.77 |
| 6" dia. | 67.13 | 77.54 | 0.55 | 6.31 | m | 83.85 |
| **Class 7** | | | | | | |
| 1/2" dia. | 3.10 | 3.58 | 0.33 | 3.79 | m | 7.37 |
| 3/4" dia. | 4.34 | 5.01 | 0.34 | 3.90 | m | 8.91 |
| 1" dia. | 6.64 | 7.67 | 0.41 | 4.70 | m | 12.37 |
| 1 1/4" dia. | 9.12 | 10.53 | 0.41 | 4.70 | m | 15.23 |
| 1 1/2" dia. | 11.29 | 13.04 | 0.42 | 4.82 | m | 17.86 |
| 2" dia. | 17.99 | 20.78 | 0.45 | 5.16 | m | 25.94 |
| **Extra Over PVC-U pipes; PVC-U fittings; solvent welded joints** | | | | | | |
| **End cap** | | | | | | |
| 1/2" dia. | 0.58 | 0.64 | 0.18 | 2.07 | nr | 2.71 |
| 3/4" dia. | 0.68 | 0.75 | 0.20 | 2.29 | nr | 3.04 |
| 1" dia. | 0.77 | 0.85 | 0.23 | 2.64 | nr | 3.49 |
| 1 1/4" dia. | 1.21 | 1.33 | 0.26 | 2.98 | nr | 4.31 |
| 1 1/2" dia. | 2.02 | 2.22 | 0.29 | 3.33 | nr | 5.55 |
| 2" dia. | 2.45 | 2.69 | 0.32 | 3.67 | nr | 6.36 |
| 3" dia. | 7.53 | 8.28 | 0.37 | 4.25 | nr | 12.53 |
| 4" dia. | 11.62 | 12.78 | 0.46 | 5.28 | nr | 18.06 |
| 6" dia. | 28.16 | 30.98 | 0.69 | 7.92 | nr | 38.90 |
| **Socket** | | | | | | |
| 1/2" dia. | 0.61 | 0.67 | 0.32 | 3.67 | nr | 4.34 |
| 3/4" dia. | 0.68 | 0.75 | 0.36 | 4.13 | nr | 4.88 |
| 1" dia. | 0.79 | 0.87 | 0.43 | 4.94 | nr | 5.81 |
| 1 1/4" dia. | 1.42 | 1.56 | 0.47 | 5.39 | nr | 6.95 |
| 1 1/2" dia. | 1.68 | 1.85 | 0.53 | 6.08 | nr | 7.93 |
| 2" dia. | 2.37 | 2.61 | 0.58 | 6.66 | nr | 9.27 |
| 3" dia. | 9.13 | 10.04 | 0.67 | 7.69 | nr | 17.73 |
| 4" dia. | 13.17 | 14.49 | 0.83 | 9.53 | nr | 24.02 |
| 6" dia. | 33.15 | 36.47 | 1.25 | 14.34 | nr | 50.81 |
| **Reducing socket** | | | | | | |
| 3/4 x 1/2" dia. | 0.73 | 0.80 | 0.32 | 3.67 | nr | 4.47 |
| 1 x 3/4" dia. | 0.90 | 0.99 | 0.36 | 4.13 | nr | 5.12 |
| 1 1/4 x 1" dia. | 1.72 | 1.89 | 0.43 | 4.94 | nr | 6.83 |
| 1 1/2 x 1 1/4" dia. | 1.94 | 2.13 | 0.47 | 5.39 | nr | 7.52 |
| 2 x 1 1/2" dia. | 2.93 | 3.22 | 0.53 | 6.08 | nr | 9.30 |
| 3 x 2" dia. | 8.87 | 9.76 | 0.58 | 6.66 | nr | 16.42 |
| 4 x 3" dia. | 13.09 | 14.40 | 0.67 | 7.69 | nr | 22.09 |
| 6 x 4" dia. | 47.79 | 52.57 | 0.83 | 9.53 | nr | 62.10 |
| 8 x 6" dia. | 77.58 | 85.34 | 1.25 | 14.34 | nr | 99.68 |

## Y:MECHANICAL AND ELECTRICAL SERVICES

| Item | Net Price £ | Material £ | Labour hours | Labour £ | Unit | Total Rate £ |
|---|---|---|---|---|---|---|
| **Y10 : PIPELINES : PLASTICS - WATER & GAS (contd)** | | | | | | |
| **Extra Over PVC-U pipes; PVC-U fittings; solvent welded joints (contd)** | | | | | | |
| Elbow 90 degree | | | | | | |
| 1/2" dia. | 0.84 | 0.92 | 0.32 | 3.67 | nr | **4.59** |
| 3/4" dia. | 0.99 | 1.09 | 0.36 | 4.13 | nr | **5.22** |
| 1" dia. | 1.33 | 1.46 | 0.43 | 4.94 | nr | **6.40** |
| 1 1/4" dia. | 2.37 | 2.61 | 0.47 | 5.39 | nr | **8.00** |
| 1 1/2" dia. | 3.06 | 3.37 | 0.47 | 5.39 | nr | **8.76** |
| 2" dia. | 4.56 | 5.02 | 0.58 | 6.66 | nr | **11.68** |
| 3" dia. | 13.09 | 14.40 | 0.67 | 7.69 | nr | **22.09** |
| 4" dia. | 19.80 | 21.78 | 0.83 | 9.53 | nr | **31.31** |
| 6" dia. | 78.27 | 86.10 | 1.25 | 14.34 | nr | **100.44** |
| Elbow 45 degree | | | | | | |
| 1/2" dia. | 1.55 | 1.71 | 0.32 | 3.67 | nr | **5.38** |
| 3/4" dia. | 1.64 | 1.80 | 0.36 | 4.13 | nr | **5.93** |
| 1" dia. | 2.02 | 2.22 | 0.47 | 5.39 | nr | **7.61** |
| 1 1/4" dia. | 2.88 | 3.17 | 0.47 | 5.39 | nr | **8.56** |
| 1 1/2" dia. | 3.62 | 3.98 | 0.53 | 6.08 | nr | **10.06** |
| 2" dia. | 5.08 | 5.59 | 0.58 | 6.66 | nr | **12.25** |
| 3" dia. | 11.97 | 13.17 | 0.67 | 7.69 | nr | **20.86** |
| 4" dia. | 24.63 | 27.09 | 0.83 | 9.53 | nr | **36.62** |
| 6" dia. | 50.80 | 55.88 | 1.25 | 14.34 | nr | **70.22** |
| Bend 90 degree (long radius) | | | | | | |
| 3" dia. | 36.59 | 40.25 | 0.67 | 7.69 | nr | **47.94** |
| 4" dia. | 73.88 | 81.27 | 0.83 | 9.53 | nr | **90.80** |
| 6" dia. | 162.31 | 178.54 | 1.25 | 14.34 | nr | **192.88** |
| Bend 45 degree (long radius) | | | | | | |
| 1" dia. | 5.08 | 5.59 | 0.47 | 5.39 | nr | **10.98** |
| 1 1/2" dia. | 8.70 | 9.57 | 0.53 | 6.08 | nr | **15.65** |
| 2" dia. | 14.21 | 15.63 | 0.58 | 6.66 | nr | **22.29** |
| 3" dia. | 30.31 | 33.34 | 0.67 | 7.69 | nr | **41.03** |
| 4" dia. | 59.07 | 64.98 | 0.83 | 9.53 | nr | **74.51** |
| 6" dia. | 136.47 | 150.12 | 1.25 | 14.34 | nr | **164.46** |
| Socket union | | | | | | |
| 1/2" dia. | 3.14 | 3.45 | 0.35 | 4.02 | nr | **7.47** |
| 3/4" dia. | 3.62 | 3.98 | 0.40 | 4.59 | nr | **8.57** |
| 1" dia. | 4.69 | 5.16 | 0.47 | 5.39 | nr | **10.55** |
| 1 1/4" dia. | 5.81 | 6.39 | 0.52 | 5.97 | nr | **12.36** |
| 1 1/2" dia. | 8.01 | 8.81 | 0.59 | 6.77 | nr | **15.58** |
| 2" dia. | 10.33 | 11.36 | 0.64 | 7.35 | nr | **18.71** |
| 3" dia. | 38.49 | 42.34 | 0.73 | 8.38 | nr | **50.72** |
| 4" dia. | 52.09 | 57.30 | 0.92 | 10.56 | nr | **67.86** |

## Y:MECHANICAL AND ELECTRICAL SERVICES

| Item | Net Price £ | Material £ | Labour hours | Labour £ | Unit | Total Rate £ |
|---|---|---|---|---|---|---|
| Saddle plain | | | | | | |
| 2" x 1 1/4" dia. | 8.37 | 9.21 | 0.44 | 5.05 | nr | 14.26 |
| 3" x 1 1/2" dia. | 11.69 | 12.86 | 0.50 | 5.74 | nr | 18.60 |
| 4" x 2" dia. | 13.28 | 14.61 | 0.70 | 8.03 | nr | 22.64 |
| 6" x 2" dia. | 15.50 | 17.05 | 0.94 | 10.79 | nr | 27.84 |
| Straight tank connector | | | | | | |
| 1/2" dia. | 2.17 | 2.39 | 0.13 | 1.49 | nr | 3.88 |
| 3/4" dia. | 2.44 | 2.68 | 0.14 | 1.61 | nr | 4.29 |
| 1" dia. | 5.18 | 5.70 | 0.15 | 1.72 | nr | 7.42 |
| 1 1/4" dia. | 13.20 | 14.52 | 0.17 | 1.95 | nr | 16.47 |
| 1 1/2" dia. | 14.44 | 15.88 | 0.19 | 2.18 | nr | 18.06 |
| 2" dia. | 17.27 | 19.00 | 0.25 | 2.87 | nr | 21.87 |
| 3" dia. | 17.71 | 19.48 | 0.30 | 3.44 | nr | 22.92 |
| Equal tee | | | | | | |
| 1/2" dia. | 0.95 | 1.04 | 0.46 | 5.28 | nr | 6.32 |
| 3/4" dia. | 1.21 | 1.33 | 0.50 | 5.74 | nr | 7.07 |
| 1" dia. | 1.81 | 1.99 | 0.56 | 6.43 | nr | 8.42 |
| 1 1/4" dia. | 2.54 | 2.79 | 0.73 | 8.38 | nr | 11.17 |
| 1 1/2" dia. | 3.70 | 4.07 | 0.77 | 8.84 | nr | 12.91 |
| 2" dia. | 5.81 | 6.39 | 0.83 | 9.53 | nr | 15.92 |
| 3" dia. | 16.88 | 18.57 | 1.08 | 12.41 | nr | 30.98 |
| 4" dia. | 24.80 | 27.28 | 1.33 | 15.30 | nr | 42.58 |
| 6" dia. | 86.28 | 94.91 | 2.00 | 22.95 | nr | 117.86 |
| **PVC-C (Chlorinated Polyvinyl Chloride) pipes; solvent welded in the running length** | | | | | | |
| Pipe; 3m long; PN25 | | | | | | |
| 16 x 2,0mm | 1.73 | 1.90 | 0.20 | 2.29 | m | 4.19 |
| 20 x 2,3mm | 2.84 | 3.12 | 0.20 | 2.29 | m | 5.41 |
| 25 x 2,8mm | 3.80 | 4.18 | 0.20 | 2.29 | m | 6.47 |
| 32 x 3,6mm | 5.52 | 6.07 | 0.20 | 2.29 | m | 8.36 |
| Pipe; 5m long; PN25 | | | | | | |
| 40 x 4,5mm | 8.86 | 9.75 | 0.20 | 2.29 | m | 12.04 |
| 50 x 5,6mm | 13.43 | 14.77 | 0.20 | 2.29 | m | 17.06 |
| 64 x 7,0mm | 20.56 | 22.62 | 0.20 | 2.29 | m | 24.91 |
| **Extra Over PVC-C pipe; PVC-C fittings; solvent welded joints** | | | | | | |
| Straight coupling; PN25 | | | | | | |
| 16mm | 0.41 | 0.45 | 0.20 | 2.29 | nr | 2.74 |
| 20mm | 0.56 | 0.62 | 0.20 | 2.29 | nr | 2.91 |
| 25mm | 0.71 | 0.78 | 0.20 | 2.29 | nr | 3.07 |
| 32mm | 2.23 | 2.45 | 0.20 | 2.29 | nr | 4.74 |
| 40mm | 2.84 | 3.12 | 0.20 | 2.29 | nr | 5.41 |
| 50mm | 3.80 | 4.18 | 0.20 | 2.29 | nr | 6.47 |
| 63mm | 6.69 | 7.36 | 0.20 | 2.29 | nr | 9.65 |

## Y:MECHANICAL AND ELECTRICAL SERVICES

| Item | Net Price £ | Material £ | Labour hours | Labour £ | Unit | Total Rate £ |
|---|---|---|---|---|---|---|
| **Y10 : PIPELINES : PLASTICS - WATER & GAS (contd)** | | | | | | |
| **Extra Over PVC-C pipe; PVC-C fittings; solvent welded joints (contd)** | | | | | | |
| Elbow; 90 degree; PN25 | | | | | | |
| 16mm | 0.66 | 0.73 | 0.20 | 2.29 | nr | 3.02 |
| 20mm | 1.01 | 1.11 | 0.20 | 2.29 | nr | 3.40 |
| 25mm | 1.27 | 1.40 | 0.20 | 2.29 | nr | 3.69 |
| 32mm | 2.63 | 2.89 | 0.20 | 2.29 | nr | 5.18 |
| 40mm | 4.06 | 4.47 | 0.20 | 2.29 | nr | 6.76 |
| 50mm | 5.63 | 6.19 | 0.20 | 2.29 | nr | 8.48 |
| 63mm | 9.62 | 10.58 | 0.20 | 2.29 | nr | 12.87 |
| Elbow; 45 degree; PN25 | | | | | | |
| 16mm | 0.66 | 0.73 | 0.20 | 2.29 | nr | 3.02 |
| 20mm | 1.01 | 1.11 | 0.20 | 2.29 | nr | 3.40 |
| 25mm | 1.27 | 1.40 | 0.20 | 2.29 | nr | 3.69 |
| 32mm | 2.63 | 2.89 | 0.20 | 2.29 | nr | 5.18 |
| 40mm | 4.06 | 4.47 | 0.20 | 2.29 | nr | 6.76 |
| 50mm | 5.63 | 6.19 | 0.20 | 2.29 | nr | 8.48 |
| 63mm | 9.62 | 10.58 | 0.20 | 2.29 | nr | 12.87 |
| Reducer fitting; single stage reduction | | | | | | |
| 20/16mm | 0.71 | 0.78 | 0.20 | 2.29 | nr | 3.07 |
| 25/20mm | 0.86 | 0.95 | 0.20 | 2.29 | nr | 3.24 |
| 32/25mm | 1.73 | 1.90 | 0.20 | 2.29 | nr | 4.19 |
| 40/32mm | 2.28 | 2.51 | 0.20 | 2.29 | nr | 4.80 |
| 50/40mm | 2.63 | 2.89 | 0.20 | 2.29 | nr | 5.18 |
| 63/50mm | 4.00 | 4.40 | 0.20 | 2.29 | nr | 6.69 |
| Equal tee; 90 degree; PN25 | | | | | | |
| 16mm | 1.11 | 1.22 | 0.20 | 2.29 | nr | 3.51 |
| 20mm | 1.52 | 1.67 | 0.20 | 2.29 | nr | 3.96 |
| 25mm | 1.92 | 2.11 | 0.20 | 2.29 | nr | 4.40 |
| 32mm | 3.14 | 3.45 | 0.20 | 2.29 | nr | 5.74 |
| 40mm | 5.42 | 5.96 | 0.20 | 2.29 | nr | 8.25 |
| 50mm | 8.10 | 8.91 | 0.20 | 2.29 | nr | 11.20 |
| 63mm | 13.68 | 15.05 | 0.20 | 2.29 | nr | 17.34 |
| Cap; PN25 | | | | | | |
| 16mm | 0.51 | 0.56 | 0.20 | 2.29 | nr | 2.85 |
| 20mm | 0.76 | 0.84 | 0.20 | 2.29 | nr | 3.13 |
| 25mm | 1.01 | 1.11 | 0.20 | 2.29 | nr | 3.40 |
| 32mm | 1.47 | 1.62 | 0.20 | 2.29 | nr | 3.91 |
| 40mm | 2.03 | 2.23 | 0.20 | 2.29 | nr | 4.52 |
| 50mm | 2.84 | 3.12 | 0.20 | 2.29 | nr | 5.41 |
| 63mm | 4.50 | 4.95 | 0.20 | 2.29 | nr | 7.24 |

## Y:MECHANICAL AND ELECTRICAL SERVICES

| Item | Net Price £ | Material £ | Labour hours | Labour £ | Unit | Total Rate £ |
|---|---|---|---|---|---|---|
| **Y10 : PIPELINES : CAST AND DUCTILE IRON** | | | | | | |
| **Cast iron soil pipe; mechanical joints; nitrile rubber gasket; fixed with holderbats; BS 416** | | | | | | |
| Pipe | | | | | | |
| 50mm dia. | 9.40 | 5.17 | 0.25 | 2.87 | m | 8.04 |
| 75mm dia. | 9.06 | 4.98 | 0.45 | 5.16 | m | 10.14 |
| 100mm dia. | 10.95 | 4.01 | 0.60 | 6.89 | m | 10.90 |
| 150mm dia. | 22.85 | 8.37 | 0.70 | 8.04 | m | 16.41 |
| **Extra Over cast iron soil pipe with mechanical joints; cast iron fittings; BS 416** | | | | | | |
| Standard coupling | | | | | | |
| 50mm dia. | 4.11 | 4.52 | 0.50 | 5.74 | nr | 10.26 |
| 75mm dia. | 4.55 | 5.00 | 0.60 | 6.89 | nr | 11.89 |
| 100mm dia. | 5.92 | 6.51 | 0.67 | 7.69 | nr | 14.20 |
| 150mm dia. | 11.84 | 13.02 | 0.83 | 9.53 | nr | 22.55 |
| Stepped; coupling | | | | | | |
| 75mm dia. | 8.15 | 8.96 | 0.60 | 6.89 | nr | 15.85 |
| 100mm dia. | 9.57 | 10.53 | 0.67 | 7.69 | nr | 18.22 |
| 150mm dia. | 11.65 | 12.81 | 0.83 | 9.53 | nr | 22.34 |
| 45 degree bend | | | | | | |
| 50mm dia. | 7.24 | 7.96 | 0.50 | 5.74 | nr | 13.70 |
| 75mm dia. | 7.24 | 7.96 | 0.60 | 6.89 | nr | 14.85 |
| 100mm dia. | 10.02 | 11.02 | 0.67 | 7.69 | nr | 18.71 |
| 150mm dia. | 17.91 | 19.70 | 0.83 | 9.53 | nr | 29.23 |
| 87.5 degree bend | | | | | | |
| 50mm dia. | 7.24 | 7.96 | 0.50 | 5.74 | nr | 13.70 |
| 75mm dia. | 7.24 | 7.96 | 0.60 | 6.89 | nr | 14.85 |
| 100mm dia. | 10.02 | 11.02 | 0.67 | 7.69 | nr | 18.71 |
| 150mm dia. | 17.91 | 19.70 | 0.83 | 9.53 | nr | 29.23 |
| 87.5 degree bend with heel rest | | | | | | |
| 100mm dia. | 19.66 | 21.63 | 0.67 | 7.70 | nr | 29.33 |
| 150mm dia. | 39.13 | 43.04 | 0.83 | 9.53 | nr | 52.57 |
| 87.5 degree bend with access | | | | | | |
| 100mm dia. | 21.20 | 23.32 | 0.67 | 7.69 | nr | 31.01 |
| 150mm dia. | 30.12 | 33.13 | 0.83 | 9.53 | nr | 42.66 |
| 87.5 degree long radius bend | | | | | | |
| 100mm dia. | 16.23 | 17.85 | 0.67 | 7.69 | nr | 25.54 |
| 150mm dia. | 29.39 | 32.33 | 0.83 | 9.53 | nr | 41.86 |
| 87.5 degree long tail bend | | | | | | |
| 100mm dia. | 12.95 | 14.24 | 0.70 | 8.04 | nr | 22.28 |

## Y:MECHANICAL AND ELECTRICAL SERVICES

| Item | Net Price £ | Material £ | Labour hours | Labour £ | Unit | Total Rate £ |
|---|---|---|---|---|---|---|
| **Y10 : PIPELINES : CAST AND DUCTILE IRON (contd)** | | | | | | |
| **Extra Over cast iron soil pipe with mechanical joints; cast iron fittings; BS 416 (contd)** | | | | | | |
| Access pipe | | | | | | |
| 100mm dia. | 36.10 | 39.71 | 0.67 | 7.70 | nr | 47.41 |
| 150mm dia. | 55.25 | 60.77 | 0.83 | 9.53 | nr | 70.30 |
| Reducer | | | | | | |
| 75mm dia. | 9.51 | 10.46 | 0.60 | 6.89 | nr | 17.35 |
| 100mm dia. | 12.65 | 13.91 | 0.67 | 7.69 | nr | 21.60 |
| 150mm dia. | 24.62 | 27.08 | 0.83 | 9.53 | nr | 36.61 |
| Branch | | | | | | |
| 50mm dia. | 10.90 | 11.99 | 0.78 | 8.95 | nr | 20.94 |
| 75mm dia. | 10.90 | 11.99 | 0.85 | 9.76 | nr | 21.75 |
| 100mm dia. | 15.50 | 17.05 | 1.00 | 11.47 | nr | 28.52 |
| 150mm dia. | 31.90 | 35.09 | 1.20 | 13.78 | nr | 48.87 |
| Access branch | | | | | | |
| 100mm dia. | 26.69 | 29.36 | 1.02 | 11.71 | nr | 41.07 |
| 150mm dia. | 44.12 | 48.53 | 1.20 | 13.78 | nr | 62.31 |
| Double branch | | | | | | |
| 100mm dia. | 19.17 | 21.09 | 1.30 | 14.92 | nr | 36.01 |
| Double access branch | | | | | | |
| 100mm dia. | 30.35 | 33.38 | 1.43 | 16.42 | nr | 49.80 |
| **Extra Over cast iron soil pipe with mechanical joints; cast iron ancillaries; BS 416** | | | | | | |
| Gully trap | | | | | | |
| 75mm dia. | 14.66 | 16.13 | 0.90 | 10.33 | nr | 26.46 |
| 100mm dia. | 22.89 | 25.18 | 1.00 | 11.47 | nr | 36.65 |
| 150mm dia. | 56.97 | 62.67 | 1.20 | 13.78 | nr | 76.45 |
| Gully trap with access | | | | | | |
| 100mm dia. | 45.66 | 50.23 | 1.16 | 13.31 | nr | 63.54 |
| Garage gully | | | | | | |
| 100mm dia. | 324.89 | 357.38 | 1.50 | 17.23 | nr | 374.61 |

## Y:MECHANICAL AND ELECTRICAL SERVICES

| Item | Net Price £ | Material £ | Labour hours | Labour £ | Unit | Total Rate £ |
|---|---|---|---|---|---|---|
| **Y10 : PIPELINES : BLACK STEEL - SCREWED** | | | | | | |
| Black steel pipes; screwed and socketed joints; BS 1387: 1985 | | | | | | |
| Varnished; light | | | | | | |
| 15mm dia. | 2.32 | 2.68 | 0.53 | 6.08 | m | 8.76 |
| 20mm dia. | 3.02 | 3.49 | 0.54 | 6.20 | m | 9.69 |
| 25mm dia. | 4.14 | 4.78 | 0.59 | 6.77 | m | 11.55 |
| 32mm dia. | 5.10 | 5.89 | 0.66 | 7.57 | m | 13.46 |
| 40mm dia. | 6.31 | 7.29 | 0.72 | 8.27 | m | 15.56 |
| 50mm dia. | 8.21 | 9.48 | 0.78 | 8.95 | m | 18.43 |
| 65mm dia. | 12.10 | 13.98 | 0.88 | 10.10 | m | 24.08 |
| 80mm dia. | 14.45 | 16.68 | 0.98 | 11.25 | m | 27.93 |
| 100mm dia. | 20.94 | 24.18 | 1.32 | 15.16 | m | 39.34 |
| Varnished; medium | | | | | | |
| 8mm dia. | 2.48 | 2.86 | 0.52 | 5.97 | m | 8.83 |
| 10mm dia. | 2.47 | 2.86 | 0.52 | 5.97 | m | 8.83 |
| 15mm dia. | 2.75 | 3.18 | 0.53 | 6.08 | m | 9.26 |
| 20mm dia. | 3.24 | 3.74 | 0.54 | 6.20 | m | 9.94 |
| 25mm dia. | 4.65 | 5.37 | 0.59 | 6.77 | m | 12.14 |
| 32mm dia. | 5.75 | 6.65 | 0.66 | 7.57 | m | 14.22 |
| 40mm dia. | 6.69 | 7.73 | 0.72 | 8.27 | m | 16.00 |
| 50mm dia. | 9.41 | 10.87 | 0.78 | 8.95 | m | 19.82 |
| 65mm dia. | 12.78 | 14.76 | 0.88 | 10.10 | m | 24.86 |
| 80mm dia. | 16.60 | 19.17 | 0.98 | 11.25 | m | 30.42 |
| 100mm dia. | 23.52 | 27.17 | 1.32 | 15.16 | m | 42.33 |
| 125mm dia. | 34.70 | 40.08 | 1.50 | 17.21 | m | 57.29 |
| 150mm dia. | 40.30 | 46.55 | 1.70 | 19.52 | m | 66.07 |
| Varnished; heavy | | | | | | |
| 15mm dia. | 3.28 | 3.79 | 0.53 | 6.08 | m | 9.87 |
| 20mm dia. | 3.89 | 4.49 | 0.54 | 6.20 | m | 10.69 |
| 25mm dia. | 5.68 | 6.56 | 0.59 | 6.77 | m | 13.33 |
| 32mm dia. | 7.05 | 8.14 | 0.66 | 7.57 | m | 15.71 |
| 40mm dia. | 8.22 | 9.49 | 0.72 | 8.27 | m | 17.76 |
| 50mm dia. | 11.42 | 13.19 | 0.78 | 8.95 | m | 22.14 |
| 65mm dia. | 15.52 | 17.93 | 0.88 | 10.10 | m | 28.03 |
| 80mm dia. | 19.76 | 22.82 | 0.98 | 11.25 | m | 34.07 |
| 100mm dia. | 27.58 | 31.85 | 1.32 | 15.16 | m | 47.01 |
| 125mm dia. | 37.00 | 42.73 | 1.50 | 17.21 | m | 59.94 |
| 150mm dia. | 43.26 | 49.97 | 1.70 | 19.52 | m | 69.49 |

## Y:MECHANICAL AND ELECTRICAL SERVICES

| Item | Net Price £ | Material £ | Labour hours | Labour £ | Unit | Total Rate £ |
|---|---|---|---|---|---|---|
| **Y10 : PIPELINES : BLACK STEEL - SCREWED (contd)** | | | | | | |
| Extra Over black steel screwed pipes; black steel flanges, screwed and drilled; metric; BS 4504 | | | | | | |
| Screwed flanges; PN6 | | | | | | |
| 15mm dia. | 6.50 | 7.15 | 0.37 | 4.25 | nr | 11.40 |
| 20mm dia. | 6.50 | 7.15 | 0.49 | 5.62 | nr | 12.77 |
| 25mm dia. | 6.88 | 7.57 | 0.56 | 6.43 | nr | 14.00 |
| 32mm dia. | 8.89 | 9.78 | 0.63 | 7.23 | nr | 17.01 |
| 40mm dia. | 8.89 | 9.78 | 0.73 | 8.38 | nr | 18.16 |
| 50mm dia. | 10.73 | 11.80 | 0.88 | 10.10 | nr | 21.90 |
| 65mm dia. | 13.80 | 15.18 | 1.08 | 12.39 | nr | 27.57 |
| 80mm dia. | 16.45 | 18.09 | 1.29 | 14.80 | nr | 32.89 |
| 100mm dia. | 19.58 | 21.54 | 1.42 | 16.29 | nr | 37.83 |
| 125mm dia. | 47.94 | 52.73 | 1.70 | 19.51 | nr | 72.24 |
| 150mm dia. | 47.94 | 52.73 | 2.08 | 23.87 | nr | 76.60 |
| Screwed flanges; PN16 | | | | | | |
| 15mm dia. | 6.20 | 6.82 | 0.37 | 4.25 | nr | 11.07 |
| 20mm dia. | 6.20 | 6.82 | 0.49 | 5.62 | nr | 12.44 |
| 25mm dia. | 6.56 | 7.22 | 0.56 | 6.43 | nr | 13.65 |
| 32mm dia. | 8.82 | 9.70 | 0.63 | 7.23 | nr | 16.93 |
| 40mm dia. | 8.82 | 9.70 | 0.73 | 8.38 | nr | 18.08 |
| 50mm dia. | 10.52 | 11.57 | 0.88 | 10.10 | nr | 21.67 |
| 65mm dia. | 13.02 | 14.32 | 1.08 | 12.39 | nr | 26.71 |
| 80mm dia. | 16.30 | 17.93 | 1.29 | 14.80 | nr | 32.73 |
| 100mm dia. | 19.03 | 20.93 | 1.42 | 16.29 | nr | 37.22 |
| 125mm dia. | 47.60 | 52.36 | 1.70 | 19.51 | nr | 71.87 |
| 150mm dia. | 47.60 | 52.36 | 2.08 | 23.87 | nr | 76.23 |
| Extra Over black steel screwed pipes; black steel flanges, screwed and drilled; imperial; BS 10 | | | | | | |
| Screwed flanges; Table E | | | | | | |
| 1/2" dia. | 7.82 | 8.60 | 0.37 | 4.25 | nr | 12.85 |
| 3/4" dia. | 7.82 | 8.60 | 0.49 | 5.62 | nr | 14.22 |
| 1" dia. | 7.82 | 8.60 | 0.56 | 6.43 | nr | 15.03 |
| 1 1/4" dia. | 7.88 | 8.67 | 0.63 | 7.23 | nr | 15.90 |
| 1 1/2" dia. | 7.88 | 8.67 | 0.73 | 8.38 | nr | 17.05 |
| 2" dia. | 7.98 | 8.78 | 0.88 | 10.10 | nr | 18.88 |
| 2 1/2" dia. | 10.08 | 11.09 | 1.08 | 12.39 | nr | 23.48 |
| 3" dia. | 10.08 | 11.09 | 1.29 | 14.80 | nr | 25.89 |
| 4" dia. | 15.32 | 16.85 | 1.42 | 16.29 | nr | 33.14 |
| 5" dia. | 29.94 | 32.93 | 1.70 | 19.51 | nr | 52.44 |
| 6" dia. | 29.94 | 32.93 | 2.08 | 23.87 | nr | 56.80 |

*Material Costs / Measured Work Prices - Mechanical Installations* 165

## Y:MECHANICAL AND ELECTRICAL SERVICES

| Item | Net Price £ | Material £ | Labour hours | Labour £ | Unit | Total Rate £ |
|---|---|---|---|---|---|---|
| **Extra Over black steel screwed pipes; black steel flange connections** | | | | | | |
| Bolted connection between pair of flanges; including gasket, bolts, nuts and washers | | | | | | |
| 50mm dia. | 3.96 | 4.36 | 0.50 | 5.74 | nr | 10.10 |
| 65mm dia. | 4.20 | 4.62 | 0.50 | 5.74 | nr | 10.36 |
| 80mm dia. | 7.50 | 8.25 | 0.50 | 5.74 | nr | 13.99 |
| 100mm dia. | 7.74 | 8.51 | 0.50 | 5.74 | nr | 14.25 |
| 125mm dia. | 8.40 | 9.24 | 0.50 | 5.74 | nr | 14.98 |
| 150mm dia. | 12.60 | 13.86 | 0.88 | 10.10 | nr | 23.96 |
| **Extra Over black steel screwed pipes; black heavy steel tubular fittings; BS 1387** | | | | | | |
| Long screw connection with socket and backnut | | | | | | |
| 15mm dia. | 2.98 | 3.28 | 0.66 | 7.57 | nr | 10.85 |
| 20mm dia. | 3.55 | 3.90 | 0.88 | 10.10 | nr | 14.00 |
| 25mm dia. | 4.91 | 5.40 | 1.00 | 11.47 | nr | 16.87 |
| 32mm dia. | 6.44 | 7.08 | 1.16 | 13.31 | nr | 20.39 |
| 40mm dia. | 7.87 | 8.66 | 1.34 | 15.38 | nr | 24.04 |
| 50mm dia. | 11.54 | 12.69 | 1.60 | 18.36 | nr | 31.05 |
| 65mm dia. | 24.91 | 27.40 | 1.96 | 22.49 | nr | 49.89 |
| 80mm dia. | 34.24 | 37.66 | 2.32 | 26.62 | nr | 64.28 |
| 100mm dia. | 55.33 | 60.86 | 3.20 | 36.72 | nr | 97.58 |
| Running nipple | | | | | | |
| 15mm dia. | 0.72 | 0.79 | 0.66 | 7.57 | nr | 8.36 |
| 20mm dia. | 0.91 | 1.00 | 0.88 | 10.10 | nr | 11.10 |
| 25mm dia. | 1.11 | 1.22 | 1.00 | 11.47 | nr | 12.69 |
| 32mm dia. | 1.56 | 1.72 | 1.16 | 13.31 | nr | 15.03 |
| 40mm dia. | 2.11 | 2.32 | 1.34 | 15.38 | nr | 17.70 |
| 50mm dia. | 3.20 | 3.52 | 1.60 | 18.36 | nr | 21.88 |
| 65mm dia. | 6.89 | 7.58 | 1.96 | 22.49 | nr | 30.07 |
| 80mm dia. | 10.74 | 11.81 | 2.32 | 26.62 | nr | 38.43 |
| 100mm dia. | 16.82 | 18.50 | 3.20 | 36.72 | nr | 55.22 |
| Barrel nipple | | | | | | |
| 15mm dia. | 0.55 | 0.60 | 0.66 | 7.57 | nr | 8.17 |
| 20mm dia. | 0.71 | 0.78 | 0.88 | 10.10 | nr | 10.88 |
| 25mm dia. | 0.92 | 1.01 | 1.00 | 11.47 | nr | 12.48 |
| 32mm dia. | 1.38 | 1.52 | 1.16 | 13.31 | nr | 14.83 |
| 40mm dia. | 1.70 | 1.87 | 1.34 | 15.38 | nr | 17.25 |
| 50mm dia. | 2.42 | 2.66 | 1.60 | 18.36 | nr | 21.02 |
| 65mm dia. | 5.20 | 5.72 | 1.96 | 22.49 | nr | 28.21 |
| 80mm dia. | 7.25 | 7.97 | 2.32 | 26.62 | nr | 34.59 |
| 100mm dia. | 13.12 | 14.43 | 3.20 | 36.72 | nr | 51.15 |
| 125mm dia. | 24.37 | 26.81 | 3.90 | 44.75 | nr | 71.56 |
| 150mm dia. | 38.38 | 42.22 | 4.60 | 52.78 | nr | 95.00 |

*Material Costs / Measured Work Prices - Mechanical Installations*

## Y:MECHANICAL AND ELECTRICAL SERVICES

| Item | Net Price £ | Material £ | Labour hours | Labour £ | Unit | Total Rate £ |
|---|---|---|---|---|---|---|
| **Y10 : PIPELINES : BLACK STEEL - SCREWED (contd)** | | | | | | |
| **Extra Over black steel screwed pipes; black heavy steel tubular fittings; BS 1387 (contd)** | | | | | | |
| Close taper nipple | | | | | | |
| 15mm dia. | 0.86 | 0.95 | 0.66 | 7.57 | nr | **8.52** |
| 20mm dia. | 1.11 | 1.22 | 0.88 | 10.10 | nr | **11.32** |
| 25mm dia. | 1.46 | 1.61 | 1.00 | 11.47 | nr | **13.08** |
| 32mm dia. | 2.18 | 2.40 | 1.16 | 13.31 | nr | **15.71** |
| 40mm dia. | 2.70 | 2.97 | 1.34 | 15.38 | nr | **18.35** |
| 50mm dia. | 4.15 | 4.57 | 1.60 | 18.36 | nr | **22.93** |
| 65mm dia. | 8.04 | 8.84 | 1.96 | 22.49 | nr | **31.33** |
| 80mm dia. | 10.70 | 11.77 | 2.32 | 26.62 | nr | **38.39** |
| 100mm dia. | 20.36 | 22.40 | 3.20 | 36.72 | nr | **59.12** |
| 125mm dia. | 36.10 | 39.71 | 3.90 | 44.75 | nr | **84.46** |
| 150mm dia. | 50.57 | 55.63 | 4.60 | 52.78 | nr | **108.41** |
| 90 degree bend with socket | | | | | | |
| 15mm dia. | 2.32 | 2.55 | 0.66 | 7.57 | nr | **10.12** |
| 20mm dia. | 3.15 | 3.46 | 0.88 | 10.10 | nr | **13.56** |
| 25mm dia. | 4.84 | 5.32 | 1.00 | 11.47 | nr | **16.79** |
| 32mm dia. | 6.94 | 7.63 | 1.16 | 13.31 | nr | **20.94** |
| 40mm dia. | 8.48 | 9.33 | 1.34 | 15.38 | nr | **24.71** |
| 50mm dia. | 13.18 | 14.50 | 1.60 | 18.36 | nr | **32.86** |
| 65mm dia. | 23.51 | 25.86 | 1.96 | 22.49 | nr | **48.35** |
| 80mm dia. | 36.22 | 39.84 | 2.32 | 26.62 | nr | **66.46** |
| 100mm dia. | 65.00 | 71.50 | 3.20 | 36.72 | nr | **108.22** |
| 125mm dia. | 108.34 | 119.17 | 3.20 | 36.72 | nr | **155.89** |
| **Extra Over black steel screwed pipes; black heavy steel fittings; BS 1740** | | | | | | |
| Plug | | | | | | |
| 15mm dia. | 0.61 | 0.67 | 0.33 | 3.79 | nr | **4.46** |
| 20mm dia. | 0.92 | 1.01 | 0.44 | 5.05 | nr | **6.06** |
| 25mm dia. | 1.66 | 1.83 | 0.50 | 5.74 | nr | **7.57** |
| 32mm dia. | 2.53 | 2.78 | 0.58 | 6.66 | nr | **9.44** |
| 40mm dia. | 2.83 | 3.11 | 0.67 | 7.69 | nr | **10.80** |
| 50mm dia. | 4.01 | 4.41 | 0.80 | 9.18 | nr | **13.59** |
| 65mm dia. | 9.63 | 10.59 | 0.98 | 11.25 | nr | **21.84** |
| 80mm dia. | 17.99 | 19.79 | 1.16 | 13.31 | nr | **33.10** |
| 100mm dia. | 34.55 | 38.01 | 1.80 | 20.66 | nr | **58.67** |
| 150mm dia. | 244.60 | 269.06 | 3.32 | 38.10 | nr | **307.16** |

# THE CONSTRUCTION NET
### Online information sources for the construction industry

by *Alan H Bridges*
*Universities of Strathclyde (Glasgow) and Delft (the Netherlands)*

You are a busy architect/engineer/contractor who has just received an Internet connection and are thinking "so what!" With slow connections it is costing you time and money to search for information - and often the information you find is less than useful. This book provides a guide to the use of tools to improve use of the Net and a comprehensive guide to where the best information sources are to be found for the building design and construction professional.

Learn where to find design images or hard technical information; where to join discussion groups for advice on CAD software; and how to join mailing lists to automatically receive up-to-date information on your chosen special subject areas.

Using this book can help the busy professional to optimize online time by determining the key sites to visit before connection to the Internet. Topics are conveniently arranged by subject showing where to find the key "index sites" together with details of many specialist sites.

- helps the user access information faster

- quick guide to specialized information

- regular online updates

---

**For further information and to order please contact**
The Marketing Dept., E & FN Spon, 2-6 Boundary Row, London SE1 8HN
Tel: 0171 865 0066  Fax: 0171 522 9621

---

Published by E & F N Spon
246 x 189mm 256 pages
October 1996
Paperback 0-419-21780-0
£24.99

# SPON'S PRICE BOOKS 1998

## SPON'S ARCHITECTS' AND BUILDERS' PRICE BOOK 1998

**123rd edition**
*Davis Langdon & Everest*

The only price book geared to market conditions that affect building tender prices. Spon's A & B provides comprehensive coverage of construction costs from small scale alterations and repairs to the largest residential or commercial developments.

*944 pages
October 1997
Hardback
0-419-23060-2
£72.50*

## SPON'S MECHANICAL AND ELECTRICAL SERVICES PRICE BOOK 1998

**29th edition**
*Davis Langdon & Everest*

"An essential reference for everybody concerned with the calculation of costs of mechanical and electrical works" - **Cost Engineer**

Outline costs for a wide variety of engineering services are followed by more detailed elemental and quantified analysis.

*560 pages
October 1997
Hardback
0-419-23080-7
£75.00*

## SPON'S LANDSCAPE AND EXTERNAL WORKS PRICE BOOK 1998
**NEW LAYOUT**

**17th edition**
*Derek Lovejoy Partnership and Davis Langdon & Everest*

Now completely revised and expanded. Every rate has been reworked and recalculated. Now includes for the first time a realistic labour rate as well as plant and material costs.

"Surely there can be no office without this publication" - **Landscape Design**

*288 pages
October 1997
Hardback
0-419-23070-X
£62.50*

## SPON'S CIVIL ENGINEERING AND HIGHWAY WORKS PRICE BOOK 1998

**12th edition**
*Davis Langdon & Everest*

"Unquestionably, this will be required reading by all estimators involved in civil engineering works. Quantity surveyors will also find it essential for their shelves." - **Civil Engineering Surveyor**

*640 pages
October 1997
Hardback
0-419-23050-5
£95.00*

FREE UPDATES every 3 months during books' currency

E & FN SPON an imprint of Thomson Professional

## reSPONse

**SPON's** price data now available in an electronic estimating system
Call **0171-522-9966** for details

## Y:MECHANICAL AND ELECTRICAL SERVICES

| Item | Net Price £ | Material £ | Labour hours | Labour £ | Unit | Total Rate £ |
|---|---|---|---|---|---|---|
| Socket | | | | | | |
| 15mm dia. | 0.60 | 0.66 | 0.66 | 7.57 | nr | 8.23 |
| 20mm dia. | 0.68 | 0.75 | 0.88 | 10.10 | nr | 10.85 |
| 25mm dia. | 0.92 | 1.01 | 1.00 | 11.47 | nr | 12.48 |
| 32mm dia. | 1.40 | 1.54 | 1.16 | 13.31 | nr | 14.85 |
| 40mm dia. | 1.69 | 1.86 | 1.34 | 15.38 | nr | 17.24 |
| 50mm dia. | 2.62 | 2.88 | 1.60 | 18.36 | nr | 21.24 |
| 65mm dia. | 4.90 | 5.39 | 1.96 | 22.49 | nr | 27.88 |
| 80mm dia. | 6.71 | 7.38 | 2.32 | 26.62 | nr | 34.00 |
| 100mm dia. | 11.95 | 13.14 | 3.20 | 36.72 | nr | 49.86 |
| 125mm dia. | 26.68 | 29.35 | 3.90 | 44.75 | nr | 74.10 |
| 150mm dia. | 38.17 | 41.99 | 4.60 | 52.78 | nr | 94.77 |
| Cone seat unions | | | | | | |
| 15mm dia. | 6.67 | 7.34 | 0.66 | 7.57 | nr | 14.91 |
| 20mm dia. | 9.06 | 9.97 | 0.88 | 10.10 | nr | 20.07 |
| 25mm dia. | 12.34 | 13.57 | 1.00 | 11.47 | nr | 25.04 |
| 32mm dia. | 24.66 | 27.13 | 1.16 | 13.31 | nr | 40.44 |
| 40mm dia. | 30.11 | 33.12 | 1.34 | 15.38 | nr | 48.50 |
| 50mm dia. | 44.32 | 48.75 | 1.60 | 18.36 | nr | 67.11 |
| Elbow, male/female | | | | | | |
| 15mm dia. | 4.16 | 4.58 | 0.66 | 7.57 | nr | 12.15 |
| 20mm dia. | 5.30 | 5.83 | 0.88 | 10.10 | nr | 15.93 |
| 25mm dia. | 8.59 | 9.45 | 1.00 | 11.47 | nr | 20.92 |
| 32mm dia. | 14.17 | 15.59 | 1.16 | 13.31 | nr | 28.90 |
| 40mm dia. | 17.09 | 18.80 | 1.34 | 15.38 | nr | 34.18 |
| 50mm dia. | 29.83 | 32.81 | 1.60 | 18.36 | nr | 51.17 |
| Elbow, female/female | | | | | | |
| 15mm dia. | 3.19 | 3.51 | 0.66 | 7.57 | nr | 11.08 |
| 20mm dia. | 4.16 | 4.58 | 0.88 | 10.10 | nr | 14.68 |
| 25mm dia. | 5.65 | 6.21 | 1.00 | 11.47 | nr | 17.68 |
| 32mm dia. | 10.51 | 11.56 | 1.16 | 13.31 | nr | 24.87 |
| 40mm dia. | 12.52 | 13.77 | 1.34 | 15.38 | nr | 29.15 |
| 50mm dia. | 20.39 | 22.43 | 1.60 | 18.36 | nr | 40.79 |
| 65mm dia. | 53.02 | 58.32 | 1.96 | 22.49 | nr | 80.81 |
| 80mm dia. | 63.21 | 69.53 | 2.32 | 26.62 | nr | 96.15 |
| 100mm dia. | 103.56 | 113.92 | 3.20 | 36.72 | nr | 150.64 |
| Equal tee | | | | | | |
| 15mm dia. | 3.96 | 4.36 | 0.95 | 10.90 | nr | 15.26 |
| 20mm dia. | 4.61 | 5.07 | 1.27 | 14.57 | nr | 19.64 |
| 25mm dia. | 6.79 | 7.47 | 1.47 | 16.87 | nr | 24.34 |
| 32mm dia. | 14.02 | 15.42 | 1.68 | 19.28 | nr | 34.70 |
| 40mm dia. | 15.26 | 16.79 | 1.94 | 22.26 | nr | 39.05 |
| 50mm dia. | 24.80 | 27.28 | 2.31 | 26.51 | nr | 53.79 |
| 65mm dia. | 63.43 | 69.77 | 2.83 | 32.47 | nr | 102.24 |
| 80mm dia. | 68.06 | 74.87 | 3.35 | 38.44 | nr | 113.31 |
| 100mm dia. | 109.54 | 120.49 | 4.63 | 53.13 | nr | 173.62 |

## Y:MECHANICAL AND ELECTRICAL SERVICES

| Item | Net Price £ | Material £ | Labour hours | Labour £ | Unit | Total Rate £ |
|---|---|---|---|---|---|---|
| **Y10 : PIPELINES : BLACK STEEL - SCREWED (contd)** | | | | | | |
| Extra Over black steel screwed pipes; black malleable iron fittings; BS 143 | | | | | | |
| Cap | | | | | | |
| 15mm dia. | 0.55 | 0.61 | 0.33 | 3.79 | nr | 4.40 |
| 20mm dia. | 0.63 | 0.69 | 0.44 | 5.05 | nr | 5.74 |
| 25mm dia. | 0.79 | 0.87 | 0.50 | 5.74 | nr | 6.61 |
| 32mm dia. | 1.14 | 1.26 | 0.58 | 6.66 | nr | 7.92 |
| 40mm dia. | 1.46 | 1.60 | 0.67 | 7.69 | nr | 9.29 |
| 50mm dia. | 2.84 | 3.12 | 0.80 | 9.18 | nr | 12.30 |
| 65mm dia. | 4.45 | 4.90 | 0.98 | 11.25 | nr | 16.15 |
| 80mm dia. | 5.04 | 5.54 | 1.16 | 13.31 | nr | 18.85 |
| 100mm dia. | 11.04 | 12.14 | 1.80 | 20.66 | nr | 32.80 |
| Plain plug, hollow | | | | | | |
| 15mm dia. | 0.43 | 0.48 | 0.33 | 3.79 | nr | 4.27 |
| 20mm dia. | 0.55 | 0.61 | 0.44 | 5.05 | nr | 5.66 |
| 25mm dia. | 0.67 | 0.74 | 0.50 | 5.74 | nr | 6.48 |
| 32mm dia. | 0.95 | 1.04 | 0.58 | 6.66 | nr | 7.70 |
| 40mm dia. | 1.58 | 1.73 | 0.67 | 7.69 | nr | 9.42 |
| 50mm dia. | 2.21 | 2.43 | 0.80 | 9.18 | nr | 11.61 |
| 65mm dia. | 3.27 | 3.60 | 0.98 | 11.25 | nr | 14.85 |
| 80mm dia. | 4.89 | 5.38 | 1.16 | 13.31 | nr | 18.69 |
| 100mm dia. | 9.01 | 9.91 | 1.80 | 20.66 | nr | 30.57 |
| 125mm dia. | 23.54 | 25.90 | 2.56 | 29.38 | nr | 55.28 |
| 150mm dia. | 28.25 | 31.08 | 3.32 | 38.10 | nr | 69.18 |
| Plain plug, solid | | | | | | |
| 15mm dia. | 1.14 | 1.26 | 0.33 | 3.79 | nr | 5.05 |
| 20mm dia. | 1.14 | 1.26 | 0.44 | 5.05 | nr | 6.31 |
| 25mm dia. | 1.70 | 1.87 | 0.50 | 5.74 | nr | 7.61 |
| 32mm dia. | 2.05 | 2.25 | 0.58 | 6.66 | nr | 8.91 |
| 40mm dia. | 2.76 | 3.04 | 0.67 | 7.69 | nr | 10.73 |
| 50mm dia. | 3.63 | 3.99 | 0.80 | 9.18 | nr | 13.17 |
| 65mm dia. | 3.75 | 4.13 | 0.98 | 11.25 | nr | 15.38 |
| 80mm dia. | 5.59 | 6.15 | 1.16 | 13.31 | nr | 19.46 |
| Elbow, male/female | | | | | | |
| 15mm dia. | 0.71 | 0.78 | 0.66 | 7.57 | nr | 8.35 |
| 20mm dia. | 0.95 | 1.04 | 0.88 | 10.10 | nr | 11.14 |
| 25mm dia. | 1.58 | 1.73 | 1.00 | 11.47 | nr | 13.20 |
| 32mm dia. | 2.60 | 2.86 | 1.16 | 13.31 | nr | 16.17 |
| 40mm dia. | 3.86 | 4.25 | 1.34 | 15.38 | nr | 19.63 |
| 50mm dia. | 4.97 | 5.46 | 1.60 | 18.36 | nr | 23.82 |
| 65mm dia. | 9.42 | 10.36 | 1.96 | 22.49 | nr | 32.85 |
| 80mm dia. | 12.88 | 14.16 | 2.32 | 26.62 | nr | 40.78 |
| 100mm dia. | 22.51 | 24.76 | 3.20 | 36.72 | nr | 61.48 |

*Material Costs / Measured Work Prices - Mechanical Installations*

## Y:MECHANICAL AND ELECTRICAL SERVICES

| Item | Net Price £ | Material £ | Labour hours | Labour £ | Unit | Total Rate £ |
|---|---|---|---|---|---|---|
| **Elbow** | | | | | | |
| 15mm dia. | 0.63 | 0.69 | 0.66 | 7.57 | nr | **8.26** |
| 20mm dia. | 0.87 | 0.95 | 0.88 | 10.10 | nr | **11.05** |
| 25mm dia. | 1.34 | 1.47 | 1.00 | 11.47 | nr | **12.94** |
| 32mm dia. | 2.21 | 2.43 | 1.16 | 13.31 | nr | **15.74** |
| 40mm dia. | 3.71 | 4.08 | 1.34 | 15.38 | nr | **19.46** |
| 50mm dia. | 4.34 | 4.77 | 1.60 | 18.36 | nr | **23.13** |
| 65mm dia. | 8.17 | 8.98 | 1.96 | 22.49 | nr | **31.47** |
| 80mm dia. | 11.99 | 13.19 | 2.32 | 26.62 | nr | **39.81** |
| 100mm dia. | 20.60 | 22.66 | 3.20 | 36.72 | nr | **59.38** |
| 125mm dia. | 44.14 | 48.56 | 4.60 | 52.78 | nr | **101.34** |
| 150mm dia. | 82.18 | 90.40 | 6.00 | 68.85 | nr | **159.25** |
| **45 degree elbow** | | | | | | |
| 15mm dia. | 1.34 | 1.47 | 0.66 | 7.57 | nr | **9.04** |
| 20mm dia. | 1.66 | 1.82 | 0.88 | 10.10 | nr | **11.92** |
| 25mm dia. | 2.45 | 2.69 | 1.00 | 11.47 | nr | **14.16** |
| 32mm dia. | 4.49 | 4.94 | 1.16 | 13.31 | nr | **18.25** |
| 40mm dia. | 5.52 | 6.07 | 1.34 | 15.38 | nr | **21.45** |
| 50mm dia. | 7.57 | 8.32 | 1.60 | 18.36 | nr | **26.68** |
| 65mm dia. | 9.71 | 10.68 | 1.96 | 22.49 | nr | **33.17** |
| 80mm dia. | 14.60 | 16.06 | 2.32 | 26.62 | nr | **42.68** |
| 100mm dia. | 28.21 | 31.04 | 3.20 | 36.72 | nr | **67.76** |
| 150mm dia. | 75.52 | 83.07 | 6.00 | 68.85 | nr | **151.92** |
| **Bend, male/female** | | | | | | |
| 15mm dia. | 1.38 | 1.52 | 0.66 | 7.57 | nr | **9.09** |
| 20mm dia. | 1.73 | 1.91 | 0.88 | 10.10 | nr | **12.01** |
| 25mm dia. | 2.60 | 2.86 | 1.00 | 11.47 | nr | **14.33** |
| 32mm dia. | 4.81 | 5.29 | 1.16 | 13.31 | nr | **18.60** |
| 40mm dia. | 5.13 | 5.64 | 1.34 | 15.38 | nr | **21.02** |
| 50mm dia. | 7.33 | 8.07 | 1.60 | 18.36 | nr | **26.43** |
| 65mm dia. | 10.59 | 11.65 | 1.96 | 22.49 | nr | **34.14** |
| 80mm dia. | 15.45 | 17.00 | 2.32 | 26.62 | nr | **43.62** |
| 100mm dia. | 36.64 | 40.30 | 3.20 | 36.72 | nr | **77.02** |
| **Bend, male** | | | | | | |
| 15mm dia. | 2.25 | 2.47 | 0.66 | 7.57 | nr | **10.04** |
| 20mm dia. | 2.52 | 2.77 | 0.88 | 10.10 | nr | **12.87** |
| 25mm dia. | 3.71 | 4.08 | 1.00 | 11.47 | nr | **15.55** |
| 32mm dia. | 7.25 | 7.98 | 1.16 | 13.31 | nr | **21.29** |
| 40mm dia. | 10.17 | 11.19 | 1.34 | 15.38 | nr | **26.57** |
| 50mm dia. | 13.60 | 14.96 | 1.60 | 18.36 | nr | **33.32** |
| **Bend, female** | | | | | | |
| 15mm dia. | 1.14 | 1.26 | 0.66 | 7.57 | nr | **8.83** |
| 20mm dia. | 1.62 | 1.78 | 0.88 | 10.10 | nr | **11.88** |
| 25mm dia. | 2.29 | 2.52 | 1.00 | 11.47 | nr | **13.99** |
| 32mm dia. | 3.90 | 4.29 | 1.16 | 13.31 | nr | **17.60** |
| 40mm dia. | 4.65 | 5.12 | 1.34 | 15.38 | nr | **20.50** |
| 50mm dia. | 7.33 | 8.07 | 1.60 | 18.36 | nr | **26.43** |
| 65mm dia. | 13.65 | 15.01 | 1.96 | 22.49 | nr | **37.50** |
| 80mm dia. | 13.84 | 15.22 | 2.32 | 26.62 | nr | **41.84** |
| 100mm dia. | 42.45 | 46.70 | 3.20 | 36.72 | nr | **83.42** |
| 125mm dia. | 108.52 | 119.37 | 4.60 | 52.78 | nr | **172.15** |
| 150mm dia. | 165.75 | 182.33 | 6.00 | 68.85 | nr | **251.18** |

## Y:MECHANICAL AND ELECTRICAL SERVICES

| Item | Net Price £ | Material £ | Labour hours | Labour £ | Unit | Total Rate £ |
|---|---|---|---|---|---|---|
| **Y10 : PIPELINES : BLACK STEEL - SCREWED (contd)** | | | | | | |
| **Extra Over black steel screwed pipes; black malleable iron fittings; BS 143 (contd)** | | | | | | |
| Return bend | | | | | | |
| 15mm dia. | 4.53 | 4.99 | 0.66 | 7.57 | nr | 12.56 |
| 20mm dia. | 7.33 | 8.07 | 0.88 | 10.10 | nr | 18.17 |
| 25mm dia. | 9.15 | 10.06 | 1.00 | 11.47 | nr | 21.53 |
| 32mm dia. | 12.77 | 14.05 | 1.16 | 13.31 | nr | 27.36 |
| 40mm dia. | 15.22 | 16.74 | 1.34 | 15.38 | nr | 32.12 |
| 50mm dia. | 23.22 | 25.54 | 1.60 | 18.36 | nr | 43.90 |
| Equal socket, parallel thread | | | | | | |
| 15mm dia. | 0.59 | 0.65 | 0.66 | 7.57 | nr | 8.22 |
| 20mm dia. | 0.71 | 0.78 | 0.88 | 10.10 | nr | 10.88 |
| 25mm dia. | 0.95 | 1.04 | 1.00 | 11.47 | nr | 12.51 |
| 32mm dia. | 1.62 | 1.78 | 1.16 | 13.31 | nr | 15.09 |
| 40mm dia. | 2.21 | 2.43 | 1.34 | 15.38 | nr | 17.81 |
| 50mm dia. | 3.31 | 3.64 | 1.60 | 18.36 | nr | 22.00 |
| 65mm dia. | 5.30 | 5.83 | 1.96 | 22.49 | nr | 28.32 |
| 80mm dia. | 7.28 | 8.01 | 2.32 | 26.62 | nr | 34.63 |
| 100mm dia. | 12.36 | 13.60 | 3.20 | 36.72 | nr | 50.32 |
| Concentric reducing socket | | | | | | |
| 20 x 15mm dia. | 0.87 | 0.95 | 0.88 | 10.10 | nr | 11.05 |
| 25 x 15mm dia. | 1.14 | 1.26 | 1.00 | 11.47 | nr | 12.73 |
| 25 x 20mm dia. | 1.06 | 1.17 | 1.00 | 11.47 | nr | 12.64 |
| 32 x 25mm dia. | 1.81 | 2.00 | 1.16 | 13.31 | nr | 15.31 |
| 40 x 25mm dia. | 2.13 | 2.34 | 1.34 | 15.38 | nr | 17.72 |
| 40 x 32mm dia. | 2.36 | 2.60 | 1.34 | 15.38 | nr | 17.98 |
| 50 x 25mm dia. | 4.10 | 4.51 | 1.60 | 18.36 | nr | 22.87 |
| 50 x 40mm dia. | 3.31 | 3.64 | 1.60 | 18.36 | nr | 22.00 |
| 65 x 50mm dia. | 5.52 | 6.07 | 1.96 | 22.49 | nr | 28.56 |
| 80 x 50mm dia. | 6.88 | 7.57 | 2.32 | 26.62 | nr | 34.19 |
| 100 x 50mm dia. | 13.72 | 15.09 | 3.20 | 36.72 | nr | 51.81 |
| 100 x 80mm dia. | 12.73 | 14.00 | 3.20 | 36.72 | nr | 50.72 |
| 150 x 100mm dia. | 33.59 | 36.94 | 4.60 | 52.78 | nr | 89.72 |
| Eccentric reducing socket | | | | | | |
| 20 x 15mm dia. | 1.54 | 1.69 | 0.88 | 10.10 | nr | 11.79 |
| 25 x 15mm dia. | 4.38 | 4.81 | 1.00 | 11.47 | nr | 16.28 |
| 25 x 20mm dia. | 4.97 | 5.46 | 1.00 | 11.47 | nr | 16.93 |
| 32 x 25mm dia. | 5.68 | 6.24 | 1.16 | 13.31 | nr | 19.55 |
| 40 x 25mm dia. | 6.50 | 7.15 | 1.34 | 15.38 | nr | 22.53 |
| 40 x 32mm dia. | 3.51 | 3.86 | 1.34 | 15.38 | nr | 19.24 |
| 50 x 25mm dia. | 4.22 | 4.64 | 1.60 | 18.36 | nr | 23.00 |
| 50 x 40mm dia. | 4.22 | 4.64 | 1.60 | 18.36 | nr | 23.00 |
| 65 x 50mm dia. | 6.88 | 7.57 | 1.96 | 22.49 | nr | 30.06 |
| 80 x 50mm dia. | 11.18 | 12.30 | 2.32 | 26.62 | nr | 38.92 |

## Y:MECHANICAL AND ELECTRICAL SERVICES

| Item | Net Price £ | Material £ | Labour hours | Labour £ | Unit | Total Rate £ |
|---|---|---|---|---|---|---|
| Hexagon bush | | | | | | |
| 20 x 15mm dia. | 0.55 | 0.61 | 0.44 | 5.05 | nr | 5.66 |
| 25 x 15mm dia. | 0.67 | 0.74 | 0.50 | 5.74 | nr | 6.48 |
| 25 x 20mm dia. | 0.71 | 0.78 | 0.50 | 5.74 | nr | 6.52 |
| 32 x 25mm dia. | 0.83 | 0.91 | 0.58 | 6.66 | nr | 7.57 |
| 40 x 25mm dia. | 1.14 | 1.26 | 0.67 | 7.69 | nr | 8.95 |
| 40 x 32mm dia. | 1.18 | 1.30 | 0.67 | 7.69 | nr | 8.99 |
| 50 x 25mm dia. | 2.36 | 2.60 | 0.80 | 9.18 | nr | 11.78 |
| 50 x 40mm dia. | 2.21 | 2.43 | 0.80 | 9.18 | nr | 11.61 |
| 65 x 50mm dia. | 3.61 | 3.97 | 0.98 | 11.25 | nr | 15.22 |
| 80 x 50mm dia. | 5.45 | 5.99 | 1.16 | 13.31 | nr | 19.30 |
| 100 x 50mm dia. | 12.06 | 13.27 | 1.80 | 20.66 | nr | 33.93 |
| 100 x 80mm dia. | 10.04 | 11.05 | 1.80 | 20.66 | nr | 31.71 |
| 150 x 100mm dia. | 31.78 | 34.96 | 3.00 | 34.42 | nr | 69.38 |
| Hexagon nipple | | | | | | |
| 15mm dia. | 0.59 | 0.65 | 0.33 | 3.79 | nr | 4.44 |
| 20mm dia. | 0.67 | 0.74 | 0.44 | 5.05 | nr | 5.79 |
| 25mm dia. | 0.95 | 1.04 | 0.50 | 5.74 | nr | 6.78 |
| 32mm dia. | 1.58 | 1.73 | 0.58 | 6.66 | nr | 8.39 |
| 40mm dia. | 1.81 | 2.00 | 0.67 | 7.69 | nr | 9.69 |
| 50mm dia. | 3.31 | 3.64 | 0.80 | 9.18 | nr | 12.82 |
| 65mm dia. | 4.71 | 5.18 | 0.98 | 11.25 | nr | 16.43 |
| 80mm dia. | 6.81 | 7.49 | 1.16 | 13.31 | nr | 20.80 |
| 100mm dia. | 11.55 | 12.71 | 1.80 | 20.66 | nr | 33.37 |
| 150mm dia. | 32.66 | 35.93 | 3.00 | 34.42 | nr | 70.35 |
| Union, male/female | | | | | | |
| 15mm dia. | 2.45 | 2.69 | 0.66 | 7.57 | nr | 10.26 |
| 20mm dia. | 3.00 | 3.30 | 0.88 | 10.10 | nr | 13.40 |
| 25mm dia. | 3.75 | 4.12 | 1.00 | 11.47 | nr | 15.59 |
| 32mm dia. | 5.20 | 5.72 | 1.16 | 13.31 | nr | 19.03 |
| 40mm dia. | 6.66 | 7.33 | 1.34 | 15.38 | nr | 22.71 |
| 50mm dia. | 10.49 | 11.54 | 1.60 | 18.36 | nr | 29.90 |
| 65mm dia. | 19.72 | 21.69 | 1.96 | 22.49 | nr | 44.18 |
| 80mm dia. | 27.11 | 29.82 | 2.32 | 26.62 | nr | 56.44 |
| Union, female | | | | | | |
| 15mm dia. | 2.52 | 2.77 | 0.66 | 7.57 | nr | 10.34 |
| 20mm dia. | 2.76 | 3.04 | 0.88 | 10.10 | nr | 13.14 |
| 25mm dia. | 3.23 | 3.55 | 1.00 | 11.47 | nr | 15.02 |
| 32mm dia. | 4.89 | 5.38 | 1.16 | 13.31 | nr | 18.69 |
| 40mm dia. | 5.52 | 6.07 | 1.34 | 15.38 | nr | 21.45 |
| 50mm dia. | 9.15 | 10.06 | 1.60 | 18.36 | nr | 28.42 |
| 65mm dia. | 17.92 | 19.71 | 1.96 | 22.49 | nr | 42.20 |
| 80mm dia. | 23.69 | 26.06 | 2.32 | 26.62 | nr | 52.68 |
| 100mm dia. | 45.10 | 49.61 | 3.20 | 36.72 | nr | 86.33 |
| Union elbow, male/female | | | | | | |
| 15mm dia. | 3.11 | 3.43 | 0.66 | 7.57 | nr | 11.00 |
| 20mm dia. | 3.90 | 4.29 | 0.88 | 10.10 | nr | 14.39 |
| 25mm dia. | 5.48 | 6.03 | 1.00 | 11.47 | nr | 17.50 |

## Y:MECHANICAL AND ELECTRICAL SERVICES

| Item | Net Price £ | Material £ | Labour hours | Labour £ | Unit | Total Rate £ |
|---|---|---|---|---|---|---|
| **Y10 : PIPELINES : BLACK STEEL - SCREWED (contd)** | | | | | | |
| Extra Over black steel screwed pipes; black malleable iron fittings; BS 143 (contd) | | | | | | |
| Twin elbow | | | | | | |
| 15mm dia. | 3.00 | 3.30 | 0.95 | 10.90 | nr | 14.20 |
| 20mm dia. | 3.31 | 3.64 | 1.27 | 14.57 | nr | 18.21 |
| 25mm dia. | 5.36 | 5.90 | 1.45 | 16.64 | nr | 22.54 |
| 32mm dia. | 9.50 | 10.45 | 1.68 | 19.28 | nr | 29.73 |
| 40mm dia. | 12.03 | 13.23 | 1.94 | 22.26 | nr | 35.49 |
| 50mm dia. | 15.45 | 17.00 | 2.31 | 26.51 | nr | 43.51 |
| 65mm dia. | 23.84 | 26.22 | 2.83 | 32.47 | nr | 58.69 |
| 80mm dia. | 40.61 | 44.67 | 3.35 | 38.44 | nr | 83.11 |
| Equal tee | | | | | | |
| 15mm dia. | 0.87 | 0.95 | 0.95 | 10.90 | nr | 11.85 |
| 20mm dia. | 1.26 | 1.39 | 1.27 | 14.57 | nr | 15.96 |
| 25mm dia. | 1.81 | 2.00 | 1.45 | 16.64 | nr | 18.64 |
| 32mm dia. | 3.00 | 3.30 | 1.68 | 19.28 | nr | 22.58 |
| 40mm dia. | 4.10 | 4.51 | 1.94 | 22.26 | nr | 26.77 |
| 50mm dia. | 5.91 | 6.50 | 2.31 | 26.51 | nr | 33.01 |
| 65mm dia. | 11.77 | 12.95 | 2.83 | 32.47 | nr | 45.42 |
| 80mm dia. | 13.72 | 15.09 | 3.35 | 38.44 | nr | 53.53 |
| 100mm dia. | 24.87 | 27.35 | 4.63 | 53.13 | nr | 80.48 |
| 125mm dia. | 60.99 | 67.09 | 5.32 | 61.05 | nr | 128.14 |
| 150mm dia. | 97.19 | 106.90 | 6.00 | 68.85 | nr | 175.75 |
| Tee reducing on branch | | | | | | |
| 20 x 15mm dia. | 1.14 | 1.26 | 1.27 | 14.57 | nr | 15.83 |
| 25 x 15mm dia. | 1.58 | 1.73 | 1.45 | 16.64 | nr | 18.37 |
| 25 x 20mm dia. | 1.66 | 1.82 | 1.45 | 16.64 | nr | 18.46 |
| 32 x 25mm dia. | 2.92 | 3.21 | 1.68 | 19.28 | nr | 22.49 |
| 40 x 25mm dia. | 3.86 | 4.25 | 1.94 | 22.26 | nr | 26.51 |
| 40 x 32mm dia. | 5.05 | 5.55 | 1.94 | 22.26 | nr | 27.81 |
| 50 x 25mm dia. | 5.13 | 5.64 | 2.31 | 26.51 | nr | 32.15 |
| 50 x 40mm dia. | 7.09 | 7.80 | 2.31 | 26.51 | nr | 34.31 |
| 65 x 50mm dia. | 10.45 | 11.49 | 2.83 | 32.47 | nr | 43.96 |
| 80 x 50mm dia. | 14.12 | 15.54 | 3.35 | 38.44 | nr | 53.98 |
| 100 x 50mm dia. | 20.60 | 22.66 | 4.63 | 53.13 | nr | 75.79 |
| 100 x 80mm dia. | 31.78 | 34.96 | 4.63 | 53.13 | nr | 88.09 |
| 150 x 100mm dia. | 71.58 | 78.74 | 6.00 | 68.85 | nr | 147.59 |
| Equal pitcher tee | | | | | | |
| 15mm dia. | 2.36 | 2.60 | 0.95 | 10.90 | nr | 13.50 |
| 20mm dia. | 2.92 | 3.21 | 1.27 | 14.57 | nr | 17.78 |
| 25mm dia. | 4.38 | 4.81 | 1.45 | 16.64 | nr | 21.45 |
| 32mm dia. | 5.99 | 6.59 | 1.68 | 19.28 | nr | 25.87 |
| 40mm dia. | 9.26 | 10.19 | 1.94 | 22.26 | nr | 32.45 |
| 50mm dia. | 13.01 | 14.31 | 2.31 | 26.51 | nr | 40.82 |
| 65mm dia. | 17.66 | 19.42 | 2.83 | 32.47 | nr | 51.89 |
| 80mm dia. | 24.28 | 26.70 | 3.35 | 38.44 | nr | 65.14 |
| 100mm dia. | 54.63 | 60.09 | 4.63 | 53.13 | nr | 113.22 |

## Y:MECHANICAL AND ELECTRICAL SERVICES

| Item | Net Price £ | Material £ | Labour hours | Labour £ | Unit | Total Rate £ |
|---|---|---|---|---|---|---|
| Cross | | | | | | |
| 15mm dia. | 2.05 | 2.25 | 1.05 | 12.05 | nr | **14.30** |
| 20mm dia. | 3.07 | 3.38 | 1.40 | 16.07 | nr | **19.45** |
| 25mm dia. | 3.90 | 4.29 | 1.60 | 18.36 | nr | **22.65** |
| 32mm dia. | 5.13 | 5.64 | 1.86 | 21.34 | nr | **26.98** |
| 40mm dia. | 6.90 | 7.59 | 2.14 | 24.56 | nr | **32.15** |
| 50mm dia. | 10.72 | 11.79 | 2.56 | 29.38 | nr | **41.17** |
| 65mm dia. | 14.60 | 16.06 | 3.14 | 36.03 | nr | **52.09** |
| 80mm dia. | 19.42 | 21.36 | 3.71 | 42.57 | nr | **63.93** |
| 100mm dia. | 35.31 | 38.85 | 5.12 | 58.75 | nr | **97.60** |

*Material Costs / Measured Work Prices - Mechanical Installations*

## Y:MECHANICAL AND ELECTRICAL SERVICES

| Item | Net Price £ | Material £ | Labour hours | Labour £ | Unit | Total Rate £ |
|---|---|---|---|---|---|---|
| **Y10 : PIPELINES : BLACK STEEL - WELDED** | | | | | | |
| **Black steel pipes; butt welded joints; BS 1387: 1985** | | | | | | |
| Varnished; light | | | | | | |
| 15mm dia. | 2.06 | 2.38 | 0.55 | 6.31 | m | 8.69 |
| 20mm dia. | 2.68 | 3.10 | 0.57 | 6.54 | m | 9.64 |
| 25mm dia. | 3.67 | 4.24 | 0.63 | 7.23 | m | 11.47 |
| 32mm dia. | 4.53 | 5.23 | 0.72 | 8.27 | m | 13.50 |
| 40mm dia. | 5.60 | 6.47 | 0.80 | 9.18 | m | 15.65 |
| 50mm dia. | 7.29 | 8.42 | 0.92 | 10.56 | m | 18.98 |
| 65mm dia. | 10.75 | 12.42 | 1.07 | 12.29 | m | 24.71 |
| 80mm dia. | 12.83 | 14.82 | 1.22 | 14.01 | m | 28.83 |
| 100mm dia. | 18.59 | 21.47 | 1.58 | 18.13 | m | 39.60 |
| Varnished; medium | | | | | | |
| 8mm dia. | 2.21 | 2.55 | 0.54 | 6.20 | m | 8.75 |
| 10mm dia. | 2.20 | 2.54 | 0.54 | 6.20 | m | 8.74 |
| 15mm dia. | 2.44 | 2.82 | 0.55 | 6.31 | m | 9.13 |
| 20mm dia. | 2.88 | 3.32 | 0.57 | 6.54 | m | 9.86 |
| 25mm dia. | 4.13 | 4.77 | 0.63 | 7.23 | m | 12.00 |
| 32mm dia. | 5.11 | 5.90 | 0.72 | 8.27 | m | 14.17 |
| 40mm dia. | 5.94 | 6.86 | 0.80 | 9.18 | m | 16.04 |
| 50mm dia. | 8.36 | 9.66 | 0.92 | 10.56 | m | 20.22 |
| 65mm dia. | 11.35 | 13.10 | 1.07 | 12.29 | m | 25.39 |
| 80mm dia. | 14.74 | 17.03 | 1.22 | 14.01 | m | 31.04 |
| 100mm dia. | 20.89 | 24.13 | 1.58 | 18.16 | m | 42.29 |
| 125mm dia. | 30.82 | 35.60 | 1.80 | 20.66 | m | 56.26 |
| 150mm dia. | 35.79 | 41.34 | 2.03 | 23.29 | m | 64.63 |
| Varnished; heavy | | | | | | |
| 15mm dia. | 2.91 | 3.20 | 0.55 | 6.31 | m | 9.51 |
| 20mm dia. | 3.45 | 3.99 | 0.57 | 6.54 | m | 10.53 |
| 25mm dia. | 5.05 | 5.83 | 0.63 | 7.23 | m | 13.06 |
| 32mm dia. | 6.26 | 7.23 | 0.72 | 8.27 | m | 15.50 |
| 40mm dia. | 7.30 | 8.43 | 0.80 | 9.18 | m | 17.61 |
| 50mm dia. | 10.14 | 11.71 | 0.92 | 10.56 | m | 22.27 |
| 65mm dia. | 13.79 | 15.92 | 1.07 | 12.29 | m | 28.21 |
| 80mm dia. | 13.79 | 15.92 | 1.22 | 14.01 | m | 29.93 |
| 100mm dia. | 24.49 | 28.29 | 1.58 | 18.16 | m | 46.45 |
| 125mm dia. | 32.86 | 37.95 | 1.80 | 20.66 | m | 58.61 |
| 150mm dia. | 38.42 | 44.38 | 2.03 | 23.29 | m | 67.67 |

## Y:MECHANICAL AND ELECTRICAL SERVICES

| Item | Net Price £ | Material £ | Labour hours | Labour £ | Unit | Total Rate £ |
|---|---|---|---|---|---|---|
| **Extra Over black steel butt welded pipes; black steel flanges, welding and drilled; metric; BS 4504** | | | | | | |
| Welded flanges; PN6 | | | | | | |
| 15mm dia. | 4.80 | 5.28 | 0.67 | 7.69 | nr | 12.97 |
| 20mm dia. | 4.80 | 5.28 | 0.77 | 8.84 | nr | 14.12 |
| 25mm dia. | 5.04 | 5.54 | 0.93 | 10.67 | nr | 16.21 |
| 32mm dia. | 5.36 | 5.90 | 1.09 | 12.51 | nr | 18.41 |
| 40mm dia. | 5.36 | 5.90 | 1.17 | 13.43 | nr | 19.33 |
| 50mm dia. | 6.13 | 6.74 | 1.44 | 16.52 | nr | 23.26 |
| 65mm dia. | 6.89 | 7.58 | 1.55 | 17.79 | nr | 25.37 |
| 80mm dia. | 10.19 | 11.21 | 1.63 | 18.70 | nr | 29.91 |
| 100mm dia. | 11.48 | 12.63 | 2.19 | 25.13 | nr | 37.76 |
| 125mm dia. | 19.10 | 21.01 | 2.52 | 28.92 | nr | 49.93 |
| 150mm dia. | 19.46 | 21.41 | 2.84 | 32.59 | nr | 54.00 |
| Welded flanges; PN16 | | | | | | |
| 15mm dia. | 4.61 | 5.07 | 0.67 | 7.69 | nr | 12.76 |
| 20mm dia. | 4.61 | 5.07 | 0.77 | 8.84 | nr | 13.91 |
| 25mm dia. | 4.86 | 5.35 | 0.93 | 10.67 | nr | 16.02 |
| 32mm dia. | 7.00 | 7.70 | 1.09 | 12.51 | nr | 20.21 |
| 40mm dia. | 7.00 | 7.70 | 1.17 | 13.43 | nr | 21.13 |
| 50mm dia. | 8.53 | 9.38 | 1.44 | 16.52 | nr | 25.90 |
| 65mm dia. | 9.92 | 10.91 | 1.55 | 17.79 | nr | 28.70 |
| 80mm dia. | 12.55 | 13.80 | 1.63 | 18.70 | nr | 32.50 |
| 100mm dia. | 14.71 | 16.18 | 2.19 | 25.13 | nr | 41.31 |
| 125mm dia. | 22.97 | 25.27 | 2.52 | 28.92 | nr | 54.19 |
| 150mm dia. | 22.97 | 25.27 | 2.84 | 32.59 | nr | 57.86 |
| Blank flanges, slip on for welding; PN6 | | | | | | |
| 15mm dia. | 3.12 | 3.43 | 0.58 | 6.66 | nr | 10.09 |
| 20mm dia. | 3.29 | 3.62 | 0.67 | 7.69 | nr | 11.31 |
| 25mm dia. | 3.78 | 4.16 | 0.80 | 9.18 | nr | 13.34 |
| 32mm dia. | 4.60 | 5.06 | 0.95 | 10.90 | nr | 15.96 |
| 40mm dia. | 5.26 | 5.79 | 1.02 | 11.70 | nr | 17.49 |
| 50mm dia. | 6.57 | 7.23 | 1.25 | 14.34 | nr | 21.57 |
| 65mm dia. | 8.54 | 9.39 | 1.35 | 15.49 | nr | 24.88 |
| 80mm dia. | 10.02 | 11.02 | 1.42 | 16.29 | nr | 27.31 |
| 100mm dia. | 11.17 | 12.29 | 1.90 | 21.80 | nr | 34.09 |
| 125mm dia. | 15.44 | 16.98 | 2.19 | 25.13 | nr | 42.11 |
| 150mm dia. | 19.38 | 21.32 | 2.48 | 28.46 | nr | 49.78 |
| Blank flanges, slip on for welding; PN16 | | | | | | |
| 15mm dia. | 3.50 | 3.85 | 0.58 | 6.66 | nr | 10.51 |
| 20mm dia. | 3.29 | 3.62 | 0.67 | 7.69 | nr | 11.31 |
| 25mm dia. | 3.78 | 4.16 | 0.80 | 9.18 | nr | 13.34 |
| 32mm dia. | 4.76 | 5.24 | 0.95 | 10.90 | nr | 16.14 |
| 40mm dia. | 5.26 | 5.79 | 1.02 | 11.70 | nr | 17.49 |
| 50mm dia. | 6.73 | 7.40 | 1.25 | 14.34 | nr | 21.74 |
| 65mm dia. | 8.71 | 9.58 | 1.35 | 15.49 | nr | 25.07 |
| 80mm dia. | 10.51 | 11.56 | 1.42 | 16.29 | nr | 27.85 |
| 100mm dia. | 12.16 | 13.38 | 1.90 | 21.80 | nr | 35.18 |
| 125mm dia. | 18.07 | 19.88 | 2.19 | 25.13 | nr | 45.01 |
| 150mm dia. | 21.02 | 23.12 | 2.48 | 28.46 | nr | 51.58 |

## Y:MECHANICAL AND ELECTRICAL SERVICES

| Item | Net Price £ | Material £ | Labour hours | Labour £ | Unit | Total Rate £ |
|---|---|---|---|---|---|---|
| **Y10 : PIPELINES : BLACK STEEL - WELDED (contd)** | | | | | | |
| **Extra Over black steel butt welded pipes; black steel flanges, welding and drilled; imperial; BS 10** | | | | | | |
| Welded flanges; Table E | | | | | | |
| 1/2" dia. | 5.87 | 6.46 | 0.67 | 7.69 | nr | 14.15 |
| 3/4" dia. | 5.87 | 6.46 | 0.77 | 8.84 | nr | 15.30 |
| 1" dia. | 5.92 | 6.51 | 0.93 | 10.67 | nr | 17.18 |
| 1 1/4" dia. | 5.92 | 6.51 | 1.09 | 12.51 | nr | 19.02 |
| 1 1/2" dia. | 5.92 | 6.51 | 1.17 | 13.43 | nr | 19.94 |
| 2" dia. | 5.99 | 6.59 | 1.44 | 16.52 | nr | 23.11 |
| 2 1/2" dia. | 7.56 | 8.32 | 1.55 | 17.79 | nr | 26.11 |
| 3" dia. | 7.56 | 8.32 | 1.63 | 18.70 | nr | 27.02 |
| 4" dia. | 11.50 | 12.65 | 2.19 | 25.13 | nr | 37.78 |
| 5" dia. | 22.46 | 24.71 | 2.52 | 28.92 | nr | 53.63 |
| 6" dia. | 22.46 | 24.71 | 2.84 | 32.59 | nr | 57.30 |
| Blank flanges, slip on for welding; Table E | | | | | | |
| 1/2" dia. | 3.85 | 4.24 | 0.58 | 6.66 | nr | 10.90 |
| 3/4" dia. | 3.85 | 4.24 | 0.67 | 7.69 | nr | 11.93 |
| 1" dia. | 3.85 | 4.24 | 0.80 | 9.18 | nr | 13.42 |
| 1 1/4" dia. | 4.43 | 4.87 | 0.95 | 10.90 | nr | 15.77 |
| 1 1/2" dia. | 5.09 | 5.60 | 1.02 | 11.70 | nr | 17.30 |
| 2" dia. | 5.12 | 5.63 | 1.25 | 14.34 | nr | 19.97 |
| 2 1/2" dia. | 7.06 | 7.77 | 1.35 | 15.49 | nr | 23.26 |
| 3" dia. | 8.05 | 8.86 | 1.42 | 16.29 | nr | 25.15 |
| 4" dia. | 10.35 | 11.38 | 1.90 | 21.80 | nr | 33.18 |
| 5" dia. | 14.45 | 15.89 | 2.19 | 25.13 | nr | 41.02 |
| 6" dia. | 20.69 | 22.76 | 2.48 | 28.46 | nr | 51.22 |
| **Extra Over black steel butt welded pipes; black steel flange connections** | | | | | | |
| Bolted connection between pair of flanges; including gasket, bolts, nuts and washers | | | | | | |
| 50mm dia. | 3.96 | 4.36 | 0.50 | 5.74 | nr | 10.10 |
| 65mm dia. | 4.20 | 4.62 | 0.50 | 5.74 | nr | 10.36 |
| 80mm dia. | 7.50 | 8.25 | 0.50 | 5.74 | nr | 13.99 |
| 100mm dia. | 7.74 | 8.51 | 0.50 | 5.74 | nr | 14.25 |
| 125mm dia. | 8.40 | 9.24 | 0.50 | 5.74 | nr | 14.98 |
| 150mm dia. | 12.60 | 13.86 | 0.88 | 10.10 | nr | 23.96 |

*Material Costs / Measured Work Prices - Mechanical Installations* 177

## Y:MECHANICAL AND ELECTRICAL SERVICES

| Item | Net Price £ | Material £ | Labour hours | Labour £ | Unit | Total Rate £ |
|---|---|---|---|---|---|---|
| **Extra Over black steel butt welded pipes; black steel welding fittings; BS 1640** | | | | | | |
| Cap | | | | | | |
| 20mm dia. | 3.78 | 4.16 | 0.35 | 4.02 | nr | 8.18 |
| 25mm dia. | 3.96 | 4.36 | 0.43 | 4.93 | nr | 9.29 |
| 32mm dia. | 3.96 | 4.36 | 0.54 | 6.20 | nr | 10.56 |
| 40mm dia. | 3.97 | 4.37 | 0.63 | 7.23 | nr | 11.60 |
| 50mm dia. | 5.00 | 5.50 | 0.93 | 10.67 | nr | 16.17 |
| 65mm dia. | 5.64 | 6.20 | 1.30 | 14.92 | nr | 21.12 |
| 80mm dia. | 6.26 | 6.89 | 1.62 | 18.59 | nr | 25.48 |
| 100mm dia. | 8.71 | 9.58 | 2.20 | 25.25 | nr | 34.83 |
| 125mm dia. | 13.21 | 14.53 | 3.08 | 35.34 | nr | 49.87 |
| 150mm dia. | 14.83 | 16.31 | 3.90 | 44.75 | nr | 61.06 |
| Concentric reducer | | | | | | |
| 20 x 15mm dia. | 3.49 | 3.84 | 0.61 | 7.00 | nr | 10.84 |
| 25 x 15mm dia. | 3.49 | 3.84 | 0.77 | 8.84 | nr | 12.68 |
| 25 x 20mm dia. | 2.89 | 3.18 | 0.77 | 8.84 | nr | 12.02 |
| 32 x 25mm dia. | 3.35 | 3.69 | 0.96 | 11.02 | nr | 14.71 |
| 40 x 25mm dia. | 3.68 | 4.05 | 1.23 | 14.11 | nr | 18.16 |
| 40 x 32mm dia. | 3.68 | 4.05 | 1.23 | 14.11 | nr | 18.16 |
| 50 x 25mm dia. | 4.12 | 4.53 | 1.66 | 19.05 | nr | 23.58 |
| 50 x 40mm dia. | 4.12 | 4.53 | 1.66 | 19.05 | nr | 23.58 |
| 65 x 50mm dia. | 5.21 | 5.73 | 2.36 | 27.08 | nr | 32.81 |
| 80 x 50mm dia. | 5.53 | 6.08 | 3.13 | 35.92 | nr | 42.00 |
| 100 x 50mm dia. | 8.41 | 9.25 | 3.91 | 44.87 | nr | 54.12 |
| 100 x 80mm dia. | 7.49 | 8.24 | 3.91 | 44.87 | nr | 53.11 |
| 125 x 80mm dia. | 14.37 | 15.81 | 4.36 | 50.03 | nr | 65.84 |
| 150 x 100mm dia. | 14.37 | 15.81 | 4.81 | 55.19 | nr | 71.00 |
| Eccentric reducer | | | | | | |
| 20 x 15mm dia. | 4.37 | 4.81 | 0.61 | 7.00 | nr | 11.81 |
| 25 x 15mm dia. | 4.51 | 4.96 | 0.77 | 8.84 | nr | 13.80 |
| 25 x 20mm dia. | 4.36 | 4.80 | 0.77 | 8.84 | nr | 13.64 |
| 32 x 25mm dia. | 4.57 | 5.03 | 0.96 | 11.02 | nr | 16.05 |
| 40 x 25mm dia. | 4.92 | 5.41 | 1.23 | 14.11 | nr | 19.52 |
| 40 x 32mm dia. | 4.92 | 5.41 | 1.23 | 14.11 | nr | 19.52 |
| 50 x 25mm dia. | 7.75 | 8.53 | 1.66 | 19.05 | nr | 27.58 |
| 50 x 40mm dia. | 7.75 | 8.53 | 1.66 | 19.05 | nr | 27.58 |
| 65 x 50mm dia. | 7.93 | 8.72 | 2.36 | 27.08 | nr | 35.80 |
| 80 x 50mm dia. | 8.28 | 9.11 | 3.13 | 35.92 | nr | 45.03 |
| 100 x 50mm dia. | 15.43 | 16.97 | 3.91 | 44.87 | nr | 61.84 |
| 100 x 80mm dia. | 10.70 | 11.77 | 3.91 | 44.87 | nr | 56.64 |
| 125 x 80mm dia. | 22.10 | 24.31 | 4.36 | 50.03 | nr | 74.34 |
| 150 x 100mm dia. | 22.10 | 24.31 | 4.81 | 55.19 | nr | 79.50 |

## Y:MECHANICAL AND ELECTRICAL SERVICES

| Item | Net Price £ | Material £ | Labour hours | Labour £ | Unit | Total Rate £ |
|---|---|---|---|---|---|---|
| **Y10 : PIPELINES : BLACK STEEL - WELDED (contd)** | | | | | | |
| **Extra Over black steel butt welded pipes; black steel welding fittings; BS 1640 (contd)** | | | | | | |
| 45 degree elbow, long radius | | | | | | |
| 15mm dia. | 1.87 | 2.06 | 0.52 | 5.97 | nr | 8.03 |
| 20mm dia. | 1.87 | 2.06 | 0.69 | 7.92 | nr | 9.98 |
| 25mm dia. | 2.12 | 2.33 | 0.85 | 9.75 | nr | 12.08 |
| 32mm dia. | 2.56 | 2.82 | 1.07 | 12.28 | nr | 15.10 |
| 40mm dia. | 2.70 | 2.97 | 1.33 | 15.26 | nr | 18.23 |
| 50mm dia. | 3.01 | 3.31 | 1.86 | 21.34 | nr | 24.65 |
| 65mm dia. | 5.54 | 6.09 | 2.60 | 29.84 | nr | 35.93 |
| 80mm dia. | 6.25 | 6.88 | 3.24 | 37.18 | nr | 44.06 |
| 100mm dia. | 7.80 | 8.58 | 3.92 | 44.98 | nr | 53.56 |
| 125mm dia. | 16.58 | 18.24 | 4.67 | 53.59 | nr | 71.83 |
| 150mm dia. | 18.77 | 20.65 | 5.41 | 62.08 | nr | 82.73 |
| 90 degree elbow, long radius | | | | | | |
| 15mm dia. | 1.87 | 2.06 | 0.52 | 5.97 | nr | 8.03 |
| 20mm dia. | 1.87 | 2.06 | 0.69 | 7.92 | nr | 9.98 |
| 25mm dia. | 2.12 | 2.33 | 0.85 | 9.75 | nr | 12.08 |
| 32mm dia. | 2.56 | 2.82 | 1.07 | 12.28 | nr | 15.10 |
| 40mm dia. | 2.70 | 2.97 | 1.33 | 15.26 | nr | 18.23 |
| 50mm dia. | 3.01 | 3.31 | 1.86 | 21.34 | nr | 24.65 |
| 65mm dia. | 5.54 | 6.09 | 2.60 | 29.84 | nr | 35.93 |
| 80mm dia. | 6.25 | 6.88 | 3.24 | 37.18 | nr | 44.06 |
| 100mm dia. | 10.40 | 11.44 | 3.92 | 44.98 | nr | 56.42 |
| 125mm dia. | 22.08 | 24.29 | 4.67 | 53.59 | nr | 77.88 |
| 150mm dia. | 25.03 | 27.53 | 5.41 | 62.08 | nr | 89.61 |
| Equal tee | | | | | | |
| 15mm dia. | 6.49 | 7.14 | 0.75 | 8.61 | nr | 15.75 |
| 20mm dia. | 6.49 | 7.14 | 1.00 | 11.47 | nr | 18.61 |
| 25mm dia. | 8.74 | 9.61 | 1.23 | 14.11 | nr | 23.72 |
| 32mm dia. | 10.88 | 11.97 | 1.44 | 16.52 | nr | 28.49 |
| 40mm dia. | 10.88 | 11.97 | 1.93 | 22.15 | nr | 34.12 |
| 50mm dia. | 11.31 | 12.44 | 2.87 | 32.93 | nr | 45.37 |
| 65mm dia. | 14.05 | 15.46 | 3.32 | 38.10 | nr | 53.56 |
| 80mm dia. | 18.32 | 20.15 | 3.79 | 43.49 | nr | 63.64 |
| 100mm dia. | 26.37 | 29.01 | 4.69 | 53.82 | nr | 82.83 |
| 125mm dia. | 46.78 | 51.46 | 6.25 | 71.72 | nr | 123.18 |
| 150mm dia. | 47.61 | 52.37 | 7.81 | 89.62 | nr | 141.99 |

## Y:MECHANICAL AND ELECTRICAL SERVICES

| Item | Net Price £ | Material £ | Labour hours | Labour £ | Unit | Total Rate £ |
|---|---|---|---|---|---|---|
| **Extra Over black steel butt welded pipes; labours** | | | | | | |
| Made bend | | | | | | |
| 15mm dia. | - | - | 0.40 | 4.59 | nr | 4.59 |
| 20mm dia. | - | - | 0.40 | 4.59 | nr | 4.59 |
| 25mm dia. | - | - | 0.48 | 5.51 | nr | 5.51 |
| 32mm dia. | - | - | 0.64 | 7.34 | nr | 7.34 |
| 40mm dia. | - | - | 0.80 | 9.18 | nr | 9.18 |
| 50mm dia. | - | - | 0.88 | 10.10 | nr | 10.10 |
| 65mm dia. | - | - | 1.00 | 11.47 | nr | 11.47 |
| 80mm dia. | - | - | 1.12 | 12.85 | nr | 12.85 |
| 100mm dia. | - | - | 3.20 | 36.72 | nr | 36.72 |
| 125mm dia. | - | - | 3.73 | 42.80 | nr | 42.80 |
| 150mm dia. | - | - | 4.25 | 48.77 | nr | 48.77 |
| Splay cut end | | | | | | |
| 15mm dia. | - | - | 0.14 | 1.61 | nr | 1.61 |
| 20mm dia. | - | - | 0.16 | 1.84 | nr | 1.84 |
| 25mm dia. | - | - | 0.18 | 2.07 | nr | 2.07 |
| 32mm dia. | - | - | 0.25 | 2.87 | nr | 2.87 |
| 40mm dia. | - | - | 0.27 | 3.10 | nr | 3.10 |
| 50mm dia. | - | - | 0.31 | 3.56 | nr | 3.56 |
| 65mm dia. | - | - | 0.35 | 4.02 | nr | 4.02 |
| 80mm dia. | - | - | 0.40 | 4.59 | nr | 4.59 |
| 100mm dia. | - | - | 0.48 | 5.51 | nr | 5.51 |
| 125mm dia. | - | - | 0.56 | 6.43 | nr | 6.43 |
| 150mm dia. | - | - | 0.64 | 7.34 | nr | 7.34 |
| Screwed joint to fitting | | | | | | |
| 15mm dia. | - | - | 0.30 | 3.44 | nr | 3.44 |
| 20mm dia. | - | - | 0.40 | 4.59 | nr | 4.59 |
| 25mm dia. | - | - | 0.46 | 5.28 | nr | 5.28 |
| 32mm dia. | - | - | 0.53 | 6.08 | nr | 6.08 |
| 40mm dia. | - | - | 0.61 | 7.00 | nr | 7.00 |
| 50mm dia. | - | - | 0.73 | 8.38 | nr | 8.38 |
| 65mm dia. | - | - | 0.89 | 10.21 | nr | 10.21 |
| 80mm dia. | - | - | 1.05 | 12.05 | nr | 12.05 |
| 100mm dia. | - | - | 1.46 | 16.75 | nr | 16.75 |
| 125mm dia. | - | - | 2.10 | 24.10 | nr | 24.10 |
| 150mm dia. | - | - | 2.73 | 31.33 | nr | 31.33 |
| Straight butt weld | | | | | | |
| 15mm dia. | - | - | 0.30 | 3.44 | nr | 3.44 |
| 20mm dia. | - | - | 0.40 | 4.59 | nr | 4.59 |
| 25mm dia. | - | - | 0.50 | 5.74 | nr | 5.74 |
| 32mm dia. | - | - | 0.66 | 7.57 | nr | 7.57 |
| 40mm dia. | - | - | 0.80 | 9.18 | nr | 9.18 |
| 50mm dia. | - | - | 1.20 | 13.77 | nr | 13.77 |
| 65mm dia. | - | - | 1.60 | 18.36 | nr | 18.36 |
| 80mm dia. | - | - | 2.00 | 22.95 | nr | 22.95 |
| 100mm dia. | - | - | 2.40 | 27.54 | nr | 27.54 |
| 125mm dia. | - | - | 2.81 | 32.24 | nr | 32.24 |
| 150mm dia. | - | - | 3.21 | 36.83 | nr | 36.83 |

## Y:MECHANICAL AND ELECTRICAL SERVICES

| Item | Net Price £ | Material £ | Labour hours | Labour £ | Unit | Total Rate £ |
|---|---|---|---|---|---|---|
| **Y10 : PIPELINES : BLACK STEEL - WELDED (contd)** | | | | | | |
| **Extra Over black steel butt welded pipes; labours (contd)** | | | | | | |
| Branch weld | | | | | | |
| 15mm dia. | - | - | 0.45 | 5.16 | nr | **5.16** |
| 20mm dia. | - | - | 0.60 | 6.88 | nr | **6.88** |
| 25mm dia. | - | - | 0.76 | 8.72 | nr | **8.72** |
| 32mm dia. | - | - | 0.99 | 11.36 | nr | **11.36** |
| 40mm dia. | - | - | 1.07 | 12.28 | nr | **12.28** |
| 50mm dia. | - | - | 1.50 | 17.21 | nr | **17.21** |
| 65mm dia. | - | - | 2.00 | 22.95 | nr | **22.95** |
| 80mm dia. | - | - | 2.50 | 28.69 | nr | **28.69** |
| 100mm dia. | - | - | 3.00 | 34.42 | nr | **34.42** |
| 125mm dia. | - | - | 3.50 | 40.16 | nr | **40.16** |
| 150mm dia. | - | - | 4.00 | 45.90 | nr | **45.90** |
| Welded reducing joint | | | | | | |
| 15mm dia. | - | - | 0.60 | 6.88 | nr | **6.88** |
| 20mm dia. | - | - | 0.80 | 9.18 | nr | **9.18** |
| 25mm dia. | - | - | 1.00 | 11.47 | nr | **11.47** |
| 32mm dia. | - | - | 1.32 | 15.15 | nr | **15.15** |
| 40mm dia. | - | - | 1.60 | 18.36 | nr | **18.36** |
| 50mm dia. | - | - | 2.40 | 27.54 | nr | **27.54** |
| 65mm dia. | - | - | 3.21 | 36.83 | nr | **36.83** |
| 80mm dia. | - | - | 4.00 | 45.90 | nr | **45.90** |
| 100mm dia. | - | - | 4.41 | 50.60 | nr | **50.60** |
| 125mm dia. | - | - | 4.81 | 55.19 | nr | **55.19** |
| 150mm dia. | - | - | 5.21 | 59.78 | nr | **59.78** |

*Material Costs / Measured Work Prices - Mechanical Installations*  181

## Y:MECHANICAL AND ELECTRICAL SERVICES

| Item | Net Price £ | Material £ | Labour hours | Labour £ | Unit | Total Rate £ |
|---|---|---|---|---|---|---|
| **Y10 : PIPELINES : GALVANISED STEEL** | | | | | | |
| **Galvanised steel pipes; screwed and socketed joints; BS 1387: 1985** | | | | | | |
| Galvanised; light | | | | | | |
| 15mm dia. | 3.60 | 4.16 | 0.53 | 6.08 | m | **10.24** |
| 20mm dia. | 4.47 | 5.16 | 0.54 | 6.20 | m | **11.36** |
| 25mm dia. | 5.97 | 6.90 | 0.59 | 6.77 | m | **13.67** |
| 32mm dia. | 7.36 | 8.50 | 0.66 | 7.57 | m | **16.07** |
| 40mm dia. | 9.06 | 10.46 | 0.72 | 8.27 | m | **18.73** |
| 50mm dia. | 11.80 | 13.63 | 0.78 | 8.95 | m | **22.58** |
| 65mm dia. | 17.30 | 19.98 | 0.88 | 10.10 | m | **30.08** |
| 80mm dia. | 20.68 | 23.88 | 0.98 | 11.25 | m | **35.13** |
| 100mm dia. | 29.88 | 34.51 | 1.32 | 15.16 | m | **49.67** |
| Galvanised; medium | | | | | | |
| 8mm dia. | 4.29 | 4.95 | 0.52 | 5.97 | m | **10.92** |
| 10mm dia. | 4.31 | 4.97 | 0.52 | 5.97 | m | **10.94** |
| 15mm dia. | 4.23 | 4.89 | 0.53 | 6.08 | m | **10.97** |
| 20mm dia. | 4.77 | 5.51 | 0.54 | 6.20 | m | **11.71** |
| 25mm dia. | 6.67 | 7.71 | 0.59 | 6.77 | m | **14.48** |
| 32mm dia. | 8.26 | 9.54 | 0.66 | 7.57 | m | **17.11** |
| 40mm dia. | 9.60 | 11.08 | 0.72 | 8.27 | m | **19.35** |
| 50mm dia. | 13.46 | 15.55 | 0.78 | 8.95 | m | **24.50** |
| 65mm dia. | 18.26 | 21.09 | 0.88 | 10.10 | m | **31.19** |
| 80mm dia. | 23.65 | 27.31 | 0.98 | 11.25 | m | **38.56** |
| 100mm dia. | 33.43 | 38.61 | 1.32 | 15.16 | m | **53.77** |
| 125mm dia. | 49.19 | 56.81 | 1.50 | 17.21 | m | **74.02** |
| 150mm dia. | 57.13 | 65.98 | 1.70 | 19.52 | m | **85.50** |
| Galvanised; heavy | | | | | | |
| 15mm dia. | 5.01 | 5.78 | 0.53 | 6.08 | m | **11.86** |
| 20mm dia. | 5.68 | 6.56 | 0.54 | 6.20 | m | **12.76** |
| 25mm dia. | 8.10 | 9.35 | 0.59 | 6.77 | m | **16.12** |
| 32mm dia. | 10.05 | 11.61 | 0.66 | 7.57 | m | **19.18** |
| 40mm dia. | 11.73 | 13.55 | 0.72 | 8.27 | m | **21.82** |
| 50mm dia. | 16.25 | 18.77 | 0.78 | 8.95 | m | **27.72** |
| 65mm dia. | 22.07 | 25.49 | 0.88 | 10.10 | m | **35.59** |
| 80mm dia. | 28.03 | 32.38 | 0.98 | 11.25 | m | **43.63** |
| 100mm dia. | 39.06 | 45.12 | 1.32 | 15.16 | m | **60.28** |
| 125mm dia. | 52.37 | 60.48 | 1.50 | 17.21 | m | **77.69** |
| 150mm dia. | 61.24 | 70.73 | 1.70 | 19.52 | m | **90.25** |
| **Galvanised steel pipes; butt welded joints; BS 1387: 1985** | | | | | | |
| Galvanised; light | | | | | | |
| 15mm dia. | 3.20 | 3.69 | 0.55 | 6.31 | m | **10.00** |
| 20mm dia. | 3.97 | 4.59 | 0.57 | 6.54 | m | **11.13** |
| 25mm dia. | 5.30 | 6.12 | 0.63 | 7.23 | m | **13.35** |
| 32mm dia. | 6.53 | 7.54 | 0.72 | 8.27 | m | **15.81** |
| 40mm dia. | 8.05 | 9.29 | 0.80 | 9.18 | m | **18.47** |
| 50mm dia. | 10.48 | 12.11 | 0.92 | 10.56 | m | **22.67** |
| 65mm dia. | 15.36 | 17.74 | 1.07 | 12.29 | m | **30.03** |
| 80mm dia. | 18.37 | 21.21 | 1.22 | 14.01 | m | **35.22** |
| 100mm dia. | 26.53 | 30.65 | 1.58 | 18.13 | m | **48.78** |

## Y:MECHANICAL AND ELECTRICAL SERVICES

| Item | Net Price £ | Material £ | Labour hours | Labour £ | Unit | Total Rate £ |
|---|---|---|---|---|---|---|
| **Y10 : PIPELINES : GALVANISED STEEL (contd)** | | | | | | |
| **Galvanised steel pipes; butt welded joints; BS 1387: 1985 (contd)** | | | | | | |
| Galvanised; medium | | | | | | |
| 8mm dia. | 3.82 | 4.41 | 0.54 | 6.20 | m | 10.61 |
| 10mm dia. | 3.82 | 4.42 | 0.54 | 6.20 | m | 10.62 |
| 15mm dia. | 3.76 | 4.34 | 0.55 | 6.31 | m | 10.65 |
| 20mm dia. | 4.24 | 4.90 | 0.57 | 6.54 | m | 11.44 |
| 25mm dia. | 5.92 | 6.84 | 0.63 | 7.23 | m | 14.07 |
| 32mm dia. | 7.33 | 8.47 | 0.72 | 8.27 | m | 16.74 |
| 40mm dia. | 8.52 | 9.84 | 0.80 | 9.18 | m | 19.02 |
| 50mm dia. | 11.96 | 13.81 | 0.92 | 10.56 | m | 24.37 |
| 65mm dia. | 16.21 | 18.73 | 1.07 | 12.29 | m | 31.02 |
| 80mm dia. | 21.00 | 24.25 | 1.22 | 14.01 | m | 38.26 |
| 100mm dia. | 29.69 | 34.29 | 1.58 | 18.16 | m | 52.45 |
| 125mm dia. | 43.68 | 50.46 | 1.80 | 20.66 | m | 71.12 |
| 150mm dia. | 50.73 | 58.60 | 2.03 | 23.29 | m | 81.89 |
| Galvanised; heavy | | | | | | |
| 15mm dia. | 4.45 | 5.14 | 0.55 | 6.31 | m | 11.45 |
| 20mm dia. | 5.05 | 5.83 | 0.57 | 6.54 | m | 12.37 |
| 25mm dia. | 7.19 | 8.31 | 0.63 | 7.23 | m | 15.54 |
| 32mm dia. | 8.93 | 10.31 | 0.72 | 8.27 | m | 18.58 |
| 40mm dia. | 10.42 | 12.03 | 0.80 | 9.18 | m | 21.21 |
| 50mm dia. | 14.43 | 16.67 | 0.92 | 10.56 | m | 27.23 |
| 65mm dia. | 19.60 | 22.64 | 1.07 | 12.29 | m | 34.93 |
| 80mm dia. | 24.90 | 28.76 | 1.22 | 14.01 | m | 42.77 |
| 100mm dia. | 34.69 | 40.07 | 1.58 | 18.16 | m | 58.23 |
| 125mm dia. | 46.51 | 53.72 | 1.80 | 20.66 | m | 74.38 |
| 150mm dia. | 54.39 | 62.82 | 2.03 | 23.29 | m | 86.11 |
| **Extra Over galvanised steel screwed pipes; galvanised steel flanges, screwed and drilled; metric; BS 4504** | | | | | | |
| Screwed flanges; PN6 | | | | | | |
| 15mm dia. | 11.38 | 12.52 | 0.37 | 4.25 | nr | 16.77 |
| 20mm dia. | 11.38 | 12.52 | 0.49 | 5.62 | nr | 18.14 |
| 25mm dia. | 12.04 | 13.24 | 0.56 | 6.43 | nr | 19.67 |
| 32mm dia. | 15.56 | 17.12 | 0.63 | 7.23 | nr | 24.35 |
| 40mm dia. | 15.56 | 17.12 | 0.73 | 8.38 | nr | 25.50 |
| 50mm dia. | 18.77 | 20.65 | 0.88 | 10.10 | nr | 30.75 |
| 65mm dia. | 24.15 | 26.57 | 1.08 | 12.39 | nr | 38.96 |
| 80mm dia. | 28.79 | 31.67 | 1.29 | 14.80 | nr | 46.47 |
| 100mm dia. | 34.27 | 37.70 | 1.42 | 16.29 | nr | 53.99 |
| 125mm dia. | 83.90 | 92.29 | 1.70 | 19.51 | nr | 111.80 |
| 150mm dia. | 83.90 | 92.29 | 2.08 | 23.87 | nr | 116.16 |

## Y:MECHANICAL AND ELECTRICAL SERVICES

| Item | Net Price £ | Material £ | Labour hours | Labour £ | Unit | Total Rate £ |
|---|---|---|---|---|---|---|
| **Screwed flanges; PN16** | | | | | | |
| 15mm dia. | 10.86 | 11.95 | 0.37 | 4.25 | nr | 16.20 |
| 20mm dia. | 10.86 | 11.95 | 0.49 | 5.62 | nr | 17.57 |
| 25mm dia. | 11.49 | 12.64 | 0.56 | 6.43 | nr | 19.07 |
| 32mm dia. | 15.44 | 16.98 | 0.63 | 7.23 | nr | 24.21 |
| 40mm dia. | 15.44 | 16.98 | 0.73 | 8.38 | nr | 25.36 |
| 50mm dia. | 18.42 | 20.26 | 0.88 | 10.10 | nr | 30.36 |
| 65mm dia. | 22.79 | 25.07 | 1.08 | 12.39 | nr | 37.46 |
| 80mm dia. | 28.52 | 31.37 | 1.29 | 14.80 | nr | 46.17 |
| 100mm dia. | 33.31 | 36.64 | 1.42 | 16.29 | nr | 52.93 |
| 125mm dia. | 83.31 | 91.64 | 1.70 | 19.51 | nr | 111.15 |
| 150mm dia. | 83.31 | 91.64 | 2.08 | 23.87 | nr | 115.51 |
| **Extra Over galvanised steel screwed pipes; galvanised steel flanges, screwed and drilled; imperial; BS 10** | | | | | | |
| Screwed flanges; Table E | | | | | | |
| 1/2" dia. | 13.69 | 15.06 | 0.37 | 4.25 | nr | 19.31 |
| 3/4" dia. | 13.69 | 15.06 | 0.49 | 5.62 | nr | 20.68 |
| 1" dia. | 13.69 | 15.06 | 0.56 | 6.43 | nr | 21.49 |
| 1 1/4" dia. | 13.80 | 15.18 | 0.63 | 7.23 | nr | 22.41 |
| 1 1/2" dia. | 13.80 | 15.18 | 0.73 | 8.38 | nr | 23.56 |
| 2" dia. | 13.80 | 15.18 | 0.88 | 10.10 | nr | 25.28 |
| 2 1/2" dia. | 17.64 | 19.40 | 1.08 | 12.39 | nr | 31.79 |
| 3" dia. | 17.64 | 19.40 | 1.29 | 14.80 | nr | 34.20 |
| 4" dia. | 26.82 | 29.50 | 1.42 | 16.29 | nr | 45.79 |
| 5" dia. | 52.40 | 57.64 | 1.70 | 19.51 | nr | 77.15 |
| 6" dia. | 52.40 | 57.64 | 2.08 | 23.87 | nr | 81.51 |
| **Extra Over galvanised steel screwed pipes; galvanised steel flange connections** | | | | | | |
| Bolted connection between pair of flanges; including gasket, bolts, nuts and washers | | | | | | |
| 50mm dia. | 4.56 | 5.02 | 0.50 | 5.74 | nr | 10.76 |
| 65mm dia. | 4.86 | 5.35 | 0.50 | 5.74 | nr | 11.09 |
| 80mm dia. | 8.83 | 9.71 | 0.50 | 5.74 | nr | 15.45 |
| 100mm dia. | 9.06 | 9.97 | 0.50 | 5.74 | nr | 15.71 |
| 125mm dia. | 9.79 | 10.77 | 0.50 | 5.74 | nr | 16.51 |
| 150mm dia. | 17.47 | 19.22 | 0.88 | 10.10 | nr | 29.32 |
| **Extra Over galvanised steel screwed pipes; galvanised heavy steel tubular fittings; BS 1387** | | | | | | |
| Long screw connection with socket and backnut | | | | | | |
| 15mm dia. | 3.98 | 4.38 | 0.66 | 7.57 | nr | 11.95 |
| 20mm dia. | 4.73 | 5.20 | 0.88 | 10.10 | nr | 15.30 |
| 25mm dia. | 6.55 | 7.21 | 1.00 | 11.47 | nr | 18.68 |
| 32mm dia. | 8.56 | 9.42 | 1.16 | 13.31 | nr | 22.73 |
| 40mm dia. | 10.42 | 11.46 | 1.34 | 15.38 | nr | 26.84 |
| 50mm dia. | 15.32 | 16.85 | 1.60 | 18.36 | nr | 35.21 |
| 65mm dia. | 32.90 | 36.19 | 1.96 | 22.49 | nr | 58.68 |
| 80mm dia. | 45.18 | 49.70 | 2.32 | 26.62 | nr | 76.32 |
| 100mm dia. | 72.88 | 80.17 | 3.20 | 36.72 | nr | 116.89 |

## Y:MECHANICAL AND ELECTRICAL SERVICES

| Item | Net Price £ | Material £ | Labour hours | Labour £ | Unit | Total Rate £ |
|---|---|---|---|---|---|---|
| **Y10 : PIPELINES : GALVANISED STEEL (contd)** | | | | | | |
| **Extra Over galvanised steel screwed pipes; galvanised heavy steel tubular fittings; BS 1387 (contd)** | | | | | | |
| Running nipple | | | | | | |
| 15mm dia. | 0.97 | 1.07 | 0.66 | 7.57 | nr | 8.64 |
| 20mm dia. | 1.22 | 1.34 | 0.88 | 10.10 | nr | 11.44 |
| 25mm dia. | 1.49 | 1.64 | 1.00 | 11.47 | nr | 13.11 |
| 32mm dia. | 2.11 | 2.32 | 1.16 | 13.31 | nr | 15.63 |
| 40mm dia. | 2.84 | 3.12 | 1.34 | 15.38 | nr | 18.50 |
| 50mm dia. | 4.33 | 4.76 | 1.60 | 18.36 | nr | 23.12 |
| 65mm dia. | 9.30 | 10.23 | 1.96 | 22.49 | nr | 32.72 |
| 80mm dia. | 14.50 | 15.95 | 2.32 | 26.62 | nr | 42.57 |
| 100mm dia. | 22.71 | 24.98 | 3.20 | 36.72 | nr | 61.70 |
| Barrel nipple | | | | | | |
| 15mm dia. | 0.74 | 0.81 | 0.66 | 7.57 | nr | 8.38 |
| 20mm dia. | 0.96 | 1.06 | 0.88 | 10.10 | nr | 11.16 |
| 25mm dia. | 1.25 | 1.38 | 1.00 | 11.47 | nr | 12.85 |
| 32mm dia. | 1.86 | 2.05 | 1.16 | 13.31 | nr | 15.36 |
| 40mm dia. | 2.30 | 2.53 | 1.34 | 15.38 | nr | 17.91 |
| 50mm dia. | 3.28 | 3.61 | 1.60 | 18.36 | nr | 21.97 |
| 65mm dia. | 7.03 | 7.73 | 1.96 | 22.49 | nr | 30.22 |
| 80mm dia. | 9.79 | 10.77 | 2.32 | 26.62 | nr | 37.39 |
| 100mm dia. | 17.71 | 19.48 | 3.20 | 36.72 | nr | 56.20 |
| 125mm dia. | 32.89 | 36.18 | 3.90 | 44.75 | nr | 80.93 |
| 150mm dia. | 51.81 | 56.99 | 4.60 | 52.78 | nr | 109.77 |
| Close taper nipple | | | | | | |
| 15mm dia. | 1.16 | 1.28 | 0.66 | 7.57 | nr | 8.85 |
| 20mm dia. | 1.51 | 1.66 | 0.88 | 10.10 | nr | 11.76 |
| 25mm dia. | 1.97 | 2.17 | 1.00 | 11.47 | nr | 13.64 |
| 32mm dia. | 2.94 | 3.23 | 1.16 | 13.31 | nr | 16.54 |
| 40mm dia. | 3.64 | 4.00 | 1.34 | 15.38 | nr | 19.38 |
| 50mm dia. | 5.59 | 6.15 | 1.60 | 18.36 | nr | 24.51 |
| 65mm dia. | 10.85 | 11.94 | 1.96 | 22.49 | nr | 34.43 |
| 80mm dia. | 14.45 | 15.89 | 2.32 | 26.62 | nr | 42.51 |
| 100mm dia. | 27.49 | 30.24 | 3.20 | 36.72 | nr | 66.96 |
| 90 degree bend with socket | | | | | | |
| 15mm dia. | 3.10 | 3.41 | 0.66 | 7.57 | nr | 10.98 |
| 20mm dia. | 4.22 | 4.64 | 0.88 | 10.10 | nr | 14.74 |
| 25mm dia. | 6.48 | 7.13 | 1.00 | 11.47 | nr | 18.60 |
| 32mm dia. | 9.29 | 10.22 | 1.16 | 13.31 | nr | 23.53 |
| 40mm dia. | 11.36 | 12.50 | 1.34 | 15.38 | nr | 27.88 |
| 50mm dia. | 17.66 | 19.43 | 1.60 | 18.36 | nr | 37.79 |
| 65mm dia. | 31.52 | 34.67 | 1.96 | 22.49 | nr | 57.16 |
| 80mm dia. | 48.56 | 53.42 | 2.32 | 26.62 | nr | 80.04 |
| 100mm dia. | 87.16 | 95.88 | 3.20 | 36.72 | nr | 132.60 |
| 125mm dia. | 303.67 | 334.04 | 3.90 | 44.75 | nr | 378.79 |
| 150mm dia. | 429.62 | 472.58 | 4.60 | 52.78 | nr | 525.36 |

## Material Costs / Measured Work Prices - Mechanical Installations

### Y:MECHANICAL AND ELECTRICAL SERVICES

| Item | Net Price £ | Material £ | Labour hours | Labour £ | Unit | Total Rate £ |
|---|---|---|---|---|---|---|
| **Extra Over galvanised steel screwed pipes; galvanised heavy steel fittings; BS 1740** | | | | | | |
| Plug | | | | | | |
| 15mm dia. | 0.79 | 0.87 | 0.33 | 3.79 | nr | 4.66 |
| 20mm dia. | 1.22 | 1.34 | 0.44 | 5.05 | nr | 6.39 |
| 25mm dia. | 2.14 | 2.35 | 0.50 | 5.74 | nr | 8.09 |
| 32mm dia. | 3.31 | 3.64 | 0.58 | 6.66 | nr | 10.30 |
| 40mm dia. | 3.66 | 4.03 | 0.67 | 7.69 | nr | 11.72 |
| 50mm dia. | 5.23 | 5.75 | 0.80 | 9.18 | nr | 14.93 |
| 65mm dia. | 12.50 | 13.75 | 0.98 | 11.25 | nr | 25.00 |
| 80mm dia. | 23.40 | 25.74 | 1.16 | 13.31 | nr | 39.05 |
| 100mm dia. | 44.93 | 49.42 | 1.80 | 20.66 | nr | 70.08 |
| Socket | | | | | | |
| 15mm dia. | 0.79 | 0.87 | 0.66 | 7.57 | nr | 8.44 |
| 20mm dia. | 0.89 | 0.98 | 0.88 | 10.10 | nr | 11.08 |
| 25mm dia. | 1.19 | 1.31 | 1.00 | 11.47 | nr | 12.78 |
| 32mm dia. | 1.82 | 2.00 | 1.16 | 13.31 | nr | 15.31 |
| 40mm dia. | 2.21 | 2.43 | 1.34 | 15.38 | nr | 17.81 |
| 50mm dia. | 3.41 | 3.75 | 1.60 | 18.36 | nr | 22.11 |
| 65mm dia. | 6.40 | 7.04 | 1.96 | 22.49 | nr | 29.53 |
| 80mm dia. | 8.72 | 9.59 | 2.32 | 26.62 | nr | 36.21 |
| 100mm dia. | 15.53 | 17.08 | 3.20 | 36.72 | nr | 53.80 |
| 150mm dia. | 49.63 | 54.59 | 4.60 | 52.78 | nr | 107.37 |
| Elbow, female/female | | | | | | |
| 15mm dia. | 4.15 | 4.57 | 0.66 | 7.57 | nr | 12.14 |
| 20mm dia. | 5.41 | 5.95 | 0.88 | 10.10 | nr | 16.05 |
| 25mm dia. | 7.33 | 8.06 | 1.00 | 11.47 | nr | 19.53 |
| 32mm dia. | 13.66 | 15.03 | 1.16 | 13.31 | nr | 28.34 |
| 40mm dia. | 16.28 | 17.91 | 1.34 | 15.38 | nr | 33.29 |
| 50mm dia. | 26.51 | 29.16 | 1.60 | 18.36 | nr | 47.52 |
| 65mm dia. | 68.96 | 75.86 | 1.96 | 22.49 | nr | 98.35 |
| 80mm dia. | 82.22 | 90.44 | 2.32 | 26.62 | nr | 117.06 |
| 100mm dia. | 134.62 | 148.08 | 3.20 | 36.72 | nr | 184.80 |
| Equal tee | | | | | | |
| 15mm dia. | 5.15 | 5.67 | 0.95 | 10.90 | nr | 16.57 |
| 20mm dia. | 5.98 | 6.58 | 1.27 | 14.57 | nr | 21.15 |
| 25mm dia. | 8.81 | 9.69 | 1.47 | 16.87 | nr | 26.56 |
| 32mm dia. | 18.22 | 20.04 | 1.68 | 19.28 | nr | 39.32 |
| 40mm dia. | 18.03 | 19.83 | 1.94 | 22.26 | nr | 42.09 |
| 50mm dia. | 32.24 | 35.46 | 2.31 | 26.51 | nr | 61.97 |
| 65mm dia. | 82.45 | 90.69 | 2.83 | 32.47 | nr | 123.16 |
| 80mm dia. | 88.48 | 97.33 | 3.35 | 38.44 | nr | 135.77 |
| 100mm dia. | 142.40 | 156.64 | 4.63 | 53.13 | nr | 209.77 |

## Y:MECHANICAL AND ELECTRICAL SERVICES

| Item | Net Price £ | Material £ | Labour hours | Labour £ | Unit | Total Rate £ |
|---|---|---|---|---|---|---|
| **Y10 : PIPELINES : GALVANISED STEEL (contd)** | | | | | | |
| **Extra Over galvanised steel screwed pipes; galvanised malleable iron fittings; BS 143** | | | | | | |
| Cap | | | | | | |
| 15mm dia. | 0.78 | 0.86 | 0.33 | 3.79 | nr | 4.65 |
| 20mm dia. | 0.89 | 0.98 | 0.44 | 5.05 | nr | 6.03 |
| 25mm dia. | 1.12 | 1.23 | 0.50 | 5.74 | nr | 6.97 |
| 32mm dia. | 1.62 | 1.78 | 0.58 | 6.66 | nr | 8.44 |
| 40mm dia. | 2.06 | 2.27 | 0.67 | 7.69 | nr | 9.96 |
| 50mm dia. | 4.01 | 4.42 | 0.80 | 9.18 | nr | 13.60 |
| 65mm dia. | 6.48 | 7.13 | 0.98 | 11.25 | nr | 18.38 |
| 80mm dia. | 7.34 | 8.08 | 1.16 | 13.31 | nr | 21.39 |
| 100mm dia. | 16.08 | 17.68 | 1.80 | 20.66 | nr | 38.34 |
| Plain plug, hollow | | | | | | |
| 15mm dia. | 0.61 | 0.67 | 0.33 | 3.79 | nr | 4.46 |
| 20mm dia. | 0.78 | 0.86 | 0.44 | 5.05 | nr | 5.91 |
| 25mm dia. | 0.95 | 1.04 | 0.50 | 5.74 | nr | 6.78 |
| 32mm dia. | 1.34 | 1.47 | 0.58 | 6.66 | nr | 8.13 |
| 40mm dia. | 2.23 | 2.45 | 0.67 | 7.69 | nr | 10.14 |
| 50mm dia. | 3.12 | 3.43 | 0.80 | 9.18 | nr | 12.61 |
| 65mm dia. | 4.77 | 5.25 | 0.98 | 11.25 | nr | 16.50 |
| 80mm dia. | 7.13 | 7.84 | 1.16 | 13.31 | nr | 21.15 |
| 100mm dia. | 13.13 | 14.44 | 1.80 | 20.66 | nr | 35.10 |
| 125mm dia. | 34.29 | 37.72 | 2.56 | 29.38 | nr | 67.10 |
| 150mm dia. | 41.15 | 45.27 | 3.32 | 38.10 | nr | 83.37 |
| Plain plug, solid | | | | | | |
| 15mm dia. | 1.62 | 1.78 | 0.33 | 3.79 | nr | 5.57 |
| 20mm dia. | 1.62 | 1.78 | 0.44 | 5.05 | nr | 6.83 |
| 25mm dia. | 2.40 | 2.64 | 0.50 | 5.74 | nr | 8.38 |
| 32mm dia. | 2.90 | 3.19 | 0.58 | 6.66 | nr | 9.85 |
| 40mm dia. | 3.90 | 4.30 | 0.67 | 7.69 | nr | 11.99 |
| 50mm dia. | 5.13 | 5.64 | 0.80 | 9.18 | nr | 14.82 |
| 65mm dia. | 5.47 | 6.01 | 0.98 | 11.25 | nr | 17.26 |
| 80mm dia. | 8.15 | 8.96 | 1.16 | 13.31 | nr | 22.27 |
| Elbow, male/female | | | | | | |
| 15mm dia. | 1.00 | 1.10 | 0.66 | 7.57 | nr | 8.67 |
| 20mm dia. | 1.34 | 1.47 | 0.88 | 10.10 | nr | 11.57 |
| 25mm dia. | 2.23 | 2.45 | 1.00 | 11.47 | nr | 13.92 |
| 32mm dia. | 3.68 | 4.05 | 1.16 | 13.31 | nr | 17.36 |
| 40mm dia. | 5.46 | 6.01 | 1.34 | 15.38 | nr | 21.39 |
| 50mm dia. | 7.03 | 7.73 | 1.60 | 18.36 | nr | 26.09 |
| 65mm dia. | 13.72 | 15.09 | 1.96 | 22.49 | nr | 37.58 |
| 80mm dia. | 18.76 | 20.63 | 2.32 | 26.62 | nr | 47.25 |
| 100mm dia. | 32.79 | 36.07 | 3.20 | 36.72 | nr | 72.79 |

## Y:MECHANICAL AND ELECTRICAL SERVICES

| Item | Net Price £ | Material £ | Labour hours | Labour £ | Unit | Total Rate £ |
|---|---|---|---|---|---|---|
| Elbow | | | | | | |
| 15mm dia. | 0.89 | 0.98 | 0.66 | 7.57 | nr | 8.55 |
| 20mm dia. | 1.23 | 1.35 | 0.88 | 10.10 | nr | 11.45 |
| 25mm dia. | 1.90 | 2.09 | 1.00 | 11.47 | nr | 13.56 |
| 32mm dia. | 3.12 | 3.43 | 1.16 | 13.31 | nr | 16.74 |
| 40mm dia. | 5.24 | 5.76 | 1.34 | 15.38 | nr | 21.14 |
| 50mm dia. | 6.13 | 6.75 | 1.60 | 18.36 | nr | 25.11 |
| 65mm dia. | 11.90 | 13.08 | 1.96 | 22.49 | nr | 35.57 |
| 80mm dia. | 17.47 | 19.22 | 2.32 | 26.62 | nr | 45.84 |
| 100mm dia. | 30.01 | 33.01 | 3.20 | 36.72 | nr | 69.73 |
| 125mm dia. | 64.30 | 70.74 | 4.60 | 52.78 | nr | 123.52 |
| 150mm dia. | 119.71 | 131.68 | 6.00 | 68.85 | nr | 200.53 |
| 45 degree elbow | | | | | | |
| 15mm dia. | 1.90 | 2.09 | 0.66 | 7.57 | nr | 9.66 |
| 20mm dia. | 2.34 | 2.58 | 0.88 | 10.10 | nr | 12.68 |
| 25mm dia. | 3.46 | 3.80 | 1.00 | 11.47 | nr | 15.27 |
| 32mm dia. | 6.36 | 6.99 | 1.16 | 13.31 | nr | 20.30 |
| 40mm dia. | 7.81 | 8.59 | 1.34 | 15.38 | nr | 23.97 |
| 50mm dia. | 10.70 | 11.77 | 1.60 | 18.36 | nr | 30.13 |
| 65mm dia. | 14.15 | 15.56 | 1.96 | 22.49 | nr | 38.05 |
| 80mm dia. | 21.27 | 23.40 | 2.32 | 26.62 | nr | 50.02 |
| 100mm dia. | 41.10 | 45.21 | 3.20 | 36.72 | nr | 81.93 |
| 150mm dia. | 110.01 | 121.01 | 6.00 | 68.85 | nr | 189.86 |
| Bend, male/female | | | | | | |
| 15mm dia. | 1.95 | 2.15 | 0.66 | 7.57 | nr | 9.72 |
| 20mm dia. | 2.45 | 2.70 | 0.88 | 10.10 | nr | 12.80 |
| 25mm dia. | 3.68 | 4.05 | 1.00 | 11.47 | nr | 15.52 |
| 32mm dia. | 6.80 | 7.48 | 1.16 | 13.31 | nr | 20.79 |
| 40mm dia. | 7.25 | 7.97 | 1.34 | 15.38 | nr | 23.35 |
| 50mm dia. | 10.37 | 11.41 | 1.60 | 18.36 | nr | 29.77 |
| 65mm dia. | 17.98 | 19.78 | 1.96 | 22.49 | nr | 42.27 |
| 80mm dia. | 26.22 | 28.84 | 2.32 | 26.62 | nr | 55.46 |
| 100mm dia. | 62.18 | 68.40 | 3.20 | 36.72 | nr | 105.12 |
| Bend, male | | | | | | |
| 15mm dia. | 3.18 | 3.50 | 0.66 | 7.57 | nr | 11.07 |
| 20mm dia. | 3.57 | 3.92 | 0.88 | 10.10 | nr | 14.02 |
| 25mm dia. | 5.24 | 5.76 | 1.00 | 11.47 | nr | 17.23 |
| 32mm dia. | 10.26 | 11.28 | 1.16 | 13.31 | nr | 24.59 |
| 40mm dia. | 14.39 | 15.82 | 1.34 | 15.38 | nr | 31.20 |
| 50mm dia. | 19.24 | 21.16 | 1.60 | 18.36 | nr | 39.52 |
| Bend, female | | | | | | |
| 15mm dia. | 1.62 | 1.78 | 0.66 | 7.57 | nr | 9.35 |
| 20mm dia. | 2.29 | 2.51 | 0.88 | 10.10 | nr | 12.61 |
| 25mm dia. | 3.24 | 3.56 | 1.00 | 11.47 | nr | 15.03 |
| 32mm dia. | 5.52 | 6.07 | 1.16 | 13.31 | nr | 19.38 |
| 40mm dia. | 6.58 | 7.24 | 1.34 | 15.38 | nr | 22.62 |
| 50mm dia. | 10.37 | 11.41 | 1.60 | 18.36 | nr | 29.77 |
| 65mm dia. | 19.88 | 21.87 | 1.96 | 22.49 | nr | 44.36 |
| 80mm dia. | 29.47 | 32.42 | 2.32 | 26.62 | nr | 59.04 |
| 100mm dia. | 61.84 | 68.02 | 3.20 | 36.72 | nr | 104.74 |
| 125mm dia. | 158.08 | 173.89 | 4.60 | 52.78 | nr | 226.67 |
| 150mm dia. | 241.46 | 265.60 | 6.00 | 68.85 | nr | 334.45 |

## Y:MECHANICAL AND ELECTRICAL SERVICES

| Item | Net Price £ | Material £ | Labour hours | Labour £ | Unit | Total Rate £ |
|---|---|---|---|---|---|---|
| **Y10 : PIPELINES : GALVANISED STEEL (contd)** | | | | | | |
| **Extra Over galvanised steel screwed pipes; galvanised malleable iron fittings; BS 143 (contd)** | | | | | | |
| Return bend | | | | | | |
| 15mm dia. | 6.41 | 7.05 | 0.66 | 7.57 | nr | 14.62 |
| 20mm dia. | 10.37 | 11.41 | 0.88 | 10.10 | nr | 21.51 |
| 25mm dia. | 12.94 | 14.23 | 1.00 | 11.47 | nr | 25.70 |
| 32mm dia. | 18.06 | 19.87 | 1.16 | 13.31 | nr | 33.18 |
| 40mm dia. | 21.52 | 23.67 | 1.34 | 15.38 | nr | 39.05 |
| 50mm dia. | 32.84 | 36.13 | 1.60 | 18.36 | nr | 54.49 |
| Equal socket, parallel thread | | | | | | |
| 15mm dia. | 0.84 | 0.92 | 0.66 | 7.57 | nr | 8.49 |
| 20mm dia. | 1.00 | 1.10 | 0.88 | 10.10 | nr | 11.20 |
| 25mm dia. | 1.34 | 1.47 | 1.00 | 11.47 | nr | 12.94 |
| 32mm dia. | 2.29 | 2.51 | 1.16 | 13.31 | nr | 15.82 |
| 40mm dia. | 3.12 | 3.43 | 1.34 | 15.38 | nr | 18.81 |
| 50mm dia. | 4.68 | 5.15 | 1.60 | 18.36 | nr | 23.51 |
| 65mm dia. | 7.72 | 8.49 | 1.96 | 22.49 | nr | 30.98 |
| 80mm dia. | 10.61 | 11.67 | 2.32 | 26.62 | nr | 38.29 |
| 100mm dia. | 18.00 | 19.80 | 3.20 | 36.72 | nr | 56.52 |
| Concentric reducing socket | | | | | | |
| 20 x 15mm dia. | 1.23 | 1.35 | 0.88 | 10.10 | nr | 11.45 |
| 25 x 15mm dia. | 1.62 | 1.78 | 1.00 | 11.47 | nr | 13.25 |
| 25 x 20mm dia. | 1.51 | 1.66 | 1.00 | 11.47 | nr | 13.13 |
| 32 x 25mm dia. | 2.57 | 2.82 | 1.16 | 13.31 | nr | 16.13 |
| 40 x 25mm dia. | 3.01 | 3.31 | 1.34 | 15.38 | nr | 18.69 |
| 40 x 32mm dia. | 3.34 | 3.68 | 1.34 | 15.38 | nr | 19.06 |
| 50 x 25mm dia. | 5.80 | 6.38 | 1.60 | 18.36 | nr | 24.74 |
| 50 x 40mm dia. | 4.68 | 5.15 | 1.60 | 18.36 | nr | 23.51 |
| 65 x 50mm dia. | 8.04 | 8.84 | 1.96 | 22.49 | nr | 31.33 |
| 80 x 50mm dia. | 10.02 | 11.02 | 2.32 | 26.62 | nr | 37.64 |
| 100 x 50mm dia. | 19.99 | 21.99 | 3.20 | 36.72 | nr | 58.71 |
| 100 x 80mm dia. | 18.54 | 20.40 | 3.20 | 36.72 | nr | 57.12 |
| 150 x 100mm dia. | 48.92 | 53.82 | 4.60 | 52.78 | nr | 106.60 |
| Eccentric reducing socket | | | | | | |
| 20 x 15mm dia. | 2.18 | 2.39 | 0.88 | 10.10 | nr | 12.49 |
| 25 x 15mm dia. | 6.19 | 6.81 | 1.00 | 11.47 | nr | 18.28 |
| 25 x 20mm dia. | 7.03 | 7.73 | 1.00 | 11.47 | nr | 19.20 |
| 32 x 25mm dia. | 8.03 | 8.83 | 1.16 | 13.31 | nr | 22.14 |
| 40 x 25mm dia. | 9.20 | 10.12 | 1.34 | 15.38 | nr | 25.50 |
| 40 x 32mm dia. | 4.96 | 5.46 | 1.34 | 15.38 | nr | 20.84 |
| 50 x 25mm dia. | 5.97 | 6.56 | 1.60 | 18.36 | nr | 24.92 |
| 50 x 40mm dia. | 5.97 | 6.56 | 1.60 | 18.36 | nr | 24.92 |
| 65 x 50mm dia. | 10.02 | 11.02 | 1.96 | 22.49 | nr | 33.51 |
| 80 x 50mm dia. | 16.29 | 17.92 | 2.32 | 26.62 | nr | 44.54 |

## Y:MECHANICAL AND ELECTRICAL SERVICES

| Item | Net Price £ | Material £ | Labour hours | Labour £ | Unit | Total Rate £ |
|---|---|---|---|---|---|---|
| Hexagon bush | | | | | | |
| 20 x 15mm dia. | 0.78 | 0.86 | 0.44 | 5.05 | nr | 5.91 |
| 25 x 15mm dia. | 0.95 | 1.04 | 0.50 | 5.74 | nr | 6.78 |
| 25 x 20mm dia. | 1.00 | 1.10 | 0.50 | 5.74 | nr | 6.84 |
| 32 x 25mm dia. | 1.17 | 1.29 | 0.58 | 6.66 | nr | 7.95 |
| 40 x 25mm dia. | 1.58 | 1.74 | 0.67 | 7.69 | nr | 9.43 |
| 40 x 32mm dia. | 1.67 | 1.84 | 0.67 | 7.69 | nr | 9.53 |
| 50 x 25mm dia. | 3.34 | 3.68 | 0.80 | 9.18 | nr | 12.86 |
| 50 x 40mm dia. | 3.12 | 3.43 | 0.80 | 9.18 | nr | 12.61 |
| 65 x 50mm dia. | 5.25 | 5.78 | 0.98 | 11.25 | nr | 17.03 |
| 80 x 50mm dia. | 7.93 | 8.73 | 1.16 | 13.31 | nr | 22.04 |
| 100 x 50mm dia. | 17.58 | 19.33 | 1.80 | 20.66 | nr | 39.99 |
| 100 x 80mm dia. | 14.63 | 16.09 | 1.80 | 20.66 | nr | 36.75 |
| 150 x 100mm dia. | 46.30 | 50.93 | 3.00 | 34.42 | nr | 85.35 |
| Hexagon nipple | | | | | | |
| 15mm dia. | 0.84 | 0.92 | 0.33 | 3.79 | nr | 4.71 |
| 20mm dia. | 0.95 | 1.04 | 0.44 | 5.05 | nr | 6.09 |
| 25mm dia. | 1.34 | 1.47 | 0.50 | 5.74 | nr | 7.21 |
| 32mm dia. | 2.23 | 2.45 | 0.58 | 6.66 | nr | 9.11 |
| 40mm dia. | 2.57 | 2.82 | 0.67 | 7.69 | nr | 10.51 |
| 50mm dia. | 4.68 | 5.15 | 0.80 | 9.18 | nr | 14.33 |
| 65mm dia. | 6.86 | 7.54 | 0.98 | 11.25 | nr | 18.79 |
| 80mm dia. | 9.91 | 10.90 | 1.16 | 13.31 | nr | 24.21 |
| 100mm dia. | 16.83 | 18.51 | 1.80 | 20.66 | nr | 39.17 |
| 150mm dia. | 47.58 | 52.34 | 3.00 | 34.42 | nr | 86.76 |
| Union, male/female | | | | | | |
| 15mm dia. | 3.46 | 3.80 | 0.66 | 7.57 | nr | 11.37 |
| 20mm dia. | 4.24 | 4.66 | 0.88 | 10.10 | nr | 14.76 |
| 25mm dia. | 5.30 | 5.83 | 1.00 | 11.47 | nr | 17.30 |
| 32mm dia. | 7.36 | 8.09 | 1.16 | 13.31 | nr | 21.40 |
| 40mm dia. | 9.42 | 10.36 | 1.34 | 15.38 | nr | 25.74 |
| 50mm dia. | 14.83 | 16.31 | 1.60 | 18.36 | nr | 34.67 |
| 65mm dia. | 28.72 | 31.59 | 1.96 | 22.49 | nr | 54.08 |
| 80mm dia. | 39.49 | 43.44 | 2.32 | 26.62 | nr | 70.06 |
| Union, female | | | | | | |
| 15mm dia. | 3.57 | 3.92 | 0.66 | 7.57 | nr | 11.49 |
| 20mm dia. | 3.90 | 4.30 | 0.88 | 10.10 | nr | 14.40 |
| 25mm dia. | 4.57 | 5.03 | 1.00 | 11.47 | nr | 16.50 |
| 32mm dia. | 6.91 | 7.60 | 1.16 | 13.31 | nr | 20.91 |
| 40mm dia. | 7.81 | 8.59 | 1.34 | 15.38 | nr | 23.97 |
| 50mm dia. | 12.94 | 14.23 | 1.60 | 18.36 | nr | 32.59 |
| 65mm dia. | 26.10 | 28.71 | 1.96 | 22.49 | nr | 51.20 |
| 80mm dia. | 34.51 | 37.96 | 2.32 | 26.62 | nr | 64.58 |
| 100mm dia. | 65.70 | 72.27 | 3.20 | 36.72 | nr | 108.99 |
| Union elbow, male/female | | | | | | |
| 15mm dia. | 4.40 | 4.84 | 0.66 | 7.57 | nr | 12.41 |
| 20mm dia. | 5.52 | 6.07 | 0.88 | 10.10 | nr | 16.17 |
| 25mm dia. | 7.75 | 8.52 | 1.00 | 11.47 | nr | 19.99 |

## Y:MECHANICAL AND ELECTRICAL SERVICES

| Item | Net Price £ | Material £ | Labour hours | Labour £ | Unit | Total Rate £ |
|---|---|---|---|---|---|---|
| **Y10 : PIPELINES : GALVANISED STEEL (contd)** | | | | | | |
| **Extra Over galvanised steel screwed pipes; galvanised malleable iron fittings; BS 143 (contd)** | | | | | | |
| Twin elbow | | | | | | |
| 15mm dia. | 4.24 | 4.66 | 0.95 | 10.90 | nr | 15.56 |
| 20mm dia. | 4.68 | 5.15 | 1.27 | 14.57 | nr | 19.72 |
| 25mm dia. | 7.58 | 8.34 | 1.45 | 16.64 | nr | 24.98 |
| 32mm dia. | 13.44 | 14.78 | 1.68 | 19.28 | nr | 34.06 |
| 40mm dia. | 17.01 | 18.71 | 1.94 | 22.26 | nr | 40.97 |
| 50mm dia. | 21.86 | 24.04 | 2.31 | 26.51 | nr | 50.55 |
| 65mm dia. | 34.72 | 38.19 | 2.83 | 32.47 | nr | 70.66 |
| 80mm dia. | 59.16 | 65.07 | 3.35 | 38.44 | nr | 103.51 |
| Equal tee | | | | | | |
| 15mm dia. | 1.23 | 1.35 | 0.95 | 10.90 | nr | 12.25 |
| 20mm dia. | 1.79 | 1.96 | 1.27 | 14.57 | nr | 16.53 |
| 25mm dia. | 2.57 | 2.82 | 1.45 | 16.64 | nr | 19.46 |
| 32mm dia. | 4.24 | 4.66 | 1.68 | 19.28 | nr | 23.94 |
| 40mm dia. | 5.80 | 6.38 | 1.94 | 22.26 | nr | 28.64 |
| 50mm dia. | 8.36 | 9.20 | 2.31 | 26.51 | nr | 35.71 |
| 65mm dia. | 17.15 | 18.86 | 2.83 | 32.47 | nr | 51.33 |
| 80mm dia. | 19.99 | 21.99 | 3.35 | 38.44 | nr | 60.43 |
| 100mm dia. | 36.23 | 39.85 | 4.63 | 53.13 | nr | 92.98 |
| 125mm dia. | 88.85 | 97.73 | 5.32 | 61.05 | nr | 158.78 |
| 150mm dia. | 141.57 | 155.73 | 6.00 | 68.85 | nr | 224.58 |
| Tee reducing on branch | | | | | | |
| 20 x 15mm dia. | 1.62 | 1.78 | 1.27 | 14.57 | nr | 16.35 |
| 25 x 15mm dia. | 2.23 | 2.45 | 1.45 | 16.64 | nr | 19.09 |
| 25 x 20mm dia. | 2.34 | 2.58 | 1.45 | 16.64 | nr | 19.22 |
| 32 x 25mm dia. | 4.12 | 4.54 | 1.68 | 19.28 | nr | 23.82 |
| 40 x 25mm dia. | 5.46 | 6.01 | 1.94 | 22.26 | nr | 28.27 |
| 40 x 32mm dia. | 7.14 | 7.85 | 1.94 | 22.26 | nr | 30.11 |
| 50 x 25mm dia. | 7.25 | 7.97 | 2.31 | 26.51 | nr | 34.48 |
| 50 x 40mm dia. | 10.03 | 11.04 | 2.31 | 26.51 | nr | 37.55 |
| 65 x 50mm dia. | 15.22 | 16.74 | 2.83 | 32.47 | nr | 49.21 |
| 80 x 50mm dia. | 20.58 | 22.63 | 3.35 | 38.44 | nr | 61.07 |
| 100 x 50mm dia. | 30.01 | 33.01 | 4.63 | 53.13 | nr | 86.14 |
| 100 x 80mm dia. | 46.30 | 50.93 | 4.63 | 53.13 | nr | 104.06 |
| 150 x 100mm dia. | 104.28 | 114.71 | 6.00 | 68.85 | nr | 183.56 |
| Equal pitcher tee | | | | | | |
| 15mm dia. | 3.34 | 3.68 | 0.95 | 10.90 | nr | 14.58 |
| 20mm dia. | 4.12 | 4.54 | 1.27 | 14.57 | nr | 19.11 |
| 25mm dia. | 6.19 | 6.81 | 1.45 | 16.64 | nr | 23.45 |
| 32mm dia. | 8.48 | 9.32 | 1.68 | 19.28 | nr | 28.60 |
| 40mm dia. | 13.10 | 14.41 | 1.94 | 22.26 | nr | 36.67 |
| 50mm dia. | 18.40 | 20.24 | 2.31 | 26.51 | nr | 46.75 |
| 65mm dia. | 25.72 | 28.29 | 2.83 | 32.47 | nr | 60.76 |
| 80mm dia. | 35.37 | 38.90 | 3.35 | 38.44 | nr | 77.34 |
| 100mm dia. | 79.58 | 87.53 | 4.63 | 53.13 | nr | 140.66 |

## Y:MECHANICAL AND ELECTRICAL SERVICES

| Item | Net Price £ | Material £ | Labour hours | Labour £ | Unit | Total Rate £ |
|---|---|---|---|---|---|---|
| Cross | | | | | | |
| 15mm dia. | 2.90 | 3.19 | 1.05 | 12.05 | nr | **15.24** |
| 20mm dia. | 4.35 | 4.78 | 1.40 | 16.07 | nr | **20.85** |
| 25mm dia. | 5.52 | 6.07 | 1.60 | 18.36 | nr | **24.43** |
| 32mm dia. | 7.25 | 7.97 | 1.86 | 21.34 | nr | **29.31** |
| 40mm dia. | 9.76 | 10.73 | 2.14 | 24.56 | nr | **35.29** |
| 50mm dia. | 15.16 | 16.68 | 2.56 | 29.38 | nr | **46.06** |
| 65mm dia. | 21.27 | 23.40 | 3.14 | 36.03 | nr | **59.43** |
| 80mm dia. | 28.29 | 31.12 | 3.71 | 42.57 | nr | **73.69** |
| 100mm dia. | 51.44 | 56.59 | 5.12 | 58.75 | nr | **115.34** |

## Y:MECHANICAL AND ELECTRICAL SERVICES

| Item | Net Price £ | Material £ | Labour hours | Labour £ | Unit | Total Rate £ |
|---|---|---|---|---|---|---|
| **Y10 : PIPELINES : MALLEABLE GROOVED JOINTING SYSTEM** | | | | | | |
| Malleable Grooved Jointing System; suitable in Hot, Cold water installations; -40C to +90C dry heat ; supplied complete with standard grade E gasket, nuts and bolts and painted finish | | | | | | |
| Grooved Joints | | | | | | |
| 20 mm | 6.09 | 6.70 | 0.50 | 5.74 | nr | **12.44** |
| 40 mm | 7.56 | 8.32 | 0.65 | 7.46 | nr | **15.78** |
| 50 mm | 7.56 | 8.32 | 0.65 | 7.46 | nr | **15.78** |
| 65 mm | 8.42 | 9.26 | 0.75 | 8.61 | nr | **17.87** |
| 80 mm | 8.80 | 9.68 | 1.00 | 11.47 | nr | **21.15** |
| 100 mm | 13.40 | 14.74 | 1.25 | 14.34 | nr | **29.08** |
| 125 mm | 20.42 | 22.46 | 1.50 | 17.21 | nr | **39.67** |
| Malleable Grooved Fittings; suitable in Hot, Cold water installations; -40C to +90C dry heat ; supplied complete with standard grade E gasket, nuts and bolts and painted finish | | | | | | |
| Mechanical tee; Threaded outlet | | | | | | |
| 32 x 15mm | 8.66 | 9.53 | 0.80 | 9.18 | nr | **18.71** |
| 32 x 20mm | 8.66 | 9.53 | 0.80 | 9.18 | nr | **18.71** |
| 32 x 25mm | 9.07 | 9.98 | 0.80 | 9.18 | nr | **19.16** |
| 40 x 15mm | 9.11 | 10.02 | 1.00 | 11.47 | nr | **21.49** |
| 40 x 25mm | 10.16 | 11.18 | 1.00 | 11.47 | nr | **22.65** |
| 50 x 15mm | 9.66 | 10.63 | 1.20 | 13.77 | nr | **24.40** |
| 50 x 20mm | 9.66 | 10.63 | 1.20 | 13.77 | nr | **24.40** |
| 50 x 25mm | 10.60 | 11.66 | 1.20 | 13.77 | nr | **25.43** |
| Grooved adaptors | | | | | | |
| 50mm | 36.85 | 40.53 | 0.60 | 6.88 | nr | **47.41** |
| 65mm | 39.22 | 43.14 | 0.75 | 8.61 | nr | **51.75** |
| 80mm | 43.05 | 47.35 | 0.75 | 8.61 | nr | **55.96** |
| 100mm | 46.17 | 50.79 | 1.35 | 15.49 | nr | **66.28** |
| 125mm | 56.64 | 62.30 | 1.37 | 15.72 | nr | **78.02** |
| Short radius elbow; 90 degree | | | | | | |
| 25mm | 11.87 | 13.06 | 0.65 | 7.46 | nr | **20.52** |
| 32mm | 12.47 | 13.72 | 0.75 | 8.61 | nr | **22.33** |
| 40mm | 13.17 | 14.49 | 0.75 | 8.61 | nr | **23.10** |
| 50mm | 14.08 | 15.49 | 0.80 | 9.18 | nr | **24.67** |
| 65mm | 17.76 | 19.54 | 0.90 | 10.33 | nr | **29.87** |
| 80mm | 18.14 | 19.95 | 1.20 | 13.77 | nr | **33.72** |
| 100mm | 24.28 | 26.71 | 1.65 | 18.93 | nr | **45.64** |
| 125mm | 40.01 | 44.01 | 1.90 | 21.80 | nr | **65.81** |

Material Costs / Measured Work Prices - Mechanical Installations

## Y:MECHANICAL AND ELECTRICAL SERVICES

| Item | Net Price £ | Material £ | Labour hours | Labour £ | Unit | Total Rate £ |
|---|---|---|---|---|---|---|
| Short radius elbow; 45 degree | | | | | | |
| 25mm | 10.32 | 11.35 | 0.65 | 7.46 | nr | 18.81 |
| 32mm | 10.83 | 11.91 | 0.75 | 8.61 | nr | 20.52 |
| 40mm | 11.45 | 12.60 | 0.75 | 8.61 | nr | 21.21 |
| 50mm | 11.95 | 13.14 | 0.80 | 9.18 | nr | 22.32 |
| 65mm | 15.24 | 16.76 | 0.90 | 10.33 | nr | 27.09 |
| 80mm | 17.11 | 18.82 | 1.20 | 13.77 | nr | 32.59 |
| 100mm | 21.24 | 23.36 | 1.65 | 18.93 | nr | 42.29 |
| 125mm | 35.91 | 39.50 | 1.90 | 21.80 | nr | 61.30 |
| Equal tee | | | | | | |
| 25mm | 16.87 | 18.56 | 0.95 | 10.90 | nr | 29.46 |
| 32mm | 19.72 | 21.69 | 1.15 | 13.20 | nr | 34.89 |
| 40mm | 21.14 | 23.25 | 1.15 | 13.20 | nr | 36.45 |
| 50mm | 28.27 | 31.10 | 1.25 | 14.34 | nr | 45.44 |
| 65mm | 31.98 | 35.18 | 1.35 | 15.49 | nr | 50.67 |
| 80mm | 33.88 | 37.27 | 1.75 | 20.08 | nr | 57.35 |
| 100mm | 37.88 | 41.67 | 2.30 | 26.39 | nr | 68.06 |
| Reducing tee; grooved outlet | | | | | | |
| 50 x 50 x 25mm | 28.55 | 31.41 | 0.95 | 10.90 | nr | 42.31 |
| 50 x 50 x 40mm | 30.75 | 33.83 | 1.15 | 13.20 | nr | 47.03 |
| 80 x 80 x 50mm | 53.61 | 58.97 | 1.25 | 14.34 | nr | 73.31 |
| 100 x 100 x 50mm | 64.85 | 71.33 | 1.25 | 14.34 | nr | 85.67 |
| 100 x 100 x 80mm | 67.42 | 74.16 | 1.75 | 20.08 | nr | 94.24 |
| Concentric reducers | | | | | | |
| 50 x 40mm | 16.90 | 18.59 | 1.00 | 11.47 | nr | 30.06 |
| 80 x 50mm | 16.96 | 18.66 | 1.10 | 12.62 | nr | 31.28 |
| 100 x 50mm | 27.41 | 30.15 | 1.40 | 16.07 | nr | 46.22 |
| 100 x 80mm | 34.03 | 37.43 | 1.50 | 17.21 | nr | 54.64 |
| 125 x 80mm | 39.40 | 43.34 | 1.70 | 19.51 | nr | 62.85 |

# Keep your figures up to date, free of charge

This section, and most of the other information in this Price Book, is brought up to date every three months, until the next annual edition, in the *Price Book Update*.

The *Update* is available free to all Price Book purchasers.

To ensure you receive your copy, simply complete the reply card from the centre of the book and return it to us.

## Y:MECHANICAL AND ELECTRICAL SERVICES

| Item | Net Price £ | Material £ | Labour hours | Labour £ | Unit | Total Rate £ |
|---|---|---|---|---|---|---|
| **Y10 : PIPELINES : CARBON STEEL** | | | | | | |
| **Hot finished seamless carbon steel pipe; API 5LB grade; butt welded joints** | | | | | | |
| 1/2 " nominal pipe size | | | | | | |
| 2.8mm pipe thickness | 1.56 | 1.80 | 0.55 | 6.31 | m | 8.11 |
| 3.7mm pipe thickness | 1.96 | 2.26 | 0.55 | 6.31 | m | 8.57 |
| 4.8mm pipe thickness | 3.58 | 4.13 | 0.55 | 6.31 | m | 10.44 |
| 7.5mm pipe thickness | 5.55 | 6.41 | 0.55 | 6.31 | m | 12.72 |
| 3/4 " nominal pipe size | | | | | | |
| 2.9mm pipe thickness | 2.04 | 2.36 | 0.57 | 6.54 | m | 8.90 |
| 3.9mm pipe thickness | 2.63 | 3.04 | 0.57 | 6.54 | m | 9.58 |
| 5.6mm pipe thickness | 4.83 | 5.58 | 0.57 | 6.54 | m | 12.12 |
| 7.8mm pipe thickness | 7.42 | 8.57 | 0.57 | 6.54 | m | 15.11 |
| 1 " nominal pipe size | | | | | | |
| 3.4mm pipe thickness | 3.03 | 3.50 | 0.63 | 7.23 | m | 10.73 |
| 4.5mm pipe thickness | 3.86 | 4.46 | 0.63 | 7.23 | m | 11.69 |
| 6.4mm pipe thickness | 5.50 | 6.35 | 0.63 | 7.23 | m | 13.58 |
| 9.1mm pipe thickness | 9.06 | 10.46 | 0.63 | 7.23 | m | 17.69 |
| 1 1/4 " nominal pipe size | | | | | | |
| 3.6mm pipe thickness | 3.80 | 4.39 | 0.72 | 8.26 | m | 12.65 |
| 4.9mm pipe thickness | 5.00 | 5.78 | 0.72 | 8.26 | m | 14.04 |
| 6.4mm pipe thickness | 6.78 | 7.83 | 0.72 | 8.26 | m | 16.09 |
| 9.7mm pipe thickness | 12.20 | 14.09 | 0.72 | 8.26 | m | 22.35 |
| 1 1/2 " nominal pipe size | | | | | | |
| 3.7mm pipe thickness | 4.13 | 4.77 | 0.80 | 9.18 | m | 13.95 |
| 5.1mm pipe thickness | 5.51 | 6.36 | 0.80 | 9.18 | m | 15.54 |
| 7.1mm pipe thickness | 7.32 | 8.45 | 0.80 | 9.18 | m | 17.63 |
| 10.2mm pipe thickness | 10.61 | 12.25 | 0.80 | 9.18 | m | 21.43 |
| 2 " nominal pipe size | | | | | | |
| 3.9mm pipe thickness | 5.63 | 6.50 | 0.92 | 10.56 | m | 17.06 |
| 5.5mm pipe thickness | 7.72 | 8.92 | 0.92 | 10.56 | m | 19.48 |
| 8.7mm pipe thickness | 12.54 | 14.48 | 0.92 | 10.56 | m | 25.04 |
| 11.1mm pipe thickness | 15.27 | 17.64 | 0.92 | 10.56 | m | 28.20 |
| 2 1/2 " nominal pipe size | | | | | | |
| 5.2mm pipe thickness | 9.03 | 10.43 | 1.07 | 12.28 | m | 22.71 |
| 7.0mm pipe thickness | 11.85 | 13.69 | 1.07 | 12.28 | m | 25.97 |
| 9.5mm pipe thickness | 16.87 | 19.48 | 1.07 | 12.28 | m | 31.76 |
| 14.0mm pipe thickness | 23.10 | 26.68 | 1.07 | 12.28 | m | 38.96 |
| 3 " nominal pipe size | | | | | | |
| 5.5mm pipe thickness | 10.69 | 12.35 | 1.22 | 14.00 | m | 26.35 |
| 7.6mm pipe thickness | 14.40 | 16.63 | 1.22 | 14.00 | m | 30.63 |
| 11.1mm pipe thickness | 22.15 | 25.58 | 1.22 | 14.00 | m | 39.58 |
| 15.2mm pipe thickness | 28.72 | 33.17 | 1.22 | 14.00 | m | 47.17 |

## Y:MECHANICAL AND ELECTRICAL SERVICES

| Item | Net Price £ | Material £ | Labour hours | Labour £ | Unit | Total Rate £ |
|---|---|---|---|---|---|---|
| 4 " nominal pipe size | | | | | | |
| 6.0mm pipe thickness | 15.15 | 17.50 | 1.58 | 18.13 | m | 35.63 |
| 8.6mm pipe thickness | 21.19 | 24.47 | 1.58 | 18.13 | m | 42.60 |
| 11.1mm pipe thickness | 29.37 | 33.92 | 1.58 | 18.13 | m | 52.05 |
| 13.5mm pipe thickness | 34.90 | 40.31 | 1.58 | 18.13 | m | 58.44 |
| 17.1mm pipe thickness | 42.61 | 49.21 | 1.58 | 18.13 | m | 67.34 |
| 5 " nominal pipe size | | | | | | |
| 6.6mm pipe thickness | 20.73 | 23.94 | 1.80 | 20.66 | m | 44.60 |
| 9.5mm pipe thickness | 29.18 | 33.70 | 1.80 | 20.66 | m | 54.36 |
| 12.7mm pipe thickness | 41.89 | 48.38 | 1.80 | 20.66 | m | 69.04 |
| 15.9mm pipe thickness | 51.11 | 59.03 | 1.80 | 20.66 | m | 79.69 |
| 19.1mm pipe thickness | 59.85 | 69.13 | 1.80 | 20.66 | m | 89.79 |
| 6 " nominal pipe size | | | | | | |
| 7.1mm pipe thickness | 29.34 | 33.89 | 2.03 | 23.29 | m | 57.18 |
| 11.0mm pipe thickness | 44.37 | 51.25 | 2.03 | 23.29 | m | 74.54 |
| 14.3mm pipe thickness | 56.47 | 65.22 | 2.03 | 23.29 | m | 88.51 |
| 18.3mm pipe thickness | 70.38 | 81.29 | 2.03 | 23.29 | m | 104.58 |
| 22.0mm pipe thickness | 82.51 | 95.30 | 2.03 | 23.29 | m | 118.59 |
| **Extra Over hot finished carbon steel pipe; forged steel welded fittings; API 3000LB** | | | | | | |
| Coupling | | | | | | |
| 1/2 " dia. | 0.60 | 0.66 | 0.52 | 5.97 | nr | 6.63 |
| 3/4 " dia. | 0.75 | 0.82 | 0.69 | 7.92 | nr | 8.74 |
| 1 " dia. | 1.00 | 1.10 | 0.85 | 9.75 | nr | 10.85 |
| 1 1/4 " dia. | 1.44 | 1.58 | 1.07 | 12.28 | nr | 13.86 |
| 1 1/2" dia. | 1.75 | 1.93 | 1.33 | 15.26 | nr | 17.19 |
| 2 " dia. | 2.23 | 2.45 | 1.86 | 21.34 | nr | 23.79 |
| Union | | | | | | |
| 1/2 " dia. | 2.46 | 2.71 | 0.52 | 5.97 | nr | 8.68 |
| 3/4 " dia. | 2.58 | 2.84 | 0.69 | 7.92 | nr | 10.76 |
| 1 " dia. | 3.24 | 3.56 | 0.85 | 9.75 | nr | 13.31 |
| 1 1/4 " dia. | 5.55 | 6.11 | 1.07 | 12.28 | nr | 18.39 |
| 1 1/2" dia. | 6.45 | 7.09 | 1.33 | 15.26 | nr | 22.35 |
| 2 " dia. | 7.92 | 8.71 | 1.86 | 21.34 | nr | 30.05 |
| 45 degree elbow | | | | | | |
| 1/2 " dia. | 0.99 | 1.09 | 0.52 | 5.97 | nr | 7.06 |
| 3/4 " dia. | 1.23 | 1.35 | 0.69 | 7.92 | nr | 9.27 |
| 1 " dia. | 1.59 | 1.75 | 0.85 | 9.75 | nr | 11.50 |
| 1 1/4 " dia. | 2.97 | 3.27 | 1.07 | 12.28 | nr | 15.55 |
| 1 1/2" dia. | 3.96 | 4.36 | 1.33 | 15.26 | nr | 19.62 |
| 2 " dia. | 6.60 | 7.26 | 1.86 | 21.34 | nr | 28.60 |

## Y:MECHANICAL AND ELECTRICAL SERVICES

| Item | Net Price £ | Material £ | Labour hours | Labour £ | Unit | Total Rate £ |
|---|---|---|---|---|---|---|
| **Y10 : PIPELINES : CARBON STEEL (contd)** | | | | | | |
| **Extra Over hot finished carbon steel pipe; forged steel welded fittings; API 3000LB (contd)** | | | | | | |
| 90 degree elbow | | | | | | |
| 1/2 " dia. | 0.90 | 0.99 | 0.52 | 5.97 | nr | **6.96** |
| 3/4 " dia. | 1.11 | 1.22 | 0.69 | 7.92 | nr | **9.14** |
| 1 " dia. | 1.44 | 1.58 | 0.85 | 9.75 | nr | **11.33** |
| 1 1/4 " dia. | 2.73 | 3.00 | 1.07 | 12.28 | nr | **15.28** |
| 1 1/2" dia. | 3.57 | 3.93 | 1.33 | 15.26 | nr | **19.19** |
| 2 " dia. | 5.79 | 6.37 | 1.86 | 21.34 | nr | **27.71** |
| Equal tee | | | | | | |
| 1/2 " dia. | 1.11 | 1.22 | 0.75 | 8.61 | nr | **9.83** |
| 3/4 " dia. | 1.38 | 1.52 | 1.00 | 11.47 | nr | **12.99** |
| 1 " dia. | 1.80 | 1.98 | 1.23 | 14.11 | nr | **16.09** |
| 1 1/4 " dia. | 3.48 | 3.83 | 1.44 | 16.52 | nr | **20.35** |
| 1 1/2" dia. | 4.80 | 5.28 | 1.93 | 22.15 | nr | **27.43** |
| 2 " dia. | 7.50 | 8.25 | 2.87 | 32.93 | nr | **41.18** |

## Y:MECHANICAL AND ELECTRICAL SERVICES

| Item | Net Price £ | Material £ | Labour hours | Labour £ | Unit | Total Rate £ |
|---|---|---|---|---|---|---|
| **Y10 : PIPELINES : STAINLESS STEEL** | | | | | | |
| Stainless steel pipes; capillary or compression joints; BS 4127 | | | | | | |
| Grade 304; satin finish | | | | | | |
| 15mm dia. | 1.13 | 1.31 | 0.41 | 4.70 | m | 6.01 |
| 22mm dia. | 1.70 | 1.96 | 0.51 | 5.85 | m | 7.81 |
| 28mm dia. | 2.40 | 2.77 | 0.58 | 6.66 | m | 9.43 |
| 35mm dia. | 3.62 | 4.18 | 0.65 | 7.46 | m | 11.64 |
| 42mm dia. | 4.58 | 5.29 | 0.71 | 8.15 | m | 13.44 |
| 54mm dia. | 6.37 | 7.36 | 0.80 | 9.18 | m | 16.54 |
| Grade 316 satin finish | | | | | | |
| 15mm dia. | 1.66 | 1.92 | 0.61 | 7.00 | m | 8.92 |
| 22mm dia. | 2.37 | 2.74 | 0.76 | 8.73 | m | 11.47 |
| 28mm dia. | 3.63 | 4.19 | 0.87 | 9.99 | m | 14.18 |
| 35mm dia. | 4.23 | 4.89 | 0.98 | 11.25 | m | 16.14 |
| 42mm dia. | 5.38 | 6.21 | 1.06 | 12.17 | m | 18.38 |
| 54mm dia. | 7.31 | 8.44 | 1.16 | 13.31 | m | 21.75 |
| Extra Over stainless steel pipes; capillary fittings | | | | | | |
| Straight coupling | | | | | | |
| 15mm dia. | 0.51 | 0.56 | 0.25 | 2.87 | nr | 3.43 |
| 22mm dia. | 0.83 | 0.91 | 0.28 | 3.21 | nr | 4.12 |
| 28mm dia. | 1.10 | 1.21 | 0.33 | 3.79 | nr | 5.00 |
| 35mm dia. | 3.30 | 3.63 | 0.37 | 4.25 | nr | 7.88 |
| 42mm dia. | 3.81 | 4.19 | 0.42 | 4.82 | nr | 9.01 |
| 54mm dia. | 5.72 | 6.29 | 0.45 | 5.16 | nr | 11.45 |
| 45 degree bend | | | | | | |
| 15mm dia. | 2.77 | 3.05 | 0.25 | 2.87 | nr | 5.92 |
| 22mm dia. | 3.63 | 3.99 | 0.30 | 3.40 | nr | 7.39 |
| 28mm dia. | 4.47 | 4.92 | 0.33 | 3.79 | nr | 8.71 |
| 35mm dia. | 7.75 | 8.53 | 0.37 | 4.25 | nr | 12.78 |
| 42mm dia. | 10.29 | 11.32 | 0.42 | 4.82 | nr | 16.14 |
| 54mm dia. | 14.45 | 15.89 | 0.45 | 5.16 | nr | 21.05 |
| 90 degree bend | | | | | | |
| 15mm dia. | 1.43 | 1.57 | 0.28 | 3.21 | nr | 4.78 |
| 22mm dia. | 1.93 | 2.12 | 0.28 | 3.21 | nr | 5.33 |
| 28mm dia. | 2.73 | 3.00 | 0.33 | 3.79 | nr | 6.79 |
| 35mm dia. | 8.64 | 9.50 | 0.37 | 4.25 | nr | 13.75 |
| 42mm dia. | 11.89 | 13.08 | 0.42 | 4.82 | nr | 17.90 |
| 54mm dia. | 16.13 | 17.74 | 0.45 | 5.16 | nr | 22.90 |
| Reducer | | | | | | |
| 22 x 15mm dia. | 3.29 | 3.62 | 0.28 | 3.21 | nr | 6.83 |
| 28 x 22mm dia. | 3.68 | 4.05 | 0.33 | 3.79 | nr | 7.84 |
| 35 x 28mm dia. | 4.50 | 4.95 | 0.37 | 4.25 | nr | 9.20 |
| 42 x 35mm dia. | 6.32 | 6.95 | 0.42 | 4.82 | nr | 11.77 |
| 54 x 42mm dia. | 18.72 | 20.59 | 0.48 | 5.52 | nr | 26.11 |

## Y:MECHANICAL AND ELECTRICAL SERVICES

| Item | Net Price £ | Material £ | Labour hours | Labour £ | Unit | Total Rate £ |
|---|---|---|---|---|---|---|
| **Y10 : PIPELINES : STAINLESS STEEL (contd)** | | | | | | |
| **Extra Over stainless steel pipes; capillary fittings (contd)** | | | | | | |
| Tap connector | | | | | | |
| 15mm dia. | 9.72 | 10.69 | 0.13 | 1.49 | nr | 12.18 |
| 22mm dia. | 12.85 | 14.13 | 0.14 | 1.61 | nr | 15.74 |
| 28mm dia. | 17.83 | 19.61 | 0.17 | 1.95 | nr | 21.56 |
| Tank connector | | | | | | |
| 15mm dia. | 12.56 | 13.82 | 0.13 | 1.49 | nr | 15.31 |
| 22mm dia. | 18.69 | 20.56 | 0.13 | 1.49 | nr | 22.05 |
| 35mm dia. | 33.26 | 36.59 | 0.18 | 2.07 | nr | 38.66 |
| 42mm dia. | 43.94 | 48.33 | 0.21 | 2.41 | nr | 50.74 |
| 54mm dia. | 66.51 | 73.16 | 0.24 | 2.75 | nr | 75.91 |
| Tee equal | | | | | | |
| 15mm dia. | 2.56 | 2.82 | 0.37 | 4.25 | nr | 7.07 |
| 22mm dia. | 3.19 | 3.51 | 0.40 | 4.59 | nr | 8.10 |
| 28mm dia. | 3.86 | 4.25 | 0.45 | 5.16 | nr | 9.41 |
| 35mm dia. | 12.06 | 13.27 | 0.59 | 6.77 | nr | 20.04 |
| 42mm dia. | 14.87 | 16.36 | 0.62 | 7.12 | nr | 23.48 |
| 54mm dia. | 30.03 | 33.03 | 0.67 | 7.69 | nr | 40.72 |
| Unequal tee | | | | | | |
| 22 x 15mm dia. | 5.87 | 6.46 | 0.37 | 4.25 | nr | 10.71 |
| 28 x 15mm dia. | 7.61 | 8.37 | 0.45 | 5.16 | nr | 13.53 |
| 28 x 22mm dia. | 7.61 | 8.37 | 0.45 | 5.17 | nr | 13.54 |
| 35 x 22mm dia. | 13.30 | 14.63 | 0.59 | 6.77 | nr | 21.40 |
| 35 x 28mm dia. | 13.30 | 14.63 | 0.59 | 6.77 | nr | 21.40 |
| 42 x 28mm dia. | 16.36 | 18.00 | 0.62 | 7.12 | nr | 25.12 |
| 42 x 35mm dia. | 16.36 | 18.00 | 0.62 | 7.12 | nr | 25.12 |
| 54 x 35mm dia. | 61.75 | 67.92 | 0.67 | 7.69 | nr | 75.61 |
| 54 x 42mm dia. | 61.75 | 67.92 | 0.67 | 7.69 | nr | 75.61 |
| Union, conical seat | | | | | | |
| 15mm dia. | 16.10 | 17.71 | 0.25 | 2.87 | nr | 20.58 |
| 22mm dia. | 25.35 | 27.89 | 0.28 | 3.21 | nr | 31.10 |
| 28mm dia. | 32.83 | 36.11 | 0.33 | 3.79 | nr | 39.90 |
| 35mm dia. | 43.02 | 47.32 | 0.37 | 4.25 | nr | 51.57 |
| 42mm dia. | 54.27 | 59.70 | 0.42 | 4.82 | nr | 64.52 |
| 54mm dia. | 71.80 | 78.98 | 0.45 | 5.16 | nr | 84.14 |
| Union, flat seat | | | | | | |
| 15mm dia. | 16.82 | 18.50 | 0.25 | 2.87 | nr | 21.37 |
| 22mm dia. | 26.18 | 28.80 | 0.28 | 3.21 | nr | 32.01 |
| 28mm dia. | 33.85 | 37.23 | 0.33 | 3.79 | nr | 41.02 |
| 35mm dia. | 44.22 | 48.64 | 0.37 | 4.25 | nr | 52.89 |
| 42mm dia. | 55.70 | 61.27 | 0.42 | 4.82 | nr | 66.09 |
| 54mm dia. | 74.75 | 82.22 | 0.45 | 5.16 | nr | 87.38 |

## Material Costs / Measured Work Prices - Mechanical Installations

### Y:MECHANICAL AND ELECTRICAL SERVICES

| Item | Net Price £ | Material £ | Labour hours | Labour £ | Unit | Total Rate £ |
|---|---|---|---|---|---|---|
| **Extra Over stainless steel pipes; compression fittings** | | | | | | |
| Straight coupling | | | | | | |
| 15mm dia. | 11.20 | 12.32 | 0.18 | 2.07 | nr | **14.39** |
| 22mm dia. | 21.32 | 23.45 | 0.22 | 2.52 | nr | **25.97** |
| 28mm dia. | 28.70 | 31.57 | 0.25 | 2.87 | nr | **34.44** |
| 35mm dia. | 44.32 | 48.75 | 0.30 | 3.44 | nr | **52.19** |
| 42mm dia. | 51.73 | 56.90 | 0.40 | 4.59 | nr | **61.49** |
| 90 degree bend | | | | | | |
| 15mm dia. | 14.11 | 15.52 | 0.18 | 2.07 | nr | **17.59** |
| 22mm dia. | 28.02 | 30.82 | 0.22 | 2.52 | nr | **33.34** |
| 28mm dia. | 38.22 | 42.04 | 0.25 | 2.87 | nr | **44.91** |
| 35mm dia. | 77.40 | 85.14 | 0.33 | 3.79 | nr | **88.93** |
| 42mm dia. | 113.12 | 124.43 | 0.35 | 4.02 | nr | **128.45** |
| Reducer | | | | | | |
| 22 x 15mm dia. | 20.32 | 22.35 | 0.28 | 3.21 | nr | **25.56** |
| 28 x 22mm dia. | 27.81 | 30.59 | 0.28 | 3.21 | nr | **33.80** |
| 35 x 28mm dia. | 40.65 | 44.72 | 0.30 | 3.44 | nr | **48.16** |
| 42 x 35mm dia. | 54.07 | 59.48 | 0.37 | 4.25 | nr | **63.73** |
| Stud coupling | | | | | | |
| 15mm dia. | 9.73 | 10.70 | 0.42 | 4.82 | nr | **15.52** |
| 22mm dia. | 15.77 | 17.35 | 0.25 | 2.87 | nr | **20.22** |
| 28mm dia. | 25.46 | 28.01 | 0.25 | 2.87 | nr | **30.88** |
| 35mm dia. | 35.65 | 39.22 | 0.37 | 4.25 | nr | **43.47** |
| 42mm dia. | 46.14 | 50.75 | 0.42 | 4.82 | nr | **55.57** |
| Equal tee | | | | | | |
| 15mm dia. | 19.87 | 21.86 | 0.37 | 4.25 | nr | **26.11** |
| 22mm dia. | 41.04 | 45.14 | 0.40 | 4.59 | nr | **49.73** |
| 28mm dia. | 56.13 | 61.74 | 0.45 | 5.16 | nr | **66.90** |
| 35mm dia. | 111.63 | 122.79 | 0.59 | 6.77 | nr | **129.56** |
| 42mm dia. | 154.79 | 170.27 | 0.62 | 7.12 | nr | **177.39** |
| Running tee | | | | | | |
| 15mm dia. | 24.45 | 26.89 | 0.37 | 4.25 | nr | **31.14** |
| 22mm dia. | 44.05 | 48.45 | 0.40 | 4.59 | nr | **53.04** |
| 28mm dia. | 74.60 | 82.06 | 0.59 | 6.77 | nr | **88.83** |

## Y:MECHANICAL AND ELECTRICAL SERVICES

| Item | Net Price £ | Material £ | Labour hours | Labour £ | Unit | Total Rate £ |
|---|---|---|---|---|---|---|
| **Y10 : PIPELINES : COPPER** | | | | | | |
| Microbore copper pipe; capillary or compression joints in the running length; BS 2871 | | | | | | |
| Table W | | | | | | |
| 6mm dia. | 0.58 | 0.67 | 0.41 | 4.70 | m | **5.37** |
| 8mm dia. | 0.60 | 0.70 | 0.41 | 4.70 | m | **5.40** |
| 10mm dia. | 0.84 | 0.97 | 0.41 | 4.70 | m | **5.67** |
| Table W; plastic coated gas and cold water service pipe for corrosive and aggresive environments | | | | | | |
| 6mm dia. | 0.77 | 0.89 | 0.45 | 5.16 | m | **6.05** |
| 8mm dia. | 0.85 | 0.98 | 0.45 | 5.16 | m | **6.14** |
| 10mm dia. | 1.13 | 1.30 | 0.50 | 5.74 | m | **7.04** |
| Table W; profiled plastic coated central heating and hot water service pipe for heat loss reduction | | | | | | |
| 8mm dia. | 0.72 | 0.84 | 0.45 | 5.16 | m | **6.00** |
| 10mm dia. | 0.97 | 1.12 | 0.50 | 5.74 | m | **6.86** |
| Microbore accessories | | | | | | |
| Manifold connectors; side entry one way flow 22mm body | | | | | | |
| 4 x 8mm connections | 8.53 | 9.38 | 0.58 | 6.66 | nr | **16.04** |
| 6 x 8mm connections | 10.15 | 11.16 | 0.87 | 9.99 | nr | **21.15** |
| 2 x 10mm connections | 5.57 | 6.13 | 0.29 | 3.33 | nr | **9.46** |
| 4 x 10mm connections | 8.92 | 9.81 | 0.58 | 6.66 | nr | **16.47** |
| Manifold connectors; linear flow 22mm body | | | | | | |
| 4 x 8mm connections | 7.33 | 8.06 | 0.58 | 6.66 | nr | **14.72** |
| 4 x 10mm connections | 9.46 | 10.41 | 0.58 | 6.66 | nr | **17.07** |
| Manifold connectors; linear flow 28mm body | | | | | | |
| 6 x 8mm connections | 10.53 | 11.58 | 0.87 | 9.99 | nr | **21.57** |

## Y:MECHANICAL AND ELECTRICAL SERVICES

| Item | Net Price £ | Material £ | Labour hours | Labour £ | Unit | Total Rate £ |
|---|---|---|---|---|---|---|
| **Copper pipe; capillary or compression joints in the running length; BS 2871** | | | | | | |
| Table X | | | | | | |
| 8mm dia. | 0.80 | 0.92 | 0.41 | 4.70 | m | 5.62 |
| 10mm dia. | 1.02 | 1.18 | 0.41 | 4.70 | m | 5.88 |
| 12mm dia. | 1.18 | 1.36 | 0.41 | 4.70 | m | 6.06 |
| 15mm dia. | 1.06 | 1.23 | 0.41 | 4.70 | m | 5.93 |
| 22mm dia. | 2.13 | 2.46 | 0.47 | 5.40 | m | 7.86 |
| 28mm dia. | 2.84 | 3.28 | 0.51 | 5.85 | m | 9.13 |
| 35mm dia. | 6.64 | 7.67 | 0.58 | 6.66 | m | 14.33 |
| 42mm dia. | 8.10 | 9.36 | 0.65 | 7.46 | m | 16.82 |
| 54mm dia. | 10.46 | 12.08 | 0.71 | 8.15 | m | 20.23 |
| 67mm dia. | 15.64 | 18.06 | 0.74 | 8.49 | m | 26.55 |
| 76mm dia. | 22.09 | 25.51 | 0.75 | 8.61 | m | 34.12 |
| 108mm dia. | 32.54 | 37.59 | 0.76 | 8.72 | m | 46.31 |
| 133mm dia. | 42.80 | 49.44 | 1.02 | 11.70 | m | 61.14 |
| 159mm dia. | 67.94 | 78.47 | 1.12 | 12.85 | m | 91.32 |
| Table Y | | | | | | |
| 8mm dia. | 1.04 | 1.20 | 0.43 | 4.94 | m | 6.14 |
| 10mm dia. | 1.25 | 1.45 | 0.43 | 4.94 | m | 6.39 |
| 12mm dia. | 1.68 | 1.94 | 0.43 | 4.94 | m | 6.88 |
| 15mm dia. | 2.47 | 2.86 | 0.43 | 4.94 | m | 7.80 |
| 22mm dia. | 4.34 | 5.01 | 0.50 | 5.74 | m | 10.75 |
| 28mm dia. | 5.72 | 6.61 | 0.54 | 6.20 | m | 12.81 |
| 35mm dia. | 8.50 | 9.82 | 0.62 | 7.11 | m | 16.93 |
| 42mm dia. | 10.25 | 11.83 | 0.70 | 8.03 | m | 19.86 |
| 54mm dia. | 17.61 | 20.34 | 0.77 | 8.84 | m | 29.18 |
| 67mm dia. | 25.63 | 29.61 | 0.81 | 9.29 | m | 38.90 |
| 76mm dia. | 29.35 | 33.90 | 0.50 | 5.74 | m | 39.64 |
| 108mm dia. | 53.83 | 62.17 | 0.87 | 9.98 | m | 72.15 |
| **Table X; plastic coated gas and cold water service pipe for corrosive and aggresive environments** | | | | | | |
| 15mm dia. | 1.76 | 2.03 | 0.60 | 6.89 | m | 8.92 |
| 22mm dia. | 3.36 | 3.88 | 0.68 | 7.81 | m | 11.69 |
| 28mm dia. | 4.18 | 4.83 | 0.74 | 8.49 | m | 13.32 |
| 35mm dia. | 8.08 | 9.33 | 0.85 | 9.76 | m | 19.09 |
| 42mm dia. | 9.63 | 11.12 | 0.96 | 11.02 | m | 22.14 |
| 54mm dia. | 12.02 | 13.89 | 1.05 | 12.05 | m | 25.94 |
| **Table Y; plastic coated gas and cold water service pipe for corrosive and aggresive environments** | | | | | | |
| 15mm dia. | 3.17 | 3.66 | 0.60 | 6.89 | m | 10.55 |
| 22mm dia. | 5.58 | 6.44 | 0.68 | 7.81 | m | 14.25 |
| 28mm dia. | 7.07 | 8.16 | 0.74 | 8.49 | m | 16.65 |
| 35mm dia. | 9.94 | 11.48 | 0.85 | 9.76 | m | 21.24 |
| 42mm dia. | 11.77 | 13.60 | 0.96 | 11.02 | m | 24.62 |
| 54mm dia. | 19.18 | 22.15 | 1.05 | 12.05 | m | 34.20 |

## Y:MECHANICAL AND ELECTRICAL SERVICES

| Item | Net Price £ | Material £ | Labour hours | Labour £ | Unit | Total Rate £ |
|---|---|---|---|---|---|---|
| **Y10 : PIPELINES : COPPER (contd)** | | | | | | |
| **Copper pipe; capillary or compression joints in the running length; BS 2871 (contd)** | | | | | | |
| Table X; profiled plastic coated central heating and hot water service pipe for heat loss reduction | | | | | | |
| 15mm dia. | 1.88 | 2.17 | 0.60 | 6.89 | m | 9.06 |
| 22mm dia. | 3.58 | 4.14 | 0.68 | 7.81 | m | 11.95 |
| 28mm dia. | 4.43 | 5.11 | 0.74 | 8.49 | m | 13.60 |
| 35mm dia. | 8.25 | 9.53 | 0.85 | 9.76 | m | 19.29 |
| 42mm dia. | 9.80 | 11.32 | 0.96 | 11.02 | m | 22.34 |
| 54mm dia. | 12.20 | 14.09 | 1.05 | 12.05 | m | 26.14 |
| Table Y; profiled plastic coated central heating and hot water service pipe for heat loss reduction | | | | | | |
| 12mm dia. | 2.01 | 2.32 | 0.60 | 6.89 | m | 9.21 |
| 15mm dia. | 3.29 | 3.80 | 0.60 | 6.89 | m | 10.69 |
| 22mm dia. | 5.79 | 6.69 | 0.68 | 7.81 | m | 14.50 |
| **Extra Over copper pipes; capillary fittings; BS 864** | | | | | | |
| Stop end | | | | | | |
| 15mm dia. | 0.91 | 1.00 | 0.13 | 1.49 | nr | 2.49 |
| 22mm dia. | 1.32 | 1.45 | 0.14 | 1.61 | nr | 3.06 |
| 28mm dia. | 2.11 | 2.32 | 0.17 | 1.95 | nr | 4.27 |
| 35mm dia. | 4.64 | 5.10 | 0.18 | 2.07 | nr | 7.17 |
| 42mm dia. | 6.53 | 7.18 | 0.21 | 2.41 | nr | 9.59 |
| 54mm dia. | 9.87 | 10.86 | 0.21 | 2.41 | nr | 13.27 |
| Straight coupling; copper to copper | | | | | | |
| 6mm dia. | 0.65 | 0.71 | 0.25 | 2.87 | nr | 3.58 |
| 8mm dia. | 0.56 | 0.62 | 0.25 | 2.87 | nr | 3.49 |
| 10mm dia. | 0.36 | 0.40 | 0.25 | 2.87 | nr | 3.27 |
| 15mm dia. | 0.49 | 0.54 | 0.25 | 2.87 | nr | 3.41 |
| 22mm dia. | 0.38 | 0.42 | 0.28 | 3.21 | nr | 3.63 |
| 28mm dia. | 0.83 | 0.91 | 0.33 | 3.79 | nr | 4.70 |
| 35mm dia. | 2.23 | 2.45 | 0.37 | 4.25 | nr | 6.70 |
| 42mm dia. | 3.21 | 3.53 | 0.42 | 4.82 | nr | 8.35 |
| 54mm dia. | 6.63 | 7.29 | 0.45 | 5.16 | nr | 12.45 |
| 67mm dia. | 18.06 | 19.87 | 0.56 | 6.43 | nr | 26.30 |
| Adaptor coupling; imperial to metric | | | | | | |
| 1/2" x 15mm dia. | 1.21 | 1.33 | 0.25 | 2.87 | nr | 4.20 |
| 3/4" x 22mm dia. | 1.23 | 1.35 | 0.28 | 3.21 | nr | 4.56 |
| 1" x 28mm dia. | 2.34 | 2.57 | 0.33 | 3.79 | nr | 6.36 |
| 1 1/4" x 35mm dia. | 4.03 | 4.43 | 0.37 | 4.25 | nr | 8.68 |
| 1 1/2" x 42mm dia. | 5.12 | 5.63 | 0.42 | 4.82 | nr | 10.45 |

## Y:MECHANICAL AND ELECTRICAL SERVICES

| Item | Net Price £ | Material £ | Labour hours | Labour £ | Unit | Total Rate £ |
|---|---|---|---|---|---|---|
| Reducing coupling | | | | | | |
| 15 x 10mm dia. | 1.01 | 1.11 | 0.25 | 2.87 | nr | 3.98 |
| 22 x 10mm dia. | 1.94 | 2.13 | 0.28 | 3.21 | nr | 5.34 |
| 22 x 15mm dia. | 1.14 | 1.25 | 0.30 | 3.44 | nr | 4.69 |
| 28 x 15mm dia. | 2.25 | 2.48 | 0.31 | 3.56 | nr | 6.04 |
| 28 x 22mm dia. | 1.78 | 1.96 | 0.33 | 3.79 | nr | 5.75 |
| 35 x 28mm dia. | 4.50 | 4.95 | 0.37 | 4.25 | nr | 9.20 |
| 42 x 35mm dia. | 6.37 | 7.01 | 0.42 | 4.82 | nr | 11.83 |
| 54 x 35mm dia. | 9.42 | 10.36 | 0.45 | 5.16 | nr | 15.52 |
| 54 x 42mm dia. | 9.30 | 10.23 | 0.45 | 5.16 | nr | 15.39 |
| Straight female connector | | | | | | |
| 15mm x 1/2" dia. | 1.55 | 1.71 | 0.25 | 2.87 | nr | 4.58 |
| 22mm x 3/4" dia. | 2.19 | 2.41 | 0.28 | 3.21 | nr | 5.62 |
| 28mm x 1" dia. | 4.47 | 4.92 | 0.33 | 3.79 | nr | 8.71 |
| 35mm x 1 1/4" dia. | 7.45 | 8.20 | 0.37 | 4.25 | nr | 12.45 |
| 42mm x 1 1/2" dia. | 9.24 | 10.16 | 0.42 | 4.82 | nr | 14.98 |
| 54mm x 2" dia. | 14.64 | 16.10 | 0.45 | 5.16 | nr | 21.26 |
| Straight male connector | | | | | | |
| 15mm x 1/2" dia. | 1.31 | 1.44 | 0.25 | 2.87 | nr | 4.31 |
| 22mm x 3/4" dia. | 2.31 | 2.54 | 0.28 | 3.21 | nr | 5.75 |
| 28mm x 1" dia. | 3.66 | 4.03 | 0.33 | 3.79 | nr | 7.82 |
| 35mm x 1 1/4" dia. | 6.80 | 7.48 | 0.37 | 4.25 | nr | 11.73 |
| 42mm x 1 1/2" dia. | 8.76 | 9.64 | 0.42 | 4.82 | nr | 14.46 |
| 54mm x 2" dia. | 13.29 | 14.62 | 0.45 | 5.16 | nr | 19.78 |
| 67mm x 2 1/2" dia. | 21.09 | 23.09 | 0.56 | 6.43 | nr | 29.52 |
| Female reducing connector | | | | | | |
| 15mm x 3/4" dia. | 3.56 | 3.92 | 0.25 | 2.87 | nr | 6.79 |
| 22mm x 1" dia. | 6.73 | 7.40 | 0.28 | 3.21 | nr | 10.61 |
| Male reducing connector | | | | | | |
| 15mm x 3/4" dia. | 3.17 | 3.49 | 0.25 | 2.87 | nr | 6.36 |
| 22mm x 1" dia. | 4.88 | 5.37 | 0.28 | 3.21 | nr | 8.58 |
| Lead connector | | | | | | |
| 15mm dia. | 1.15 | 1.26 | 0.25 | 2.87 | nr | 4.13 |
| 22mm dia. | 1.73 | 1.90 | 0.28 | 3.21 | nr | 5.11 |
| 28mm dia. | 2.34 | 2.57 | 0.33 | 3.79 | nr | 6.36 |
| Flanged connector | | | | | | |
| 28mm dia. | 23.73 | 26.10 | 0.33 | 3.79 | nr | 29.89 |
| 35mm dia. | 30.04 | 33.04 | 0.37 | 4.25 | nr | 37.29 |
| 42mm dia. | 35.90 | 39.49 | 0.42 | 4.82 | nr | 44.31 |
| 54mm dia. | 54.27 | 59.70 | 0.45 | 5.16 | nr | 64.86 |
| 67mm dia. | 67.02 | 73.72 | 0.56 | 6.43 | nr | 80.15 |

Y:MECHANICAL AND ELECTRICAL SERVICES

| Item | Net Price £ | Material £ | Labour hours | Labour £ | Unit | Total Rate £ |
|---|---|---|---|---|---|---|
| **Y10 : PIPELINES : COPPER (contd)** | | | | | | |
| **Extra Over copper pipes; capillary fittings; BS 864 (contd)** | | | | | | |
| Tank connector | | | | | | |
| 15mm x 1/2" dia. | 3.22 | 3.54 | 0.25 | 2.87 | nr | 6.41 |
| 22mm x 3/4" dia. | 4.92 | 5.41 | 0.28 | 3.21 | nr | 8.62 |
| 28mm x 1" dia. | 6.76 | 7.44 | 0.33 | 3.79 | nr | 11.23 |
| 35mm x 1 1/4" dia. | 8.17 | 8.99 | 0.37 | 4.25 | nr | 13.24 |
| 42mm x 1 1/2" dia. | 10.71 | 11.78 | 0.42 | 4.82 | nr | 16.60 |
| 54mm x 2" dia. | 16.35 | 17.98 | 0.45 | 5.16 | nr | 23.14 |
| Tank connector with long thread | | | | | | |
| 15mm x 1/2" dia. | 4.22 | 4.64 | 0.32 | 3.67 | nr | 8.31 |
| 22mm x 3/4" dia. | 6.00 | 6.60 | 0.35 | 4.02 | nr | 10.62 |
| 28mm x 1" dia. | 7.41 | 8.15 | 0.42 | 4.82 | nr | 12.97 |
| Reducer | | | | | | |
| 15 x 10mm dia. | 0.62 | 0.68 | 0.25 | 2.87 | nr | 3.55 |
| 22 x 15mm dia. | 0.71 | 0.78 | 0.28 | 3.21 | nr | 3.99 |
| 28 x 15mm dia. | 1.63 | 1.79 | 0.30 | 3.44 | nr | 5.23 |
| 28 x 22mm dia. | 1.07 | 1.18 | 0.33 | 3.79 | nr | 4.97 |
| 35 x 22mm dia. | 3.11 | 3.42 | 0.37 | 4.25 | nr | 7.67 |
| 42 x 22mm dia. | 5.34 | 5.87 | 0.39 | 4.48 | nr | 10.35 |
| 42 x 35mm dia. | 4.74 | 5.21 | 0.42 | 4.82 | nr | 10.03 |
| 54 x 35mm dia. | 8.66 | 9.53 | 0.43 | 4.94 | nr | 14.47 |
| 54 x 42mm dia. | 8.43 | 9.27 | 0.45 | 5.16 | nr | 14.43 |
| 67 x 54mm dia. | 13.57 | 14.93 | 0.56 | 6.43 | nr | 21.36 |
| Adaptor; copper to female iron | | | | | | |
| 15mm x 1/2" dia. | 2.47 | 2.72 | 0.25 | 2.87 | nr | 5.59 |
| 22mm x 3/4" dia. | 3.30 | 3.63 | 0.28 | 3.21 | nr | 6.84 |
| 28mm x 1" dia. | 4.65 | 5.12 | 0.33 | 3.79 | nr | 8.91 |
| 35mm x 1 1/4" dia. | 8.80 | 9.68 | 0.37 | 4.25 | nr | 13.93 |
| 42mm x 1 1/2" dia. | 11.09 | 12.20 | 0.42 | 4.82 | nr | 17.02 |
| 54mm x 2" dia. | 13.35 | 14.69 | 0.45 | 5.16 | nr | 19.85 |
| Adaptor; copper to male iron | | | | | | |
| 15mm x 1/2" dia. | 2.51 | 2.76 | 0.25 | 2.87 | nr | 5.63 |
| 22mm x 3/4" dia. | 3.21 | 3.53 | 0.28 | 3.21 | nr | 6.74 |
| 28mm x 1" dia. | 4.91 | 5.40 | 0.33 | 3.79 | nr | 9.19 |
| 35mm x 1 1/4" dia. | 7.50 | 8.25 | 0.37 | 4.25 | nr | 12.50 |
| 42mm x 1 1/2" dia. | 11.29 | 12.42 | 0.42 | 4.82 | nr | 17.24 |
| 54mm x 2" dia. | 13.35 | 14.69 | 0.45 | 5.16 | nr | 19.85 |
| Union coupling | | | | | | |
| 15mm dia. | 3.69 | 4.06 | 0.37 | 4.25 | nr | 8.31 |
| 22mm dia. | 5.91 | 6.50 | 0.40 | 4.59 | nr | 11.09 |
| 28mm dia. | 8.63 | 9.49 | 0.45 | 5.16 | nr | 14.65 |
| 35mm dia. | 11.30 | 12.43 | 0.59 | 6.77 | nr | 19.20 |
| 42mm dia. | 16.53 | 18.18 | 0.62 | 7.12 | nr | 25.30 |
| 54mm dia. | 26.65 | 29.32 | 0.67 | 7.69 | nr | 37.01 |
| 67mm dia. | 49.05 | 53.95 | 0.83 | 9.53 | nr | 63.48 |

*Material Costs / Measured Work Prices - Mechanical Installations*

## Y:MECHANICAL AND ELECTRICAL SERVICES

| Item | Net Price £ | Material £ | Labour hours | Labour £ | Unit | Total Rate £ |
|---|---|---|---|---|---|---|
| Elbow | | | | | | |
| 15mm dia. | 0.28 | 0.31 | 0.25 | 2.87 | nr | 3.18 |
| 22mm dia. | 0.62 | 0.68 | 0.28 | 3.21 | nr | 3.89 |
| 28mm dia. | 1.33 | 1.46 | 0.33 | 3.79 | nr | 5.25 |
| 35mm dia. | 4.38 | 4.82 | 0.37 | 4.25 | nr | 9.07 |
| 42mm dia. | 7.88 | 8.67 | 0.42 | 4.82 | nr | 13.49 |
| 54mm dia. | 16.21 | 17.83 | 0.45 | 5.16 | nr | 22.99 |
| 67mm dia. | 39.28 | 43.21 | 0.56 | 6.43 | nr | 49.64 |
| Backplate elbow | | | | | | |
| 15mm dia. | 2.88 | 3.17 | 0.50 | 5.74 | nr | 8.91 |
| 22mm dia. | 5.95 | 6.54 | 0.53 | 6.08 | nr | 12.62 |
| Overflow bend | | | | | | |
| 15mm dia. | 6.46 | 7.11 | 0.25 | 2.87 | nr | 9.98 |
| 22mm dia. | 7.40 | 8.14 | 0.28 | 3.21 | nr | 11.35 |
| Return bend | | | | | | |
| 15mm dia. | 3.25 | 3.58 | 0.25 | 2.87 | nr | 6.45 |
| 22mm dia. | 5.22 | 5.74 | 0.28 | 3.21 | nr | 8.95 |
| 28mm dia. | 9.39 | 10.33 | 0.33 | 3.79 | nr | 14.12 |
| Obtuse elbow | | | | | | |
| 15mm dia. | 0.88 | 0.97 | 0.25 | 2.87 | nr | 3.84 |
| 22mm dia. | 1.75 | 1.93 | 0.28 | 3.21 | nr | 5.14 |
| 28mm dia. | 3.32 | 3.65 | 0.33 | 3.79 | nr | 7.44 |
| 35mm dia. | 6.31 | 6.94 | 0.37 | 4.25 | nr | 11.19 |
| 42mm dia. | 11.09 | 12.20 | 0.42 | 4.82 | nr | 17.02 |
| 54mm dia. | 18.87 | 20.76 | 0.45 | 5.16 | nr | 25.92 |
| 67mm dia. | 37.32 | 41.05 | 0.56 | 6.43 | nr | 47.48 |
| Straight tap connector | | | | | | |
| 15mm x 1/2" dia. | 1.25 | 1.38 | 0.13 | 1.49 | nr | 2.87 |
| 22mm x 3/4" dia. | 1.75 | 1.93 | 0.14 | 1.61 | nr | 3.54 |
| Bent tap connector | | | | | | |
| 15mm x 1/2" dia. | 1.60 | 1.76 | 0.13 | 1.49 | nr | 3.25 |
| 22mm x 3/4" dia. | 2.81 | 3.09 | 0.14 | 1.61 | nr | 4.70 |
| Bent male union connector | | | | | | |
| 15mm x 1/2" dia. | 5.42 | 5.96 | 0.37 | 4.25 | nr | 10.21 |
| 22mm x 3/4" dia. | 6.93 | 7.62 | 0.40 | 4.59 | nr | 12.21 |
| 28mm x 1" dia. | 9.66 | 10.63 | 0.45 | 5.16 | nr | 15.79 |
| 35mm x 1 1/4" dia. | 14.65 | 16.11 | 0.59 | 6.77 | nr | 22.88 |
| 42mm x 1 1/2" dia. | 25.32 | 27.85 | 0.62 | 7.12 | nr | 34.97 |
| 54mm x 2" dia. | 40.34 | 44.37 | 0.67 | 7.69 | nr | 52.06 |

## Y:MECHANICAL AND ELECTRICAL SERVICES

| Item | Net Price £ | Material £ | Labour hours | Labour £ | Unit | Total Rate £ |
|---|---|---|---|---|---|---|
| **Y10 : PIPELINES : COPPER (contd)** | | | | | | |
| **Extra Over copper pipes; capillary fittings; BS 864 (contd)** | | | | | | |
| Bent female union connector | | | | | | |
| 15mm dia. | 5.43 | 5.97 | 0.37 | 4.25 | nr | 10.22 |
| 22mm x 3/4" dia. | 6.84 | 7.52 | 0.40 | 4.59 | nr | 12.11 |
| 28mm x 1" dia. | 10.25 | 11.28 | 0.45 | 5.16 | nr | 16.44 |
| 35mm x 1 1/4" dia. | 16.06 | 17.67 | 0.59 | 6.77 | nr | 24.44 |
| 42mm x 1 1/2" dia. | 27.74 | 30.51 | 0.62 | 7.12 | nr | 37.63 |
| 54mm x 2" dia. | 44.20 | 48.62 | 0.67 | 7.69 | nr | 56.31 |
| Straight union adaptor | | | | | | |
| 15mm x 3/4" dia. | 2.45 | 2.69 | 0.37 | 4.25 | nr | 6.94 |
| 22mm x 1" dia. | 3.37 | 3.71 | 0.40 | 4.59 | nr | 8.30 |
| 28mm x 1 1/4" dia. | 5.12 | 5.63 | 0.45 | 5.16 | nr | 10.79 |
| 35mm x 1 1/2" dia. | 7.86 | 8.65 | 0.59 | 6.77 | nr | 15.42 |
| 42mm x 2" dia. | 10.49 | 11.54 | 0.62 | 7.12 | nr | 18.66 |
| 54mm x 2 1/2" dia. | 15.32 | 16.85 | 0.67 | 7.69 | nr | 24.54 |
| 67mm x 3" dia. | 27.48 | 30.23 | 0.83 | 9.53 | nr | 39.76 |
| Straight male union connector | | | | | | |
| 15mm x 1/2" dia. | 4.97 | 5.47 | 0.37 | 4.25 | nr | 9.72 |
| 22mm x 3/4" dia. | 6.46 | 7.11 | 0.40 | 4.59 | nr | 11.70 |
| 28mm x 1" dia. | 9.63 | 10.59 | 0.45 | 5.16 | nr | 15.75 |
| 35mm x 1 1/4" dia. | 12.33 | 13.56 | 0.59 | 6.77 | nr | 20.33 |
| 42mm x 1 1/2" dia. | 19.14 | 21.05 | 0.62 | 7.12 | nr | 28.17 |
| 54mm x 2" dia. | 30.56 | 33.62 | 0.67 | 7.69 | nr | 41.31 |
| Straight female union connector | | | | | | |
| 15mm x 1/2" dia. | 5.08 | 5.59 | 0.37 | 4.25 | nr | 9.84 |
| 22mm x 3/4" dia. | 6.30 | 6.93 | 0.40 | 4.59 | nr | 11.52 |
| 28mm x 1" dia. | 9.46 | 10.41 | 0.45 | 5.16 | nr | 15.57 |
| 35mm x 1 1/4" dia. | 11.99 | 13.19 | 0.59 | 6.77 | nr | 19.96 |
| 42mm x 1 1/2" dia. | 18.97 | 20.87 | 0.62 | 7.11 | nr | 27.98 |
| 54mm x 2" dia. | 27.87 | 30.66 | 0.67 | 7.69 | nr | 38.35 |
| Male reducing nipple | | | | | | |
| 3/4 x 1/2" dia. | 2.30 | 2.53 | 0.28 | 3.21 | nr | 5.74 |
| 1 x 3/4" dia | 2.66 | 2.93 | 0.33 | 3.79 | nr | 6.72 |
| 1 1/4 x 1" dia | 3.72 | 4.09 | 0.37 | 4.25 | nr | 8.34 |
| 1 1/2 x 1 1/4" dia. | 5.42 | 5.96 | 0.42 | 4.82 | nr | 10.78 |
| 2 x 1 1/2" dia. | 10.99 | 12.09 | 0.45 | 5.16 | nr | 17.25 |
| 2 1/2 x 2" dia. | 14.68 | 16.15 | 0.56 | 6.43 | nr | 22.58 |
| Female reducing nipple | | | | | | |
| 3/4 x 1/2" dia. | 2.30 | 2.53 | 0.28 | 3.21 | nr | 5.74 |
| 1 x 3/4" dia. | 2.60 | 2.93 | 0.33 | 3.79 | nr | 6.72 |
| 1 1/4 x 1" dia. | 3.72 | 4.09 | 0.37 | 4.25 | nr | 8.34 |
| 1 1/2 x 1 1/4" dia. | 5.42 | 5.96 | 0.42 | 4.82 | nr | 10.78 |
| 2 x 1 1/2" dia. | 10.99 | 12.09 | 0.45 | 5.16 | nr | 17.25 |
| 2 1/2 x 2" dia. | 14.68 | 16.15 | 0.56 | 6.43 | nr | 22.58 |

*Material Costs / Measured Work Prices - Mechanical Installations*

## Y:MECHANICAL AND ELECTRICAL SERVICES

| Item | Net Price £ | Material £ | Labour hours | Labour £ | Unit | Total Rate £ |
|---|---|---|---|---|---|---|
| Equal tee | | | | | | |
| 10mm dia. | 1.15 | 1.26 | 0.23 | 2.64 | nr | 3.90 |
| 15mm dia. | 0.47 | 0.52 | 0.37 | 4.25 | nr | 4.77 |
| 22mm dia. | 1.18 | 1.30 | 0.40 | 4.59 | nr | 5.89 |
| 28mm dia. | 2.91 | 3.20 | 0.45 | 5.16 | nr | 8.36 |
| 35mm dia. | 7.59 | 8.35 | 0.59 | 6.77 | nr | 15.12 |
| 42mm dia. | 11.49 | 12.64 | 0.62 | 7.12 | nr | 19.76 |
| 54mm dia. | 22.04 | 24.24 | 0.67 | 7.69 | nr | 31.93 |
| 67mm dia. | 50.85 | 55.94 | 0.83 | 9.52 | nr | 65.46 |
| Female tee, reducing branch FI | | | | | | |
| 15 x 15mm x 1/4" dia. | 3.46 | 3.81 | 0.37 | 4.25 | nr | 8.06 |
| 22 x 22mm x 1/2" dia. | 2.41 | 2.65 | 0.40 | 4.59 | nr | 7.24 |
| 28 x 28mm x 3/4" dia. | 8.29 | 9.12 | 0.45 | 5.16 | nr | 14.28 |
| 35 x 35mm x 3/4" dia. | 12.17 | 13.39 | 0.45 | 5.16 | nr | 18.55 |
| 42 x 42mm x 1/2" dia. | 14.52 | 15.97 | 0.62 | 7.11 | nr | 23.08 |
| Backplate tee | | | | | | |
| 15 x 15mm x 1/2" dia. | 6.46 | 7.11 | 0.62 | 7.11 | nr | 14.22 |
| Heater tee | | | | | | |
| 1/2 x 1/2" x 15mm dia. | 5.80 | 6.38 | 0.37 | 4.25 | nr | 10.63 |
| Union heater tee | | | | | | |
| 1/2 x 1/2" x 15mm dia. | 7.80 | 8.58 | 0.37 | 4.25 | nr | 12.83 |
| Sweep tee - equal | | | | | | |
| 15mm dia. | 4.69 | 5.16 | 0.37 | 4.25 | nr | 9.41 |
| 22mm dia. | 6.03 | 6.63 | 0.40 | 4.59 | nr | 11.22 |
| 28mm dia. | 11.44 | 12.58 | 0.45 | 5.16 | nr | 17.74 |
| 35mm dia. | 15.24 | 16.76 | 0.59 | 6.77 | nr | 23.53 |
| 42mm dia. | 22.59 | 24.85 | 0.62 | 7.11 | nr | 31.96 |
| 54mm dia. | 25.03 | 27.53 | 0.67 | 7.69 | nr | 35.22 |
| 67mm dia. | 43.57 | 47.93 | 0.83 | 9.53 | nr | 57.46 |
| Sweep tee - reducing | | | | | | |
| 22 x 22 x 15mm dia. | 5.06 | 5.57 | 0.40 | 4.59 | nr | 10.16 |
| 28 x 28 x 22mm dia. | 9.09 | 10.00 | 0.45 | 5.16 | nr | 15.16 |
| 35 x 35 x 22mm dia. | 14.98 | 16.48 | 0.59 | 6.77 | nr | 23.25 |
| Sweep tee - double | | | | | | |
| 15mm dia. | 5.30 | 5.83 | 0.37 | 4.25 | nr | 10.08 |
| 22mm dia. | 7.20 | 7.92 | 0.40 | 4.59 | nr | 12.51 |
| 28mm dia. | 10.95 | 12.04 | 0.45 | 5.16 | nr | 17.20 |
| Cross | | | | | | |
| 15mm dia. | 7.00 | 7.70 | 0.49 | 5.62 | nr | 13.32 |
| 22mm dia. | 7.83 | 8.61 | 0.53 | 6.08 | nr | 14.69 |
| 28mm dia. | 11.22 | 12.34 | 0.59 | 6.77 | nr | 19.11 |

## Y:MECHANICAL AND ELECTRICAL SERVICES

| Item | Net Price £ | Material £ | Labour hours | Labour £ | Unit | Total Rate £ |
|---|---|---|---|---|---|---|
| **Y10 : PIPELINES : COPPER (contd)** | | | | | | |
| **Extra Over copper pipes; high duty capillary fittings; BS 864** | | | | | | |
| Stop end | | | | | | |
| 15mm dia. | 3.52 | 3.87 | 0.13 | 1.49 | nr | 5.36 |
| Straight coupling; copper to copper | | | | | | |
| 15mm dia. | 1.62 | 1.78 | 0.25 | 2.87 | nr | 4.65 |
| 22mm dia. | 2.58 | 2.84 | 0.28 | 3.21 | nr | 6.05 |
| 28mm dia. | 3.66 | 4.03 | 0.33 | 3.79 | nr | 7.82 |
| 35mm dia. | 6.44 | 7.08 | 0.37 | 4.25 | nr | 11.33 |
| 42mm dia. | 7.05 | 7.75 | 0.42 | 4.82 | nr | 12.57 |
| 54mm dia. | 10.38 | 11.42 | 0.45 | 5.16 | nr | 16.58 |
| Reducing coupling | | | | | | |
| 15 x 12mm dia. | 3.04 | 3.34 | 0.25 | 2.87 | nr | 6.21 |
| 22 x 15mm dia. | 3.52 | 3.87 | 0.28 | 3.21 | nr | 7.08 |
| 28 x 22mm dia. | 4.85 | 5.33 | 0.33 | 3.79 | nr | 9.12 |
| Straight female connector | | | | | | |
| 15mm x 1/2" dia. | 3.96 | 4.36 | 0.25 | 2.87 | nr | 7.23 |
| 22mm x 3/4" dia. | 4.48 | 4.93 | 0.28 | 3.21 | nr | 8.14 |
| 28mm x 1" dia. | 6.60 | 7.26 | 0.33 | 3.79 | nr | 11.05 |
| Straight male connector | | | | | | |
| 15mm x 1/2" dia. | 3.86 | 4.25 | 0.25 | 2.87 | nr | 7.12 |
| 22mm x 3/4" dia. | 4.48 | 4.93 | 0.28 | 3.21 | nr | 8.14 |
| 28mm x 1" dia. | 6.60 | 7.26 | 0.33 | 3.79 | nr | 11.05 |
| 42mm x 1 1/2" dia. | 12.88 | 14.17 | 0.42 | 4.82 | nr | 18.99 |
| 54mm x 2" dia. | 20.91 | 23.00 | 0.45 | 5.16 | nr | 28.16 |
| Reducer | | | | | | |
| 15 x 12mm dia. | 2.00 | 2.20 | 0.25 | 2.87 | nr | 5.07 |
| 22 x 15mm dia. | 1.96 | 2.16 | 0.28 | 3.21 | nr | 5.37 |
| 28 x 22mm dia. | 3.52 | 3.87 | 0.33 | 3.79 | nr | 7.66 |
| 35 x 28mm dia. | 4.48 | 4.93 | 0.37 | 4.25 | nr | 9.18 |
| 42 x 35mm dia. | 5.77 | 6.35 | 0.42 | 4.82 | nr | 11.17 |
| 54 x 42mm dia. | 9.30 | 10.23 | 0.25 | 2.87 | nr | 13.10 |
| Straight union adaptor | | | | | | |
| 15mm x 3/4" dia. | 3.18 | 3.50 | 0.25 | 2.87 | nr | 6.37 |
| 22mm x 1" dia. | 4.37 | 4.81 | 0.28 | 3.21 | nr | 8.02 |
| 28mm x 1 1/4" dia. | 5.77 | 6.35 | 0.33 | 3.79 | nr | 10.14 |
| 35mm x 1 1/2" dia. | 10.46 | 11.51 | 0.37 | 4.25 | nr | 15.76 |
| 42mm x 2" dia. | 13.22 | 14.54 | 0.42 | 4.82 | nr | 19.36 |
| Bent union adaptor | | | | | | |
| 15mm x 3/4" dia. | 8.38 | 9.22 | 0.25 | 2.87 | nr | 12.09 |
| 22mm x 1" dia. | 11.30 | 12.43 | 0.28 | 3.21 | nr | 15.64 |
| 28mm x 1 1/4" dia. | 15.22 | 16.74 | 0.33 | 3.79 | nr | 20.53 |

## Y:MECHANICAL AND ELECTRICAL SERVICES

| Item | Net Price £ | Material £ | Labour hours | Labour £ | Unit | Total Rate £ |
|---|---|---|---|---|---|---|
| Adaptor; male copper to Fl | | | | | | |
| 15mm x 1/2" dia. | 5.97 | 6.57 | 0.25 | 2.87 | nr | 9.44 |
| 22mm x 3/4" dia. | 6.08 | 6.69 | 0.28 | 3.21 | nr | 9.90 |
| Union coupling | | | | | | |
| 15mm dia. | 7.26 | 7.99 | 0.37 | 4.25 | nr | 12.24 |
| 22mm dia. | 9.30 | 10.23 | 0.40 | 4.59 | nr | 14.82 |
| 28mm dia. | 12.92 | 14.21 | 0.45 | 5.16 | nr | 19.37 |
| 35mm dia. | 22.52 | 24.77 | 0.59 | 6.77 | nr | 31.54 |
| 42mm dia. | 26.52 | 29.17 | 0.62 | 7.11 | nr | 36.28 |
| Elbow | | | | | | |
| 15mm dia. | 4.67 | 5.14 | 0.25 | 2.87 | nr | 8.01 |
| 22mm dia. | 4.99 | 5.49 | 0.28 | 3.21 | nr | 8.70 |
| 28mm dia. | 7.42 | 8.16 | 0.33 | 3.79 | nr | 11.95 |
| 35mm dia. | 11.61 | 12.77 | 0.37 | 4.25 | nr | 17.02 |
| 42mm dia. | 14.45 | 15.89 | 0.42 | 4.82 | nr | 20.71 |
| 54mm dia. | 25.14 | 27.65 | 0.42 | 4.82 | nr | 32.47 |
| Return bend | | | | | | |
| 28mm dia. | 15.22 | 16.74 | 0.33 | 3.79 | nr | 20.53 |
| 35mm dia. | 17.68 | 19.45 | 0.37 | 4.25 | nr | 23.70 |
| Bent male union connector | | | | | | |
| 15mm x 1/2" dia. | 10.84 | 11.92 | 0.37 | 4.25 | nr | 16.17 |
| 22mm x 3/4" dia. | 14.61 | 16.07 | 0.40 | 4.59 | nr | 20.66 |
| 28mm x 1" dia. | 26.52 | 29.17 | 0.45 | 5.16 | nr | 34.33 |
| Composite flange | | | | | | |
| 22mm dia. | 17.91 | 19.70 | 0.35 | 4.02 | nr | 23.72 |
| 28mm dia. | 19.91 | 21.90 | 0.37 | 4.25 | nr | 26.15 |
| 35mm dia. | 25.91 | 28.50 | 0.38 | 4.36 | nr | 32.86 |
| 42mm dia. | 29.83 | 32.81 | 0.41 | 4.70 | nr | 37.51 |
| 54mm dia. | 41.82 | 46.00 | 0.43 | 4.94 | nr | 50.94 |
| Equal tee | | | | | | |
| 15mm dia. | 5.37 | 5.91 | 0.37 | 4.25 | nr | 10.16 |
| 22mm dia. | 6.77 | 7.45 | 0.40 | 4.59 | nr | 12.04 |
| 28mm dia. | 8.92 | 9.81 | 0.45 | 5.16 | nr | 14.97 |
| 35mm dia. | 15.22 | 16.74 | 0.59 | 6.77 | nr | 23.51 |
| 42mm dia. | 19.37 | 21.31 | 0.76 | 8.73 | nr | 30.04 |
| 54mm dia. | 30.52 | 33.57 | 0.67 | 7.69 | nr | 41.26 |
| Reducing tee | | | | | | |
| 15 x 12mm dia. | 7.36 | 8.10 | 0.37 | 4.25 | nr | 12.35 |
| 22 x 15mm dia. | 8.73 | 9.60 | 0.40 | 4.59 | nr | 14.19 |
| 28 x 22mm dia. | 12.44 | 13.68 | 0.45 | 5.16 | nr | 18.84 |
| 35 x 28mm dia. | 19.70 | 21.67 | 0.63 | 7.23 | nr | 28.90 |
| 42 x 28mm dia. | 25.21 | 27.73 | 0.76 | 8.73 | nr | 36.46 |
| 54 x 28mm dia. | 39.82 | 43.80 | 0.96 | 11.02 | nr | 54.82 |

## Y:MECHANICAL AND ELECTRICAL SERVICES

| Item | Net Price £ | Material £ | Labour hours | Labour £ | Unit | Total Rate £ |
|---|---|---|---|---|---|---|
| **Y10 : PIPELINES : COPPER (contd)** | | | | | | |
| **Extra Over copper pipes; compression fittings; BS 864** | | | | | | |
| Stop end | | | | | | |
| 15mm dia. | 1.09 | 1.20 | 0.10 | 1.15 | nr | **2.35** |
| 22mm dia. | 1.34 | 1.47 | 0.33 | 3.79 | nr | **5.26** |
| 28mm dia. | 3.67 | 4.04 | 0.14 | 1.61 | nr | **5.65** |
| Straight connector; copper to copper | | | | | | |
| 15mm dia. | 0.71 | 0.78 | 0.18 | 2.07 | nr | **2.85** |
| 22mm dia. | 1.24 | 1.36 | 0.22 | 2.52 | nr | **3.88** |
| 28mm dia. | 3.31 | 3.64 | 0.25 | 2.87 | nr | **6.51** |
| Straight connector; copper to imperial copper | | | | | | |
| 22mm dia. | 3.11 | 3.42 | 0.22 | 2.52 | nr | **5.94** |
| Male coupling; copper to MI (BSP) | | | | | | |
| 15mm dia. | 0.65 | 0.71 | 0.18 | 2.07 | nr | **2.78** |
| 22mm dia. | 1.04 | 1.14 | 0.22 | 2.52 | nr | **3.66** |
| 28mm dia. | 1.97 | 2.17 | 0.25 | 2.87 | nr | **5.04** |
| Male coupling with long thread and backnut | | | | | | |
| 15mm dia. | 3.44 | 3.78 | 0.18 | 2.07 | nr | **5.85** |
| 22mm dia. | 4.37 | 4.81 | 0.22 | 2.52 | nr | **7.33** |
| Female coupling; copper to FI (BSP) | | | | | | |
| 15mm dia. | 0.80 | 0.88 | 0.18 | 2.07 | nr | **2.95** |
| 22mm dia. | 1.15 | 1.26 | 0.22 | 2.52 | nr | **3.78** |
| 28mm dia. | 2.74 | 3.01 | 0.25 | 2.87 | nr | **5.88** |
| Elbow | | | | | | |
| 15mm dia. | 0.87 | 0.96 | 0.18 | 2.07 | nr | **3.03** |
| 22mm dia. | 1.48 | 1.63 | 0.22 | 2.52 | nr | **4.15** |
| 28mm dia. | 4.04 | 4.44 | 0.25 | 2.87 | nr | **7.31** |
| Male elbow; copper to FI (BSP) | | | | | | |
| 15mm x 1/2" dia. | 1.35 | 1.49 | 0.18 | 2.07 | nr | **3.56** |
| 22mm x 3/4" dia. | 1.74 | 1.91 | 0.22 | 2.52 | nr | **4.43** |
| 28mm x 1" dia. | 4.23 | 4.65 | 0.25 | 2.87 | nr | **7.52** |
| Female elbow; copper to FI (BSP) | | | | | | |
| 15mm x 1/2" dia. | 2.07 | 2.28 | 0.18 | 2.07 | nr | **4.35** |
| 22mm x 3/4" dia. | 2.77 | 3.05 | 0.22 | 2.52 | nr | **5.57** |
| 28mm x 1" dia. | 4.76 | 5.24 | 0.25 | 2.87 | nr | **8.11** |
| Backplate elbow | | | | | | |
| 15mm x 1/2" dia. | 3.00 | 3.30 | 0.48 | 5.51 | nr | **8.81** |

## Material Costs / Measured Work Prices - Mechanical Installations

### Y:MECHANICAL AND ELECTRICAL SERVICES

| Item | Net Price £ | Material £ | Labour hours | Labour £ | Unit | Total Rate £ |
|---|---|---|---|---|---|---|
| Reducing set; internal | | | | | | |
| 15mm dia. | 1.11 | 1.22 | 0.18 | 2.07 | nr | 3.29 |
| Tank coupling; long thread | | | | | | |
| 22mm dia. | 5.49 | 6.04 | 0.53 | 6.08 | nr | 12.12 |
| Tee equal | | | | | | |
| 15mm dia. | 1.23 | 1.35 | 0.29 | 3.33 | nr | 4.68 |
| 22mm dia. | 2.07 | 2.28 | 0.30 | 3.44 | nr | 5.72 |
| 28mm dia. | 7.61 | 8.37 | 0.34 | 3.90 | nr | 12.27 |
| Tee reducing | | | | | | |
| 22mm dia. | 4.30 | 4.73 | 0.30 | 3.44 | nr | 8.17 |
| Backplate tee | | | | | | |
| 15mm dia. | 5.68 | 6.25 | 0.59 | 6.77 | nr | 13.02 |
| **Extra Over copper pipes; dezincification resistant compression fittings; BS 864** | | | | | | |
| Stop end | | | | | | |
| 15mm dia. | 1.49 | 1.64 | 0.10 | 1.15 | nr | 2.79 |
| 22mm dia. | 2.17 | 2.39 | 0.12 | 1.38 | nr | 3.77 |
| 28mm dia. | 4.02 | 4.42 | 0.14 | 1.61 | nr | 6.03 |
| 35mm dia. | 6.20 | 6.82 | 0.17 | 1.95 | nr | 8.77 |
| 42mm dia. | 10.30 | 11.33 | 0.19 | 2.18 | nr | 13.51 |
| Straight coupling; copper to copper | | | | | | |
| 15mm dia. | 1.20 | 1.32 | 0.18 | 2.07 | nr | 3.39 |
| 22mm dia. | 1.95 | 2.15 | 0.22 | 2.52 | nr | 4.67 |
| 28mm dia. | 3.84 | 4.22 | 0.25 | 2.87 | nr | 7.09 |
| 35mm dia. | 8.00 | 8.80 | 0.30 | 3.44 | nr | 12.24 |
| 42mm dia. | 10.51 | 11.56 | 0.35 | 4.02 | nr | 15.58 |
| 54mm dia. | 15.72 | 17.29 | 0.40 | 4.59 | nr | 21.88 |
| Straight swivel connector; copper to imperial copper | | | | | | |
| 22mm dia. | 3.74 | 4.11 | 0.22 | 2.52 | nr | 6.63 |
| Male coupling; copper to MI (BSP) | | | | | | |
| 15mm x 1/2" dia. | 1.06 | 1.17 | 0.18 | 2.07 | nr | 3.24 |
| 22mm x 3/4" dia. | 1.63 | 1.79 | 0.22 | 2.52 | nr | 4.31 |
| 28mm x 1" dia. | 2.81 | 3.09 | 0.25 | 2.87 | nr | 5.96 |
| 35mm x 1 1/4" dia. | 6.07 | 6.68 | 0.30 | 3.44 | nr | 10.12 |
| 42mm x 1 1/2" dia. | 9.11 | 10.02 | 0.35 | 4.02 | nr | 14.04 |
| 54mm x 2" dia. | 13.45 | 14.79 | 0.60 | 6.89 | nr | 21.68 |
| Male coupling with long thread and backnuts | | | | | | |
| 22mm dia. | 5.26 | 5.79 | 0.22 | 2.52 | nr | 8.31 |
| 28mm dia. | 5.84 | 6.42 | 0.22 | 2.52 | nr | 8.94 |

## Y:MECHANICAL AND ELECTRICAL SERVICES

| Item | Net Price £ | Material £ | Labour hours | Labour £ | Unit | Total Rate £ |
|---|---|---|---|---|---|---|
| **Y10 : PIPELINES : COPPER (contd)** | | | | | | |
| **Extra Over copper pipes; dezincification resistant compression fittings; BS 864 (contd)** | | | | | | |
| Female coupling; copper to FI (BSP) | | | | | | |
| 15mm x 1/2" dia. | 1.28 | 1.41 | 0.18 | 2.07 | nr | 3.48 |
| 22mm x 3/4" dia. | 1.89 | 2.08 | 0.22 | 2.52 | nr | 4.60 |
| 28mm x 1" dia. | 3.54 | 3.89 | 0.25 | 2.87 | nr | 6.76 |
| 35mm x 1 1/4" dia. | 7.30 | 8.03 | 0.30 | 3.44 | nr | 11.47 |
| 42mm x 1 1/2" dia. | 9.80 | 10.78 | 0.35 | 4.02 | nr | 14.80 |
| 54mm x 2" dia. | 14.40 | 15.84 | 0.40 | 4.59 | nr | 20.43 |
| Elbow | | | | | | |
| 15mm dia. | 1.45 | 1.59 | 0.18 | 2.07 | nr | 3.66 |
| 22mm dia. | 2.31 | 2.54 | 0.22 | 2.52 | nr | 5.06 |
| 28mm dia. | 4.97 | 5.47 | 0.25 | 2.87 | nr | 8.34 |
| 35mm dia. | 10.04 | 11.04 | 0.30 | 3.44 | nr | 14.48 |
| 42mm dia. | 14.61 | 16.07 | 0.35 | 4.02 | nr | 20.09 |
| 54mm dia. | 25.15 | 27.66 | 0.40 | 4.59 | nr | 32.25 |
| Male elbow; copper to MI (BSP) | | | | | | |
| 15mm x 1/2" dia. | 1.35 | 1.49 | 0.18 | 2.07 | nr | 3.56 |
| 22mm x 3/4" dia. | 1.74 | 1.91 | 0.22 | 2.52 | nr | 4.43 |
| 28mm x 1" dia. | 4.23 | 4.65 | 0.25 | 2.87 | nr | 7.52 |
| Female elbow; copper to FI (BSP) | | | | | | |
| 15mm x 1/2" dia. | 2.07 | 2.28 | 0.18 | 2.07 | nr | 4.35 |
| 22mm x 3/4" dia. | 2.77 | 3.05 | 0.22 | 2.52 | nr | 5.57 |
| 28mm x 1" dia. | 4.76 | 5.24 | 0.25 | 2.87 | nr | 8.11 |
| Backplate elbow | | | | | | |
| 15mm x 1/2" dia. | 3.00 | 3.30 | 0.48 | 5.51 | nr | 8.81 |
| Straight tap connector | | | | | | |
| 15mm dia. | 3.11 | 3.42 | 0.14 | 1.61 | nr | 5.03 |
| 22mm dia. | 3.74 | 4.11 | 0.15 | 1.72 | nr | 5.83 |
| Tank coupling | | | | | | |
| 15mm dia. | 3.35 | 3.69 | 0.18 | 2.07 | nr | 5.76 |
| 22mm dia. | 3.57 | 3.93 | 0.22 | 2.52 | nr | 6.45 |
| 28mm dia. | 6.15 | 6.76 | 0.25 | 2.87 | nr | 9.63 |
| 35mm dia. | 7.32 | 8.05 | 0.30 | 3.44 | nr | 11.49 |
| 42mm dia. | 9.82 | 10.80 | 0.35 | 4.02 | nr | 14.82 |
| 54mm dia. | 14.63 | 16.09 | 0.25 | 2.87 | nr | 18.96 |
| Reducing set; internal | | | | | | |
| 15mm dia. | 1.21 | 1.33 | 0.18 | 2.07 | nr | 3.40 |
| 22mm dia. | 1.31 | 1.44 | 0.22 | 2.52 | nr | 3.96 |

*Material Costs / Measured Work Prices - Mechanical Installations*

## Y:MECHANICAL AND ELECTRICAL SERVICES

| Item | Net Price £ | Material £ | Labour hours | Labour £ | Unit | Total Rate £ |
|---|---|---|---|---|---|---|
| Tee equal | | | | | | |
| 15mm dia. | 2.03 | 2.23 | 0.29 | 3.33 | nr | 5.56 |
| 22mm dia. | 3.37 | 3.71 | 0.30 | 3.44 | nr | 7.15 |
| 28mm dia. | 7.92 | 8.71 | 0.34 | 3.90 | nr | 12.61 |
| 35mm dia. | 14.04 | 15.44 | 0.45 | 5.16 | nr | 20.60 |
| 42mm dia. | 22.06 | 24.27 | 0.47 | 5.39 | nr | 29.66 |
| 54mm dia. | 35.45 | 38.99 | 0.56 | 6.43 | nr | 45.42 |
| Tee reducing | | | | | | |
| 22mm dia. | 4.64 | 5.10 | 0.30 | 3.44 | nr | 8.54 |
| 28mm dia. | 7.65 | 8.41 | 0.34 | 3.90 | nr | 12.31 |
| 35mm dia. | 13.71 | 15.08 | 0.45 | 5.16 | nr | 20.24 |
| 42mm dia. | 21.19 | 23.31 | 0.47 | 5.39 | nr | 28.70 |
| 54mm dia. | 35.45 | 38.99 | 0.56 | 6.43 | nr | 45.42 |
| **Extra Over copper pipes; bronze one piece brazing flanges; metric** | | | | | | |
| Bronze flange; PN6 | | | | | | |
| 15mm dia. | 12.09 | 13.30 | 0.25 | 2.87 | nr | 16.17 |
| 22mm dia. | 14.35 | 15.79 | 0.28 | 3.21 | nr | 19.00 |
| 28mm dia. | 16.46 | 18.11 | 0.30 | 3.44 | nr | 21.55 |
| 35mm dia. | 23.18 | 25.50 | 0.37 | 4.25 | nr | 29.75 |
| 42mm dia. | 28.36 | 31.20 | 0.42 | 4.82 | nr | 36.02 |
| 54mm dia. | 40.38 | 44.42 | 0.45 | 5.16 | nr | 49.58 |
| 67mm dia. | 46.59 | 51.25 | 0.56 | 6.43 | nr | 57.68 |
| 76mm dia. | 53.03 | 58.33 | 0.67 | 7.69 | nr | 66.02 |
| 108mm dia. | 71.26 | 78.39 | 0.75 | 8.61 | nr | 87.00 |
| 133mm dia. | 86.50 | 95.15 | 0.92 | 10.56 | nr | 105.71 |
| 159mm dia. | 122.59 | 134.85 | 1.17 | 13.44 | nr | 148.29 |
| Bronze flange; PN10 | | | | | | |
| 15mm dia. | 15.80 | 17.38 | 0.25 | 2.87 | nr | 20.25 |
| 22mm dia. | 18.41 | 20.25 | 0.28 | 3.21 | nr | 23.46 |
| 28mm dia. | 18.52 | 20.37 | 0.33 | 3.82 | nr | 24.19 |
| 35mm dia. | 25.28 | 27.81 | 0.37 | 4.25 | nr | 32.06 |
| 42mm dia. | 30.18 | 33.20 | 0.42 | 4.82 | nr | 38.02 |
| 54mm dia. | 42.92 | 47.21 | 0.45 | 5.16 | nr | 52.37 |
| 67mm dia. | 46.59 | 51.25 | 0.56 | 6.42 | nr | 57.67 |
| 76mm dia. | 58.97 | 64.87 | 0.67 | 7.69 | nr | 72.56 |
| 108mm dia. | 87.32 | 96.05 | 0.75 | 8.61 | nr | 104.66 |
| 133mm dia. | 100.36 | 110.40 | 0.92 | 10.56 | nr | 120.96 |
| 159mm dia. | 153.52 | 168.87 | 1.17 | 13.44 | nr | 182.31 |
| Bronze flange; PN16 | | | | | | |
| 15mm dia. | 15.80 | 17.38 | 0.25 | 2.87 | nr | 20.25 |
| 22mm dia. | 18.41 | 20.25 | 0.28 | 3.21 | nr | 23.46 |
| 28mm dia. | 19.09 | 21.00 | 0.33 | 3.82 | nr | 24.82 |
| 35mm dia. | 25.28 | 27.81 | 0.37 | 4.25 | nr | 32.06 |
| 42mm dia. | 30.18 | 33.20 | 0.42 | 4.82 | nr | 38.02 |
| 54mm dia. | 45.35 | 49.88 | 0.45 | 5.16 | nr | 55.04 |
| 67mm dia. | 55.23 | 60.75 | 0.56 | 6.42 | nr | 67.17 |
| 76mm dia. | 70.17 | 77.19 | 0.67 | 7.69 | nr | 84.88 |
| 108mm dia. | 90.18 | 99.20 | 0.75 | 8.61 | nr | 107.81 |
| 133mm dia. | 149.22 | 164.14 | 0.92 | 10.56 | nr | 174.70 |
| 159mm dia. | 185.82 | 204.40 | 1.17 | 13.44 | nr | 217.84 |

## Y:MECHANICAL AND ELECTRICAL SERVICES

| Item | Net Price £ | Material £ | Labour hours | Labour £ | Unit | Total Rate £ |
|---|---|---|---|---|---|---|
| Y10 : PIPELINES : COPPER (contd) | | | | | | |
| Extra Over copper pipes; bronze blank flanges; metric | | | | | | |
| Bronze blank flange; PN6 | | | | | | |
| 15mm dia. | 10.74 | 11.81 | 0.25 | 2.87 | nr | 14.68 |
| 22mm dia. | 13.71 | 15.08 | 0.25 | 2.87 | nr | 17.95 |
| 28mm dia. | 14.09 | 15.50 | 0.25 | 2.87 | nr | 18.37 |
| 35mm dia. | 23.05 | 25.36 | 0.30 | 3.44 | nr | 28.80 |
| 42mm dia. | 31.09 | 34.20 | 0.30 | 3.44 | nr | 37.64 |
| 54mm dia. | 33.76 | 37.14 | 0.30 | 3.44 | nr | 40.58 |
| 67mm dia. | 42.20 | 46.42 | 0.33 | 3.79 | nr | 50.21 |
| 76mm dia. | 54.35 | 59.78 | 0.33 | 3.79 | nr | 63.57 |
| 108mm dia. | 86.36 | 95.00 | 0.38 | 4.36 | nr | 99.36 |
| 133mm dia. | 101.88 | 112.07 | 0.52 | 5.97 | nr | 118.04 |
| 159mm dia. | 128.35 | 141.19 | 0.57 | 6.54 | nr | 147.73 |
| Bronze blank flange; PN10 | | | | | | |
| 15mm dia. | 13.00 | 14.30 | 0.25 | 2.87 | nr | 17.17 |
| 22mm dia. | 16.81 | 18.49 | 0.25 | 2.87 | nr | 21.36 |
| 28mm dia. | 18.61 | 20.47 | 0.25 | 2.87 | nr | 23.34 |
| 35mm dia. | 23.05 | 25.36 | 0.30 | 3.44 | nr | 28.80 |
| 42mm dia. | 43.21 | 47.53 | 0.30 | 3.44 | nr | 50.97 |
| 54mm dia. | 49.26 | 54.19 | 0.30 | 3.44 | nr | 57.63 |
| 67mm dia. | 52.84 | 58.12 | 0.47 | 5.39 | nr | 63.51 |
| 76mm dia. | 67.90 | 74.69 | 0.47 | 5.39 | nr | 80.08 |
| 108mm dia. | 103.70 | 114.07 | 0.52 | 5.97 | nr | 120.04 |
| 133mm dia. | 109.48 | 120.43 | 0.52 | 5.97 | nr | 126.40 |
| 159mm dia. | 200.22 | 220.24 | 0.70 | 8.04 | nr | 228.28 |
| Bronze blank flange; PN16 | | | | | | |
| 15mm dia. | 13.00 | 14.30 | 0.25 | 2.87 | nr | 17.17 |
| 22mm dia. | 17.07 | 18.78 | 0.25 | 2.87 | nr | 21.65 |
| 28mm dia. | 18.61 | 20.47 | 0.25 | 2.87 | nr | 23.34 |
| 35mm dia. | 23.05 | 25.36 | 0.30 | 3.44 | nr | 28.80 |
| 42mm dia. | 43.21 | 47.53 | 0.30 | 3.44 | nr | 50.97 |
| 54mm dia. | 49.26 | 54.19 | 0.30 | 3.44 | nr | 57.63 |
| 67mm dia. | 76.60 | 84.26 | 0.47 | 5.39 | nr | 89.65 |
| 76mm dia. | 90.73 | 99.80 | 0.47 | 5.39 | nr | 105.19 |
| 108mm dia. | 115.34 | 126.87 | 0.52 | 5.97 | nr | 132.84 |
| 133mm dia. | 189.29 | 208.22 | 0.52 | 5.97 | nr | 214.19 |
| 159mm dia. | 242.29 | 266.52 | 0.70 | 8.04 | nr | 274.56 |

## Y:MECHANICAL AND ELECTRICAL SERVICES

| Item | Net Price £ | Material £ | Labour hours | Labour £ | Unit | Total Rate £ |
|---|---|---|---|---|---|---|
| **Extra Over copper pipes; bronze screwed flanges; metric** | | | | | | |
| Bronze screwed flange; 6 BSP | | | | | | |
| 15mm dia. | 9.94 | 10.93 | 0.37 | 4.25 | nr | 15.18 |
| 22mm dia. | 11.50 | 12.65 | 0.49 | 5.62 | nr | 18.27 |
| 28mm dia. | 11.98 | 13.18 | 0.55 | 6.31 | nr | 19.49 |
| 35mm dia. | 16.15 | 17.77 | 0.63 | 7.23 | nr | 25.00 |
| 42mm dia. | 19.40 | 21.34 | 0.73 | 8.38 | nr | 29.72 |
| 54mm dia. | 26.35 | 28.98 | 0.88 | 10.10 | nr | 39.08 |
| 67mm dia. | 33.09 | 36.40 | 1.08 | 12.41 | nr | 48.81 |
| 76mm dia. | 39.92 | 43.91 | 1.28 | 14.69 | nr | 58.60 |
| 108mm dia. | 63.22 | 69.54 | 1.42 | 16.30 | nr | 85.84 |
| 133mm dia. | 74.95 | 82.44 | 1.67 | 19.19 | nr | 101.63 |
| 159mm dia. | 95.41 | 104.95 | 2.08 | 23.91 | nr | 128.86 |
| Bronze screwed flange; 10 BSP | | | | | | |
| 15mm dia. | 12.05 | 13.26 | 0.37 | 4.25 | nr | 17.51 |
| 22mm dia. | 14.04 | 15.44 | 0.49 | 5.63 | nr | 21.07 |
| 28mm dia. | 15.50 | 17.05 | 0.55 | 6.31 | nr | 23.36 |
| 35mm dia. | 22.07 | 24.28 | 0.63 | 7.23 | nr | 31.51 |
| 42mm dia. | 26.95 | 29.64 | 0.73 | 8.38 | nr | 38.02 |
| 54mm dia. | 38.33 | 42.16 | 0.88 | 10.10 | nr | 52.26 |
| 67mm dia. | 44.87 | 49.36 | 1.08 | 12.41 | nr | 61.77 |
| 76mm dia. | 50.65 | 55.72 | 1.28 | 14.69 | nr | 70.41 |
| 108mm dia. | 67.25 | 73.97 | 1.42 | 16.30 | nr | 90.27 |
| 133mm dia. | 81.49 | 89.64 | 1.67 | 19.19 | nr | 108.83 |
| 159mm dia. | 144.61 | 159.07 | 2.08 | 23.91 | nr | 182.98 |
| Bronze screwed flange; 16 BSP | | | | | | |
| 15mm dia. | 12.05 | 13.26 | 0.37 | 4.25 | nr | 17.51 |
| 22mm dia. | 14.04 | 15.44 | 0.49 | 5.63 | nr | 21.07 |
| 28mm dia. | 15.50 | 17.05 | 0.55 | 6.31 | nr | 23.36 |
| 35mm dia. | 22.07 | 24.28 | 0.64 | 7.30 | nr | 31.58 |
| 42mm dia. | 26.95 | 29.64 | 0.73 | 8.38 | nr | 38.02 |
| 54mm dia. | 38.33 | 42.16 | 0.88 | 10.10 | nr | 52.26 |
| 67mm dia. | 52.83 | 58.11 | 1.08 | 12.41 | nr | 70.52 |
| 76mm dia. | 66.69 | 73.36 | 1.28 | 14.69 | nr | 88.05 |
| 108mm dia. | 85.17 | 93.69 | 1.42 | 16.30 | nr | 109.99 |
| 133mm dia. | 140.56 | 154.62 | 1.67 | 19.19 | nr | 173.81 |
| 159mm dia. | 174.25 | 191.68 | 2.08 | 23.91 | nr | 215.59 |
| **Extra Over copper pipes; labours** | | | | | | |
| Made bend | | | | | | |
| 15mm dia. | - | - | 0.30 | 3.44 | nr | 3.44 |
| 22mm dia. | - | - | 0.33 | 3.79 | nr | 3.79 |
| 28mm dia. | - | - | 0.36 | 4.13 | nr | 4.13 |
| 35mm dia. | - | - | 0.48 | 5.51 | nr | 5.51 |
| 42mm dia. | - | - | 0.60 | 6.89 | nr | 6.89 |
| 54mm dia. | - | - | 0.66 | 7.57 | nr | 7.57 |
| 67mm dia. | - | - | 0.76 | 8.73 | nr | 8.73 |
| 76mm dia. | - | - | 0.84 | 9.64 | nr | 9.64 |

*Material Costs / Measured Work Prices - Mechanical Installations*

## Y:MECHANICAL AND ELECTRICAL SERVICES

| Item | Net Price £ | Material £ | Labour hours | Labour £ | Unit | Total Rate £ |
|---|---|---|---|---|---|---|
| **Y10 : PIPELINES : COPPER (contd)** | | | | | | |
| **Extra Over copper pipes; labours (contd)** | | | | | | |
| Bronze butt weld | | | | | | |
| 15mm dia. | - | - | 0.22 | 2.52 | nr | **2.52** |
| 22mm dia. | - | - | 0.25 | 2.87 | nr | **2.87** |
| 28mm dia. | - | - | 0.28 | 3.21 | nr | **3.21** |
| 35mm dia. | - | - | 0.35 | 4.02 | nr | **4.02** |
| 42mm dia. | - | - | 0.42 | 4.82 | nr | **4.82** |
| 54mm dia. | - | - | 0.50 | 5.74 | nr | **5.74** |
| 67mm dia. | - | - | 0.63 | 7.23 | nr | **7.23** |
| 76mm dia. | - | - | 0.75 | 8.61 | nr | **8.61** |
| 108mm dia. | - | - | 0.91 | 10.45 | nr | **10.45** |
| 133mm dia. | - | - | 1.15 | 13.20 | nr | **13.20** |
| 159mm dia. | - | - | 1.32 | 15.16 | nr | **15.16** |

## Y:MECHANICAL AND ELECTRICAL SERVICES

| Item | Net Price £ | Material £ | Labour hours | Labour £ | Unit | Total Rate £ |
|---|---|---|---|---|---|---|
| **Y10 : PIPELINES : PIPE FIXINGS** | | | | | | |
| **For steel pipes; black malleable iron** | | | | | | |
| Single pipe bracket, screw on, black malleable iron; screwed to wood | | | | | | |
| 15mm dia. | 0.70 | 0.77 | 0.15 | 1.72 | nr | 2.49 |
| 20mm dia. | 0.78 | 0.86 | 0.15 | 1.72 | nr | 2.58 |
| 25mm dia. | 0.91 | 1.00 | 0.20 | 2.29 | nr | 3.29 |
| 32mm dia. | 1.24 | 1.36 | 0.20 | 2.29 | nr | 3.65 |
| 40mm dia. | 1.65 | 1.82 | 0.25 | 2.87 | nr | 4.69 |
| 50mm dia. | 2.19 | 2.41 | 0.25 | 2.87 | nr | 5.28 |
| 65mm dia. | 2.90 | 3.18 | 0.30 | 3.44 | nr | 6.62 |
| 80mm dia. | 3.97 | 4.37 | 0.35 | 4.02 | nr | 8.39 |
| 100mm dia. | 5.79 | 6.37 | 0.40 | 4.59 | nr | 10.96 |
| Single pipe bracket, screw on, black malleable iron; plugged and screwed | | | | | | |
| 15mm dia. | 0.70 | 0.77 | 0.30 | 3.44 | nr | 4.21 |
| 20mm dia. | 0.78 | 0.86 | 0.30 | 3.44 | nr | 4.30 |
| 25mm dia. | 0.91 | 1.00 | 0.38 | 4.36 | nr | 5.36 |
| 32mm dia. | 1.24 | 1.36 | 0.38 | 4.36 | nr | 5.72 |
| 40mm dia. | 1.65 | 1.82 | 0.38 | 4.36 | nr | 6.18 |
| 50mm dia. | 2.19 | 2.41 | 0.38 | 4.36 | nr | 6.77 |
| 65mm dia. | 2.89 | 3.18 | 0.40 | 4.59 | nr | 7.77 |
| 80mm dia. | 3.96 | 4.37 | 0.50 | 5.74 | nr | 10.11 |
| 100mm dia. | 5.78 | 6.37 | 0.50 | 5.74 | nr | 12.11 |
| Single pipe bracket for building in, black malleable iron | | | | | | |
| 15mm dia. | 0.70 | 0.77 | 0.11 | 1.26 | nr | 2.03 |
| 20mm dia. | 0.78 | 0.86 | 0.12 | 1.38 | nr | 2.24 |
| 25mm dia. | 0.95 | 1.05 | 0.13 | 1.49 | nr | 2.54 |
| 32mm dia. | 1.24 | 1.36 | 0.14 | 1.61 | nr | 2.97 |
| 40mm dia. | 1.65 | 1.82 | 0.15 | 1.72 | nr | 3.54 |
| 50mm dia. | 1.94 | 2.14 | 0.16 | 1.84 | nr | 3.98 |
| Pipe ring, single socket, black malleable iron | | | | | | |
| 15mm dia. | 0.78 | 0.86 | 0.11 | 1.26 | nr | 2.12 |
| 20mm dia. | 0.91 | 1.00 | 0.12 | 1.38 | nr | 2.38 |
| 25mm dia. | 0.99 | 1.09 | 0.13 | 1.49 | nr | 2.58 |
| 32mm dia. | 1.03 | 1.14 | 0.14 | 1.61 | nr | 2.75 |
| 40mm dia. | 1.32 | 1.46 | 0.15 | 1.72 | nr | 3.18 |
| 50mm dia. | 1.69 | 1.86 | 0.16 | 1.84 | nr | 3.70 |
| 65mm dia. | 2.44 | 2.68 | 0.30 | 3.44 | nr | 6.12 |
| 80mm dia. | 2.94 | 3.23 | 0.35 | 4.02 | nr | 7.25 |
| 100mm dia. | 4.47 | 4.91 | 0.40 | 4.59 | nr | 9.50 |
| 125mm dia. | 8.97 | 9.87 | 0.60 | 6.89 | nr | 16.76 |
| 150mm dia. | 10.05 | 11.05 | 0.80 | 9.18 | nr | 20.23 |

*Material Costs / Measured Work Prices - Mechanical Installations*

## Y:MECHANICAL AND ELECTRICAL SERVICES

| Item | Net Price £ | Material £ | Labour hours | Labour £ | Unit | Total Rate £ |
|---|---|---|---|---|---|---|
| **Y10 : PIPELINES : PIPE FIXINGS (contd)** | | | | | | |
| **For steel pipes; black malleable iron (contd)** | | | | | | |
| Pipe ring, double socket, black malleable iron | | | | | | |
| 15mm dia. | 0.95 | 1.05 | 0.11 | 1.26 | nr | 2.31 |
| 20mm dia. | 1.07 | 1.18 | 0.12 | 1.38 | nr | 2.56 |
| 25mm dia. | 1.20 | 1.32 | 0.13 | 1.49 | nr | 2.81 |
| 32mm dia. | 1.40 | 1.55 | 0.14 | 1.61 | nr | 3.16 |
| 40mm dia. | 1.65 | 1.82 | 0.15 | 1.72 | nr | 3.54 |
| 50mm dia. | 1.86 | 2.05 | 0.16 | 1.84 | nr | 3.89 |
| Screw on backplate, black malleable iron; screwed to wood | | | | | | |
| 15mm dia. | 0.95 | 1.05 | 0.15 | 1.72 | nr | 2.77 |
| Screw on backplate, black malleable iron; plugged and screwed | | | | | | |
| 15mm dia. | 0.95 | 1.05 | 0.30 | 3.44 | nr | 4.49 |
| **For steel pipes; galvanised iron** | | | | | | |
| Single pipe bracket, screw on, galvanised iron; screwed to wood | | | | | | |
| 15mm dia. | 0.95 | 1.04 | 0.15 | 1.72 | nr | 2.76 |
| 20mm dia. | 1.06 | 1.16 | 0.15 | 1.72 | nr | 2.88 |
| 25mm dia. | 1.23 | 1.35 | 0.20 | 2.29 | nr | 3.64 |
| 32mm dia. | 1.67 | 1.84 | 0.20 | 2.29 | nr | 4.13 |
| 40mm dia. | 2.23 | 2.45 | 0.25 | 2.87 | nr | 5.32 |
| 50mm dia. | 2.95 | 3.25 | 0.25 | 2.87 | nr | 6.12 |
| 65mm dia. | 3.90 | 4.29 | 0.30 | 3.44 | nr | 7.73 |
| 80mm dia. | 5.34 | 5.88 | 0.35 | 4.02 | nr | 9.90 |
| 100mm dia. | 7.79 | 8.57 | 0.40 | 4.59 | nr | 13.16 |
| Single pipe bracket, screw on, galvanised iron; plugged and screwed | | | | | | |
| 15mm dia. | 0.95 | 1.04 | 0.30 | 3.44 | nr | 4.48 |
| 20mm dia. | 1.06 | 1.16 | 0.30 | 3.44 | nr | 4.60 |
| 25mm dia. | 1.23 | 1.35 | 0.38 | 4.36 | nr | 5.71 |
| 32mm dia. | 1.67 | 1.84 | 0.38 | 4.36 | nr | 6.20 |
| 40mm dia. | 2.27 | 2.45 | 0.38 | 4.36 | nr | 6.81 |
| 50mm dia. | 2.95 | 3.25 | 0.38 | 4.36 | nr | 7.61 |
| 65mm dia. | 3.89 | 4.29 | 0.40 | 4.59 | nr | 8.88 |
| 80mm dia. | 5.34 | 5.88 | 0.50 | 5.74 | nr | 11.62 |
| 100mm dia. | 7.79 | 8.57 | 0.50 | 5.74 | nr | 14.31 |
| Single pipe bracket for building in, galvanised iron | | | | | | |
| 15mm dia. | 0.95 | 1.04 | 0.11 | 1.26 | nr | 2.30 |
| 20mm dia. | 1.06 | 1.16 | 0.12 | 1.38 | nr | 2.54 |
| 25mm dia. | 1.28 | 1.41 | 0.13 | 1.49 | nr | 2.90 |
| 32mm dia. | 1.67 | 1.84 | 0.14 | 1.61 | nr | 3.45 |
| 40mm dia. | 2.23 | 2.45 | 0.15 | 1.72 | nr | 4.17 |
| 50mm dia. | 2.62 | 2.88 | 0.16 | 1.84 | nr | 4.72 |

*Material Costs / Measured Work Prices - Mechanical Installations*

## Y:MECHANICAL AND ELECTRICAL SERVICES

| Item | Net Price £ | Material £ | Labour hours | Labour £ | Unit | Total Rate £ |
|---|---|---|---|---|---|---|
| Pipe ring, single socket, galvanised iron | | | | | | |
| 15mm dia. | 1.06 | 1.16 | 0.11 | 1.26 | nr | 2.42 |
| 20mm dia. | 1.23 | 1.35 | 0.12 | 1.38 | nr | 2.73 |
| 25mm dia. | 1.33 | 1.47 | 0.13 | 1.49 | nr | 2.96 |
| 32mm dia. | 1.39 | 1.53 | 0.15 | 1.72 | nr | 3.25 |
| 40mm dia. | 1.78 | 1.96 | 0.15 | 1.72 | nr | 3.68 |
| 50mm dia. | 2.28 | 2.51 | 0.16 | 1.84 | nr | 4.35 |
| 65mm dia. | 3.28 | 3.61 | 0.30 | 3.44 | nr | 7.05 |
| 80mm dia. | 3.95 | 4.35 | 0.35 | 4.02 | nr | 8.37 |
| 100mm dia. | 6.01 | 6.61 | 0.40 | 4.59 | nr | 11.20 |
| 125mm dia. | 12.08 | 13.28 | 0.60 | 6.89 | nr | 20.17 |
| 150mm dia. | 12.08 | 13.28 | 0.80 | 9.18 | nr | 22.46 |
| Pipe ring, double socket, galvanised iron | | | | | | |
| 15mm dia. | 0.95 | 1.05 | 0.11 | 1.26 | nr | 2.31 |
| 20mm dia. | 1.07 | 1.18 | 0.12 | 1.38 | nr | 2.56 |
| 25mm dia. | 1.20 | 1.32 | 0.13 | 1.49 | nr | 2.81 |
| 32mm dia. | 1.40 | 1.55 | 0.14 | 1.61 | nr | 3.16 |
| 40mm dia. | 1.65 | 1.82 | 0.15 | 1.72 | nr | 3.54 |
| 50mm dia. | 1.86 | 2.05 | 0.16 | 1.84 | nr | 3.89 |
| Screw on backplate, galvanised iron; screwed to wood | | | | | | |
| 15mm dia. | 1.28 | 1.41 | 0.15 | 1.72 | nr | 3.13 |
| Screw on backplate, galvanised iron; plugged and screwed | | | | | | |
| 15mm dia. | 1.28 | 1.41 | 0.30 | 3.44 | nr | 4.85 |
| **For copper pipes** | | | | | | |
| Saddle band | | | | | | |
| 6mm dia. | 0.07 | 0.08 | 0.12 | 1.38 | nr | 1.46 |
| 8mm dia. | 0.07 | 0.08 | 0.13 | 1.43 | nr | 1.51 |
| 10mm dia. | 0.07 | 0.08 | 0.13 | 1.49 | nr | 1.57 |
| 12mm dia. | 0.07 | 0.08 | 0.14 | 1.55 | nr | 1.63 |
| 15mm dia. | 0.08 | 0.09 | 0.15 | 1.72 | nr | 1.81 |
| 22mm dia. | 0.08 | 0.09 | 0.15 | 1.72 | nr | 1.81 |
| 28mm dia. | 0.09 | 0.10 | 0.20 | 2.29 | nr | 2.39 |
| 35mm dia. | 0.14 | 0.15 | 0.20 | 2.29 | nr | 2.44 |
| 42mm dia. | 0.30 | 0.33 | 0.25 | 2.87 | nr | 3.20 |
| 54mm dia. | 0.42 | 0.46 | 0.25 | 2.87 | nr | 3.33 |
| Single spacing clip | | | | | | |
| 15mm dia. | 0.09 | 0.10 | 0.15 | 1.72 | nr | 1.82 |
| 22mm dia. | 0.09 | 0.10 | 0.16 | 1.84 | nr | 1.94 |
| 28mm dia. | 0.42 | 0.46 | 0.20 | 2.29 | nr | 2.75 |

## Y:MECHANICAL AND ELECTRICAL SERVICES

| Item | Net Price £ | Material £ | Labour hours | Labour £ | Unit | Total Rate £ |
|---|---|---|---|---|---|---|
| **Y10 : PIPELINES : PIPE FIXINGS (contd)** | | | | | | |
| **For copper pipes (contd)** | | | | | | |
| Two piece spacing clip | | | | | | |
| 8mm dia. Bottom | 0.07 | 0.08 | 0.12 | 1.38 | nr | 1.46 |
| 8mm dia. Top | 0.07 | 0.08 | 0.12 | 1.38 | nr | 1.46 |
| 12mm dia. Bottom | 0.07 | 0.08 | 0.15 | 1.72 | nr | 1.80 |
| 12mm dia. Top | 0.07 | 0.08 | 0.15 | 1.72 | nr | 1.80 |
| 15mm dia. Bottom | 0.08 | 0.09 | 0.15 | 1.72 | nr | 1.81 |
| 15mm dia. Top | 0.08 | 0.09 | 0.15 | 1.72 | nr | 1.81 |
| 22mm dia. Bottom | 0.08 | 0.09 | 0.17 | 2.01 | nr | 2.10 |
| 22mm dia. Top | 0.08 | 0.09 | 0.17 | 2.01 | nr | 2.10 |
| 28mm dia. Bottom | 0.08 | 0.09 | 0.20 | 2.29 | nr | 2.38 |
| 28mm dia. Top | 0.14 | 0.15 | 0.20 | 2.29 | nr | 2.44 |
| 35mm dia. Bottom | 0.09 | 0.10 | 0.25 | 2.87 | nr | 2.97 |
| 35mm dia. Top | 0.22 | 0.24 | 0.25 | 2.87 | nr | 3.11 |
| 42mm dia. Bottom | 0.22 | 0.24 | 0.25 | 2.87 | nr | 3.11 |
| 42mm dia. Top | 0.37 | 0.41 | 0.25 | 2.87 | nr | 3.28 |
| 54mm dia. Bottom | 0.25 | 0.28 | 0.25 | 2.87 | nr | 3.15 |
| 54mm dia. Top | 0.51 | 0.56 | 0.25 | 2.87 | nr | 3.43 |
| Single pipe bracket | | | | | | |
| 15mm dia. | 1.00 | 1.10 | 0.15 | 1.72 | nr | 2.82 |
| 22mm dia. | 1.15 | 1.26 | 0.15 | 1.72 | nr | 2.98 |
| 28mm dia. | 1.37 | 1.51 | 0.20 | 2.29 | nr | 3.80 |
| Single pipe bracket for building in | | | | | | |
| 22mm dia. | 1.68 | 1.85 | 0.05 | 0.57 | nr | 2.42 |
| 28mm dia. | 1.79 | 1.97 | 0.05 | 0.57 | nr | 2.54 |
| Single pipe ring | | | | | | |
| 15mm dia. | 1.78 | 1.96 | 0.30 | 3.44 | nr | 5.40 |
| 22mm dia. | 1.90 | 2.09 | 0.30 | 3.44 | nr | 5.53 |
| 28mm dia. | 2.26 | 2.49 | 0.38 | 4.36 | nr | 6.85 |
| 35mm dia. | 2.43 | 2.67 | 0.38 | 4.36 | nr | 7.03 |
| 42mm dia. | 2.64 | 2.90 | 0.38 | 4.36 | nr | 7.26 |
| 54mm dia. | 3.18 | 3.50 | 0.40 | 4.59 | nr | 8.09 |
| 67mm dia. | 7.43 | 8.17 | 0.40 | 4.59 | nr | 12.76 |
| 76mm dia. | 9.38 | 10.32 | 0.50 | 5.74 | nr | 16.06 |
| 108mm dia. | 14.53 | 15.98 | 0.50 | 5.74 | nr | 21.72 |
| Double pipe ring | | | | | | |
| 15mm dia. | 2.02 | 2.22 | 0.30 | 3.44 | nr | 5.66 |
| 22mm dia. | 2.15 | 2.37 | 0.30 | 3.44 | nr | 5.81 |
| 28mm dia. | 2.88 | 3.17 | 0.38 | 4.36 | nr | 7.53 |
| 35mm dia. | 2.99 | 3.29 | 0.38 | 4.36 | nr | 7.65 |
| 42mm dia. | 3.27 | 3.60 | 0.38 | 4.36 | nr | 7.96 |
| 54mm dia. | 3.92 | 4.31 | 0.40 | 4.59 | nr | 8.90 |
| 67mm dia. | 8.38 | 9.22 | 0.40 | 4.59 | nr | 13.81 |
| 76mm dia. | 10.46 | 11.51 | 0.50 | 5.74 | nr | 17.25 |
| 108mm dia. | 17.60 | 19.36 | 0.50 | 5.74 | nr | 25.10 |

## Y: MECHANICAL AND ELECTRICAL SERVICES

| Item | Net Price £ | Material £ | Labour hours | Labour £ | Unit | Total Rate £ |
|---|---|---|---|---|---|---|
| Wall bracket | | | | | | |
| 15mm dia. | 2.54 | 2.79 | 0.05 | 0.57 | nr | 3.36 |
| 22mm dia. | 2.99 | 3.29 | 0.05 | 0.57 | nr | 3.86 |
| 28mm dia. | 3.64 | 4.00 | 0.05 | 0.57 | nr | 4.57 |
| 35mm dia. | 4.70 | 5.17 | 0.05 | 0.57 | nr | 5.74 |
| 42mm dia. | 6.20 | 6.82 | 0.05 | 0.57 | nr | 7.39 |
| 54mm dia. | 7.84 | 8.62 | 0.05 | 0.57 | nr | 9.19 |
| Hospital bracket | | | | | | |
| 15mm dia. | 2.15 | 2.37 | 0.30 | 3.44 | nr | 5.81 |
| 22mm dia. | 2.27 | 2.50 | 0.30 | 3.44 | nr | 5.94 |
| 28mm dia. | 2.78 | 3.06 | 0.38 | 4.36 | nr | 7.42 |
| 35mm dia. | 2.99 | 3.29 | 0.38 | 4.36 | nr | 7.65 |
| 42mm dia. | 4.19 | 4.61 | 0.38 | 4.36 | nr | 8.97 |
| 54mm dia. | 5.70 | 6.27 | 0.40 | 4.59 | nr | 10.86 |
| Screw on backplate, female | | | | | | |
| 15mm dia. | 0.95 | 1.04 | 0.30 | 3.44 | nr | 4.48 |
| 22mm dia. | 0.95 | 1.04 | 0.30 | 3.44 | nr | 4.48 |
| 28mm dia. | 0.95 | 1.04 | 0.38 | 4.36 | nr | 5.40 |
| 35mm dia. | 0.95 | 1.04 | 0.38 | 4.36 | nr | 5.40 |
| 42mm dia. | 0.95 | 1.04 | 0.38 | 4.36 | nr | 5.40 |
| 54mm dia. | 0.95 | 1.04 | 0.40 | 4.59 | nr | 5.63 |
| 67mm dia. | 0.95 | 1.04 | 0.40 | 4.59 | nr | 5.63 |
| 76mm dia. | 0.95 | 1.04 | 0.50 | 5.74 | nr | 6.78 |
| 108mm dia. | 0.95 | 1.04 | 0.50 | 5.74 | nr | 6.78 |
| Screw on backplate, male | | | | | | |
| 15mm dia. | 0.85 | 0.94 | 0.30 | 3.44 | nr | 4.38 |
| 22mm dia. | 0.85 | 0.94 | 0.30 | 3.44 | nr | 4.38 |
| 28mm dia. | 0.85 | 0.94 | 0.38 | 4.36 | nr | 5.30 |
| 35mm dia. | 0.85 | 0.94 | 0.38 | 4.36 | nr | 5.30 |
| 42mm dia. | 0.85 | 0.94 | 0.38 | 4.36 | nr | 5.30 |
| 54mm dia. | 0.85 | 0.94 | 0.40 | 4.59 | nr | 5.53 |
| 67mm dia. | 0.85 | 0.94 | 0.40 | 4.59 | nr | 5.53 |
| 76mm dia. | 0.85 | 0.94 | 0.50 | 5.74 | nr | 6.68 |
| 108mm dia. | 0.95 | 1.04 | 0.50 | 5.74 | nr | 6.78 |
| Pipe joist clips, single | | | | | | |
| 15mm dia. | 0.45 | 0.49 | 0.08 | 0.92 | nr | 1.41 |
| 22mm dia. | 0.45 | 0.49 | 0.08 | 0.92 | nr | 1.41 |
| Pipe joist clips, double | | | | | | |
| 15mm dia. | 0.62 | 0.68 | 0.08 | 0.92 | nr | 1.60 |
| 22mm dia. | 0.62 | 0.68 | 0.08 | 0.92 | nr | 1.60 |

**Material Costs / Measured Work Prices - Mechanical Installations**

## Y:MECHANICAL AND ELECTRICAL SERVICES

| Item | Net Price £ | Material £ | Labour hours | Labour £ | Unit | Total Rate £ |
|---|---|---|---|---|---|---|
| **Y10 : PIPELINES : PIPE FIXINGS (contd)** | | | | | | |
| **Fabricated hangers and brackets (Note: It has been assumed there would be sufficient quantities required to gain the benefit of bulk purchase)** | | | | | | |
| Various sizes cut to length; including handling and fixing | | | | | | |
| Mild steel flats | 9.83 | 10.81 | 0.13 | 1.49 | kg | 12.30 |
| Mild steel angle | 6.70 | 7.37 | 0.17 | 1.95 | kg | 9.32 |
| Mild steel channel | 24.57 | 27.03 | 0.17 | 1.95 | kg | 28.98 |
| Mild steel rods up to 10mm | 1.66 | 1.83 | 0.17 | 1.95 | kg | 3.78 |
| **Labours on fabricated hangers and brackets** | | | | | | |
| Cut other than initial cut on metal of thickness shown | | | | | | |
| 6mm thick | - | - | 0.04 | 0.46 | nr | 0.46 |
| 8mm thick | - | - | 0.05 | 0.57 | nr | 0.57 |
| 10mm thick | - | - | 0.06 | 0.69 | nr | 0.69 |
| 13mm thick | - | - | 0.07 | 0.80 | nr | 0.80 |
| 16mm thick | - | - | 0.08 | 0.92 | nr | 0.92 |
| Ragged ends of flat | | | | | | |
| 6mm thick | - | - | 0.12 | 1.38 | nr | 1.38 |
| 8mm thick | - | - | 0.13 | 1.49 | nr | 1.49 |
| 10mm thick | - | - | 0.14 | 1.61 | nr | 1.61 |
| Ragged ends of angle | | | | | | |
| 5mm thick | - | - | 0.18 | 2.07 | nr | 2.07 |
| 6mm thick | - | - | 0.18 | 2.07 | nr | 2.07 |
| 8mm thick | - | - | 0.19 | 2.18 | nr | 2.18 |
| 10mm thick | - | - | 0.20 | 2.29 | nr | 2.29 |
| Bend on flat | | | | | | |
| 6mm thick | - | - | 0.08 | 0.92 | nr | 0.92 |
| 8mm thick | - | - | 0.09 | 1.03 | nr | 1.03 |
| 10mm thick | - | - | 0.09 | 1.03 | nr | 1.03 |
| Twist on flat | | | | | | |
| 6mm thick | - | - | 0.08 | 0.92 | nr | 0.92 |
| 8mm thick | - | - | 0.09 | 1.03 | nr | 1.03 |
| 10mm thick | - | - | 0.09 | 1.03 | nr | 1.03 |
| Drill hole not exceeding 6mm dia. in steel, not exceeding thickness shown | | | | | | |
| 6mm thick | - | - | 0.04 | 0.46 | nr | 0.46 |
| 13mm thick | - | - | 0.05 | 0.57 | nr | 0.57 |
| Drill hole not exceeding 13mm dia. in steel, not exceeding thickness shown | | | | | | |
| 6mm thick | - | - | 0.05 | 0.57 | nr | 0.57 |

## Y:MECHANICAL AND ELECTRICAL SERVICES

| Item | Net Price £ | Material £ | Labour hours | Labour £ | Unit | Total Rate £ |
|---|---|---|---|---|---|---|
| 13mm thick | - | - | 0.06 | 0.69 | nr | 0.69 |
| Drill hole not exceeding 25mm dia. in steel, not exceeding thickness shown | | | | | | |
| 6mm thick | - | - | 0.08 | 0.92 | nr | 0.92 |
| 13mm thick | - | - | 0.09 | 1.03 | nr | 1.03 |
| Thread end of rod | | | | | | |
| 6mm thick | - | - | 0.07 | 0.80 | nr | 0.80 |
| 13mm thick | - | - | 0.08 | 0.92 | nr | 0.92 |
| 10mm thick | - | - | 0.09 | 1.03 | nr | 1.03 |
| Bend in rod | | | | | | |
| 3mm thick | - | - | 0.06 | 0.69 | nr | 0.69 |
| 6mm thick | - | - | 0.06 | 0.69 | nr | 0.69 |
| 10mm thick | - | - | 0.06 | 0.69 | nr | 0.69 |
| Form eye in rod | | | | | | |
| 3mm thick | - | - | 0.06 | 0.69 | nr | 0.69 |
| 6mm thick | - | - | 0.06 | 0.69 | nr | 0.69 |
| 10mm thick | - | - | 0.06 | 0.69 | nr | 0.69 |
| Butt welding mild steel | | | | | | |
| 5mm thick | - | - | 0.10 | 1.15 | nr | 1.15 |
| 6mm thick | - | - | 0.13 | 1.49 | nr | 1.49 |
| 10mm thick | - | - | 0.19 | 2.18 | nr | 2.18 |
| 13mm thick | - | - | 0.26 | 2.99 | nr | 2.99 |
| Welding ends of rods to steel | | | | | | |
| 3mm thick | - | - | 0.08 | 0.92 | nr | 0.92 |
| 6mm thick | - | - | 0.11 | 1.26 | nr | 1.26 |
| 10mm thick | - | - | 0.16 | 1.84 | nr | 1.84 |
| **Floor or ceiling cover plates** | | | | | | |
| Plastic | | | | | | |
| 15mm dia. | 0.29 | 0.32 | 0.15 | 1.72 | nr | 2.04 |
| 20mm dia. | 0.31 | 0.34 | 0.21 | 2.41 | nr | 2.75 |
| 25mm dia. | 0.35 | 0.39 | 0.21 | 2.41 | nr | 2.80 |
| 32mm dia. | 0.51 | 0.56 | 0.23 | 2.64 | nr | 3.20 |
| 40mm dia. | 0.78 | 0.86 | 0.25 | 2.87 | nr | 3.73 |
| 50mm dia. | 0.86 | 0.95 | 0.25 | 2.87 | nr | 3.82 |
| Chromium plated | | | | | | |
| 15mm dia. | 2.11 | 2.32 | 0.15 | 1.72 | nr | 4.04 |
| 20mm dia. | 2.25 | 2.48 | 0.15 | 1.72 | nr | 4.20 |
| 25mm dia. | 2.34 | 2.57 | 0.20 | 2.29 | nr | 4.86 |
| 32mm dia. | 2.39 | 2.63 | 0.20 | 2.29 | nr | 4.92 |
| 40mm dia. | 2.69 | 2.96 | 0.25 | 2.87 | nr | 5.83 |
| 50mm dia. | 3.23 | 3.55 | 0.25 | 2.87 | nr | 6.42 |

## Material Costs / Measured Work Prices - Mechanical Installations

### Y:MECHANICAL AND ELECTRICAL SERVICES

| Item | Net Price £ | Material £ | Labour hours | Labour £ | Unit | Total Rate £ |
|---|---|---|---|---|---|---|
| **Y10 : PIPELINES : PIPE FIXINGS (contd)** | | | | | | |
| **Pipe roller and chair** | | | | | | |
| Roller and chair; black malleable | | | | | | |
| Up to 50mm dia. | 2.35 | 2.58 | 0.20 | 2.29 | nr | 4.87 |
| 65mm dia. | 2.84 | 3.13 | 0.20 | 2.29 | nr | 5.42 |
| 80mm dia. | 4.06 | 4.47 | 0.20 | 2.29 | nr | 6.76 |
| 100mm dia. | 4.37 | 4.81 | 0.20 | 2.29 | nr | 7.10 |
| 125mm dia. | 4.77 | 5.25 | 0.20 | 2.29 | nr | 7.54 |
| 150mm dia. | 5.30 | 5.83 | 0.30 | 3.44 | nr | 9.27 |
| 175mm dia. | 13.29 | 14.62 | 0.30 | 3.44 | nr | 18.06 |
| 200mm dia. | 13.29 | 14.62 | 0.30 | 3.44 | nr | 18.06 |
| 250mm dia. | 18.99 | 20.89 | 0.30 | 3.44 | nr | 24.33 |
| 300mm dia. | 19.94 | 21.93 | 0.30 | 3.44 | nr | 25.37 |
| Roller and chair; galvanised | | | | | | |
| Up to 50mm dia. | 3.52 | 3.88 | 0.20 | 2.29 | nr | 6.17 |
| 65mm dia. | 4.27 | 4.69 | 0.20 | 2.29 | nr | 6.98 |
| 80mm dia. | 6.10 | 6.71 | 0.20 | 2.29 | nr | 9.00 |
| 100mm dia. | 6.55 | 7.21 | 0.20 | 2.29 | nr | 9.50 |
| 125mm dia. | 7.16 | 7.88 | 0.20 | 2.29 | nr | 10.17 |
| 150mm dia. | 7.95 | 8.75 | 0.30 | 3.44 | nr | 12.19 |
| 175mm dia. | 19.94 | 21.93 | 0.30 | 3.44 | nr | 25.37 |
| 200mm dia. | 19.94 | 21.93 | 0.30 | 3.44 | nr | 25.37 |
| 250mm dia. | 28.48 | 31.33 | 0.30 | 3.44 | nr | 34.77 |
| 300mm dia. | 29.91 | 32.90 | 0.30 | 3.44 | nr | 36.34 |
| Roller bracket; black malleable | | | | | | |
| 25mm dia. | 1.73 | 1.90 | 0.20 | 2.29 | nr | 4.19 |
| 32mm dia. | 1.82 | 2.00 | 0.20 | 2.29 | nr | 4.29 |
| 40mm dia. | 1.95 | 2.14 | 0.20 | 2.29 | nr | 4.43 |
| 50mm dia. | 2.05 | 2.26 | 0.20 | 2.29 | nr | 4.55 |
| 65mm dia. | 2.70 | 2.97 | 0.20 | 2.29 | nr | 5.26 |
| 80mm dia. | 3.89 | 4.28 | 0.20 | 2.29 | nr | 6.57 |
| 100mm dia. | 4.32 | 4.76 | 0.20 | 2.29 | nr | 7.05 |
| 125mm dia. | 7.14 | 7.85 | 0.20 | 2.29 | nr | 10.14 |
| 150mm dia. | 7.14 | 7.85 | 0.30 | 3.44 | nr | 11.29 |
| 175mm dia. | 15.90 | 17.49 | 0.30 | 3.44 | nr | 20.93 |
| 200mm dia. | 15.90 | 17.49 | 0.30 | 3.44 | nr | 20.93 |
| 250mm dia. | 21.14 | 23.25 | 0.30 | 3.44 | nr | 26.69 |
| 300mm dia. | 26.19 | 28.81 | 0.30 | 3.44 | nr | 32.25 |
| Roller bracket; galvanised | | | | | | |
| 25mm dia. | 2.59 | 2.85 | 0.20 | 2.29 | nr | 5.14 |
| 32mm dia. | 2.72 | 3.00 | 0.20 | 2.29 | nr | 5.29 |
| 40mm dia. | 2.92 | 3.21 | 0.20 | 2.29 | nr | 5.50 |
| 50mm dia. | 3.08 | 3.39 | 0.20 | 2.29 | nr | 5.68 |
| 65mm dia. | 4.05 | 4.46 | 0.20 | 2.29 | nr | 6.75 |
| 80mm dia. | 5.84 | 6.42 | 0.20 | 2.29 | nr | 8.71 |
| 100mm dia. | 6.49 | 7.13 | 0.20 | 2.29 | nr | 9.42 |
| 125mm dia. | 10.70 | 11.78 | 0.20 | 2.29 | nr | 14.07 |
| 150mm dia. | 10.70 | 11.78 | 0.30 | 3.44 | nr | 15.22 |
| 175mm dia. | 23.85 | 26.24 | 0.30 | 3.44 | nr | 29.68 |
| 200mm dia. | 23.85 | 26.24 | 0.30 | 3.44 | nr | 29.68 |
| 250mm dia. | 31.71 | 34.88 | 0.30 | 3.44 | nr | 38.32 |
| 300mm dia. | 39.29 | 43.22 | 0.30 | 3.44 | nr | 46.66 |

## Y:MECHANICAL AND ELECTRICAL SERVICES

| Item | Net Price £ | Material £ | Labour hours | Labour £ | Unit | Total Rate £ |
|---|---|---|---|---|---|---|
| **Y11 : PIPELINE ANCILLARIES : REGULATING VALVES** | | | | | | |
| **Regulators** | | | | | | |
| Gunmetal; self-acting two port thermostatic regulator; single seat; water or steam; normally closed or normally open; screwed ends; complete with sensing element; 2m long capillary tube | | | | | | |
| 15mm dia. | 284.00 | 312.40 | 1.50 | 17.23 | nr | 329.63 |
| 20mm dia. | 292.00 | 321.20 | 1.80 | 20.68 | nr | 341.88 |
| 25mm dia. | 300.80 | 330.88 | 1.94 | 22.28 | nr | 353.16 |
| Gunmetal; self-acting two port thermostatic regulator; double seat; water or steam; flanged ends (BS 4504 PN25); with sensing element; 2m long capillary tube; steel body | | | | | | |
| 65mm dia. | 964.80 | 1061.28 | 1.00 | 13.60 | nr | 1074.88 |
| 80mm dia. | 1133.60 | 1246.96 | 1.20 | 16.33 | nr | 1263.29 |
| Gunmetal; self-acting two port thermostat; single seat; screwed; normally closed; with adjustable or fixed bleed device | | | | | | |
| 25mm dia. | 219.20 | 241.12 | 1.70 | 19.52 | nr | 260.64 |
| 32mm dia. | 224.80 | 247.28 | 1.50 | 17.23 | nr | 264.51 |
| 40mm dia. | 240.80 | 264.88 | 1.55 | 17.79 | nr | 282.67 |
| 50mm dia. | 289.60 | 318.56 | 1.60 | 18.36 | nr | 336.92 |
| Self acting temperature regulator for storage calorifier; integral sensing element and pocket; screwed ends | | | | | | |
| 15mm dia. | 261.36 | 287.50 | 1.40 | 16.07 | nr | 303.57 |
| 25mm dia. | 286.88 | 315.57 | 1.70 | 19.52 | nr | 335.09 |
| 32mm dia. | 370.48 | 407.53 | 2.06 | 23.66 | nr | 431.19 |
| 40mm dia. | 453.20 | 498.52 | 2.30 | 26.44 | nr | 524.96 |
| 50mm dia. | 529.76 | 582.74 | 2.65 | 30.36 | nr | 613.10 |
| Self acting temperature regulator for storage calorifier; integral sensing element and pocket; flanged ends; bolted connection | | | | | | |
| 15mm dia. | 383.68 | 422.05 | 0.50 | 5.74 | nr | 427.79 |
| 25mm dia. | 439.12 | 483.03 | 0.61 | 7.00 | nr | 490.03 |
| 32mm dia. | 553.52 | 608.87 | 0.74 | 8.49 | nr | 617.36 |
| 40mm dia. | 655.60 | 721.16 | 0.83 | 9.53 | nr | 730.69 |
| 50mm dia. | 761.20 | 837.32 | 0.95 | 10.91 | nr | 848.23 |
| Chrome plated thermostatic mixing valves including non-return valves and inlet swivel connections with strainers; copper compression fittings | | | | | | |
| 15mm dia. | 96.80 | 106.48 | 0.50 | 5.74 | nr | 112.22 |

## Y:MECHANICAL AND ELECTRICAL SERVICES

| Item | Net Price £ | Material £ | Labour hours | Labour £ | Unit | Total Rate £ |
|---|---|---|---|---|---|---|
| **Y11 : PIPELINE ANCILLARIES : REGULATING VALVES (contd)** | | | | | | |
| **Regulators (contd)** | | | | | | |
| Chrome plated thermostatic mixing valves including non-return valves and inlet swivel connections with angle pattern combined isolating valves and strainers; copper compression fittings | | | | | | |
| 15mm dia. | 101.20 | 111.32 | 0.50 | 5.74 | nr | **117.06** |
| Gunmetal thermostatic mixing valves including non-return valves and inlet swivel connections with strainers; copper compression fittings | | | | | | |
| 15mm dia. | 87.88 | 96.67 | 0.50 | 5.74 | nr | **102.41** |
| Gunmetal thermostatic mixing valves including non-return valves and inlet swivel connections with angle pattern combined isolating valves and strainers; copper compression fittings | | | | | | |
| 15mm dia. | 92.40 | 101.64 | 0.50 | 5.74 | nr | **107.38** |
| **Pipeline ancillaries; steam traps and accessories** | | | | | | |
| Cast iron; inverted bucket type; steam trap pressure range up to 17 bar at 210 degree celsius; screwed ends | | | | | | |
| 1/2" dia. | 71.69 | 78.86 | 0.84 | 9.64 | nr | **88.50** |
| 3/4" dia. | 104.65 | 115.11 | 1.14 | 13.08 | nr | **128.19** |
| 1" dia. | 164.80 | 181.28 | 1.30 | 14.92 | nr | **196.20** |
| 11/2" dia. | 304.06 | 334.47 | 1.71 | 19.65 | nr | **354.12** |
| 2" dia. | 469.68 | 516.65 | 2.08 | 23.91 | nr | **540.56** |
| Cast iron; inverted bucket type; steam trap pressure range up to 17 bar at 210 degree celsius; flanged ends (BS 10 table H); bolted connections | | | | | | |
| 15mm dia. | 169.74 | 186.71 | 1.24 | 14.24 | nr | **200.95** |
| 20mm dia. | 200.23 | 220.25 | 1.34 | 15.38 | nr | **235.63** |
| 25mm dia. | 306.53 | 337.18 | 1.40 | 16.07 | nr | **353.25** |
| 40mm dia. | 474.62 | 522.08 | 1.54 | 17.68 | nr | **539.76** |
| 50mm dia. | 578.45 | 636.29 | 1.64 | 18.84 | nr | **655.13** |

## Y:MECHANICAL AND ELECTRICAL SERVICES

| Item | Net Price £ | Material £ | Labour hours | Labour £ | Unit | Total Rate £ |
|---|---|---|---|---|---|---|
| **Y11 : PIPELINE ANCILLARIES : RADIATOR VALVES** | | | | | | |
| Radiator valves | | | | | | |
| Bronze, straight pattern; wheelhead or lockshield; matt chromium plated finish; screwed joints to steel | | | | | | |
| 15mm dia. | 21.50 | 23.65 | 0.84 | 9.64 | nr | 33.29 |
| 22mm dia. | 28.19 | 31.01 | 1.14 | 13.08 | nr | 44.09 |
| 25mm dia. | 35.40 | 38.94 | 1.30 | 14.92 | nr | 53.86 |
| Bronze, angle pattern; wheelhead or lockshield; matt chromium plated finish; screwed joints to steel | | | | | | |
| 15mm dia. | 13.90 | 15.29 | 0.84 | 9.64 | nr | 24.93 |
| 22mm dia. | 17.65 | 19.41 | 1.14 | 13.08 | nr | 32.49 |
| 25mm dia. | 25.35 | 27.89 | 1.30 | 14.92 | nr | 42.81 |
| Bronze, straight pattern; polished chromium plated; compression joint to copper | | | | | | |
| 15mm dia. | 28.19 | 31.01 | 0.46 | 5.28 | nr | 36.29 |
| 22mm dia. | 36.71 | 40.38 | 0.54 | 6.20 | nr | 46.58 |
| 25mm dia. | 45.89 | 50.48 | 0.54 | 6.20 | nr | 56.68 |
| Bronze, angle pattern; polished chromium plated; compression joint to copper | | | | | | |
| 15mm dia. | 20.06 | 22.07 | 0.46 | 5.28 | nr | 27.35 |
| 22mm dia. | 26.44 | 29.08 | 0.54 | 6.20 | nr | 35.28 |
| 25mm dia. | 34.09 | 37.50 | 0.54 | 6.20 | nr | 43.70 |
| Thermostatic angle valve with built in sensor | | | | | | |
| 10mm dia. | 11.79 | 13.10 | 0.50 | 5.74 | nr | 18.84 |
| 15mm dia. | 11.79 | 13.10 | 0.50 | 5.74 | nr | 18.84 |
| Thermostatic straight valve with built in sensor | | | | | | |
| 15mm dia. | 11.79 | 13.10 | 0.50 | 5.74 | nr | 18.84 |
| Microbore, single entry; brass; wheelhead/lockshield | | | | | | |
| 8mm dia. | 4.74 | 5.27 | 0.50 | 5.74 | nr | 11.01 |
| 10mm dia. | 4.74 | 5.27 | 0.50 | 5.74 | nr | 11.01 |
| 15mm dia. | 4.74 | 5.27 | 0.50 | 5.74 | nr | 11.01 |
| Microbore, single entry; chromium plated; wheelhead/lockshield | | | | | | |
| 8mm dia. | 5.05 | 5.61 | 0.50 | 5.74 | nr | 11.35 |
| 10mm dia. | 5.05 | 5.61 | 0.50 | 5.74 | nr | 11.35 |
| 15mm dia. | 5.05 | 5.61 | 0.50 | 5.74 | nr | 11.35 |

**Material Costs / Measured Work Prices - Mechanical Installations**

Y:MECHANICAL AND ELECTRICAL SERVICES

| Item | Net Price £ | Material £ | Labour hours | Labour £ | Unit | Total Rate £ |
|---|---|---|---|---|---|---|
| **Y11 : PIPELINE ANCILLARIES : RADIATOR VALVES (contd)** | | | | | | |
| **Radiator valves (contd)** | | | | | | |
| Microbore, twin entry; brass; wheelhead/lockshield | | | | | | |
| 8mm dia. | 12.15 | 13.50 | 0.50 | 5.74 | nr | **19.24** |
| 10mm dia. | 13.22 | 14.69 | 0.50 | 5.74 | nr | **20.43** |
| Microbore, twin entry; chromium plated; wheelhead/lockshield | | | | | | |
| 8mm dia. | 14.76 | 16.40 | 0.50 | 5.74 | nr | **22.14** |
| 10mm dia. | 16.22 | 18.02 | 0.50 | 5.74 | nr | **23.76** |

## Material Costs / Measured Work Prices - Mechanical Installations

### Y:MECHANICAL AND ELECTRICAL SERVICES

| Item | Net Price £ | Material £ | Labour hours | Labour £ | Unit | Total Rate £ |
|---|---|---|---|---|---|---|
| **Y11 : PIPELINE ANCILLARIES : BALL FLOAT VALVES** | | | | | | |
| **Ball float valves** | | | | | | |
| Bronze, equilibrium; copper float; working pressure cold services up to 16 bar; flanged ends; BS 4504 Table 16/21; bolted connections | | | | | | |
| 25mm dia. | 319.01 | 350.91 | 1.00 | 11.47 | nr | **362.38** |
| 32mm dia. | 463.22 | 509.54 | 1.16 | 13.31 | nr | **522.85** |
| 40mm dia. | 520.03 | 572.03 | 1.34 | 15.38 | nr | **587.41** |
| 50mm dia. | 834.67 | 918.14 | 1.64 | 18.84 | nr | **936.98** |
| 65mm dia. | 878.37 | 966.21 | 1.80 | 20.68 | nr | **986.89** |
| 80mm dia. | 1066.28 | 1172.91 | 1.92 | 22.07 | nr | **1194.98** |
| Heavy, equilibrium; with long tail and backnut; copper float; screwed for iron | | | | | | |
| 25mm dia. | 101.29 | 106.35 | 1.64 | 18.84 | nr | **125.19** |
| 32mm dia. | 138.82 | 145.76 | 1.80 | 20.68 | nr | **166.44** |
| 40mm dia. | 175.55 | 184.33 | 1.92 | 22.07 | nr | **206.40** |
| 50mm dia. | 367.77 | 386.16 | 2.80 | 32.14 | nr | **418.30** |
| Brass, ball valve; BS 1212; copper float; screwed | | | | | | |
| 15mm dia. | 4.75 | 4.99 | 0.24 | 2.75 | nr | **7.74** |

RESPONSE - SPON's price data now available in an electronic estimating system call us on 0171-522-9966 for details

Y:MECHANICAL AND ELECTRICAL SERVICES

| Item | Net Price £ | Material £ | Labour hours | Labour £ | Unit | Total Rate £ |
|---|---|---|---|---|---|---|
| **Y11 : PIPELINE ANCILLARIES : BALL VALVES** | | | | | | |
| **Ball valves** | | | | | | |
| Malleable iron body; lever operated stainless steel ball and stem; Class 125; cold working pressure up to 12 bar; flanged ends (BS 4504 16/11); bolted connections | | | | | | |
| 40mm dia. | 92.77 | 102.05 | 1.54 | 17.67 | nr | 119.72 |
| 50mm dia. | 116.71 | 128.38 | 1.64 | 18.82 | nr | 147.20 |
| 80mm dia. | 195.19 | 214.71 | 1.92 | 22.03 | nr | 236.74 |
| 100mm dia. | 360.79 | 396.87 | 2.80 | 32.13 | nr | 429.00 |
| 150mm dia. | 490.82 | 539.90 | 12.05 | 138.27 | nr | 678.17 |
| Malleable iron body; lever operated stainless steel ball and stem; cold working pressure up to 16 bar; screwed ends to steel | | | | | | |
| 20mm dia. | 20.43 | 22.47 | 1.34 | 15.38 | nr | 37.85 |
| 25mm dia. | 21.00 | 23.10 | 1.40 | 16.07 | nr | 39.17 |
| 32mm dia. | 28.90 | 31.79 | 1.46 | 16.78 | nr | 48.57 |
| 40mm dia. | 28.90 | 31.79 | 1.54 | 17.68 | nr | 49.47 |
| 50mm dia. | 34.58 | 38.04 | 1.64 | 18.84 | nr | 56.88 |
| Carbon steel body; lever operated stainless steel ball and stem; Class 150; cold working pressure up to 19 bar; screwed ends to steel | | | | | | |
| 15mm dia. | 14.99 | 16.49 | 0.84 | 9.64 | nr | 26.13 |
| 20mm dia. | 15.75 | 17.32 | 1.14 | 13.08 | nr | 30.40 |
| 25mm dia. | 18.02 | 19.82 | 1.30 | 14.92 | nr | 34.74 |

## Y:MECHANICAL AND ELECTRICAL SERVICES

| Item | Net Price £ | Material £ | Labour hours | Labour £ | Unit | Total Rate £ |
|---|---|---|---|---|---|---|
| **Y11 : PIPELINE ANCILLARIES : CHECK VALVES** | | | | | | |
| Check valves | | | | | | |
| DZR copper alloy and bronze, WRc approved cartridge double check valve; BS 6282; working pressure cold services up to 10 bar at 65 degrees Celsius; screwed ends | | | | | | |
| 32mm dia. | 66.29 | 72.92 | 1.50 | 17.23 | nr | 90.15 |
| 40mm dia. | 98.54 | 108.39 | 1.74 | 19.99 | nr | 128.38 |
| 50mm dia. | 116.01 | 127.61 | 2.08 | 23.91 | nr | 151.52 |
| Bronze, swing pattern; BS 5154 series B, PN 25; working pressure saturated steam up to 10.5 bar, cold services up to 25 bar; screwed ends | | | | | | |
| 15mm dia. | 15.63 | 17.19 | 0.84 | 9.64 | nr | 26.83 |
| 20mm dia. | 18.59 | 20.45 | 1.14 | 13.08 | nr | 33.53 |
| 25mm dia. | 25.76 | 28.34 | 1.30 | 14.92 | nr | 43.26 |
| 32mm dia. | 38.52 | 42.37 | 1.50 | 17.23 | nr | 59.60 |
| 40mm dia. | 47.93 | 52.72 | 1.74 | 19.99 | nr | 72.71 |
| 50mm dia. | 75.25 | 80.33 | 2.08 | 23.91 | nr | 104.24 |
| 65mm dia. | 135.58 | 149.14 | 2.60 | 29.88 | nr | 179.02 |
| 80mm dia. | 204.30 | 224.73 | 3.00 | 34.46 | nr | 259.19 |
| Bronze, horizontal lift pattern; BS 5154 series B, PN32; working pressure saturated steam up to 14 bar, cold services up to 32 bar; screwed connections to steel | | | | | | |
| 15mm dia. | 15.32 | 16.85 | 1.24 | 14.24 | nr | 31.09 |
| 20mm dia. | 22.84 | 25.12 | 1.34 | 15.38 | nr | 40.50 |
| 25mm dia. | 32.47 | 35.72 | 1.40 | 16.07 | nr | 51.79 |
| 32mm dia. | 43.67 | 45.38 | 1.54 | 17.68 | nr | 63.06 |
| 40mm dia. | 51.06 | 56.17 | 1.54 | 17.68 | nr | 73.85 |
| 50mm dia. | 82.42 | 90.66 | 1.64 | 18.84 | nr | 109.50 |
| 65mm dia. | 294.39 | 323.83 | 1.80 | 20.68 | nr | 344.51 |
| 80mm dia. | 442.67 | 486.94 | 1.92 | 22.07 | nr | 509.01 |
| Bronze, oblique lift pattern; BS 5154 series B, PN32; working pressure saturated steam up to 14 bar, cold services up to 32 bar; screwed connections to steel | | | | | | |
| 15mm dia. | 34.27 | 37.70 | 1.24 | 14.24 | nr | 51.94 |
| 20mm dia. | 37.63 | 41.39 | 1.34 | 15.38 | nr | 56.77 |
| 25mm dia. | 57.33 | 63.06 | 1.40 | 16.07 | nr | 79.13 |
| 32mm dia. | 82.87 | 91.16 | 1.54 | 17.68 | nr | 108.84 |
| 40mm dia. | 91.82 | 101.00 | 1.54 | 17.68 | nr | 118.68 |
| 50mm dia. | 131.69 | 144.86 | 1.64 | 18.84 | nr | 163.70 |

## Y:MECHANICAL AND ELECTRICAL SERVICES

| Item | Net Price £ | Material £ | Labour hours | Labour £ | Unit | Total Rate £ |
|---|---|---|---|---|---|---|
| **Y11 : PIPELINE ANCILLARIES : CHECK VALVES (contd)** | | | | | | |
| Check valves (contd) | | | | | | |
| Cast iron, swing pattern; BS 5153 PN6; working pressure for cold services up to 6 bar; BS 4504 flanged ends; bolted connections | | | | | | |
| 50mm dia. | 166.27 | 182.90 | 1.64 | 18.84 | nr | 201.74 |
| 65mm dia. | 188.64 | 207.50 | 1.80 | 20.68 | nr | 228.18 |
| 80mm dia. | 212.33 | 233.56 | 1.92 | 22.07 | nr | 255.63 |
| 100mm dia. | 271.99 | 299.19 | 2.80 | 32.14 | nr | 331.33 |
| 125mm dia. | 405.80 | 446.38 | 9.01 | 103.38 | nr | 549.76 |
| 150mm dia. | 456.25 | 501.88 | 12.05 | 138.25 | nr | 640.13 |
| 200mm dia. | 1009.01 | 1109.91 | 13.16 | 151.01 | nr | 1260.92 |
| 250mm dia. | 1535.45 | 1688.99 | 14.08 | 161.57 | nr | 1850.56 |
| 300mm dia. | 2039.95 | 2243.95 | 15.15 | 173.85 | nr | 2417.80 |
| Cast iron, horizontal lift pattern; BS 5153 PN16; working pressure for saturated steam 13 bar, cold services up to 16 bar; BS 4504 flanged ends; bolted connections | | | | | | |
| 50mm dia. | 296.12 | 325.73 | 1.64 | 18.84 | nr | 344.57 |
| 65mm dia. | 333.41 | 366.75 | 1.80 | 20.68 | nr | 387.43 |
| 80mm dia. | 423.35 | 465.69 | 1.92 | 22.07 | nr | 487.76 |
| 100mm dia. | 557.15 | 612.87 | 2.80 | 32.14 | nr | 645.01 |
| 125mm dia. | 899.33 | 989.26 | 9.01 | 103.38 | nr | 1092.64 |
| 150mm dia. | 960.75 | 1056.83 | 12.05 | 138.25 | nr | 1195.08 |

# Keep your figures up to date, free of charge

This section, and most of the other information in this Price Book, is brought up to date every three months, until the next annual edition, in the *Price Book Update*.

The *Update* is available free to all Price Book purchasers.

To ensure you receive your copy, simply complete the reply card from the centre of the book and return it to us.

## Y:MECHANICAL AND ELECTRICAL SERVICES

| Item | Net Price £ | Material £ | Labour hours | Labour £ | Unit | Total Rate £ |
|---|---|---|---|---|---|---|
| **Y11 : PIPELINE ANCILLARIES : COMMISSIONING VALVES** | | | | | | |
| **Commissioning valves** | | | | | | |
| Bronze commissioning set; LTHW, MTHW and chilled water; metering station; double regulating valve; BS5154 PN20 Series B; working pressure 20 bar at 100 degrees Celsius; screwed ends to steel | | | | | | |
| 15mm dia. | 47.05 | 51.76 | 1.26 | 14.47 | nr | 66.23 |
| 20mm dia. | 77.21 | 84.93 | 1.71 | 19.65 | nr | 104.58 |
| 25mm dia. | 93.80 | 103.18 | 1.95 | 22.41 | nr | 125.59 |
| 32mm dia. | 123.57 | 135.93 | 2.25 | 25.85 | nr | 161.78 |
| 40mm dia. | 164.19 | 180.61 | 2.61 | 29.96 | nr | 210.57 |
| 50mm dia. | 237.36 | 261.10 | 3.13 | 35.86 | nr | 296.96 |
| Cast iron, double regulating valve; LTHW, MTHW and chilled water; BS5152 PN16; working pressure 16 bar at 120 degrees Celsius; flanged ends (BS 4504, Part 1, Table 16), bolted connections | | | | | | |
| 65mm dia. | 524.35 | 576.78 | 1.80 | 20.68 | nr | 597.46 |
| 80mm dia. | 586.77 | 645.45 | 1.92 | 22.07 | nr | 667.52 |
| 100mm dia. | 711.62 | 782.78 | 2.80 | 32.14 | nr | 814.92 |
| 125mm dia. | 936.34 | 1029.97 | 9.01 | 103.38 | nr | 1133.35 |
| 150mm dia. | 1111.12 | 1222.23 | 12.05 | 138.25 | nr | 1360.48 |
| Cast iron, orifice valve; LTHW, MTHW and chilled water; complete with controlled test points; BS5152 PN16; working pressure 16 bar at 120 degrees Celsius; flanged ends (BS 4504, Part 1, Table 16); bolted connections | | | | | | |
| 65mm dia. | 478.57 | 526.43 | 1.80 | 20.68 | nr | 547.11 |
| 80mm dia. | 511.86 | 563.05 | 1.92 | 22.07 | nr | 585.12 |
| 100mm dia. | 632.55 | 695.80 | 2.80 | 32.14 | nr | 727.94 |
| 125mm dia. | 898.88 | 988.77 | 9.01 | 103.38 | nr | 1092.15 |
| 150mm dia. | 1077.83 | 1185.61 | 12.05 | 138.25 | nr | 1323.86 |
| Cast iron commissioning set; LTHW, MTHW and chilled water; metering station; double regulating valve; BS5152 PN16; woking pressure 16 bar at 120 degrees Celsius; flanged ends (BS 4504, Part 1, Table 16); bolted connections | | | | | | |
| 65mm dia. | 649.20 | 714.12 | 1.80 | 20.68 | nr | 734.80 |
| 80mm dia. | 733.67 | 807.04 | 1.92 | 22.07 | nr | 829.11 |
| 100mm dia. | 887.24 | 975.96 | 1.92 | 22.07 | nr | 998.03 |
| 125mm dia | 1248.45 | 1373.30 | 1.92 | 22.07 | nr | 1395.37 |
| 150mm dia. | 147.22 | 1622.78 | 1.92 | 22.07 | nr | 1644.85 |

## Y:MECHANICAL AND ELECTRICAL SERVICES

| Item | Net Price £ | Material £ | Labour hours | Labour £ | Unit | Total Rate £ |
|---|---|---|---|---|---|---|
| **Y11 : PIPELINE ANCILLARIES : CONTROL VALVES** | | | | | | |
| **Control valves; manually operated** | | | | | | |
| Pressure reducing valve for steam; maximum range of 17 bar and 232 degree celsius; screwed ends to steel | | | | | | |
| 15mm dia. | 268.14 | 294.95 | 0.84 | 9.64 | nr | 304.59 |
| 20mm dia. | 289.46 | 318.41 | 0.88 | 10.07 | nr | 328.48 |
| 25mm dia. | 314.06 | 345.47 | 1.30 | 14.92 | nr | 360.39 |
| Pressure reducing valve for steam; maximum range of 17 bar and 232 degree celsius; flanged ends (BS 10, Table H or BS 4504, Tables 25 and 16) | | | | | | |
| 25mm dia. | 376.38 | 414.02 | 1.64 | 18.84 | nr | 432.86 |
| 32mm dia. | 428.04 | 470.84 | 1.80 | 20.68 | nr | 491.52 |
| 40mm dia. | 511.68 | 562.85 | 2.04 | 23.42 | nr | 586.27 |
| 50mm dia. | 590.40 | 649.44 | 2.48 | 28.47 | nr | 677.91 |
| **Control valves; electrically operated (electrical work elsewhere)** | | | | | | |
| Cast iron; butterfly type; two position electrically controlled 240V motor and linkage mechanism; for low pressure hot water; max 6 bar 120 degree celsius; flanged ends; counter flanges | | | | | | |
| 25mm dia. | 105.55 | 294.36 | 1.40 | 16.07 | nr | 310.43 |
| 32mm dia. | 109.09 | 298.25 | 1.46 | 16.78 | nr | 315.03 |
| 40mm dia. | 116.77 | 306.70 | 1.54 | 17.68 | nr | 324.38 |
| 50mm dia. | 129.38 | 320.57 | 1.64 | 18.84 | nr | 339.41 |
| 65mm dia. | 142.60 | 335.12 | 2.68 | 30.76 | nr | 365.88 |
| 80mm dia. | 153.83 | 347.47 | 2.68 | 30.76 | nr | 378.23 |
| 100mm dia. | 176.27 | 372.15 | 2.80 | 32.14 | nr | 404.29 |
| 125mm dia. | 207.02 | 663.89 | 2.80 | 32.14 | nr | 696.03 |
| 150mm dia. | 278.29 | 742.29 | 2.80 | 32.14 | nr | 774.43 |
| 200mm dia. | 481.08 | 965.36 | 2.80 | 32.14 | nr | 997.50 |
| Cast iron; three way 240V motorized; for low pressure hot water; max 6 bar 120 degree celsius; flanged ends and drilled (BS 10, Table F) | | | | | | |
| 25mm dia. | 151.06 | 427.98 | 1.90 | 21.82 | nr | 449.80 |
| 40mm dia. | 174.89 | 454.19 | 2.04 | 23.42 | nr | 477.61 |
| 50mm dia. | 215.40 | 498.75 | 3.45 | 39.57 | nr | 538.32 |
| 65mm dia. | 279.75 | 569.54 | 3.45 | 39.57 | nr | 609.11 |
| 80mm dia. | 457.33 | 764.87 | 3.57 | 40.98 | nr | 805.85 |
| Two port normally closed motorised valve; electric actuator; spring return; domestic usage | | | | | | |
| 22mm dia. | 36.63 | 40.29 | 1.14 | 13.08 | nr | 53.37 |
| 28mm dia. | 43.68 | 48.05 | 1.30 | 14.92 | nr | 62.97 |

## Y:MECHANICAL AND ELECTRICAL SERVICES

| Item | Net Price £ | Material £ | Labour hours | Labour £ | Unit | Total Rate £ |
|---|---|---|---|---|---|---|
| Two port on/off motorised valve; electric actuator; spring return; domestic usage | | | | | | |
| 22mm dia. | 51.25 | 56.38 | 1.14 | 13.08 | nr | **69.46** |
| Three port motorised valve; electric actuator; spring return; domestic usage | | | | | | |
| 22mm dia. | 45.95 | 50.55 | 1.14 | 13.08 | nr | **63.63** |

## Y:MECHANICAL AND ELECTRICAL SERVICES

| Item | Net Price £ | Material £ | Labour hours | Labour £ | Unit | Total Rate £ |
|---|---|---|---|---|---|---|
| **Y11 : PIPELINE ANCILLARIES : LUBRICATED PLUG VALVES** | | | | | | |
| **Lubricated plug valves** | | | | | | |
| Cast iron; lubricated plug valve; PN16; working pressure 10.8 bar at 260 degrees Celsius, cold services up to 16 bar; wrench operated with lever; screwed connections | | | | | | |
| 15mm dia. | 56.62 | 62.90 | 0.94 | 10.79 | nr | **73.69** |
| 20mm dia. | 57.97 | 64.40 | 1.12 | 12.86 | nr | **77.26** |
| 25mm dia. | 68.31 | 75.89 | 1.30 | 14.92 | nr | **90.81** |
| 32mm dia. | 80.44 | 89.37 | 1.50 | 17.23 | nr | **106.60** |
| 40mm dia. | 80.44 | 89.37 | 1.74 | 19.99 | nr | **109.36** |
| 50mm dia. | 126.28 | 140.30 | 2.08 | 23.91 | nr | **164.21** |
| Cast iron; lubricated plug valve; PN16; working pressure 11.2 bar at 250 degrees Celsius, cold services up to 16 bar; wrench operated with lever; flanged ends to BS 4504, Table 16/11; bolted connections | | | | | | |
| 65mm dia. | 181.56 | 201.71 | 1.80 | 20.68 | nr | **222.39** |
| 80mm dia. | 219.31 | 243.65 | 1.92 | 22.07 | nr | **265.72** |
| 100mm dia. | 361.77 | 401.93 | 2.80 | 32.14 | nr | **434.07** |
| Cast iron; lubricated plug valve; PN16; working pressure 11.2 bar at 250 degrees Celsius, cold service up to 16 bar; gear operated with lever; flanged ends to BS 4504, Table 16/11; bolted connections | | | | | | |
| 100mm dia. | 1127.99 | 1253.20 | 7.04 | 80.81 | nr | **1334.01** |
| 150mm dia. | 1923.43 | 2136.93 | 12.05 | 138.25 | nr | **2275.18** |

*Material Costs / Measured Work Prices - Mechanical Installations*

**Y:MECHANICAL AND ELECTRICAL SERVICES**

| Item | Net Price £ | Material £ | Labour hours | Labour £ | Unit | Total Rate £ |
|---|---|---|---|---|---|---|
| **Y11 : PIPELINE ANCILLARIES : GATE VALVES** | | | | | | |
| Gate valves | | | | | | |
| Bronze wedge non-rising stem; BS 5154, series B, PN 32; wheelhead operated; working pressure saturated steam up to 14 bar; cold services up to 32 bar; screwed ends to steel | | | | | | |
| 15mm dia. | 18.86 | 20.75 | 0.94 | 10.79 | nr | 31.54 |
| 20mm dia. | 24.31 | 26.74 | 1.12 | 12.86 | nr | 39.60 |
| 25mm dia. | 32.95 | 36.24 | 1.30 | 14.92 | nr | 51.16 |
| 32mm dia. | 48.17 | 52.99 | 1.50 | 17.23 | nr | 70.22 |
| 40mm dia. | 62.26 | 68.49 | 1.74 | 19.99 | nr | 88.48 |
| 50mm dia. | 89.90 | 98.99 | 2.08 | 23.91 | nr | 122.90 |
| Bronze wedge non-rising stem; BS 5154, series B, PN 20; wheelhead operated; working pressure saturated steam up to 9 bar; cold services up to 20 bar; screwed ends to steel | | | | | | |
| 15mm dia. | 12.33 | 13.56 | 0.94 | 10.79 | nr | 24.35 |
| 20mm dia. | 12.33 | 13.56 | 1.12 | 12.86 | nr | 26.42 |
| 25mm dia. | 22.64 | 24.90 | 1.30 | 14.92 | nr | 39.82 |
| 32mm dia. | 32.28 | 35.51 | 1.50 | 17.23 | nr | 52.74 |
| 40mm dia. | 44.16 | 48.58 | 1.74 | 19.99 | nr | 68.57 |
| 50mm dia. | 63.22 | 69.54 | 2.08 | 23.91 | nr | 93.45 |
| DZR copper alloy wedge non-rising stem; BS 5154, PN 20; wheelhead operated; working pressure saturated steam up to 9 bar; cold services up to 20 bar; screwed ends to steel | | | | | | |
| 15mm dia. | 7.73 | 8.50 | 0.94 | 10.79 | nr | 19.29 |
| 20mm dia. | 9.13 | 10.04 | 1.12 | 12.86 | nr | 22.90 |
| 25mm dia. | 12.63 | 13.89 | 1.30 | 14.92 | nr | 28.81 |
| 32mm dia. | 18.92 | 20.81 | 1.50 | 17.23 | nr | 38.04 |
| 40mm dia. | 26.66 | 29.33 | 1.74 | 19.99 | nr | 49.32 |
| 50mm dia. | 36.93 | 40.62 | 2.08 | 23.91 | nr | 64.53 |
| Bronze wedge non-rising stem; BS 5154, PN 16; wheelhead operated; working pressure saturated steam up to 7 bar; cold services up to 16 bar; BS4504 flanged ends; bolted connections | | | | | | |
| 15mm dia. | 50.26 | 55.29 | 1.36 | 15.61 | nr | 70.90 |
| 20mm dia. | 64.96 | 71.46 | 1.40 | 16.07 | nr | 87.53 |
| 25mm dia. | 85.20 | 93.72 | 1.46 | 16.78 | nr | 110.50 |
| 32mm dia. | 111.00 | 122.10 | 1.54 | 17.68 | nr | 139.78 |
| 40mm dia. | 132.97 | 146.27 | 1.64 | 18.84 | nr | 165.11 |
| 50mm dia. | 185.44 | 203.98 | 1.64 | 18.84 | nr | 222.82 |
| 65mm dia. | 281.87 | 310.06 | 1.80 | 20.68 | nr | 330.74 |
| 80mm dia. | 408.60 | 449.46 | 1.92 | 22.07 | nr | 471.53 |
| 100mm dia. | 760.38 | 836.42 | 2.80 | 32.14 | nr | 868.56 |

## Y:MECHANICAL AND ELECTRICAL SERVICES

| Item | Net Price £ | Material £ | Labour hours | Labour £ | Unit | Total Rate £ |
|---|---|---|---|---|---|---|
| **Y11 : PIPELINE ANCILLARIES : GATE VALVES (contd)** | | | | | | |
| **Gate valves (contd)** | | | | | | |
| Cast iron, trim material bronze; non rising stem, inside screwed; BS 5150, PN6; wheelhead operated; working pressure up to 6 bar at 120 degrees C; BS4504 flanged ends; bolted connections | | | | | | |
| 40mm dia. | 129.26 | 142.19 | 1.64 | 18.84 | nr | 161.03 |
| 50mm dia. | 130.54 | 143.59 | 1.64 | 18.84 | nr | 162.43 |
| 65mm dia. | 142.95 | 157.25 | 1.80 | 20.68 | nr | 177.93 |
| 80mm dia. | 165.21 | 181.73 | 1.92 | 22.07 | nr | 203.80 |
| 100mm dia. | 218.28 | 240.11 | 2.80 | 32.14 | nr | 272.25 |
| 125mm dia. | 312.44 | 343.68 | 9.01 | 103.38 | nr | 447.06 |
| 150mm dia. | 359.52 | 395.47 | 12.05 | 138.25 | nr | 533.72 |
| 200mm dia. | 684.80 | 753.28 | 13.16 | 150.99 | nr | 904.27 |
| 250mm dia. | 1005.80 | 1106.38 | 14.08 | 161.62 | nr | 1268.00 |
| 300mm dia. | 1249.76 | 1374.74 | 15.15 | 173.86 | nr | 1548.60 |
| Cast iron, trim material bronze; non rising stem, inside screwed; BS 5150, PN10; wheelhead operated; working pressure saturated steam up to 8.4 bar, cold services up to 10 bar; BS4504 flanged ends; bolted connections | | | | | | |
| 50mm dia. | 138.67 | 152.54 | 1.64 | 18.84 | nr | 171.38 |
| 65mm dia. | 155.79 | 171.37 | 1.80 | 20.68 | nr | 192.05 |
| 80mm dia. | 180.62 | 198.68 | 1.92 | 22.07 | nr | 220.75 |
| 100mm dia. | 239.68 | 263.65 | 2.80 | 32.14 | nr | 295.79 |
| 125mm dia. | 338.12 | 371.93 | 9.01 | 103.38 | nr | 475.31 |
| 150mm dia. | 389.48 | 428.43 | 12.05 | 138.25 | nr | 566.68 |
| 200mm dia. | 706.20 | 776.82 | 13.16 | 150.99 | nr | 927.81 |
| 250mm dia. | 1052.88 | 1158.17 | 14.08 | 161.62 | nr | 1319.79 |
| 300mm dia. | 1339.64 | 1473.60 | 15.15 | 173.86 | nr | 1647.46 |
| 350mm dia. | 2251.28 | 2476.41 | 15.15 | 173.86 | nr | 2650.27 |
| Cast iron, trim material bronze; non rising stem, inside screwed; BS 5150, PN16; wheelhead operated; working pressure saturated steam up to 12.8 bar, cold services up to 16 bar; BS4504 flanged ends; bolted connections | | | | | | |
| 40mm dia. | 154.51 | 169.96 | 1.64 | 18.84 | nr | 188.80 |
| 50mm dia. | 154.51 | 169.96 | 1.64 | 18.84 | nr | 188.80 |
| 65mm dia. | 177.62 | 195.38 | 1.80 | 20.68 | nr | 216.06 |
| 80mm dia. | 197.74 | 217.51 | 1.92 | 22.07 | nr | 239.58 |
| 100mm dia. | 263.22 | 289.54 | 2.80 | 32.14 | nr | 321.68 |
| 125mm dia. | 380.92 | 419.01 | 9.01 | 103.38 | nr | 522.39 |
| 150mm dia. | 449.40 | 494.34 | 12.05 | 138.25 | nr | 632.59 |
| 200mm dia. | 778.96 | 856.86 | 13.16 | 150.99 | nr | 1007.85 |
| 250mm dia. | 1168.44 | 1285.28 | 14.08 | 161.62 | nr | 1446.90 |
| 300mm dia. | 1326.80 | 1459.48 | 15.15 | 173.86 | nr | 1633.34 |
| 350mm dia. | 2606.52 | 2867.17 | 15.15 | 173.86 | nr | 3041.03 |
| 400mm dia. | 3269.92 | 3596.91 | 16.13 | 185.08 | nr | 3781.99 |
| 450mm dia. | 4609.56 | 5070.52 | 16.13 | 185.08 | nr | 5255.60 |

## Y:MECHANICAL AND ELECTRICAL SERVICES

| Item | Net Price £ | Material £ | Labour hours | Labour £ | Unit | Total Rate £ |
|---|---|---|---|---|---|---|
| Cast iron, trim material bronze; non rising stem; BS 5163 series A, PN16; wheelhead operated; working pressure cold services up to 16 bar; BS4504 flanged ends; bolted connections | | | | | | |
| 50mm dia. | 276.06 | 303.67 | 1.64 | 18.84 | nr | **322.51** |
| 65mm dia. | 284.62 | 313.08 | 1.80 | 20.68 | nr | **333.76** |
| 80mm dia. | 293.18 | 322.50 | 1.92 | 22.07 | nr | **344.57** |
| 100mm dia. | 383.06 | 421.37 | 2.80 | 32.14 | nr | **453.51** |
| 125mm dia. | 500.76 | 550.84 | 9.01 | 103.38 | nr | **654.22** |
| 150mm dia. | 607.76 | 668.54 | 12.05 | 138.25 | nr | **806.79** |

## Y:MECHANICAL AND ELECTRICAL SERVICES

| Item | Net Price £ | Material £ | Labour hours | Labour £ | Unit | Total Rate £ |
|---|---|---|---|---|---|---|
| **Y11 : PIPELINE ANCILLARIES : GLOBE VALVES** | | | | | | |
| Globe valves | | | | | | |
| Bronze; renewable disc; BS 5154 series B, PN32; working pressure saturated steam up to 14 bar; cold services up to 32 bar; screwed ends to steel | | | | | | |
| 15mm dia. | 18.41 | 20.25 | 0.84 | 9.64 | nr | **29.89** |
| 20mm dia. | 25.76 | 28.34 | 1.14 | 13.08 | nr | **41.42** |
| 25mm dia. | 36.28 | 39.91 | 1.30 | 14.92 | nr | **54.83** |
| 32mm dia. | 50.62 | 55.68 | 1.50 | 17.23 | nr | **72.91** |
| 40mm dia. | 70.32 | 77.35 | 1.74 | 19.99 | nr | **97.34** |
| 50mm dia. | 107.95 | 118.75 | 1.64 | 18.84 | nr | **137.59** |
| Bronze; needle valve; BS 5154, series B, PN32; working pressure saturated steam up to 14 bar; cold services up to 32 bar; BS4504 screwed ends to steel | | | | | | |
| 15mm dia. | 19.71 | 21.68 | 1.24 | 14.24 | nr | **35.92** |
| 20mm dia. | 33.37 | 36.71 | 1.34 | 15.38 | nr | **52.09** |
| 25mm dia. | 47.03 | 51.73 | 1.40 | 16.07 | nr | **67.80** |
| 32mm dia. | 98.10 | 107.91 | 1.46 | 16.78 | nr | **124.69** |
| 40mm dia. | 154.53 | 169.98 | 1.54 | 17.68 | nr | **187.66** |
| 50mm dia. | 195.74 | 215.31 | 1.64 | 18.84 | nr | **234.15** |
| Bronze; renewable disc; BS 5154, series B, PN16; working pressure saturated steam up to 7 bar; cold services up to 16 bar; BS4504 flanged ends; bolted connections | | | | | | |
| 15mm dia. | 48.82 | 53.70 | 1.24 | 14.24 | nr | **67.94** |
| 20mm dia. | 56.44 | 62.08 | 1.34 | 15.38 | nr | **77.46** |
| 25mm dia. | 98.99 | 108.89 | 1.40 | 16.07 | nr | **124.96** |
| 32mm dia. | 123.63 | 135.99 | 1.46 | 16.78 | nr | **152.77** |
| 40mm dia. | 146.02 | 160.62 | 1.54 | 17.68 | nr | **178.30** |
| 50mm dia. | 203.81 | 224.19 | 1.64 | 18.84 | nr | **243.03** |
| Bronze; renewable disc; BS 2060, class 250; working pressure saturated steam up to 24 bar; cold services up to 38 bar; flanged ends (BS 10 table H); bolted connections | | | | | | |
| 15mm dia. | 124.52 | 136.97 | 1.24 | 14.24 | nr | **151.21** |
| 20mm dia. | 145.13 | 159.64 | 1.34 | 15.38 | nr | **175.02** |
| 25mm dia. | 198.88 | 218.77 | 1.40 | 16.07 | nr | **234.84** |
| 32mm dia. | 259.80 | 285.78 | 1.46 | 16.78 | nr | **302.56** |
| 40mm dia. | 329.22 | 362.14 | 1.54 | 17.68 | nr | **379.82** |
| 50mm dia. | 483.76 | 532.14 | 1.64 | 18.84 | nr | **550.98** |
| 65mm dia. | 671.89 | 739.08 | 1.80 | 20.68 | nr | **759.76** |
| 80mm dia. | 958.56 | 1054.42 | 1.92 | 22.07 | nr | **1076.49** |

## Y:MECHANICAL AND ELECTRICAL SERVICES

| Item | Net Price £ | Material £ | Labour hours | Labour £ | Unit | Total Rate £ |
|---|---|---|---|---|---|---|
| **Y11 : PIPELINE ANCILLARIES : SAFETY AND RELIEF VALVES** | | | | | | |
| **Safety and relief valves** | | | | | | |
| Bronze safety valve; 'pop' type; side outlet; including easing lever; working pressure saturated steam up to 20.7 bar; screwed ends to steel | | | | | | |
| 15mm dia. | 85.22 | 96.55 | 0.24 | 2.75 | nr | 99.30 |
| 20mm dia. | 98.33 | 111.41 | 0.34 | 3.90 | nr | 115.31 |
| Bronze safety valve; 'pop' type; side outlet; including easing lever; working pressure saturated steam up to 17.2 bar; screwed ends to steel | | | | | | |
| 25mm dia. | 121.49 | 137.65 | 0.34 | 3.90 | nr | 141.55 |
| 32mm dia. | 157.76 | 178.74 | 0.46 | 5.28 | nr | 184.02 |
| Bronze safety valve; 'pop' type; side outlet; including easing lever; working pressure saturated steam up to 13.8 bar; screwed ends to steel | | | | | | |
| 40mm dia. | 206.70 | 234.19 | 0.60 | 6.89 | nr | 241.08 |
| 50mm dia. | 281.87 | 319.36 | 0.72 | 8.27 | nr | 327.63 |
| 65mm dia. | 502.55 | 569.39 | 0.90 | 10.33 | nr | 579.72 |
| 80mm dia. | 563.73 | 638.71 | 1.06 | 12.17 | nr | 650.88 |
| Bronze relief valve; spring type; side outlet; working pressure up to 20.7 bar at 120 degrees Celsius; screwed ends to steel | | | | | | |
| 15mm dia. | 47.63 | 53.96 | 0.24 | 2.75 | nr | 56.71 |
| 20mm dia. | 58.56 | 66.35 | 0.34 | 3.90 | nr | 70.25 |
| Bronze relief valve; spring type; side outlet; working pressure up to 17.2 bar at 120 degrees Celsius; screwed ends to steel | | | | | | |
| 25mm dia. | 79.10 | 89.62 | 0.34 | 3.90 | nr | 93.52 |
| 32mm dia. | 115.81 | 131.21 | 0.46 | 5.28 | nr | 136.49 |
| Bronze relief valve; spring type; side outlet; working pressure up to 13.8 bar at 120 degrees Celsius; screwed ends to steel | | | | | | |
| 40mm dia. | 138.97 | 157.45 | 0.60 | 6.89 | nr | 164.34 |
| 50mm dia. | 194.03 | 219.84 | 0.72 | 8.27 | nr | 228.11 |
| 65mm dia. | 308.09 | 349.07 | 0.90 | 10.33 | nr | 359.40 |
| 80mm dia. | 404.23 | 457.99 | 1.06 | 12.17 | nr | 470.16 |

## Y:MECHANICAL AND ELECTRICAL SERVICES

| Item | Net Price £ | Material £ | Labour hours | Labour £ | Unit | Total Rate £ |
|---|---|---|---|---|---|---|
| **Y11 : PIPELINE ANCILLARIES : DRAIN COCKS AND VENT COCKS** | | | | | | |
| Cocks; screwed joints to steel | | | | | | |
| Bronze gland cock; complete with malleable iron lever; working pressure cold services up to 10 bar; screwed ends | | | | | | |
| 15mm dia. | 27.19 | 29.91 | 0.84 | 9.64 | nr | 39.55 |
| 20mm dia. | 38.23 | 42.05 | 1.14 | 13.08 | nr | 55.13 |
| 25mm dia. | 54.63 | 60.09 | 1.30 | 14.92 | nr | 75.01 |
| 32mm dia. | 106.70 | 117.37 | 1.50 | 17.23 | nr | 134.60 |
| 40mm dia. | 145.17 | 159.69 | 1.74 | 19.99 | nr | 179.68 |
| 50mm dia. | 211.51 | 232.66 | 2.08 | 23.91 | nr | 256.57 |
| Bronze three-way plug cock; complete with malleable iron lever; working pressure cold services up to 10 bar; screwed ends | | | | | | |
| 15mm dia. | 54.29 | 59.72 | 0.84 | 9.64 | nr | 69.36 |
| 20mm dia. | 63.25 | 69.58 | 1.14 | 13.08 | nr | 82.66 |
| 25mm dia. | 76.19 | 83.81 | 1.30 | 14.92 | nr | 98.73 |
| 32mm dia. | 125.33 | 137.86 | 1.50 | 17.23 | nr | 155.09 |
| 40mm dia. | 150.14 | 165.15 | 1.74 | 19.99 | nr | 185.14 |
| Cocks; capillary joints to copper | | | | | | |
| Stopcock; brass head with gun metal body | | | | | | |
| 15mm dia. | 2.97 | 3.27 | 0.50 | 5.74 | nr | 9.01 |
| 22mm dia. | 5.54 | 6.09 | 0.50 | 5.74 | nr | 11.83 |
| 28mm dia. | 15.76 | 17.34 | 0.53 | 6.08 | nr | 23.42 |
| Lockshield stop cocks; brass head with gun metal body | | | | | | |
| 15mm dia. | 7.27 | 8.00 | 0.50 | 5.74 | nr | 13.74 |
| 22mm dia. | 10.46 | 11.51 | 0.50 | 5.74 | nr | 17.25 |
| 28mm dia. | 18.53 | 20.38 | 0.53 | 6.08 | nr | 26.46 |
| DZR stopcock; brass head with gun metal body | | | | | | |
| 15mm dia. | 7.45 | 8.20 | 0.50 | 5.74 | nr | 13.94 |
| 22mm dia. | 12.92 | 14.21 | 0.50 | 5.74 | nr | 19.95 |
| 28mm dia. | 21.52 | 23.67 | 0.53 | 6.08 | nr | 29.75 |
| Gunmetal stopcock | | | | | | |
| 35mm dia. | 33.75 | 37.13 | 0.75 | 8.61 | nr | 45.74 |
| 42mm dia. | 44.82 | 49.30 | 0.75 | 8.61 | nr | 57.91 |
| 54mm dia. | 66.96 | 73.66 | 0.83 | 9.53 | nr | 83.19 |
| Lockshield DZR stopcock; brass head gun metal body | | | | | | |
| 15mm dia. | 9.84 | 10.82 | 0.50 | 5.74 | nr | 16.56 |
| 22mm dia. | 14.61 | 16.07 | 0.50 | 5.74 | nr | 21.81 |
| 28mm dia. | 24.29 | 26.72 | 0.53 | 6.08 | nr | 32.80 |

## Y:MECHANICAL AND ELECTRICAL SERVICES

| Item | Net Price £ | Material £ | Labour hours | Labour £ | Unit | Total Rate £ |
|---|---|---|---|---|---|---|
| Double union stopcock | | | | | | |
| 15mm dia. | 9.61 | 10.57 | 0.66 | 7.57 | nr | 18.14 |
| 22mm dia. | 13.53 | 14.88 | 0.66 | 7.57 | nr | 22.45 |
| 28mm dia. | 24.06 | 26.47 | 0.70 | 8.04 | nr | 34.51 |
| Double union lockshield stopcock | | | | | | |
| 15mm dia. | 10.76 | 11.84 | 0.67 | 7.64 | nr | 19.48 |
| 22mm dia. | 15.07 | 16.58 | 0.67 | 7.64 | nr | 24.22 |
| 28mm dia. | 26.91 | 29.60 | 0.70 | 8.04 | nr | 37.64 |
| Double union DZR stopcock | | | | | | |
| 15mm dia. | 15.68 | 17.25 | 0.67 | 7.64 | nr | 24.89 |
| 22mm dia. | 19.30 | 21.23 | 0.67 | 7.64 | nr | 28.87 |
| 28mm dia. | 35.67 | 39.24 | 0.70 | 8.04 | nr | 47.28 |
| Double union lockshield DZR stopcock | | | | | | |
| 15mm dia. | 16.61 | 18.27 | 0.67 | 7.64 | nr | 25.91 |
| 22mm dia. | 20.83 | 22.91 | 0.67 | 7.64 | nr | 30.55 |
| 28mm dia. | 38.44 | 42.28 | 0.70 | 8.04 | nr | 50.32 |
| Double union gun metal stopcock | | | | | | |
| 35mm dia. | 59.42 | 65.36 | 0.70 | 8.04 | nr | 73.40 |
| 42mm dia. | 81.56 | 89.72 | 0.75 | 8.61 | nr | 98.33 |
| 54mm dia. | 128.30 | 141.13 | 0.91 | 10.39 | nr | 151.52 |
| Stopcock with easy clean cover | | | | | | |
| 15mm dia. | 7.65 | 8.41 | 0.70 | 8.04 | nr | 16.45 |
| 22mm dia. | 10.84 | 11.92 | 0.67 | 7.64 | nr | 19.56 |
| 28mm dia. | 19.07 | 20.98 | 0.70 | 8.04 | nr | 29.02 |
| Union stopcock with easy clean cover | | | | | | |
| 15mm dia. | 9.30 | 10.23 | 0.67 | 7.64 | nr | 17.87 |
| 22mm dia. | 14.22 | 15.64 | 0.67 | 7.64 | nr | 23.28 |
| 28mm dia. | 22.52 | 24.77 | 0.70 | 8.04 | nr | 32.81 |
| Double union stopcock with easy clean cover | | | | | | |
| 15mm dia. | 11.15 | 12.27 | 0.67 | 7.64 | nr | 19.91 |
| 22mm dia. | 15.45 | 17.00 | 0.67 | 7.64 | nr | 24.64 |
| 28mm dia. | 27.14 | 29.85 | 0.70 | 8.04 | nr | 37.89 |
| Combined stopcock and drain | | | | | | |
| 15mm dia. | 16.37 | 18.01 | 0.72 | 8.27 | nr | 26.28 |
| 22mm dia. | 20.29 | 22.32 | 0.72 | 8.27 | nr | 30.59 |
| Combined DZR stopcock and drain | | | | | | |
| 15mm dia. | 21.83 | 24.01 | 0.72 | 8.27 | nr | 32.28 |

## Y:MECHANICAL AND ELECTRICAL SERVICES

| Item | Net Price £ | Material £ | Labour hours | Labour £ | Unit | Total Rate £ |
|---|---|---|---|---|---|---|
| **Y11 : PIPELINE ANCILLARIES : DRAIN COCKS AND VENT COCKS** | | | | | | |
| **Cocks; compression joints to copper (contd)** | | | | | | |
| Stopcock; brass head gun metal body | | | | | | |
| 15mm dia. | 3.96 | 4.36 | 0.50 | 5.74 | nr | **10.10** |
| 22mm dia. | 6.96 | 7.66 | 0.50 | 5.74 | nr | **13.40** |
| 28mm dia. | 18.14 | 19.95 | 0.53 | 6.08 | nr | **26.03** |
| Lockshield stopcock; brass head gun metal body | | | | | | |
| 15mm dia. | 8.61 | 9.47 | 0.50 | 5.74 | nr | **15.21** |
| 22mm dia. | 12.15 | 13.37 | 0.50 | 5.74 | nr | **19.11** |
| 28mm dia. | 23.37 | 25.71 | 0.53 | 6.08 | nr | **31.79** |
| DZR Stopcock | | | | | | |
| 15mm dia. | 9.22 | 10.14 | 0.45 | 5.16 | nr | **15.30** |
| 22mm dia. | 15.14 | 16.65 | 0.45 | 5.16 | nr | **21.81** |
| 28mm dia. | 25.06 | 27.57 | 0.47 | 5.37 | nr | **32.94** |
| 35mm dia. | 46.13 | 50.74 | 0.59 | 6.77 | nr | **57.51** |
| 42mm dia. | 66.04 | 72.64 | 0.59 | 6.77 | nr | **79.41** |
| 54mm dia. | 89.94 | 98.93 | 0.67 | 7.69 | nr | **106.62** |
| DZR Lockshield stopcock | | | | | | |
| 15mm dia. | 10.61 | 11.67 | 0.45 | 5.16 | nr | **16.83** |
| 22mm dia. | 16.61 | 18.27 | 0.45 | 5.16 | nr | **23.43** |
| Stopcock with easy clean cover | | | | | | |
| 15mm dia. | 9.53 | 10.48 | 0.50 | 5.74 | nr | **16.22** |
| 22mm dia. | 13.38 | 14.72 | 0.50 | 5.74 | nr | **20.46** |
| 28mm dia. | 23.98 | 26.38 | 0.53 | 6.08 | nr | **32.46** |
| Combined stop/draincock | | | | | | |
| 15mm dia. | 13.30 | 14.63 | 0.20 | 2.29 | nr | **16.92** |
| 22mm dia. | 20.45 | 22.50 | 0.50 | 5.74 | nr | **28.24** |
| DZR Combined stop/draincock | | | | | | |
| 15mm dia. | 17.14 | 18.85 | 0.45 | 5.16 | nr | **24.01** |
| 22mm dia. | 25.14 | 27.65 | 0.45 | 5.16 | nr | **32.81** |
| Stopcock to polyethylene | | | | | | |
| 15mm dia. | 10.84 | 11.92 | 0.45 | 5.16 | nr | **17.08** |
| 20mm dia. | 16.30 | 17.93 | 0.45 | 5.16 | nr | **23.09** |
| 25mm dia. | 23.60 | 25.96 | 0.47 | 5.39 | nr | **31.35** |
| Draw off coupling | | | | | | |
| 15mm dia. | 5.74 | 6.31 | 0.45 | 5.16 | nr | **11.47** |

## Y:MECHANICAL AND ELECTRICAL SERVICES

| Item | Net Price £ | Material £ | Labour hours | Labour £ | Unit | Total Rate £ |
|---|---|---|---|---|---|---|
| DZR Draw off coupling | | | | | | |
|   15mm dia. | 7.92 | 8.71 | 0.45 | 5.16 | nr | **13.87** |
|   22mm dia. | 9.17 | 10.09 | 0.45 | 5.16 | nr | **15.25** |
| Draw off elbow | | | | | | |
|   15mm dia. | 8.70 | 9.57 | 0.45 | 5.16 | nr | **14.73** |
|   22mm dia. | 10.72 | 11.79 | 0.45 | 5.16 | nr | **16.95** |
| Lockshield drain cock | | | | | | |
|   15mm dia. | 4.03 | 4.43 | 0.45 | 5.16 | nr | **9.59** |

## Y:MECHANICAL AND ELECTRICAL SERVICES

| Item | Net Price £ | Material £ | Labour hours | Labour £ | Unit | Total Rate £ |
|---|---|---|---|---|---|---|
| **Y11 : PIPELINE ANCILLARIES : AUTOMATIC AIR VENTS** | | | | | | |
| **Air vents; including regulating, adjusting and testing** | | | | | | |
| Automatic air vent; pressures up to 7 bar; 93 degree celsius; screwed ends to steel | | | | | | |
| 15mm dia. | 52.58 | 57.84 | 0.80 | 9.18 | nr | **67.02** |
| Automatic air vent; pressures up to 7 bar; 93 degree celsius; lockhead isolating valve; screwed ends to steel | | | | | | |
| 15mm dia. | 65.04 | 71.54 | 0.80 | 9.18 | nr | **80.72** |
| Automatic air vent; pressures up to 17 bar; 204 degree celsius; flanged end (BS10, Table H); bolted connections to counter flange (measured separately) | | | | | | |
| 15mm dia. | 372.69 | 409.96 | 0.80 | 9.18 | nr | **419.14** |

## Y:MECHANICAL AND ELECTRICAL SERVICES

| Item | Net Price £ | Material £ | Labour hours | Labour £ | Unit | Total Rate £ |
|---|---|---|---|---|---|---|
| **Y11 : PIPELINE ANCILLARIES : STEAM TRAPS AND STRAINERS** | | | | | | |
| **Steam traps and strainers** | | | | | | |
| Stainless steel; thermodynamic trap with pressure range up to 42 bar; temperature range to 400 degree celsius; screwed ends to steel | | | | | | |
| 15mm dia. | 50.26 | 55.29 | 0.84 | 9.64 | nr | 64.93 |
| 20mm dia. | 73.75 | 81.13 | 1.14 | 13.08 | nr | 94.21 |
| Stainless steel; thermodynamic trap with pressure range up to 24 bar; temperature range to 288 degree celsius; flanged ends (BS 10, Table H); bolted connections | | | | | | |
| 15mm dia. | 191.17 | 210.29 | 1.24 | 14.24 | nr | 224.53 |
| 20mm dia. | 196.11 | 215.72 | 1.34 | 15.38 | nr | 231.10 |
| 25mm dia. | 211.77 | 232.95 | 1.40 | 16.07 | nr | 249.02 |
| Malleable iron pipeline strainer; max steam working pressure 14 bar and temperature range to 230 degree celsius; screwed ends to steel | | | | | | |
| 1/2" dia. | 7.36 | 8.10 | 0.84 | 9.64 | nr | 17.74 |
| 3/4" dia. | 9.85 | 10.84 | 1.14 | 13.08 | nr | 23.92 |
| 1" dia. | 14.54 | 15.99 | 1.30 | 14.92 | nr | 30.91 |
| 11/2" dia. | 24.10 | 26.51 | 1.50 | 17.23 | nr | 43.74 |
| 2" dia. | 42.85 | 47.13 | 1.74 | 19.99 | nr | 67.12 |
| Bronze pipeline strainer; max steam working pressure 25 bar; flanged ends (BS 19, Table H); bolted connections | | | | | | |
| 15mm dia. | 79.93 | 87.92 | 1.24 | 14.24 | nr | 102.16 |
| 20mm dia. | 83.22 | 91.54 | 1.34 | 15.38 | nr | 106.92 |
| 25mm dia. | 101.35 | 111.48 | 1.40 | 16.07 | nr | 127.55 |
| 32mm dia. | 148.32 | 163.15 | 1.46 | 16.78 | nr | 179.93 |
| 40mm dia. | 185.40 | 203.94 | 1.54 | 17.68 | nr | 221.62 |
| 50mm dia. | 277.69 | 305.46 | 1.64 | 18.84 | nr | 324.30 |
| 65mm dia. | 354.32 | 389.75 | 2.50 | 28.69 | nr | 418.44 |
| 80mm dia. | 444.14 | 488.55 | 2.91 | 33.36 | nr | 521.91 |
| 100mm dia. | 794.34 | 873.77 | 3.51 | 40.26 | nr | 914.03 |
| Balanced pressure thermostatic steam trap and strainer; max working pressure up to 13 bar; screwed ends to steel | | | | | | |
| 1/2" dia. | 34.77 | 38.25 | 1.26 | 14.47 | nr | 52.72 |
| 3/4" dia. | 40.21 | 44.23 | 1.71 | 19.65 | nr | 63.88 |
| Bimetallic thermostatic steam trap and strainer; max working pressure up to 21 bar; flanged ends | | | | | | |
| 15mm | 106.30 | 116.93 | 1.24 | 14.24 | nr | 131.17 |
| 20mm | 106.30 | 116.93 | 1.34 | 15.38 | nr | 132.31 |

## Y:MECHANICAL AND ELECTRICAL SERVICES

| Item | Net Price £ | Material £ | Labour hours | Labour £ | Unit | Total Rate £ |
|---|---|---|---|---|---|---|
| **Y11 : PIPELINE ANCILLARIES : SIGHT GLASSES** | | | | | | |
| Sight glasses | | | | | | |
| Pressed brass; straight; single window; screwed ends to steel | | | | | | |
| 15mm dia. | 21.58 | 23.74 | 0.84 | 9.64 | nr | 33.38 |
| 20mm dia. | 24.11 | 26.52 | 1.14 | 13.08 | nr | 39.60 |
| 25mm dia. | 29.95 | 32.95 | 1.30 | 14.92 | nr | 47.87 |
| Gunmetal; straight; double window; screwed ends to steel | | | | | | |
| 15mm dia. | 35.05 | 38.55 | 0.84 | 9.64 | nr | 48.19 |
| 20mm dia. | 38.27 | 42.10 | 1.14 | 13.08 | nr | 55.18 |
| 25mm dia. | 47.53 | 52.28 | 1.30 | 14.92 | nr | 67.20 |
| 32mm dia. | 77.93 | 85.72 | 1.35 | 15.51 | nr | 101.23 |
| 40mm dia. | 77.93 | 85.72 | 1.74 | 19.99 | nr | 105.71 |
| 50mm dia. | 94.25 | 103.67 | 2.08 | 23.91 | nr | 127.58 |
| SG Iron flanged; BS 4504, PN 25 | | | | | | |
| 15mm dia. | 71.81 | 78.99 | 1.00 | 11.47 | nr | 90.46 |
| 20mm dia. | 84.05 | 92.45 | 1.25 | 14.34 | nr | 106.79 |
| 25mm dia. | 106.90 | 117.59 | 1.50 | 17.23 | nr | 134.82 |
| 32mm dia. | 118.32 | 130.15 | 1.70 | 19.52 | nr | 149.67 |
| 40mm dia. | 155.04 | 170.54 | 2.00 | 22.95 | nr | 193.49 |
| 50mm dia. | 186.86 | 205.55 | 2.30 | 26.44 | nr | 231.99 |
| Check valve and sight glass; gun metal; screwed | | | | | | |
| 15mm dia. | 35.21 | 38.73 | 0.84 | 9.64 | nr | 48.37 |
| 20mm dia. | 37.09 | 40.80 | 1.14 | 13.08 | nr | 53.88 |
| 25mm dia. | 62.83 | 69.11 | 1.30 | 14.92 | nr | 84.03 |

## Y:MECHANICAL AND ELECTRICAL SERVICES

| Item | Net Price £ | Material £ | Labour hours | Labour £ | Unit | Total Rate £ |
|---|---|---|---|---|---|---|
| **Y11 : PIPELINE ANCILLARIES : EXPANSION JOINTS** | | | | | | |
| **Axial movement bellows expansion joints; stainless steel** | | | | | | |
| Screwed ends for steel pipework; up to 6 bar G at 100 degrees Celsius | | | | | | |
| 15mm dia. | 37.92 | 41.71 | 0.48 | 5.51 | nr | **47.22** |
| 20mm dia. | 46.13 | 50.74 | 0.60 | 6.89 | nr | **57.63** |
| 25mm dia. | 46.13 | 50.74 | 0.72 | 8.27 | nr | **59.01** |
| 32mm dia. | 54.33 | 59.76 | 0.78 | 8.95 | nr | **68.71** |
| 40mm dia. | 65.60 | 72.16 | 0.84 | 9.64 | nr | **81.80** |
| 50mm dia. | 90.20 | 99.22 | 0.84 | 9.64 | nr | **108.86** |
| Screwed ends for steel pipework; aluminium and steel outer sleeves; up to 16 bar G at 120 degrees Celsius | | | | | | |
| 20mm dia. | 65.60 | 72.16 | 1.28 | 14.69 | nr | **86.85** |
| 25mm dia. | 69.70 | 76.67 | 1.50 | 17.23 | nr | **93.90** |
| 32mm dia. | 98.40 | 108.24 | 1.76 | 20.20 | nr | **128.44** |
| 40mm dia. | 107.63 | 118.39 | 2.00 | 22.95 | nr | **141.34** |
| 50mm dia. | 116.85 | 128.53 | 2.28 | 26.20 | nr | **154.73** |
| Flanged ends for steel pipework; aluminium and steel outer sleeves; up to 16 bar G at 120 degrees Celsius | | | | | | |
| 20mm dia. | 67.65 | 74.42 | 0.48 | 5.51 | nr | **79.93** |
| 25mm dia. | 73.80 | 81.18 | 0.60 | 6.89 | nr | **88.07** |
| 32mm dia. | 101.47 | 111.62 | 0.72 | 8.27 | nr | **119.89** |
| 40mm dia. | 107.63 | 118.39 | 0.78 | 8.95 | nr | **127.34** |
| 50mm dia. | 131.20 | 144.32 | 0.84 | 9.64 | nr | **153.96** |
| Flanged ends for steel pipework; up to 16 bar G at 120 degrees Celsius | | | | | | |
| 65mm dia. | 153.75 | 169.13 | 1.08 | 12.41 | nr | **181.54** |
| 80mm dia. | 169.13 | 186.04 | 1.32 | 15.16 | nr | **201.20** |
| 100mm dia. | 179.38 | 197.32 | 1.92 | 22.07 | nr | **219.39** |
| 150mm dia. | 266.50 | 293.15 | 3.61 | 41.43 | nr | **334.58** |
| Screwed ends for non-ferrous pipework; up to 6 bar G at 100 degrees Celsius | | | | | | |
| 20mm dia. | 58.42 | 64.26 | 0.48 | 5.51 | nr | **69.77** |
| 25mm dia. | 59.45 | 65.39 | 0.60 | 6.89 | nr | **72.28** |
| 32mm dia. | 69.70 | 76.67 | 0.72 | 8.27 | nr | **84.94** |
| 40mm dia. | 83.03 | 91.33 | 0.78 | 8.95 | nr | **100.28** |
| 50mm dia. | 117.88 | 129.67 | 0.84 | 9.64 | nr | **139.31** |
| Flanged ends for steel, copper or non-ferrous pipework; up to 16 bar G at 120 degrees Celsius | | | | | | |
| 65mm dia. | 176.30 | 193.93 | 0.78 | 8.95 | nr | **202.88** |
| 80mm dia. | 183.47 | 201.82 | 0.84 | 9.64 | nr | **211.46** |
| 100mm dia. | 210.13 | 231.14 | 1.08 | 12.41 | nr | **243.55** |
| 150mm dia. | 343.38 | 377.72 | 1.32 | 15.16 | nr | **392.88** |

## Y:MECHANICAL AND ELECTRICAL SERVICES

| Item | Net Price £ | Material £ | Labour hours | Labour £ | Unit | Total Rate £ |
|---|---|---|---|---|---|---|
| **Y11 : PIPELINE ANCILLARIES : EXPANSION JOINTS (contd)** | | | | | | |
| **Angular movement bellows expansion joints; stainless steel** | | | | | | |
| Flanged ends for steel pipework; up to 16 bar G at 120 degrees Celsius | | | | | | |
| 50mm dia. | 348.50 | 383.35 | 0.60 | 6.89 | nr | **390.24** |
| 65mm dia. | 389.50 | 428.45 | 0.72 | 8.27 | nr | **436.72** |
| 80mm dia. | 410.00 | 451.00 | 0.78 | 8.95 | nr | **459.95** |
| 100mm dia. | 527.88 | 580.67 | 0.84 | 9.64 | nr | **590.31** |
| 125mm dia. | 543.25 | 597.58 | 1.08 | 12.41 | nr | **609.99** |
| 150mm dia. | 579.13 | 637.04 | 1.08 | 12.41 | nr | **649.45** |
| **Universal lateral movement bellows expansion joints; stainless steel** | | | | | | |
| Flanged ends for steel pipework; up to 16 bar G at 120 degrees Celsius | | | | | | |
| 50mm dia. | 341.32 | 375.45 | 0.84 | 9.64 | nr | **385.09** |
| 65mm dia. | 365.93 | 402.52 | 1.08 | 12.41 | nr | **414.93** |
| 80mm dia. | 393.60 | 432.96 | 1.32 | 15.16 | nr | **448.12** |
| 100mm dia. | 457.15 | 502.87 | 1.92 | 22.07 | nr | **524.94** |
| 125mm dia. | 520.70 | 572.77 | 3.61 | 41.43 | nr | **614.20** |
| 150mm dia. | 567.85 | 624.63 | 3.61 | 41.43 | nr | **666.06** |
| **Universal movement expansion joints; reinforced neoprene flexible connector** | | | | | | |
| Spherical expansion joints; flanged to BS 10, Table E; up to 10 bar at 100 degrees Celsius | | | | | | |
| 40mm dia. | 152.72 | 167.99 | 0.78 | 8.95 | nr | **176.94** |
| 50mm dia. | 164.00 | 180.40 | 0.84 | 9.64 | nr | **190.04** |
| 65mm dia. | 182.45 | 200.69 | 1.08 | 12.41 | nr | **213.10** |
| 80mm dia. | 215.25 | 236.78 | 1.32 | 15.16 | nr | **251.94** |
| 100mm dia. | 246.00 | 270.60 | 1.92 | 22.07 | nr | **292.67** |
| 150mm dia. | 351.57 | 386.73 | 3.61 | 41.43 | nr | **428.16** |
| Hose connector; BSP threaded union ends; up to 8 bar at 100 degrees Celsius | | | | | | |
| 20mm dia. | 35.88 | 39.47 | 1.28 | 14.69 | nr | **54.16** |
| 25mm dia. | 41.00 | 45.10 | 1.50 | 17.23 | nr | **62.33** |
| 32mm dia. | 46.13 | 50.74 | 1.76 | 20.20 | nr | **70.94** |
| 40mm dia. | 56.38 | 62.02 | 2.00 | 22.95 | nr | **84.97** |
| 50mm dia. | 66.63 | 73.29 | 2.28 | 26.20 | nr | **99.49** |

## Y:MECHANICAL AND ELECTRICAL SERVICES

| Item | Net Price £ | Material £ | Labour hours | Labour £ | Unit | Total Rate £ |
|---|---|---|---|---|---|---|
| **Y11 : PIPELINE ANCILLARIES : GAUGES** | | | | | | |
| **Thermometers and pressure gauges** | | | | | | |
| Dial thermometer; Class 2 bimetal operation; plastic case and dial; acrylic window; brass pocket; temperature range 0-120 degrees Celsius in metric and imperial scales; screwed end | | | | | | |
| 65mm dia. face; horizontal pocket length 50mm | 8.71 | 9.58 | 0.81 | 9.30 | nr | 18.88 |
| 65mm dia. face; horizontal pocket length 100mm | 12.67 | 13.94 | 0.81 | 9.30 | nr | 23.24 |
| 65mm dia. face; horizontal pocket length 150mm | 15.87 | 17.46 | 0.81 | 9.30 | nr | 26.76 |
| 65mm dia. face; vertical pocket length 50mm | 9.67 | 10.64 | 0.81 | 9.30 | nr | 19.94 |
| 65mm dia. face; vertical pocket length 100mm | 14.05 | 15.46 | 0.81 | 9.30 | nr | 24.76 |
| 65mm dia. face; vertical pocket length 150mm | 17.61 | 19.37 | 0.81 | 9.30 | nr | 28.67 |
| Dial thermometer; Class 2 bimetal operation; coated steel case and dial; acrylic window; brass pocket; temperature range 0-120 degrees Celsius in metric and imperial scales; screwed end | | | | | | |
| 80mm dia. face; horizontal pocket length 50mm | 26.69 | 29.36 | 0.81 | 9.30 | nr | 38.66 |
| 80mm dia. face; horizontal pocket length 100mm | 31.16 | 34.28 | 0.81 | 9.30 | nr | 43.58 |
| 100mm dia. face; horizontal pocket length 50mm | 29.82 | 32.80 | 0.81 | 9.30 | nr | 42.10 |
| 100mm dia. face; horizontal pocket length 100mm | 34.81 | 38.29 | 0.81 | 9.30 | nr | 47.59 |
| 80mm dia. face; vertical pocket length 50mm | 30.04 | 33.04 | 0.81 | 9.30 | nr | 42.34 |
| 80mm dia. face; vertical pocket length 100mm | 34.52 | 37.97 | 0.81 | 9.30 | nr | 47.27 |
| 100mm dia. face; vertical pocket length 50mm | 33.56 | 36.92 | 0.81 | 9.30 | nr | 46.22 |
| 100mm dia. face; vertical pocket length 100mm | 38.56 | 42.42 | 0.81 | 9.30 | nr | 51.72 |
| Dial thermometer; Class 2 bimetal operation; plastic case and dial; acrylic window; brass pocket; temperature range 0-160 degrees Celsius in metric and imperial scales; screwed end | | | | | | |
| 65mm dia. face; horizontal pocket length 50mm | 13.96 | 15.36 | 0.81 | 9.30 | nr | 24.66 |
| 65mm dia. face; horizontal pocket length 100mm | 16.19 | 17.81 | 0.81 | 9.30 | nr | 27.11 |
| 65mm dia. face; horizontal pocket length 150mm | 18.82 | 20.70 | 0.81 | 9.30 | nr | 30.00 |
| 65mm dia. face; vertical pocket length 50mm | 15.49 | 17.04 | 0.81 | 9.30 | nr | 26.34 |
| 65mm dia. face; vertical pocket length 100mm | 17.97 | 19.77 | 0.81 | 9.30 | nr | 29.07 |
| 65mm dia. face; vertical pocket length 150mm | 20.89 | 22.98 | 0.81 | 9.30 | nr | 32.28 |

## Y:MECHANICAL AND ELECTRICAL SERVICES

| Item | Net Price £ | Material £ | Labour hours | Labour £ | Unit | Total Rate £ |
|---|---|---|---|---|---|---|
| **Y11 : PIPELINE ANCILLARIES : GAUGES (contd)** | | | | | | |
| **Thermometers and pressure gauges (contd)** | | | | | | |
| Dial thermometer; Class 2 bimetal operation; coated steel case and dial; acrylic window; brass pocket; temperature range 0-160 degrees Celsius in metric and imperial scales; screwed end | | | | | | |
| 80mm dia. face; horizontal pocket length 50mm | 27.76 | 30.54 | 0.81 | 9.30 | nr | 39.84 |
| 80mm dia. face; horizontal pocket length 100mm | 32.81 | 36.09 | 0.81 | 9.30 | nr | 45.39 |
| 80mm dia. face; horizontal pocket length 150mm | 34.65 | 38.12 | 0.81 | 9.30 | nr | 47.42 |
| 100mm dia. face; horizontal pocket length 50mm | 31.01 | 34.11 | 0.81 | 9.30 | nr | 43.41 |
| 100mm dia. face; horizontal pocket length 100mm | 36.66 | 40.33 | 0.81 | 9.30 | nr | 49.63 |
| 100mm dia. face; horizontal pocket length 150mm | 38.71 | 42.58 | 0.81 | 9.30 | nr | 51.88 |
| 150mm dia. face; horizontal pocket length 50mm | 40.89 | 44.98 | 0.81 | 9.30 | nr | 54.28 |
| 150mm dia. face; horizontal pocket length 100mm | 54.78 | 60.26 | 0.81 | 9.30 | nr | 69.56 |
| 150mm dia. face; horizontal pocket length 150mm | 57.83 | 63.61 | 0.81 | 9.30 | nr | 72.91 |
| 80mm dia. face; vertical pocket length 50mm | 30.79 | 33.87 | 0.81 | 9.30 | nr | 43.17 |
| 80mm dia. face; vertical pocket length 100mm | 35.59 | 39.15 | 0.81 | 9.30 | nr | 48.45 |
| 80mm dia. face; vertical pocket length 150mm | 37.61 | 41.37 | 0.81 | 9.30 | nr | 50.67 |
| 100mm dia. face; vertical pocket length 50mm | 34.41 | 37.85 | 0.81 | 9.30 | nr | 47.15 |
| 100mm dia. face; vertical pocket length 100mm | 39.77 | 43.75 | 0.81 | 9.30 | nr | 53.05 |
| 100mm dia. face; vertical pocket length 150mm | 42.02 | 46.22 | 0.81 | 9.30 | nr | 55.52 |
| 150mm dia. face; vertical pocket length 50mm | 56.30 | 61.93 | 0.81 | 9.30 | nr | 71.23 |
| 150mm dia. face; vertical pocket length 100mm | 59.67 | 65.64 | 0.81 | 9.30 | nr | 74.94 |
| 150mm dia. face; vertical pocket length 150mm | 62.76 | 69.04 | 0.81 | 9.30 | nr | 78.34 |
| Dial thermometer; Class 1 bimetal operation; stainless steel case, dial and probe; acrylic window; temperature range 0-400 degrees Celsius; fitted to and including machined stainless steel pocket; maximum working pressure 150 bar; screwed end | | | | | | |
| 100mm dia. face; horizontal pocket length 63mm | 50.67 | 55.74 | 1.62 | 18.60 | nr | 74.34 |
| 100mm dia. face; horizontal pocket length 100mm | 58.05 | 63.85 | 1.62 | 18.60 | nr | 82.45 |
| 100mm dia. face; horizontal pocket length 250mm | 87.43 | 96.17 | 1.62 | 18.60 | nr | 114.77 |
| 100mm dia. face; vertical pocket length 63mm | 55.10 | 60.61 | 1.62 | 18.60 | nr | 79.21 |
| 100mm dia. face; vertical pocket length 100mm | 62.47 | 68.72 | 1.62 | 18.60 | nr | 87.32 |
| 100mm dia. face; vertical pocket length 250mm | 91.85 | 101.03 | 1.62 | 18.60 | nr | 119.63 |

## Y:MECHANICAL AND ELECTRICAL SERVICES

| Item | Net Price £ | Material £ | Labour hours | Labour £ | Unit | Total Rate £ |
|---|---|---|---|---|---|---|
| Dial thermometer; Class 1 bimetal operation; stainless steel case, dial and probe; acrylic window; temperature range 0-400 degrees Celsius; fitted to and including fabricated stainless steel pocket; maximum working pressure 40 bar; screwed end | | | | | | |
| 100mm dia. face; horizontal pocket length 250mm | 75.92 | 83.51 | 1.62 | 18.60 | nr | 102.11 |
| 100mm dia. face; vertical pocket length 250mm | 80.34 | 88.37 | 1.62 | 18.60 | nr | 106.97 |
| Dial thermometer; Class 1 bimetal operation; stainless steel case, dial and probe; acrylic window; temperature range 0-400 degrees Celsius; fitted to and including machined brass pocket; maximum working pressure 64 bar; maximum working temperature 160 degrees Celsius; screwed end | | | | | | |
| 100mm dia. face; horizontal pocket length 63mm | 41.73 | 45.90 | 1.62 | 18.60 | nr | 64.50 |
| 100mm dia. face; horizontal pocket length 100mm | 47.97 | 52.77 | 1.62 | 18.60 | nr | 71.37 |
| 100mm dia. face; horizontal pocket length 250mm | 63.88 | 70.27 | 1.62 | 18.60 | nr | 88.87 |
| 100mm dia. face; vertical pocket length 63mm | 46.16 | 50.78 | 1.62 | 18.60 | nr | 69.38 |
| 100mm dia. face; vertical pocket length 100mm | 52.39 | 57.63 | 1.62 | 18.60 | nr | 76.23 |
| 100mm dia. face; vertical pocket length 250mm | 68.30 | 75.13 | 1.62 | 18.60 | nr | 93.73 |
| Dial pressure gauge; Class 2.5 copper alloy direct measurement; coated steel case and dial; acrylic window; 0-40 bar / psi scales; screwed end | | | | | | |
| 50mm dia. face; horizontal connection | 4.04 | 4.44 | 0.81 | 9.30 | nr | 13.74 |
| 63mm dia. face; horizontal connection | 4.74 | 5.21 | 0.81 | 9.30 | nr | 14.51 |
| 80mm dia. face; horizontal connection | 19.92 | 21.91 | 0.81 | 9.30 | nr | 31.21 |
| 100mm dia. face; horizontal connection | 21.19 | 23.31 | 0.81 | 9.30 | nr | 32.61 |
| 50mm dia. face; vertical connection | 3.76 | 4.14 | 0.81 | 9.30 | nr | 13.44 |
| 63mm dia. face; vertical connection | 4.51 | 4.96 | 0.81 | 9.30 | nr | 14.26 |
| 80mm dia. face; vertical connection | 18.26 | 20.09 | 0.81 | 9.30 | nr | 29.39 |
| 100mm dia. face; vertical connection | 19.43 | 21.37 | 0.81 | 9.30 | nr | 30.67 |
| 150mm dia. face; vertical connection | 33.75 | 37.13 | 0.81 | 9.30 | nr | 46.43 |
| Dial pressure gauge; Class 1.6 copper alloy direct measurement; liquid filled stainless steel case and dial; acrylic window; 0-400/600 bar / psi scales; screwed end | | | | | | |
| 50mm dia. face; vertical connection | 7.67 | 8.44 | 0.81 | 9.30 | nr | 17.74 |
| 63mm dia. face; horizontal connection | 10.13 | 11.14 | 0.81 | 9.30 | nr | 20.44 |
| 63mm dia. face; vertical connection | 9.25 | 10.18 | 0.81 | 9.30 | nr | 19.48 |
| 100mm dia. face; horizontal connection | 25.84 | 28.42 | 0.81 | 9.30 | nr | 37.72 |
| 100mm dia. face; vertical connection | 24.91 | 27.40 | 0.81 | 9.30 | nr | 36.70 |

## Y:MECHANICAL AND ELECTRICAL SERVICES

| Item | Net Price £ | Material £ | Labour hours | Labour £ | Unit | Total Rate £ |
|---|---|---|---|---|---|---|
| **Y11 : PIPELINE ANCILLARIES : GAUGES (contd)** | | | | | | |
| **Thermometers and pressure gauges (contd)** | | | | | | |
| Dial pressure gauge; flanged ends; direct 0/4 bar | | | | | | |
| 100mm | 26.16 | 28.78 | 1.30 | 14.92 | nr | **43.70** |
| 150mm | 41.63 | 45.79 | 1.30 | 14.92 | nr | **60.71** |
| Dial pressure gauge; flanged ends; direct 0/20 bar | | | | | | |
| 100mm | 26.16 | 28.78 | 1.30 | 14.92 | nr | **43.70** |
| 150mm | 41.63 | 45.79 | 1.30 | 14.92 | nr | **60.71** |
| Combined altitude gauge and thermometer; 20/120 degrees C; range 0/20m; 50mm tail; screwed ends; back entry | | | | | | |
| 100mm | 20.88 | 22.97 | 1.00 | 11.47 | nr | **34.44** |
| Combined altitude gauge and thermometer; 20/120 degrees C; range 0/20m; 50mm tail; screwed ends; bottom entry | | | | | | |
| 100mm | 30.11 | 33.12 | 1.00 | 11.47 | nr | **44.59** |
| Dial pattern; mercury-in-steel thermometer; 15mm screwed tail; brass pockets; 20/60 degrees C; screwed ends to steel; back entry | | | | | | |
| 100mm dia. | 93.23 | 102.55 | 0.81 | 9.30 | nr | **111.85** |
| 150mm dia. | 106.55 | 117.20 | 0.81 | 9.30 | nr | **126.50** |
| Dial pattern; mercury-in-steel thermometer; 15mm screwed tail; brass pockets; 20/60 degrees C; screwed ends to steel; bottom entry | | | | | | |
| 100mm dia. | 93.23 | 102.55 | 0.81 | 9.30 | nr | **111.85** |
| 150mm dia. | 106.55 | 117.20 | 0.81 | 9.30 | nr | **126.50** |

## Y:MECHANICAL AND ELECTRICAL SERVICES

| Item | Net Price £ | Material £ | Labour hours | Labour £ | Unit | Total Rate £ |
|---|---|---|---|---|---|---|
| **Y20 : PUMPS : PUMPS, CIRCULATORS AND ACCELERATORS** | | | | | | |
| Centrifugal heating pump; belt drive; 3 phase, 1450 rpm motor; max. pressure 400kN/m2; max. temperature 125 degree C; bed plate; coupling guard bolted connections; supply only mating flanges; includes fixing on prepared base; electrical work elsewhere | | | | | | |
| Heating pumps | | | | | | |
| 40mm pump size; 4.0 litre/sec maximum delivery; 70 kPa maximum head; 0.55kW maximum motor rating | 682.43 | 750.67 | 9.01 | 103.38 | nr | 854.05 |
| 40mm pump size; 4.0 litre/sec maximum delivery; 130 kPa maximum head; 1.1kW maximum motor rating | 731.09 | 804.20 | 9.01 | 103.38 | nr | 907.58 |
| 50mm pump size; 8.5 litre/sec maximum delivery; 90 kPa maximum head; 1.1kW maximum motor rating | 780.84 | 858.92 | 9.01 | 103.38 | nr | 962.30 |
| 50mm pump size; 8.5 litre/sec maximum delivery; 190 kPa maximum head; 2.2kW maximum motor rating | 899.81 | 989.79 | 14.08 | 161.62 | nr | **1151.41** |
| 50mm pump size; 8.5 litre/sec maximum delivery; 215 kPa maximum head; 3.0 kW maximum motor rating | 958.21 | 1054.03 | 14.08 | 161.62 | nr | **1215.65** |
| 65mm pump size; 14.0 litre/sec maximum delivery; 90 kPa maximum head; 1.50kW maximum motor rating | 881.42 | 969.56 | 14.08 | 161.62 | nr | **1131.18** |
| 65mm pump size; 14.0 litre/sec maximum delivery; 160 kPa maximum head; 3.0 kW maximum motor rating | 949.56 | 1044.52 | 14.08 | 161.62 | nr | **1206.14** |
| 65mm pump size; 14.5 litre/sec maximum delivery; 210 kPa maximum head; 4.0 kW maximum motor rating | 1007.96 | 1108.76 | 14.08 | 161.62 | nr | **1270.38** |
| 80mm pump size; 22.0 litre/sec maximum delivery; 130 kPa maximum head; 4.0 kW maximum motor rating | 1210.20 | 1331.22 | 15.15 | 173.86 | nr | **1505.08** |
| 80mm pump size; 22.0 litre/sec maximum delivery; 200 kPa maximum head; 5.5 kW maximum motor rating | 1534.65 | 1688.12 | 15.15 | 173.86 | nr | **1861.98** |
| 80mm pump size; 22.0 litre/sec maximum delivery; 250 kPa maximum head; 7.50kW maximum motor rating | 1696.87 | 1866.56 | 15.15 | 173.86 | nr | **2040.42** |
| 100mm pump size; 30.0 litre/sec maximum delivery; 100 kPa maximum head; 4.0 kW maximum motor rating | 1471.92 | 1619.11 | 22.22 | 255.00 | nr | **1874.11** |

Y:MECHANICAL AND ELECTRICAL SERVICES

| Item | Net Price £ | Material £ | Labour hours | Labour £ | Unit | Total Rate £ |
|---|---|---|---|---|---|---|
| **Y20 : PUMPS : PUMPS, CIRCULATORS AND ACCELERATORS (contd)** | | | | | | |
| Centrifugal heating pump; belt drive; 3 phase, 1450 rpm motor; max. pressure 400kN/m2; max. temperature 125 degree C; bed plate; coupling guard bolted connections; supply only mating flanges; includes fixing on prepared base; electrical work elsewhere (contd) | | | | | | |
| Heating pumps (contd) | | | | | | |
| 100mm pump size; 36.0 litre/sec maximum delivery; 25 kPa maximum head; 11.0kW maximum motor rating | 2253.85 | 2479.24 | 22.22 | 255.00 | nr | 2734.24 |
| 100mm pump size; 36.0 litre/sec maximum delivery; 550 kPa maximum head; 30.0 kW maximum motor rating | 3532.18 | 3885.40 | 22.22 | 255.00 | nr | 4140.40 |
| Extra for single phase motor | | | | | | |
| 0.37 kW | 154.18 | 169.60 | 6.02 | 69.13 | nr | 238.73 |
| 0.55 kW | 154.18 | 169.60 | 6.02 | 69.13 | nr | 238.73 |
| 0.75 kW | 154.18 | 169.60 | 6.02 | 69.13 | nr | 238.73 |
| 1.10 kW | 154.18 | 169.60 | 6.02 | 69.13 | nr | 238.73 |
| Centrifugal heating pump; close coupled; 3 phase, 1450 rpm motor; max. pressure 400kN/m2; max. temperature 110 degree C; bed plate; coupling guard; bolted connections; supply only mating flanges; includes fixing on prepared base; electrical work elsewhere | | | | | | |
| Heating pumps | | | | | | |
| 40mm pump size; 4.0 litre/sec maximum delivery; 4 kPa maximum head; 0.55kW maximum motor rating | 473.70 | 521.07 | 9.01 | 103.38 | nr | 624.45 |
| 40mm pump size; 4.0 litre/sec maximum delivery; 23 kPa maximum head; 2.20kW maximum motor rating | 543.99 | 598.39 | 9.01 | 103.38 | nr | 701.77 |
| 40mm pump size; 4.0 litre/sec maximum delivery; 75 kPa maximum head; 11 kW maximum motor rating | 1626.58 | 1789.24 | 9.01 | 103.38 | nr | 1892.62 |
| 50mm pump size; 7.0 litre/sec maximum delivery; 65 kPa maximum head; 0.75 kW maximum motor rating | 508.31 | 559.14 | 9.01 | 103.38 | nr | 662.52 |
| 50mm pump size; 10.0 litre/sec maximum delivery; 33 kPa maximum head; 5.5kW maximum motor rating | 966.86 | 1063.55 | 9.01 | 103.38 | nr | 1166.93 |
| 50mm pump size; 4.0 litre/sec maximum delivery; 120 kPa maximum head; 1.1 kW maximum motor rating | 682.43 | 750.67 | 9.01 | 103.38 | nr | 854.05 |

## Y:MECHANICAL AND ELECTRICAL SERVICES

| Item | Net Price £ | Material £ | Labour hours | Labour £ | Unit | Total Rate £ |
|---|---|---|---|---|---|---|
| 80mm pump size; 16.0 litre/sec maximum delivery; 80 kPa maximum head; 2.2 kW maximum motor rating | 883.59 | 971.95 | 15.15 | 173.86 | nr | 1145.81 |
| 80mm pump size; 16.0 litre/sec maximum delivery; 120 kPa maximum head; 3.0 kW maximum motor rating | 908.46 | 999.31 | 15.15 | 173.86 | nr | 1173.17 |
| 100mm pump size; 28.0 litre/sec maximum delivery; 40 kPa maximum head; 2.2 kW maximum motor rating | 872.77 | 960.05 | 22.22 | 255.00 | nr | 1215.05 |
| 100mm pump size; 28.0 litre/sec maximum delivery; 90 kPa maximum head; 4.0 kW maximum motor rating | 962.53 | 1058.78 | 22.22 | 255.00 | nr | 1313.78 |
| 125mm pump size; 40.0 litre/sec maximum delivery; 50 kPa maximum head; 3.0 kW maximum motor rating | 910.62 | 1001.68 | 30.30 | 347.73 | nr | 1349.41 |
| 125mm pump size; 40.0 litre/sec maximum delivery; 120 kPa maximum head; 7.5 kW maximum motor rating | 1124.76 | 1237.24 | 30.30 | 347.73 | nr | 1584.97 |
| 150mm pump size; 70.0 litre/sec maximum delivery; 75 kPa maximum head; 7.5 kW maximum motor rating | 1107.46 | 1218.21 | 38.46 | 441.35 | nr | 1659.56 |
| 150mm pump size; 70.0 litre/sec maximum delivery; 120 kPa maximum head; 11.0 kW maximum motor rating | 1455.70 | 1601.27 | 38.46 | 441.35 | nr | 2042.62 |
| 150mm pump size; 70.0 litre/sec maximum delivery; 150 kPa maximum head; 15.0 kW maximum motor rating | 1556.28 | 1711.91 | 38.46 | 441.35 | nr | 2153.26 |
| Glandless domestic heating pump; for low pressure domestic hot water heating systems; 240 volt; 50Hz electric motor; max working pressure 1000N/m2 and max temperature of 110 degree C; includes fixing; electrical work elsewhere | | | | | | |
| Single speed | | | | | | |
| 20mm BSP unions | 98.36 | 108.20 | 1.50 | 17.23 | nr | 125.43 |
| Three speed; pump only | | | | | | |
| 40mm | 231.66 | 254.83 | 1.50 | 17.23 | nr | 272.06 |
| Automatic sump pump; self contained, totally enclosed electric motor, float switch and gear; includes fixing; electrical work elsewhere | | | | | | |
| Small portable | | | | | | |
| Single phase | 153.57 | 168.93 | 1.50 | 17.23 | nr | 186.16 |

## Material Costs / Measured Work Prices - Mechanical Installations
### Y:MECHANICAL AND ELECTRICAL SERVICES

| Item | Net Price £ | Material £ | Labour hours | Labour £ | Unit | Total Rate £ |
|---|---|---|---|---|---|---|
| **Y20 : PUMPS : PUMPS, CIRCULATORS AND ACCELERATORS (contd)** | | | | | | |
| Automatic sump pump; self contained, totally enclosed electric motor, float switch and gear; includes fixing; electrical work elsewhere (contd) | | | | | | |
| Fixed installation | | | | | | |
| Single phase | 302.82 | 333.10 | 2.00 | 22.95 | nr | 356.05 |
| Three phase | 316.88 | 348.57 | 2.00 | 22.95 | nr | 371.52 |
| Pipeline mounted circulator; for low and medium pressure heating and hot water services; silent running; 3 phase; 1450 rpm motor; max pressure 600 kN/m2; max temperature 110 degree C; bolted connections; supply only mating flanges; includes fixing; electrical work elsewhere | | | | | | |
| Circulator | | | | | | |
| 32mm pump size; 2.0 litre/sec maximum delivery; 17 kPa maximum head; 0.2kW maximum motor rating | 246.17 | 270.79 | 9.01 | 103.38 | nr | 374.17 |
| 50mm pump size; 3.0 litre/sec maximum delivery; 20 kPa maximum head; 0.2kW maximum motor rating | 312.09 | 343.30 | 9.01 | 103.38 | nr | 446.68 |
| 65mm pump size; 5.0 litre/sec maximum delivery; 30 kPa maximum head; 0.37 kW maximum motor rating | 501.61 | 551.77 | 10.00 | 114.75 | nr | 666.52 |
| 65mm pump size; 8.0 litre/sec maximum delivery; 37 kPa maximum head; 0.75 kW maximum motor rating | 558.26 | 614.09 | 10.00 | 114.75 | nr | 728.84 |
| 80mm pump size; 12.0 litre/sec maximum delivery; 42 kPa maximum head; 1.1 kW maximum motor rating | 677.74 | 745.51 | 10.00 | 114.75 | nr | 860.26 |
| 100mm pump size; 25.0 litre/sec maximum delivery; 37 kPa maximum head; 2.2 kW maximum motor rating | 853.87 | 939.26 | 11.11 | 127.50 | nr | 1066.76 |
| Dual pipeline mounted circulator; for low and medium pressure heating and hot water services; silent running; 3 phase; 1450 rpm motor; max pressure 600 kN/m2; max temperature 110 degree C; bolted connections; supply only mating flanges; includes fixing; electrical work elsewhere | | | | | | |
| Circulator | | | | | | |
| 40mm pump size; 2.0 litre/sec maximum delivery; 17 kPa maximum head; 0.8kW maximum motor rating | 568.56 | 625.42 | 9.01 | 103.38 | nr | 728.80 |

## Material Costs / Measured Work Prices - Mechanical Installations

### Y:MECHANICAL AND ELECTRICAL SERVICES

| Item | Net Price £ | Material £ | Labour hours | Labour £ | Unit | Total Rate £ |
|---|---|---|---|---|---|---|
| 50mm pump size; 3.0 litre/sec maximum delivery; 20 kPa maximum head; 0.2 kW maximum motor rating | 581.95 | 640.14 | 9.01 | 103.38 | nr | 743.52 |
| 65mm pump size; 5.0 litre/sec maximum delivery; 30 kPa maximum head; 0.37 kW maximum motor rating | 1000.13 | 1100.14 | 10.00 | 114.75 | nr | 1214.89 |
| 65mm pump size; 8.0 litre/sec maximum delivery; 37 kPa maximum head; 0.75 kW maximum motor rating | 1147.42 | 1262.16 | 10.00 | 114.75 | nr | 1376.91 |
| 80mm pump size; 12.0 litre/sec maximum delivery; 42 kPa maximum head; 1.1 kW maximum motor rating | 1188.62 | 1307.48 | 10.00 | 114.75 | nr | 1422.23 |
| Glandless accelerator pumps; for low and medium pressure heating and hot water services; silent running; 3 phase; 1450 rpm motor; max pressure 600 kN/m2; max temperature 120 degree C; bolted connections; supply only mating flanges; includes fixing; electrical work elsewhere | | | | | | |
| Accelerator pump | | | | | | |
| 40mm pump size; 4.0 litre/sec maximum delivery; 15 kPa maximum head; 0.35kW maximum motor rating | 326.61 | 359.27 | 10.00 | 114.75 | nr | 474.02 |
| 50mm pump size; 6.0 litre/sec maximum delivery; 20 kPa maximum head; 0.45kW maximum motor rating | 410.97 | 452.07 | 10.00 | 114.75 | nr | 566.82 |
| 80mm pump size; 13.0 litre/sec maximum delivery; 28 kPa maximum head; 0.58kW maximum motor rating | 483.43 | 531.77 | 10.00 | 114.75 | nr | 646.52 |

## Y:MECHANICAL AND ELECTRICAL SERVICES

| Item | Net Price £ | Material £ | Labour hours | Labour £ | Unit | Total Rate £ |
|---|---|---|---|---|---|---|
| **Y20 : PUMPS : PRESSURISED COLD WATER SUPPLY SET** | | | | | | |
| **Pressurised cold water supply set; packaged; fully automatic; pumps with 3 phase motors; membrane tank; valves; control panel; inter connecting pipework; fixed on steel frame; includes fixing; electrical work elsewhere** | | | | | | |
| Cold water supply set | | | | | | |
| 3 litre/sec maximum delivery; 600 kN/m2 maximum head | 4167.45 | 4584.19 | 10.00 | 114.75 | nr | **4698.94** |
| 7 litre/sec maximum delivery; 600 kN/m2 maximum head | 5670.00 | 6237.00 | 10.00 | 114.75 | nr | **6351.75** |
| 16 litre/sec maximum delivery; 600 kN/m2 maximum head | 8013.60 | 8814.96 | 10.00 | 114.75 | nr | **8929.71** |

# ENERGY MANAGEMENT AND OPERATING COSTS IN BUILDINGS

# Keith J. Moss

Managing the consumption and conservation of energy in buildings must now become the concern of both building managers and occupants. The provision of lighting, hot water supply, communications, cooking, space heating and cooling accounts for 45 per cent of UK energy consumption.

**Energy Management and Operating Costs in Buildings** introduces the reader to the principles of managing and conserving energy consumption in buildings people use for work or leisure. Energy consumption is considered for the provision of space heating, hot water supply ventilation and air conditioning. The author introduces the use of standard performance indicators and energy consumption yardsticks, and discusses the use and application of Degree Days. Following an introduction to the preparation of the energy audit, monitoring and targeting techniques are investigated and analysed.

Readers are not expected to have prior knowledge of the design of building services. Each chapter of the book is set out with:

- nomenclature:
- introduction;
- worked examples and case studies;
- data, text and illustrations appropriate to each topic.

This is a key text for undergraduates on building services engineering, energy management, environmental engineering and related construction courses. It will also be an invaluable work for professional energy managers and building services engineers.

Preface. Acknowledgements. Introduction. The economics of space heating plants. Estimating energy consumption - space heating. Intermittent space heating. Estimating the annual cost for the provision of a hot water supply. Energy consumption for cooling loads. Performance indicators. Energy conservation strategies. Cost benefit analysis. Energy audits. Monitoring and targeting. Appendices. Bibliography. Index

**Keith Moss** is a consultant engineer and a visiting lecturer at the University of Bath and City of Bath College.

246x189mm, 200 pages
25 line illustrations
Paperback ISBN 0-419-21770-3
£24.99 June 1997

# New From

## E & FN SPON
An Imprint of Thomson Professional
2-6 Boundary Row
London SE1 8HN

# Site Management of Building Services Contractors

**Jim Wild, Building Services Management Consultant, Berkshire, UK**

There is an increasingly wide range of building engineering services. When installed, these must sustain and protect a specified internal environment to the satisfaction of the client, designers, insurers and the relevant authorities.

Managing building services contractors can prove to be a minefield. The most successful jobs will always be those where building site managers have first built teams focused on tackling issues that might cause adversarial attitudes later on and jeopardize the project.

The author shows how a simple common management approach can improve site managers' competency in overseeing building services contractors, sub traders and specialists, and maximize the effectiveness of time spent on building services. By providing an account of building services from the site management rather than the corporate viewpoint, this book breaks new ground.

**Site Management of Building Services Contractors:**
- provides step by step guidance from pre award to post handover;
- emphasises system based nature of building services contracts;
- covers risk management throughout the process.

**Contents:**
Overview of building services. Basic planning strategy. Project plans for quality, safety and the environment. Planning and programming. Schedules. Supervision and inspection. Assessing construction progress. Commissioning and its management. The management of defects. Handover. Getting help. Summary. Index

January 1997: 246x189mm
384 pp: 32 line illustrations
Hardback: 0 419 20450 4: £35.00

# E & FN SPON

An Imprint of
Thomson Professional

## Material Costs / Measured Work Prices - Mechanical Installations

### Y:MECHANICAL AND ELECTRICAL SERVICES

| Item | Net Price £ | Material £ | Labour hours | Labour £ | Unit | Total Rate £ |
|---|---|---|---|---|---|---|
| **Y21: WATER TANKS/CISTERNS : SECTIONAL GLASS FIBRE TANKS** | | | | | | |
| **Combination hot and cold water tanks** | | | | | | |
| Super seven high performance insulated copper storage units; direct pattern; standard connections; BS 3198; includes placing in position | | | | | | |
| 450mm dia.; 1200mm height | 176.56 | 194.22 | 1.50 | 17.23 | nr | **211.45** |
| 450mm dia.; 1850mm height | 273.99 | 301.39 | 1.50 | 17.23 | nr | **318.62** |

## Y:MECHANICAL AND ELECTRICAL SERVICES

| Item | Net Price £ | Material £ | Labour hours | Labour £ | Unit | Total Rate £ |
|---|---|---|---|---|---|---|
| **Y21 : WATER TANKS/CISTERNS : MOULDED FIBREGLASS CISTERNS** | | | | | | |
| Cisterns; fibreglass; complete with ball valve, fixing plate and fitted covers; complies to byelaw 30 | | | | | | |
| Rectangular | | | | | | |
| 70 litres capacity | 209.00 | 229.90 | 1.00 | 11.47 | nr | **241.37** |
| 110 litres capacity | 229.00 | 251.90 | 1.00 | 11.47 | nr | **263.37** |
| 170 litres capacity | 273.00 | 300.30 | 1.00 | 11.47 | nr | **311.77** |
| 280 litres capacity | 324.00 | 356.40 | 1.00 | 11.47 | nr | **367.87** |
| 420 litres capacity | 418.00 | 459.80 | 1.50 | 17.23 | nr | **477.03** |
| 710 litres capacity | 615.00 | 676.50 | 2.50 | 28.69 | nr | **705.19** |
| 840 litres capacity | 669.00 | 735.90 | 2.50 | 28.69 | nr | **764.59** |
| 1590 litres capacity | 986.00 | 1084.60 | 13.16 | 150.99 | nr | **1235.59** |
| 2275 litres capacity | 1205.00 | 1325.50 | 20.00 | 229.50 | nr | **1555.00** |
| 3365 litres capacity | 1693.00 | 1862.30 | 25.00 | 286.88 | nr | **2149.18** |
| 4545 litres capacity | 2066.00 | 2272.60 | 30.30 | 347.73 | nr | **2620.33** |

## Y:MECHANICAL AND ELECTRICAL SERVICES

| Item | Net Price £ | Material £ | Labour hours | Labour £ | Unit | Total Rate £ |
|---|---|---|---|---|---|---|
| **Y21 : WATER TANKS/CISTERNS : MOULDED PLASTIC CISTERNS** | | | | | | |
| Cisterns; polypropylene; complete with ball valve, fixing plate and cover; includes hoisting and placing in position | | | | | | |
| Rectangular | | | | | | |
| 18 litres capacity | 3.44 | 3.78 | 1.00 | 11.47 | nr | 15.25 |
| 68 litres capacity | 15.42 | 16.96 | 1.00 | 11.47 | nr | 28.43 |
| 91 litres capacity | 15.71 | 17.28 | 1.00 | 11.47 | nr | 28.75 |
| 114 litres capacity | 21.69 | 23.86 | 1.00 | 11.47 | nr | 35.33 |
| 182 litres capacity | 37.37 | 41.11 | 1.00 | 11.47 | nr | 52.58 |
| 227 litres capacity | 37.69 | 41.46 | 1.00 | 11.47 | nr | 52.93 |
| Circular | | | | | | |
| 114 litres capacity | 17.58 | 19.34 | 1.00 | 11.47 | nr | 30.81 |
| 227 litres capacity | 27.42 | 30.16 | 1.00 | 11.47 | nr | 41.63 |
| 318 litres capacity | 71.59 | 78.75 | 1.00 | 11.47 | nr | 90.22 |
| 455 litres capacity | 81.36 | 89.50 | 1.00 | 11.47 | nr | 100.97 |
| Extra over rectangular cisterns; byelaw 30 kit | | | | | | |
| 18 litres capacity | 9.27 | 10.20 | 0.50 | 5.74 | nr | 15.94 |
| 68 litres capacity | 16.24 | 17.86 | 0.50 | 5.74 | nr | 23.60 |
| 91 litres capacity | 18.10 | 19.91 | 0.50 | 5.74 | nr | 25.65 |
| 114 litres capacity | 19.24 | 21.16 | 0.50 | 5.74 | nr | 26.90 |
| 182 litres capacity | 27.35 | 30.09 | 0.75 | 8.61 | nr | 38.70 |
| 227 litres capacity | 28.70 | 31.57 | 0.75 | 8.61 | nr | 40.18 |
| Extra over circular cisterns; byelaw 30 kit | | | | | | |
| 114 litres capacity | 19.09 | 21.00 | 0.50 | 5.74 | nr | 26.74 |
| 127 litres capacity | 26.93 | 29.62 | 0.75 | 8.61 | nr | 38.23 |
| 318 litres capacity | 43.22 | 47.54 | 0.50 | 5.74 | nr | 53.28 |
| 455 litres capacity | 51.01 | 56.11 | 0.75 | 8.61 | nr | 64.72 |

## Material Costs / Measured Work Prices - Mechanical Installations

### Y:MECHANICAL AND ELECTRICAL SERVICES

| Item | Net Price £ | Material £ | Labour hours | Labour £ | Unit | Total Rate £ |
|---|---|---|---|---|---|---|
| **Y22 : HEAT EXCHANGERS : ROTARY AIR TO AIR WHEEL** | | | | | | |
| **Thermal wheels; fixed in operational location; electrical work elsewhere** | | | | | | |
| Thermal wheels; fixed speed drive; rotary air to air energy exchanger comprising supporting frame, rotor wheel and drive motor; wheel constructed of alternate flat and corrugated foils of inorganic material glued together to form axial flutes coated with a dessicant or similar treatment; placing in ductwork system | | | | | | |
| 1 m3/sec. flow rate | 1740.90 | 1914.99 | 18.18 | 208.64 | nr | **2123.63** |
| 2 m3/sec. flow rate | 2307.90 | 2538.69 | 25.00 | 286.88 | nr | **2825.57** |
| Thermal wheels; variable speed drive; rotary air to air energy exchanger comprising supporting frame, rotor wheel and drive motor; wheel constructed of alternate flat and corrugated foils of inorganic material glued together to form axial flutes coated with a dessicant or similar treatment; placing in ductwork system | | | | | | |
| 4 m3/sec. flow rate | 3742.20 | 4116.42 | 35.71 | 409.82 | nr | **4526.24** |
| 7 m/3sec. flow rate | 4860.45 | 5346.49 | 50.00 | 573.75 | nr | **5920.24** |
| 12 m3/sec. flow rate | 8516.55 | 9368.20 | 76.92 | 882.69 | nr | **10250.89** |

## Y:MECHANICAL AND ELECTRICAL SERVICES

| Item | Net Price £ | Material £ | Labour hours | Labour £ | Unit | Total Rate £ |
|---|---|---|---|---|---|---|
| **Y23 : STORAGE CYLINDERS/CALORIFERS : COPPER DIRECT CYLINDERS** | | | | | | |
| **Insulated copper storage cylinders; BS 699; includes placing in position** | | | | | | |
| Grade 3 (maximum 10m working head) | | | | | | |
| BS size 6; 115 litres capacity; 400mm dia.; 1050mm height | 58.11 | 63.92 | 1.50 | 17.23 | nr | 81.15 |
| BS size 7; 120 litres capacity; 450mm dia.; 900mm height | 68.71 | 75.58 | 2.00 | 22.95 | nr | 98.53 |
| BS size 8; 144 litres capacity; 450mm dia.; 1050mm height | 64.34 | 70.77 | 2.80 | 32.14 | nr | 102.91 |
| Grade 4 (maximum 6m working head) | | | | | | |
| BS size 2; 96 litres capacity; 400mm dia.; 900mm height | 54.37 | 59.81 | 1.50 | 17.23 | nr | 77.04 |
| BS size 7; 120 litres capacity; 450mm dia.; 900mm height | 44.99 | 49.49 | 1.50 | 17.23 | nr | 66.72 |
| BS size 8; 144 litres capacity; 450mm dia.; 1050mm height | 56.45 | 62.09 | 1.50 | 17.23 | nr | 79.32 |
| BS size 9; 166 litres capacity; 450mm dia.; 1200mm height | 82.53 | 90.78 | 1.50 | 17.23 | nr | 108.01 |
| **Storage cylinders; brazed copper construction; to BS 699; screwed bosses; includes placing in position** | | | | | | |
| Tested to 2.2 bar, 15m maximum head | | | | | | |
| 144 litres | 245.18 | 269.70 | 3.00 | 34.46 | nr | 304.16 |
| 160 litres | 277.16 | 304.88 | 3.00 | 34.46 | nr | 339.34 |
| 200 litres | 285.69 | 314.26 | 3.76 | 43.14 | nr | 357.40 |
| 255 litres | 325.13 | 357.64 | 3.76 | 43.14 | nr | 400.78 |
| 290 litres | 439.19 | 483.11 | 3.76 | 43.14 | nr | 526.25 |
| 370 litres | 503.15 | 553.47 | 4.50 | 51.69 | nr | 605.16 |
| 450 litres | 669.45 | 736.39 | 5.00 | 57.38 | nr | 793.77 |
| Tested to 2.55 bar, 17m maximum head | | | | | | |
| 550 litres | 724.88 | 797.37 | 5.00 | 57.38 | nr | 854.75 |
| 700 litres | 847.47 | 932.22 | 6.02 | 69.13 | nr | 1001.35 |
| 800 litres | 1012.70 | 1113.97 | 6.54 | 75.00 | nr | 1188.97 |
| 900 litres | 1060.67 | 1166.74 | 8.00 | 91.80 | nr | 1258.54 |
| 1000 litres | 1119.30 | 1231.23 | 8.00 | 91.80 | nr | 1323.03 |
| 1250 litres | 1225.90 | 1348.49 | 13.16 | 150.99 | nr | 1499.48 |
| 1500 litres | 1865.50 | 2052.05 | 15.15 | 173.86 | nr | 2225.91 |
| 2000 litres | 2238.60 | 2462.46 | 17.24 | 197.85 | nr | 2660.31 |
| 3000 litres | 3144.70 | 3459.17 | 24.39 | 279.88 | nr | 3739.05 |

## Y:MECHANICAL AND ELECTRICAL SERVICES

| Item | Net Price £ | Material £ | Labour hours | Labour £ | Unit | Total Rate £ |
|---|---|---|---|---|---|---|
| **Y23 : STORAGE CYLINDERS/CALORIFIERS : STEEL INDIRECT CYLINDERS** | | | | | | |
| **Indirect cylinders; mild steel, welded throughout, galvanised; with bolted connections; includes placing in position** | | | | | | |
| 3.2mm plate | | | | | | |
| 136 litres capacity | 346.45 | 381.10 | 2.50 | 28.69 | nr | **409.79** |
| 159 litres capacity | 373.10 | 410.41 | 2.80 | 32.14 | nr | **442.55** |
| 182 litres capacity | 458.38 | 504.22 | 3.00 | 34.46 | nr | **538.68** |
| 227 litres capacity | 564.98 | 621.48 | 3.00 | 34.46 | nr | **655.94** |
| 273 litres capacity | 666.25 | 732.88 | 4.00 | 45.90 | nr | **778.78** |
| 364 litres capacity | 767.52 | 844.27 | 4.50 | 51.69 | nr | **895.96** |
| 455 litres capacity | 884.78 | 973.26 | 5.00 | 57.38 | nr | **1030.64** |
| 683 litres capacity | 1353.82 | 1489.20 | 6.02 | 69.13 | nr | **1558.33** |
| 910 litres capacity | 1662.96 | 1829.26 | 7.04 | 80.81 | nr | **1910.07** |
| **Welded cylinders; mild steel, galvanised; bolted top; BS 417; includes hoisting and placing in position** | | | | | | |
| 3.2mm gauge; 30m working head | | | | | | |
| YM141-123 litres capacity | 170.56 | 187.62 | 2.00 | 22.95 | nr | **210.57** |
| YM150-136 litres capacity | 181.22 | 199.34 | 2.50 | 28.69 | nr | **228.03** |
| YM177-159 litres capacity | 213.20 | 234.52 | 2.50 | 28.69 | nr | **263.21** |

## Material Costs / Measured Work Prices - Mechanical Installations

### Y: MECHANICAL AND ELECTRICAL SERVICES

| Item | Net Price £ | Material £ | Labour hours | Labour £ | Unit | Total Rate £ |
|---|---|---|---|---|---|---|
| **Y23 : STORAGE CYLINDERS/CALORIFERS : COPPER INDIRECT CYLINDERS** | | | | | | |
| Indirect cylinders; copper; bolted top; up to 5 tappings for connections; BS 1586; includes placing in position | | | | | | |
| Grade 3, tested to 1.45 bar, 10m maximum head | | | | | | |
| 74 litres capacity | 117.26 | 128.99 | 1.50 | 17.23 | nr | 146.22 |
| 96 litres capacity | 119.39 | 131.33 | 1.50 | 17.23 | nr | 148.56 |
| 114 litres capacity | 122.59 | 134.85 | 1.50 | 17.23 | nr | 152.08 |
| 117 litres capacity | 127.92 | 140.71 | 2.00 | 22.95 | nr | 163.66 |
| 140 litres capacity | 131.12 | 144.23 | 2.50 | 28.69 | nr | 172.92 |
| 162 litres capacity | 183.35 | 201.69 | 3.00 | 34.46 | nr | 236.15 |
| 190 litres capacity | 200.41 | 220.45 | 3.51 | 40.26 | nr | 260.71 |
| 245 litres capacity | 234.52 | 257.97 | 3.80 | 43.63 | nr | 301.60 |
| 280 litres capacity | 415.74 | 457.31 | 4.00 | 45.90 | nr | 503.21 |
| 360 litres capacity | 449.85 | 494.83 | 4.50 | 51.69 | nr | 546.52 |
| 440 litres capacity | 522.34 | 574.57 | 4.50 | 51.69 | nr | 626.26 |
| Grade 2, tested to 2.2 bar, 15m maximum head | | | | | | |
| 117 litres capacity | 169.49 | 186.44 | 2.00 | 22.95 | nr | 209.39 |
| 140 litres capacity | 184.42 | 202.86 | 2.50 | 28.69 | nr | 231.55 |
| 162 litres capacity | 211.07 | 232.18 | 2.80 | 32.14 | nr | 264.32 |
| 190 litres capacity | 245.18 | 269.70 | 3.00 | 34.46 | nr | 304.16 |
| 245 litres capacity | 296.35 | 325.99 | 4.00 | 45.90 | nr | 371.89 |
| 280 litres capacity | 473.30 | 520.63 | 4.00 | 45.90 | nr | 566.53 |
| 360 litres capacity | 522.34 | 574.57 | 4.50 | 51.69 | nr | 626.26 |
| 440 litres capacity | 618.28 | 680.11 | 4.50 | 51.69 | nr | 731.80 |
| Grade 1, tested 3.65 bar, 25m maximum head | | | | | | |
| 190 litres capacity | 364.57 | 401.03 | 3.00 | 34.46 | nr | 435.49 |
| 245 litres capacity | 414.67 | 456.14 | 3.00 | 34.46 | nr | 490.60 |
| 280 litres capacity | 586.30 | 644.93 | 4.00 | 45.90 | nr | 690.83 |
| 360 litres capacity | 741.94 | 816.13 | 4.50 | 51.69 | nr | 867.82 |
| 440 litres capacity | 900.77 | 990.85 | 4.50 | 51.69 | nr | 1042.54 |
| Indirect cylinders, including manhole; BS 853 | | | | | | |
| Grade 3, tested to 1.5 bar, 10m maximum head | | | | | | |
| 550 litres capacity | 746.20 | 820.82 | 5.21 | 59.77 | nr | 880.59 |
| 700 litres capacity | 826.15 | 908.76 | 6.02 | 69.13 | nr | 977.89 |
| 800 litres capacity | 959.40 | 1055.34 | 6.54 | 75.00 | nr | 1130.34 |
| 1000 litres capacity | 1199.25 | 1319.17 | 7.04 | 80.81 | nr | 1399.98 |
| 1500 litres capacity | 1385.80 | 1524.38 | 10.00 | 114.75 | nr | 1639.13 |
| 2000 litres capacity | 1918.80 | 2110.68 | 16.13 | 185.08 | nr | 2295.76 |

## Y:MECHANICAL AND ELECTRICAL SERVICES

| Item | Net Price £ | Material £ | Labour hours | Labour £ | Unit | Total Rate £ |
|---|---|---|---|---|---|---|
| **Y23 : STORAGE CYLINDERS/CALORIFERS : COPPER INDIRECT CYLINDERS (contd)** | | | | | | |
| Indirect cylinders, including manhole; BS 853 (contd) | | | | | | |
| Grade 2, tested to 2.55 bar, 15m maximum head | | | | | | |
| 550 litres capacity | 847.47 | 932.22 | 5.21 | 59.77 | nr | 991.99 |
| 700 litres capacity | 1060.67 | 1166.74 | 6.02 | 69.13 | nr | 1235.87 |
| 800 litres capacity | 1119.30 | 1231.23 | 6.54 | 75.00 | nr | 1306.23 |
| 1000 litres capacity | 1385.80 | 1524.38 | 7.04 | 80.81 | nr | 1605.19 |
| 1500 litres capacity | 1705.60 | 1876.16 | 10.00 | 114.75 | nr | 1990.91 |
| 2000 litres capacity | 2132.00 | 2345.20 | 16.13 | 185.08 | nr | 2530.28 |
| Grade 1, tested to 4 bar, 25m maximum head | | | | | | |
| 550 litres capacity | 986.05 | 1084.65 | 5.21 | 59.77 | nr | 1144.42 |
| 700 litres capacity | 1119.30 | 1231.23 | 6.02 | 69.13 | nr | 1300.36 |
| 800 litres capacity | 1199.25 | 1319.17 | 6.54 | 75.00 | nr | 1394.17 |
| 1000 litres capacity | 1599.00 | 1758.90 | 7.04 | 80.81 | nr | 1839.71 |
| 1500 litres capacity | 1918.80 | 2110.68 | 10.00 | 114.75 | nr | 2225.43 |
| 2000 litres capacity | 2345.20 | 2579.72 | 16.13 | 185.08 | nr | 2764.80 |

# Keep your figures up to date, free of charge

This section, and most of the other information in this Price Book, is brought up to date every three months, until the next annual edition, in the *Price Book Update*.

The *Update* is available free to all Price Book purchasers.

To ensure you receive your copy, simply complete the reply card from the centre of the book and return it to us.

## Y:MECHANICAL AND ELECTRICAL SERVICES

| Item | Net Price £ | Material £ | Labour hours | Labour £ | Unit | Total Rate £ |
|---|---|---|---|---|---|---|
| **Y23 : STORAGE CYLINDERS/CALORIFIERS : STEEL NON STORAGE CALORIFIER** | | | | | | |
| Non-storage calorifiers; mild steel; heater battery duty 82.71 degrees C to BS 853, maximum test on shell 11.55 bar, tubes 26.25 bar | | | | | | |
| Horizontal; steam at 3.2 bar | | | | | | |
| 88 kW capacity | 373.10 | 410.41 | 8.00 | 91.80 | nr | 502.21 |
| 176 kW capacity | 559.65 | 615.62 | 12.05 | 138.25 | nr | 753.87 |
| 293 kW capacity | 586.30 | 644.93 | 14.08 | 161.62 | nr | 806.55 |
| 586 kW capacity | 767.52 | 844.27 | 37.04 | 425.00 | nr | 1269.27 |
| 879 kW capacity | 954.07 | 1049.48 | 40.00 | 459.00 | nr | 1508.48 |
| 1465 kW capacity | 1519.05 | 1670.95 | 45.45 | 521.59 | nr | 2192.54 |
| Horizontal; steam at 4.8 bar | | | | | | |
| 88 kW capacity | 373.10 | 410.41 | 5.00 | 57.38 | nr | 467.79 |
| 176 kW capacity | 559.65 | 615.62 | 8.00 | 91.80 | nr | 707.42 |
| 293 kW capacity | 580.97 | 639.07 | 12.05 | 138.25 | nr | 777.32 |
| 586 kW capacity | 756.86 | 832.55 | 22.22 | 255.00 | nr | 1087.55 |
| 879 kW capacity | 943.41 | 1037.75 | 28.57 | 327.86 | nr | 1365.61 |
| 1465 kW capacity | 1519.05 | 1670.95 | 40.00 | 459.00 | nr | 2129.95 |
| Vertical; steam at 3.2 bar | | | | | | |
| 88 kW capacity | 383.76 | 422.14 | 8.00 | 91.80 | nr | 513.94 |
| 176 kW capacity | 575.64 | 633.20 | 12.05 | 138.25 | nr | 771.45 |
| 293 kW capacity | 607.62 | 668.38 | 14.08 | 161.62 | nr | 830.00 |
| 586 kW capacity | 772.85 | 850.13 | 37.04 | 425.00 | nr | 1275.13 |
| 879 kW capacity | 959.40 | 1055.34 | 40.00 | 459.00 | nr | 1514.34 |
| 1465 kW capacity | 1545.70 | 1700.27 | 45.45 | 521.59 | nr | 2221.86 |
| Vertical; steam at 4.8 bar | | | | | | |
| 88 kW capacity | 383.76 | 422.14 | 5.00 | 57.38 | nr | 479.52 |
| 176 kW capacity | 575.64 | 633.20 | 12.05 | 138.25 | nr | 771.45 |
| 293 kW capacity | 607.62 | 668.38 | 12.05 | 138.25 | nr | 806.63 |
| 586 kW capacity | 772.85 | 850.13 | 22.22 | 255.00 | nr | 1105.13 |
| 879 kW capacity | 959.40 | 1055.34 | 28.57 | 327.86 | nr | 1383.20 |
| 1465 kW capacity | 1545.70 | 1700.27 | 40.00 | 459.00 | nr | 2159.27 |
| Horizontal; primary water at 116 degree C (on) 90 degree C (off) | | | | | | |
| 88 kW capacity | 405.08 | 445.59 | 5.00 | 57.38 | nr | 502.97 |
| 176 kW capacity | 692.90 | 762.19 | 7.04 | 80.81 | nr | 843.00 |
| 293 kW capacity | 852.80 | 938.08 | 9.01 | 103.38 | nr | 1041.46 |
| 586 kW capacity | 1439.10 | 1583.01 | 22.22 | 255.00 | nr | 1838.01 |
| 879 kW capacity | 1684.28 | 1852.71 | 28.57 | 327.86 | nr | 2180.57 |
| 1465 kW capacity | 2345.20 | 2579.72 | 50.00 | 573.75 | nr | 3153.47 |

## Y:MECHANICAL AND ELECTRICAL SERVICES

| Item | Net Price £ | Material £ | Labour hours | Labour £ | Unit | Total Rate £ |
|---|---|---|---|---|---|---|
| **Y23 : STORAGE CYLINDERS/CALORIFIERS : STEEL NON STORAGE CALORIFIER (contd)** | | | | | | |
| Non-storage calorifiers; mild steel; heater battery duty 82.71 degrees C to BS 853, maximum test on shell 11.55 bar, tubes 26.25 bar (contd) | | | | | | |
| Vertical; primary water at 116 degree C (on) 93 degree C (off) | | | | | | |
| 88 kW capacity | 415.74 | 457.31 | 5.00 | 57.38 | nr | **514.69** |
| 176 kW capacity | 708.89 | 779.78 | 7.04 | 80.81 | nr | **860.59** |
| 293 kW capacity | 895.44 | 984.98 | 9.01 | 103.38 | nr | **1088.36** |
| 586 kW capacity | 1439.10 | 1583.01 | 22.22 | 255.00 | nr | **1838.01** |
| 879 kW capacity | 1705.60 | 1876.16 | 28.57 | 327.86 | nr | **2204.02** |
| 1465 kW capacity | 2398.50 | 2638.35 | 50.00 | 573.75 | nr | **3212.10** |

## Y:MECHANICAL AND ELECTRICAL SERVICES

| Item | Net Price £ | Material £ | Labour hours | Labour £ | Unit | Total Rate £ |
|---|---|---|---|---|---|---|
| **Y23 : STORAGE CYLINDERS/CALORIFIERS : STEEL STORAGE CALORIFIER** | | | | | | |
| Storage calorifiers; galvanised mild steel; heater battery capable of raising temperature of contents from 10 degree C to 65 degree C in one hour; static head not exceeding 1.35 bar; includes fixing in position on cradles or legs | | | | | | |
| Horizontal; primary LPHW at 82 degree C (on) 71 degree C (off) | | | | | | |
| 400 litres capacity | 986.05 | 1084.65 | 6.71 | 77.01 | nr | 1161.66 |
| 1000 litres capacity | 1449.76 | 1594.74 | 8.00 | 91.80 | nr | 1686.54 |
| 2000 litres capacity | 2345.20 | 2579.72 | 14.08 | 161.62 | nr | 2741.34 |
| 3000 litres capacity | 3731.00 | 4104.10 | 25.00 | 286.88 | nr | 4390.98 |
| 4000 litres capacity | 4317.30 | 4749.03 | 40.00 | 459.00 | nr | 5208.03 |
| 4500 litres capacity | 4477.20 | 4924.92 | 35.71 | 409.82 | nr | 5334.74 |
| Horizontal; primary steam at 3.2 bar | | | | | | |
| 400 litres capacity | 868.79 | 955.67 | 6.71 | 77.01 | nr | 1032.68 |
| 1000 litres capacity | 1279.20 | 1407.12 | 8.00 | 91.80 | nr | 1498.92 |
| 2000 litres capacity | 2078.70 | 2286.57 | 14.08 | 161.62 | nr | 2448.19 |
| 3000 litres capacity | 3283.28 | 3611.61 | 25.00 | 286.88 | nr | 3898.49 |
| 4000 litres capacity | 3784.30 | 4162.73 | 40.00 | 459.00 | nr | 4621.73 |
| 4500 litres capacity | 3944.20 | 4338.62 | 50.00 | 573.75 | nr | 4912.37 |
| Vertical; primary LPHW at 82 degree C (on) 71 degree C (off) | | | | | | |
| 400 litres capacity | 959.40 | 1055.34 | 7.04 | 80.81 | nr | 1136.15 |
| 1000 litres capacity | 1423.11 | 1565.42 | 8.00 | 91.80 | nr | 1657.22 |
| 2000 litres capacity | 2318.55 | 2550.41 | 14.08 | 161.62 | nr | 2712.03 |
| 3000 litres capacity | 3704.35 | 4074.78 | 25.00 | 286.88 | nr | 4361.66 |
| 4000 litres capacity | 4290.65 | 4719.72 | 40.00 | 459.00 | nr | 5178.72 |
| 4500 litres capacity | 4450.55 | 4895.60 | 50.00 | 573.75 | nr | 5469.35 |
| Vertical; primary steam at 3.2 bar | | | | | | |
| 400 litres capacity | 842.14 | 926.35 | 7.04 | 80.81 | nr | 1007.16 |
| 1000 litres capacity | 1252.55 | 1377.81 | 8.00 | 91.80 | nr | 1469.61 |
| 2000 litres capacity | 2052.05 | 2257.26 | 14.08 | 161.62 | nr | 2418.88 |
| 3000 litres capacity | 3256.63 | 3582.29 | 25.00 | 286.88 | nr | 3869.17 |
| 4000 litres capacity | 3757.65 | 4133.41 | 40.00 | 459.00 | nr | 4592.41 |
| 4500 litres capacity | 3917.55 | 4309.31 | 50.00 | 573.75 | nr | 4883.06 |

## Y:MECHANICAL AND ELECTRICAL SERVICES

| Item | Net Price £ | Material £ | Labour hours | Labour £ | Unit | Total Rate £ |
|---|---|---|---|---|---|---|
| **Y23 : STORAGE CYLINDERS/CALORIFIERS : COPPER STORAGE CALORIFIER** | | | | | | |
| Storage calorifiers; copper; heater battery capable of raising temperature of contents from 10 degree C to 65 degree C in one hour; static head not exceeding 1.35 bar; BS 853; includes fixing in position on cradles or legs | | | | | | |
| Horizontal; primary LPHW at 82 degree C (on) 71 degree C (off) | | | | | | |
| 400 litres capacity | 1066.00 | 1172.60 | 7.04 | 80.81 | nr | **1253.41** |
| 1000 litres capacity | 1705.60 | 1876.16 | 8.00 | 91.80 | nr | **1967.96** |
| 2000 litres capacity | 3411.20 | 3752.32 | 14.08 | 161.62 | nr | **3913.94** |
| 3000 litres capacity | 4210.70 | 4631.77 | 25.00 | 286.88 | nr | **4918.65** |
| 4000 litres capacity | 5116.80 | 5628.48 | 40.00 | 459.00 | nr | **6087.48** |
| 4500 litres capacity | 5756.40 | 6332.04 | 50.00 | 573.75 | nr | **6905.79** |
| Horizontal; primary steam at 3.2 bar | | | | | | |
| 400 litres capacity | 980.72 | 1078.79 | 7.04 | 80.81 | nr | **1159.60** |
| 1000 litres capacity | 1439.10 | 1583.01 | 8.00 | 91.80 | nr | **1674.81** |
| 2000 litres capacity | 3198.00 | 3517.80 | 14.08 | 161.62 | nr | **3679.42** |
| 3000 litres capacity | 3954.86 | 4350.35 | 25.00 | 286.88 | nr | **4637.23** |
| 4000 litres capacity | 4903.60 | 5393.96 | 40.00 | 459.00 | nr | **5852.96** |
| 4500 litres capacity | 5543.20 | 6097.52 | 50.00 | 573.75 | nr | **6671.27** |
| Vertical; primary LPHW at 82 degree C (on) 71 degree C (off) | | | | | | |
| 400 litres capacity | 1044.68 | 1149.15 | 7.04 | 80.81 | nr | **1229.96** |
| 1000 litres capacity | 1678.95 | 1846.85 | 8.00 | 91.80 | nr | **1938.65** |
| 2000 litres capacity | 3198.00 | 3517.80 | 14.08 | 161.62 | nr | **3679.42** |
| 3000 litres capacity | 3997.50 | 4397.25 | 25.00 | 286.88 | nr | **4684.13** |
| 4000 litres capacity | 4903.60 | 5393.96 | 40.00 | 459.00 | nr | **5852.96** |
| 4500 litres capacity | 5543.20 | 6097.52 | 50.00 | 573.75 | nr | **6671.27** |
| Vertical; primary steam at 3.2 bar | | | | | | |
| 400 litres capacity | 959.40 | 1055.34 | 7.04 | 80.81 | nr | **1136.15** |
| 1000 litres capacity | 1412.45 | 1553.69 | 8.00 | 91.80 | nr | **1645.49** |
| 2000 litres capacity | 2984.80 | 3283.28 | 14.08 | 161.62 | nr | **3444.90** |
| 3000 litres capacity | 3731.00 | 4104.10 | 25.00 | 286.88 | nr | **4390.98** |
| 4000 litres capacity | 4690.40 | 5159.44 | 40.00 | 459.00 | nr | **5618.44** |
| 4500 litres capacity | 5330.00 | 5863.00 | 50.00 | 573.75 | nr | **6436.75** |

## Y:MECHANICAL AND ELECTRICAL SERVICES

| Item | Net Price £ | Material £ | Labour hours | Labour £ | Unit | Total Rate £ |
|---|---|---|---|---|---|---|
| **Y30 : AIR DUCTLINES : MILD STEEL DUCTWORK** | | | | | | |
| Galvanised sheet metal DW142 class A rectangular section ductwork; including all necessary stiffeners, joints, couplers in the running length and duct supports | | | | | | |
| Straight duct | | | | | | |
| Sum of two sides 200mm | 6.10 | 7.72 | 1.19 | 18.87 | m | 26.59 |
| Sum of two sides 250mm | 6.93 | 8.77 | 1.19 | 18.87 | m | 27.64 |
| Sum of two sides 300mm | 7.90 | 9.99 | 1.19 | 18.87 | m | 28.86 |
| Sum of two sides 350mm | 8.85 | 11.20 | 1.19 | 18.87 | m | 30.07 |
| Sum of two sides 400mm | 9.76 | 12.35 | 1.16 | 18.40 | m | 30.75 |
| Sum of two sides 450mm | 10.65 | 13.47 | 1.16 | 18.40 | m | 31.87 |
| Sum of two sides 500mm | 11.54 | 14.60 | 1.16 | 18.40 | m | 33.00 |
| Sum of two sides 550mm | 12.47 | 15.77 | 1.27 | 20.14 | m | 35.91 |
| Sum of two sides 600mm | 13.35 | 16.89 | 1.27 | 20.14 | m | 37.03 |
| Sum of two sides 650mm | 14.26 | 18.04 | 1.27 | 20.14 | m | 38.18 |
| Sum of two sides 700mm | 15.13 | 19.14 | 1.27 | 20.14 | m | 39.28 |
| Sum of two sides 750mm | 16.08 | 20.34 | 1.27 | 20.14 | m | 40.48 |
| Sum of two sides 800mm | 16.95 | 21.44 | 1.27 | 20.14 | m | 41.58 |
| Sum of two sides 850mm | 17.90 | 22.64 | 1.27 | 20.14 | m | 42.78 |
| Sum of two sides 900mm | 18.79 | 23.77 | 1.27 | 20.14 | m | 43.91 |
| Sum of two sides 950mm | 19.86 | 25.12 | 1.37 | 21.73 | m | 46.85 |
| Sum of two sides 1000mm | 20.58 | 26.03 | 1.37 | 21.73 | m | 47.76 |
| Sum of two sides 1100mm | 22.36 | 28.29 | 1.37 | 21.73 | m | 50.02 |
| Sum of two sides 1200mm | 24.21 | 30.63 | 1.37 | 21.73 | m | 52.36 |
| Sum of two sides 1300mm | 25.99 | 32.88 | 1.40 | 22.20 | m | 55.08 |
| Sum of two sides 1400mm | 27.81 | 35.18 | 1.40 | 22.20 | m | 57.38 |
| Sum of two sides 1500mm | 29.63 | 37.48 | 1.48 | 23.47 | m | 60.95 |
| Sum of two sides 1600mm | 31.45 | 39.78 | 1.55 | 24.58 | m | 64.36 |
| Sum of two sides 1700mm | 36.53 | 46.21 | 1.55 | 24.58 | m | 70.79 |
| Sum of two sides 1800mm | 38.49 | 48.69 | 1.61 | 25.54 | m | 74.23 |
| Sum of two sides 1900mm | 40.42 | 51.13 | 1.61 | 25.54 | m | 76.67 |
| Sum of two sides 2000mm | 42.37 | 53.60 | 1.61 | 25.54 | m | 79.14 |
| Sum of two sides 2100mm | 55.94 | 70.76 | 2.17 | 34.42 | m | 105.18 |
| Sum of two sides 2200mm | 58.38 | 73.85 | 2.19 | 34.73 | m | 108.58 |
| Sum of two sides 2300mm | 60.89 | 77.03 | 2.19 | 34.73 | m | 111.76 |
| Sum of two sides 2400mm | 63.39 | 80.19 | 2.38 | 37.75 | m | 117.94 |
| Sum of two sides 2500mm | 65.84 | 83.29 | 2.38 | 37.75 | m | 121.04 |
| Sum of two sides 2600mm | 73.32 | 92.75 | 2.64 | 41.87 | m | 134.62 |
| Sum of two sides 2700mm | 76.00 | 96.14 | 2.66 | 42.19 | m | 138.33 |
| Sum of two sides 2800mm | 78.65 | 99.49 | 2.95 | 46.79 | m | 146.28 |
| Sum of two sides 2900mm | 81.33 | 102.88 | 2.96 | 46.95 | m | 149.83 |
| Sum of two sides 3000mm | 84.01 | 106.27 | 3.15 | 49.96 | m | 156.23 |
| Sum of two sides 3100mm | 86.65 | 109.61 | 3.15 | 49.96 | m | 159.57 |
| Sum of two sides 3200mm | 89.34 | 113.02 | 3.18 | 50.44 | m | 163.46 |

## Y:MECHANICAL AND ELECTRICAL SERVICES

| Item | Net Price £ | Material £ | Labour hours | Labour £ | Unit | Total Rate £ |
|---|---|---|---|---|---|---|
| **Y30 : AIR DUCTLINES : MILD STEEL DUCTWORK (contd)** | | | | | | |
| **Extra Over galvanised sheet metal DW142 class A rectangular section ductwork; fittings; including all necessary stiffeners, joints etc.** | | | | | | |
| Stopped end | | | | | | |
| Sum of two sides 200mm | 0.14 | 0.18 | 0.38 | 6.03 | nr | 6.21 |
| Sum of two sides 250mm | 0.21 | 0.27 | 0.38 | 6.03 | nr | 6.30 |
| Sum of two sides 300mm | 0.21 | 0.27 | 0.38 | 6.03 | nr | 6.30 |
| Sum of two sides 350mm | 0.29 | 0.37 | 0.38 | 6.03 | nr | 6.40 |
| Sum of two sides 400mm | 0.29 | 0.37 | 0.38 | 6.03 | nr | 6.40 |
| Sum of two sides 450mm | 0.34 | 0.43 | 0.38 | 6.03 | nr | 6.46 |
| Sum of two sides 500mm | 0.34 | 0.43 | 0.38 | 6.03 | nr | 6.46 |
| Sum of two sides 550mm | 0.34 | 0.43 | 0.38 | 6.03 | nr | 6.46 |
| Sum of two sides 600mm | 0.42 | 0.53 | 0.38 | 6.03 | nr | 6.56 |
| Sum of two sides 650mm | 0.42 | 0.53 | 0.38 | 6.03 | nr | 6.56 |
| Sum of two sides 700mm | 0.48 | 0.61 | 0.38 | 6.03 | nr | 6.64 |
| Sum of two sides 750mm | 0.56 | 0.71 | 0.38 | 6.03 | nr | 6.74 |
| Sum of two sides 800mm | 0.56 | 0.71 | 0.38 | 6.03 | nr | 6.74 |
| Sum of two sides 850mm | 0.65 | 0.82 | 0.58 | 9.20 | nr | 10.02 |
| Sum of two sides 900mm | 0.69 | 0.87 | 0.58 | 9.20 | nr | 10.07 |
| Sum of two sides 950mm | 0.73 | 0.92 | 0.58 | 9.20 | nr | 10.12 |
| Sum of two sides 1000mm | 0.77 | 0.97 | 0.58 | 9.20 | nr | 10.17 |
| Sum of two sides 1100mm | 0.81 | 1.02 | 0.58 | 9.20 | nr | 10.22 |
| Sum of two sides 1200mm | 0.87 | 1.10 | 0.58 | 9.20 | nr | 10.30 |
| Sum of two sides 1300mm | 0.92 | 1.16 | 0.58 | 9.20 | nr | 10.36 |
| Sum of two sides 1400mm | 0.97 | 1.23 | 0.58 | 9.20 | nr | 10.43 |
| Sum of two sides 1500mm | 1.04 | 1.32 | 0.58 | 9.20 | nr | 10.52 |
| Sum of two sides 1600mm | 1.09 | 1.38 | 0.58 | 9.20 | nr | 10.58 |
| Sum of two sides 1700mm | 1.17 | 1.48 | 0.58 | 9.20 | nr | 10.68 |
| Sum of two sides 1800mm | 1.25 | 1.58 | 0.58 | 9.20 | nr | 10.78 |
| Sum of two sides 1900mm | 1.31 | 1.66 | 0.58 | 9.20 | nr | 10.86 |
| Sum of two sides 2000mm | 1.36 | 1.72 | 0.58 | 9.20 | nr | 10.92 |
| Sum of two sides 2100mm | 1.82 | 2.30 | 0.87 | 13.80 | nr | 16.10 |
| Sum of two sides 2200mm | 1.89 | 2.39 | 0.87 | 13.80 | nr | 16.19 |
| Sum of two sides 2300mm | 1.98 | 2.50 | 0.87 | 13.80 | nr | 16.30 |
| Sum of two sides 2400mm | 2.04 | 2.58 | 0.87 | 13.80 | nr | 16.38 |
| Sum of two sides 2500mm | 2.16 | 2.73 | 0.87 | 13.80 | nr | 16.53 |
| Sum of two sides 2600mm | 2.25 | 2.85 | 0.87 | 13.80 | nr | 16.65 |
| Sum of two sides 2700mm | 2.34 | 2.96 | 0.87 | 13.80 | nr | 16.76 |
| Sum of two sides 2800mm | 2.43 | 3.07 | 1.16 | 18.40 | nr | 21.47 |
| Sum of two sides 2900mm | 2.50 | 3.16 | 1.16 | 18.40 | nr | 21.56 |
| Sum of two sides 3000mm | 2.59 | 3.28 | 1.73 | 27.44 | nr | 30.72 |
| Sum of two sides 3100mm | 2.69 | 3.40 | 1.80 | 28.55 | nr | 31.95 |
| Sum of two sides 3200mm | 2.74 | 3.47 | 1.80 | 28.55 | nr | 32.02 |
| Taper; large end measured | | | | | | |
| Sum of two sides 200mm | 4.47 | 5.65 | 1.42 | 22.52 | nr | 28.17 |
| Sum of two sides 250mm | 4.63 | 5.86 | 1.42 | 22.52 | nr | 28.38 |
| Sum of two sides 300mm | 5.68 | 7.19 | 1.42 | 22.52 | nr | 29.71 |
| Sum of two sides 350mm | 6.30 | 7.97 | 1.40 | 22.20 | nr | 30.17 |
| Sum of two sides 400mm | 6.88 | 8.70 | 1.40 | 22.20 | nr | 30.90 |
| Sum of two sides 450mm | 7.49 | 9.47 | 1.40 | 22.20 | nr | 31.67 |
| Sum of two sides 500mm | 8.13 | 10.28 | 1.40 | 22.20 | nr | 32.48 |
| Sum of two sides 550mm | 8.71 | 11.02 | 1.69 | 26.80 | nr | 37.82 |
| Sum of two sides 600mm | 9.28 | 11.74 | 1.69 | 26.80 | nr | 38.54 |

*Material Costs / Measured Work Prices - Mechanical Installations*

## Y:MECHANICAL AND ELECTRICAL SERVICES

| Item | Net Price £ | Material £ | Labour hours | Labour £ | Unit | Total Rate £ |
|---|---|---|---|---|---|---|
| Sum of two sides 650mm | 9.91 | 12.54 | 1.69 | 26.80 | nr | 39.34 |
| Sum of two sides 700mm | 10.53 | 13.32 | 1.69 | 26.80 | nr | 40.12 |
| Sum of two sides 750mm | 11.13 | 14.08 | 1.92 | 30.45 | nr | 44.53 |
| Sum of two sides 800mm | 11.73 | 14.84 | 1.92 | 30.45 | nr | 45.29 |
| Sum of two sides 850mm | 12.32 | 15.58 | 1.92 | 30.45 | nr | 46.03 |
| Sum of two sides 900mm | 12.91 | 16.33 | 1.92 | 30.45 | nr | 46.78 |
| Sum of two sides 950mm | 13.53 | 17.12 | 1.95 | 30.93 | nr | 48.05 |
| Sum of two sides 1000mm | 14.14 | 17.89 | 2.18 | 34.58 | nr | 52.47 |
| Sum of two sides 1100mm | 15.34 | 19.41 | 2.18 | 34.58 | nr | 53.99 |
| Sum of two sides 1200mm | 16.55 | 20.94 | 2.18 | 34.58 | nr | 55.52 |
| Sum of two sides 1300mm | 17.77 | 22.48 | 2.30 | 36.48 | nr | 58.96 |
| Sum of two sides 1400mm | 18.94 | 23.96 | 2.30 | 36.48 | nr | 60.44 |
| Sum of two sides 1500mm | 20.19 | 25.54 | 2.47 | 39.17 | nr | 64.71 |
| Sum of two sides 1600mm | 21.37 | 27.03 | 2.47 | 39.17 | nr | 66.20 |
| Sum of two sides 1700mm | 26.71 | 33.79 | 2.47 | 39.17 | nr | 72.96 |
| Sum of two sides 1800mm | 28.07 | 35.51 | 2.59 | 41.08 | nr | 76.59 |
| Sum of two sides 1900mm | 29.46 | 37.27 | 2.71 | 42.98 | nr | 80.25 |
| Sum of two sides 2000mm | 30.81 | 38.97 | 2.59 | 41.08 | nr | 80.05 |
| Sum of two sides 2100mm | 35.81 | 45.30 | 2.92 | 46.31 | nr | 91.61 |
| Sum of two sides 2200mm | 37.22 | 47.08 | 2.92 | 46.31 | nr | 93.39 |
| Sum of two sides 2300mm | 38.91 | 49.22 | 2.92 | 46.31 | nr | 95.53 |
| Sum of two sides 2400mm | 40.43 | 51.14 | 3.12 | 49.48 | nr | 100.62 |
| Sum of two sides 2500mm | 42.00 | 53.13 | 3.12 | 49.48 | nr | 102.61 |
| Sum of two sides 2600mm | 43.54 | 55.08 | 3.12 | 49.48 | nr | 104.56 |
| Sum of two sides 2700mm | 45.11 | 57.06 | 3.12 | 49.48 | nr | 106.54 |
| Sum of two sides 2800mm | 46.63 | 58.99 | 3.95 | 62.65 | nr | 121.64 |
| Sum of two sides 2900mm | 48.22 | 61.00 | 3.97 | 62.97 | nr | 123.97 |
| Sum of two sides 3000mm | 49.71 | 62.88 | 4.52 | 71.69 | nr | 134.57 |
| Sum of two sides 3100mm | 51.21 | 64.78 | 4.52 | 71.69 | nr | 136.47 |
| Sum of two sides 3200mm | 52.84 | 66.84 | 4.52 | 71.69 | nr | 138.53 |
| Set piece | | | | | | |
| Sum of two sides 200mm | 3.91 | 4.95 | 1.65 | 26.17 | nr | 31.12 |
| Sum of two sides 250mm | 4.45 | 5.63 | 1.65 | 26.17 | nr | 31.80 |
| Sum of two sides 300mm | 5.06 | 6.40 | 1.65 | 26.17 | nr | 32.57 |
| Sum of two sides 350mm | 5.68 | 7.19 | 1.63 | 25.85 | nr | 33.04 |
| Sum of two sides 400mm | 6.32 | 7.99 | 1.63 | 25.85 | nr | 33.84 |
| Sum of two sides 450mm | 7.01 | 8.87 | 1.63 | 25.85 | nr | 34.72 |
| Sum of two sides 500mm | 7.74 | 9.79 | 1.63 | 25.85 | nr | 35.64 |
| Sum of two sides 550mm | 8.48 | 10.73 | 1.92 | 30.45 | nr | 41.18 |
| Sum of two sides 600mm | 9.28 | 11.74 | 1.92 | 30.45 | nr | 42.19 |
| Sum of two sides 650mm | 10.13 | 12.81 | 1.92 | 30.45 | nr | 43.26 |
| Sum of two sides 700mm | 11.01 | 13.93 | 1.92 | 30.45 | nr | 44.38 |
| Sum of two sides 750mm | 11.87 | 15.02 | 1.92 | 30.45 | nr | 45.47 |
| Sum of two sides 800mm | 12.82 | 16.22 | 1.92 | 30.45 | nr | 46.67 |
| Sum of two sides 850mm | 13.77 | 17.42 | 1.92 | 30.45 | nr | 47.87 |
| Sum of two sides 900mm | 15.18 | 19.20 | 1.92 | 30.45 | nr | 49.65 |
| Sum of two sides 950mm | 15.81 | 20.00 | 2.18 | 34.58 | nr | 54.58 |
| Sum of two sides 1000mm | 16.89 | 21.37 | 2.18 | 34.58 | nr | 55.95 |
| Sum of two sides 1100mm | 19.15 | 24.22 | 2.18 | 34.58 | nr | 58.80 |
| Sum of two sides 1200mm | 21.52 | 27.22 | 2.18 | 34.58 | nr | 61.80 |
| Sum of two sides 1300mm | 24.04 | 30.41 | 2.30 | 36.48 | nr | 66.89 |
| Sum of two sides 1400mm | 26.69 | 33.76 | 2.30 | 36.48 | nr | 70.24 |
| Sum of two sides 1500mm | 29.47 | 37.28 | 2.47 | 39.17 | nr | 76.45 |
| Sum of two sides 1600mm | 32.42 | 41.01 | 2.47 | 39.17 | nr | 80.18 |
| Sum of two sides 1700mm | 39.59 | 50.08 | 2.59 | 41.08 | nr | 91.16 |
| Sum of two sides 1800mm | 42.96 | 54.34 | 2.61 | 41.40 | nr | 95.74 |
| Sum of two sides 1900mm | 46.46 | 58.77 | 2.71 | 42.98 | nr | 101.75 |
| Sum of two sides 2000mm | 50.14 | 63.43 | 2.71 | 42.98 | nr | 106.41 |

## Y:MECHANICAL AND ELECTRICAL SERVICES

| Item | Net Price £ | Material £ | Labour hours | Labour £ | Unit | Total Rate £ |
|---|---|---|---|---|---|---|
| **Y30 : AIR DUCTLINES : MILD STEEL DUCTWORK (contd)** | | | | | | |
| Extra Over galvanised sheet metal DW142 class A rectangular section ductwork; fittings; including all necessary stiffeners, joints etc. (contd) | | | | | | |
| Set piece (contd) | | | | | | |
| Sum of two sides 2100mm | 62.97 | 79.66 | 2.92 | 46.31 | nr | 125.97 |
| Sum of two sides 2200mm | 73.74 | 93.28 | 3.26 | 51.70 | nr | 144.98 |
| Sum of two sides 2300mm | 78.91 | 99.82 | 3.26 | 51.70 | nr | 151.52 |
| Sum of two sides 2400mm | 84.27 | 106.60 | 3.47 | 55.04 | nr | 161.64 |
| Sum of two sides 2500mm | 89.78 | 113.57 | 3.48 | 55.19 | nr | 168.76 |
| Sum of two sides 2600mm | 95.51 | 120.82 | 3.49 | 55.35 | nr | 176.17 |
| Sum of two sides 2700mm | 101.37 | 128.23 | 3.50 | 55.51 | nr | 183.74 |
| Sum of two sides 2800mm | 107.39 | 135.85 | 4.34 | 68.83 | nr | 204.68 |
| Sum of two sides 2900mm | 121.58 | 153.80 | 4.76 | 75.50 | nr | 229.30 |
| Sum of two sides 3000mm | 128.25 | 162.24 | 5.32 | 84.38 | nr | 246.62 |
| Sum of two sides 3100mm | 135.10 | 170.90 | 5.35 | 84.85 | nr | 255.75 |
| Sum of two sides 3200mm | 142.09 | 179.74 | 5.35 | 84.85 | nr | 264.59 |
| 90 degree radius bend | | | | | | |
| Sum of two sides 200mm | 3.91 | 4.95 | 1.25 | 19.83 | nr | 24.78 |
| Sum of two sides 250mm | 4.45 | 5.63 | 1.25 | 19.83 | nr | 25.46 |
| Sum of two sides 300mm | 5.06 | 6.40 | 1.25 | 19.83 | nr | 26.23 |
| Sum of two sides 350mm | 5.68 | 7.19 | 1.22 | 19.35 | nr | 26.54 |
| Sum of two sides 400mm | 6.32 | 7.99 | 1.22 | 19.35 | nr | 27.34 |
| Sum of two sides 450mm | 7.01 | 8.87 | 1.22 | 19.35 | nr | 28.22 |
| Sum of two sides 500mm | 7.74 | 9.79 | 1.22 | 19.35 | nr | 29.14 |
| Sum of two sides 550mm | 8.48 | 10.73 | 1.33 | 21.09 | nr | 31.82 |
| Sum of two sides 600mm | 9.28 | 11.74 | 1.33 | 21.09 | nr | 32.83 |
| Sum of two sides 650mm | 10.13 | 12.81 | 1.33 | 21.09 | nr | 33.90 |
| Sum of two sides 700mm | 11.01 | 13.93 | 1.33 | 21.09 | nr | 35.02 |
| Sum of two sides 750mm | 11.87 | 15.02 | 1.40 | 22.20 | nr | 37.22 |
| Sum of two sides 800mm | 12.82 | 16.22 | 1.40 | 22.20 | nr | 38.42 |
| Sum of two sides 850mm | 13.77 | 17.42 | 1.40 | 22.20 | nr | 39.62 |
| Sum of two sides 900mm | 14.78 | 18.70 | 1.40 | 22.20 | nr | 40.90 |
| Sum of two sides 950mm | 15.81 | 20.00 | 1.82 | 28.87 | nr | 48.87 |
| Sum of two sides 1000mm | 16.89 | 21.37 | 1.86 | 29.50 | nr | 50.87 |
| Sum of two sides 1100mm | 19.15 | 24.22 | 1.82 | 20.88 | nr | 45.10 |
| Sum of two sides 1200mm | 21.52 | 27.22 | 1.82 | 28.87 | nr | 56.09 |
| Sum of two sides 1300mm | 24.04 | 30.41 | 2.15 | 34.10 | nr | 64.51 |
| Sum of two sides 1400mm | 26.69 | 33.76 | 2.15 | 34.10 | nr | 67.86 |
| Sum of two sides 1500mm | 29.47 | 37.28 | 2.38 | 37.75 | nr | 75.03 |
| Sum of two sides 1600mm | 32.42 | 41.01 | 2.38 | 37.75 | nr | 78.76 |
| Sum of two sides 1700mm | 39.59 | 50.08 | 2.55 | 40.44 | nr | 90.52 |
| Sum of two sides 1800mm | 42.96 | 54.34 | 2.55 | 40.44 | nr | 94.78 |
| Sum of two sides 1900mm | 46.46 | 58.77 | 2.83 | 44.88 | nr | 103.65 |
| Sum of two sides 2000mm | 50.14 | 63.43 | 2.83 | 44.88 | nr | 108.31 |
| Sum of two sides 2100mm | 62.97 | 79.66 | 3.85 | 61.06 | nr | 140.72 |
| Sum of two sides 2200mm | 73.74 | 93.28 | 4.18 | 66.30 | nr | 159.58 |
| Sum of two sides 2300mm | 78.91 | 99.82 | 4.22 | 66.93 | nr | 166.75 |
| Sum of two sides 2400mm | 84.27 | 106.60 | 4.68 | 74.23 | nr | 180.83 |
| Sum of two sides 2500mm | 89.78 | 113.57 | 4.70 | 74.54 | nr | 188.11 |
| Sum of two sides 2600mm | 95.51 | 120.82 | 4.70 | 74.54 | nr | 195.36 |
| Sum of two sides 2700mm | 101.37 | 128.23 | 4.71 | 74.70 | nr | 202.93 |
| Sum of two sides 2800mm | 107.39 | 135.85 | 8.19 | 129.90 | nr | 265.75 |

*Material Costs / Measured Work Prices - Mechanical Installations* 277

## Y:MECHANICAL AND ELECTRICAL SERVICES

| Item | Net Price £ | Material £ | Labour hours | Labour £ | Unit | Total Rate £ |
|---|---|---|---|---|---|---|
| Sum of two sides 2900mm | 121.58 | 153.80 | 8.62 | 136.72 | nr | 290.52 |
| Sum of two sides 3000mm | 128.25 | 162.24 | 8.75 | 138.78 | nr | 301.02 |
| Sum of two sides 3100mm | 135.10 | 170.90 | 8.75 | 138.78 | nr | 309.68 |
| Sum of two sides 3200mm | 142.09 | 179.74 | 8.75 | 138.78 | nr | 318.52 |
| 45 degree radius bend | | | | | | |
| Sum of two sides 200mm | 3.38 | 4.28 | 0.89 | 14.12 | nr | 18.40 |
| Sum of two sides 250mm | 3.80 | 4.81 | 1.12 | 17.76 | nr | 22.57 |
| Sum of two sides 300mm | 4.25 | 5.38 | 1.12 | 17.76 | nr | 23.14 |
| Sum of two sides 350mm | 4.67 | 5.91 | 1.12 | 17.76 | nr | 23.67 |
| Sum of two sides 400mm | 5.23 | 6.62 | 1.10 | 17.45 | nr | 24.07 |
| Sum of two sides 450mm | 5.77 | 7.30 | 1.10 | 17.45 | nr | 24.75 |
| Sum of two sides 500mm | 6.37 | 8.06 | 1.10 | 17.45 | nr | 25.51 |
| Sum of two sides 550mm | 6.98 | 8.83 | 1.16 | 18.40 | nr | 27.23 |
| Sum of two sides 600mm | 6.77 | 8.56 | 1.16 | 18.40 | nr | 26.96 |
| Sum of two sides 650mm | 8.35 | 10.56 | 1.16 | 18.40 | nr | 28.96 |
| Sum of two sides 700mm | 9.05 | 11.45 | 1.16 | 18.40 | nr | 29.85 |
| Sum of two sides 750mm | 9.76 | 12.35 | 1.22 | 19.35 | nr | 31.70 |
| Sum of two sides 800mm | 10.63 | 13.45 | 1.22 | 19.35 | nr | 32.80 |
| Sum of two sides 850mm | 11.44 | 14.47 | 1.22 | 19.35 | nr | 33.82 |
| Sum of two sides 900mm | 12.32 | 15.58 | 1.22 | 19.35 | nr | 34.93 |
| Sum of two sides 950mm | 13.20 | 16.70 | 1.40 | 22.20 | nr | 38.90 |
| Sum of two sides 1000mm | 14.14 | 17.89 | 1.40 | 22.20 | nr | 40.09 |
| Sum of two sides 1100mm | 16.09 | 20.35 | 1.40 | 22.20 | nr | 42.55 |
| Sum of two sides 1200mm | 18.21 | 23.04 | 1.40 | 22.20 | nr | 45.24 |
| Sum of two sides 1300mm | 20.44 | 25.86 | 1.91 | 30.29 | nr | 56.15 |
| Sum of two sides 1400mm | 22.82 | 28.87 | 1.91 | 30.29 | nr | 59.16 |
| Sum of two sides 1500mm | 25.36 | 32.08 | 2.11 | 33.47 | nr | 65.55 |
| Sum of two sides 1600mm | 28.00 | 35.42 | 2.11 | 33.47 | nr | 68.89 |
| Sum of two sides 1700mm | 34.89 | 44.14 | 2.26 | 35.84 | nr | 79.98 |
| Sum of two sides 1800mm | 37.99 | 48.06 | 2.26 | 35.84 | nr | 83.90 |
| Sum of two sides 1900mm | 41.24 | 52.17 | 2.48 | 39.33 | nr | 91.50 |
| Sum of two sides 2000mm | 44.62 | 56.44 | 2.48 | 39.33 | nr | 95.77 |
| Sum of two sides 2100mm | 55.69 | 70.45 | 2.19 | 34.73 | nr | 105.18 |
| Sum of two sides 2200mm | 60.10 | 76.03 | 2.19 | 34.73 | nr | 110.76 |
| Sum of two sides 2300mm | 64.69 | 81.83 | 2.19 | 34.73 | nr | 116.56 |
| Sum of two sides 2400mm | 69.43 | 87.83 | 3.90 | 61.86 | nr | 149.69 |
| Sum of two sides 2500mm | 74.31 | 94.00 | 3.90 | 61.86 | nr | 155.86 |
| Sum of two sides 2600mm | 86.54 | 109.47 | 4.26 | 67.56 | nr | 177.03 |
| Sum of two sides 2700mm | 92.08 | 116.48 | 4.55 | 72.16 | nr | 188.64 |
| Sum of two sides 2800mm | 97.75 | 123.65 | 4.55 | 72.16 | nr | 195.81 |
| Sum of two sides 2900mm | 103.62 | 131.08 | 6.87 | 108.96 | nr | 240.04 |
| Sum of two sides 3000mm | 109.67 | 138.73 | 7.00 | 111.02 | nr | 249.75 |
| Sum of two sides 3100mm | 115.90 | 146.61 | 7.00 | 111.02 | nr | 257.63 |
| Sum of two sides 3200mm | 122.28 | 154.68 | 7.00 | 111.02 | nr | 265.70 |
| 90 degree bend with turning vanes | | | | | | |
| Sum of two sides 200mm | 4.32 | 5.46 | 1.29 | 20.46 | nr | 25.92 |
| Sum of two sides 250mm | 5.40 | 6.83 | 1.29 | 20.46 | nr | 27.29 |
| Sum of two sides 300mm | 6.05 | 7.65 | 1.29 | 20.46 | nr | 28.11 |
| Sum of two sides 350mm | 7.08 | 8.96 | 1.29 | 20.46 | nr | 29.42 |
| Sum of two sides 400mm | 8.36 | 10.58 | 1.29 | 20.46 | nr | 31.04 |
| Sum of two sides 450mm | 9.44 | 11.94 | 1.29 | 20.46 | nr | 32.40 |
| Sum of two sides 500mm | 10.62 | 13.43 | 1.29 | 20.46 | nr | 33.89 |
| Sum of two sides 550mm | 12.21 | 15.45 | 1.39 | 22.05 | nr | 37.50 |
| Sum of two sides 600mm | 13.70 | 17.33 | 1.39 | 22.05 | nr | 39.38 |
| Sum of two sides 650mm | 15.35 | 19.42 | 1.39 | 22.05 | nr | 41.47 |
| Sum of two sides 700mm | 17.08 | 21.61 | 1.39 | 22.05 | nr | 43.66 |

## Y:MECHANICAL AND ELECTRICAL SERVICES

| Item | Net Price £ | Material £ | Labour hours | Labour £ | Unit | Total Rate £ |
|---|---|---|---|---|---|---|
| **Y30 : AIR DUCTLINES : MILD STEEL DUCTWORK (contd)** | | | | | | |
| **Extra Over galvanised sheet metal DW142 class A rectangular section ductwork; fittings; including all necessary stiffeners, joints etc. (contd)** | | | | | | |
| 90 degree bend with turning vanes (contd) | | | | | | |
| Sum of two sides 750mm | 18.93 | 23.95 | 1.46 | 23.16 | nr | 47.11 |
| Sum of two sides 800mm | 20.83 | 26.35 | 1.46 | 23.16 | nr | 49.51 |
| Sum of two sides 850mm | 22.87 | 28.93 | 1.46 | 23.16 | nr | 52.09 |
| Sum of two sides 900mm | 25.01 | 31.64 | 1.46 | 23.16 | nr | 54.80 |
| Sum of two sides 950mm | 27.27 | 34.50 | 1.88 | 29.82 | nr | 64.32 |
| Sum of two sides 1000mm | 29.59 | 37.43 | 1.88 | 29.82 | nr | 67.25 |
| Sum of two sides 1100mm | 34.58 | 43.74 | 1.88 | 29.82 | nr | 73.56 |
| Sum of two sides 1200mm | 39.92 | 50.50 | 1.88 | 29.82 | nr | 80.32 |
| Sum of two sides 1300mm | 45.71 | 57.82 | 2.26 | 35.84 | nr | 93.66 |
| Sum of two sides 1400mm | 51.92 | 65.68 | 2.26 | 35.84 | nr | 101.52 |
| Sum of two sides 1500mm | 58.51 | 74.02 | 2.49 | 39.49 | nr | 113.51 |
| Sum of two sides 1600mm | 65.49 | 82.84 | 2.49 | 39.49 | nr | 122.33 |
| Sum of two sides 1700mm | 77.05 | 97.47 | 2.67 | 42.35 | nr | 139.82 |
| Sum of two sides 1800mm | 85.06 | 107.60 | 2.67 | 42.35 | nr | 149.95 |
| Sum of two sides 1900mm | 93.43 | 118.19 | 3.06 | 48.53 | nr | 166.72 |
| Sum of two sides 2000mm | 102.22 | 129.31 | 3.06 | 48.53 | nr | 177.84 |
| Sum of two sides 2100mm | 120.00 | 151.80 | 4.05 | 64.23 | nr | 216.03 |
| Sum of two sides 2200mm | 130.49 | 165.07 | 4.05 | 64.23 | nr | 229.30 |
| Sum of two sides 2300mm | 147.63 | 186.75 | 4.39 | 69.63 | nr | 256.38 |
| Sum of two sides 2400mm | 159.17 | 201.35 | 4.85 | 76.92 | nr | 278.27 |
| Sum of two sides 2500mm | 171.17 | 216.53 | 4.85 | 76.92 | nr | 293.45 |
| Sum of two sides 2600mm | 183.61 | 232.27 | 4.87 | 77.24 | nr | 309.51 |
| Sum of two sides 2700mm | 196.47 | 248.53 | 4.87 | 77.24 | nr | 325.77 |
| Sum of two sides 2800mm | 209.77 | 265.36 | 8.81 | 139.73 | nr | 405.09 |
| Sum of two sides 2900mm | 223.53 | 282.77 | 8.81 | 139.73 | nr | 422.50 |
| Sum of two sides 3000mm | 245.97 | 311.15 | 9.31 | 147.66 | nr | 458.81 |
| Sum of two sides 3100mm | 260.88 | 330.01 | 9.31 | 147.66 | nr | 477.67 |
| Sum of two sides 3200mm | 276.20 | 349.39 | 9.39 | 148.93 | nr | 498.32 |
| Equal twin square bend with turning vanes | | | | | | |
| Sum of two sides 200mm | 6.42 | 8.12 | 2.09 | 33.15 | nr | 41.27 |
| Sum of two sides 250mm | 7.67 | 9.70 | 2.09 | 33.15 | nr | 42.85 |
| Sum of two sides 300mm | 9.16 | 11.59 | 2.09 | 33.15 | nr | 44.74 |
| Sum of two sides 350mm | 10.82 | 13.69 | 2.04 | 32.35 | nr | 46.04 |
| Sum of two sides 400mm | 12.68 | 16.04 | 2.04 | 32.35 | nr | 48.39 |
| Sum of two sides 450mm | 14.73 | 18.63 | 2.04 | 32.35 | nr | 50.98 |
| Sum of two sides 500mm | 16.90 | 21.38 | 2.04 | 32.35 | nr | 53.73 |
| Sum of two sides 550mm | 19.29 | 24.40 | 2.15 | 34.10 | nr | 58.50 |
| Sum of two sides 600mm | 21.94 | 27.75 | 2.15 | 34.10 | nr | 61.85 |
| Sum of two sides 650mm | 24.69 | 31.23 | 2.16 | 34.26 | nr | 65.49 |
| Sum of two sides 700mm | 27.69 | 35.03 | 2.16 | 34.26 | nr | 69.29 |
| Sum of two sides 750mm | 30.81 | 38.97 | 2.26 | 35.84 | nr | 74.81 |
| Sum of two sides 800mm | 34.15 | 43.20 | 2.26 | 35.84 | nr | 79.04 |
| Sum of two sides 850mm | 37.63 | 47.60 | 2.26 | 35.84 | nr | 83.44 |
| Sum of two sides 900mm | 41.33 | 52.28 | 2.26 | 35.84 | nr | 88.12 |
| Sum of two sides 950mm | 45.21 | 57.19 | 3.01 | 47.74 | nr | 104.93 |
| Sum of two sides 1000mm | 49.32 | 62.39 | 3.01 | 47.74 | nr | 110.13 |
| Sum of two sides 1100mm | 57.99 | 73.36 | 3.01 | 47.74 | nr | 121.10 |
| Sum of two sides 1200mm | 67.44 | 85.31 | 3.01 | 47.74 | nr | 133.05 |

## Y:MECHANICAL AND ELECTRICAL SERVICES

| Item | Net Price £ | Material £ | Labour hours | Labour £ | Unit | Total Rate £ |
|---|---|---|---|---|---|---|
| Sum of two sides 1300mm | 77.65 | 98.23 | 3.67 | 58.21 | nr | 156.44 |
| Sum of two sides 1400mm | 88.57 | 112.04 | 3.67 | 58.21 | nr | 170.25 |
| Sum of two sides 1500mm | 100.19 | 126.74 | 4.07 | 64.55 | nr | 191.29 |
| Sum of two sides 1600mm | 112.65 | 142.50 | 4.07 | 64.55 | nr | 207.05 |
| Sum of two sides 1700mm | 131.96 | 166.93 | 2.80 | 44.41 | nr | 211.34 |
| Sum of two sides 1800mm | 146.09 | 184.80 | 2.67 | 42.35 | nr | 227.15 |
| Sum of two sides 1900mm | 161.02 | 203.69 | 2.95 | 46.79 | nr | 250.48 |
| Sum of two sides 2000mm | 176.60 | 223.40 | 2.95 | 46.79 | nr | 270.19 |
| Sum of two sides 2100mm | 210.83 | 266.70 | 4.05 | 64.23 | nr | 330.93 |
| Sum of two sides 2200mm | 229.40 | 290.19 | 4.05 | 64.23 | nr | 354.42 |
| Sum of two sides 2300mm | 248.73 | 314.64 | 4.39 | 69.63 | nr | 384.27 |
| Sum of two sides 2400mm | 268.80 | 340.03 | 4.85 | 76.92 | nr | 416.95 |
| Sum of two sides 2500mm | 289.80 | 366.60 | 4.85 | 76.92 | nr | 443.52 |
| Sum of two sides 2600mm | 318.59 | 403.02 | 4.87 | 77.24 | nr | 480.26 |
| Sum of two sides 2700mm | 341.39 | 431.86 | 4.87 | 77.24 | nr | 509.10 |
| Sum of two sides 2800mm | 364.90 | 461.60 | 8.81 | 139.73 | nr | 601.33 |
| Sum of two sides 2900mm | 389.21 | 492.35 | 14.81 | 234.89 | nr | 727.24 |
| Sum of two sides 3000mm | 414.40 | 524.22 | 15.20 | 241.08 | nr | 765.30 |
| Sum of two sides 3100mm | 448.81 | 567.74 | 15.60 | 247.42 | nr | 815.16 |
| Sum of two sides 3200mm | 475.87 | 601.98 | 15.60 | 247.42 | nr | 849.40 |
| **Spigot branch** | | | | | | |
| Sum of two sides 200mm | 0.29 | 0.37 | 0.95 | 15.07 | nr | 15.44 |
| Sum of two sides 250mm | 0.34 | 0.43 | 0.95 | 15.07 | nr | 15.50 |
| Sum of two sides 300mm | 0.42 | 0.53 | 0.95 | 15.07 | nr | 15.60 |
| Sum of two sides 350mm | 0.48 | 0.61 | 0.92 | 14.59 | nr | 15.20 |
| Sum of two sides 400mm | 0.56 | 0.71 | 0.92 | 14.59 | nr | 15.30 |
| Sum of two sides 450mm | 0.61 | 0.77 | 0.92 | 14.59 | nr | 15.36 |
| Sum of two sides 500mm | 0.66 | 0.83 | 0.92 | 14.59 | nr | 15.42 |
| Sum of two sides 550mm | 0.75 | 0.95 | 1.03 | 16.34 | nr | 17.29 |
| Sum of two sides 600mm | 0.81 | 1.02 | 1.03 | 16.34 | nr | 17.36 |
| Sum of two sides 650mm | 0.90 | 1.14 | 1.03 | 16.34 | nr | 17.48 |
| Sum of two sides 700mm | 0.95 | 1.20 | 1.03 | 16.34 | nr | 17.54 |
| Sum of two sides 750mm | 1.04 | 1.32 | 1.03 | 16.34 | nr | 17.66 |
| Sum of two sides 800mm | 1.12 | 1.42 | 1.03 | 16.34 | nr | 17.76 |
| Sum of two sides 850mm | 1.17 | 1.48 | 1.03 | 16.34 | nr | 17.82 |
| Sum of two sides 900mm | 1.25 | 1.58 | 1.03 | 16.34 | nr | 17.92 |
| Sum of two sides 950mm | 1.31 | 1.66 | 1.29 | 20.46 | nr | 22.12 |
| Sum of two sides 1000mm | 1.36 | 1.72 | 1.29 | 20.46 | nr | 22.18 |
| Sum of two sides 1100mm | 1.52 | 1.92 | 1.29 | 20.46 | nr | 22.38 |
| Sum of two sides 1200mm | 1.65 | 2.09 | 1.29 | 20.46 | nr | 22.55 |
| Sum of two sides 1300mm | 1.81 | 2.29 | 1.39 | 22.05 | nr | 24.34 |
| Sum of two sides 1400mm | 1.92 | 2.43 | 1.39 | 22.05 | nr | 24.48 |
| Sum of two sides 1500mm | 2.04 | 2.58 | 1.64 | 26.01 | nr | 28.59 |
| Sum of two sides 1600mm | 2.20 | 2.78 | 1.64 | 26.01 | nr | 28.79 |
| Sum of two sides 1700mm | 2.34 | 2.96 | 1.69 | 26.80 | nr | 29.76 |
| Sum of two sides 1800mm | 2.46 | 3.11 | 1.69 | 26.80 | nr | 29.91 |
| Sum of two sides 1900mm | 2.61 | 3.30 | 1.85 | 29.34 | nr | 32.64 |
| Sum of two sides 2000mm | 2.74 | 3.47 | 1.85 | 29.34 | nr | 32.81 |
| Sum of two sides 2100mm | 3.63 | 4.59 | 2.61 | 41.40 | nr | 45.99 |
| Sum of two sides 2200mm | 3.81 | 4.82 | 2.61 | 41.40 | nr | 46.22 |
| Sum of two sides 2300mm | 3.93 | 4.97 | 2.61 | 41.40 | nr | 46.37 |
| Sum of two sides 2400mm | 4.11 | 5.20 | 2.88 | 45.68 | nr | 50.88 |
| Sum of two sides 2500mm | 4.30 | 5.44 | 2.88 | 45.68 | nr | 51.12 |
| Sum of two sides 2600mm | 4.30 | 5.44 | 2.88 | 45.68 | nr | 51.12 |
| Sum of two sides 2700mm | 4.67 | 5.91 | 2.88 | 45.68 | nr | 51.59 |
| Sum of two sides 2800mm | 4.84 | 6.12 | 3.94 | 62.49 | nr | 68.61 |
| Sum of two sides 2900mm | 4.99 | 6.31 | 3.94 | 62.49 | nr | 68.80 |
| Sum of two sides 3000mm | 5.17 | 6.54 | 4.83 | 76.61 | nr | 83.15 |

## Y:MECHANICAL AND ELECTRICAL SERVICES

| Item | Net Price £ | Material £ | Labour hours | Labour £ | Unit | Total Rate £ |
|---|---|---|---|---|---|---|
| **Y30 : AIR DUCTLINES : MILD STEEL DUCTWORK (contd)** | | | | | | |
| **Extra Over galvanised sheet metal DW142 class A rectangular section ductwork; fittings; including all necessary stiffeners, joints etc. (contd)** | | | | | | |
| Spigot branch (contd) | | | | | | |
| Sum of two sides 3100mm | 5.32 | 6.73 | 4.83 | 76.61 | nr | 83.34 |
| Sum of two sides 3200mm | 5.51 | 6.97 | 4.83 | 76.61 | nr | 83.58 |
| Branch; based on branch size | | | | | | |
| Sum of two sides 200mm | 2.38 | 3.01 | 1.10 | 17.45 | nr | 20.46 |
| Sum of two sides 250mm | 2.65 | 3.35 | 1.10 | 17.45 | nr | 20.80 |
| Sum of two sides 300mm | 3.06 | 3.87 | 1.10 | 17.45 | nr | 21.32 |
| Sum of two sides 350mm | 3.38 | 4.28 | 1.16 | 18.40 | nr | 22.68 |
| Sum of two sides 400mm | 3.72 | 4.71 | 1.16 | 18.40 | nr | 23.11 |
| Sum of two sides 450mm | 4.06 | 5.14 | 1.16 | 18.40 | nr | 23.54 |
| Sum of two sides 500mm | 4.38 | 5.54 | 1.16 | 18.40 | nr | 23.94 |
| Sum of two sides 550mm | 4.73 | 5.98 | 1.18 | 18.72 | nr | 24.70 |
| Sum of two sides 600mm | 5.06 | 6.40 | 1.18 | 18.72 | nr | 25.12 |
| Sum of two sides 650mm | 5.45 | 6.89 | 1.18 | 18.72 | nr | 25.61 |
| Sum of two sides 700mm | 5.75 | 7.27 | 1.18 | 18.72 | nr | 25.99 |
| Sum of two sides 750mm | 6.07 | 7.68 | 1.18 | 18.72 | nr | 26.40 |
| Sum of two sides 800mm | 6.42 | 8.12 | 1.18 | 18.72 | nr | 26.84 |
| Sum of two sides 850mm | 6.77 | 8.56 | 1.18 | 18.72 | nr | 27.28 |
| Sum of two sides 900mm | 7.08 | 8.96 | 1.18 | 18.72 | nr | 27.68 |
| Sum of two sides 950mm | 7.42 | 9.39 | 1.44 | 22.84 | nr | 32.23 |
| Sum of two sides 1000mm | 7.74 | 9.79 | 1.44 | 22.84 | nr | 32.63 |
| Sum of two sides 1100mm | 8.44 | 10.68 | 1.44 | 22.84 | nr | 33.52 |
| Sum of two sides 1200mm | 9.10 | 11.51 | 1.44 | 22.84 | nr | 34.35 |
| Sum of two sides 1300mm | 9.78 | 12.37 | 1.69 | 26.80 | nr | 39.17 |
| Sum of two sides 1400mm | 10.45 | 13.22 | 1.69 | 26.80 | nr | 40.02 |
| Sum of two sides 1500mm | 11.13 | 14.08 | 1.79 | 28.39 | nr | 42.47 |
| Sum of two sides 1600mm | 11.78 | 14.90 | 1.79 | 28.39 | nr | 43.29 |
| Sum of two sides 1700mm | 14.51 | 18.36 | 1.86 | 29.50 | nr | 47.86 |
| Sum of two sides 1800mm | 15.29 | 19.34 | 2.02 | 32.04 | nr | 51.38 |
| Sum of two sides 1900mm | 16.03 | 20.28 | 2.02 | 32.04 | nr | 52.32 |
| Sum of two sides 2000mm | 16.77 | 21.21 | 2.02 | 32.04 | nr | 53.25 |
| Sum of two sides 2100mm | 19.71 | 24.93 | 2.61 | 41.40 | nr | 66.33 |
| Sum of two sides 2200mm | 20.58 | 26.03 | 2.61 | 41.40 | nr | 67.43 |
| Sum of two sides 2300mm | 21.42 | 27.10 | 2.61 | 41.40 | nr | 68.50 |
| Sum of two sides 2400mm | 22.29 | 28.20 | 2.88 | 45.68 | nr | 73.88 |
| Sum of two sides 2500mm | 23.15 | 29.28 | 2.88 | 45.68 | nr | 74.96 |
| Sum of two sides 2600mm | 24.01 | 30.37 | 2.88 | 45.68 | nr | 76.05 |
| Sum of two sides 2700mm | 24.86 | 31.45 | 2.88 | 45.68 | nr | 77.13 |
| Sum of two sides 2800mm | 25.75 | 32.57 | 3.94 | 62.49 | nr | 95.06 |
| Sum of two sides 2900mm | 26.59 | 33.64 | 4.12 | 65.34 | nr | 98.98 |
| Sum of two sides 3000mm | 27.47 | 34.75 | 5.00 | 79.30 | nr | 114.05 |
| Sum of two sides 3100mm | 28.31 | 35.81 | 5.00 | 79.30 | nr | 115.11 |
| Sum of two sides 3200mm | 29.16 | 36.89 | 5.00 | 79.30 | nr | 116.19 |

## Y:MECHANICAL AND ELECTRICAL SERVICES

| Item | Net Price £ | Material £ | Labour hours | Labour £ | Unit | Total Rate £ |
|---|---|---|---|---|---|---|
| **Galvanised sheet metal DW142 class C rectangular section ductwork; including all necessary stiffeners, joints, couplers in the running length and duct supports** | | | | | | |
| Straight duct | | | | | | |
| Sum of two sides 200mm | 4.54 | 5.74 | 1.19 | 18.87 | m | 24.61 |
| Sum of two sides 250mm | 5.21 | 6.59 | 1.19 | 18.87 | m | 25.46 |
| Sum of two sides 300mm | 5.85 | 7.40 | 1.19 | 18.87 | m | 26.27 |
| Sum of two sides 350mm | 6.56 | 8.30 | 1.16 | 18.40 | m | 26.70 |
| Sum of two sides 400mm | 7.21 | 9.12 | 1.16 | 18.40 | m | 27.52 |
| Sum of two sides 450mm | 7.89 | 9.98 | 1.16 | 18.40 | m | 28.38 |
| Sum of two sides 500mm | 8.54 | 10.80 | 1.16 | 18.40 | m | 29.20 |
| Sum of two sides 550mm | 9.23 | 11.68 | 1.17 | 18.56 | m | 30.24 |
| Sum of two sides 600mm | 9.89 | 12.51 | 1.17 | 18.56 | m | 31.07 |
| Sum of two sides 650mm | 10.57 | 13.37 | 1.17 | 18.56 | m | 31.93 |
| Sum of two sides 700mm | 11.22 | 14.19 | 1.17 | 18.56 | m | 32.75 |
| Sum of two sides 750mm | 11.92 | 15.08 | 1.17 | 18.56 | m | 33.64 |
| Sum of two sides 800mm | 12.56 | 15.89 | 1.17 | 18.56 | m | 34.45 |
| Sum of two sides 850mm | 14.53 | 18.38 | 1.26 | 19.98 | m | 38.36 |
| Sum of two sides 900mm | 15.28 | 19.33 | 1.27 | 20.14 | m | 39.47 |
| Sum of two sides 950mm | 16.02 | 20.27 | 1.48 | 23.47 | m | 43.74 |
| Sum of two sides 1000mm | 16.73 | 21.16 | 1.48 | 23.47 | m | 44.63 |
| Sum of two sides 1100mm | 18.20 | 23.02 | 1.49 | 23.63 | m | 46.65 |
| Sum of two sides 1200mm | 19.68 | 24.90 | 1.49 | 23.63 | m | 48.53 |
| Sum of two sides 1300mm | 21.14 | 26.74 | 1.51 | 23.95 | m | 50.69 |
| Sum of two sides 1400mm | 22.63 | 28.63 | 1.55 | 24.58 | m | 53.21 |
| Sum of two sides 1500mm | 24.09 | 30.47 | 1.61 | 25.54 | m | 56.01 |
| Sum of two sides 1600mm | 25.58 | 32.36 | 1.62 | 25.69 | m | 58.05 |
| Sum of two sides 1700mm | 29.77 | 37.66 | 1.74 | 27.60 | m | 65.26 |
| Sum of two sides 1800mm | 31.34 | 39.65 | 1.76 | 27.91 | m | 67.56 |
| Sum of two sides 1900mm | 32.93 | 41.66 | 1.81 | 28.71 | m | 70.37 |
| Sum of two sides 2000mm | 34.49 | 43.63 | 1.82 | 28.87 | m | 72.50 |
| Sum of two sides 2100mm | 46.33 | 58.61 | 2.53 | 40.13 | m | 98.74 |
| Sum of two sides 2200mm | 48.39 | 61.21 | 2.55 | 40.44 | m | 101.65 |
| Sum of two sides 2300mm | 50.47 | 63.84 | 2.56 | 40.60 | m | 104.44 |
| Sum of two sides 2400mm | 52.53 | 66.45 | 2.76 | 43.77 | m | 110.22 |
| Sum of two sides 2500mm | 54.61 | 69.08 | 2.77 | 43.93 | m | 113.01 |
| Sum of two sides 2600mm | 69.32 | 87.69 | 2.97 | 47.11 | m | 134.80 |
| Sum of two sides 2700mm | 71.73 | 90.74 | 2.99 | 47.42 | m | 138.16 |
| Sum of two sides 2800mm | 74.15 | 93.80 | 3.30 | 52.34 | m | 146.14 |
| Sum of two sides 2900mm | 76.67 | 96.99 | 3.31 | 52.50 | m | 149.49 |
| Sum of two sides 3000mm | 79.20 | 100.19 | 3.53 | 55.99 | m | 156.18 |
| Sum of two sides 3100mm | 81.69 | 103.34 | 3.55 | 56.30 | m | 159.64 |
| Sum of two sides 3200mm | 84.21 | 106.53 | 3.56 | 56.46 | m | 162.99 |
| **Extra Over galvanised sheet metal DW142 class C rectangular section ductwork; fittings; including all necessary stiffeners, joints etc.** | | | | | | |
| Stopped end | | | | | | |
| Sum of two sides 200mm | 0.10 | 0.13 | 0.38 | 6.03 | nr | 6.16 |
| Sum of two sides 250mm | 0.16 | 0.20 | 0.38 | 6.03 | nr | 6.23 |
| Sum of two sides 300mm | 0.16 | 0.20 | 0.38 | 6.03 | nr | 6.23 |
| Sum of two sides 350mm | 0.21 | 0.27 | 0.38 | 6.03 | nr | 6.30 |
| Sum of two sides 400mm | 0.21 | 0.27 | 0.38 | 6.03 | nr | 6.30 |
| Sum of two sides 450mm | 0.25 | 0.32 | 0.38 | 6.03 | nr | 6.35 |

## Y:MECHANICAL AND ELECTRICAL SERVICES

| Item | Net Price £ | Material £ | Labour hours | Labour £ | Unit | Total Rate £ |
|---|---|---|---|---|---|---|
| **Y30 : AIR DUCTLINES : MILD STEEL DUCTWORK (contd)** | | | | | | |
| **Extra Over galvanised sheet metal DW142 class C rectangular section ductwork; fittings; including all necessary stiffeners, joints etc. (contd)** | | | | | | |
| Stopped end | | | | | | |
| Sum of two sides 500mm | 0.25 | 0.32 | 0.38 | 6.03 | nr | 6.35 |
| Sum of two sides 550mm | 0.25 | 0.32 | 0.38 | 6.03 | nr | 6.35 |
| Sum of two sides 600mm | 0.31 | 0.39 | 0.38 | 6.03 | nr | 6.42 |
| Sum of two sides 650mm | 0.31 | 0.39 | 0.38 | 6.03 | nr | 6.42 |
| Sum of two sides 700mm | 0.35 | 0.44 | 0.38 | 6.03 | nr | 6.47 |
| Sum of two sides 750mm | 0.42 | 0.53 | 0.38 | 6.03 | nr | 6.56 |
| Sum of two sides 800mm | 0.42 | 0.53 | 0.38 | 6.03 | nr | 6.56 |
| Sum of two sides 850mm | 0.45 | 0.57 | 0.38 | 6.03 | nr | 6.60 |
| Sum of two sides 900mm | 0.45 | 0.57 | 0.38 | 6.03 | nr | 6.60 |
| Sum of two sides 950mm | 0.45 | 0.57 | 0.38 | 6.03 | nr | 6.60 |
| Sum of two sides 1000mm | 0.51 | 0.65 | 0.38 | 6.03 | nr | 6.68 |
| Sum of two sides 1100mm | 0.56 | 0.71 | 0.38 | 6.03 | nr | 6.74 |
| Sum of two sides 1200mm | 0.60 | 0.76 | 0.38 | 6.03 | nr | 6.79 |
| Sum of two sides 1300mm | 0.66 | 0.83 | 0.38 | 6.03 | nr | 6.86 |
| Sum of two sides 1400mm | 0.72 | 0.91 | 0.38 | 6.03 | nr | 6.94 |
| Sum of two sides 1500mm | 0.77 | 0.97 | 0.38 | 6.03 | nr | 7.00 |
| Sum of two sides 1600mm | 0.81 | 1.02 | 0.38 | 6.03 | nr | 7.05 |
| Sum of two sides 1700mm | 0.87 | 1.10 | 0.58 | 9.20 | nr | 10.30 |
| Sum of two sides 1800mm | 0.91 | 1.15 | 0.58 | 9.20 | nr | 10.35 |
| Sum of two sides 1900mm | 0.97 | 1.23 | 0.58 | 9.20 | nr | 10.43 |
| Sum of two sides 2000mm | 1.03 | 1.30 | 0.58 | 9.20 | nr | 10.50 |
| Sum of two sides 2100mm | 1.34 | 1.70 | 0.87 | 13.80 | nr | 15.50 |
| Sum of two sides 2200mm | 1.40 | 1.77 | 0.87 | 13.80 | nr | 15.57 |
| Sum of two sides 2300mm | 1.46 | 1.85 | 0.87 | 13.80 | nr | 15.65 |
| Sum of two sides 2400mm | 1.52 | 1.92 | 0.87 | 13.80 | nr | 15.72 |
| Sum of two sides 2500mm | 1.60 | 2.02 | 0.87 | 13.80 | nr | 15.82 |
| Sum of two sides 2600mm | 1.66 | 2.10 | 0.87 | 13.80 | nr | 15.90 |
| Sum of two sides 2700mm | 1.72 | 2.18 | 0.87 | 13.80 | nr | 15.98 |
| Sum of two sides 2800mm | 1.81 | 2.29 | 1.16 | 18.40 | nr | 20.69 |
| Sum of two sides 2900mm | 2.50 | 3.16 | 1.16 | 18.40 | nr | 21.56 |
| Sum of two sides 3000mm | 2.59 | 3.28 | 1.73 | 27.44 | nr | 30.72 |
| Sum of two sides 3100mm | 2.69 | 3.40 | 1.73 | 27.44 | nr | 30.84 |
| Sum of two sides 3200mm | 2.74 | 3.47 | 1.73 | 27.44 | nr | 30.91 |
| Taper; large end measured | | | | | | |
| Sum of two sides 200mm | 3.31 | 4.19 | 1.42 | 22.52 | nr | 26.71 |
| Sum of two sides 250mm | 3.76 | 4.76 | 1.42 | 22.52 | nr | 27.28 |
| Sum of two sides 300mm | 4.20 | 5.31 | 1.42 | 22.52 | nr | 27.83 |
| Sum of two sides 350mm | 4.67 | 5.91 | 1.40 | 22.20 | nr | 28.11 |
| Sum of two sides 400mm | 5.11 | 6.46 | 1.40 | 22.20 | nr | 28.66 |
| Sum of two sides 450mm | 5.54 | 7.01 | 1.40 | 22.20 | nr | 29.21 |
| Sum of two sides 500mm | 6.02 | 7.62 | 1.40 | 22.20 | nr | 29.82 |
| Sum of two sides 550mm | 6.46 | 8.17 | 1.69 | 26.80 | nr | 34.97 |
| Sum of two sides 600mm | 6.88 | 8.70 | 1.69 | 26.80 | nr | 35.50 |
| Sum of two sides 650mm | 7.35 | 9.30 | 1.69 | 26.80 | nr | 36.10 |
| Sum of two sides 700mm | 7.80 | 9.87 | 1.69 | 26.80 | nr | 36.67 |
| Sum of two sides 750mm | 8.24 | 10.42 | 1.92 | 30.45 | nr | 40.87 |
| Sum of two sides 800mm | 8.68 | 10.98 | 1.92 | 30.45 | nr | 41.43 |
| Sum of two sides 850mm | 9.13 | 11.55 | 1.92 | 30.45 | nr | 42.00 |
| Sum of two sides 900mm | 9.58 | 12.12 | 1.92 | 30.45 | nr | 42.57 |

*Material Costs / Measured Work Prices - Mechanical Installations*

## Y:MECHANICAL AND ELECTRICAL SERVICES

| Item | Net Price £ | Material £ | Labour hours | Labour £ | Unit | Total Rate £ |
|---|---|---|---|---|---|---|
| Sum of two sides 950mm | 10.02 | 12.68 | 2.18 | 34.58 | nr | 47.26 |
| Sum of two sides 1000mm | 10.47 | 13.24 | 2.18 | 34.58 | nr | 47.82 |
| Sum of two sides 1100mm | 11.38 | 14.40 | 2.18 | 34.58 | nr | 48.98 |
| Sum of two sides 1200mm | 12.25 | 15.50 | 2.18 | 34.58 | nr | 50.08 |
| Sum of two sides 1300mm | 13.16 | 16.65 | 2.30 | 36.48 | nr | 53.13 |
| Sum of two sides 1400mm | 14.03 | 17.75 | 2.30 | 36.48 | nr | 54.23 |
| Sum of two sides 1500mm | 14.95 | 18.91 | 2.47 | 39.17 | nr | 58.08 |
| Sum of two sides 1600mm | 15.83 | 20.02 | 2.47 | 39.17 | nr | 59.19 |
| Sum of two sides 1700mm | 19.77 | 25.01 | 2.59 | 41.08 | nr | 66.09 |
| Sum of two sides 1800mm | 20.79 | 26.30 | 2.59 | 41.08 | nr | 67.38 |
| Sum of two sides 1900mm | 21.81 | 27.59 | 2.71 | 42.98 | nr | 70.57 |
| Sum of two sides 2000mm | 22.83 | 28.88 | 2.71 | 42.98 | nr | 71.86 |
| Sum of two sides 2100mm | 26.53 | 33.56 | 2.92 | 46.31 | nr | 79.87 |
| Sum of two sides 2200mm | 27.66 | 34.99 | 2.92 | 46.31 | nr | 81.30 |
| Sum of two sides 2300mm | 28.82 | 36.46 | 2.92 | 46.31 | nr | 82.77 |
| Sum of two sides 2400mm | 29.96 | 37.90 | 3.12 | 49.48 | nr | 87.38 |
| Sum of two sides 2500mm | 31.11 | 39.35 | 3.12 | 49.48 | nr | 88.83 |
| Sum of two sides 2600mm | 47.72 | 60.37 | 3.16 | 50.12 | nr | 110.49 |
| Sum of two sides 2700mm | 49.26 | 62.31 | 3.16 | 50.12 | nr | 112.43 |
| Sum of two sides 2800mm | 50.80 | 64.26 | 4.00 | 63.44 | nr | 127.70 |
| Sum of two sides 2900mm | 52.51 | 66.43 | 4.01 | 63.60 | nr | 130.03 |
| Sum of two sides 3000mm | 54.16 | 68.51 | 4.56 | 72.32 | nr | 140.83 |
| Sum of two sides 3100mm | 55.85 | 70.65 | 4.56 | 72.32 | nr | 142.97 |
| Sum of two sides 3200mm | 57.56 | 72.81 | 4.56 | 72.32 | nr | 145.13 |
| Set piece | | | | | | |
| Sum of two sides 200mm | 2.91 | 3.68 | 1.65 | 26.17 | nr | 29.85 |
| Sum of two sides 250mm | 3.30 | 4.17 | 1.65 | 26.17 | nr | 30.34 |
| Sum of two sides 300mm | 3.74 | 4.73 | 1.65 | 26.17 | nr | 30.90 |
| Sum of two sides 350mm | 4.20 | 5.31 | 1.63 | 25.85 | nr | 31.16 |
| Sum of two sides 400mm | 4.68 | 5.92 | 1.63 | 25.85 | nr | 31.77 |
| Sum of two sides 450mm | 5.19 | 6.57 | 1.63 | 25.85 | nr | 32.42 |
| Sum of two sides 500mm | 5.73 | 7.25 | 1.63 | 25.85 | nr | 33.10 |
| Sum of two sides 550mm | 6.28 | 7.94 | 1.92 | 30.45 | nr | 38.39 |
| Sum of two sides 600mm | 6.88 | 8.70 | 1.92 | 30.45 | nr | 39.15 |
| Sum of two sides 650mm | 7.51 | 9.50 | 1.92 | 30.45 | nr | 39.95 |
| Sum of two sides 700mm | 8.15 | 10.31 | 1.92 | 30.45 | nr | 40.76 |
| Sum of two sides 750mm | 8.80 | 11.13 | 1.92 | 30.45 | nr | 41.58 |
| Sum of two sides 800mm | 9.50 | 12.02 | 1.92 | 30.45 | nr | 42.47 |
| Sum of two sides 850mm | 10.19 | 12.89 | 1.92 | 30.45 | nr | 43.34 |
| Sum of two sides 900mm | 10.95 | 13.85 | 1.92 | 30.45 | nr | 44.30 |
| Sum of two sides 950mm | 11.71 | 14.81 | 2.18 | 34.58 | nr | 49.39 |
| Sum of two sides 1000mm | 12.51 | 15.83 | 2.18 | 34.58 | nr | 50.41 |
| Sum of two sides 1100mm | 14.18 | 17.94 | 2.18 | 34.58 | nr | 52.52 |
| Sum of two sides 1200mm | 15.93 | 20.15 | 2.18 | 34.58 | nr | 54.73 |
| Sum of two sides 1300mm | 20.11 | 25.44 | 2.47 | 39.17 | nr | 64.61 |
| Sum of two sides 1400mm | 22.26 | 28.16 | 2.47 | 39.17 | nr | 67.33 |
| Sum of two sides 1500mm | 24.49 | 30.98 | 2.47 | 39.17 | nr | 70.15 |
| Sum of two sides 1600mm | 26.84 | 33.95 | 2.47 | 39.17 | nr | 73.12 |
| Sum of two sides 1700mm | 32.79 | 41.48 | 2.61 | 41.40 | nr | 82.88 |
| Sum of two sides 1800mm | 35.49 | 44.89 | 2.61 | 41.40 | nr | 86.29 |
| Sum of two sides 1900mm | 38.28 | 48.42 | 2.71 | 42.98 | nr | 91.40 |
| Sum of two sides 2000mm | 41.21 | 52.13 | 2.71 | 42.98 | nr | 95.11 |
| Sum of two sides 2100mm | 52.47 | 66.37 | 2.92 | 46.31 | nr | 112.68 |
| Sum of two sides 2200mm | 55.59 | 70.32 | 3.26 | 51.70 | nr | 122.02 |
| Sum of two sides 2300mm | 60.14 | 76.08 | 3.26 | 51.70 | nr | 127.78 |
| Sum of two sides 2400mm | 64.18 | 81.19 | 3.47 | 55.04 | nr | 136.23 |
| Sum of two sides 2500mm | 68.34 | 86.45 | 3.48 | 55.19 | nr | 141.64 |
| Sum of two sides 2600mm | 89.43 | 113.13 | 3.49 | 55.35 | nr | 168.48 |
| Sum of two sides 2700mm | 94.28 | 119.26 | 3.50 | 55.51 | nr | 174.77 |

## Y:MECHANICAL AND ELECTRICAL SERVICES

| Item | Net Price £ | Material £ | Labour hours | Labour £ | Unit | Total Rate £ |
|---|---|---|---|---|---|---|
| **Y30 : AIR DUCTLINES : MILD STEEL DUCTWORK (contd)** | | | | | | |
| **Extra Over galvanised sheet metal DW142 class C rectangular section ductwork; fittings; including all necessary stiffeners, joints etc. (contd)** | | | | | | |
| Set piece (contd) | | | | | | |
| Sum of two sides 2800mm | 99.29 | 125.60 | 4.33 | 68.68 | nr | **194.28** |
| Sum of two sides 2900mm | 112.40 | 142.19 | 4.74 | 75.18 | nr | **217.37** |
| Sum of two sides 3000mm | 118.56 | 149.98 | 5.31 | 84.22 | nr | **234.20** |
| Sum of two sides 3100mm | 124.90 | 158.00 | 5.34 | 84.69 | nr | **242.69** |
| Sum of two sides 3200mm | 131.38 | 166.20 | 5.35 | 84.85 | nr | **251.05** |
| 90 degree radius bend | | | | | | |
| Sum of two sides 200mm | 2.91 | 3.68 | 1.25 | 19.83 | nr | **23.51** |
| Sum of two sides 250mm | 3.30 | 4.17 | 1.25 | 19.83 | nr | **24.00** |
| Sum of two sides 300mm | 3.74 | 4.73 | 1.25 | 19.83 | nr | **24.56** |
| Sum of two sides 350mm | 4.20 | 5.31 | 1.22 | 19.35 | nr | **24.66** |
| Sum of two sides 400mm | 4.68 | 5.92 | 1.22 | 19.35 | nr | **25.27** |
| Sum of two sides 450mm | 5.19 | 6.57 | 1.22 | 19.35 | nr | **25.92** |
| Sum of two sides 500mm | 5.73 | 7.25 | 1.22 | 19.35 | nr | **26.60** |
| Sum of two sides 550mm | 6.28 | 7.94 | 1.33 | 21.09 | nr | **29.03** |
| Sum of two sides 600mm | 6.88 | 8.70 | 1.33 | 21.09 | nr | **29.79** |
| Sum of two sides 650mm | 7.51 | 9.50 | 1.33 | 21.09 | nr | **30.59** |
| Sum of two sides 700mm | 8.15 | 10.31 | 1.33 | 21.09 | nr | **31.40** |
| Sum of two sides 750mm | 8.80 | 11.13 | 1.40 | 22.20 | nr | **33.33** |
| Sum of two sides 800mm | 9.50 | 12.02 | 1.40 | 22.20 | nr | **34.22** |
| Sum of two sides 850mm | 10.19 | 12.89 | 1.40 | 22.20 | nr | **35.09** |
| Sum of two sides 900mm | 10.95 | 13.85 | 1.40 | 22.20 | nr | **36.05** |
| Sum of two sides 950mm | 11.71 | 14.81 | 1.82 | 28.87 | nr | **43.68** |
| Sum of two sides 1000mm | 12.51 | 15.83 | 1.82 | 28.87 | nr | **44.70** |
| Sum of two sides 1100mm | 14.18 | 17.94 | 1.82 | 28.87 | nr | **46.81** |
| Sum of two sides 1200mm | 15.93 | 20.15 | 1.82 | 28.87 | nr | **49.02** |
| Sum of two sides 1300mm | 20.11 | 25.44 | 2.32 | 36.80 | nr | **62.24** |
| Sum of two sides 1400mm | 22.26 | 28.16 | 2.32 | 36.80 | nr | **64.96** |
| Sum of two sides 1500mm | 24.49 | 30.98 | 2.56 | 40.60 | nr | **71.58** |
| Sum of two sides 1600mm | 26.84 | 33.95 | 2.58 | 40.92 | nr | **74.87** |
| Sum of two sides 1700mm | 32.79 | 41.48 | 2.84 | 45.04 | nr | **86.52** |
| Sum of two sides 1800mm | 35.49 | 44.89 | 2.84 | 45.04 | nr | **89.93** |
| Sum of two sides 1900mm | 38.28 | 48.42 | 3.13 | 49.64 | nr | **98.06** |
| Sum of two sides 2000mm | 41.21 | 52.13 | 3.13 | 49.64 | nr | **101.77** |
| Sum of two sides 2100mm | 52.47 | 66.37 | 4.26 | 67.56 | nr | **133.93** |
| Sum of two sides 2200mm | 56.24 | 71.14 | 4.27 | 67.72 | nr | **138.86** |
| Sum of two sides 2300mm | 60.14 | 76.08 | 4.30 | 68.20 | nr | **144.28** |
| Sum of two sides 2400mm | 64.18 | 81.19 | 4.77 | 75.65 | nr | **156.84** |
| Sum of two sides 2500mm | 68.34 | 86.45 | 4.79 | 75.97 | nr | **162.42** |
| Sum of two sides 2600mm | 89.43 | 113.13 | 4.95 | 78.51 | nr | **191.64** |
| Sum of two sides 2700mm | 94.28 | 119.26 | 4.95 | 78.51 | nr | **197.77** |
| Sum of two sides 2800mm | 99.29 | 125.60 | 8.49 | 134.65 | nr | **260.25** |
| Sum of two sides 2900mm | 112.40 | 142.19 | 8.88 | 140.84 | nr | **283.03** |
| Sum of two sides 3000mm | 118.56 | 149.98 | 9.02 | 143.06 | nr | **293.04** |
| Sum of two sides 3100mm | 124.90 | 158.00 | 9.02 | 143.06 | nr | **301.06** |
| Sum of two sides 3200mm | 177.35 | 224.35 | 9.09 | 144.17 | nr | **368.52** |

## Y:MECHANICAL AND ELECTRICAL SERVICES

| Item | Net Price £ | Material £ | Labour hours | Labour £ | Unit | Total Rate £ |
|---|---|---|---|---|---|---|
| **45 degree radius bend** | | | | | | |
| Sum of two sides 200mm | 2.51 | 3.18 | 1.12 | 17.76 | nr | 20.94 |
| Sum of two sides 250mm | 2.79 | 3.53 | 1.12 | 17.76 | nr | 21.29 |
| Sum of two sides 300mm | 3.15 | 3.98 | 1.12 | 17.76 | nr | 21.74 |
| Sum of two sides 350mm | 3.48 | 4.40 | 1.10 | 17.45 | nr | 21.85 |
| Sum of two sides 400mm | 3.87 | 4.90 | 1.10 | 17.45 | nr | 22.35 |
| Sum of two sides 450mm | 4.28 | 5.41 | 1.10 | 17.45 | nr | 22.86 |
| Sum of two sides 500mm | 4.72 | 5.97 | 1.10 | 17.45 | nr | 23.42 |
| Sum of two sides 550mm | 5.16 | 6.53 | 1.16 | 18.40 | nr | 24.93 |
| Sum of two sides 600mm | 5.67 | 7.17 | 1.16 | 18.40 | nr | 25.57 |
| Sum of two sides 650mm | 6.19 | 7.83 | 1.16 | 18.40 | nr | 26.23 |
| Sum of two sides 700mm | 6.71 | 8.49 | 1.16 | 18.40 | nr | 26.89 |
| Sum of two sides 750mm | 7.25 | 9.17 | 1.16 | 18.40 | nr | 27.57 |
| Sum of two sides 800mm | 7.88 | 9.97 | 1.22 | 19.35 | nr | 29.32 |
| Sum of two sides 850mm | 8.48 | 10.73 | 1.22 | 19.35 | nr | 30.08 |
| Sum of two sides 900mm | 9.13 | 11.55 | 1.22 | 19.35 | nr | 30.90 |
| Sum of two sides 950mm | 9.78 | 12.37 | 1.40 | 22.20 | nr | 34.57 |
| Sum of two sides 1000mm | 10.47 | 13.24 | 1.40 | 22.20 | nr | 35.44 |
| Sum of two sides 1100mm | 11.93 | 15.09 | 1.40 | 22.20 | nr | 37.29 |
| Sum of two sides 1200mm | 13.48 | 17.05 | 1.40 | 22.20 | nr | 39.25 |
| Sum of two sides 1300mm | 15.13 | 19.14 | 1.91 | 30.29 | nr | 49.43 |
| Sum of two sides 1400mm | 16.90 | 21.38 | 1.91 | 30.29 | nr | 51.67 |
| Sum of two sides 1500mm | 18.79 | 23.77 | 2.11 | 33.47 | nr | 57.24 |
| Sum of two sides 1600mm | 20.75 | 26.25 | 2.11 | 33.47 | nr | 59.72 |
| Sum of two sides 1700mm | 29.30 | 37.06 | 2.55 | 40.44 | nr | 77.50 |
| Sum of two sides 1800mm | 31.80 | 40.23 | 2.55 | 40.44 | nr | 80.67 |
| Sum of two sides 1900mm | 34.41 | 43.53 | 2.80 | 44.41 | nr | 87.94 |
| Sum of two sides 2000mm | 37.12 | 46.96 | 2.80 | 44.41 | nr | 91.37 |
| Sum of two sides 2100mm | 47.10 | 59.58 | 2.61 | 41.40 | nr | 100.98 |
| Sum of two sides 2200mm | 50.62 | 64.03 | 2.62 | 41.55 | nr | 105.58 |
| Sum of two sides 2300mm | 54.29 | 68.68 | 2.63 | 41.71 | nr | 110.39 |
| Sum of two sides 2400mm | 58.07 | 73.46 | 4.34 | 68.83 | nr | 142.29 |
| Sum of two sides 2500mm | 61.97 | 78.39 | 4.35 | 68.99 | nr | 147.38 |
| Sum of two sides 2600mm | 82.77 | 104.70 | 4.53 | 71.85 | nr | 176.55 |
| Sum of two sides 2700mm | 87.39 | 110.55 | 4.53 | 71.85 | nr | 182.40 |
| Sum of two sides 2800mm | 92.16 | 116.58 | 7.13 | 113.08 | nr | 229.66 |
| Sum of two sides 2900mm | 97.68 | 123.57 | 7.17 | 113.72 | nr | 237.29 |
| Sum of two sides 3000mm | 103.38 | 130.78 | 7.26 | 115.15 | nr | 245.93 |
| Sum of two sides 3100mm | 109.24 | 138.19 | 7.26 | 115.15 | nr | 253.34 |
| Sum of two sides 3200mm | 115.27 | 145.82 | 7.31 | 115.94 | nr | 261.76 |
| **90 degree square bend with turning vanes** | | | | | | |
| Sum of two sides 200mm | 3.20 | 4.05 | 1.29 | 20.46 | nr | 24.51 |
| Sum of two sides 250mm | 3.82 | 4.83 | 1.29 | 20.46 | nr | 25.29 |
| Sum of two sides 300mm | 4.56 | 5.77 | 1.29 | 20.46 | nr | 26.23 |
| Sum of two sides 350mm | 5.24 | 6.63 | 1.29 | 20.46 | nr | 27.09 |
| Sum of two sides 400mm | 6.07 | 7.68 | 1.29 | 20.46 | nr | 28.14 |
| Sum of two sides 450mm | 6.99 | 8.84 | 1.29 | 20.46 | nr | 29.30 |
| Sum of two sides 500mm | 7.96 | 10.07 | 1.29 | 20.46 | nr | 30.53 |
| Sum of two sides 550mm | 9.04 | 11.44 | 1.39 | 22.05 | nr | 33.49 |
| Sum of two sides 600mm | 10.15 | 12.84 | 1.39 | 22.05 | nr | 34.89 |
| Sum of two sides 650mm | 11.39 | 14.41 | 1.39 | 22.05 | nr | 36.46 |
| Sum of two sides 700mm | 12.65 | 16.00 | 1.39 | 22.05 | nr | 38.05 |
| Sum of two sides 750mm | 14.01 | 17.72 | 1.46 | 23.16 | nr | 40.88 |
| Sum of two sides 800mm | 15.44 | 19.53 | 1.46 | 23.16 | nr | 42.69 |
| Sum of two sides 850mm | 16.94 | 21.43 | 1.46 | 23.16 | nr | 44.59 |
| Sum of two sides 900mm | 18.51 | 23.42 | 1.46 | 23.16 | nr | 46.58 |
| Sum of two sides 950mm | 20.20 | 25.55 | 1.88 | 29.82 | nr | 55.37 |

## Y:MECHANICAL AND ELECTRICAL SERVICES

| Item | Net Price £ | Material £ | Labour hours | Labour £ | Unit | Total Rate £ |
|---|---|---|---|---|---|---|
| **Y30 : AIR DUCTLINES : MILD STEEL DUCTWORK (contd)** | | | | | | |
| **Extra Over galvanised sheet metal DW142 class C rectangular section ductwork; fittings; including all necessary stiffeners, joints etc. (contd)** | | | | | | |
| 90 degree square bend with turning vanes (contd) | | | | | | |
| Sum of two sides 1000mm | 21.92 | 27.73 | 1.88 | 29.82 | nr | 57.55 |
| Sum of two sides 1100mm | 25.61 | 32.40 | 1.88 | 29.82 | nr | 62.22 |
| Sum of two sides 1200mm | 29.56 | 37.39 | 1.88 | 29.82 | nr | 67.21 |
| Sum of two sides 1300mm | 33.87 | 42.85 | 2.26 | 35.84 | nr | 78.69 |
| Sum of two sides 1400mm | 40.94 | 51.79 | 2.44 | 38.70 | nr | 90.49 |
| Sum of two sides 1500mm | 46.01 | 58.20 | 2.68 | 42.51 | nr | 100.71 |
| Sum of two sides 1600mm | 51.35 | 64.96 | 2.69 | 42.66 | nr | 107.62 |
| Sum of two sides 1700mm | 60.55 | 76.60 | 2.96 | 46.95 | nr | 123.55 |
| Sum of two sides 1800mm | 66.65 | 84.31 | 2.96 | 46.95 | nr | 131.26 |
| Sum of two sides 1900mm | 72.81 | 92.10 | 3.26 | 51.70 | nr | 143.80 |
| Sum of two sides 2000mm | 79.83 | 100.98 | 3.26 | 51.70 | nr | 152.68 |
| Sum of two sides 2100mm | 94.77 | 119.88 | 7.50 | 118.95 | nr | 238.83 |
| Sum of two sides 2200mm | 102.78 | 130.02 | 7.50 | 118.95 | nr | 248.97 |
| Sum of two sides 2300mm | 111.06 | 140.49 | 7.55 | 119.75 | nr | 260.24 |
| Sum of two sides 2400mm | 119.69 | 151.41 | 8.13 | 128.94 | nr | 280.35 |
| Sum of two sides 2500mm | 128.63 | 162.72 | 8.30 | 131.64 | nr | 294.36 |
| Sum of two sides 2600mm | 154.69 | 195.68 | 8.56 | 135.76 | nr | 331.44 |
| Sum of two sides 2700mm | 164.71 | 208.36 | 8.62 | 136.72 | nr | 345.08 |
| Sum of two sides 2800mm | 175.14 | 221.55 | 9.09 | 144.17 | nr | 365.72 |
| Sum of two sides 2900mm | 186.62 | 236.07 | 9.09 | 144.17 | nr | 380.24 |
| Sum of two sides 3000mm | 205.33 | 259.74 | 9.62 | 152.58 | nr | 412.32 |
| Sum of two sides 3100mm | 217.79 | 275.50 | 9.62 | 152.58 | nr | 428.08 |
| Sum of two sides 3200mm | 230.58 | 291.68 | 9.62 | 152.58 | nr | 444.26 |
| Equal twin square bend with turning vanes | | | | | | |
| Sum of two sides 200mm | 4.75 | 6.01 | 2.09 | 33.15 | nr | 39.16 |
| Sum of two sides 250mm | 5.68 | 7.19 | 2.09 | 33.15 | nr | 40.34 |
| Sum of two sides 300mm | 6.79 | 8.59 | 2.09 | 33.15 | nr | 41.74 |
| Sum of two sides 350mm | 8.00 | 10.12 | 2.04 | 32.35 | nr | 42.47 |
| Sum of two sides 400mm | 9.39 | 11.88 | 2.04 | 32.35 | nr | 44.23 |
| Sum of two sides 450mm | 10.91 | 13.80 | 2.04 | 32.35 | nr | 46.15 |
| Sum of two sides 500mm | 12.53 | 15.85 | 2.04 | 32.35 | nr | 48.20 |
| Sum of two sides 550mm | 14.29 | 18.08 | 2.15 | 34.10 | nr | 52.18 |
| Sum of two sides 600mm | 16.65 | 21.06 | 2.15 | 34.10 | nr | 55.16 |
| Sum of two sides 650mm | 18.29 | 23.14 | 2.15 | 34.10 | nr | 57.24 |
| Sum of two sides 700mm | 20.51 | 25.95 | 2.15 | 34.10 | nr | 60.05 |
| Sum of two sides 750mm | 22.83 | 28.88 | 2.26 | 35.84 | nr | 64.72 |
| Sum of two sides 800mm | 25.30 | 32.00 | 2.26 | 35.84 | nr | 67.84 |
| Sum of two sides 850mm | 27.87 | 35.26 | 2.26 | 35.84 | nr | 71.10 |
| Sum of two sides 900mm | 30.62 | 38.73 | 2.26 | 35.84 | nr | 74.57 |
| Sum of two sides 950mm | 33.51 | 42.39 | 3.03 | 48.06 | nr | 90.45 |
| Sum of two sides 1000mm | 36.53 | 46.21 | 3.03 | 48.06 | nr | 94.27 |
| Sum of two sides 1100mm | 42.98 | 54.37 | 3.03 | 48.06 | nr | 102.43 |
| Sum of two sides 1200mm | 49.96 | 63.20 | 3.03 | 48.06 | nr | 111.26 |
| Sum of two sides 1300mm | 59.83 | 75.69 | 3.85 | 61.06 | nr | 136.75 |
| Sum of two sides 1400mm | 68.07 | 86.11 | 3.85 | 61.06 | nr | 147.17 |
| Sum of two sides 1500mm | 76.87 | 97.24 | 4.25 | 67.41 | nr | 164.65 |
| Sum of two sides 1600mm | 86.27 | 109.13 | 4.26 | 67.56 | nr | 176.69 |
| Sum of two sides 1700mm | 101.22 | 128.04 | 4.68 | 74.23 | nr | 202.27 |

## Y:MECHANICAL AND ELECTRICAL SERVICES

| Item | Net Price £ | Material £ | Labour hours | Labour £ | Unit | Total Rate £ |
|---|---|---|---|---|---|---|
| Sum of two sides 1800mm | 111.88 | 141.53 | 4.68 | 74.23 | nr | 215.76 |
| Sum of two sides 1900mm | 123.12 | 155.75 | 4.87 | 77.24 | nr | 232.99 |
| Sum of two sides 2000mm | 134.89 | 170.64 | 4.87 | 77.24 | nr | 247.88 |
| Sum of two sides 2100mm | 163.55 | 206.89 | 7.50 | 118.95 | nr | 325.84 |
| Sum of two sides 2200mm | 177.65 | 224.73 | 7.50 | 118.95 | nr | 343.68 |
| Sum of two sides 2300mm | 192.32 | 243.28 | 7.55 | 119.75 | nr | 363.03 |
| Sum of two sides 2400mm | 207.52 | 262.51 | 8.13 | 128.94 | nr | 391.45 |
| Sum of two sides 2500mm | 223.42 | 282.63 | 8.30 | 131.64 | nr | 414.27 |
| Sum of two sides 2600mm | 265.60 | 335.98 | 8.56 | 135.76 | nr | 471.74 |
| Sum of two sides 2700mm | 283.36 | 358.45 | 8.62 | 136.72 | nr | 495.17 |
| Sum of two sides 2800mm | 301.60 | 381.52 | 15.20 | 241.08 | nr | 622.60 |
| Sum of two sides 2900mm | 321.67 | 406.91 | 15.20 | 241.08 | nr | 647.99 |
| Sum of two sides 3000mm | 342.50 | 433.26 | 15.60 | 247.42 | nr | 680.68 |
| Sum of two sides 3100mm | 370.94 | 469.24 | 16.04 | 254.40 | nr | 723.64 |
| Sum of two sides 3200mm | 393.30 | 497.52 | 16.04 | 254.40 | nr | 751.92 |
| Spigot branch | | | | | | |
| Sum of two sides 200mm | 0.21 | 0.27 | 0.95 | 15.07 | nr | 15.34 |
| Sum of two sides 250mm | 0.25 | 0.32 | 0.95 | 15.07 | nr | 15.39 |
| Sum of two sides 300mm | 0.31 | 0.39 | 0.95 | 15.07 | nr | 15.46 |
| Sum of two sides 350mm | 0.35 | 0.44 | 0.92 | 14.59 | nr | 15.03 |
| Sum of two sides 400mm | 0.42 | 0.53 | 0.92 | 14.59 | nr | 15.12 |
| Sum of two sides 450mm | 0.45 | 0.57 | 0.92 | 14.59 | nr | 15.16 |
| Sum of two sides 500mm | 0.51 | 0.65 | 0.92 | 14.59 | nr | 15.24 |
| Sum of two sides 550mm | 0.56 | 0.71 | 1.03 | 16.34 | nr | 17.05 |
| Sum of two sides 600mm | 0.60 | 0.76 | 1.03 | 16.34 | nr | 17.10 |
| Sum of two sides 650mm | 0.66 | 0.83 | 1.03 | 16.34 | nr | 17.17 |
| Sum of two sides 700mm | 0.72 | 0.91 | 1.03 | 16.34 | nr | 17.25 |
| Sum of two sides 750mm | 0.77 | 0.97 | 1.03 | 16.34 | nr | 17.31 |
| Sum of two sides 800mm | 0.81 | 1.02 | 1.03 | 16.34 | nr | 17.36 |
| Sum of two sides 850mm | 0.87 | 1.10 | 1.03 | 16.34 | nr | 17.44 |
| Sum of two sides 900mm | 0.91 | 1.15 | 1.03 | 16.34 | nr | 17.49 |
| Sum of two sides 950mm | 0.97 | 1.23 | 1.29 | 20.46 | nr | 21.69 |
| Sum of two sides 1000mm | 1.03 | 1.30 | 1.29 | 20.46 | nr | 21.76 |
| Sum of two sides 1100mm | 1.12 | 1.42 | 1.29 | 20.46 | nr | 21.88 |
| Sum of two sides 1200mm | 1.23 | 1.56 | 1.29 | 20.46 | nr | 22.02 |
| Sum of two sides 1300mm | 1.33 | 1.68 | 1.39 | 22.05 | nr | 23.73 |
| Sum of two sides 1400mm | 1.43 | 1.81 | 1.39 | 22.05 | nr | 23.86 |
| Sum of two sides 1500mm | 1.52 | 1.92 | 1.64 | 26.01 | nr | 27.93 |
| Sum of two sides 1600mm | 1.63 | 2.06 | 1.64 | 26.01 | nr | 28.07 |
| Sum of two sides 1700mm | 1.72 | 2.18 | 1.69 | 26.80 | nr | 28.98 |
| Sum of two sides 1800mm | 1.83 | 2.31 | 1.69 | 26.80 | nr | 29.11 |
| Sum of two sides 1900mm | 1.92 | 2.43 | 1.85 | 29.34 | nr | 31.77 |
| Sum of two sides 2000mm | 2.03 | 2.57 | 1.85 | 29.34 | nr | 31.91 |
| Sum of two sides 2100mm | 2.69 | 3.40 | 2.61 | 41.40 | nr | 44.80 |
| Sum of two sides 2200mm | 2.81 | 3.55 | 2.61 | 41.40 | nr | 44.95 |
| Sum of two sides 2300mm | 2.94 | 3.72 | 2.61 | 41.40 | nr | 45.12 |
| Sum of two sides 2400mm | 3.06 | 3.87 | 2.88 | 45.68 | nr | 49.55 |
| Sum of two sides 2500mm | 3.19 | 4.04 | 2.88 | 45.68 | nr | 49.72 |
| Sum of two sides 2600mm | 3.31 | 4.19 | 2.88 | 45.68 | nr | 49.87 |
| Sum of two sides 2700mm | 3.47 | 4.39 | 2.88 | 45.68 | nr | 50.07 |
| Sum of two sides 2800mm | 3.58 | 4.53 | 3.94 | 62.49 | nr | 67.02 |
| Sum of two sides 2900mm | 3.71 | 4.69 | 3.94 | 62.49 | nr | 67.18 |
| Sum of two sides 3000mm | 3.84 | 4.86 | 4.83 | 76.61 | nr | 81.47 |
| Sum of two sides 3100mm | 3.94 | 4.98 | 4.83 | 76.61 | nr | 81.59 |
| Sum of two sides 3200mm | 4.08 | 5.16 | 4.83 | 76.61 | nr | 81.77 |

Y:MECHANICAL AND ELECTRICAL SERVICES

| Item | Net Price £ | Material £ | Labour hours | Labour £ | Unit | Total Rate £ |
|---|---|---|---|---|---|---|
| **Y30 : AIR DUCTLINES : MILD STEEL DUCTWORK (contd)** | | | | | | |
| **Extra Over galvanised sheet metal DW142 class C rectangular section ductwork; fittings; including all necessary stiffeners, joints etc. (contd)** | | | | | | |
| Branch; based on branch size (contd) | | | | | | |
| Sum of two sides 200mm | 1.90 | 2.40 | 1.10 | 17.45 | nr | 19.85 |
| Sum of two sides 250mm | 2.00 | 2.53 | 1.10 | 17.45 | nr | 19.98 |
| Sum of two sides 300mm | 2.25 | 2.85 | 1.10 | 17.45 | nr | 20.30 |
| Sum of two sides 350mm | 2.51 | 3.18 | 1.16 | 18.40 | nr | 21.58 |
| Sum of two sides 400mm | 2.76 | 3.49 | 1.16 | 18.40 | nr | 21.89 |
| Sum of two sides 450mm | 3.00 | 3.79 | 1.16 | 18.40 | nr | 22.19 |
| Sum of two sides 500mm | 3.25 | 4.11 | 1.16 | 18.40 | nr | 22.51 |
| Sum of two sides 550mm | 3.48 | 4.40 | 1.18 | 18.72 | nr | 23.12 |
| Sum of two sides 600mm | 3.74 | 4.73 | 1.18 | 18.72 | nr | 23.45 |
| Sum of two sides 650mm | 4.03 | 5.10 | 1.18 | 18.72 | nr | 23.82 |
| Sum of two sides 700mm | 4.26 | 5.39 | 1.18 | 18.72 | nr | 24.11 |
| Sum of two sides 750mm | 4.49 | 5.68 | 1.18 | 18.72 | nr | 24.40 |
| Sum of two sides 800mm | 4.75 | 6.01 | 1.18 | 18.72 | nr | 24.73 |
| Sum of two sides 850mm | 5.00 | 6.33 | 1.18 | 18.72 | nr | 25.05 |
| Sum of two sides 900mm | 5.24 | 6.63 | 1.18 | 18.72 | nr | 25.35 |
| Sum of two sides 950mm | 5.50 | 6.96 | 1.44 | 22.84 | nr | 29.80 |
| Sum of two sides 1000mm | 5.73 | 7.25 | 1.44 | 22.84 | nr | 30.09 |
| Sum of two sides 1100mm | 6.25 | 7.91 | 1.44 | 22.84 | nr | 30.75 |
| Sum of two sides 1200mm | 6.76 | 8.55 | 1.44 | 22.84 | nr | 31.39 |
| Sum of two sides 1300mm | 7.23 | 9.15 | 1.69 | 26.80 | nr | 35.95 |
| Sum of two sides 1400mm | 7.74 | 9.79 | 1.69 | 26.80 | nr | 36.59 |
| Sum of two sides 1500mm | 8.24 | 10.42 | 1.79 | 28.39 | nr | 38.81 |
| Sum of two sides 1600mm | 8.72 | 11.03 | 1.79 | 28.39 | nr | 39.42 |
| Sum of two sides 1700mm | 10.74 | 13.59 | 1.86 | 29.50 | nr | 43.09 |
| Sum of two sides 1800mm | 11.31 | 14.31 | 1.86 | 29.50 | nr | 43.81 |
| Sum of two sides 1900mm | 11.86 | 15.00 | 2.02 | 32.04 | nr | 47.04 |
| Sum of two sides 2000mm | 12.42 | 15.71 | 2.02 | 32.04 | nr | 47.75 |
| Sum of two sides 2100mm | 14.59 | 18.46 | 2.80 | 44.41 | nr | 62.87 |
| Sum of two sides 2200mm | 15.25 | 19.29 | 2.80 | 44.41 | nr | 63.70 |
| Sum of two sides 2300mm | 15.87 | 20.08 | 2.80 | 44.41 | nr | 64.49 |
| Sum of two sides 2400mm | 16.52 | 20.90 | 3.06 | 48.53 | nr | 69.43 |
| Sum of two sides 2500mm | 17.15 | 21.69 | 3.06 | 48.53 | nr | 70.22 |
| Sum of two sides 2600mm | 25.54 | 32.31 | 3.08 | 48.85 | nr | 81.16 |
| Sum of two sides 2700mm | 26.35 | 33.33 | 3.08 | 48.85 | nr | 82.18 |
| Sum of two sides 2800mm | 27.21 | 34.42 | 4.13 | 65.50 | nr | 99.92 |
| Sum of two sides 2900mm | 28.07 | 35.51 | 4.13 | 65.50 | nr | 101.01 |
| Sum of two sides 3000mm | 29.00 | 36.69 | 5.02 | 79.62 | nr | 116.31 |
| Sum of two sides 3100mm | 29.87 | 37.79 | 5.02 | 79.62 | nr | 117.41 |
| Sum of two sides 3200mm | 30.80 | 38.96 | 5.02 | 79.62 | nr | 118.58 |
| **Galvanised sheet metal DW142 class A spirally wound circular section ductwork; including all necessary stiffeners, joints, couplers in the running length and duct supports** | | | | | | |
| Straight duct | | | | | | |
| 80mm dia. | 4.49 | 5.19 | 0.15 | 2.38 | m | 7.57 |
| 100mm dia. | 4.49 | 5.19 | 0.87 | 13.80 | m | 18.99 |

*Material Costs / Measured Work Prices - Mechanical Installations*

## Y:MECHANICAL AND ELECTRICAL SERVICES

| Item | Net Price £ | Material £ | Labour hours | Labour £ | Unit | Total Rate £ |
|---|---|---|---|---|---|---|
| 150mm dia. | 6.77 | 7.82 | 0.87 | 13.80 | m | 21.62 |
| 200mm dia. | 11.70 | 13.51 | 0.87 | 13.80 | m | 27.31 |
| 250mm dia. | 14.16 | 16.35 | 1.21 | 19.19 | m | 35.54 |
| 300mm dia. | 14.16 | 16.35 | 1.21 | 19.19 | m | 35.54 |
| 355mm dia. | 20.06 | 23.17 | 1.21 | 19.19 | m | 42.36 |
| 400mm dia. | 20.06 | 23.17 | 1.21 | 19.19 | m | 42.36 |
| 450mm dia. | 22.74 | 26.26 | 1.21 | 19.19 | m | 45.45 |
| 500mm dia. | 24.88 | 28.74 | 1.21 | 19.19 | m | 47.93 |
| 630mm dia. | 34.37 | 39.70 | 1.39 | 22.05 | m | 61.75 |
| 710mm dia. | 37.61 | 43.44 | 1.39 | 22.05 | m | 65.49 |
| 800mm dia. | 50.31 | 58.11 | 1.44 | 22.84 | m | 80.95 |
| 900mm dia. | 56.20 | 64.91 | 1.46 | 23.16 | m | 88.07 |
| 1000mm dia. | 62.45 | 72.13 | 1.65 | 26.17 | m | 98.30 |
| 1120mm dia. | 80.46 | 92.93 | 2.43 | 38.54 | m | 131.47 |
| 1250mm dia. | 89.56 | 103.44 | 2.43 | 38.54 | m | 141.98 |
| 1400mm dia. | 102.11 | 117.94 | 2.77 | 43.93 | m | 161.87 |
| 1600mm dia. | 114.82 | 132.62 | 3.06 | 48.53 | m | 181.15 |
| **Extra Over galvanised sheet metal DW142 class A spirally wound circular section ductwork; fittings; including all necessary stiffeners, joints, etc.** | | | | | | |
| Stopped end | | | | | | |
| 80mm dia. | 1.77 | 1.95 | 0.15 | 2.38 | nr | 4.33 |
| 100mm dia. | 1.77 | 1.95 | 0.15 | 2.38 | nr | 4.33 |
| 150mm dia. | 3.33 | 3.66 | 0.15 | 2.38 | nr | 6.04 |
| 200mm dia. | 4.15 | 4.57 | 0.20 | 3.17 | nr | 7.74 |
| 250mm dia. | 6.51 | 7.16 | 0.29 | 4.60 | nr | 11.76 |
| 300mm dia. | 6.51 | 7.16 | 0.29 | 4.60 | nr | 11.76 |
| 355mm dia. | 8.76 | 9.64 | 0.44 | 6.98 | nr | 16.62 |
| 400mm dia. | 8.76 | 9.64 | 0.44 | 6.98 | nr | 16.62 |
| 450mm dia. | 10.66 | 11.73 | 0.44 | 6.98 | nr | 18.71 |
| 500mm dia. | 11.73 | 12.90 | 0.44 | 6.98 | nr | 19.88 |
| 630mm dia. | 18.00 | 19.80 | 0.58 | 9.20 | nr | 29.00 |
| 710mm dia. | 52.60 | 57.86 | 0.69 | 10.94 | nr | 68.80 |
| 800mm dia. | 54.83 | 60.31 | 0.81 | 12.85 | nr | 73.16 |
| 900mm dia. | 63.01 | 69.31 | 0.92 | 14.59 | nr | 83.90 |
| 1000mm dia. | 63.01 | 69.31 | 1.04 | 16.49 | nr | 85.80 |
| 1120mm dia. | 93.82 | 103.20 | 1.16 | 18.40 | nr | 121.60 |
| 1250mm dia. | 93.82 | 103.20 | 1.16 | 18.40 | nr | 121.60 |
| 1400mm dia. | 105.30 | 115.83 | 1.16 | 18.40 | nr | 134.23 |
| 1600mm dia. | 119.34 | 131.27 | 1.16 | 18.40 | nr | 149.67 |
| Taper; large end measured | | | | | | |
| 100mm dia. | 7.10 | 7.81 | 0.29 | 4.60 | nr | 12.41 |
| 150mm dia. | 8.05 | 8.86 | 0.29 | 4.60 | nr | 13.46 |
| 200mm dia. | 14.60 | 16.06 | 0.44 | 6.98 | nr | 23.04 |
| 250mm dia. | 15.86 | 17.45 | 0.58 | 9.20 | nr | 26.65 |
| 300mm dia. | 22.27 | 24.50 | 0.58 | 9.20 | nr | 33.70 |
| 355mm dia. | 31.75 | 34.92 | 0.87 | 13.80 | nr | 48.72 |
| 400mm dia. | 31.75 | 34.92 | 0.87 | 13.80 | nr | 48.72 |
| 450mm dia. | 34.35 | 37.78 | 0.87 | 13.80 | nr | 51.58 |
| 500mm dia. | 35.41 | 38.95 | 0.87 | 13.80 | nr | 52.75 |
| 630mm dia. | 46.67 | 51.34 | 0.87 | 13.80 | nr | 65.14 |
| 710mm dia. | 79.00 | 86.90 | 0.96 | 15.23 | nr | 102.13 |
| 800mm dia. | 87.65 | 96.42 | 1.06 | 16.81 | nr | 113.23 |
| 900mm dia. | 126.75 | 139.43 | 1.16 | 18.40 | nr | 157.83 |

## Y:MECHANICAL AND ELECTRICAL SERVICES

| Item | Net Price £ | Material £ | Labour hours | Labour £ | Unit | Total Rate £ |
|---|---|---|---|---|---|---|
| **Y30 : AIR DUCTLINES : MILD STEEL DUCTWORK (contd)** | | | | | | |
| Extra Over galvanised sheet metal DW142 class A spirally wound circular section ductwork; fittings; including all necessary stiffeners, joints, etc. (contd) | | | | | | |
| Taper; large end measured (contd) | | | | | | |
| 1000mm dia. | 126.75 | 139.43 | 1.25 | 19.83 | nr | 159.26 |
| 1120mm dia. | 160.39 | 176.43 | 3.47 | 55.04 | nr | 231.47 |
| 1250mm dia. | 160.39 | 176.43 | 3.47 | 55.04 | nr | 231.47 |
| 1400mm dia. | 175.50 | 193.05 | 4.05 | 64.23 | nr | 257.28 |
| 1600mm dia. | 189.54 | 208.49 | 4.62 | 73.27 | nr | 281.76 |
| 90 degree segmented radius bend | | | | | | |
| 80mm dia. | 6.28 | 6.91 | 0.29 | 4.60 | nr | 11.51 |
| 100mm dia. | 11.48 | 12.63 | 0.29 | 4.60 | nr | 17.23 |
| 150mm dia. | 11.48 | 12.63 | 0.29 | 4.60 | nr | 17.23 |
| 200mm dia. | 15.95 | 17.55 | 0.44 | 6.98 | nr | 24.53 |
| 250mm dia. | 22.04 | 24.24 | 0.58 | 9.20 | nr | 33.44 |
| 300mm dia. | 23.56 | 25.92 | 0.58 | 9.20 | nr | 35.12 |
| 355mm dia. | 34.94 | 38.43 | 0.87 | 13.80 | nr | 52.23 |
| 400mm dia. | 34.94 | 38.43 | 0.87 | 13.80 | nr | 52.23 |
| 450mm dia. | 41.70 | 45.87 | 0.87 | 13.80 | nr | 59.67 |
| 500mm dia. | 51.29 | 56.42 | 0.87 | 13.80 | nr | 70.22 |
| 630mm dia. | 73.20 | 80.52 | 0.87 | 13.80 | nr | 94.32 |
| 710mm dia. | 157.66 | 173.43 | 0.96 | 15.23 | nr | 188.66 |
| 800mm dia. | 178.76 | 196.64 | 1.06 | 16.81 | nr | 213.45 |
| 900mm dia. | 210.61 | 231.67 | 1.16 | 18.40 | nr | 250.07 |
| 1000mm dia. | 230.88 | 253.97 | 1.25 | 19.83 | nr | 273.80 |
| 1120mm dia. | 301.13 | 331.24 | 3.47 | 55.04 | nr | 386.28 |
| 1250mm dia. | 334.54 | 367.99 | 3.47 | 55.04 | nr | 423.03 |
| 1400mm dia. | 582.00 | 640.20 | 4.05 | 64.23 | nr | 704.43 |
| 1600mm dia. | 797.37 | 877.11 | 4.62 | 73.27 | nr | 950.38 |
| 45 degree radius bend | | | | | | |
| 80mm dia. | 4.97 | 5.47 | 0.29 | 4.60 | nr | 10.07 |
| 100mm dia. | 4.97 | 5.47 | 0.29 | 4.60 | nr | 10.07 |
| 150mm dia. | 8.28 | 9.11 | 0.29 | 4.60 | nr | 13.71 |
| 200mm dia. | 14.92 | 16.41 | 0.40 | 6.34 | nr | 22.75 |
| 250mm dia. | 15.63 | 17.19 | 0.58 | 9.20 | nr | 26.39 |
| 300mm dia. | 20.79 | 22.87 | 0.58 | 9.20 | nr | 32.07 |
| 355mm dia. | 23.57 | 25.93 | 0.87 | 13.80 | nr | 39.73 |
| 400mm dia. | 23.57 | 25.93 | 0.87 | 13.80 | nr | 39.73 |
| 450mm dia. | 27.13 | 29.84 | 0.87 | 13.80 | nr | 43.64 |
| 500mm dia. | 33.16 | 36.48 | 0.87 | 13.80 | nr | 50.28 |
| 630mm dia. | 48.22 | 53.04 | 0.87 | 13.80 | nr | 66.84 |
| 710mm dia. | 52.60 | 57.86 | 0.96 | 15.23 | nr | 73.09 |
| 800mm dia. | 114.43 | 125.87 | 1.06 | 16.81 | nr | 142.68 |
| 900mm dia. | 149.47 | 164.42 | 1.16 | 18.40 | nr | 182.82 |
| 1000mm dia. | 166.43 | 183.07 | 1.25 | 19.83 | nr | 202.90 |
| 1120mm dia. | 216.42 | 238.06 | 3.47 | 55.04 | nr | 293.10 |
| 1250mm dia. | 232.41 | 255.65 | 3.47 | 55.04 | nr | 310.69 |
| 1400mm dia. | 330.75 | 363.82 | 4.05 | 64.23 | nr | 428.05 |
| 1600mm dia. | 797.37 | 877.11 | 4.62 | 73.27 | nr | 950.38 |

*Material Costs / Measured Work Prices - Mechanical Installations* 291

## Y:MECHANICAL AND ELECTRICAL SERVICES

| Item | Net Price £ | Material £ | Labour hours | Labour £ | Unit | Total Rate £ |
|---|---|---|---|---|---|---|
| **90 degree equal twin bend** | | | | | | |
| 80mm dia. | 19.18 | 21.10 | 0.58 | 9.20 | nr | 30.30 |
| 100mm dia. | 19.18 | 21.10 | 0.58 | 9.20 | nr | 30.30 |
| 150mm dia. | 32.23 | 35.45 | 0.58 | 9.20 | nr | 44.65 |
| 200mm dia. | 71.19 | 78.31 | 0.87 | 13.80 | nr | 92.11 |
| 250mm dia. | 99.27 | 109.20 | 1.16 | 18.40 | nr | 127.60 |
| 300mm dia. | 102.82 | 113.10 | 1.16 | 18.40 | nr | 131.50 |
| 355mm dia. | 125.55 | 138.10 | 1.73 | 27.44 | nr | 165.54 |
| 400mm dia. | 125.55 | 138.10 | 1.73 | 27.44 | nr | 165.54 |
| 450mm dia. | 148.07 | 162.88 | 1.73 | 27.44 | nr | 190.32 |
| 500mm dia. | 167.02 | 183.72 | 1.73 | 27.44 | nr | 211.16 |
| 630mm dia. | 256.23 | 281.85 | 1.73 | 27.44 | nr | 309.29 |
| 710mm dia. | 482.26 | 530.49 | 1.82 | 28.87 | nr | 559.36 |
| 800mm dia. | 580.45 | 638.50 | 1.93 | 30.61 | nr | 669.11 |
| 900mm dia. | 665.51 | 732.06 | 2.02 | 32.04 | nr | 764.10 |
| 1000mm dia. | 710.54 | 781.59 | 2.11 | 33.47 | nr | 815.06 |
| 1120mm dia. | 975.77 | 1073.35 | 4.62 | 73.27 | nr | 1146.62 |
| 1250mm dia. | 1062.84 | 1169.12 | 4.62 | 73.27 | nr | 1242.39 |
| **Conical branch** | | | | | | |
| 80mm dia. | 12.91 | 14.20 | 0.58 | 9.20 | nr | 23.40 |
| 100mm dia. | 12.91 | 14.20 | 0.58 | 9.20 | nr | 23.40 |
| 150mm dia. | 17.41 | 19.15 | 0.58 | 9.20 | nr | 28.35 |
| 200mm dia. | 28.90 | 31.79 | 0.87 | 13.80 | nr | 45.59 |
| 250mm dia. | 38.02 | 41.82 | 1.16 | 18.40 | nr | 60.22 |
| 300mm dia. | 38.02 | 41.82 | 1.16 | 18.40 | nr | 60.22 |
| 355mm dia. | 52.00 | 57.20 | 1.73 | 27.44 | nr | 84.64 |
| 400mm dia. | 52.00 | 57.20 | 1.73 | 27.44 | nr | 84.64 |
| 450mm dia. | 73.44 | 80.78 | 1.73 | 27.44 | nr | 108.22 |
| 500mm dia. | 73.44 | 80.78 | 1.73 | 27.44 | nr | 108.22 |
| 630mm dia. | 105.78 | 116.36 | 1.73 | 27.44 | nr | 143.80 |
| 710mm dia. | 109.56 | 120.52 | 1.82 | 28.87 | nr | 149.39 |
| 800mm dia. | 109.56 | 120.52 | 1.93 | 30.61 | nr | 151.13 |
| 900mm dia. | 179.11 | 197.02 | 2.02 | 32.04 | nr | 229.06 |
| 1000mm dia. | 179.11 | 197.02 | 2.11 | 33.47 | nr | 230.49 |
| 1120mm dia. | 250.55 | 275.61 | 4.62 | 73.27 | nr | 348.88 |
| 1250mm dia. | 250.55 | 275.61 | 5.20 | 82.47 | nr | 358.08 |
| **45 degree branch** | | | | | | |
| 80mm dia. | 16.82 | 18.50 | 0.58 | 9.20 | nr | 27.70 |
| 100mm dia. | 16.82 | 18.50 | 0.58 | 9.20 | nr | 27.70 |
| 150mm dia. | 20.49 | 22.54 | 0.58 | 9.20 | nr | 31.74 |
| 200mm dia. | 30.20 | 33.22 | 1.16 | 18.40 | nr | 51.62 |
| 250mm dia. | 71.55 | 78.70 | 1.16 | 18.40 | nr | 97.10 |
| 300mm dia. | 71.66 | 78.83 | 1.16 | 18.40 | nr | 97.23 |
| 355mm dia. | 71.66 | 78.83 | 1.73 | 27.44 | nr | 106.27 |
| 400mm dia. | 71.66 | 78.83 | 1.73 | 27.44 | nr | 106.27 |
| 450mm dia. | 105.30 | 115.83 | 1.73 | 27.44 | nr | 143.27 |
| 500mm dia. | 105.30 | 115.83 | 1.73 | 27.44 | nr | 143.27 |
| 630mm dia. | 105.30 | 115.83 | 1.73 | 27.44 | nr | 143.27 |
| 710mm dia. | 179.11 | 197.02 | 1.82 | 28.87 | nr | 225.89 |
| 800mm dia. | 250.55 | 275.61 | 1.93 | 30.61 | nr | 306.22 |
| 900mm dia. | 250.55 | 275.61 | 2.31 | 36.64 | nr | 312.25 |
| 1000mm dia. | 250.55 | 275.61 | 2.31 | 36.64 | nr | 312.25 |
| 1120mm dia. | 250.55 | 275.61 | 4.62 | 73.27 | nr | 348.88 |
| 1250mm dia. | 250.55 | 275.61 | 4.62 | 73.27 | nr | 348.88 |

## Y:MECHANICAL AND ELECTRICAL SERVICES

| Item | Net Price £ | Material £ | Labour hours | Labour £ | Unit | Total Rate £ |
|---|---|---|---|---|---|---|
| **Y30 : AIR DUCTLINES : MILD STEEL DUCTWORK (contd)** | | | | | | |
| Galvanised sheet metal DW142 class C spirally wound circular section ductwork; including all necessary stiffeners, joints, couplers in the running length and duct supports | | | | | | |
| Straight duct | | | | | | |
| 80mm dia. | 4.49 | 5.19 | 0.87 | 13.80 | m | **18.99** |
| 100mm dia. | 4.49 | 5.19 | 0.87 | 13.80 | m | **18.99** |
| 150mm dia. | 6.77 | 7.82 | 0.87 | 13.80 | m | **21.62** |
| 200mm dia. | 11.70 | 13.51 | 0.87 | 13.80 | m | **27.31** |
| 250mm dia. | 14.16 | 16.35 | 1.21 | 19.19 | m | **35.54** |
| 300mm dia. | 15.77 | 18.21 | 1.21 | 19.19 | m | **37.40** |
| 355mm dia. | 20.06 | 23.17 | 1.21 | 19.19 | m | **42.36** |
| 400mm dia. | 20.06 | 23.17 | 1.21 | 19.19 | m | **42.36** |
| 450mm dia. | 22.74 | 26.26 | 1.21 | 19.19 | m | **45.45** |
| 500mm dia. | 24.88 | 28.74 | 1.21 | 19.19 | m | **47.93** |
| 630mm dia. | 34.37 | 39.70 | 1.39 | 22.05 | m | **61.75** |
| 710mm dia. | 37.61 | 43.44 | 1.39 | 22.05 | m | **65.49** |
| 800mm dia. | 50.31 | 58.11 | 1.44 | 22.84 | m | **80.95** |
| 900mm dia. | 56.20 | 64.91 | 1.46 | 23.16 | m | **88.07** |
| 1000mm dia. | 62.45 | 72.13 | 1.65 | 26.17 | m | **98.30** |
| 1120mm dia. | 80.46 | 92.93 | 2.43 | 38.54 | m | **131.47** |
| 1250mm dia. | 89.56 | 103.44 | 2.43 | 38.54 | m | **141.98** |
| 1400mm dia. | 102.11 | 117.94 | 2.77 | 43.93 | m | **161.87** |
| 1600mm dia. | 114.82 | 132.62 | 3.06 | 48.53 | m | **181.15** |
| Extra Over galvanised sheet metal DW142 class C spirally wound circular section ductwork; fittings; including all necessary stiffeners, joints, etc. | | | | | | |
| Stopped end | | | | | | |
| 80mm dia. | 1.77 | 1.95 | 0.15 | 2.38 | nr | **4.33** |
| 100mm dia. | 1.77 | 1.95 | 0.15 | 2.38 | nr | **4.33** |
| 150mm dia. | 3.33 | 3.66 | 0.15 | 2.38 | nr | **6.04** |
| 200mm dia. | 4.15 | 4.57 | 0.20 | 3.17 | nr | **7.74** |
| 250mm dia. | 4.15 | 4.57 | 0.29 | 4.60 | nr | **9.17** |
| 300mm dia. | 6.51 | 7.16 | 0.29 | 4.60 | nr | **11.76** |
| 355mm dia. | 8.76 | 9.64 | 0.44 | 6.98 | nr | **16.62** |
| 400mm dia. | 8.76 | 9.64 | 0.44 | 6.98 | nr | **16.62** |
| 450mm dia. | 10.66 | 11.73 | 0.44 | 6.98 | nr | **18.71** |
| 500mm dia. | 11.73 | 12.90 | 0.44 | 6.98 | nr | **19.88** |
| 630mm dia. | 18.00 | 19.80 | 0.58 | 9.20 | nr | **29.00** |
| 710mm dia. | 52.60 | 57.86 | 0.69 | 10.94 | nr | **68.80** |
| 800mm dia. | 54.83 | 60.31 | 0.81 | 12.85 | nr | **73.16** |
| 900mm dia. | 63.01 | 69.31 | 0.92 | 14.59 | nr | **83.90** |
| 1000mm dia. | 63.01 | 69.31 | 1.04 | 16.49 | nr | **85.80** |
| 1120mm dia. | 93.82 | 103.20 | 1.16 | 18.40 | nr | **121.60** |
| 1250mm dia. | 93.82 | 103.20 | 1.16 | 18.40 | nr | **121.60** |
| 1400mm dia. | 105.30 | 115.83 | 1.16 | 18.40 | nr | **134.23** |
| 1600mm dia. | 105.30 | 115.83 | 1.16 | 18.40 | nr | **134.23** |

## Y:MECHANICAL AND ELECTRICAL SERVICES

| Item | Net Price £ | Material £ | Labour hours | Labour £ | Unit | Total Rate £ |
|---|---|---|---|---|---|---|
| Taper; large end measured | | | | | | |
| 100mm dia. | 7.10 | 7.81 | 0.29 | 4.60 | nr | 12.41 |
| 150mm dia. | 8.05 | 8.86 | 0.29 | 4.60 | nr | 13.46 |
| 200mm dia. | 14.09 | 15.50 | 0.44 | 6.98 | nr | 22.48 |
| 250mm dia. | 15.86 | 17.45 | 0.58 | 9.20 | nr | 26.65 |
| 300mm dia. | 22.27 | 24.50 | 0.58 | 9.20 | nr | 33.70 |
| 355mm dia. | 31.75 | 34.92 | 0.87 | 13.80 | nr | 48.72 |
| 400mm dia. | 31.75 | 34.92 | 0.87 | 13.80 | nr | 48.72 |
| 450mm dia. | 34.35 | 37.78 | 0.87 | 13.80 | nr | 51.58 |
| 500mm dia. | 35.41 | 38.95 | 0.87 | 13.80 | nr | 52.75 |
| 630mm dia. | 46.67 | 51.34 | 0.87 | 13.80 | nr | 65.14 |
| 710mm dia. | 79.00 | 86.90 | 0.96 | 15.23 | nr | 102.13 |
| 800mm dia. | 87.65 | 96.42 | 1.06 | 16.81 | nr | 113.23 |
| 900mm dia. | 126.75 | 139.43 | 1.16 | 18.40 | nr | 157.83 |
| 1000mm dia. | 126.75 | 139.43 | 1.25 | 19.83 | nr | 159.26 |
| 1120mm dia. | 160.39 | 176.43 | 3.47 | 55.04 | nr | 231.47 |
| 1250mm dia. | 160.39 | 176.43 | 3.47 | 55.04 | nr | 231.47 |
| 1400mm dia. | 175.50 | 193.05 | 4.05 | 64.23 | nr | 257.28 |
| 1600mm dia. | 189.54 | 208.49 | 4.62 | 73.27 | nr | 281.76 |
| 90 degree segmented radius bend | | | | | | |
| 80mm dia. | 6.28 | 6.91 | 0.29 | 4.60 | nr | 11.51 |
| 100mm dia. | 6.28 | 6.91 | 0.29 | 4.60 | nr | 11.51 |
| 150mm dia. | 11.48 | 12.63 | 0.29 | 4.60 | nr | 17.23 |
| 200mm dia. | 15.95 | 17.55 | 0.44 | 6.98 | nr | 24.53 |
| 250mm dia. | 22.04 | 24.24 | 0.58 | 9.20 | nr | 33.44 |
| 300mm dia. | 23.56 | 25.92 | 0.58 | 9.20 | nr | 35.12 |
| 355mm dia. | 34.94 | 38.43 | 0.87 | 13.80 | nr | 52.23 |
| 400mm dia. | 34.94 | 38.43 | 0.87 | 13.80 | nr | 52.23 |
| 450mm dia. | 41.70 | 45.87 | 0.87 | 13.80 | nr | 59.67 |
| 500mm dia. | 51.29 | 56.42 | 0.87 | 13.80 | nr | 70.22 |
| 630mm dia. | 73.20 | 80.52 | 0.87 | 13.80 | nr | 94.32 |
| 710mm dia. | 157.66 | 173.43 | 0.96 | 15.23 | nr | 188.66 |
| 800mm dia. | 178.76 | 196.64 | 1.06 | 16.81 | nr | 213.45 |
| 900mm dia. | 210.61 | 231.67 | 1.16 | 18.40 | nr | 250.07 |
| 1000mm dia. | 230.88 | 253.97 | 1.25 | 19.83 | nr | 273.80 |
| 1120mm dia. | 301.13 | 331.24 | 3.47 | 55.04 | nr | 386.28 |
| 1250mm dia. | 334.54 | 367.99 | 3.47 | 55.04 | nr | 423.03 |
| 1400mm dia. | 582.00 | 640.20 | 4.05 | 64.23 | nr | 704.43 |
| 1600mm dia. | 797.37 | 877.11 | 4.62 | 73.27 | nr | 950.38 |
| 45 degree radius bend | | | | | | |
| 80mm dia. | 4.97 | 5.47 | 0.29 | 4.60 | nr | 10.07 |
| 100mm dia. | 4.97 | 5.47 | 0.29 | 4.60 | nr | 10.07 |
| 150mm dia. | 8.28 | 9.11 | 0.29 | 4.60 | nr | 13.71 |
| 200mm dia. | 14.92 | 16.41 | 0.44 | 6.98 | nr | 23.39 |
| 250mm dia. | 15.63 | 17.19 | 0.58 | 9.20 | nr | 26.39 |
| 300mm dia. | 20.79 | 22.87 | 0.58 | 9.20 | nr | 32.07 |
| 355mm dia. | 23.57 | 25.93 | 0.87 | 13.80 | nr | 39.73 |
| 400mm dia. | 23.57 | 25.93 | 0.87 | 13.80 | nr | 39.73 |
| 450mm dia. | 27.13 | 29.84 | 0.87 | 13.80 | nr | 43.64 |
| 500mm dia. | 33.16 | 36.48 | 0.87 | 13.80 | nr | 50.28 |
| 630mm dia. | 48.22 | 53.04 | 0.87 | 13.80 | nr | 66.84 |
| 710mm dia. | 89.44 | 98.38 | 0.96 | 15.23 | nr | 113.61 |
| 800mm dia. | 114.43 | 125.87 | 1.06 | 16.81 | nr | 142.68 |
| 900mm dia. | 149.47 | 164.42 | 1.16 | 18.40 | nr | 182.82 |
| 1000mm dia. | 166.43 | 183.07 | 1.25 | 19.83 | nr | 202.90 |

## Y:MECHANICAL AND ELECTRICAL SERVICES

| Item | Net Price £ | Material £ | Labour hours | Labour £ | Unit | Total Rate £ |
|---|---|---|---|---|---|---|
| **Y30 : AIR DUCTLINES : MILD STEEL DUCTWORK (contd)** | | | | | | |
| **Extra Over galvanised sheet metal DW142 class C spirally wound circular section ductwork; fittings; including all necessary stiffeners, joints, etc. (contd)** | | | | | | |
| 45 degree radius bend (contd) | | | | | | |
| 1120mm dia. | 216.42 | 238.06 | 3.47 | 55.04 | nr | 293.10 |
| 1250mm dia. | 232.41 | 255.65 | 3.47 | 55.04 | nr | 310.69 |
| 1400mm dia. | 330.75 | 363.82 | 5.20 | 82.47 | nr | 446.29 |
| 1600mm dia. | 434.88 | 478.37 | 4.62 | 73.27 | nr | 551.64 |
| 90 degree equal twin bend | | | | | | |
| 80mm dia. | 19.18 | 21.10 | 0.58 | 9.20 | nr | 30.30 |
| 100mm dia. | 19.18 | 21.10 | 0.58 | 9.20 | nr | 30.30 |
| 150mm dia. | 32.23 | 35.45 | 0.58 | 9.20 | nr | 44.65 |
| 200mm dia. | 71.19 | 78.31 | 0.87 | 13.80 | nr | 92.11 |
| 250mm dia. | 99.27 | 109.20 | 1.16 | 18.40 | nr | 127.60 |
| 300mm dia. | 102.82 | 113.10 | 1.16 | 18.40 | nr | 131.50 |
| 355mm dia. | 125.55 | 138.10 | 1.73 | 27.44 | nr | 165.54 |
| 400mm dia. | 125.55 | 138.10 | 1.73 | 27.44 | nr | 165.54 |
| 450mm dia. | 148.07 | 162.88 | 1.73 | 27.44 | nr | 190.32 |
| 500mm dia. | 167.02 | 183.72 | 1.73 | 27.44 | nr | 211.16 |
| 630mm dia. | 256.23 | 281.85 | 1.73 | 27.44 | nr | 309.29 |
| 710mm dia. | 482.26 | 530.49 | 1.82 | 28.87 | nr | 559.36 |
| 800mm dia. | 580.45 | 638.50 | 2.02 | 32.04 | nr | 670.54 |
| 900mm dia. | 665.51 | 732.06 | 2.02 | 32.04 | nr | 764.10 |
| 1000mm dia. | 710.54 | 781.59 | 2.11 | 33.47 | nr | 815.06 |
| 1120mm dia. | 975.77 | 1073.35 | 4.62 | 73.27 | nr | 1146.62 |
| 1250mm dia. | 1062.84 | 1169.12 | 4.62 | 73.27 | nr | 1242.39 |
| Conical branch | | | | | | |
| 80mm dia. | 12.91 | 14.20 | 0.58 | 9.20 | nr | 23.40 |
| 100mm dia. | 12.91 | 14.20 | 0.58 | 9.20 | nr | 23.40 |
| 150mm dia. | 17.41 | 19.15 | 0.58 | 9.20 | nr | 28.35 |
| 200mm dia. | 28.90 | 31.79 | 0.87 | 13.80 | nr | 45.59 |
| 250mm dia. | 38.02 | 41.82 | 1.16 | 18.40 | nr | 60.22 |
| 300mm dia. | 38.02 | 41.82 | 1.16 | 18.40 | nr | 60.22 |
| 355mm dia. | 52.00 | 57.20 | 1.73 | 27.44 | nr | 84.64 |
| 400mm dia. | 52.00 | 57.20 | 1.73 | 27.44 | nr | 84.64 |
| 450mm dia. | 73.44 | 80.78 | 1.73 | 27.44 | nr | 108.22 |
| 500mm dia. | 73.44 | 80.78 | 1.73 | 27.44 | nr | 108.22 |
| 630mm dia. | 105.30 | 115.83 | 1.73 | 27.44 | nr | 143.27 |
| 710mm dia. | 109.56 | 120.52 | 1.82 | 28.87 | nr | 149.39 |
| 800mm dia. | 109.56 | 120.52 | 1.93 | 30.61 | nr | 151.13 |
| 900mm dia. | 179.11 | 197.02 | 2.02 | 32.04 | nr | 229.06 |
| 1000mm dia. | 179.11 | 197.02 | 2.11 | 33.47 | nr | 230.49 |
| 1120mm dia. | 194.62 | 214.08 | 4.62 | 73.27 | nr | 287.35 |
| 1250mm dia. | 250.55 | 275.61 | 4.62 | 73.27 | nr | 348.88 |
| 45 degree branch | | | | | | |
| 80mm dia. | 16.82 | 18.50 | 0.58 | 9.20 | nr | 27.70 |
| 100mm dia. | 16.82 | 18.50 | 0.58 | 9.20 | nr | 27.70 |
| 150mm dia. | 20.49 | 22.54 | 0.58 | 9.20 | nr | 31.74 |

## Y: MECHANICAL AND ELECTRICAL SERVICES

| Item | Net Price £ | Material £ | Labour hours | Labour £ | Unit | Total Rate £ |
|---|---|---|---|---|---|---|
| 200mm dia. | 30.20 | 33.22 | 0.87 | 13.80 | nr | 47.02 |
| 250mm dia. | 71.55 | 78.70 | 1.16 | 18.40 | nr | 97.10 |
| 300mm dia. | 71.66 | 78.83 | 1.16 | 18.40 | nr | 97.23 |
| 355mm dia. | 71.66 | 78.83 | 1.73 | 27.44 | nr | 106.27 |
| 400mm dia. | 71.66 | 78.83 | 1.73 | 27.44 | nr | 106.27 |
| 450mm dia. | 105.30 | 115.83 | 1.73 | 27.44 | nr | 143.27 |
| 500mm dia. | 105.30 | 115.83 | 1.73 | 27.44 | nr | 143.27 |
| 630mm dia. | 105.30 | 115.83 | 1.73 | 27.44 | nr | 143.27 |
| 710mm dia. | 179.11 | 197.02 | 1.82 | 28.87 | nr | 225.89 |
| 800mm dia. | 250.55 | 275.61 | 2.13 | 33.78 | nr | 309.39 |
| 900mm dia. | 250.55 | 275.61 | 2.31 | 36.64 | nr | 312.25 |
| 1000mm dia. | 250.55 | 275.61 | 2.31 | 36.64 | nr | 312.25 |
| 1120mm dia. | 250.55 | 275.61 | 4.62 | 73.27 | nr | 348.88 |
| 1250mm dia. | 250.55 | 275.61 | 4.62 | 73.27 | nr | 348.88 |

**Galvanised sheet metal DW142 class C spirally wound flat oval section ductwork; including all necessary stiffeners, joints, couplers in the running length and duct supports**

Straight duct

| Item | Net Price £ | Material £ | Labour hours | Labour £ | Unit | Total Rate £ |
|---|---|---|---|---|---|---|
| 345 x 102mm | 21.83 | 25.21 | 2.71 | 42.98 | m | 68.19 |
| 427 x 102mm | 25.86 | 29.87 | 2.99 | 47.42 | m | 77.29 |
| 508 x 102mm | 28.99 | 33.48 | 3.14 | 49.80 | m | 83.28 |
| 559 x 152mm | 28.99 | 33.48 | 3.43 | 54.40 | m | 87.88 |
| 531 x 203mm | 40.18 | 46.41 | 3.43 | 54.40 | m | 100.81 |
| 851 x 203mm | 53.17 | 61.41 | 5.72 | 90.72 | m | 152.13 |
| 582 x 254mm | 44.41 | 51.29 | 3.62 | 57.41 | m | 108.70 |
| 1303 x 254mm | 111.67 | 128.98 | 8.13 | 128.94 | m | 257.92 |
| 632 x 305mm | 49.28 | 56.92 | 3.93 | 62.33 | m | 119.25 |
| 1275 x 305mm | 100.62 | 116.22 | 8.13 | 128.94 | m | 245.16 |
| 765 x 356mm | 58.30 | 67.34 | 5.72 | 90.72 | m | 158.06 |
| 1247 x 356mm | 102.11 | 117.94 | 8.13 | 128.94 | m | 246.88 |
| 1727 x 356mm | 143.22 | 165.42 | 10.41 | 165.11 | m | 330.53 |
| 737 x 406mm | 59.07 | 68.23 | 5.72 | 90.72 | m | 158.95 |
| 818 x 406mm | 62.41 | 72.08 | 6.21 | 98.49 | m | 170.57 |
| 978 x 406mm | 69.32 | 80.06 | 6.92 | 109.75 | m | 189.81 |
| 1379 x 406mm | 113.33 | 130.90 | 8.75 | 138.78 | m | 269.68 |
| 1699 x 406mm | 143.22 | 165.42 | 10.41 | 165.11 | m | 330.53 |
| 709 x 457mm | 59.07 | 68.23 | 5.72 | 90.72 | m | 158.95 |
| 1671 x 457mm | 143.22 | 165.42 | 10.31 | 163.52 | m | 328.94 |
| 678 x 508mm | 59.07 | 68.23 | 5.72 | 90.72 | m | 158.95 |

**Extra Over galvanised sheet metal DW142 class C spirally wound flat oval section ductwork; fittings; including all necessary stiffeners, joints, etc.**

Plug end

| Item | Net Price £ | Material £ | Labour hours | Labour £ | Unit | Total Rate £ |
|---|---|---|---|---|---|---|
| 345 x 102mm | 29.41 | 32.35 | 0.20 | 3.17 | nr | 35.52 |
| 427 x 102mm | 29.90 | 32.89 | 0.20 | 3.17 | nr | 36.06 |
| 508 x 102mm | 30.43 | 33.47 | 0.29 | 4.60 | nr | 38.07 |
| 559 x 152mm | 37.91 | 41.70 | 0.29 | 4.60 | nr | 46.30 |
| 531 x 203mm | 38.45 | 42.30 | 0.44 | 6.98 | nr | 49.28 |
| 851 x 203mm | 41.60 | 45.76 | 0.44 | 6.98 | nr | 52.74 |
| 582 x 254mm | 39.39 | 43.33 | 0.44 | 6.98 | nr | 50.31 |
| 1303 x 254mm | 79.72 | 87.69 | 0.69 | 10.94 | nr | 98.63 |

## Y:MECHANICAL AND ELECTRICAL SERVICES

| Item | Net Price £ | Material £ | Labour hours | Labour £ | Unit | Total Rate £ |
|---|---|---|---|---|---|---|
| **Y30 : AIR DUCTLINES : MILD STEEL DUCTWORK (contd)** | | | | | | |
| **Extra Over galvanised sheet metal DW142 class C spirally wound flat oval section ductwork; fittings; including all necessary stiffeners, joints, etc. (contd)** | | | | | | |
| Plug end (contd) | | | | | | |
| 632 x 305mm | 39.55 | 43.51 | 0.44 | 6.98 | nr | 50.49 |
| 1275 x 305mm | 66.04 | 72.64 | 0.69 | 10.94 | nr | 83.58 |
| 765 x 356mm | 50.61 | 55.67 | 0.69 | 10.94 | nr | 66.61 |
| 1727 x 356mm | 107.13 | 117.84 | 1.04 | 16.49 | nr | 134.33 |
| 737 x 406mm | 42.65 | 46.91 | 0.44 | 6.98 | nr | 53.89 |
| 818 x 406mm | 43.39 | 47.73 | 0.44 | 6.98 | nr | 54.71 |
| 978 x 406mm | 45.32 | 49.85 | 0.44 | 6.98 | nr | 56.83 |
| 1379 x 406mm | 105.16 | 115.68 | 0.69 | 10.94 | nr | 126.62 |
| 1699 x 406mm | 108.28 | 119.11 | 1.04 | 16.49 | nr | 135.60 |
| 709 x 457mm | 41.55 | 45.70 | 0.44 | 6.98 | nr | 52.68 |
| 1671 x 457mm | 108.11 | 118.92 | 1.04 | 16.49 | nr | 135.41 |
| 678 x 508mm | 56.52 | 62.17 | 0.44 | 6.98 | nr | 69.15 |
| Taper; large end measured | | | | | | |
| 345 x 102mm | 59.94 | 65.93 | 0.95 | 15.07 | nr | 81.00 |
| 427 x 102mm | 62.27 | 68.50 | 1.06 | 16.81 | nr | 85.31 |
| 508 x 102mm | 64.96 | 71.46 | 1.13 | 17.92 | nr | 89.38 |
| 559 x 152mm | 78.87 | 86.76 | 1.26 | 19.98 | nr | 106.74 |
| 531 x 203mm | 76.93 | 84.62 | 1.26 | 19.98 | nr | 104.60 |
| 851 x 203mm | 108.88 | 119.77 | 1.16 | 18.40 | nr | 138.17 |
| 582 x 254mm | 82.86 | 91.15 | 1.34 | 21.25 | nr | 112.40 |
| 1303 x 254mm | 183.05 | 201.35 | 1.16 | 18.40 | nr | 219.75 |
| 632 x 305mm | 96.76 | 106.44 | 0.70 | 11.10 | nr | 117.54 |
| 1275 x 305mm | 185.89 | 204.48 | 1.16 | 18.40 | nr | 222.88 |
| 765 x 356mm | 109.10 | 120.01 | 1.16 | 18.40 | nr | 138.41 |
| 1247 x 356mm | 188.88 | 207.77 | 1.16 | 18.40 | nr | 226.17 |
| 1727 x 356mm | 240.99 | 265.09 | 1.25 | 19.83 | nr | 284.92 |
| 737 x 406mm | 118.13 | 129.94 | 1.16 | 18.40 | nr | 148.34 |
| 818 x 406mm | 123.03 | 135.33 | 1.27 | 20.14 | nr | 155.47 |
| 978 x 406mm | 139.72 | 153.69 | 1.44 | 22.84 | nr | 176.53 |
| 1379 x 406mm | 205.50 | 226.05 | 1.44 | 22.84 | nr | 248.89 |
| 1699 x 406mm | 237.39 | 261.13 | 1.44 | 22.84 | nr | 283.97 |
| 709 x 457mm | 118.42 | 130.26 | 1.16 | 18.40 | nr | 148.66 |
| 1671 x 457mm | 238.04 | 261.84 | 1.44 | 22.84 | nr | 284.68 |
| 678 x 508mm | 123.53 | 135.88 | 1.16 | 18.40 | nr | 154.28 |
| 90 degree radius bend | | | | | | |
| 345 x 102mm | 95.77 | 105.35 | 0.29 | 4.60 | nr | 109.95 |
| 427 x 102mm | 96.80 | 106.48 | 0.58 | 9.20 | nr | 115.68 |
| 508 x 102mm | 97.81 | 107.59 | 0.58 | 9.20 | nr | 116.79 |
| 559 x 152mm | 96.58 | 106.24 | 0.58 | 9.20 | nr | 115.44 |
| 531 x 203mm | 98.80 | 108.68 | 0.87 | 13.80 | nr | 122.48 |
| 851 x 203mm | 111.24 | 122.36 | 0.87 | 13.80 | nr | 136.16 |
| 582 x 254mm | 106.77 | 117.45 | 0.87 | 13.80 | nr | 131.25 |
| 1303 x 254mm | 146.93 | 161.62 | 0.96 | 15.23 | nr | 176.85 |
| 632 x 305mm | 120.91 | 133.00 | 0.87 | 13.80 | nr | 146.80 |
| 1275 x 305mm | 167.66 | 184.43 | 0.96 | 15.23 | nr | 199.66 |
| 765 x 356mm | 138.31 | 152.14 | 0.87 | 13.80 | nr | 165.94 |

## Material Costs / Measured Work Prices - Mechanical Installations

### Y: MECHANICAL AND ELECTRICAL SERVICES

| Item | Net Price £ | Material £ | Labour hours | Labour £ | Unit | Total Rate £ |
|---|---|---|---|---|---|---|
| 1247 x 356mm | 190.66 | 209.73 | 0.96 | 15.23 | nr | **224.96** |
| 1727 x 356mm | 237.68 | 261.45 | 1.25 | 19.83 | nr | **281.28** |
| 737 x 406mm | 157.66 | 173.43 | 0.96 | 15.23 | nr | **188.66** |
| 818 x 406mm | 160.47 | 176.52 | 0.87 | 13.80 | nr | **190.32** |
| 978 x 406mm | 192.32 | 211.55 | 0.96 | 15.23 | nr | **226.78** |
| 1379 x 406mm | 227.60 | 250.36 | 1.16 | 18.40 | nr | **268.76** |
| 1699 x 406mm | 246.39 | 271.03 | 1.25 | 19.83 | nr | **290.86** |
| 709 x 457mm | 172.07 | 189.28 | 0.87 | 13.80 | nr | **203.08** |
| 1671 x 457mm | 284.23 | 312.65 | 1.25 | 19.83 | nr | **332.48** |
| 678 x 508mm | 179.69 | 197.66 | 0.87 | 13.80 | nr | **211.46** |
| **45 degree radius bend** | | | | | | |
| 345 x 102mm | 86.66 | 95.33 | 1.03 | 16.34 | nr | **111.67** |
| 427 x 102mm | 87.46 | 96.21 | 0.95 | 15.07 | nr | **111.28** |
| 508 x 102mm | 88.23 | 97.05 | 0.82 | 13.01 | nr | **110.06** |
| 559 x 152mm | 81.68 | 89.85 | 0.79 | 12.53 | nr | **102.38** |
| 531 x 203mm | 83.08 | 91.39 | 0.85 | 13.48 | nr | **104.87** |
| 851 x 203mm | 134.19 | 147.61 | 0.18 | 2.85 | nr | **150.46** |
| 582 x 254mm | 85.47 | 94.02 | 0.76 | 12.05 | nr | **106.07** |
| 1303 x 254mm | 124.81 | 137.29 | 1.16 | 18.40 | nr | **155.69** |
| 632 x 305mm | 87.26 | 95.99 | 0.58 | 9.20 | nr | **105.19** |
| 1275 x 305mm | 140.23 | 154.25 | 1.16 | 18.40 | nr | **172.65** |
| 765 x 356mm | 108.08 | 118.89 | 0.87 | 13.80 | nr | **132.69** |
| 1247 x 356mm | 146.26 | 160.89 | 1.16 | 18.40 | nr | **179.29** |
| 1727 x 356mm | 181.43 | 199.57 | 1.26 | 19.98 | nr | **219.55** |
| 737 x 406mm | 121.52 | 133.67 | 0.69 | 10.94 | nr | **144.61** |
| 818 x 406mm | 123.21 | 135.53 | 0.40 | 6.34 | nr | **141.87** |
| 978 x 406mm | 147.95 | 162.75 | 0.87 | 13.80 | nr | **176.55** |
| 1379 x 406mm | 173.68 | 191.05 | 1.16 | 18.40 | nr | **209.45** |
| 1699 x 406mm | 186.84 | 205.52 | 1.27 | 20.14 | nr | **225.66** |
| 709 x 457mm | 132.08 | 145.29 | 0.81 | 12.85 | nr | **158.14** |
| 1671 x 457mm | 215.49 | 237.04 | 1.26 | 19.98 | nr | **257.02** |
| 678 x 508mm | 138.93 | 152.82 | 0.92 | 14.59 | nr | **167.41** |
| **90 degree hard bend with turning vanes** | | | | | | |
| 345 x 102mm | 88.60 | 97.46 | 0.55 | 8.72 | nr | **106.18** |
| 427 x 102mm | 91.78 | 100.96 | 1.16 | 18.40 | nr | **119.36** |
| 508 x 102mm | 95.28 | 104.81 | 1.16 | 18.40 | nr | **123.21** |
| 559 x 152mm | 117.83 | 129.61 | 1.16 | 18.40 | nr | **148.01** |
| 531 x 203mm | 118.00 | 129.80 | 1.73 | 27.44 | nr | **157.24** |
| 851 x 203mm | 187.20 | 205.92 | 1.73 | 27.44 | nr | **233.36** |
| 582 x 254mm | 121.65 | 133.81 | 1.73 | 27.44 | nr | **161.25** |
| 1303 x 254mm | 445.15 | 489.67 | 1.82 | 28.87 | nr | **518.54** |
| 632 x 305mm | 136.53 | 150.18 | 1.73 | 27.44 | nr | **177.62** |
| 1275 x 305mm | 489.01 | 537.91 | 1.82 | 28.87 | nr | **566.78** |
| 765 x 356mm | 156.82 | 172.50 | 1.73 | 27.44 | nr | **199.94** |
| 1247 x 356mm | 488.14 | 536.95 | 1.82 | 28.87 | nr | **565.82** |
| 1727 x 356mm | 775.03 | 852.53 | 1.82 | 28.87 | nr | **881.40** |
| 737 x 406mm | 162.59 | 178.85 | 1.73 | 27.44 | nr | **206.29** |
| 818 x 406mm | 213.89 | 235.28 | 1.73 | 27.44 | nr | **262.72** |
| 978 x 406mm | 227.99 | 250.79 | 1.73 | 27.44 | nr | **278.23** |
| 1379 x 406mm | 506.31 | 556.94 | 1.82 | 28.87 | nr | **585.81** |
| 1699 x 406mm | 773.28 | 850.61 | 2.11 | 33.47 | nr | **884.08** |
| 709 x 457mm | 173.16 | 190.48 | 1.73 | 27.44 | nr | **217.92** |
| 1671 x 457mm | 861.20 | 947.32 | 2.11 | 33.47 | nr | **980.79** |
| 678 x 508mm | 166.14 | 182.75 | 1.82 | 28.87 | nr | **211.62** |

## Y:MECHANICAL AND ELECTRICAL SERVICES

| Item | Net Price £ | Material £ | Labour hours | Labour £ | Unit | Total Rate £ |
|---|---|---|---|---|---|---|
| **Y30 : AIR DUCTLINES : MILD STEEL DUCTWORK (contd)** | | | | | | |
| Extra Over galvanised sheet metal DW142 class C spirally wound flat oval section ductwork; fittings; including all necessary stiffeners, joints, etc. (contd) | | | | | | |
| 90 degree branch | | | | | | |
| 345 x 102mm | 37.61 | 41.37 | 0.58 | 9.20 | nr | 50.57 |
| 427 x 102mm | 38.36 | 42.20 | 0.58 | 9.20 | nr | 51.40 |
| 508 x 102mm | 39.06 | 42.97 | 1.16 | 18.40 | nr | 61.37 |
| 559 x 152mm | 47.70 | 52.47 | 1.16 | 18.40 | nr | 70.87 |
| 531 x 203mm | 48.69 | 53.56 | 1.16 | 18.40 | nr | 71.96 |
| 851 x 203mm | 61.98 | 68.18 | 1.73 | 27.44 | nr | 95.62 |
| 582 x 254mm | 55.04 | 60.54 | 1.73 | 27.44 | nr | 87.98 |
| 1303 x 254mm | 102.67 | 112.94 | 1.82 | 28.87 | nr | 141.81 |
| 632 x 305mm | 58.80 | 64.68 | 1.73 | 27.44 | nr | 92.12 |
| 1275 x 305mm | 103.86 | 114.25 | 1.82 | 28.87 | nr | 143.12 |
| 765 x 356mm | 68.41 | 75.25 | 1.73 | 27.44 | nr | 102.69 |
| 1247 x 356mm | 109.97 | 120.97 | 1.82 | 28.87 | nr | 149.84 |
| 1727 x 356mm | 137.97 | 151.77 | 2.11 | 33.47 | nr | 185.24 |
| 737 x 406mm | 71.54 | 78.69 | 1.73 | 27.44 | nr | 106.13 |
| 818 x 406mm | 74.76 | 82.24 | 1.73 | 27.44 | nr | 109.68 |
| 978 x 406mm | 81.08 | 89.19 | 0.96 | 15.23 | nr | 104.42 |
| 1379 x 406mm | 119.50 | 131.45 | 1.93 | 30.61 | nr | 162.06 |
| 1699 x 406mm | 133.97 | 147.37 | 2.11 | 33.47 | nr | 180.84 |
| 709 x 457mm | 74.52 | 81.97 | 1.73 | 27.44 | nr | 109.41 |
| 1671 x 457mm | 140.10 | 154.11 | 2.11 | 33.47 | nr | 187.58 |
| 678 x 508mm | 79.89 | 87.88 | 1.73 | 27.44 | nr | 115.32 |
| 45 degree branch | | | | | | |
| 345 x 102mm | 63.19 | 69.51 | 0.58 | 9.20 | nr | 78.71 |
| 427 x 102mm | 66.35 | 72.98 | 0.58 | 9.20 | nr | 82.18 |
| 508 x 102mm | 69.82 | 76.80 | 1.16 | 18.40 | nr | 95.20 |
| 599 x 152mm | 83.60 | 91.96 | 1.73 | 27.44 | nr | 119.40 |
| 531 x 203mm | 84.60 | 93.06 | 1.73 | 27.44 | nr | 120.50 |
| 851 x 203mm | 124.88 | 137.37 | 1.73 | 27.44 | nr | 164.81 |
| 582 x 254mm | 90.97 | 100.07 | 1.73 | 27.44 | nr | 127.51 |
| 1303 x 254mm | 228.40 | 251.24 | 0.96 | 15.23 | nr | 266.47 |
| 632 x 305mm | 102.39 | 112.63 | 1.73 | 27.44 | nr | 140.07 |
| 1275 x 305mm | 224.44 | 246.88 | 1.82 | 28.87 | nr | 275.75 |
| 765 x 356mm | 119.73 | 131.70 | 1.73 | 27.44 | nr | 159.14 |
| 1247 x 356mm | 225.45 | 248.00 | 1.82 | 28.87 | nr | 276.87 |
| 1727 x 356mm | 325.27 | 357.80 | 1.82 | 28.87 | nr | 386.67 |
| 737 x 406mm | 130.53 | 143.58 | 1.73 | 27.44 | nr | 171.02 |
| 818 x 406mm | 141.49 | 155.64 | 1.73 | 27.44 | nr | 183.08 |
| 978 x 406mm | 163.23 | 179.55 | 1.73 | 27.44 | nr | 206.99 |
| 1379 x 406mm | 252.94 | 278.23 | 1.93 | 30.61 | nr | 308.84 |
| 1699 x 406mm | 326.44 | 359.08 | 2.19 | 34.73 | nr | 393.81 |
| 709 x 457mm | 130.94 | 144.03 | 1.73 | 27.44 | nr | 171.47 |
| 1671 x 457mm | 301.76 | 331.94 | 2.11 | 33.47 | nr | 365.41 |
| 678 x 508mm | 136.36 | 150.00 | 1.73 | 27.44 | nr | 177.44 |

*Material Costs / Measured Work Prices - Mechanical Installations*

## Y:MECHANICAL AND ELECTRICAL SERVICES

| Item | Net Price £ | Material £ | Labour hours | Labour £ | Unit | Total Rate £ |
|---|---|---|---|---|---|---|
| **Y31 : AIR DUCTLINE ANCILLARIES : DAMPERS** | | | | | | |
| **Dampers for rectangular section ductwork; including all necessary joints** | | | | | | |
| Volume control damper; galvanised steel casing; aluminium aerofoil blades; manually operated | | | | | | |
| Sum of two sides 200mm | 21.35 | 23.48 | 1.60 | 25.38 | nr | 48.86 |
| Sum of two sides 250mm | 21.90 | 24.09 | 1.60 | 25.38 | nr | 49.47 |
| Sum of two sides 300mm | 24.27 | 26.70 | 1.60 | 25.38 | nr | 52.08 |
| Sum of two sides 350mm | 24.90 | 27.39 | 1.60 | 25.38 | nr | 52.77 |
| Sum of two sides 400mm | 27.27 | 30.00 | 1.60 | 25.38 | nr | 55.38 |
| Sum of two sides 450mm | 27.89 | 30.68 | 1.60 | 25.38 | nr | 56.06 |
| Sum of two sides 500mm | 30.41 | 33.45 | 1.60 | 25.38 | nr | 58.83 |
| Sum of two sides 550mm | 31.13 | 34.24 | 1.60 | 25.38 | nr | 59.62 |
| Sum of two sides 600mm | 33.65 | 37.02 | 1.70 | 26.97 | nr | 63.99 |
| Sum of two sides 650mm | 34.36 | 37.80 | 2.00 | 31.72 | nr | 69.52 |
| Sum of two sides 700mm | 36.95 | 40.65 | 2.10 | 33.32 | nr | 73.97 |
| Sum of two sides 750mm | 37.82 | 41.60 | 2.10 | 33.32 | nr | 74.92 |
| Sum of two sides 800mm | 40.43 | 44.47 | 2.15 | 34.10 | nr | 78.57 |
| Sum of two sides 850mm | 41.29 | 45.42 | 2.20 | 34.89 | nr | 80.31 |
| Sum of two sides 900mm | 43.97 | 48.37 | 2.30 | 36.48 | nr | 84.85 |
| Sum of two sides 950mm | 44.92 | 49.41 | 2.30 | 36.48 | nr | 85.89 |
| Sum of two sides 1000mm | 47.67 | 52.44 | 2.40 | 38.06 | nr | 90.50 |
| Sum of two sides 1100mm | 51.45 | 56.59 | 2.60 | 41.24 | nr | 97.83 |
| Sum of two sides 1200mm | 55.39 | 60.93 | 2.80 | 44.41 | nr | 105.34 |
| Sum of two sides 1300mm | 59.41 | 65.35 | 3.10 | 49.17 | nr | 114.52 |
| Sum of two sides 1400mm | 63.51 | 69.86 | 3.25 | 51.55 | nr | 121.41 |
| Sum of two sides 1500mm | 67.77 | 74.55 | 3.40 | 53.92 | nr | 128.47 |
| Sum of two sides 1600mm | 72.10 | 79.31 | 3.45 | 54.72 | nr | 134.03 |
| Sum of two sides 1700mm | 76.59 | 84.25 | 3.60 | 57.10 | nr | 141.35 |
| Sum of two sides 1800mm | 81.16 | 89.28 | 3.90 | 61.86 | nr | 151.14 |
| Sum of two sides 1900mm | 85.82 | 94.40 | 4.20 | 66.64 | nr | 161.04 |
| Sum of two sides 2000mm | 90.62 | 99.68 | 4.33 | 68.68 | nr | 168.36 |
| Sum of two sides 2100mm | 105.14 | 115.65 | 4.43 | 70.26 | nr | 185.91 |
| Sum of two sides 2200mm | 110.77 | 121.85 | 4.55 | 72.16 | nr | 194.01 |
| Sum of two sides 2300mm | 116.57 | 128.23 | 4.70 | 74.54 | nr | 202.77 |
| Sum of two sides 2400mm | 122.55 | 134.81 | 4.88 | 77.40 | nr | 212.21 |
| Sum of two sides 2500mm | 128.62 | 141.48 | 5.00 | 79.30 | nr | 220.78 |
| Sum of two sides 2600mm | 134.98 | 148.48 | 5.10 | 80.92 | nr | 229.40 |
| Sum of two sides 2700mm | 141.45 | 155.59 | 5.30 | 84.06 | nr | 239.65 |
| Sum of two sides 2800mm | 148.08 | 162.89 | 5.50 | 87.23 | nr | 250.12 |
| Sum of two sides 2900mm | 154.86 | 170.35 | 5.60 | 88.82 | nr | 259.17 |
| Sum of two sides 3000mm | 161.79 | 177.97 | 5.75 | 91.20 | nr | 269.17 |
| Sum of two sides 3100mm | 168.88 | 185.77 | 6.00 | 95.16 | nr | 280.93 |
| Sum of two sides 3200mm | 176.11 | 193.72 | 6.15 | 97.54 | nr | 291.26 |
| Fire damper; galvanised steel casing; stainless steel folding shutter; fusible link and 24V d.c. electro-magnetic shutter release mechanisms; spring operated; BS 476 4-hour fire rating | | | | | | |
| Sum of two sides 200mm | 74.59 | 82.06 | 1.60 | 25.38 | nr | 107.44 |
| Sum of two sides 250mm | 74.59 | 82.06 | 1.60 | 25.38 | nr | 107.44 |
| Sum of two sides 300mm | 74.59 | 82.06 | 1.60 | 25.38 | nr | 107.44 |
| Sum of two sides 350mm | 74.59 | 82.06 | 1.60 | 25.38 | nr | 107.44 |
| Sum of two sides 400mm | 75.46 | 83.02 | 1.60 | 25.38 | nr | 108.40 |

## Y:MECHANICAL AND ELECTRICAL SERVICES

| Item | Net Price £ | Material £ | Labour hours | Labour £ | Unit | Total Rate £ |
|---|---|---|---|---|---|---|
| **Y31 : AIR DUCTLINE ANCILLARIES : DAMPERS (contd)** | | | | | | |
| **Dampers for rectangular section ductwork; including all necessary joints (contd)** | | | | | | |
| Fire damper; galvanised steel casing; stainless steel folding shutter; fusible link and 24V d.c. electro-magnetic shutter release mechanisms; spring operated; BS 476 4-hour fire rating (contd) | | | | | | |
| Sum of two sides 450mm | 76.33 | 83.97 | 1.60 | 25.38 | nr | 109.35 |
| Sum of two sides 500mm | 77.28 | 85.01 | 1.60 | 25.38 | nr | 110.39 |
| Sum of two sides 550mm | 78.14 | 85.95 | 1.60 | 25.38 | nr | 111.33 |
| Sum of two sides 600mm | 79.40 | 87.34 | 1.70 | 26.97 | nr | 114.31 |
| Sum of two sides 650mm | 80.34 | 88.39 | 2.00 | 31.72 | nr | 120.11 |
| Sum of two sides 700mm | 81.69 | 89.86 | 2.10 | 33.32 | nr | 123.18 |
| Sum of two sides 750mm | 82.64 | 90.90 | 2.10 | 33.32 | nr | 124.22 |
| Sum of two sides 800mm | 83.97 | 92.38 | 2.15 | 34.11 | nr | 126.49 |
| Sum of two sides 850mm | 85.00 | 93.50 | 2.20 | 34.93 | nr | 128.43 |
| Sum of two sides 900mm | 86.42 | 95.06 | 2.30 | 36.54 | nr | 131.60 |
| Sum of two sides 950mm | 87.52 | 96.27 | 2.30 | 36.54 | nr | 132.81 |
| Sum of two sides 1000mm | 88.94 | 97.83 | 2.40 | 38.13 | nr | 135.96 |
| Sum of two sides 1100mm | 91.54 | 100.69 | 2.60 | 41.30 | nr | 141.99 |
| Sum of two sides 1200mm | 94.21 | 103.64 | 2.80 | 44.43 | nr | 148.07 |
| Sum of two sides 1300mm | 97.05 | 106.75 | 3.10 | 49.17 | nr | 155.92 |
| Sum of two sides 1400mm | 99.89 | 109.88 | 3.25 | 51.55 | nr | 161.43 |
| Sum of two sides 1500mm | 102.88 | 113.18 | 3.40 | 53.95 | nr | 167.13 |
| Sum of two sides 1600mm | 105.96 | 116.56 | 3.45 | 54.72 | nr | 171.28 |
| Sum of two sides 1700mm | 109.11 | 120.03 | 3.60 | 57.10 | nr | 177.13 |
| Sum of two sides 1800mm | 112.34 | 123.57 | 3.90 | 61.86 | nr | 185.43 |
| Sum of two sides 1900mm | 115.65 | 127.22 | 4.20 | 66.64 | nr | 193.86 |
| Sum of two sides 2000mm | 119.11 | 131.03 | 4.33 | 68.68 | nr | 199.71 |
| Sum of two sides 2100mm | 129.14 | 142.05 | 4.43 | 70.26 | nr | 212.31 |
| Sum of two sides 2200mm | 133.30 | 146.64 | 4.55 | 72.16 | nr | 218.80 |
| Sum of two sides 2300mm | 137.68 | 151.45 | 4.70 | 74.54 | nr | 225.99 |
| Sum of two sides 2400mm | 142.24 | 156.46 | 4.88 | 77.40 | nr | 233.86 |
| Sum of two sides 2500mm | 147.00 | 161.70 | 5.00 | 79.30 | nr | 241.00 |
| Sum of two sides 2600mm | 151.95 | 167.16 | 5.10 | 80.92 | nr | 248.08 |
| Sum of two sides 2700mm | 157.13 | 172.84 | 5.30 | 84.06 | nr | 256.90 |
| Sum of two sides 2800mm | 162.51 | 178.76 | 5.50 | 87.23 | nr | 265.99 |
| Sum of two sides 2900mm | 168.11 | 184.92 | 5.60 | 88.82 | nr | 273.74 |
| Sum of two sides 3000mm | 173.93 | 191.33 | 5.75 | 91.20 | nr | 282.53 |
| Sum of two sides 3100mm | 179.99 | 197.99 | 6.00 | 95.16 | nr | 293.15 |
| Sum of two sides 3200mm | 186.28 | 204.91 | 6.15 | 97.54 | nr | 302.45 |
| **Dampers for circular section ductwork; including all necessary joints** | | | | | | |
| Volume control damper; galvanised steel casing; aluminium aerofoil blades; manually operated | | | | | | |
| 100mm dia. | 32.23 | 35.45 | 0.80 | 12.69 | nr | 48.14 |
| 150mm dia. | 33.80 | 37.18 | 0.90 | 14.27 | nr | 51.45 |
| 200mm dia. | 36.01 | 39.61 | 1.05 | 16.66 | nr | 56.27 |
| 250mm dia. | 38.84 | 42.72 | 1.20 | 19.04 | nr | 61.76 |
| 300mm dia. | 42.24 | 46.46 | 1.35 | 21.43 | nr | 67.89 |
| 355mm dia. | 46.18 | 50.80 | 1.65 | 26.17 | nr | 76.97 |

## Y:MECHANICAL AND ELECTRICAL SERVICES

| Item | Net Price £ | Material £ | Labour hours | Labour £ | Unit | Total Rate £ |
|---|---|---|---|---|---|---|
| 400mm dia. | 50.75 | 55.83 | 1.90 | 30.15 | nr | 85.98 |
| 450mm dia. | 55.94 | 61.53 | 2.10 | 33.32 | nr | 94.85 |
| 500mm dia. | 61.62 | 67.78 | 2.95 | 46.79 | nr | 114.57 |
| 630mm dia. | 82.34 | 90.57 | 4.55 | 72.16 | nr | 162.73 |
| 710mm dia. | 99.13 | 109.04 | 5.20 | 82.47 | nr | 191.51 |
| 800mm dia. | 108.35 | 119.19 | 5.80 | 91.99 | nr | 211.18 |
| 900mm dia. | 128.67 | 141.54 | 6.40 | 101.51 | nr | 243.05 |
| 1000mm dia. | 151.29 | 166.42 | 7.00 | 111.02 | nr | 277.44 |
| Fire damper; galvanised steel casing; stainless steel folding shutter; fusible link and 24V d.c. electro-magnetic shutter release mechanisms; spring operated; BS 476 4-hour fire rating | | | | | | |
| 100mm dia. | 76.72 | 84.40 | 0.80 | 12.69 | nr | 97.09 |
| 150mm dia. | 76.72 | 84.40 | 0.90 | 14.27 | nr | 98.67 |
| 200mm dia. | 78.22 | 86.05 | 1.05 | 16.65 | nr | 102.70 |
| 250mm dia. | 81.14 | 89.25 | 1.20 | 19.04 | nr | 108.29 |
| 300mm dia. | 84.21 | 92.64 | 1.35 | 21.43 | nr | 114.07 |
| 355mm dia. | 90.67 | 99.74 | 1.65 | 26.17 | nr | 125.91 |
| 400mm dia. | 90.67 | 99.74 | 1.90 | 30.15 | nr | 129.89 |
| 450mm dia. | 94.14 | 103.55 | 2.10 | 33.32 | nr | 136.87 |
| 500mm dia. | 97.68 | 107.45 | 2.95 | 46.79 | nr | 154.24 |
| 630mm dia. | 109.11 | 120.03 | 4.55 | 72.16 | nr | 192.19 |
| 710mm dia. | 117.38 | 129.13 | 5.20 | 82.47 | nr | 211.60 |
| 800mm dia. | 121.63 | 133.80 | 5.80 | 91.99 | nr | 225.79 |
| 900mm dia. | 130.70 | 143.77 | 6.40 | 101.51 | nr | 245.28 |
| 1000mm dia. | 140.15 | 154.18 | 7.00 | 111.02 | nr | 265.20 |
| **Dampers for flat oval section ductwork; including all necessary joints** | | | | | | |
| Volume control damper; galvanised steel casing; aluminium aerofoil blades; manually operated | | | | | | |
| 345 x 102mm | 46.33 | 50.96 | 1.20 | 19.04 | nr | 70.00 |
| 427 x 102mm | 48.39 | 53.23 | 1.35 | 21.43 | nr | 74.66 |
| 508 x 102mm | 50.43 | 55.47 | 1.60 | 25.38 | nr | 80.85 |
| 559 x 152mm | 56.42 | 62.06 | 1.90 | 30.15 | nr | 92.21 |
| 531 x 203mm | 60.20 | 66.22 | 1.90 | 30.15 | nr | 96.37 |
| 851 x 203mm | 68.71 | 75.58 | 4.55 | 72.16 | nr | 147.74 |
| 582 x 254mm | 66.43 | 73.07 | 2.10 | 33.32 | nr | 106.39 |
| 1303 x 254mm | 86.33 | 94.96 | 6.40 | 101.51 | nr | 196.47 |
| 632 x 305mm | 72.89 | 80.18 | 2.95 | 46.79 | nr | 126.97 |
| 1275 x 305mm | 91.75 | 100.92 | 6.40 | 101.51 | nr | 202.43 |
| 765 x 356mm | 82.50 | 90.75 | 4.55 | 72.16 | nr | 162.91 |
| 1247 x 356mm | 98.82 | 108.70 | 6.40 | 101.51 | nr | 210.21 |
| 1727 x 356mm | 112.85 | 124.14 | 8.20 | 130.00 | nr | 254.14 |
| 737 x 406mm | 86.28 | 94.91 | 4.55 | 72.16 | nr | 167.07 |
| 818 x 406mm | 89.44 | 98.38 | 5.20 | 82.47 | nr | 180.85 |
| 978 x 406mm | 94.24 | 103.66 | 5.50 | 87.23 | nr | 190.89 |
| 1379 x 406mm | 110.52 | 121.57 | 7.00 | 111.02 | nr | 232.59 |
| 1699 x 406mm | 120.39 | 132.43 | 8.20 | 130.05 | nr | 262.48 |
| 709 x 457mm | 91.57 | 100.73 | 4.50 | 71.44 | nr | 172.17 |
| 1671 x 457mm | 126.54 | 139.19 | 8.10 | 128.47 | nr | 267.66 |
| 678 x 508mm | 89.83 | 98.81 | 4.55 | 72.16 | nr | 170.97 |

## Y:MECHANICAL AND ELECTRICAL SERVICES

| Item | Net Price £ | Material £ | Labour hours | Labour £ | Unit | Total Rate £ |
|---|---|---|---|---|---|---|
| **Y31 : AIR DUCTLINE ANCILLARIES : DAMPERS (contd)** | | | | | | |
| **Dampers for flat oval section ductwork; including all necessary joints (contd)** | | | | | | |
| Fire damper; galvanised steel casing; stainless steel folding shutter; fusible link and 24V d.c. electro-magnetic shutter release mechanisms; BS 476 4-hour fire rating | | | | | | |
| 345 x 102mm | 89.02 | 97.92 | 1.20 | 19.04 | nr | **116.96** |
| 427 x 102mm | 91.78 | 100.96 | 1.35 | 21.43 | nr | **122.39** |
| 508 x 102mm | 94.53 | 103.98 | 1.60 | 25.38 | nr | **129.36** |
| 559 x 152mm | 98.16 | 107.98 | 1.90 | 30.15 | nr | **138.13** |
| 531 x 203mm | 99.03 | 108.93 | 1.90 | 30.15 | nr | **139.08** |
| 851 x 203mm | 108.87 | 119.77 | 4.55 | 72.16 | nr | **191.93** |
| 582 x 254mm | 102.72 | 113.00 | 2.10 | 33.32 | nr | **146.32** |
| 1303 x 254mm | 126.27 | 138.90 | 6.40 | 101.51 | nr | **240.41** |
| 632 x 305mm | 106.43 | 117.07 | 2.95 | 46.79 | nr | **163.86** |
| 1275 x 305mm | 121.09 | 133.20 | 6.40 | 101.51 | nr | **234.71** |
| 765 x 356mm | 112.97 | 124.28 | 4.55 | 72.16 | nr | **196.44** |
| 1247 x 356mm | 125.82 | 138.40 | 6.40 | 101.51 | nr | **239.91** |
| 1727 x 356mm | 143.95 | 158.36 | 8.20 | 130.00 | nr | **288.36** |
| 737 x 406mm | 113.84 | 125.22 | 4.55 | 72.16 | nr | **197.38** |
| 818 x 406mm | 116.68 | 128.35 | 5.20 | 82.47 | nr | **210.82** |
| 978 x 406mm | 121.00 | 133.11 | 5.50 | 87.23 | nr | **220.34** |
| 1379 x 406mm | 137.19 | 150.92 | 7.00 | 111.02 | nr | **261.94** |
| 1699 x 406mm | 146.55 | 161.20 | 8.20 | 130.00 | nr | **291.20** |
| 709 x 457mm | 116.12 | 127.74 | 4.50 | 71.44 | nr | **199.18** |
| 1671 x 457mm | 147.59 | 162.36 | 8.10 | 128.47 | nr | **290.83** |
| 678 x 508mm | 116.91 | 128.60 | 4.55 | 72.16 | nr | **200.76** |

## Y:MECHANICAL AND ELECTRICAL SERVICES

| Item | Net Price £ | Material £ | Labour hours | Labour £ | Unit | Total Rate £ |
|---|---|---|---|---|---|---|
| **Y31 : AIR DUCTLINE ANCILLARIES : ACCESS OPENINGS, DOORS AND COVERS** | | | | | | |
| Access doors, hollow steel construction; 25mm mineral wool insulation; removeable; fixed with cams; including sub-frame and integral sealing gaskets | | | | | | |
| Rectangular duct | | | | | | |
| 150 x 150mm | 8.67 | 9.54 | 1.25 | 19.83 | nr | **29.37** |
| 200 x 200mm | 9.62 | 10.58 | 1.25 | 19.83 | nr | **30.41** |
| 300 x 150mm | 9.85 | 10.84 | 1.25 | 19.83 | nr | **30.67** |
| 300 x 300mm | 11.27 | 12.40 | 1.25 | 19.83 | nr | **32.23** |
| 400 x 400mm | 13.08 | 14.39 | 1.35 | 21.43 | nr | **35.82** |
| 450 x 300mm | 12.85 | 14.13 | 1.50 | 23.81 | nr | **37.94** |
| 450 x 450mm | 14.18 | 15.60 | 1.50 | 23.81 | nr | **39.41** |
| Access doors, hollow steel construction; 25mm mineral wool insulation; hinged; locked with cams; including sub-frame and integral sealing gaskets | | | | | | |
| Rectangular duct | | | | | | |
| 150 x 150mm | 8.67 | 9.54 | 1.25 | 19.83 | nr | **29.37** |
| 200 x 200mm | 9.62 | 10.58 | 1.25 | 19.83 | nr | **30.41** |
| 300 x 150mm | 9.85 | 10.84 | 1.25 | 19.83 | nr | **30.67** |
| 300 x 300mm | 11.27 | 12.40 | 1.25 | 19.83 | nr | **32.23** |
| 400 x 400mm | 13.08 | 14.39 | 1.35 | 21.43 | nr | **35.82** |
| 450 x 300mm | 12.85 | 14.13 | 1.50 | 23.81 | nr | **37.94** |
| 450 x 450mm | 14.18 | 15.60 | 1.50 | 23.81 | nr | **39.41** |

RESPONSE — SPON's price data now available in an electronic estimating system call us on 0171-522-9966 for details

## Y:MECHANICAL AND ELECTRICAL SERVICES

| Item | Net Price £ | Material £ | Labour hours | Labour £ | Unit | Total Rate £ |
|---|---|---|---|---|---|---|
| **Y40 : AIR HANDLING UNITS : SOUND ATTENUATORS** | | | | | | |
| Attenuators; shaped acoustic foam; self securing; fitted to ductwork | | | | | | |
| To suit rectangular ducts; unit length 600mm | | | | | | |
| 100 x 100mm | 4.80 | 5.28 | 0.05 | 0.57 | nr | 5.85 |
| 100 x 150mm | 6.36 | 7.00 | 0.05 | 0.57 | nr | 7.57 |
| 100 x 200mm | 8.30 | 9.13 | 0.07 | 0.77 | nr | 9.90 |
| 100 x 400mm | 15.48 | 17.03 | 0.07 | 0.77 | nr | 17.80 |
| 150 x 150mm | 8.54 | 9.39 | 0.07 | 0.77 | nr | 10.16 |
| 150 x 200mm | 11.21 | 12.33 | 0.07 | 0.77 | nr | 13.10 |
| 150 x 300mm | 16.28 | 17.91 | 0.07 | 0.77 | nr | 18.68 |
| 200 x 200mm | 14.80 | 16.28 | 0.07 | 0.77 | nr | 17.05 |
| 200 x 300mm | 21.16 | 23.28 | 0.07 | 0.77 | nr | 24.05 |
| 200 x 400mm | 27.96 | 30.76 | 0.10 | 1.15 | nr | 31.91 |
| 300 x 150mm | 16.28 | 17.91 | 0.07 | 0.77 | nr | 18.68 |
| 300 x 200mm | 21.16 | 23.28 | 0.07 | 0.77 | nr | 24.05 |
| 300 x 300mm | 32.06 | 35.27 | 0.10 | 1.15 | nr | 36.42 |
| To suit rectangular ducts; unit length 1200mm | | | | | | |
| 100 x 100mm | 9.31 | 10.24 | 0.08 | 0.92 | nr | 11.16 |
| 100 x 150mm | 12.34 | 13.57 | 0.08 | 0.92 | nr | 14.49 |
| 100 x 200mm | 16.12 | 17.73 | 0.11 | 1.21 | nr | 18.94 |
| 100 x 400mm | 30.09 | 33.10 | 0.11 | 1.21 | nr | 34.31 |
| 150 x 150mm | 16.61 | 18.27 | 0.11 | 1.21 | nr | 19.48 |
| 150 x 200mm | 21.81 | 23.99 | 0.11 | 1.21 | nr | 25.20 |
| 150 x 300mm | 31.63 | 34.79 | 0.11 | 1.21 | nr | 36.00 |
| 200 x 200mm | 28.74 | 31.61 | 0.11 | 1.21 | nr | 32.82 |
| 200 x 300mm | 41.10 | 45.21 | 0.11 | 1.21 | nr | 46.42 |
| 200 x 400mm | 54.33 | 59.76 | 0.15 | 1.76 | nr | 61.52 |
| 300 x 150mm | 16.28 | 17.91 | 0.11 | 1.21 | nr | 19.12 |
| 300 x 200mm | 41.10 | 45.21 | 0.11 | 1.21 | nr | 46.42 |
| 300 x 300mm | 62.32 | 68.55 | 0.15 | 1.76 | nr | 70.31 |
| To suit circular ducts; unit length 600mm | | | | | | |
| 100mm dia. | 4.40 | 4.84 | 0.04 | 0.41 | nr | 5.25 |
| 125mm dia. | 5.62 | 6.18 | 0.04 | 0.41 | nr | 6.59 |
| 160mm dia. | 8.86 | 9.75 | 0.04 | 0.41 | nr | 10.16 |
| 200mm dia. | 12.55 | 13.80 | 0.04 | 0.41 | nr | 14.21 |
| 224mm dia. | 15.20 | 16.72 | 0.04 | 0.41 | nr | 17.13 |
| 250mm dia. | 18.46 | 20.31 | 0.05 | 0.60 | nr | 20.91 |
| 315mm dia. | 25.98 | 28.58 | 0.05 | 0.60 | nr | 29.18 |
| 355mm dia. | 33.98 | 37.38 | 0.05 | 0.60 | nr | 37.98 |
| 400mm dia. | 45.45 | 49.99 | 0.07 | 0.82 | nr | 50.81 |
| 450mm dia. | 50.20 | 55.22 | 0.07 | 0.82 | nr | 56.04 |

## Y:MECHANICAL AND ELECTRICAL SERVICES

| Item | Net Price £ | Material £ | Labour hours | Labour £ | Unit | Total Rate £ |
|---|---|---|---|---|---|---|
| To suit circular ducts; unit length 1200mm | | | | | | |
| 100mm dia. | 8.75 | 9.63 | 0.06 | 0.64 | nr | **10.27** |
| 125mm dia. | 11.04 | 12.14 | 0.06 | 0.64 | nr | **12.78** |
| 160mm dia. | 17.54 | 19.29 | 0.06 | 0.64 | nr | **19.93** |
| 200mm dia. | 24.87 | 27.36 | 0.06 | 0.64 | nr | **28.00** |
| 224mm dia. | 30.08 | 33.09 | 0.06 | 0.64 | nr | **33.73** |
| 250mm dia. | 36.64 | 40.30 | 0.08 | 0.96 | nr | **41.26** |
| 315mm dia. | 51.66 | 56.83 | 0.08 | 0.96 | nr | **57.79** |
| 355mm dia. | 67.66 | 74.43 | 0.08 | 0.96 | nr | **75.39** |
| 400mm dia. | 90.25 | 99.28 | 0.11 | 1.27 | nr | **100.55** |
| 450mm dia. | 99.60 | 109.56 | 0.11 | 1.27 | nr | **110.83** |

## Y:MECHANICAL AND ELECTRICAL SERVICES

| Item | Net Price £ | Material £ | Labour hours | Labour £ | Unit | Total Rate £ |
|---|---|---|---|---|---|---|
| **Y40 : AIR HANDLING UNITS : AIR DISTRIBUTION EQUIPMENT** | | | | | | |
| **One piece package; ceiling void unit; electrical work elsewhere** | | | | | | |
| Low profile; heat pump units; four stage remote control; indoor or outdoor | | | | | | |
| Cooling 13.7kW, heating 15.5kW | 5417.80 | 5959.58 | 40.00 | 459.00 | nr | 6418.58 |
| Cooling 16.4kW, heating 18.8kW | 5547.58 | 6102.34 | 40.00 | 459.00 | nr | 6561.34 |
| Cooling 21.7kW, heating 24.9kW | 6888.64 | 7577.50 | 40.00 | 459.00 | nr | 8036.50 |
| Low profile; cooling only units; four stage remote control; indoor or outdoor | | | | | | |
| Cooling 7.20kW, 24,500(BTU/hr) | 1358.17 | 1493.99 | 40.00 | 459.00 | nr | 1952.99 |
| Cooling 10.6kW, 36,000(BTU/hr) | 1680.90 | 1848.99 | 40.00 | 459.00 | nr | 2307.99 |
| Cooling 12.7kW, 43,300(BTU/hr) | 1889.55 | 2078.51 | 40.00 | 459.00 | nr | 2537.51 |
| Cooling 20.50kW, 70,000(BTU/hr) | 3219.45 | 3541.39 | 40.00 | 459.00 | nr | 4000.39 |
| Cooling 25.00kW, 85,300(BTU/hr) | 3908.78 | 4299.66 | 40.00 | 459.00 | nr | 4758.66 |
| **Split systems; ceiling void units; electrical work elsewhere** | | | | | | |
| Low profile heat pump units; four stage remote control; indoor | | | | | | |
| Cooling 13.7kW, heating 15.5kW | 1842.67 | 2026.94 | 16.00 | 183.60 | nr | 2210.54 |
| Cooling 16.4kW, heating 18.8kW | 1886.96 | 2075.66 | 16.00 | 183.60 | nr | 2259.26 |
| Cooling 21.7kW, heating 24.9kW | 2342.22 | 2576.44 | 16.00 | 183.60 | nr | 2760.04 |
| Low profile heat pump units; four stage remote control; outdoor | | | | | | |
| Cooling 13.7kW, heating 15.5kW | 3575.13 | 3932.64 | 16.00 | 183.60 | nr | 4116.24 |
| Cooling 16.4kW, heating 18.8kW | 3660.62 | 4026.68 | 16.00 | 183.60 | nr | 4210.28 |
| Cooling 21.7kW, heating 24.9kW | 4546.42 | 5001.06 | 16.00 | 183.60 | nr | 5184.66 |
| Low profile cooling only units; four stage remote control; indoor | | | | | | |
| Cooling 2.4kW, 8,200(BTU/hr) | 516.75 | 568.42 | 16.00 | 183.60 | nr | 752.02 |
| Cooling 3.5kW, 11,900(BTU/hr) | 724.42 | 796.86 | 16.00 | 183.60 | nr | 980.46 |
| Cooling 5.1kW, 17,400(BTU/hr) | 913.58 | 1004.94 | 16.00 | 183.60 | nr | 1188.54 |
| Cooling 6.5kW, 22,100(BTU/hr) | 1103.70 | 1214.07 | 16.00 | 183.60 | nr | 1397.67 |
| Low profile cooling only units; four stage remote control; outdoor | | | | | | |
| Cooling 13.7kW | 2955.07 | 3250.58 | 16.00 | 183.60 | nr | 3434.18 |
| Cooling 16.4kW | 3007.60 | 3308.36 | 16.00 | 183.60 | nr | 3491.96 |
| Cooling 21.7kW | 4002.58 | 4402.84 | 16.00 | 183.60 | nr | 4586.44 |

## Y:MECHANICAL AND ELECTRICAL SERVICES

| Item | Net Price £ | Material £ | Labour hours | Labour £ | Unit | Total Rate £ |
|---|---|---|---|---|---|---|
| **Chilled water cassette** | | | | | | |
| Standard model with remote thermostat / fan speed controller; suitable for 2,3 or 4 way blow; including fascia | | | | | | |
| Cooling duty 3.25kW | 1037.21 | 1140.93 | 4.00 | 45.90 | nr | **1186.83** |
| Cooling duty 3.75kW | 1084.59 | 1193.05 | 4.00 | 45.90 | nr | **1238.95** |
| Cooling duty 5.80kW | 1493.50 | 1642.85 | 4.00 | 45.90 | nr | **1688.75** |
| Cooling duty 7.6kW | 1596.50 | 1756.15 | 4.00 | 45.90 | nr | **1802.05** |

## Y:MECHANICAL AND ELECTRICAL SERVICES

| Item | Net Price £ | Material £ | Labour hours | Labour £ | Unit | Total Rate £ |
|---|---|---|---|---|---|---|
| **Y41 : FANS : AXIAL FLOW UNIT** | | | | | | |
| Axial flow fan; including ancillaries, anti vibration mountings, mounting feet, matching flanges, flex connectors and clips; 415V, 3 phase, 50Hz motor; includes fixing in position; electrical work elsewhere | | | | | | |
| Aerofoil blade fan unit; short duct case | | | | | | |
| 315mm dia.; 0.47 m3/s duty; 147 Pa | 297.17 | 393.99 | 4.50 | 51.64 | nr | 445.63 |
| 500mm dia.; 1.89 m3/s duty; 500 Pa | 594.91 | 728.88 | 5.00 | 57.38 | nr | 786.26 |
| 560mm dia.; 2.36 m3/s duty; 147 Pa | 526.63 | 661.17 | 5.50 | 63.11 | nr | 724.28 |
| 710mm dia.; 5.67 m3/s duty; 245 Pa | 709.07 | 866.78 | 6.00 | 68.85 | nr | 935.63 |
| Aerofoil blade fan unit; long duct case | | | | | | |
| 315mm dia.; 0.47 m3/s duty; 147 Pa | 311.16 | 416.15 | 4.50 | 51.64 | nr | 467.79 |
| 500mm dia.; 1.89 m3/s duty; 500 Pa | 566.36 | 704.25 | 5.00 | 57.38 | nr | 761.63 |
| 560mm dia.; 2.36 m3/s duty; 147 Pa | 464.51 | 599.60 | 5.50 | 63.11 | nr | 662.71 |
| 710mm dia.; 5.67 m3/s duty; 245 Pa | 726.43 | 892.64 | 6.00 | 68.85 | nr | 961.49 |
| Aerofoil blade fan unit; two stage; long duct case | | | | | | |
| 315mm; 0.47m3/s duty; 500 Pa | 615.06 | 764.59 | 4.50 | 51.64 | nr | 816.23 |
| 355mm; 0.83m3/s duty; 147 Pa | 655.34 | 808.91 | 4.75 | 54.51 | nr | 863.42 |
| 710mm; 3.77m3/s duty; 431 Pa | 1671.67 | 1951.49 | 6.00 | 68.85 | nr | 2020.34 |
| 710mm; 6.61m3/s duty; 500 Pa | 1671.67 | 1951.49 | 6.00 | 68.85 | nr | 2020.34 |
| Roof mounted extract fan; including ancillaries, fibreglass cowling, fitted shutters and bird guard; 415V, 3 phase, 50Hz motor; includes fixing in position; electrical work elsewhere | | | | | | |
| Flat roof installation, fixed to curb | | | | | | |
| 315mm; 900rpm | 351.46 | 386.61 | 4.50 | 51.64 | nr | 438.25 |
| 315mm; 1380rpm | 294.38 | 363.22 | 4.50 | 51.64 | nr | 414.86 |
| 400mm; 900rpm | 383.92 | 422.31 | 5.50 | 63.11 | nr | 485.42 |
| 400mm; 1360rpm | 402.95 | 443.25 | 5.50 | 63.11 | nr | 506.36 |
| 800mm; 530rpm | 1040.95 | 1145.05 | 7.00 | 80.33 | nr | 1225.38 |
| 800mm; 700rpm | 1040.95 | 1145.05 | 7.00 | 80.33 | nr | 1225.38 |
| 800mm; 920rpm | 1102.51 | 1212.76 | 7.00 | 80.33 | nr | 1293.09 |
| 1000mm; 470rpm | 1742.75 | 1917.01 | 8.00 | 91.80 | nr | 2008.81 |
| 1000mm; 570rpm | 1589.41 | 1748.34 | 8.00 | 91.80 | nr | 1840.14 |
| 1000mm; 710rpm | 1552.47 | 1707.71 | 8.00 | 91.80 | nr | 1799.51 |

## Y:MECHANICAL AND ELECTRICAL SERVICES

| Item | Net Price £ | Material £ | Labour hours | Labour £ | Unit | Total Rate £ |
|---|---|---|---|---|---|---|
| Pitched roof installation; including purlin mounting box | | | | | | |
| 315mm; 900rpm | 397.35 | 437.10 | 4.50 | 51.64 | nr | **488.74** |
| 315mm; 1380rpm | 376.09 | 413.71 | 4.50 | 51.64 | nr | **465.35** |
| 400mm; 900rpm | 439.89 | 483.87 | 5.50 | 63.11 | nr | **546.98** |
| 400mm; 1360rpm | 458.91 | 504.80 | 5.50 | 63.11 | nr | **567.91** |
| 800mm; 530rpm | 1150.64 | 1265.70 | 7.00 | 80.33 | nr | **1346.03** |
| 800mm; 700rpm | 1150.64 | 1265.70 | 7.00 | 80.33 | nr | **1346.03** |
| 800mm; 920rpm | 1212.20 | 1333.42 | 7.00 | 80.33 | nr | **1413.75** |
| 1000mm; 470rpm | 1880.42 | 2068.46 | 8.00 | 91.80 | nr | **2160.26** |
| 1000mm; 570rpm | 1727.08 | 1899.79 | 8.00 | 91.80 | nr | **1991.59** |
| 1000mm; 710rpm | 1690.14 | 1859.15 | 8.00 | 91.80 | nr | **1950.95** |

# Keep your figures up to date, free of charge

This section, and most of the other information in this Price Book, is brought up to date every three months, until the next annual edition, in the *Price Book Update*.

The *Update* is available free to all Price Book purchasers.

To ensure you receive your copy, simply complete the reply card from the centre of the book and return it to us.

## Y:MECHANICAL AND ELECTRICAL SERVICES

| Item | Net Price £ | Material £ | Labour hours | Labour £ | Unit | Total Rate £ |
|---|---|---|---|---|---|---|
| **Y41 : FANS : TOILET EXTRACT UNITS** | | | | | | |
| Toilet extract units; centrifugal fan; various speeds for internal domestic bathrooms/ W.C's with built-in filter; complete with housing; includes placing in position; electrical work elsewhere | | | | | | |
| Fan unit; fixed to wall; including shutter | | | | | | |
|    Single speed 85m3/hr | 55.70 | 61.27 | 0.75 | 8.61 | nr | **69.88** |
|    Two speed 60-85m3/hr | 65.64 | 72.20 | 0.83 | 9.56 | nr | **81.76** |
| Humidity controlled; autospeed; fixed to wall; including shutter | | | | | | |
|    30-60-85m3/hr | 109.54 | 120.49 | 1.00 | 11.47 | nr | **131.96** |

## Y:MECHANICAL AND ELECTRICAL SERVICES

| Item | Net Price £ | Material £ | Labour hours | Labour £ | Unit | Total Rate £ |
|---|---|---|---|---|---|---|
| **Y41 : FANS : KITCHEN EXTRACT UNITS** | | | | | | |
| Centrifugal fan; single speed for internal domestic kitchens/ utility rooms; fitted with standard overload protection; complete with housing; includes placing in position; electrical work elsewhere | | | | | | |
| Window mounted | | | | | | |
| 245m3/hr | 56.67 | 62.34 | 0.50 | 5.74 | nr | 68.08 |
| 500m3/hr | 86.30 | 94.93 | 0.50 | 5.74 | nr | 100.67 |
| Wall mounted | | | | | | |
| 245m3/hr | 68.65 | 75.52 | 0.83 | 9.56 | nr | 85.08 |
| 500m3/hr | 98.28 | 108.11 | 0.83 | 9.56 | nr | 117.67 |

Material Costs / Measured Work Prices - Mechanical Installations

## Y:MECHANICAL AND ELECTRICAL SERVICES

| Item | Net Price £ | Material £ | Labour hours | Labour £ | Unit | Total Rate £ |
|---|---|---|---|---|---|---|
| **Y41 : FANS : MULTI VENT UNIT** | | | | | | |
| Centrifugal fan; various speeds, simultaneous ventilation from seperate areas fitted with standard overload protection; complete with housing; includes placing in position; ducting and electrical work elsewhere | | | | | | |
| Fan unit | | | | | | |
| 147-300m3/hr | 148.42 | 163.26 | 1.00 | 11.47 | nr | **174.73** |
| 175-411m3/hr | 162.05 | 178.25 | 1.00 | 11.47 | nr | **189.72** |

## Material Costs / Measured Work Prices - Mechanical Installations

### Y:MECHANICAL AND ELECTRICAL SERVICES

| Item | Net Price £ | Material £ | Labour hours | Labour £ | Unit | Total Rate £ |
|---|---|---|---|---|---|---|
| **Y42 : AIR FILTRATION : FILTERS** | | | | | | |
| **High efficiency filters; 99.997% EU13; tested to BS 3928** | | | | | | |
| Standard; 1700m3/ hr air volume; continuous rating up to 70 C; sealed wood case, aluminium spacers, neoprene gaskets; water repellant filter media; including placing in position | | | | | | |
| 609 x 609 x 298mm | 254.31 | 279.74 | 2.00 | 22.95 | nr | 302.69 |
| High capacity; 3400m3/hr air volume; continuous rating up to 70 C; anti-corrosion coated mild steel frame, polyurethane sealant and neoprene gaskets; water repellant filter media; including fixing in position | | | | | | |
| 609 x 609 x 298mm | 351.33 | 386.46 | 2.00 | 22.95 | nr | 409.41 |
| Chemical resistant; 1700m3/hr air volume; continuous rating up to 66 C; anti-corrosion coated mild steel frame, polyurethane sealant and silicone rubber gaskets; water repellant filter media; including fixing in position | | | | | | |
| 609 x 609 x 298mm | 285.71 | 314.28 | 2.00 | 22.95 | nr | 337.23 |
| **Bag Filters; 40/60% EU5; tested to BS 6540** | | | | | | |
| Bag filter; continuous rating up to 60 C; rigid filter assembly; sealed into one piece coated mild steel header with sealed pocket separators; including placing in position | | | | | | |
| 6 pocket, 592 x 592 x 25mm frame; pockets 350mm long; 1350m3/hr | 25.17 | 27.69 | 2.00 | 22.95 | nr | 50.64 |
| 6 pocket, 592 x 592 x 25mm frame; pockets 500mm long; 1900m3/hr | 29.43 | 32.37 | 2.50 | 28.69 | nr | 61.06 |
| 8 pocket, 592 x 592 x 25mm frame; pockets 900mm long; 3400m3/hr | 58.20 | 64.02 | 3.00 | 34.42 | nr | 98.44 |
| **Grease filters, washable; minimum 65%** | | | | | | |
| Double sided extract unit; lightweight stainless steel construction; demountable composite filter media of woven metal mat and expanded metal mesh supports; for mounting on hood and extract systems; including placing in position | | | | | | |
| 500 x 686 x 565mm, 4080m3/hr | 327.38 | 360.12 | 2.00 | 22.95 | nr | 383.07 |
| 1000 x 686 x 565mm, 8160m3/hr; | 499.21 | 549.13 | 3.00 | 34.42 | nr | 583.55 |
| 1500 x 686 x 565mm, 12240m3/hr; | 684.75 | 753.23 | 3.50 | 40.16 | nr | 793.39 |

Material Costs / Measured Work Prices - Mechanical Installations

Y:MECHANICAL AND ELECTRICAL SERVICES

| Item | Net Price £ | Material £ | Labour hours | Labour £ | Unit | Total Rate £ |
|---|---|---|---|---|---|---|
| **Y42 : AIR FILTRATION : FILTERS (contd)** | | | | | | |
| **Panel filters; 82% EU3/G3; tested to BS EN779** | | | | | | |
| Modular filter panels; continuous rating up to 100 C; graduated density media; rigid cardboard frame; including placing in position | | | | | | |
| 594 x 594 x 50mm; 2380m3/hr | 5.30 | 5.83 | 2.50 | 28.69 | nr | 34.52 |
| 594 x 296 x 50mm; 1190m3/hr | 4.03 | 4.43 | 2.50 | 28.69 | nr | 33.12 |
| **Panel filters; 90% EU4; tested to BS 6540** | | | | | | |
| Modular filter panels; continuous rating up to 100 C; pleated media with wire support; rigid cardboard frame; including placing in position | | | | | | |
| 594 x 594 x 50mm; 2380m3/hr | 10.63 | 11.69 | 3.00 | 34.42 | nr | 46.11 |
| 594 x 296 x 50mm; 1190m3/hr | 8.74 | 9.61 | 3.00 | 34.42 | nr | 44.03 |
| **Carbon filters; standard duty discarb filters; steel frame with bonded carbon panels; for fixing to ductwork; including placing in position** | | | | | | |
| 12 panels | | | | | | |
| 597 x 597 x 298mm, 1460m3/hr | 255.00 | 280.50 | 0.33 | 3.79 | nr | 284.29 |
| 597 x 597 x 451mm, 2200m3/hr | 313.50 | 344.85 | 0.33 | 3.79 | nr | 348.64 |
| 597 x 597 x 597mm, 2930m3/hr | 370.50 | 407.55 | 0.33 | 3.79 | nr | 411.34 |
| 8 panels | | | | | | |
| 451 x 451 x 298mm, 740m3/hr | 162.00 | 178.20 | 0.29 | 3.33 | nr | 181.53 |
| 451 x 451 x 451mm, 1105m3/hr | 195.00 | 214.50 | 0.29 | 3.33 | nr | 217.83 |
| 451 x 451 x 597mm, 1460m3/hr | 226.50 | 249.15 | 0.29 | 3.33 | nr | 252.48 |
| 6 panels | | | | | | |
| 298 x 298 x 298mm, 365m3/hr | 108.00 | 118.80 | 0.25 | 2.87 | nr | 121.67 |
| 298 x 298 x 451mm, 550m3/hr | 127.50 | 140.25 | 0.25 | 2.87 | nr | 143.12 |
| 298 x 298 x 597mm, 780m3/hr | 147.00 | 161.70 | 0.25 | 2.87 | nr | 164.57 |

## Y:MECHANICAL AND ELECTRICAL SERVICES

| Item | Net Price £ | Material £ | Labour hours | Labour £ | Unit | Total Rate £ |
|---|---|---|---|---|---|---|
| **Y45 : SILENCERS/ACOUSTIC TREATMENT : ACOUSTIC LOUVRES** | | | | | | |
| Acoustic louvres; opening mounted; 300mm deep steel louvres with blades packed with acoustic infill; 12mm galvanised mesh birdscreen; screw fixing in opening | | | | | | |
| Louvre units; self finished galvanised steel | | | | | | |
| 900 high x 600 wide | 127.41 | 140.15 | 3.00 | 34.42 | nr | 174.57 |
| 900 high x 900 wide | 175.81 | 193.39 | 3.00 | 34.42 | nr | 227.81 |
| 900 high x 1200 wide | 195.55 | 215.10 | 3.34 | 38.33 | nr | 253.43 |
| 900 high x 1500 wide | 303.22 | 333.54 | 3.34 | 38.33 | nr | 371.87 |
| 900 high x 1800 wide | 325.92 | 358.51 | 3.34 | 38.33 | nr | 396.84 |
| 900 high x 2100 wide | 370.35 | 407.38 | 3.34 | 38.33 | nr | 445.71 |
| 900 high x 2400 wide | 390.13 | 429.14 | 3.68 | 42.23 | nr | 471.37 |
| 900 high x 2700 wide | 498.79 | 548.67 | 3.68 | 42.23 | nr | 590.90 |
| 900 high x 3000 wide | 518.53 | 570.38 | 3.68 | 42.23 | nr | 612.61 |
| 1200 high x 600 wide | 176.80 | 194.48 | 3.00 | 34.42 | nr | 228.90 |
| 1200 high x 900 wide | 250.83 | 275.91 | 3.34 | 38.33 | nr | 314.24 |
| 1200 high x 1200 wide | 273.52 | 300.87 | 3.34 | 38.33 | nr | 339.20 |
| 1200 high x 1500 wide | 431.61 | 474.77 | 3.34 | 38.33 | nr | 513.10 |
| 1200 high x 1800 wide | 463.17 | 509.49 | 3.68 | 42.23 | nr | 551.72 |
| 1200 high x 2100 wide | 524.43 | 576.87 | 3.68 | 42.23 | nr | 619.10 |
| 1200 high x 2400 wide | 546.16 | 600.78 | 3.68 | 42.23 | nr | 643.01 |
| 1500 high x 600 wide | 208.37 | 229.21 | 3.00 | 34.42 | nr | 263.63 |
| 1500 high x 900 wide | 293.33 | 322.66 | 3.34 | 38.33 | nr | 360.99 |
| 1500 high x 1200 wide | 318.00 | 349.80 | 3.34 | 38.33 | nr | 388.13 |
| 1500 high x 1500 wide | 503.69 | 554.06 | 3.68 | 42.23 | nr | 596.29 |
| 1500 high x 1800 wide | 539.27 | 593.20 | 3.68 | 42.23 | nr | 635.43 |
| 1500 high x 2100 wide | 612.35 | 673.59 | 4.00 | 45.90 | nr | 719.49 |
| 1800 high x 600 wide | 238.01 | 261.81 | 3.34 | 38.33 | nr | 300.14 |
| 1800 high x 900 wide | 334.79 | 368.27 | 3.34 | 38.33 | nr | 406.60 |
| 1800 high x 1200 wide | 362.40 | 398.64 | 3.68 | 42.23 | nr | 440.87 |
| 1800 high x 1500 wide | 573.79 | 631.17 | 3.68 | 42.23 | nr | 673.40 |
| Louvre units; polyester powder coated steel | | | | | | |
| 900 high x 600 wide | 158.98 | 174.88 | 3.00 | 34.42 | nr | 209.30 |
| 900 high x 900 wide | 223.20 | 245.52 | 3.00 | 34.42 | nr | 279.94 |
| 900 high x 1200 wide | 258.75 | 284.63 | 3.34 | 38.33 | nr | 322.96 |
| 900 high x 1500 wide | 383.15 | 421.46 | 3.34 | 38.33 | nr | 459.79 |
| 900 high x 1800 wide | 421.69 | 463.86 | 3.34 | 38.33 | nr | 502.19 |
| 900 high x 2100 wide | 481.93 | 530.12 | 3.34 | 38.33 | nr | 568.45 |
| 900 high x 2400 wide | 517.54 | 569.29 | 3.68 | 42.23 | nr | 611.52 |
| 900 high x 2700 wide | 641.94 | 706.13 | 3.68 | 42.23 | nr | 748.36 |
| 900 high x 3000 wide | 661.68 | 727.85 | 3.68 | 42.23 | nr | 770.08 |
| 1200 high x 600 wide | 219.25 | 241.18 | 3.00 | 34.42 | nr | 275.60 |
| 1200 high x 900 wide | 314.06 | 345.47 | 3.34 | 38.33 | nr | 383.80 |
| 1200 high x 1200 wide | 358.47 | 394.32 | 3.34 | 38.33 | nr | 432.65 |
| 1200 high x 1500 wide | 538.21 | 592.03 | 3.34 | 38.33 | nr | 630.36 |
| 1200 high x 1800 wide | 590.58 | 649.64 | 3.68 | 42.23 | nr | 691.87 |
| 1200 high x 2100 wide | 673.57 | 740.93 | 3.68 | 42.23 | nr | 783.16 |
| 1200 high x 2400 wide | 716.06 | 787.67 | 3.68 | 42.23 | nr | 829.90 |
| 1500 high x 600 wide | 261.71 | 287.88 | 3.00 | 34.42 | nr | 322.30 |
| 1500 high x 900 wide | 373.30 | 410.63 | 3.34 | 38.33 | nr | 448.96 |
| 1500 high x 1200 wide | 424.65 | 467.12 | 3.34 | 38.33 | nr | 505.45 |
| 1500 high x 1500 wide | 637.00 | 700.70 | 3.68 | 42.23 | nr | 742.93 |
| 1500 high x 1800 wide | 699.22 | 769.14 | 3.68 | 42.23 | nr | 811.37 |

## Y:MECHANICAL AND ELECTRICAL SERVICES

| Item | Net Price £ | Material £ | Labour hours | Labour £ | Unit | Total Rate £ |
|---|---|---|---|---|---|---|
| **Y45 : SILENCERS/ACOUSTIC TREATMENT : ACOUSTIC LOUVRES (contd)** | | | | | | |
| Acoustic louvres; opening mounted; 300mm deep steel louvres with blades packed with acoustic infill; 12mm galvanised mesh birdscreen; screw fixing in opening (contd) | | | | | | |
| Louvre units; polyester powder coated steel (contd) | | | | | | |
| 1500 high x 2100 wide | 798.93 | 878.82 | 4.00 | 45.90 | nr | **924.72** |
| 1800 high x 600 wide | 301.25 | 331.38 | 3.34 | 38.33 | nr | **369.71** |
| 1800 high x 900 wide | 430.62 | 473.68 | 3.34 | 38.33 | nr | **512.01** |
| 1800 high x 1200 wide | 489.82 | 538.80 | 3.68 | 42.23 | nr | **581.03** |
| 1800 high x 1500 wide | 733.77 | 807.15 | 3.68 | 42.23 | nr | **849.38** |

## Y:MECHANICAL AND ELECTRICAL SERVICES

| Item | Net Price £ | Material £ | Labour hours | Labour £ | Unit | Total Rate £ |
|---|---|---|---|---|---|---|
| **Y46 : GRILLES/DIFFUSERS/LOUVRES : GRILLES** | | | | | | |
| Supply grilles; single deflection; extruded aluminium alloy frame and adjustable horizontal vanes; silver grey polyester powder coated; screw fixed | | | | | | |
| Rectangular; for duct, ceiling and sidewall applications | | | | | | |
| 100 x 100mm | 7.96 | 8.76 | 0.60 | 6.88 | nr | 15.64 |
| 150 x 150mm | 9.46 | 10.41 | 0.60 | 6.88 | nr | 17.29 |
| 200 x 150mm | 10.40 | 11.44 | 0.65 | 7.46 | nr | 18.90 |
| 200 x 200mm | 11.66 | 12.83 | 0.72 | 8.26 | nr | 21.09 |
| 300 x 100mm | 10.40 | 11.44 | 0.72 | 8.26 | nr | 19.70 |
| 300 x 150mm | 11.99 | 13.19 | 0.80 | 9.18 | nr | 22.37 |
| 300 x 200mm | 14.11 | 15.52 | 0.88 | 10.10 | nr | 25.62 |
| 300 x 300mm | 17.81 | 19.59 | 1.04 | 11.93 | nr | 31.52 |
| 400 x 100mm | 11.66 | 12.83 | 0.88 | 10.10 | nr | 22.93 |
| 400 x 150mm | 14.11 | 15.52 | 0.94 | 10.79 | nr | 26.31 |
| 400 x 200mm | 16.63 | 18.29 | 1.04 | 11.93 | nr | 30.22 |
| 400 x 300mm | 21.51 | 23.66 | 1.12 | 12.85 | nr | 36.51 |
| 600 x 200mm | 21.51 | 23.66 | 1.26 | 14.46 | nr | 38.12 |
| 600 x 300mm | 29.00 | 31.90 | 1.40 | 16.07 | nr | 47.97 |
| 600 x 400mm | 36.41 | 40.05 | 1.61 | 18.47 | nr | 58.52 |
| 600 x 500mm | 43.81 | 48.19 | 1.76 | 20.20 | nr | 68.39 |
| 600 x 600mm | 51.30 | 56.43 | 2.17 | 24.90 | nr | 81.33 |
| 800 x 300mm | 36.41 | 40.05 | 1.76 | 20.20 | nr | 60.25 |
| 800 x 400mm | 46.33 | 50.96 | 2.17 | 24.90 | nr | 75.86 |
| 800 x 600mm | 66.11 | 72.72 | 3.00 | 34.42 | nr | 107.14 |
| 1000 x 300mm | 43.81 | 48.19 | 2.60 | 29.84 | nr | 78.03 |
| 1000 x 400mm | 56.26 | 61.89 | 3.00 | 34.42 | nr | 96.31 |
| 1000 x 600mm | 81.00 | 89.10 | 3.80 | 43.60 | nr | 132.70 |
| 1000 x 800mm | 105.75 | 116.33 | 4.61 | 52.90 | nr | 169.23 |
| 1200 x 600mm | 97.24 | 106.96 | 4.61 | 52.90 | nr | 159.86 |
| 1200 x 800mm | 126.87 | 139.56 | 6.02 | 69.08 | nr | 208.64 |
| 1200 x 1000mm | 156.65 | 172.31 | 8.00 | 91.80 | nr | 264.11 |
| Rectangular; for duct, ceiling and sidewall applications; including opposed blade damper volume regulator | | | | | | |
| 100 x 100mm | 14.81 | 16.29 | 0.72 | 8.27 | nr | 24.56 |
| 150 x 150mm | 17.50 | 19.25 | 0.72 | 8.27 | nr | 27.52 |
| 200 x 150mm | 19.15 | 21.07 | 0.83 | 9.53 | nr | 30.60 |
| 200 x 200mm | 21.36 | 23.50 | 0.90 | 10.33 | nr | 33.83 |
| 300 x 100mm | 19.15 | 21.07 | 0.90 | 10.33 | nr | 31.40 |
| 300 x 150mm | 22.46 | 24.71 | 0.98 | 11.25 | nr | 35.96 |
| 300 x 200mm | 25.69 | 28.26 | 1.06 | 12.17 | nr | 40.43 |
| 300 x 300mm | 32.31 | 35.54 | 1.20 | 13.78 | nr | 49.32 |
| 400 x 100mm | 21.36 | 23.50 | 1.06 | 12.17 | nr | 35.67 |
| 400 x 150mm | 25.69 | 28.26 | 1.13 | 12.97 | nr | 41.23 |
| 400 x 200mm | 30.18 | 33.20 | 1.20 | 13.78 | nr | 46.98 |
| 400 x 300mm | 38.84 | 42.72 | 1.34 | 15.38 | nr | 58.10 |
| 600 x 200mm | 38.84 | 42.72 | 1.50 | 17.23 | nr | 59.95 |
| 600 x 300mm | 52.08 | 57.29 | 1.66 | 19.06 | nr | 76.35 |
| 600 x 400mm | 66.73 | 73.40 | 1.80 | 20.68 | nr | 94.08 |
| 600 x 500mm | 78.40 | 86.24 | 2.00 | 22.95 | nr | 109.19 |
| 600 x 600mm | 91.64 | 100.80 | 2.60 | 29.88 | nr | 130.68 |

## Y:MECHANICAL AND ELECTRICAL SERVICES

| Item | Net Price £ | Material £ | Labour hours | Labour £ | Unit | Total Rate £ |
|---|---|---|---|---|---|---|
| **Y46 : GRILLES/DIFFUSERS/LOUVRES : GRILLES (contd)** | | | | | | |
| **Supply grilles; single deflection; extruded aluminium alloy frame and adjustable horizontal vanes; silver grey polyester powder coated; screw fixed (contd)** | | | | | | |
| Rectangular; for duct, ceiling and sidewall applications; including opposed blade damper volume regulator (contd) | | | | | | |
| 800 x 300mm | 65.24 | 71.76 | 2.00 | 22.95 | nr | 94.71 |
| 800 x 400mm | 82.82 | 91.10 | 2.60 | 29.88 | nr | 120.98 |
| 800 x 600mm | 117.96 | 129.76 | 3.61 | 41.43 | nr | 171.19 |
| 1000 x 300mm | 78.40 | 86.24 | 3.00 | 34.46 | nr | 120.70 |
| 1000 x 400mm | 100.39 | 110.43 | 3.61 | 41.43 | nr | 151.86 |
| 1000 x 600mm | 144.28 | 158.71 | 4.61 | 52.88 | nr | 211.59 |
| 1000 x 800mm | 188.17 | 206.99 | 5.62 | 64.47 | nr | 271.46 |
| 1200 x 600mm | 173.20 | 190.52 | 5.62 | 64.47 | nr | 254.99 |
| 1200 x 800mm | 225.75 | 248.32 | 7.02 | 80.55 | nr | 328.87 |
| 1200 x 1000mm | 278.55 | 306.40 | 9.01 | 103.38 | nr | 409.78 |
| **Supply grilles; double deflection; extruded aluminium alloy frame and adjustable horizontal and vertical vanes; silver grey polyester powder coated; screw fixed** | | | | | | |
| Rectangular; for duct, ceiling and sidewall applications | | | | | | |
| 100 x 100mm | 9.06 | 9.97 | 0.88 | 10.10 | nr | 20.07 |
| 150 x 150mm | 11.19 | 12.31 | 0.88 | 10.10 | nr | 22.41 |
| 200 x 150mm | 12.45 | 13.70 | 1.08 | 12.39 | nr | 26.09 |
| 200 x 200mm | 14.18 | 15.60 | 1.25 | 14.34 | nr | 29.94 |
| 300 x 100mm | 12.45 | 13.70 | 1.25 | 14.34 | nr | 28.04 |
| 300 x 150mm | 15.05 | 16.55 | 1.50 | 17.21 | nr | 33.76 |
| 300 x 200mm | 17.57 | 19.33 | 1.75 | 20.08 | nr | 39.41 |
| 300 x 300mm | 22.69 | 24.96 | 2.15 | 24.67 | nr | 49.63 |
| 400 x 100mm | 14.18 | 15.60 | 1.75 | 20.08 | nr | 35.68 |
| 400 x 150mm | 17.57 | 19.33 | 1.95 | 22.38 | nr | 41.71 |
| 400 x 200mm | 20.45 | 22.50 | 2.15 | 24.67 | nr | 47.17 |
| 400 x 300mm | 27.74 | 30.51 | 2.55 | 29.26 | nr | 59.77 |
| 600 x 200mm | 27.74 | 30.51 | 3.01 | 34.54 | nr | 65.05 |
| 600 x 300mm | 37.90 | 41.69 | 3.36 | 38.56 | nr | 80.25 |
| 600 x 400mm | 48.15 | 52.97 | 3.80 | 43.60 | nr | 96.57 |
| 600 x 500mm | 58.31 | 64.14 | 4.20 | 48.20 | nr | 112.34 |
| 600 x 600mm | 68.47 | 75.32 | 4.51 | 51.75 | nr | 127.07 |
| 800 x 300mm | 48.15 | 52.97 | 4.20 | 48.20 | nr | 101.17 |
| 800 x 400mm | 61.70 | 67.87 | 4.51 | 51.75 | nr | 119.62 |
| 800 x 600mm | 88.81 | 97.69 | 5.10 | 58.52 | nr | 156.21 |
| 1000 x 300mm | 58.31 | 64.14 | 4.80 | 55.08 | nr | 119.22 |
| 1000 x 400mm | 75.25 | 82.78 | 5.10 | 58.52 | nr | 141.30 |
| 1000 x 600mm | 109.22 | 120.14 | 5.72 | 65.64 | nr | 185.78 |
| 1000 x 800mm | 143.10 | 157.41 | 6.33 | 72.64 | nr | 230.05 |
| 1200 x 600mm | 131.04 | 144.14 | 6.33 | 72.64 | nr | 216.78 |
| 1200 x 800mm | 171.70 | 188.87 | 8.19 | 93.98 | nr | 282.85 |
| 1200 x 1000mm | 212.44 | 233.68 | 10.04 | 115.21 | nr | 348.89 |

## Y:MECHANICAL AND ELECTRICAL SERVICES

| Item | Net Price £ | Material £ | Labour hours | Labour £ | Unit | Total Rate £ |
|---|---|---|---|---|---|---|
| Rectangular; for duct, ceiling and sidewall applications; including opposed blade damper volume regulator | | | | | | |
| 100 x 100mm | 15.92- | 17.51 | 1.00 | 11.47 | nr | 28.98 |
| 150 x 150mm | 19.23 | 21.15 | 1.00 | 11.47 | nr | 32.62 |
| 200 x 150mm | 21.19 | 23.31 | 1.26 | 14.46 | nr | 37.77 |
| 200 x 200mm | 23.88 | 26.27 | 1.43 | 16.41 | nr | 42.68 |
| 300 x 100mm | 21.19 | 23.31 | 1.43 | 16.41 | nr | 39.72 |
| 300 x 150mm | 25.21 | 27.73 | 1.68 | 19.28 | nr | 47.01 |
| 300 x 200mm | 29.16 | 32.08 | 1.93 | 22.15 | nr | 54.23 |
| 300 x 300mm | 37.19 | 40.91 | 2.31 | 26.51 | nr | 67.42 |
| 400 x 100mm | 23.88 | 26.27 | 1.93 | 22.15 | nr | 48.42 |
| 400 x 150mm | 29.16 | 32.08 | 2.14 | 24.56 | nr | 56.64 |
| 400 x 200mm | 34.52 | 37.97 | 2.31 | 26.51 | nr | 64.48 |
| 400 x 300mm | 45.07 | 49.58 | 2.77 | 31.79 | nr | 81.37 |
| 600 x 200mm | 45.07 | 49.58 | 3.25 | 37.29 | nr | 86.87 |
| 600 x 300mm | 60.99 | 67.09 | 3.62 | 41.54 | nr | 108.63 |
| 600 x 400mm | 76.98 | 84.68 | 3.99 | 45.79 | nr | 130.47 |
| 600 x 500mm | 92.90 | 102.19 | 4.44 | 50.95 | nr | 153.14 |
| 600 x 600mm | 108.82 | 119.70 | 4.94 | 56.69 | nr | 176.39 |
| 800 x 300mm | 76.98 | 84.68 | 4.44 | 50.95 | nr | 135.63 |
| 800 x 400mm | 98.18 | 108.00 | 4.94 | 56.69 | nr | 164.69 |
| 800 x 600mm | 140.65 | 154.72 | 5.71 | 65.52 | nr | 220.24 |
| 1000 x 300mm | 92.90 | 102.19 | 5.20 | 59.67 | nr | 161.86 |
| 1000 x 400mm | 119.38 | 131.32 | 5.71 | 65.52 | nr | 196.84 |
| 1000 x 600mm | 172.48 | 189.73 | 6.53 | 74.93 | nr | 264.66 |
| 1000 x 800mm | 225.52 | 248.07 | 7.34 | 84.23 | nr | 332.30 |
| 1200 x 600mm | 207.00 | 227.70 | 7.34 | 84.23 | nr | 311.93 |
| 1200 x 800mm | 270.59 | 297.65 | 9.19 | 105.46 | nr | 403.11 |
| 1200 x 1000mm | 334.34 | 367.77 | 11.05 | 126.80 | nr | 494.57 |
| **Exhaust grilles; aluminium** | | | | | | |
| 0 degree fixed blade core | | | | | | |
| 150 x 150mm | 12.82 | 14.10 | 0.62 | 7.12 | nr | 21.22 |
| 200 x 200mm | 14.44 | 15.88 | 0.72 | 8.27 | nr | 24.15 |
| 250 x 250mm | 16.47 | 18.12 | 0.80 | 9.18 | nr | 27.30 |
| 300 x 300mm | 18.95 | 20.84 | 1.00 | 11.47 | nr | 32.31 |
| 350 x 350mm | 21.82 | 24.00 | 1.20 | 13.77 | nr | 37.77 |
| 0 degree fixed blade core; including opposed blade damper volume regulator | | | | | | |
| 150 x 150mm | 20.73 | 22.80 | 0.62 | 7.12 | nr | 29.92 |
| 200 x 200mm | 24.00 | 26.40 | 0.72 | 8.27 | nr | 34.67 |
| 250 x 250mm | 28.19 | 31.01 | 0.80 | 9.18 | nr | 40.19 |
| 300 x 300mm | 33.24 | 36.56 | 1.00 | 11.47 | nr | 48.03 |
| 350 x 350mm | 39.14 | 43.05 | 1.20 | 13.77 | nr | 56.82 |
| 45 degree fixed blade core | | | | | | |
| 150 x 150mm | 12.82 | 14.10 | 0.62 | 7.12 | nr | 21.22 |
| 200 x 200mm | 14.44 | 15.88 | 0.72 | 8.27 | nr | 24.15 |
| 250 x 250mm | 16.47 | 18.12 | 0.80 | 9.18 | nr | 27.30 |
| 300 x 300mm | 18.95 | 20.84 | 1.00 | 11.47 | nr | 32.31 |
| 350 x 350mm | 21.82 | 24.00 | 1.20 | 13.77 | nr | 37.77 |

## Y:MECHANICAL AND ELECTRICAL SERVICES

| Item | Net Price £ | Material £ | Labour hours | Labour £ | Unit | Total Rate £ |
|---|---|---|---|---|---|---|
| **Y46 : GRILLES/DIFFUSERS/LOUVRES : GRILLES (contd)** | | | | | | |
| **Exhaust grilles; aluminium (contd)** | | | | | | |
| 45 degree fixed blade core; including opposed blade damper volume regulator | | | | | | |
| 150 x 150mm | 20.73 | 22.80 | 0.62 | 7.12 | nr | 29.92 |
| 200 x 200mm | 24.00 | 26.40 | 0.72 | 8.27 | nr | 34.67 |
| 250 x 250mm | 28.19 | 31.01 | 0.80 | 9.18 | nr | 40.19 |
| 300 x 300mm | 33.24 | 36.56 | 1.00 | 11.47 | nr | 48.03 |
| 350 x 350mm | 39.14 | 43.05 | 1.20 | 13.77 | nr | 56.82 |
| Eggcrate core | | | | | | |
| 150 x 150mm | 8.46 | 9.31 | 0.62 | 7.12 | nr | 16.43 |
| 200 x 200mm | 11.49 | 12.64 | 0.72 | 8.27 | nr | 20.91 |
| 250 x 250mm | 14.60 | 16.06 | 0.80 | 9.18 | nr | 25.24 |
| 300 x 300mm | 17.78 | 19.56 | 1.00 | 11.47 | nr | 31.03 |
| 350 x 350mm | 20.97 | 23.07 | 1.20 | 13.77 | nr | 36.84 |
| Eggcrate core; including opposed blade damper volume regulator | | | | | | |
| 150 x 150mm | 16.39 | 18.03 | 0.62 | 7.12 | nr | 25.15 |
| 200 x 200mm | 21.05 | 23.16 | 0.72 | 8.27 | nr | 31.43 |
| 250 x 250mm | 26.33 | 28.96 | 0.80 | 9.18 | nr | 38.14 |
| 300 x 300mm | 32.07 | 35.28 | 1.00 | 11.47 | nr | 46.75 |
| 350 x 350mm | 38.28 | 42.11 | 1.20 | 13.77 | nr | 55.88 |
| Mesh/perforated plate core | | | | | | |
| 150 x 150mm | 12.27 | 13.50 | 0.62 | 7.11 | nr | 20.61 |
| 200 x 200mm | 16.39 | 18.03 | 0.72 | 8.26 | nr | 26.29 |
| 250 x 250mm | 20.43 | 22.47 | 0.80 | 9.18 | nr | 31.65 |
| 300 x 300mm | 24.46 | 26.91 | 1.00 | 11.47 | nr | 38.38 |
| 350 x 350mm | 28.50 | 31.35 | 1.20 | 13.77 | nr | 45.12 |
| Mesh/perforated plate core; including opposed blade damper volume regulator | | | | | | |
| 150 x 150mm | 20.20 | 22.22 | 0.62 | 7.11 | nr | 29.33 |
| 200 x 200mm | 25.94 | 28.53 | 0.72 | 8.26 | nr | 36.79 |
| 250 x 250mm | 32.15 | 35.37 | 0.80 | 9.18 | nr | 44.55 |
| 300 x 300mm | 38.75 | 42.63 | 0.80 | 9.18 | nr | 51.81 |
| 350 x 350mm | 45.82 | 50.40 | 1.20 | 13.77 | nr | 64.17 |
| **Plastic air diffusion system** | | | | | | |
| Eggcrate grilles | | | | | | |
| 150 x 150mm | 5.05 | 5.55 | 0.62 | 7.12 | nr | 12.67 |
| 200 x 200mm | 6.68 | 7.35 | 0.72 | 8.27 | nr | 15.62 |
| 250 x 250mm | 7.84 | 8.62 | 0.80 | 9.18 | nr | 17.80 |
| 300 x 300mm | 10.17 | 11.19 | 1.00 | 11.47 | nr | 22.66 |
| 310 x 315mm | 7.38 | 8.12 | 1.05 | 12.05 | nr | 20.17 |

## Y:MECHANICAL AND ELECTRICAL SERVICES

| Item | Net Price £ | Material £ | Labour hours | Labour £ | Unit | Total Rate £ |
|---|---|---|---|---|---|---|
| Single deflection grilles | | | | | | |
| 150 x 150mm | 4.66 | 5.13 | 0.62 | 7.12 | nr | 12.25 |
| 200 x 200mm | 6.21 | 6.83 | 0.72 | 8.27 | nr | 15.10 |
| 250 x 250mm | 6.68 | 7.35 | 0.80 | 9.18 | nr | 16.53 |
| 300 x 300mm | 8.62 | 9.48 | 1.00 | 11.47 | nr | 20.95 |
| 315 x 315mm | 6.68 | 7.35 | 1.05 | 12.05 | nr | 19.40 |
| Double deflection grilles | | | | | | |
| 150 x 150mm | 5.51 | 6.06 | 0.62 | 7.12 | nr | 13.18 |
| 200 x 200mm | 7.38 | 8.12 | 0.72 | 8.27 | nr | 16.39 |
| 250 x 250mm | 9.39 | 10.33 | 0.80 | 9.18 | nr | 19.51 |
| 300 x 300mm | 12.97 | 14.27 | 1.00 | 11.47 | nr | 25.74 |
| 315 x 315mm | 9.39 | 10.33 | 1.05 | 12.05 | nr | 22.38 |
| Door transfer grilles | | | | | | |
| 150 x 150mm | 9.39 | 10.33 | 0.62 | 7.12 | nr | 17.45 |
| 200 x 200mm | 12.42 | 13.66 | 0.72 | 8.27 | nr | 21.93 |
| 250 x 250mm | 13.28 | 14.61 | 0.80 | 9.18 | nr | 23.79 |
| 300 x 300mm | 17.17 | 18.89 | 1.00 | 11.47 | nr | 30.36 |
| 315 x 315mm | 13.28 | 14.61 | 1.05 | 12.05 | nr | 26.66 |
| Opposed blade dampers | | | | | | |
| 150 x 150mm | 3.57 | 3.93 | 0.62 | 7.12 | nr | 11.05 |
| 200 x 200mm | 4.66 | 5.13 | 0.72 | 8.27 | nr | 13.40 |
| 250 x 250mm | 5.51 | 6.06 | 0.80 | 9.18 | nr | 15.24 |
| 300 x 300mm | 6.68 | 7.35 | 1.00 | 11.47 | nr | 18.82 |
| 315 x 315mm | 5.51 | 6.06 | 1.05 | 12.05 | nr | 18.11 |
| Neck reducers | | | | | | |
| 150 x 150mm | 4.27 | 4.70 | 0.62 | 7.12 | nr | 11.82 |
| 200 x 200mm | 5.05 | 5.55 | 0.72 | 8.27 | nr | 13.82 |
| 250 x 250mm | 5.90 | 6.49 | 0.80 | 9.18 | nr | 15.67 |
| 300 x 300mm | 6.68 | 7.35 | 1.00 | 11.47 | nr | 18.82 |
| 350 x 350mm | 5.51 | 6.06 | 1.05 | 12.05 | nr | 18.11 |

## Y:MECHANICAL AND ELECTRICAL SERVICES

| Item | Net Price £ | Material £ | Labour hours | Labour £ | Unit | Total Rate £ |
|---|---|---|---|---|---|---|
| **Y46 : GRILLES/DIFFUSERS/LOUVRES : DIFFUSERS** | | | | | | |
| **Ceiling mounted diffusers; circular aluminium multi-core diffuser** | | | | | | |
| Circular; for ceiling mounting | | | | | | |
| 152mm dia. neck | 58.86 | 64.75 | 0.80 | 9.18 | nr | 73.93 |
| 203mm dia. neck | 70.44 | 77.48 | 1.10 | 12.62 | nr | 90.10 |
| 305mm dia. neck | 99.56 | 109.52 | 1.40 | 16.07 | nr | 125.59 |
| 381mm dia. neck | 125.65 | 138.22 | 1.50 | 17.23 | nr | 155.45 |
| 457mm dia. neck | 216.05 | 237.66 | 2.00 | 22.95 | nr | 260.61 |
| Circular; for ceiling mounting; including louvre damper volume control | | | | | | |
| 152mm dia. neck | 78.90 | 86.79 | 1.00 | 11.47 | nr | 98.26 |
| 203mm dia. neck | 89.85 | 98.83 | 1.20 | 13.78 | nr | 112.61 |
| 305mm dia. neck | 121.69 | 133.86 | 1.60 | 18.36 | nr | 152.22 |
| 381mm dia. neck | 150.97 | 166.07 | 1.90 | 21.82 | nr | 187.89 |
| 457mm dia. neck | 243.85 | 268.24 | 2.40 | 27.58 | nr | 295.82 |
| **Ceiling mounted diffusers; rectangular aluminium multi cone diffuser; four way flow** | | | | | | |
| Rectangular; for ceiling mounting | | | | | | |
| 150 x 150 mm neck | 28.89 | 31.78 | 1.80 | 20.68 | nr | 52.46 |
| 300 x 150 mm neck | 33.16 | 36.48 | 2.35 | 27.00 | nr | 63.48 |
| 300 x 300 mm neck | 41.86 | 46.05 | 2.80 | 32.14 | nr | 78.19 |
| 450 x 150 mm neck | 37.51 | 41.26 | 2.80 | 32.14 | nr | 73.40 |
| 450 x 300 mm neck | 50.55 | 55.60 | 3.11 | 35.64 | nr | 91.24 |
| 450 x 450 mm neck | 63.52 | 69.87 | 3.40 | 39.03 | nr | 108.90 |
| 600 x 150 mm neck | 41.86 | 46.05 | 3.11 | 35.64 | nr | 81.69 |
| 600 x 300 mm neck | 59.18 | 65.10 | 3.40 | 39.03 | nr | 104.13 |
| 600 x 600 mm neck | 93.81 | 103.19 | 4.00 | 45.90 | nr | 149.09 |
| 900 x 300 mm neck | 76.49 | 84.14 | 4.00 | 45.90 | nr | 130.04 |
| Rectangular; for ceiling mounting; including opposed blade damper volume regulator | | | | | | |
| 150 x 150 mm neck | 36.81 | 40.49 | 1.80 | 20.68 | nr | 61.17 |
| 300 x 150 mm neck | 43.18 | 47.50 | 2.35 | 27.00 | nr | 74.50 |
| 300 x 300 mm neck | 56.15 | 61.77 | 2.80 | 32.14 | nr | 93.91 |
| 450 x 150 mm neck | 49.70 | 54.67 | 2.80 | 32.14 | nr | 86.81 |
| 450 x 300 mm neck | 69.12 | 76.03 | 3.11 | 35.64 | nr | 111.67 |
| 450 x 450 mm neck | 88.45 | 97.30 | 3.51 | 40.26 | nr | 137.56 |
| 600 x 150 mm neck | 56.15 | 61.77 | 3.11 | 35.64 | nr | 97.41 |
| 600 x 300 mm neck | 81.93 | 90.12 | 3.51 | 40.26 | nr | 130.38 |
| 600 x 600 mm neck | 133.57 | 146.93 | 5.62 | 64.47 | nr | 211.40 |
| 900 x 300 mm neck | 107.79 | 118.57 | 5.62 | 64.47 | nr | 183.04 |

## Y:MECHANICAL AND ELECTRICAL SERVICES

| Item | Net Price £ | Material £ | Labour hours | Labour £ | Unit | Total Rate £ |
|---|---|---|---|---|---|---|
| **Slot diffusers; continuous aluminium slot diffuser with flanged frame** | | | | | | |
| Diffuser | | | | | | |
| 1 slot | 37.66 | 41.43 | 3.76 | 43.14 | m | **84.57** |
| 2 slot | 51.41 | 56.55 | 3.76 | 43.14 | m | **99.69** |
| 3 slot | 65.24 | 71.76 | 3.76 | 43.14 | m | **114.90** |
| 4 slot | 78.98 | 86.88 | 4.50 | 51.69 | m | **138.57** |
| 6 slot | 106.55 | 117.20 | 4.50 | 51.69 | m | **168.89** |
| 8 slot | 134.04 | 147.44 | 4.50 | 51.69 | m | **199.13** |
| Diffuser; including equalizing deflector | | | | | | |
| 1 slot | 56.07 | 61.68 | 5.26 | 60.40 | m | **122.08** |
| 2 slot | 78.98 | 86.88 | 5.26 | 60.40 | m | **147.28** |
| 3 slot | 94.59 | 104.05 | 5.26 | 60.40 | m | **164.45** |
| 4 slot | 110.20 | 121.22 | 6.33 | 72.63 | m | **193.85** |
| 6 slot | 141.42 | 155.56 | 6.33 | 72.63 | m | **228.19** |
| 8 slot | 172.64 | 189.90 | 6.33 | 72.63 | m | **262.53** |
| Ends | | | | | | |
| 1 slot | 2.88 | 3.17 | 1.00 | 11.47 | nr | **14.64** |
| 2 slot | 3.03 | 3.33 | 1.00 | 11.47 | nr | **14.80** |
| 3 slot | 3.18 | 3.50 | 1.00 | 11.47 | nr | **14.97** |
| 4 slot | 3.41 | 3.75 | 1.30 | 14.92 | nr | **18.67** |
| 6 slot | 3.73 | 4.10 | 1.40 | 16.07 | nr | **20.17** |
| 8 slot | 4.11 | 4.52 | 1.50 | 17.23 | nr | **21.75** |
| Plenum boxes; 1.0m long; circular spigot; including cord operated flap damper | | | | | | |
| 1 slot | 35.33 | 38.86 | 2.75 | 31.61 | nr | **70.47** |
| 2 slot | 35.33 | 38.86 | 2.75 | 31.61 | nr | **70.47** |
| 3 slot | 38.68 | 42.55 | 2.75 | 31.61 | nr | **74.16** |
| 4 slot | 38.68 | 42.55 | 3.51 | 40.26 | nr | **82.81** |
| 6 slot | 42.33 | 46.56 | 3.51 | 40.26 | nr | **86.82** |
| 8 slot | 42.33 | 46.56 | 3.51 | 40.26 | nr | **86.82** |
| Plenum boxes; 2.0m long; circular spigot; including cord operated flap damper | | | | | | |
| 1 slot | 35.33 | 38.86 | 3.26 | 37.38 | nr | **76.24** |
| 2 slot | 42.87 | 47.16 | 3.26 | 37.38 | nr | **84.54** |
| 3 slot | 47.92 | 52.71 | 3.26 | 37.38 | nr | **90.09** |
| 4 slot | 47.92 | 52.71 | 3.76 | 43.14 | nr | **95.85** |
| 6 slot | 51.57 | 56.73 | 3.76 | 43.14 | nr | **99.87** |
| 8 slot | 51.57 | 56.73 | 3.76 | 43.14 | nr | **99.87** |
| **Perforated diffusers; rectangular face aluminium perforated diffuser; quick release face plate; for integration with rectangular ceiling tiles** | | | | | | |
| Circular spigot; rectangular diffuser | | | | | | |
| 150mm dia. spigot; 300 x 300 diffuser | 47.14 | 51.85 | 0.80 | 9.18 | nr | **61.03** |
| 300mm dia. spigot; 600 x 600 diffuser | 87.21 | 95.93 | 1.40 | 16.07 | nr | **112.00** |

## Y:MECHANICAL AND ELECTRICAL SERVICES

| Item | Net Price £ | Material £ | Labour hours | Labour £ | Unit | Total Rate £ |
|---|---|---|---|---|---|---|
| **Y46 : GRILLES/DIFFUSERS/LOUVRES : DIFFUSERS (contd)** | | | | | | |
| **Perforated diffusers; rectangular face aluminium perforated diffuser; quick release face plate; for integration with rectangular ceiling tiles (contd)** | | | | | | |
| Circular spigot; rectangilar diffuser; including louvre damper volume regulator | | | | | | |
| 150mm dia. spigot; 300 x 300 diffuser | 65.94 | 72.53 | 1.00 | 11.47 | nr | 84.00 |
| 300mm dia. spigot; 600 x 600 diffuser | 106.32 | 116.95 | 1.60 | 18.36 | nr | 135.31 |
| Rectangular spigot; rectangular diffuser | | | | | | |
| 150 x 150mm dia. spigot; 300 x 300mm diffuser | 49.62 | 54.58 | 1.00 | 11.47 | nr | 66.05 |
| 300 x 150mm dia. spigot; 600 x 300mm diffuser | 57.39 | 63.13 | 1.20 | 13.77 | nr | 76.90 |
| 300 x 300mm dia. spigot; 600 x 600mm diffuser | 84.65 | 93.11 | 1.40 | 16.07 | nr | 109.18 |
| 600 x 300mm dia. spigot; 1200 x 600mm diffuser | 150.34 | 165.37 | 1.60 | 18.36 | nr | 183.73 |
| Rectangular spigot; rectangular diffuser; including opposed blade damper volume regulator | | | | | | |
| 150 x 150mm dia. spigot; 300 x 300mm diffuser | 64.30 | 70.73 | 1.20 | 13.77 | nr | 84.50 |
| 300 x 150mm dia. spigot; 600 x 300mm diffuser | 67.72 | 104.29 | 1.40 | 16.07 | nr | 120.36 |
| 300 x 300mm dia. spigot; 600 x 600mm diffuser | 102.82 | 113.10 | 1.60 | 18.36 | nr | 131.46 |
| 600 x 300mm dia. spigot; 1200 x 600mm diffuser | 173.65 | 191.01 | 1.80 | 20.66 | nr | 211.67 |
| **Plastic air diffusion system** | | | | | | |
| Cellular diffusers | | | | | | |
| 300 x 300mm | 16.54 | 18.19 | 2.80 | 32.14 | nr | 50.33 |
| 600 x 600mm | 35.88 | 39.47 | 4.00 | 45.90 | nr | 85.37 |
| Multicone diffusers | | | | | | |
| 300 x 300mm | 16.54 | 18.19 | 2.80 | 32.14 | nr | 50.33 |
| 450 x 450mm | 24.78 | 27.26 | 3.40 | 39.03 | nr | 66.29 |
| 500 x 500mm | 24.78 | 27.26 | 3.80 | 43.63 | nr | 70.89 |
| 600 x 600mm | 35.88 | 39.47 | 4.00 | 45.90 | nr | 85.37 |
| 625 x 625mm | 35.88 | 39.47 | 4.26 | 48.83 | nr | 88.30 |
| Opposed blade dampers | | | | | | |
| 300 x 300mm | 4.66 | 5.13 | 1.20 | 13.78 | nr | 18.91 |
| 450 x 450mm | 7.38 | 8.12 | 1.50 | 17.23 | nr | 25.35 |
| 600 x 600mm | 14.75 | 16.23 | 2.60 | 29.88 | nr | 46.11 |

## Y:MECHANICAL AND ELECTRICAL SERVICES

| Item | Net Price £ | Material £ | Labour hours | Labour £ | Unit | Total Rate £ |
|---|---|---|---|---|---|---|
| Plenum boxes | | | | | | |
| 300mm | 7.38 | 8.12 | 2.80 | 32.14 | nr | **40.26** |
| 450mm | 11.02 | 12.12 | 3.40 | 39.03 | nr | **51.15** |
| 600mm | 14.75 | 16.23 | 4.00 | 45.90 | nr | **62.13** |
| Plenum spigot reducer | | | | | | |
| 600mm | 5.51 | 6.06 | 1.00 | 11.47 | nr | **17.53** |
| Blanking kits for cellular diffusers | | | | | | |
| 300mm | 3.65 | 4.01 | 0.88 | 10.10 | nr | **14.11** |
| 600mm | 5.51 | 6.06 | 1.10 | 12.62 | nr | **18.68** |
| Blanking kits for multicone diffusers | | | | | | |
| 300mm | 3.65 | 4.01 | 0.88 | 10.10 | nr | **14.11** |
| 450mm | 4.66 | 5.13 | 0.90 | 10.33 | nr | **15.46** |
| 600mm | 5.51 | 6.06 | 1.10 | 12.62 | nr | **18.68** |

## Y:MECHANICAL AND ELECTRICAL SERVICES

| Item | Net Price £ | Material £ | Labour hours | Labour £ | Unit | Total Rate £ |
|---|---|---|---|---|---|---|
| **Y46 : GRILLES/DIFFUSERS/LOUVRES : EXTERNAL LOUVRES** | | | | | | |
| Weather louvres; opening mounted; 300mm deep galvanised steel louvres; screw fixing in position | | | | | | |
| Louvre units; including 12mm galvanised mesh birdscreen | | | | | | |
| 900 x 600mm | 159.28 | 175.21 | 2.25 | 25.85 | nr | **201.06** |
| 900 x 900mm | 215.50 | 237.05 | 2.25 | 25.85 | nr | **262.90** |
| 900 x 1200mm | 267.07 | 293.78 | 2.50 | 28.69 | nr | **322.47** |
| 900 x 1500mm | 315.61 | 347.17 | 2.50 | 28.69 | nr | **375.86** |
| 900 x 1800mm | 361.74 | 397.91 | 2.50 | 28.69 | nr | **426.60** |
| 900 x 2100mm | 498.34 | 548.17 | 2.50 | 28.69 | nr | **576.86** |
| 900 x 2400mm | 534.06 | 587.47 | 2.76 | 31.70 | nr | **619.17** |
| 900 x 2700mm | 597.20 | 656.92 | 2.76 | 31.70 | nr | **688.62** |
| 900 x 3000mm | 631.21 | 694.33 | 2.76 | 31.70 | nr | **726.03** |
| 1200 x 600mm | 193.92 | 213.31 | 2.25 | 25.85 | nr | **239.16** |
| 1200 x 900mm | 262.57 | 288.83 | 2.50 | 28.69 | nr | **317.52** |
| 1200 x 1200mm | 325.63 | 358.19 | 2.50 | 28.69 | nr | **386.88** |
| 1200 x 1500mm | 385.04 | 423.54 | 2.50 | 28.69 | nr | **452.23** |
| 1200 x 1800mm | 441.57 | 485.73 | 2.76 | 31.70 | nr | **517.43** |
| 1200 x 2100mm | 607.30 | 668.03 | 2.76 | 31.70 | nr | **699.73** |
| 1200 x 2400mm | 651.25 | 716.38 | 2.76 | 31.70 | nr | **748.08** |
| 1500 x 600mm | 226.07 | 248.68 | 2.25 | 25.85 | nr | **274.53** |
| 1500 x 900mm | 306.21 | 336.83 | 2.50 | 28.69 | nr | **365.52** |
| 1500 x 1200mm | 379.99 | 417.99 | 2.50 | 28.69 | nr | **446.68** |
| 1500 x 1500mm | 449.49 | 494.44 | 2.76 | 31.70 | nr | **526.14** |
| 1500 x 1800mm | 515.89 | 567.48 | 2.76 | 31.70 | nr | **599.18** |
| 1500 x 2100mm | 708.25 | 779.08 | 3.00 | 34.46 | nr | **813.54** |
| 1800 x 600mm | 256.28 | 281.91 | 2.50 | 28.69 | nr | **310.60** |
| 1800 x 900mm | 347.37 | 382.11 | 2.50 | 28.69 | nr | **410.80** |
| 1800 x 1200mm | 431.32 | 474.45 | 2.76 | 31.70 | nr | **506.15** |
| 1800 x 1500mm | 510.53 | 561.58 | 3.00 | 34.46 | nr | **596.04** |

*Material Costs / Measured Work Prices - Mechanical Installations*

## Y:MECHANICAL AND ELECTRICAL SERVICES

| Item | Net Price £ | Material £ | Labour hours | Labour £ | Unit | Total Rate £ |
|---|---|---|---|---|---|---|
| **Y50 : THERMAL INSULATION : PREFORMED RIGID SECTIONS AND SLABS** | | | | | | |
| Rigid mineral fibre, preformed sections, wire fixings, aluminium sheeting; horizontal joints sealed with water resistant sealant; in plant rooms | | | | | | |
| Plant rooms; pipework dia. 3/4 inch; insulation thickness :- | | | | | | |
| 25mm | 3.33 | 3.67 | 0.44 | 5.05 | m | 8.72 |
| 40mm | 5.08 | 5.60 | 0.49 | 5.62 | m | 11.22 |
| 50mm | 7.10 | 7.83 | 0.52 | 5.97 | m | 13.80 |
| 80mm | 14.84 | 16.35 | 0.75 | 8.61 | m | 24.96 |
| Plant rooms; pipework dia. 1 inch; insulation thickness :- | | | | | | |
| 25mm | 3.61 | 3.98 | 0.45 | 5.16 | m | 9.14 |
| 40mm | 5.47 | 6.02 | 0.49 | 5.62 | m | 11.64 |
| 50mm | 7.53 | 8.30 | 0.53 | 6.08 | m | 14.38 |
| 80mm | 15.28 | 16.84 | 0.76 | 8.72 | m | 25.56 |
| Plant rooms; pipework dia. 1 1/2 inch; insulation thickness :- | | | | | | |
| 25mm | 4.20 | 4.62 | 0.46 | 5.28 | m | 9.90 |
| 40mm | 6.10 | 6.73 | 0.51 | 5.85 | m | 12.58 |
| 50mm | 8.28 | 9.13 | 0.54 | 6.20 | m | 15.33 |
| 80mm | 16.51 | 18.18 | 0.78 | 8.95 | m | 27.13 |
| 100mm | 19.27 | 21.24 | 0.87 | 9.98 | m | 31.22 |
| Plant rooms; pipework dia. 2 inch; insulation thickness :- | | | | | | |
| 25mm | 4.79 | 5.28 | 0.46 | 5.28 | m | 10.56 |
| 40mm | 6.87 | 7.59 | 0.52 | 5.97 | m | 13.56 |
| 50mm | 9.23 | 10.17 | 0.55 | 6.31 | m | 16.48 |
| 80mm | 17.73 | 19.54 | 0.80 | 9.18 | m | 28.72 |
| 100mm | 21.88 | 24.10 | 0.89 | 10.21 | m | 34.31 |
| Plant rooms; pipework dia. 2 1/2 inch; insulation thickness :- | | | | | | |
| 25mm | 5.44 | 6.01 | 0.52 | 5.97 | m | 11.98 |
| 40mm | 7.71 | 8.50 | 0.58 | 6.66 | m | 15.16 |
| 50mm | 10.77 | 11.87 | 0.62 | 7.11 | m | 18.98 |
| 80mm | 19.17 | 21.11 | 0.87 | 9.98 | m | 31.09 |
| 100mm | 25.54 | 28.12 | 0.97 | 11.13 | m | 39.25 |
| Plant rooms; pipework dia. 3 inch; insulation thickness :- | | | | | | |
| 25mm | 6.08 | 6.71 | 0.53 | 6.08 | m | 12.79 |
| 40mm | 8.36 | 9.22 | 0.60 | 6.88 | m | 16.10 |
| 50mm | 10.84 | 11.95 | 0.63 | 7.23 | m | 19.18 |
| 80mm | 20.64 | 22.73 | 0.88 | 10.10 | m | 32.83 |
| 100mm | 28.56 | 31.45 | 0.98 | 11.25 | m | 42.70 |
| 120mm | 35.02 | 38.56 | 1.19 | 13.66 | m | 52.22 |

## Y:MECHANICAL AND ELECTRICAL SERVICES

| Item | Net Price £ | Material £ | Labour hours | Labour £ | Unit | Total Rate £ |
|---|---|---|---|---|---|---|
| **Y50 : THERMAL INSULATION : PREFORMED RIGID SECTIONS AND SLABS (contd)** | | | | | | |
| **Rigid mineral fibre, preformed sections, wire fixings, aluminium sheeting; horizontal joints sealed with water resistant sealant; in plant rooms (contd)** | | | | | | |
| Plant rooms; pipework dia. 4 inch; insulation thickness :- | | | | | | |
| 25mm | 7.77 | 8.57 | 0.55 | 6.31 | m | **14.88** |
| 40mm | 10.69 | 11.78 | 0.61 | 7.00 | m | **18.78** |
| 50mm | 13.75 | 15.15 | 0.65 | 7.46 | m | **22.61** |
| 80mm | 25.25 | 27.79 | 0.91 | 10.44 | m | **38.23** |
| 100mm | 33.41 | 36.80 | 1.01 | 11.59 | m | **48.39** |
| 120mm | 38.94 | 42.87 | 1.21 | 13.88 | m | **56.75** |
| Plant rooms; pipework dia. 6 inch; insulation thickness :- | | | | | | |
| 25mm | 10.76 | 11.86 | 0.70 | 8.03 | m | **19.89** |
| 40mm | 14.02 | 15.44 | 0.76 | 8.72 | m | **24.16** |
| 50mm | 17.26 | 19.00 | 0.81 | 9.29 | m | **28.29** |
| 80mm | 30.46 | 33.54 | 1.10 | 12.62 | m | **46.16** |
| 100mm | 40.73 | 44.84 | 1.21 | 13.88 | m | **58.72** |
| 120mm | 45.43 | 50.01 | 1.40 | 16.07 | m | **66.08** |
| **Extra Over plant room pipework insulation; working around fittings to include elbows, tees, reducers and the like** | | | | | | |
| Plant rooms; fittings dia. 3/4 inch; insulation thickness :- | | | | | | |
| 25mm | - | - | 0.54 | 6.20 | nr | **6.20** |
| 40mm | - | - | 0.52 | 5.97 | nr | **5.97** |
| 50mm | - | - | 0.52 | 5.97 | nr | **5.97** |
| 80mm | - | - | 0.65 | 7.46 | nr | **7.46** |
| Plant rooms; fittings dia. 1 inch; insulation thickness :- | | | | | | |
| 25mm | - | - | 0.54 | 6.20 | nr | **6.20** |
| 40mm | - | - | 0.53 | 6.08 | nr | **6.08** |
| 50mm | - | - | 0.52 | 5.97 | nr | **5.97** |
| 80mm | - | - | 0.65 | 7.46 | nr | **7.46** |
| Plant rooms; fittings dia. 1 1/2 inch; insulation thickness :- | | | | | | |
| 25mm | - | - | 0.54 | 6.20 | nr | **6.20** |
| 40mm | - | - | 0.52 | 5.97 | nr | **5.97** |
| 50mm | - | - | 0.52 | 5.97 | nr | **5.97** |
| 80mm | - | - | 0.65 | 7.46 | nr | **7.46** |
| 100mm | - | - | 0.66 | 7.57 | nr | **7.57** |

## Y:MECHANICAL AND ELECTRICAL SERVICES

| Item | Net Price £ | Material £ | Labour hours | Labour £ | Unit | Total Rate £ |
|---|---|---|---|---|---|---|
| Plant rooms; fittings dia. 2 inch; insulation thickness :- | | | | | | |
| 25mm | - | - | 0.57 | 6.54 | nr | 6.54 |
| 40mm | - | - | 0.55 | 6.31 | nr | 6.31 |
| 50mm | - | - | 0.55 | 6.31 | nr | 6.31 |
| 80mm | - | - | 0.69 | 7.92 | nr | 7.92 |
| 100mm | - | - | 0.70 | 8.03 | nr | 8.03 |
| Plant rooms; fittings dia. 2 1/2 inch; insulation thickness :- | | | | | | |
| 25mm | - | - | 0.72 | 8.26 | nr | 8.26 |
| 40mm | - | - | 0.70 | 8.03 | nr | 8.03 |
| 50mm | - | - | 0.70 | 8.03 | nr | 8.03 |
| 80mm | - | - | 0.83 | 9.52 | nr | 9.52 |
| 100mm | - | - | 0.84 | 9.64 | nr | 9.64 |
| Plant rooms; fittings dia. 3 inch; insulation thickness :- | | | | | | |
| 25mm | - | - | 0.72 | 8.26 | nr | 8.26 |
| 40mm | - | - | 0.70 | 8.03 | nr | 8.03 |
| 50mm | - | - | 0.70 | 8.03 | nr | 8.03 |
| 80mm | - | - | 0.84 | 9.64 | nr | 9.64 |
| 100mm | - | - | 0.85 | 9.75 | nr | 9.75 |
| 120mm | - | - | 1.16 | 13.31 | nr | 13.31 |
| Plant rooms; fittings dia. 4 inch; insulation thickness :- | | | | | | |
| 25mm | - | - | 0.74 | 8.49 | nr | 8.49 |
| 40mm | - | - | 0.74 | 8.49 | nr | 8.49 |
| 50mm | - | - | 0.73 | 8.38 | nr | 8.38 |
| 80mm | - | - | 0.87 | 9.98 | nr | 9.98 |
| 100mm | - | - | 0.91 | 10.44 | nr | 10.44 |
| 120mm | - | - | 1.24 | 14.23 | nr | 14.23 |
| Plant rooms; fittings dia. 6 inch; insulation thickness :- | | | | | | |
| 25mm | - | - | 0.94 | 10.79 | nr | 10.79 |
| 40mm | - | - | 0.95 | 10.90 | nr | 10.90 |
| 50mm | - | - | 0.95 | 10.90 | nr | 10.90 |
| 80mm | - | - | 1.09 | 12.51 | nr | 12.51 |
| 100mm | - | - | 1.13 | 12.97 | nr | 12.97 |
| 120mm | - | - | 1.54 | 17.67 | nr | 17.67 |
| **Rigid mineral fibre, preformed sections, wire fixings, aluminium sheeting; horizontal joints sealed with water resistant sealant; in non-plant rooms** | | | | | | |
| Non-plant rooms; pipework dia. 3/4 inch; insulation thickness :- | | | | | | |
| 25mm | 3.33 | 3.67 | 0.32 | 3.67 | m | 7.34 |
| 40mm | 5.08 | 5.60 | 0.35 | 4.02 | m | 9.62 |
| 50mm | 7.10 | 7.83 | 0.37 | 4.25 | m | 12.08 |
| 80mm | 14.84 | 16.35 | 0.54 | 6.20 | m | 22.55 |

Y:MECHANICAL AND ELECTRICAL SERVICES

| Item | Net Price £ | Material £ | Labour hours | Labour £ | Unit | Total Rate £ |
|---|---|---|---|---|---|---|
| **Y50 : THERMAL INSULATION : PREFORMED RIGID SECTIONS AND SLABS (contd)** | | | | | | |
| Rigid mineral fibre, preformed sections, wire fixings, aluminium sheeting; horizontal joints sealed with water resistant sealant; in non-plant rooms (contd) | | | | | | |
| Non-plant rooms; pipework dia. 1 inch; insulation thickness :- | | | | | | |
| 25mm | 3.61 | 3.98 | 0.32 | 3.67 | m | **7.65** |
| 40mm | 5.47 | 6.02 | 0.35 | 4.02 | m | **10.04** |
| 50mm | 7.53 | 8.30 | 0.38 | 4.36 | m | **12.66** |
| 80mm | 15.28 | 16.84 | 0.55 | 6.31 | m | **23.15** |
| Non-plant rooms; pipework dia. 1 1/2 inch; insulation thickness :- | | | | | | |
| 25mm | 4.20 | 4.62 | 0.33 | 3.79 | m | **8.41** |
| 40mm | 6.10 | 6.73 | 0.36 | 4.13 | m | **10.86** |
| 50mm | 8.28 | 9.13 | 0.39 | 4.48 | m | **13.61** |
| 80mm | 16.51 | 18.18 | 0.56 | 6.43 | m | **24.61** |
| 100mm | 19.27 | 21.24 | 0.62 | 7.11 | m | **28.35** |
| Non-plant rooms; pipework dia. 2 inch; insulation thickness :- | | | | | | |
| 25mm | 4.79 | 5.28 | 0.33 | 3.79 | m | **9.07** |
| 40mm | 6.87 | 7.59 | 0.37 | 4.25 | m | **11.84** |
| 50mm | 9.23 | 10.17 | 0.40 | 4.59 | m | **14.76** |
| 80mm | 17.73 | 19.54 | 0.57 | 6.54 | m | **26.08** |
| 100mm | 21.88 | 24.10 | 0.63 | 7.23 | m | **31.33** |
| Non-plant rooms; pipework dia. 2 1/2 inch; insulation thickness :- | | | | | | |
| 25mm | 5.44 | 6.01 | 0.37 | 4.25 | m | **10.26** |
| 40mm | 7.71 | 8.50 | 0.42 | 4.82 | m | **13.32** |
| 50mm | 10.77 | 11.87 | 0.44 | 5.05 | m | **16.92** |
| 80mm | 19.17 | 21.11 | 0.62 | 7.11 | m | **28.22** |
| 100mm | 25.54 | 28.12 | 0.69 | 7.92 | m | **36.04** |
| Non-plant rooms; pipework dia. 3 inch; insulation thickness :- | | | | | | |
| 25mm | 6.08 | 6.71 | 0.38 | 4.36 | m | **11.07** |
| 40mm | 8.36 | 9.22 | 0.43 | 4.93 | m | **14.15** |
| 50mm | 10.84 | 11.95 | 0.45 | 5.16 | m | **17.11** |
| 80mm | 20.64 | 22.73 | 0.63 | 7.23 | m | **29.96** |
| 100mm | 28.56 | 31.45 | 0.70 | 8.03 | m | **39.48** |
| 120mm | 35.02 | 38.56 | 0.85 | 9.75 | m | **48.31** |

Material Costs / Measured Work Prices - Mechanical Installations

## Y: MECHANICAL AND ELECTRICAL SERVICES

| Item | Net Price £ | Material £ | Labour hours | Labour £ | Unit | Total Rate £ |
|---|---|---|---|---|---|---|
| Non-plant rooms; pipework dia. 4 inch; insulation thickness :- | | | | | | |
| 25mm | 7.77 | 8.57 | 0.39 | 4.48 | m | **13.05** |
| 40mm | 10.69 | 11.78 | 0.44 | 5.05 | m | **16.83** |
| 50mm | 13.75 | 15.15 | 0.47 | 5.39 | m | **20.54** |
| 80mm | 25.25 | 27.79 | 0.65 | 7.46 | m | **35.25** |
| 100mm | 33.41 | 36.80 | 0.72 | 8.26 | m | **45.06** |
| 120mm | 38.94 | 42.87 | 0.87 | 9.98 | m | **52.85** |
| Non-plant rooms; pipework dia. 6 inch; insulation thickness :- | | | | | | |
| 25mm | 10.76 | 11.86 | 0.50 | 5.74 | m | **17.60** |
| 40mm | 14.02 | 15.44 | 0.54 | 6.20 | m | **21.64** |
| 50mm | 17.26 | 19.00 | 0.58 | 6.66 | m | **25.66** |
| 80mm | 30.46 | 33.54 | 0.79 | 9.07 | m | **42.61** |
| 100mm | 40.73 | 44.84 | 0.86 | 9.87 | m | **54.71** |
| 120mm | 45.43 | 50.01 | 1.00 | 11.47 | m | **61.48** |
| **Extra Over non-plant room pipework insulation; working around fittings to include elbows, tees, reducers and the like** | | | | | | |
| Non-plant rooms; fittings dia. 3/4 inch; insulation thickness :- | | | | | | |
| 25mm | - | - | 0.39 | 4.48 | nr | **4.48** |
| 40mm | - | - | 0.37 | 4.25 | nr | **4.25** |
| 50mm | - | - | 0.37 | 4.25 | nr | **4.25** |
| 80mm | - | - | 0.46 | 5.28 | nr | **5.28** |
| Non-plant rooms; fittings dia. 1 inch; insulation thickness :- | | | | | | |
| 25mm | - | - | 0.39 | 4.48 | nr | **4.48** |
| 40mm | - | - | 0.38 | 4.36 | nr | **4.36** |
| 50mm | - | - | 0.37 | 4.25 | nr | **4.25** |
| 80mm | - | - | 0.46 | 5.28 | nr | **5.28** |
| Non-plant rooms; fittings dia. 1 1/2 inch; insulation thickness :- | | | | | | |
| 25mm | - | - | 0.38 | 4.36 | nr | **4.36** |
| 40mm | - | - | 0.37 | 4.25 | nr | **4.25** |
| 50mm | - | - | 0.37 | 4.25 | nr | **4.25** |
| 80mm | - | - | 0.46 | 5.28 | nr | **5.28** |
| 100mm | - | - | 0.47 | 5.39 | nr | **5.39** |
| Non-plant rooms; fittings dia. 2 inch; insulation thickness :- | | | | | | |
| 25mm | - | - | 0.41 | 4.70 | nr | **4.70** |
| 40mm | - | - | 0.39 | 4.48 | nr | **4.48** |
| 50mm | - | - | 0.39 | 4.48 | nr | **4.48** |
| 80mm | - | - | 0.49 | 5.62 | nr | **5.62** |
| 100mm | - | - | 0.50 | 5.74 | nr | **5.74** |

Y:MECHANICAL AND ELECTRICAL SERVICES

| Item | Net Price £ | Material £ | Labour hours | Labour £ | Unit | Total Rate £ |
|---|---|---|---|---|---|---|
| **Y50 : THERMAL INSULATION : PREFORMED RIGID SECTIONS AND SLABS (contd)** | | | | | | |
| Extra Over non-plant room pipework insulation; working around fittings to include elbows, tees, reducers and the like (contd) | | | | | | |
| Non-plant rooms; fittings dia. 2 1/2 inch; insulation thickness :- | | | | | | |
| 25mm | - | - | 0.52 | 5.97 | nr | **5.97** |
| 40mm | - | - | 0.50 | 5.74 | nr | **5.74** |
| 50mm | - | - | 0.50 | 5.74 | nr | **5.74** |
| 80mm | - | - | 0.59 | 6.77 | nr | **6.77** |
| 100mm | - | - | 0.60 | 6.88 | nr | **6.88** |
| Non-plant rooms; fittings dia. 3 inch; insulation thickness :- | | | | | | |
| 25mm | - | - | 0.51 | 5.85 | nr | **5.85** |
| 40mm | - | - | 0.50 | 5.74 | nr | **5.74** |
| 50mm | - | - | 0.50 | 5.74 | nr | **5.74** |
| 80mm | - | - | 0.60 | 6.88 | nr | **6.88** |
| 100mm | - | - | 0.61 | 7.00 | nr | **7.00** |
| 120mm | - | - | 0.83 | 9.52 | nr | **9.52** |
| Non-plant rooms; fittings dia. 4 inch; insulation thickness :- | | | | | | |
| 25mm | - | - | 0.53 | 6.08 | nr | **6.08** |
| 40mm | - | - | 0.53 | 6.08 | nr | **6.08** |
| 50mm | - | - | 0.52 | 5.97 | nr | **5.97** |
| 80mm | - | - | 0.62 | 7.11 | nr | **7.11** |
| 100mm | - | - | 0.65 | 7.46 | nr | **7.46** |
| 120mm | - | - | 0.89 | 10.21 | nr | **10.21** |
| Non-plant rooms; fittings dia. 6 inch; insulation thickness :- | | | | | | |
| 25mm | - | - | 0.67 | 7.69 | nr | **7.69** |
| 40mm | - | - | 0.68 | 7.80 | nr | **7.80** |
| 50mm | - | - | 0.68 | 7.80 | nr | **7.80** |
| 80mm | - | - | 0.78 | 8.95 | nr | **8.95** |
| 100mm | - | - | 0.81 | 9.29 | nr | **9.29** |
| 120mm | - | - | 1.10 | 12.62 | nr | **12.62** |

## Y:MECHANICAL AND ELECTRICAL SERVICES

| Item | Net Price £ | Material £ | Labour hours | Labour £ | Unit | Total Rate £ |
|---|---|---|---|---|---|---|
| **Y50 : THERMAL INSULATION : VALVE BOXES** | | | | | | |
| Rigid mineral fibre, preformed sections, wire fixings, aluminium sheeting; horizontal joints sealed with water resistant sealant; in plant rooms | | | | | | |
| Plant rooms; valve box dia. 3/4 inch; insulation thickness :- | | | | | | |
| 25mm | 3.33 | 3.85 | 0.60 | 6.88 | m | 10.73 |
| 40mm | 5.08 | 5.87 | 0.62 | 7.11 | m | 12.98 |
| 50mm | 7.10 | 8.20 | 0.64 | 7.34 | m | 15.54 |
| 80mm | 14.84 | 17.14 | 0.79 | 9.07 | m | 26.21 |
| Plant rooms; valve box dia. 1 inch; insulation thickness :- | | | | | | |
| 25mm | 3.61 | 4.17 | 0.61 | 7.00 | m | 11.17 |
| 40mm | 5.47 | 6.31 | 0.63 | 7.23 | m | 13.54 |
| 50mm | 7.53 | 8.70 | 0.65 | 7.46 | m | 16.16 |
| 80mm | 15.28 | 17.66 | 0.80 | 9.18 | m | 26.84 |
| Plant rooms; valve box dia. 1 1/2 inch; insulation thickness :- | | | | | | |
| 25mm | 4.20 | 4.84 | 0.61 | 7.00 | m | 11.84 |
| 40mm | 6.10 | 7.05 | 0.63 | 7.23 | m | 14.28 |
| 50mm | 8.28 | 9.56 | 0.65 | 7.46 | m | 17.02 |
| 80mm | 16.51 | 19.07 | 0.81 | 9.29 | m | 28.36 |
| 100mm | 19.27 | 22.27 | 0.88 | 10.10 | m | 32.37 |
| Plant rooms; valve box dia. 2 inch; insulation thickness :- | | | | | | |
| 25mm | 4.79 | 5.53 | 0.63 | 7.23 | m | 12.76 |
| 40mm | 6.87 | 7.95 | 0.66 | 7.57 | m | 15.52 |
| 50mm | 9.23 | 10.66 | 0.68 | 7.80 | m | 18.46 |
| 80mm | 17.73 | 20.49 | 0.85 | 9.75 | m | 30.24 |
| 100mm | 21.88 | 25.27 | 0.91 | 10.44 | m | 35.71 |
| Plant rooms; valve box dia. 2 1/2 inch; insulation thickness :- | | | | | | |
| 25mm | 5.44 | 6.29 | 0.69 | 7.92 | m | 14.21 |
| 40mm | 7.71 | 8.91 | 0.72 | 8.26 | m | 17.17 |
| 50mm | 10.08 | 11.64 | 0.74 | 8.49 | m | 20.13 |
| 80mm | 19.17 | 22.14 | 0.97 | 11.13 | m | 33.27 |
| 100mm | 25.54 | 29.50 | 1.04 | 11.93 | m | 41.43 |
| Plant rooms; valve box dia. 3 inch; insulation thickness :- | | | | | | |
| 25mm | 6.08 | 7.02 | 0.70 | 8.03 | m | 15.05 |
| 40mm | 8.36 | 9.66 | 0.73 | 8.38 | m | 18.04 |
| 50mm | 10.84 | 12.52 | 0.75 | 8.61 | m | 21.13 |
| 80mm | 20.64 | 23.84 | 0.98 | 11.25 | m | 35.09 |
| 100mm | 28.56 | 32.99 | 1.07 | 12.28 | m | 45.27 |
| 120mm | 35.02 | 40.45 | 1.41 | 16.18 | m | 56.63 |

## Y:MECHANICAL AND ELECTRICAL SERVICES

| Item | Net Price £ | Material £ | Labour hours | Labour £ | Unit | Total Rate £ |
|---|---|---|---|---|---|---|
| **Y50 : THERMAL INSULATION : VALVE BOXES (contd)** | | | | | | |
| **Rigid mineral fibre, preformed sections, wire fixings, aluminium sheeting; horizontal joints sealed with water resistant sealant; in plant rooms (contd)** | | | | | | |
| Plant rooms; valve box dia. 4 inch; insulation thickness :- | | | | | | |
| 25mm | 7.77 | 8.97 | 1.00 | 11.47 | m | **20.44** |
| 40mm | 10.69 | 12.35 | 1.03 | 11.82 | m | **24.17** |
| 50mm | 13.75 | 15.88 | 1.06 | 12.16 | m | **28.04** |
| 80mm | 25.25 | 29.15 | 1.28 | 14.69 | m | **43.84** |
| 100mm | 33.41 | 38.60 | 1.36 | 15.61 | m | **54.21** |
| 120mm | 38.94 | 44.98 | 1.79 | 20.54 | m | **65.52** |
| Plant rooms; valve box dia. 6 inch; insulation thickness :- | | | | | | |
| 25mm | 10.76 | 12.43 | 1.05 | 12.05 | m | **24.48** |
| 40mm | 14.02 | 16.19 | 1.09 | 12.51 | m | **28.70** |
| 50mm | 17.26 | 19.94 | 1.13 | 12.97 | m | **32.91** |
| 80mm | 30.46 | 35.18 | 1.37 | 15.72 | m | **50.90** |
| 100mm | 40.73 | 47.04 | 1.46 | 16.75 | m | **63.79** |
| 120mm | 45.43 | 52.47 | 1.92 | 22.03 | m | **74.50** |
| **Rigid mineral fibre, preformed sections, wire fixings, aluminium sheeting; horizontal joints sealed with water resistant sealant; in non-plant rooms** | | | | | | |
| Non-plant rooms; valve box dia. 3/4 inch; insulation thickness :- | | | | | | |
| 25mm | 3.33 | 3.85 | 0.52 | 5.97 | m | **9.82** |
| 40mm | 5.08 | 5.87 | 0.54 | 6.20 | m | **12.07** |
| 50mm | 7.10 | 8.20 | 0.56 | 6.43 | m | **14.63** |
| 80mm | 14.84 | 17.14 | 0.69 | 7.92 | m | **25.06** |
| Non-plant rooms; valve box dia. 1 inch; insulation thickness :- | | | | | | |
| 25mm | 3.61 | 4.17 | 0.53 | 6.08 | m | **10.25** |
| 40mm | 5.47 | 6.31 | 0.55 | 6.31 | m | **12.62** |
| 50mm | 7.53 | 8.70 | 0.56 | 6.43 | m | **15.13** |
| 80mm | 15.28 | 17.66 | 0.69 | 7.92 | m | **25.58** |
| Non-plant rooms; valve box dia. 1 1/2 inch; insulation thickness :- | | | | | | |
| 25mm | 3.42 | 3.95 | 0.53 | 6.08 | m | **10.03** |
| 40mm | 6.10 | 7.05 | 0.55 | 6.31 | m | **13.36** |
| 50mm | 8.28 | 9.56 | 0.57 | 6.54 | m | **16.10** |
| 80mm | 16.51 | 19.07 | 0.71 | 8.15 | m | **27.22** |
| 100mm | 19.27 | 22.27 | 0.76 | 8.72 | m | **30.99** |

*Material Costs / Measured Work Prices - Mechanical Installations* 335

## Y:MECHANICAL AND ELECTRICAL SERVICES

| Item | Net Price £ | Material £ | Labour hours | Labour £ | Unit | Total Rate £ |
|---|---|---|---|---|---|---|
| Non-plant rooms; valve box dia. 2 inch; insulation thickness :- | | | | | | |
| 25mm | 4.79 | 5.53 | 0.55 | 6.31 | m | 11.84 |
| 40mm | 6.87 | 7.95 | 0.57 | 6.54 | m | 14.49 |
| 50mm | 9.23 | 10.66 | 0.59 | 6.77 | m | 17.43 |
| 80mm | 17.73 | 20.49 | 0.74 | 8.49 | m | 28.98 |
| 100mm | 21.88 | 25.27 | 0.79 | 9.07 | m | 34.34 |
| Non-plant rooms; valve box dia. 2 1/2 inch; insulation thickness :- | | | | | | |
| 25mm | 5.44 | 6.29 | 0.60 | 6.88 | m | 13.17 |
| 40mm | 7.71 | 8.91 | 0.63 | 7.23 | m | 16.14 |
| 50mm | 10.08 | 11.64 | 0.65 | 7.46 | m | 19.10 |
| 80mm | 19.17 | 22.14 | 0.84 | 9.64 | m | 31.78 |
| 100mm | 25.54 | 29.50 | 0.90 | 10.33 | m | 39.83 |
| Non-plant rooms; valve box dia. 3 inch; insulation thickness :- | | | | | | |
| 25mm | 6.08 | 7.02 | 0.61 | 7.00 | m | 14.02 |
| 40mm | 8.36 | 9.66 | 0.63 | 7.23 | m | 16.89 |
| 50mm | 10.84 | 12.52 | 0.65 | 7.46 | m | 19.98 |
| 80mm | 20.64 | 23.84 | 0.86 | 9.87 | m | 33.71 |
| 100mm | 28.56 | 32.99 | 0.93 | 10.67 | m | 43.66 |
| 120mm | 35.02 | 40.45 | 1.22 | 14.00 | m | 54.45 |
| Non-plant rooms; valve box dia. 4 inch; insulation thickness :- | | | | | | |
| 25mm | 7.77 | 8.97 | 0.87 | 9.98 | m | 18.95 |
| 40mm | 10.69 | 12.35 | 0.90 | 10.33 | m | 22.68 |
| 50mm | 13.75 | 15.88 | 0.92 | 10.56 | m | 26.44 |
| 80mm | 25.25 | 29.15 | 1.11 | 12.74 | m | 41.89 |
| 100mm | 33.41 | 38.60 | 1.18 | 13.54 | m | 52.14 |
| 120mm | 38.94 | 44.98 | 1.56 | 17.90 | m | 62.88 |
| Non-plant rooms; valve box dia. 6 inch; insulation thickness :- | | | | | | |
| 25mm | 10.76 | 12.43 | 0.92 | 10.56 | m | 22.99 |
| 40mm | 14.02 | 16.19 | 0.95 | 10.90 | m | 27.09 |
| 50mm | 17.26 | 19.94 | 0.98 | 11.25 | m | 31.19 |
| 80mm | 30.46 | 35.18 | 1.19 | 13.66 | m | 48.84 |
| 100mm | 40.73 | 47.04 | 1.27 | 14.57 | m | 61.61 |
| 120mm | 45.43 | 52.47 | 1.67 | 19.16 | m | 71.63 |

## Y:MECHANICAL AND ELECTRICAL SERVICES

| Item | Net Price £ | Material £ | Labour hours | Labour £ | Unit | Total Rate £ |
|---|---|---|---|---|---|---|
| **Y50 : THERMAL INSULATION : INSULATION OF DUCTWORK** | | | | | | |
| Rigid mineral fibre slabs; foil faced; adhesive and tape fixings; with wire netting; plain sheet aluminium cladding; coloured identification markings | | | | | | |
| Plant rooms; concealed ductwork; plain sheet aluminium cladding | | | | | | |
|    40mm thick | 4.85 | 5.33 | 1.21 | 13.89 | m | **19.22** |
|    50mm thick | 5.57 | 6.13 | 1.21 | 13.89 | m | **20.02** |
| Non-plant rooms; concealed ductwork; plain sheet aluminium cladding | | | | | | |
|    40mm thick | 4.85 | 5.33 | 1.10 | 12.62 | m | **17.95** |
|    50mm thick | 5.57 | 6.13 | 1.10 | 12.62 | m | **18.75** |

*Material Costs / Measured Work Prices - Mechanical Installations*

## Y:MECHANICAL AND ELECTRICAL SERVICES

| Item | Net Price £ | Material £ | Labour hours | Labour £ | Unit | Total Rate £ |
|---|---|---|---|---|---|---|
| **Y52 : VIBRATION ISOLATION MOUNTINGS** | | | | | | |
| **Anti-vibration mountings; installed with fans or other items of plant** | | | | | | |
| Suspended fans; 305-601mm dia. | | | | | | |
|    loads up to 23 kg; four fixing points | 6.16 | 6.77 | 2.00 | 22.95 | Set | **29.72** |
|    loads up to 41 kg; four fixing points | 6.16 | 6.77 | 2.00 | 22.95 | Set | **29.72** |
|    loads up to 41 kg; six fixing points; for two-stage units | 6.16 | 10.16 | 3.00 | 34.42 | Set | **44.58** |
| Suspended fans; 610-1219mm dia. | | | | | | |
|    loads up to 115 kg; four fixing points | 20.43 | 22.47 | 2.00 | 22.95 | Set | **45.42** |
|    loads up to 115 kg; six fixing points; for two-stage units | 20.43 | 33.71 | 3.00 | 34.42 | Set | **68.13** |
| Suspended fans; 762-1219mm dia. | | | | | | |
|    loads up to 220 Kg; six fixing points; for two-stage units | 24.63 | 40.63 | 3.00 | 34.42 | Set | **75.05** |
| **Butterfly type damper; horizontal aerofoil blades; installed with fans** | | | | | | |
| Dampers; manual operation; to suit fan size | | | | | | |
|    305mm dia. | 192.52 | 211.77 | 2.00 | 22.95 | nr | **234.72** |
|    483mm dia. | 244.01 | 268.41 | 2.00 | 22.95 | nr | **291.36** |
|    615mm dia. | 304.45 | 334.89 | 2.00 | 22.95 | nr | **357.84** |
|    770mm dia. | 340.26 | 374.29 | 2.00 | 22.95 | nr | **397.24** |
| Dampers; pneumatic operation; to suit fan size | | | | | | |
|    315mm dia. | 127.60 | 140.36 | 2.00 | 22.95 | nr | **163.31** |
|    500mm dia. | 162.30 | 178.53 | 2.00 | 22.95 | nr | **201.48** |
|    560mm dia. | 189.16 | 208.08 | 2.00 | 22.95 | nr | **231.03** |
|    710mm dia. | 205.95 | 226.54 | 2.00 | 22.95 | nr | **249.49** |

## Y:MECHANICAL AND ELECTRICAL SERVICES

| Item | Net Price £ | Material £ | Labour hours | Labour £ | Unit | Total Rate £ |
|---|---|---|---|---|---|---|
| **Y53 : CONTROL COMPONENTS- MECHANICAL** | | | | | | |
| **Room thermostats; light and medium duty; installed and connected to provide system control** | | | | | | |
| Range 3 to 27 degree C; 240 Volt | | | | | | |
| 1 amp; on/off type | 18.40 | 20.24 | 0.30 | 3.44 | nr | **23.68** |
| Range 0 to +15 degree C; 240 Volt | | | | | | |
| 6 amp; frost thermostat | 12.41 | 13.65 | 0.30 | 3.44 | nr | **17.09** |
| Range 3 to 27 degree C; 250 Volt | | | | | | |
| 2 amp; changeover type; dead zone | 30.77 | 33.85 | 0.30 | 3.44 | nr | **37.29** |
| 2 amp; changeover type | 14.27 | 15.70 | 0.30 | 3.44 | nr | **19.14** |
| 2 amp; changeover type; concealed setting | 18.00 | 19.80 | 0.30 | 3.44 | nr | **23.24** |
| 6 amp; on/off type | 11.12 | 12.23 | 0.30 | 3.44 | nr | **15.67** |
| 6 amp; temperature set-back | 23.02 | 25.32 | 0.30 | 3.44 | nr | **28.76** |
| 16 amp; on/off type | 17.11 | 18.82 | 0.30 | 3.44 | nr | **22.26** |
| 16 amp; on/off type; concealed setting | 18.67 | 20.54 | 0.30 | 3.44 | nr | **23.98** |
| 20 amp; on/off type; concealed setting; industrial non-ventilated cover | 20.38 | 22.42 | 0.30 | 3.44 | nr | **25.86** |
| 20 amp; indicated "off" position | 20.03 | 22.03 | 0.30 | 3.44 | nr | **25.48** |
| 20 amp; manual; double pole on/off and neon indicator | 37.57 | 41.33 | 0.30 | 3.44 | nr | **44.77** |
| 20 amp; indicated "off" position | 24.58 | 27.04 | 0.30 | 3.44 | nr | **30.48** |
| Range 10 to 40 degree C; 240 Volt | | | | | | |
| 20 amp; changeover contacts | 21.75 | 23.93 | 0.30 | 3.44 | nr | **27.37** |
| 2 amp; 'Heating-Cooling' switch | 47.04 | 51.74 | 0.30 | 3.44 | nr | **55.18** |
| **Surface thermostats** | | | | | | |
| Cylinder thermostat | | | | | | |
| 6 amp; changeover type; with cable | 13.39 | 15.10 | 0.25 | 2.87 | nr | **17.97** |
| **Electrical thermostats; installed and connected to provide system control** | | | | | | |
| Range 5 to 30 degree C; 230 Volt Standard Port Single Time | | | | | | |
| 10 amp with sensor | 13.78 | 15.16 | 0.30 | 3.44 | nr | **18.60** |
| Range 5 to 30 degree C; 230 Volt Standard Port Double Time | | | | | | |
| 10 amp with sensor | 15.61 | 17.17 | 0.30 | 3.44 | nr | **20.61** |
| 10 amp with sensor and on/off switch | 23.82 | 26.20 | 0.30 | 3.44 | nr | **29.64** |

## Y:MECHANICAL AND ELECTRICAL SERVICES

| Item | Net Price £ | Material £ | Labour hours | Labour £ | Unit | Total Rate £ |
|---|---|---|---|---|---|---|
| **Radiator thermostats** | | | | | | |
| Angled valve body; thermostatic head; built in sensor | | | | | | |
| 15mm; liquid filled | 10.75 | 11.82 | 0.84 | 9.64 | nr | **21.46** |
| 15mm; wax filled | 10.75 | 11.82 | 0.84 | 9.64 | nr | **21.46** |
| **Immersion thermostats; stem type; domestic water boilers; fitted; electrical work elsewhere** | | | | | | |
| Temperature range 0 to 40 degree C | | | | | | |
| Non standard; 280mm stem | 6.77 | 7.45 | 0.25 | 2.87 | nr | **10.32** |
| Temperature range 18 to 88 degree C | | | | | | |
| 13 amp; 178mm stem | 4.63 | 5.09 | 0.25 | 2.87 | nr | **7.96** |
| 20 amp; 178mm stem | 6.99 | 7.69 | 0.25 | 2.87 | nr | **10.56** |
| Non standard; pocket clip; 280mm stem | 6.48 | 7.13 | 0.25 | 2.87 | nr | **10.00** |
| Temperature range 40 to 80 degree C | | | | | | |
| 13 amp; 178mm stem | 3.01 | 3.31 | 0.25 | 2.87 | nr | **6.18** |
| 20 amp; 178mm stem | 4.80 | 5.28 | 0.25 | 2.87 | nr | **8.15** |
| Non standard; pocket clip; 280mm stem | 7.26 | 7.99 | 0.25 | 2.87 | nr | **10.86** |
| 13 amp; 457mm stem | 3.51 | 3.86 | 0.25 | 2.87 | nr | **6.73** |
| 20 amp; 457mm stem | 5.26 | 5.79 | 0.25 | 2.87 | nr | **8.66** |
| Temperature range 50 to 100 degree C | | | | | | |
| Non standard; 1780mm stem | 6.11 | 6.72 | 0.25 | 2.87 | nr | **9.59** |
| Non standard; 280mm stem | 6.37 | 7.01 | 0.25 | 2.87 | nr | **9.88** |
| Pockets for thermostats | | | | | | |
| For 178mm stem | 8.00 | 8.80 | 0.25 | 2.87 | nr | **11.67** |
| For 280mm stem | 8.16 | 8.98 | 0.25 | 2.87 | nr | **11.85** |
| **Immersion thermostats; stem type; industrial installations; fitted; electrical work elsewhere** | | | | | | |
| Temperature range 5 to 105 degree C | | | | | | |
| For 305mm stem | 103.44 | 113.78 | 0.50 | 5.74 | nr | **119.52** |

# DAVIS LANGDON & EVEREST
## CHARTERED QUANTITY SURVEYORS
## CONSTRUCTION COST CONSULTANTS

**LONDON**
Princes House
39 Kingsway
London
WC2B 6TP
Tel : (0171) 497 9000
Fax : (0171) 497 8858

**GATESHEAD**
11 Regent Terrace
Gateshead
Tyne and Wear
NE8 1LU
Tel : (0191) 477 3844
Fax : (0191) 490 1742

**MILTON KEYNES**
6 Bassett Court
Newport Pagnell
Buckinghamshire
MK16 OJN
Tel : (01908) 613777
Fax : (01908) 210642

**SOUTHAMPTON**
Clifford House
New Road
Southampton
SO14 OAB
Tel : (01703) 333438
Fax : (01703) 226099

**BRISTOL**
St Lawrence House
29/31 Broad Street
Bristol BS1 2HF
Tel : (0117) 927 7832
Fax : (0117) 925 1350

**GLASGOW**
Cumbrae House
15 Carlton Court
Glasgow G5 9JP
Tel : (0141) 429 6677
Fax : (0141) 429 2255

**NEWPORT**
34 Godfrey Road
Newport
Gwent NP9 4PE
Tel : (01633) 259712
Fax : (01633) 215694

**DAVIS LANGDON CONSULTANCY**
Princes House
39 Kingsway
London WC2B 6TP
Tel : (0171) 379 3322
Fax : (0171) 379 3030

**CAMBRIDGE**
36 Storey's Way
Cambridge CB3 ODT
Tel : (01223) 351258
Fax : (01223) 321002

**IPSWICH**
17 St Helens Street
Ipswich IP4 1HE
Tel : (01473) 253405
Fax : (01473) 231215

**NORWICH**
63 Thorpe Road
Norwich NR1 1UD
Tel : (01603) 628194
Fax : (01603) 615928

**CARDIFF**
3 Raleigh Walk
Brigantine Place
Atlantic Wharf
Cardiff CF1 5LN
Tel : (01222) 471306
Fax : (01222) 471465

**LEEDS**
Duncan House
14 Duncan Street
Leeds
LS1 6DL
Tel : (0113) 243 2481
Fax: (0113) 242 4601

**OXFORD**
Avalon House
Marcham Road
Abingdon
Oxford OX14 1TZ
Tel : (01235) 555025
Fax : (01235) 554909

**CHESTER**
Ford Lane Farm
Lower Lane
Aldford
Chester CH3 6HP
Tel : (01244) 620222
Fax : (01244) 620303

**LIVERPOOL**
Cunard Building
Water Street
Liverpool
L3 1JR
Tel : (0151) 236 1992
Fax : (0151) 227 5401

**PLYMOUTH**
3 Russell Court
St Andrews Street
Plymouth
Devon PL6 2AX
Tel : (01752) 668372
Fax : (01752) 221219

**EDINBURGH**
74 Great King Street
Edinburgh
EH3 6QU
Tel : (0131) 557 5306
Fax : (0131) 557 5704

**MANCHESTER**
Boulton House
Chorlton Street
Manchester M1 3HY
Tel : (0161) 228 2011
Fax : (0161) 228 6317

**PORTSMOUTH**
Kings House
4 Kings Road
Portsmouth PO5 3BQ
Tel : (01705) 815218
Fax : (01705) 827156

A MEMBER OF
## DAVIS LANGDON & SEAH INTERNATIONAL

# Electrical Installations
## MATERIAL COSTS/MEASURED WORK PRICES

**DIRECTIONS**

The following explanations are given for each of the column headings and letter codes. It should be noted that not only are full material costs per item declared but also the published list price, the latest published price increase (update).

| | |
|---|---|
| Unit | Prices for each unit are given as singular (1 metre, 1 nr). |
| Net price | Manufacturer's latest issued material/component price list, plus nominal allowance for fixings, plus any percentage uplift advised by manufacturer, plus where appropriate waste less percentage discount. The net price also reflects the applicable quantity for the minimum purchase of each batch of material and against which discounts etc. apply. |
| Material cost | Net price plus percentage allowance for overheads and profit. |
| Labour constant | Gang norm (in manhours) for each operation. |
| Labour cost | Labour constant multiplied by the appropriate all-in manhour cost.(See also relevant Rates of Wages Section) |
| Measured work price | Material cost plus Labour cost. |

MATERIAL COSTS
The Material Costs given includes for delivery to sites in the London area at April/May 1997 with an allowance for waste, overheads and profit, and represents the price paid by contractors after the deduction of all trade discounts but excludes any charges in respect of VAT.

MEASURED WORK PRICES
These prices are intended to apply to new work in the London area and include allowances for all charges, preliminary items and profit. The prices are for reasonable quantities of work and the user should make suitable adjustments if the quantities are especially small or especially large. Adjustments may also be required for locality (eg outside London) and for the market conditions (eg volume of work on hand or on offer) at the time of use.

## DIRECTIONS

**LABOUR RATE**
The labour rate on which these prices have been based is £10.72 per man hour which is the London Standard Rate at 4 January 1997 plus allowances for all other emoluments and expenses. To this rate has been added 25% to cover site and head office overheads and preliminary items together with a further 2½% for profit, resulting in an inclusive rate of **£13.67** per man hour. The rate of £10.72 per man hour has been calculated on a working year of 1732.50 hours; a detailed build-up of the rate is given at the end of these Directions.

In calculating the 'Measured Work Prices' the following assumptions have been made:
- (a) That the work is carried out as a sub-contract under the Standard Form of Building Contract and that such facilities as are usual would be afforded by the main contractor.
- (b) That, unless otherwise stated, the work is being carried out in open areas at a height which would not require more than simple scaffolding.
- (c) That the building in which the work is being carried out is no more than six storey's high.

Where these assumptions are not valid, as for example where work is carried out in ducts and similar confined spaces or in multi-storey structures when additional time is needed to get to and from upper floors, then an appropriate adjustment must be made to the prices. Such adjustment will normally be to the labour element only.

No allowance has been made in the prices for any cash discount to the main contractor.

## DIRECTIONS

## LABOUR RATE - ELECTRICAL

The following detail shows how the labour rate of £10.72 per man hour has been calculated.

**The annual cost of notional eleven man gang.**

|  | TECHNICIAN<br>1 NR | APPROVED ELECTRICIANS<br>4 NR | ELECTRICIANS<br>6 NR | SUB-TOTALS |
|---|---|---|---|---|
| **Hourly Rate from 04/01/1997** | 8.11 | 7.06 | 6.55 | |
| Working hours/annum | 1,732.50 | 6,930.00 | 10,395.00 | |
| x Hourly rate = £/annum | 14,050.58 | 48,925.80 | 68,087.25 | 131,063.63 |
| Daily travel rate/day | 8.02 | 6.99 | 6.48 | |
| Days/annum | 231 | 924 | 1386 | |
| £/annum | 1,852.62 | 6,458.76 | 8,981.28 | 17,292.66 |
| Daily travel fare | 3.43 | 3.43 | 3.43 | |
| Days/annum | 226 | 904 | 1356 | |
| £/annum | 775.18 | 3,100.72 | 4,651.08 | 8,526.98 |
| JIB Combined Benefits Scheme(nr of weeks) | 52 | 208 | 312 | |
| Benefit Credit | 29.57 | 26.39 | 24.85 | |
| £/annum | 1,537.64 | 5,489.12 | 7,753.20 | 14,779.96 |
| National insurance contributions | | | | |
| Weekly gross pay EACH | 332.70 | 289.67 | 268.72 | |
| % Contributions | 0.100 | 0.100 | 0.100 | |
| £ Contributions EACH | 33.27 | 28.97 | 26.87 | |
| x Nr of men = £ contributions/week | 33.27 | 115.88 | 161.22 | |
| £ Contributions/annum | 1,590.31 | 5,539.06 | 7,706.32 | 14,835.69 |
| SUB-TOTAL | | | | 186,498.92 |
| TRAINING (INCLUDING ANY TRADE REGISTRATIONS) - SAY 1% | | | | 1,864.99 |
| SEVERANCE PAY AND SUNDRY COSTS - SAY 1.5% | | | | 2,825.46 |
| EMPLOYER'S LIABILITY AND THIRD PARTY INSURANCE - SAY 2% | | | | 3,823.79 |
| ANNUAL COST OF NOTIONAL 11 MAN GANG (10.5 MEN ACTUALLY WORKING) | | | | 195,013.16 |
| ANNUAL COST PER PRODUCTIVE MAN (ie÷10.5) | | | | 18,572.68 |
| ALL IN MAN HOUR (BASED ON 1732.6 HOURS) | | | | 10.72 |

Notes:
(1) The following assumptions have been made in the above calculations:-
    (a) The working week of 37.5 hours is made up of 7.5 hours Monday to Friday.
    (b) The actual hours worked are five days of 8.5 hours each.
    (c) Five days in the year are lost through sickness or similar reasons.
    (d) A working year of 1732.50 hours.
(2) National insurance contributions are those effective from April 1997.
(3) Weekly Holiday Credit/Welfare Stamp values are those effective from 30 September 1996.

## Y:MECHANICAL AND ELECTRICAL SERVICES

| Item | Net Price £ | Material £ | Labour hours | Labour £ | Unit | Total Rate £ |
|---|---|---|---|---|---|---|
| **T32 : LOW TEMPERATURE HOT WATER HEATING (SMALL SCALE)** | | | | | | |
| **Control Components** | | | | | | |
| Connector unit : moulded white plastic cover and block; galvanised steel back box; to immersion heaters | | | | | | |
| 3Kw up to 915mm long; fitted to thermostat | 14.79 | 16.27 | 0.75 | 10.25 | nr | **26.52** |
| Water heater switch : 20 Amp; switched with neon indicator | | | | | | |
| DP Switched with neon indicator | 7.15 | 7.87 | 0.45 | 6.15 | nr | **14.02** |

## V:ELECTRICAL SUPPLY/POWER/LIGHTING SYSTEMS

| Item | Net Price £ | Material £ | Labour hours | Labour £ | Unit | Total Rate £ |
|---|---|---|---|---|---|---|
| **V10 : ELECTRICITY GENERATION PLANT : GENERATING PLANT** | | | | | | |
| Standby Diesel Generating Sets; Supply and installation fixing to base; all supports and fixings (Raw bolts or equal); all necessary connections to equipment ( Provision and fixing of plates, discs etc. for identification is included | | | | | | |
| Three phase, 440 Volt, four wire 50 Hz packaged standby diesel generating set, complete with radio and television suppressors, daily service fuel tank and associated piping, 4 metres of exhaust pipe and primary exhaust silencer, control panel, mains failure relay, starting battery with charger, all internal wiring, interconnections, earthing and labels rated for standby duty; delivery, installation and commissioning is included | | | | | | |
| 50 kVA | 11466.49 | 12613.14 | - | - | nr | 12613.14 |
| 100 kVA | 14916.70 | 16408.37 | - | - | nr | 16408.37 |
| 150 kVA | 19243.68 | 21168.05 | - | - | nr | 21168.05 |
| 300 kVA | 25725.02 | 28297.52 | - | - | nr | 28297.52 |
| 500 kVA | 42570.75 | 46827.82 | - | - | nr | 46827.82 |
| 750 kVA | 62423.25 | 68665.57 | - | - | nr | 68665.57 |
| 1000 kVA | 77421.59 | 85163.75 | - | - | nr | 85163.75 |
| **EXTRA FOR:** | | | | | | |
| Residential Silencer | | | | | | |
| 300 kVA | 760.14 | 836.15 | - | - | nr | 836.15 |
| 500 kVA | 950.44 | 1045.48 | - | - | nr | 1045.48 |
| 750 kVA | 1075.46 | 1183.01 | - | - | nr | 1183.01 |
| Synchronisation Panel | | | | | | |
| 2 x 100 kVA | 7665.59 | 8432.15 | - | - | nr | 8432.15 |
| 2 x 300 kVA | 8304.39 | 9134.83 | - | - | nr | 9134.83 |
| 2 x 500 kVA | 8678.03 | 9545.83 | - | - | nr | 9545.83 |

Note: Prices excludes cost for bulk storage tank, acoustic treatment or engine room ventilation and have been priced as a total package, i.e. supplied, delivered, installed and commisioned by the manufacturer.

## V:ELECTRICAL SUPPLY/POWER/LIGHTING SYSTEMS

| Item | Net Price £ | Material £ | Labour hours | Labour £ | Unit | Total Rate £ |
|---|---|---|---|---|---|---|
| **V11 : HV SUPPLY/INSTALLATION/PUBLIC UTILITY : TRANSFORMERS** | | | | | | |
| Step Down Transformers; mounted on skids; provision for inserting axles and wheels; all necessary connections to equipment. (Provision and fixing of plates, discs, etc., for identification is included ). | | | | | | |
| 500 kVA; | | | | | | |
| Three Phase 11 kV/433 Volt 50 Hz and LV cable boxes, with glands and silica gel breathers. | | | | | | |
| Naturally cooled, oil filled type ODGM; offload tap changer excluding conservator, dial thermometer or Bucholz relay. | 4200.00 | 4620.00 | 38.46 | 525.69 | nr | 5145.69 |
| Naturally cooled, silicon filled, type ODGM in ventilated steel tank. | 6050.00 | 6655.00 | 38.46 | 525.69 | nr | 7180.69 |
| Dry type; class C insulated air cooled; type AN in ventilated steel tank; temperature indicator. | 10712.00 | 11783.20 | 38.46 | 525.69 | nr | 12308.89 |
| Dry type; cast resin filled, type AN in ventilated steel tank; temperature indicator. | 11433.00 | 12576.30 | 38.46 | 525.69 | nr | 13101.99 |
| Extra For: | | | | | | |
| Conservator | 181.28 | 199.41 | 8.00 | 109.34 | nr | 308.75 |
| Dial thermometer | 294.58 | 324.04 | 2.00 | 27.34 | nr | 351.38 |
| Bucholz relay | 152.44 | 167.68 | 3.00 | 41.05 | nr | 208.73 |
| Pressure Relief Valve | 276.04 | 303.64 | 2.00 | 27.34 | nr | 330.98 |
| 800 kVA; | | | | | | |
| Three Phase 11 kV/433 Volt 50 Hz and LV cable boxes, with glands and silica gel breathers. | | | | | | |
| Naturally cooled, oil filled type ODGM; offload tap changer excluding conservator, dial thermometer or Bucholz relay | 5090.00 | 5599.00 | 45.45 | 621.27 | nr | 6220.27 |
| Naturally cooled, silicon filled, type ODGM in ventilated steel tank. | 7850.00 | 8635.00 | 45.45 | 621.27 | nr | 9256.27 |
| Dry type; class C insulated air cooled; type AN in ventilated steel tank; temperature indicator. | 12154.00 | 13369.40 | 45.45 | 621.27 | nr | 13990.67 |
| Dry type; cast resin filled, type AN in ventilated steel tank; temperature indicator. | 13441.50 | 14785.65 | 45.45 | 621.27 | nr | 15406.92 |
| Extra For: | | | | | | |
| Conservator | 183.34 | 201.67 | 10.00 | 136.68 | nr | 338.35 |
| Dial thermometer | 294.58 | 324.04 | 2.00 | 27.34 | nr | 351.38 |
| Bucholz relay | 152.44 | 167.68 | 3.00 | 41.05 | nr | 208.73 |
| Pressure Relief Valve | 276.04 | 303.64 | 2.00 | 27.34 | nr | 330.98 |

*Material Costs / Measured Work Prices - Electrical Installations* 347

## V:ELECTRICAL SUPPLY/POWER/LIGHTING SYSTEMS

| Item | Net Price £ | Material £ | Labour hours | Labour £ | Unit | Total Rate £ |
|---|---|---|---|---|---|---|
| 1000 kVA | | | | | | |
| Three Phase 11KV/433 volt 50 Hz; HV and LV cable boxes, with glands and silica gel breathers. | | | | | | |
| Naturally cooled, oil filled type ODGM; offload tap changer excluding conservator, dial thermometer or Bucholz relay | 5455.00 | 6000.50 | 83.33 | 1139.00 | nr | 7139.50 |
| Naturally cooled, silicon filled, type ODGM in ventilated steel tank. | 9100.00 | 10010.00 | 83.33 | 1139.00 | nr | 11149.00 |
| Dry type; class C insulated air cooled; type AN in ventilated steel tank; temperature indicator. | 13596.00 | 14955.60 | 83.33 | 1139.00 | nr | 16094.60 |
| Dry type; cast resin filled, type AN in ventilated steel tank; temperature indicator. | 15759.00 | 17334.90 | 83.33 | 1139.00 | nr | 18473.90 |
| Extra For: | | | | | | |
| Conservator | 186.43 | 205.07 | 10.00 | 136.68 | nr | 341.75 |
| Dial thermometer | 294.58 | 324.04 | 2.00 | 27.34 | nr | 351.38 |
| Bucholz relay | 152.44 | 167.68 | 3.00 | 41.05 | nr | 208.73 |
| Pressure Relief Valve | 276.04 | 303.64 | 2.00 | 27.34 | nr | 330.98 |
| 1500 kVA; | | | | | | |
| Three Phase 11KV/433 Volt 50 Hz; HV and LV cable boxes, with glands and silica gel breathers. | | | | | | |
| Naturally cooled, oil filled type ODGM; offload tap changer excluding conservator, dial thermometer or Bucholz relay | 7735.00 | 8508.50 | 100.00 | 1366.80 | nr | 9875.30 |
| Naturally cooled, silicon filled, type ODGM in ventilated steel tank. | 12150.00 | 13365.00 | 100.00 | 1366.80 | nr | 14731.80 |
| Dry type; class C insulated air cooled; type AN in ventilated steel tank; temperature indicator. | 18128.00 | 19940.80 | 100.00 | 1366.80 | nr | 21307.60 |
| Dry type; cast resin filled; type AN in ventilated steel tank; temperature indicator. | 19776.00 | 21753.60 | 100.00 | 1366.80 | nr | 23120.40 |
| Extra For: | | | | | | |
| Conservator | 189.52 | 208.47 | 12.05 | 164.67 | nr | 373.14 |
| Dial thermometer | 294.58 | 324.04 | 2.00 | 27.34 | nr | 351.38 |
| Bucholz relay | 152.44 | 163.87 | 3.00 | 41.05 | nr | 204.92 |
| Pressure Relief Valve | 276.04 | 303.64 | 2.00 | 27.34 | nr | 330.98 |
| 2000 kVA; | | | | | | |
| Three Phase 11KV/433 volt 50Hz; HV and LV cable boxes, with glands and silica gel breathers. | | | | | | |
| Naturally cooled, oil filled type ODGM; offload tap changer excluding conservator, dial thermometer or Bucholz relay | 9955.00 | 10950.50 | 125.00 | 1708.50 | nr | 12659.00 |

## V:ELECTRICAL SUPPLY/POWER/LIGHTING SYSTEMS

| Item | Net Price £ | Material £ | Labour hours | Labour £ | Unit | Total Rate £ |
|---|---|---|---|---|---|---|
| **V11 : HV SUPPLY/INSTALLATION/PUBLIC UTILITY : TRANSFORMERS (contd)** | | | | | | |
| Step Down Transformers; mounted on skids; provision for inserting axles and wheels; all necessary connections to equipment. (Provision and fixing of plates, discs, etc., for identification is included ). (contd) | | | | | | |
| 2000 kVA; (contd) | | | | | | |
| Naturally cooled, silicon filled, type ODGM in ventilated steel tank. | 14950.00 | 16445.00 | 125.00 | 1708.50 | nr | 18153.50 |
| Dry type; class C insulated air cooled; type AN in ventilated steel tank; temperature indicator | 23690.00 | 26059.00 | 125.00 | 1708.50 | nr | 27767.50 |
| Dry type; cast resin filled, type AN in ventilated steel tank; temperature indicator | 25235.00 | 27758.50 | 125.00 | 1708.50 | nr | 29467.00 |
| Extra For: | | | | | | |
| Conservator | 191.58 | 210.74 | 12.05 | 164.67 | nr | 375.41 |
| Dial thermometer | 294.58 | 324.04 | 2.00 | 27.34 | nr | 351.38 |
| Bucholz relay | 152.44 | 167.68 | 3.00 | 41.05 | nr | 208.73 |
| Pressure Relief Valve | 276.04 | 303.64 | 2.00 | 27.34 | nr | 330.98 |

# Keep your figures up to date, free of charge

This section, and most of the other information in this Price Book, is brought up to date every three months, until the next annual edition, in the *Price Book Update*.

The *Update* is available free to all Price Book purchasers.

To ensure you receive your copy, simply complete the reply card from the centre of the book and return it to us.

## V:ELECTRICAL SUPPLY/POWER/LIGHTING SYSTEMS

| Item | Net Price £ | Material £ | Labour hours | Labour £ | Unit | Total Rate £ |
|---|---|---|---|---|---|---|
| **V32 : UNINTERRUPED POWER SUPPLY : U.P.S SYSTEMS/INVERTERS/BATTERIES/ CHARGERS** | | | | | | |
| Uninterruptable Power Supply; 240 Volt AC input and output; standard 13 Amp socket outlet connection; sheet steel enclosure; installed in office environs;self contained battery pack. ( Includes postioning on site and commissioning ). | | | | | | |
| Single Phase; I/P and O/P | | | | | | |
| 1. 2kVA (7 minute supply) | 1293.00 | 1422.30 | 0.31 | 4.27 | nr | 1426.57 |
| 1.2kVA (18 minute supply) | 1479.00 | 1626.90 | 3.21 | 43.81 | nr | 1670.71 |
| 1.6kVA (7 minute supply) | 1376.00 | 1513.60 | 3.40 | 46.49 | nr | 1560.09 |
| 1.6kVA (15 minute supply) | 1596.00 | 1755.60 | 3.40 | 46.49 | nr | 1802.09 |
| 2.2kVA (10 minute supply) | 1760.00 | 1936.00 | 3.61 | 49.34 | nr | 1985.34 |
| 2.2kVA (15 minute supply) | 1923.00 | 2115.30 | 3.61 | 49.34 | nr | 2164.64 |
| 3.0kVA (7 minute supply) | 1950.00 | 2145.00 | 4.00 | 54.67 | nr | 2199.67 |
| 3.0kVA (24 minute supply) | 2340.00 | 2574.00 | 4.00 | 54.67 | nr | 2628.67 |
| 5.0kVA (5 minute supply) | 2654.00 | 2919.40 | 4.50 | 61.57 | nr | 2980.97 |
| 5.0kVA (15 minute supply) | 3156.00 | 3471.60 | 4.50 | 61.57 | nr | 3533.17 |
| 7.5kVA (6 minute supply) | 3168.00 | 3484.80 | 4.90 | 67.00 | nr | 3551.80 |
| 7.5kVA (18 minute supply) | 4119.00 | 4530.90 | 4.90 | 67.00 | nr | 4597.90 |
| 10.0kVA (5 minute supply) | 4519.00 | 4970.90 | 5.00 | 68.34 | nr | 5039.24 |
| 10.0kVA (12 minute supply) | 5688.00 | 6256.80 | 5.00 | 68.34 | nr | 6325.14 |
| Uninterruptable Power Supply; positioned on site; final connections and commissioning. ( Includes delivery up to75 miles to a road/ level site ). | | | | | | |
| Medium size static; three phase input, single phase output; integral sealed battery. | | | | | | |
| 7.5 kVA (32 minutes supply) | 5333.00 | 5866.30 | 30.30 | 414.18 | nr | 6280.48 |
| 10.0 kVA (30 minutes supply) | 5609.00 | 6169.90 | 30.30 | 414.18 | nr | 6584.08 |
| 15.0 kVA (20 minutes supply) | 6344.00 | 6978.40 | 35.71 | 488.14 | nr | 7466.54 |
| 20.0 kVA (12 minutes supply) | 6741.00 | 7415.10 | 40.00 | 546.72 | nr | 7961.82 |
| 25.0 kVA (6 minutes supply) | 7295.00 | 8024.50 | 40.00 | 546.72 | nr | 8571.22 |
| Medium size static; three phase input, three phase output; integral sealed battery; | | | | | | |
| 10.0 kVA (30 minutes supply) | 6959.00 | 7654.90 | 35.71 | 488.14 | nr | 8143.04 |
| 15.0 kVA (20 minutes supply) | 7072.00 | 7779.20 | 35.71 | 488.14 | nr | 8267.34 |
| 20.0 kVA (12 minutes supply) | 7463.00 | 8209.30 | 40.00 | 546.72 | nr | 8756.02 |
| 30.0 kVA (6 minutes supply) | 7738.00 | 8511.80 | 30.30 | 414.18 | nr | 8925.98 |
| Large size static; three phase input/output; lead acid batteries on separate open racks. | | | | | | |
| 40 kVA (16 minutes supply) | 13033.00 | 14336.30 | 43.48 | 594.26 | nr | 14930.56 |
| 60 kVA (9 minutes supply) | 14358.00 | 15793.80 | 45.45 | 621.27 | nr | 16415.07 |
| 80 kVA (16 minutes supply) | 18630.00 | 20493.00 | 45.45 | 621.27 | nr | 21114.27 |
| 100 kVA (12 minutes supply) | 19522.00 | 21474.20 | 50.00 | 683.40 | nr | 22157.60 |
| 200 kVA (10 minutes supply) | 31901.00 | 35091.10 | 55.56 | 759.33 | nr | 35850.43 |
| 300 kVA (10 minutes supply) | 39189.00 | 43107.90 | 62.50 | 854.25 | nr | 43962.15 |

Material Costs / Measured Work Prices - Electrical Installations

## V:ELECTRICAL SUPPLY/POWER/LIGHTING SYSTEMS

| Item | Net Price £ | Material £ | Labour hours | Labour £ | Unit | Total Rate £ |
|---|---|---|---|---|---|---|
| **V32: UNINTERRUPED POWER SUPPLY : U.P.S SYSTEMS/INVERTERS/BATTERIES/ CHARGERS (contd)** | | | | | | |
| **Uninterruptable Power Supply; 240 Volt AC input and output; standard 13 Amp socket outlet connection; sheet steel enclosure; installed in office environs;self contained battery pack. ( Includes postioning on site and commissioning ). (contd)** | | | | | | |
| Integral diesel rotary; three phase input/output; no break supply including ventilation and accoustic attenuation | | | | | | |
| 75 kVA | 73647.00 | 81011.70 | - | - | nr | 81011.70 |
| 100 kVA | 75852.00 | 83437.20 | - | - | nr | 83437.20 |
| 150 kVA | 85995.00 | 94594.50 | - | - | nr | 94594.50 |
| 200 kVA | 147082.95 | 161791.24 | - | - | nr | 161791.24 |
| 250 kVA | 170134.65 | 187148.11 | - | - | nr | 187148.11 |
| 330 kVA | 189308.70 | 208239.57 | - | - | nr | 208239.57 |
| 400 kVA | 217929.60 | 239722.56 | - | - | nr | 239722.56 |
| 500 kVA | 227080.35 | 249788.39 | - | - | nr | 249788.39 |
| 630 kVA | 265006.35 | 291506.98 | - | - | nr | 291506.98 |
| 800 kVA | 303417.45 | 333759.20 | - | - | nr | 333759.20 |
| 1000 kVA | 358123.50 | 393935.85 | - | - | nr | 393935.85 |
| Note: The integral and integrated Diesel Rotary Systems are each priced as a total package, i.e. supplied, delivered, installed and commisioned by the manufacturer. | | | | | | |

*Material Costs / Measured Work Prices - Electrical Installations*

## W:COMMUNICATIONS/SECURITY/CONTROL SYSTEMS

| Item | Net Price £ | Material £ | Labour hours | Labour £ | Unit | Total Rate £ |
|---|---|---|---|---|---|---|
| **W10 : TELECOMMUNICATIONS : TELECOMMUNICATION WIRING** | | | | | | |
| **Multipair Internal Telephone cable; BS 6746; loose laid on tray or drawn in conduit or trunking.( Including cable sleeves ).** | | | | | | |
| 0.5 millimeter diameter conductor p.v.c.insulated and sheathed multipair cables; BT specification CW 1308 | | | | | | |
| 3 pair | 0.12 | 0.00 | 0.05 | 0.68 | m | **0.68** |
| 4 pair | 0.14 | 0.00 | 0.06 | 0.82 | m | **0.82** |
| 6 pair | 0.20 | 0.00 | 0.06 | 0.82 | m | **0.82** |
| 10 pair | 0.39 | 0.43 | 0.07 | 0.96 | m | **1.39** |
| 12 pair | 0.41 | 0.48 | 0.07 | 0.96 | m | **1.44** |
| 15 pair | 0.52 | 0.01 | 0.07 | 0.96 | m | **0.97** |
| 20 pair + 1 wire | 0.70 | 0.01 | 0.09 | 1.23 | m | **1.24** |
| 25 pair | 0.83 | 0.01 | 0.10 | 1.37 | m | **1.38** |
| 40 pair + earth | 1.19 | 0.01 | 0.13 | 1.78 | m | **1.79** |
| 60 pair + earth | 2.00 | 0.02 | 0.15 | 2.05 | m | **2.07** |
| 80 pair + earth | 2.21 | 0.03 | 0.20 | 2.73 | m | **2.76** |
| **Telephone Undercarpet Cable; Low Profile; laid loose** | | | | | | |
| 0.5 millimetre; PVC Insulated; PVC Sheathed multicore cable; BT CW 1316 | | | | | | |
| 6 Core | 0.31 | 0.00 | 0.05 | 0.68 | m | **0.68** |
| **Telephone Drop Wire Cable; drawn in conduit or trunking** | | | | | | |
| 0.5 millimetre conductor; PVC insulate twisted pairs; Polyehylene sheathed; BT CW 1378 | | | | | | |
| Drop Wire 10 | 0.18 | 0.00 | 0.06 | 0.82 | m | **0.82** |

## W: COMMUNICATIONS/SECURITY/CONTROL SYSTEMS

| Item | Net Price £ | Material £ | Labour hours | Labour £ | Unit | Total Rate £ |
|---|---|---|---|---|---|---|
| **W20 : RADIO/TV/CCTV : COAXIAL CABLES** | | | | | | |
| **Television Aerial Cable; coaxial; PVC sheathed; fixed to backgrounds** | | | | | | |
| General purpose TV aerial downlead; copper stranded inner conductor; cellular polythene insulation; copper braid outer conductor; 75 ohm impedance | | | | | | |
| 7/0.25mm | 0.18 | 0.00 | 0.06 | 0.82 | m | 0.82 |
| Low loss TV aerial downlead; solid copper inner conductor; cellular polythene insulation; copper braid outer; conductor; 75 ohm impedance | | | | | | |
| 1/1.12mm | 0.29 | 0.00 | 0.06 | 0.82 | m | 0.82 |
| Low loss air spaced; solid copper inner conductor; air spaced polythene insulation; copper braid outer conductor; 75 ohm impedance | | | | | | |
| 1/1.00mm | 0.18 | 0.00 | 0.06 | 0.82 | m | 0.82 |
| Satelite aerial downlead; solid copper inner conductor; air spaced polythene insulation; copper tape and braid outer conductor; 75 ohm impedance | | | | | | |
| 1/1.00mm | 0.36 | 0.00 | 0.07 | 0.96 | m | 0.96 |
| Satelite TV coaxial; solid copper inner conductor; semi air spaced polyethylene dielectric insulation; plain annealed copper foil and copper braid screen in outer conductor; PVC sheath; 75 ohm impedance | | | | | | |
| 1/1.25mm | 0.49 | 0.01 | 0.08 | 1.09 | m | 1.10 |
| Satelite TV coaxial; solid copper inner conductor; air spaced polyethylene dielectric insulation; plain annealed copper foil and copper braid screen in outer conductor; PVC sheath; 75 ohm impedance | | | | | | |
| 1/1.67mm | 0.81 | 0.01 | 0.09 | 1.23 | m | 1.24 |
| **Radio Frequency Cable; BS 2316 ; PVC sheathed; laid loose** | | | | | | |
| 7/0.41mm tinned copper inner conductor; solid polyethylene dielectric insulation; bare copper wire braid; PVC sheath; 75 ohm impedance | | | | | | |
| Cable | 0.69 | 0.01 | 0.05 | 0.68 | m | 0.69 |

## W:COMMUNICATIONS/SECURITY/CONTROL SYSTEMS

| Item | Net Price £ | Material £ | Labour hours | Labour £ | Unit | Total Rate £ |
|---|---|---|---|---|---|---|
| Twin 1/0.58mm copper covered steel solid core wire conductor; solid polyethylene dielectric insulation; bare copper wire braid; PVC sheath; 75 ohm impedance | | | | | | |
| Cable | 1.00 | 0.05 | 0.05 | 0.68 | m | 0.73 |
| **Video Cable; PVC Flame Retardant sheath; laid loose** | | | | | | |
| 7/0.1mm silver coated copper covered annealed steel wire conductor; polyethylene dielectric insulation with tin coated copper wire braid; 75 ohms impedance | | | | | | |
| Cable | 0.46 | 0.01 | 0.05 | 0.68 | m | 0.69 |

## W: COMMUNICATIONS/SECURITY/CONTROL SYSTEMS

| Item | Net Price £ | Material £ | Labour hours | Labour £ | Unit | Total Rate £ |
|---|---|---|---|---|---|---|
| **W23 : CLOCKS : CLOCK SYSTEMS** | | | | | | |
| **Clock Timing Systems; Master and Slave Units; fixed to background; having battery standby power reserve; 24 DC; ( Including supports and fixings ).** | | | | | | |
| Quartz master clock with solid state digital readout for parallel loop operation; one minute, half minute and one second pulse; maximum of 80 clocks per loop | | | | | | |
|   Two outputs of 1 Amp per loop | 486.88 | 535.57 | 4.00 | 54.67 | nr | 590.24 |
|   Four outputs of 1 Amp per loop | 527.88 | 580.67 | 6.02 | 82.34 | nr | 663.01 |
| Power supplies for above, giving 24 hours power reserve | | | | | | |
|   2 6 Amp hour batteries | 256.25 | 281.88 | 3.00 | 41.05 | nr | 322.93 |
|   2 15 Amp hour batteries | 358.75 | 394.63 | 5.00 | 68.34 | nr | 462.97 |
| Larger quartz master clock with solid state digital readout for parallel loop operation, one minute, half minute and one second pulse maximum of 500 impulse clocks over 6 circuits. | | | | | | |
|   Facility for BST & GMT programmed changeover | 369.00 | 405.90 | 8.00 | 109.34 | nr | 515.24 |
|   Power supply for larger master clock with 24 hours reserve power, with 2 Nr 38A hour batteries | 563.75 | 620.13 | 7.04 | 96.25 | nr | 716.38 |
| Radio receiver to accept BBC Rugby Transmitter MSF signal | | | | | | |
|   To synchronise time of above larger master clock | 210.13 | 231.14 | 7.04 | 96.25 | nr | 327.39 |
| Wall clocks for slave (impulse) systems;24V DC, white dial with black numerals fitted with axxis polycarbonate disc; BS 467.7 Class O | | | | | | |
|   227mm diameter 1 minute impulse | 48.38 | 53.22 | 0.50 | 6.83 | nr | 60.05 |
|   305mm diameter 1 minute impulse | 51.06 | 56.17 | 2.00 | 27.34 | nr | 83.51 |
|   227mm diameter 1/2 minute impulse | 51.06 | 56.17 | 2.00 | 27.34 | nr | 83.51 |
|   305mm diameter 1/2 minute impulse | 50.74 | 55.81 | 2.00 | 27.34 | nr | 83.15 |
|   227mm diameter 1 second impulse | 72.56 | 79.82 | 2.00 | 27.34 | nr | 107.16 |
|   305mm diameter 1 second impulse | 75.25 | 82.78 | 2.00 | 27.34 | nr | 110.12 |
| Quartz Battery Movement; BS 467.7 Class O; white dial with black numerals and sweep second hand; fitted with axxis polycarbonate disc; stove enamel case | | | | | | |
|   227mm diameter | 22.55 | 24.80 | 0.80 | 10.93 | nr | 35.73 |
|   305mm diameter | 30.24 | 33.26 | 0.80 | 10.93 | nr | 44.19 |

*Material Costs / Measured Work Prices - Electrical Installations*

## W:COMMUNICATIONS/SECURITY/CONTROL SYSTEMS

| Item | Net Price £ | Material £ | Labour hours | Labour £ | Unit | Total Rate £ |
|---|---|---|---|---|---|---|
| Large wall clock; BS 467.7 Class O; synchronous mains movement; white dial with black numerals; fitted with axis polycarbonate disc; stove enamel case | | | | | | |
| 458mm diameter | 133.25 | 146.57 | 1.50 | 20.52 | nr | 167.09 |
| 610mm diameter | 150.68 | 165.75 | 1.50 | 20.52 | nr | 186.27 |
| Weather resistant clock; synchronous mains movement; white dial with black numerals; fitted with axxis polycarbonate disc; stove enamel case and heavy duty hanging bracket and extended handset | | | | | | |
| 610mm diameter | 281.88 | 310.07 | 2.00 | 27.34 | nr | 337.41 |
| Elapsed time clock; BS 467.7 Class O; 240V AC, 50 Hz mains supply; 12 hour duration; dial with 0-55 and 1-12 duration; remote control facility; IP66; axxis polycarbonate disc; spun metal movement cover; 6 point fixing bezel | | | | | | |
| 227mm diameter | 563.75 | 620.13 | 0.50 | 6.83 | nr | 626.96 |
| Matching clock; BS 467.7 Class O; 240V AC, 50/60 Hz mains supply; 12 hour duration; dial with 1-12; dial with 1-12duration; IP 66; axxis polycarbonate disc; spun metal movement cover; semi flush mount on 6 point fixing bezel | | | | | | |
| 227mm diameter | 169.13 | 186.04 | 0.50 | 6.83 | nr | 192.87 |
| Remote Control Unit; BS 1363; integral Stop/Start and Reset; 2 gang; flush or surface mounted | | | | | | |
| Remote Control Unit | 61.50 | 67.65 | 0.50 | 6.83 | nr | 74.48 |
| Digital clocks; 240V, 50 hz supply; with/without synchronisation from Masterclock;12/24 hour display; stand alone operation; multifunctional display; o/a size 144 x 72 x 200mm | | | | | | |
| Type hours/minutes/seconds or minutes/seconds/10th seconds | 199.88 | 219.87 | 0.40 | 5.47 | nr | 225.34 |
| Type hours/minutes or minutes/seconds | 179.38 | 197.32 | 0.40 | 5.47 | nr | 202.79 |
| Digital clocks; 240V, 50 hz supply; with/without synchronisation from Masterclock;12/24 hour display; stand alone operation; multifunctional display; o/a size 240 x 90 x 56mm | | | | | | |
| Type hours/minutes or minutes/seconds | 179.38 | 197.32 | 0.60 | 8.20 | nr | 205.52 |

# Material Costs / Measured Work Prices - Electrical Installations
## W: COMMUNICATIONS/SECURITY/CONTROL SYSTEMS

Guidelines for network design:

IEEE 802.3 ETHERNET systems

These systems are laid out in a BUS topology - devices (workstations/printers/signal repeaters) are connected to a single long main cable. This main cable is called a trunk. The extreme ends of the trunk are fitted with terminators, at least one of of which should be grounded. A long trunk is made up of several individual trunks joined with repeaters.

Thin Ethernet systems use network interface cards with the transceivers built in. Connection to the network is via a T-connector joined directly to the trunk cable segment. Thick ethernet systems use transceivers which are permanently fixed to the trunk cable. Connection between the workstation NIC and the transceiver is via a patch cable.

|  | 10 Base 5 'Thick-net' | 10 Base 2 'Thin-net' | 1 Base 5 | 10 Base T | 100 Base T | 100 Base VG |
|---|---|---|---|---|---|---|
| Length of individual trunk segments | 500m | 185m | 500m | 100m | 100m | 500m |
| Maximum number of trunk segments: | 5nr | 5nr |  |  |  |  |
| Devices / Segment (including repeaters): | 100nr | 30nr |  |  |  | 1,024nr |
| Minimum separation between devices: | 2.4m | 0.5m |  |  |  |  |
| Maximum length of patch cables: | 50m | n/a |  |  |  | 200m |

IEEE 802.5 TOKEN RING systems

These systems are connected in a RING topology electrically, but are star-wired physically. Each device is connected to a wiring hub/concentrator with data travelling from the sending device by way of the hub to the next device, where it is re-transmitted on to the next device. Each device contains a network interface card (NIC) and a 2.4/3.0m adaptor cable (fly-lead). The fly-lead is connected directly, or via extension cables (patch cables) to a wiring concentrator / hub (multiple access / MAU).
The length of cable between the fly-lead and the MAU is called the lobe. The MAU's are wired together to form the ring. The original IEEE 802.5 standard applies to passive hubs; modern hardware contains signal repeaters and re-timers which allow transmission over greater distances and/or through lower quality media.

| SUMMARY: | passive MAU (IEEE 802.5) | passive MAU | | | active MAU *** | | |
|---|---|---|---|---|---|---|---|
|  |  | Type 3 UTP 4Mbs (16Mbs) | Cat 5 UTP 4Mbs (16Mbs) | Type 1 STP 4Mbs (16Mbs) | Type 3 UTP 4Mbs (16Mbs) | Cat 5 UTP 4Mbs (16Mbs) | Type 1 STP 4Mbs (16Mbs) |
| Devices per ring: | 96Nr | 72Nr | 72Nr | 260Nr * | 260Nr | 260Nr | 260Nr |
| Devices per MAU: | 8Nr | 8/16Nr | 8/16Nr | 8/16Nr | 10/16Nr | 10/16Nr | 10/16Nr |
| MAUs per ring: | 12Nr | 9/5Nr | 9/5Nr | 33/17Nr * | 26/17Nr | 26/17Nr | 26/17Nr |
| Lobe length: | 45m | 150m (60m) | 150m (60m) | 385m (173m) | 300m (150m) | 400m (300m) | 600m (400m) |
| Length between MAUs: | 45m | 300m (100m) | 300m (100m) | 770m (346m) | 300m (150m) | 400m (300m) | 600m (400m) |
| Drive Distance **: | 120m | 300m (100m) | 300m (100m) | 770m (346m) | 7.5Km(3.75Km) | 10Km (7.5Km) | 15Km (10Km) |

NB The distances quoted above assume the simplest ring - two widely spaced MAUs with 16 separate devices. More MAUs and devices will increase signal attenuation and will reduce the distances unless signal repeators/lobe extenders are fitted.
NB * Use of discrete signal re-timers required for enhanced connection parameters shown for Type 1 STP.
 ** Maximum Drive Distance consists of longest lobe length + ( main ring length - shortest ring segment ).
 Device fly leads and inter-MAU patch cables up to 3m long do not form part of the maximum distance calculations.
NB *** Assumes use of signal boosting/re-timing hardware within MAUs but not discrete signal repeaters/lobe extenders which would increase these figures further.

NB Unassisted 16Mbs networks have very short transmission distances - typically 160m maximum, assuming high-quality cable as Type 1 STP.

Cable qualities:

| Level 1 | voice grade | Capable of 20Kbs (Kbits/sec), an example is telephone cable. Rarely used for networks but modern hardware can make use of it. |
| Level 2 | data-grade | Capable of 4Mbs. Token Ring (4Mbs) / 1Base5 supported. Unshielded twisted-pair cable. |
| IBM Type 3 | data-grade | Capable of of 4Mbs (16Mbs with certain hardware). Token Ring (4Mbs/16Mbs) Shielded twisted-pair cable. |
| Level 3 | high-speed data grade | Capable of 10Mbs and 16 Mbs. Token Ring (4Mbs/16 Mbs) / 10BaseT supported. Unshielded twisted-pair cable. |
| IBM Type 9 | high-speed data grade | Capable of 16Mbs. Length limited to 100m. Token Ring (4Mbs/16Mbs) Shielded twisted-pair. |
| IBM Type 6 | high-speed data grade | Capable of 16Mbs. Token Ring (4Mbs/16Mbs). Shielded twisted-pair (extension/patch cables rather than trunks) |
| Category 4 | extended distance high-speed data grade | Capable of 20 Mbs. Token Ring (4Mbs/16 Mbs) / 10BaseT supported. UTP/STP. |
| Category 5 | low loss ext. distance high-speed data grade | Capable of 100Mbs. Token Ring (4Mbs/16Mbs)/ Ethernet / Fast Ethernet supported. UTP/STP. |
| IBM Type 1 | high-speed data grade | Capable of 16 Mbs. Token Ring (4Mbs/16 Mbs) Shielded twisted-pair. |
| IEEE 802.3 | Thin Ethernet | Capable of 10Mbs over 200m. Shielded coaxial cable. |
| IEEE 802.3 | Thin Ethernet | Capable of 10Mbs over 500m. Shielded coaxial cable. |

**Material Costs / Measured Work Prices - Electrical Installations**

## W: COMMUNICATIONS/SECURITY/CONTROL SYSTEMS

| | IEEE 802.3 ETHERNET-based systems | | | | | IEE802.12 | IEEE 802.5 IBM TOKEN RING systems | | | | | | | | |
|---|---|---|---|---|---|---|---|---|---|---|---|---|---|---|---|
| 10Base5 Ethernet | 10Base5 Ethernet | 10Base2 Thin Ethernet | Thicknet | 1Base5 Starlan | 10BaseT | 100BaseT Fast Ethernet | 100BaseVG Compatible: Fast Ethernet Token Ring | 4Mbs systems | | | | 16Mbs systems | | | |
| Thicknet 10A.Bs 500m | | Thinnet 10A.Bs 185m | | 1A.Bs 500m | 10A.Bs 100m | 100A.Bs 100m | 100A.Bs 500m | 4A.Bs 100m | 4A.Bs 300m | 4A.Bs 600m | 4A.Bs TP (Drops) | 16A.Bs 100m | 16A.Bs 300m | 16A.Bs TP (Drops) | |
| Thick coax (Trunk) | TP (Drops) | Thin coax | | TP | TP | TP | | | | | | | | | |
| Thick coax | . | . | . | . | . | . | . | . | . | . | . | . | . | . | FES 10Base5-Trunk COAX | 50 Ω |
| Thick coax | . | . | . | . | . | . | . | . | . | . | . | . | . | . | Belden 9880 COAX | 50 Ω |
| Thick coax | . | . | . | . | . | . | . | . | . | . | . | . | . | . | Belden 9880NH COAX | 50 Ω |
| | STP | . | . | . | . | . | . | . | . | . | . | . | . | . | FES 10Base5, 4pr FTP | 78 Ω |
| | STP | . | . | . | . | . | . | . | . | . | . | . | . | . | FES 10Base5, 4pr LSF FTP | 78 Ω |
| | STP | . | . | . | . | . | . | . | . | . | . | . | . | . | FES 10Base5, miniature 4pr FTP | 78 Ω |
| | STP | . | . | . | . | . | . | . | . | . | . | . | . | . | Belden 9901 4pr FTP | 78 Ω |
| | STP | . | . | . | . | . | . | . | . | . | . | . | . | . | Belden 9901NH 4pr FTP | 78 Ω |
| | STP | . | . | . | . | . | . | . | . | . | . | . | . | . | Belden 9902 5pr FTP | 78 Ω |
| | STP | . | . | . | . | . | . | . | . | . | . | . | . | . | Belden 9903 4pr FTP | 78 Ω |
| | . | Thin coax | . | . | . | . | . | . | . | . | . | . | . | . | FES 10Base2 COAX | 50 Ω |
| | . | Thin coax | . | . | . | . | . | . | . | . | . | . | . | . | FES 10Base2 LSF COAX | 50 Ω |
| | . | Thin coax | . | . | . | . | . | . | . | . | . | . | . | . | FES 10Base2 Dual COAX | 50 Ω |
| | . | Thin coax | . | . | . | . | . | . | . | . | . | . | . | . | Belden 9907 COAX | 50 Ω |
| | . | Thin coax | . | . | . | . | . | . | . | . | . | . | . | . | Belden 9907NH COAX | 50 Ω |
| | . | . | . | UTP | UTP | UTP | UTP | UTP | UTP | . | . | UTP | . | . | AT&T 1024 UTP | 105 Ω |
| | . | . | . | UTP | UTP | UTP | UTP | UTP | UTP | . | . | UTP | . | . | IBM Type 3 UTP | 105 Ω |
| | . | . | . | UTP | UTP | UTP | UTP | UTP | UTP | . | . | UTP | . | . | Belden 1154A UTP | 105 Ω |
| | . | . | . | UTP | UTP | UTP | UTP | UTP | UTP | . | . | UTP | . | . | AT&T 1010 UTP | 100 Ω |
| | . | . | . | UTP | UTP | UTP | UTP | UTP | UTP | . | . | UTP | . | . | FES UTP-2PR-LEV3 | 100 Ω |
| | . | . | . | UTP | UTP | UTP | UTP | UTP | UTP | . | . | UTP | . | . | FES UTP-4PR-LEV3 | 100 Ω |
| | . | . | . | UTP | UTP | UTP | UTP | UTP | UTP | . | . | UTP | . | . | FES UTP-12PR-LEV3 | 100 Ω |
| | . | . | . | UTP | UTP | UTP | UTP | UTP | UTP | . | . | UTP | . | . | FES UTP-25PR-LEV3 | 100 Ω |
| | . | . | . | UTP | UTP | UTP | UTP | UTP | UTP | . | . | UTP | . | . | Belden 1455A UTP | 100 Ω |
| | . | . | . | UTP | UTP | UTP | UTP | UTP | UTP | . | . | UTP | . | . | AT&T 1061 UTP | 100 Ω |
| | . | . | . | UTP | UTP | UTP | UTP | UTP | UTP | . | . | UTP | UTP | UTP | FES UTP-4PR-CAT5 | 100 Ω |
| | . | . | . | UTP | UTP | UTP | UTP | UTP | UTP | . | . | UTP | UTP | UTP | FES UTP-4PR-CAT5 LSF | 100 Ω |
| | . | . | . | STP | STP | STP | STP | STP | STP | . | . | STP | STP | STP | Belden 1583A UTP | 100 Ω |
| | . | . | . | STP | STP | STP | STP | STP | STP | . | . | STP | STP | STP | FES FTP-4PR-CAT5 | 100 Ω |
| | . | . | . | STP | STP | STP | STP | STP | STP | . | . | STP | STP | STP | Belden 1584-A FTP | 100 Ω |
| | . | . | . | STP | STP | STP | STP | STP | STP | . | . | STP | STP | UTP | FES UTP-PATCH26-CAT5 | 100 Ω |
| | . | . | . | STP | STP | STP | STP | STP | STP | . | . | STP | STP | UTP | FES UTP-PATCH24-CAT5 | 100 Ω |
| | . | . | . | STP | STP | STP | STP | STP | STP | . | . | STP | STP | STP | FES FTP-PATCH-CAT5 | 100 Ω |
| | . | . | . | UTP | UTP | UTP | UTP | UTP | UTP | . | . | UTP | UTP | . | Belden Datatwist 350 UTP | 100 Ω |
| | . | . | . | UTP | UTP | UTP | UTP | UTP | UTP | . | . | UTP | UTP | . | BICC H9682 UTP | 100 Ω |
| | . | . | . | STP | . | . | . | STP | STP | STP | STP | STP | STP | . | IBM Type 9 FTP | 150Ω |
| | . | . | . | STP | . | . | . | STP | STP | STP | STP | STP | STP | . | IBM Type 1 FTP | 150Ω |
| | . | . | . | STP | . | . | . | STP | STP | STP | STP | STP | STP | . | IBM Type 1A FTP | 150Ω |
| | . | . | . | STP | . | . | . | STP | STP | STP | STP | STP | STP | . | IBM Type 1A LSF FTP | 150Ω |
| | . | . | . | STP | . | . | . | STP | STP | STP | STP | STP | STP | STP | Belden Type 1A FTP | 150Ω |
| | . | . | . | STP | . | . | . | STP | STP | STP | STP | STP | STP | STP | IBM Type 6A FTP | 150Ω |
| | . | . | . | STP | . | . | . | STP | STP | STP | STP | STP | STP | STP | IBM Type 6A LSF FTP | 150Ω |
| | . | . | . | STP | . | . | . | STP | STP | STP | STP | STP | STP | STP | Belden Type 6A FTP | 150Ω |

**Material Costs / Measured Work Prices - Electrical Installations**

## W: COMMUNICATIONS/SECURITY/CONTROL SYSTEMS

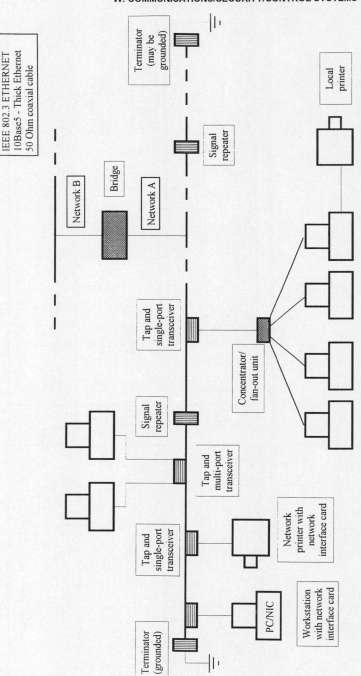

# W: COMMUNICATIONS/SECURITY/CONTROL SYSTEMS

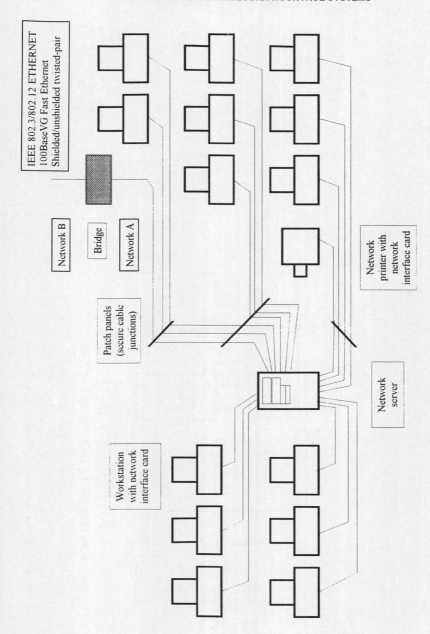

## W: COMMUNICATIONS/SECURITY/CONTROL SYSTEMS

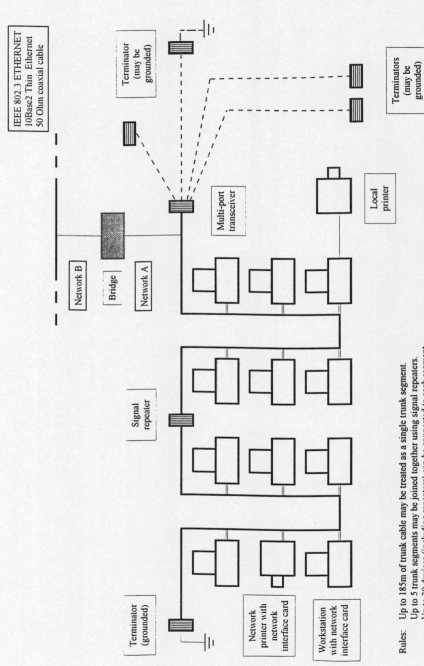

## W: COMMUNICATIONS/SECURITY/CONTROL SYSTEMS

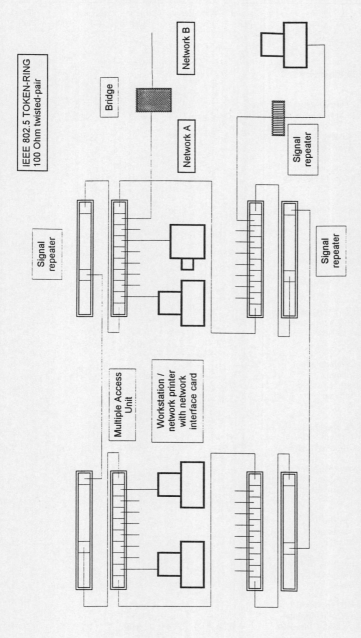

Rules: Maximum transmission distance (longest lobe + total ring - shortest ring segment) is 770m (346m). Self-repeating MAUs increase this to 15Km (10Km). Devices can be up to 150m (60m) from a MAU. If repeater/retiming equipment is used this maximum lobe length is extended to 600m (400m). MAUs can normally be up to 300m (100m) from the next. This distance is extended to 600m (400m) with the extra equipment. Normally up to 72 devices (including repeaters) can be connected to each ring.. Repeaters/retimers increase this to 260 devices per ring.

N.B. Figures are for 4Mbit/sec networks (16Mbs in brackets) over Type 1 shielded twisted-pair. Type 3/Category 5 cables are capable of 40-45% of these distances.

## W: COMMUNICATIONS/SECURITY/CONTROL SYSTEMS

| Item | Net Price £ | Material £ | Labour hours | Labour £ | Unit | Total Rate £ |
|---|---|---|---|---|---|---|
| **W30 : DATA TRANSMISSION : DATA COMMUNICATIONS** | | | | | | |
| Local area networking; IEEE Ethernet 802.3 Systems | | | | | | |
| Data Transmission Cables; PVC sheathed; fixed to backgrounds or laid in trunking. | | | | | | |
| Trunk/backbone data cable; coaxial; tinned copper inner conductor; cellular polyethene insulation; polyester / aluminium foil shield; tinned copper braid screen; second foil shield and second braid screen; nominal impedance 50 Ohms | | | | | | |
|   1 core 2.17mm; PVC sheathed; nominal outside diameter 10.3mm; nominal attenuation 13.8dB/100m at 100Mhz (10Base5 100BaseVG) | 1.16 | 1.27 | 0.03 | 0.41 | m | 1.68 |
|   1 core 2.47mm; PVC sheathed; nominal outside diameter 10.27mm; nominal attenuation 13.8dB/100m at 100Mhz (10Base5 100BaseVG) | 1.78 | 1.95 | 0.03 | 0.41 | m | 2.36 |
|   1 core 2.47mm; flame resistant PVC sheathed; nominal outside diameter 10.27mm; nominal attenuation 13.8dB/100mat 100Mhz (10Base5 100BaseVG) | 2.02 | 2.22 | 0.03 | 0.41 | m | 2.63 |
| Segment data cable; coaxial; tinned copper inner conductors; cellularpolyethene insulation; polyester / aluminium foil shield; tinned copper braid screen; nominal impedance 50 Ohms | | | | | | |
|   19 32AWG strand core; PVC sheathed; nominal outside diameter 4.62mm; nominal attenuation 4.3dB/100m at 100Mhz(10Base2) | 0.27 | 0.30 | 0.09 | 1.23 | m | 1.53 |
|   19 32AWG strand core; flame resistant PVC sheathed; nominal outside diameter 4.62mm; nominal attenuation 4.3dB/100m at 100Mhz (10Base2) | 0.30 | 0.33 | 0.09 | 1.23 | m | 1.56 |
|   Multicore; PVC sheathed; nominal outside diameter 4.62mm; nominal attenuation 4.3dB/100m at 10Mhz (10Base2) | 0.34 | 0.37 | 0.09 | 1.23 | m | 1.60 |
|   Multicore; coloured PVC sheathed; nominal outside diameter 4.62mm; nominal attenuation 4.3dB/100m at 10Mhz (10Base2) | 0.57 | 0.63 | 0.09 | 1.23 | m | 1.86 |
|   Multicore; flame resistant PVC sheathed; nominal outside diameter 4.62mm; nominal attenuation 4.3dB/100m at 10Mhz (10Base2) | 0.59 | 0.64 | 0.09 | 1.23 | m | 1.87 |

## W:COMMUNICATIONS/SECURITY/CONTROL SYSTEMS

| Item | Net Price £ | Material £ | Labour hours | Labour £ | Unit | Total Rate £ |
|---|---|---|---|---|---|---|
| Outlet drop data cable; dual coaxial; parallel tinned copper inner conductors; cellular polyethene insulation; polyester / aluminium foil shield; tinned copper braid screen; nominal impedance 50 Ohms | | | | | | |
| 2 x 19 core 32AWG; PVC sheathed; nominal outside dimensions 4.62 x 9.74mm; nominal attenuation 4.3dB/100m at 10Mhz.(10Base2) | 0.58 | 0.64 | 0.06 | 0.82 | m | **1.46** |
| 1.0m length; PVC sheathed; terminated with BNC and make-before- break safety tap connectors. (10Base2) | 18.04 | 19.84 | 0.06 | 0.82 | m | **20.66** |
| 2.0m length; PVC sheathed; terminated with BNC and make-before- break safety tap connectors. (10Base2) | 21.19 | 23.31 | 0.06 | 0.82 | m | **24.13** |
| 2.5m length; PVC sheathed; terminated with BNC and make-before- break safety tap connectors. (10Base2) | 19.94 | 21.93 | 0.06 | 0.82 | m | **22.75** |
| 3.0m length; PVC sheathed; terminated with BNC and make-before- break safety tap connectors. (10Base2) | 23.22 | 25.54 | 0.06 | 0.82 | m | **26.36** |
| 5.0m length; PVC sheathed; terminated with BNC and make-before- break safety tap connectors. (10Base2) | 30.79 | 33.87 | 0.06 | 0.82 | m | **34.69** |
| Outlet drop data cable; shielded twisted pair; tinned copper inner conductors with polyolefin insulation and polyester/ aluminium foil screen to each; tinned copper braid shield; nominal impedance 78 Ohm | | | | | | |
| 4 pair 28AWG; PVC sheathed; nominal outside diameter 6.6mm (10Base5) | 1.12 | 1.24 | 0.08 | 1.09 | m | **2.33** |
| 4 pair 3,28AWG and 1,24AWG; PVC sheathed; nominal outside diameter 6.6mm(10Base5) | 0.85 | 0.94 | 0.08 | 1.09 | m | **2.03** |
| 4 pair 20AWG; PVC sheathed; nominal outside diameter 10.54mm (10Base5) | 2.26 | 2.48 | 0.08 | 1.09 | m | **3.57** |
| 4 pair 20AWG; flame resistant; PVC sheathed; nominal outside diameter 10.54mm. (10Base5) | 1.40 | 1.54 | 0.08 | 1.09 | m | **2.63** |
| 5 pair 20AWG; PVC sheathed; nominal outside diameter 13.16mm (10Base5) | 3.10 | 3.41 | 0.08 | 1.09 | m | **4.50** |
| Data cable; unshielded twisted pair; solid copper conductors; PVC insulation; nomimal impedance 100 Ohm; Level 2 media/Data grade/IBM type 3 | | | | | | |
| 4 pair 24AWG; PVC sheathed; nominal outside diameter 4.8mm. (1Base5 10BaseT 100BaseVG TR4/300 TR 16/100) | 0.12 | 0.13 | 0.08 | 1.09 | m | **1.22** |

## W: COMMUNICATIONS/SECURITY/CONTROL SYSTEMS

| Item | Net Price £ | Material £ | Labour hours | Labour £ | Unit | Total Rate £ |
|---|---|---|---|---|---|---|
| **W30 : DATA TRANSMISSION : DATA COMMUNICATIONS (contd)** | | | | | | |
| **Local area networking; IEEE Ethernet 802.3 Systems (contd)** | | | | | | |
| **Data Transmission Cables; PVC sheathed; fixed to backgrounds or laid in trunking. (contd)** | | | | | | |
| Data cable; unshielded twisted pair; solid copper conductors; PVC insulation; nomimal impedance 105 Ohm; Level 2 media/Data grade/IBM type 3 | | | | | | |
| 4 pair 24AWG; PVC sheathed; nominal outside diameter 4.7mm; nominal attenuation 2.1dB/100m at 1 Mhz. (1Base5 10BaseT 100BaseVG TR4/300 TR 16/100) | 0.18 | 0.19 | 0.08 | 1.09 | m | **1.28** |
| Data cable; unshielded twisted pair; solid copper conductors; PVC insulation; nomimal impedance 100 Ohm; Level 3 media/High Speed/Data grade/IBM type 3 | | | | | | |
| 4 pair 24AWG; PVC sheathed; nominal outside diameter 4.6mm; nominal attenuation 10dB/100m at 10 Mhz. (1Base5 10BaseT 100BaseVG TR4/300 TR 16/100) | 0.14 | 0.15 | 0.08 | 1.09 | m | **1.24** |
| 2 pair 24AWG; PVC sheathed; nominal outside diameter 3.9mm; nominal attenuation 10dB/100m at 10 Mhz. (1Base5 10BaseT 100BaseVG TR4/300 TR 16/100) | 0.14 | 0.15 | 0.06 | 0.82 | m | **0.97** |
| 4 pair 24AWG; PVC sheathed; nominal outside diameter 4.9mm; nominal attenuation 10dB/100m at 10 Mhz. (1Base5 10BaseT 100BaseVG TR4/300 TR 16/100) | 0.18 | 0.20 | 0.08 | 1.09 | m | **1.29** |
| 12 pair 24AWG; PVC sheathed; nominal outside diameter 7.6mm; nominal attenuation 10dB/100m at 10 Mhz. (1Base5 10BaseT 100BaseVG TR4/300 TR 16/100) | 0.59 | 0.65 | 0.08 | 1.09 | m | **1.74** |
| 25 pair 24AWG; PVC sheathed; nominal outside diameter 10.4mm; nominal attenuation 10dB/100m at 10 Mhz. (1Base5 10BaseT 100BaseVG TR4/300 TR 16/100) | 1.11 | 1.22 | 0.10 | 1.37 | m | **2.59** |
| Data cable; unshielded twisted pair; solid copper conductors; polyolefin insulation; nomimal impedance100 Ohm; Category 4 media/Extended Distance High-speed Data-grade | | | | | | |
| 4 pair 24AWG; PVC sheathed; nominal outside diameter 4.93mm; nominal attenuation 7.2dB/100m at 16 Mhz. (1Base5 10BaseT 100BaseVG TR4/300 TR 16/100) | 0.33 | 0.36 | 0.08 | 1.09 | m | **1.45** |

# Building Services Engineering Spreadsheets

**D.V. Chadderton**, Chartered building services engineer, Victoria, Australia, formerly at the Southampton Institute of Higher Education, Southampton, UK

Building Services Engineering Spreadsheets is a versatile, user friendly tool for design calculations. Spreadsheet application software is readily understandable since each formula is readable in the location where it is used. Each step in the development of these engineering solutions is fully explained.

- fills the gap between manual calculation methods using a calculator and specifically engineered software costing thousands of pounds

### Contents

Contents. Preface. Acknowledgements. Introduction. Units and constants. Symbols. Chapter 1. Computer and spreadsheet use. Chapter 2. Thermal transmittance. Chapter 3. Heat gain. Chapter 4. Combustion of a fuel. Chapter 5. Building heat loss. Chapter 6. Fan and system selection. Chapter 7. Air duct network. Chapter 8. Water pipe sizing. Chapter 9. Lighting. Chapter 10. Electrical cable sizing. References. Index. Answers.

*E & F N Spon*
*most major IBM compatible spreadsheet applications*
*246x189: approx. 300pp: 63 line illus, 3 halftone illus: September 1997*
*Paperback: 0-419-22620-6: c. £29.95*

# The Technology of Building Defects

**J. Hinks**, Reader in Facilities Management, Department of Building Engineering and Surveying, Heriot-Watt University, Edinburgh, UK
**G. Cook**, Senior Lecturer, Department of Construction Management and Engineering, University of Reading, UK

The Technology of Building Defects has been developed to provide a unique stand alone review of building defects. It gives the reader a comprehensive understanding of how and why building defects occur. Defects are considered as part of the whole building rather than in isolation. General education objectives are set out which offer the reader the opportunity of self-assessment and build up an understanding of a range of technical topics concerned with building defects. This is a one stop resource which dispenses with the need to consult a mass of different information sources.

*E & F N Spon*
*246x189: c. 304pp: 95 line illus, 46 halftone illus: October 1997*
*Paperback: 0-419-19780-X: £24.99*

## W:COMMUNICATIONS/SECURITY/CONTROL SYSTEMS

| Item | Net Price £ | Material £ | Labour hours | Labour £ | Unit | Total Rate £ |
|---|---|---|---|---|---|---|
| Data cable; shielded twisted pair; solid copper conductors; polyolefin insulation; nomimal impedance100 Ohm; Category 4 media/Extended Distance High-speed Data-grade | | | | | | |
| 4 pair 24AWG; PVC sheathed; nominal outside diameter 5.25mm; nominal attenuation 9.8dB/100m at 16 Mhz. (1Base5 10BaseT 100BaseVG TR4/300 TR 16/100) | 0.40 | 0.44 | 0.08 | 1.09 | m | **1.53** |
| Data cable; unshielded twisted pair; solid copper conductors; polythene insulation; nomimal impedance 100 Ohm; Category 5 media/Low loss Extended-Distance High-speed Data-grade | | | | | | |
| 4 pair 24AWG; PVC sheathed; nominal outside diameter 4.8mm; (1Base5 10BaseT 100BaseVG TR4/300 TR 16/300) | 0.45 | 0.50 | 0.08 | 1.09 | m | **1.59** |
| 4 pair 24AWG; PVC sheathed; nominal outside diameter 5.51mm; nominal attenuation 8.2dB/100m at 16Mhz (1Base5 10BaseT 100BaseVG TR4/300 TR 16/300) | 0.21 | 0.23 | 0.08 | 1.09 | m | **1.32** |
| 4 pair 24AWG; PVC sheathed; nominal outside diameter 5.2mm; nominal attenuation 19.1dB/100m at 100Mhz (1Base5 10BaseT 100BaseT 100BaseVG TR4/300 TR 16/300) | 0.26 | 0.29 | 0.08 | 1.09 | m | **1.38** |
| 4 pair 24AWG; PVC sheathed; nominal outside diameter 4.8mm; nominal attenuation 21.9dB/100m at 100Mhz (1Base5 10BaseT 100BaseT 100BaseVG TR4/300 TR 16/300) | 0.19 | 0.20 | 0.08 | 1.09 | m | **1.29** |
| Data cable; shielded twisted pairs; solid copper conductors; polethene insulation; aluminium foil shield nomimal impedance 100 Ohm; Category 5 media/Low loss Extended-Distance High-speed Data-grade | | | | | | |
| 4 pair 24AWG; PVC sheathed; nominal outside diameter 5.3mm; nominal attenuation 21.98dB/100m at 100Mhz (1Base5 10BaseT 100BaseT 100BaseVG TR4/300 TR 16/300) | 0.32 | 0.35 | 0.08 | 1.09 | m | **1.44** |
| 4 pair 24AWG; PVC sheathed; nominal outside diameter 6.1mm; nominal attenuation 8.2dB/100m at 16Mhz (1Base5 10BaseT 100BaseT 100BaseVG TR4/300 TR 16/300) | 0.33 | 0.36 | 0.08 | 1.09 | m | **1.45** |
| **Accessories for Ethernet Systems; Fixed to backgrounds** | | | | | | |
| Surge Protectors; shielded; BNC connectors | | | | | | |
| 10KA surge capacity (10Base2) | 51.67 | 56.83 | 0.42 | 5.74 | nr | **62.57** |
| Line terminator; nominal impedance 50Ohm | | | | | | |

## Material Costs / Measured Work Prices - Electrical Installations

### W: COMMUNICATIONS/SECURITY/CONTROL SYSTEMS

| Item | Net Price £ | Material £ | Labour hours | Labour £ | Unit | Total Rate £ |
|---|---|---|---|---|---|---|
| **W30 : DATA TRANSMISSION : DATA COMMUNICATIONS (contd)** | | | | | | |
| **Local area networking; IEEE Ethernet 802.3 Systems (contd)** | | | | | | |
| **Accessories for Ethernet Systems; Fixed to backgrounds (contd)** | | | | | | |
| fitted to make-before-break safety line tap (10Base2) | 3.20 | 3.52 | 0.18 | 2.50 | nr | **6.02** |
| fitted to make-before-break safety line tap; including ground connector (10Base2) | 3.97 | 4.37 | 0.18 | 2.50 | nr | **6.87** |
| fitted to BNC jack (10Base2) | 1.91 | 2.10 | 0.17 | 2.27 | nr | **4.37** |
| Connectors; BNC push-connector; cable secured by twist-on barrel; nominal impedance 50 Ohm | | | | | | |
| end socket; straight (10Base2) | 1.55 | 1.70 | 0.17 | 2.27 | nr | **3.97** |
| end socket; right-angled (10Base2) | 3.80 | 4.18 | 0.17 | 2.27 | nr | **6.45** |
| in-line jack (10Base2) | 1.42 | 1.56 | 0.17 | 2.27 | nr | **3.83** |
| bulkhead jack (for fascia plates) (10Base2) | 2.40 | 2.63 | 0.17 | 2.27 | nr | **4.90** |
| Connectors; N series BNC twist-connector; cable secured by twist-on barrel; nominal impedance 50 Ohm | | | | | | |
| end socket; straight (10Base5) | 1.68 | 1.85 | 0.17 | 2.27 | nr | **4.12** |
| end socket; right-angled (10Base5) | 3.44 | 3.78 | 0.17 | 2.27 | nr | **6.05** |
| bulkhead jack (for fascia plates) (10Base5) | 2.77 | 3.04 | 0.17 | 2.27 | nr | **5.31** |
| Line tap; passive; insulation-displacement connection through sheath of cable; plate phosphor-bronze contacts | | | | | | |
| make-before-break safety tap connection;switching time 500ms (10Base2) | 8.75 | 9.63 | 0.10 | 1.37 | nr | **11.00** |
| Line tap; for fixing into transceiver body | | | | | | |
| insulation-displacment connection (vampire/bee-sting); low-profile; accepts cable diameters between 9.53mm and 10.41mm (10Base5) | 11.19 | 12.31 | 0.17 | 2.27 | nr | **14.58** |
| N-series BNC twist-connector; single (10Base5) | 12.70 | 13.97 | 0.10 | 1.37 | nr | **15.34** |
| BNC twist-connector; single (10Base2) | 10.03 | 11.03 | 0.10 | 1.37 | nr | **12.40** |
| BNC push-connector; tee (10Base2) | 9.40 | 10.34 | 0.10 | 1.37 | nr | **11.71** |
| Transceiver body; including indicator LEDs for power, signal quality error/SEQ, transmission/reception/collision errors | | | | | | |
| 1-port; for connection to separate line tap/adaptor (10Base5 10Base2) | 55.89 | 61.48 | 0.17 | 2.27 | nr | **63.75** |
| 2-port; for connection to separate line tap/adaptor (10Base5 10Base2) | 135.93 | 149.52 | 0.25 | 3.42 | nr | **152.94** |

## W:COMMUNICATIONS/SECURITY/CONTROL SYSTEMS

| Item | Net Price £ | Material £ | Labour hours | Labour £ | Unit | Total Rate £ |
|---|---|---|---|---|---|---|
| 4-port; for connection to separate line tap/adaptor (10Base5 10Base2) | 204.51 | 224.96 | 0.60 | 8.20 | nr | **233.16** |
| 1-port; including line tap; insulation-displacement connection (vampire/bee-sting); accepts cable diameters between 9.53mm and 10.41mm (10Base5) | 106.14 | 116.75 | 0.17 | 2.27 | nr | **119.02** |
| 2-port; including line tap; insulation-displacement connection (vampire/bee-sting) (10Base5) | 239.95 | 263.94 | 0.33 | 4.55 | nr | **268.49** |
| 1-port; including N-series BNC-type connector (10Base5) | 106.14 | 116.75 | 0.17 | 2.27 | nr | **119.02** |
| 2-port; including N-series BNC-type connector (10Base5) | 239.95 | 263.94 | 0.33 | 4.55 | nr | **268.49** |
| 1-port; including BNC-type connector (10Base5) | 131.07 | 144.18 | 0.17 | 2.27 | nr | **146.45** |
| 2-port; including BNC-type connector (10Base5) | 239.95 | 263.94 | 0.33 | 4.55 | nr | **268.49** |
| Transceiver body; no status indicators | | | | | | |
| 1-port; including integral line tap; insulation-displacement connection (vampire/bee-sting); accepts cable diameters between 9.53mm and 10.41mm (10Base5) | 97.37 | 107.11 | 0.17 | 2.27 | nr | **109.38** |
| 1-port; including N-series BNC-type connector (10Base5) | 97.37 | 107.11 | 0.17 | 2.27 | nr | **109.38** |
| Repeaters/retimers; fixing between extended segments or at junctions between media; including status indicators (contd) | | | | | | |
| 2, AUI connectors (10Base5 10Base2) | 196.89 | 216.58 | 0.83 | 11.35 | nr | **227.93** |
| 1, AUI/1 BNC connectors (10Base5 10Base2) | 196.89 | 216.58 | 0.67 | 9.11 | nr | **225.69** |
| 1, AUI/1 RJ45 connectors (10Base5 10Base2) | 173.39 | 190.73 | 0.67 | 9.11 | nr | **199.84** |
| 2, BNC connectors (10Base5 10Base2) | 214.04 | 235.44 | 0.50 | 6.83 | nr | **242.27** |
| 1, BNC/1, RJ45 connectors (10Base5 10Base2) | 179.11 | 197.02 | 0.50 | 6.83 | nr | **203.85** |
| Multiport transceiver; unpowered AUI connection to LAN; status indicators | | | | | | |
| 2-port; AUI connectors (10Base5 10Base2) | 134.01 | 147.41 | 0.50 | 6.83 | nr | **154.24** |
| 4-port; AUI connectors (10Base5 10Base2) | 182.92 | 201.21 | 1.17 | 15.95 | nr | **217.16** |
| Multiport transceiver; mains powered; single AUI connection to LAN; status indicators | | | | | | |
| 8-port; AUI connectors (10Base5 10Base2) | 370.91 | 408.00 | 1.50 | 20.52 | nr | **428.52** |
| Multiport transceiver; mains powered; electrically stackable; single AUI connection to LAN; status indicators | | | | | | |
| 8-port; AUI connectors (10Base5 10Base2) | 517.06 | 568.77 | 1.50 | 20.52 | nr | **589.29** |

## W: COMMUNICATIONS/SECURITY/CONTROL SYSTEMS

| Item | Net Price £ | Material £ | Labour hours | Labour £ | Unit | Total Rate £ |
|---|---|---|---|---|---|---|
| **W30 : DATA TRANSMISSION : DATA COMMUNICATIONS (contd)** | | | | | | |
| **Local area networking; IEEE Ethernet 802.3 Systems (contd)** | | | | | | |
| **Accessories for Ethernet Systems; Fixed to backgrounds (contd)** | | | | | | |
| Multiport repeater/hubs; mains powered; electrically stackable; single AUI connection to LAN; fixing between segments and/or connected to several segments; including status indicators | | | | | | |
| 8-port; AUI connectors (10BaseT) | 437.21 | 480.93 | 1.50 | 20.52 | nr | 501.45 |
| 2-port; BNC connectors (10Base2) | 583.28 | 641.61 | 0.50 | 6.83 | nr | 648.44 |
| 4-port; BNC connectors (10Base2) | 924.09 | 1016.50 | 0.83 | 11.39 | nr | 1027.89 |
| 4-port; BNC connectors; slim line case (10Base2) | 511.22 | 562.34 | 0.83 | 11.39 | nr | 573.73 |
| 8-port; BNC connectors (10Base2) | 1167.53 | 1284.28 | 1.50 | 20.52 | nr | 1304.80 |
| 8-port; BNC connectors; slim line case (10Base2) | 704.02 | 774.42 | 1.50 | 20.52 | nr | 794.94 |
| 8-port; BNC connectors (10BaseT) | 875.40 | 962.94 | 1.50 | 20.52 | nr | 983.46 |
| 4-port; RJ45 connectors (10BaseT) | 91.20 | 100.32 | 0.83 | 11.39 | nr | 111.71 |
| 8-port; RJ45 connectors (10BaseT) | 95.24 | 104.76 | 1.50 | 20.52 | nr | 125.28 |
| 12-port; RJ45 connectors (10BaseT) | 167.87 | 184.66 | 1.67 | 22.78 | nr | 207.44 |
| Multiport repeater/hubs; mains powered; electrically stackable; single AUI connection to LAN; fixing between segments and/or connected to several segments; including status indicators | | | | | | |
| 8-port; RJ45 connectors (10BaseT) | 222.99 | 245.29 | 1.50 | 20.52 | nr | 265.81 |
| 12-port; RJ45 connectors/1, RS-232 connector/1, Inter-bus connector; SNMP software management (10BaseT) | 483.06 | 531.37 | 2.17 | 29.65 | nr | 561.02 |
| Multiport repeater/hubs; unpowered; electrically stackable; single AUI connection to LAN; fixing between segments and/or connected to several segments; including status indicators | | | | | | |
| 8-port; RJ45 connectors (10BaseT) | 213.25 | 234.57 | 1.50 | 20.52 | nr | 255.09 |
| Local bridges/filters; mains powered; spanning tree-type bridges; fixing between adjacent LANs; BNC and AUI connectors; filtering speed 25000 packets per second (pps); forwarding at 12500pps; storage for 2000 MAC addresses; includes status indicators | | | | | | |
| enhanced unit; SNMP software management (10Base5 10Base2) | 1184.89 | 1303.38 | 0.33 | 4.55 | nr | 1307.93 |
| standard unit (10Base5 10Base2) | 987.72 | 1086.49 | 0.33 | 4.55 | nr | 1091.04 |

*Material Costs / Measured Work Prices - Electrical Installations*

## W:COMMUNICATIONS/SECURITY/CONTROL SYSTEMS

| Item | Net Price £ | Material £ | Labour hours | Labour £ | Unit | Total Rate £ |
|---|---|---|---|---|---|---|
| Patch panel; passive multi-port connectors; mechanical connections | | | | | | |
| 16-port, unkeyed unshielded RJ45 connectors (1Base5 10Base5 10Base2 10BaseT 100BaseT 100BaseVG) | 116.79 | 128.47 | 0.50 | 6.83 | nr | **135.30** |
| 16-port, unkeyed shielded RJ45 connectors (1Base5 10Base5 10Base2 10BaseT 100BaseT 100BaseVG) | 134.96 | 148.46 | 0.50 | 6.83 | nr | **155.29** |
| 48-port, unkeyed unshielded RJ45 connectors (1Base5 10Base5 10Base2 10BaseT 100BaseT 100BaseVG) | 267.78 | 294.56 | 1.37 | 18.67 | nr | **313.23** |
| 48-port, unkeyed shielded RJ45 connectors (1Base5 10Base5 10Base2 10BaseT 100BaseT 100BaseVG) | 335.26 | 368.79 | 1.37 | 18.67 | nr | **387.46** |
| Patch panel; passive multi-port connectors; insulation-displacement connections | | | | | | |
| 16-port, unkeyed unshielded RJ45 connectors (1Base5 10Base5 10Base2 10BaseT 100BaseT 100BaseVG) | 78.87 | 86.76 | 0.33 | 4.55 | nr | **91.31** |
| 16-port, keyed unshielded RJ45 connectors (1Base5 10Base5 10Base2 10BaseT 100BaseT 100BaseVG) | 87.63 | 96.39 | 0.33 | 4.55 | nr | **100.94** |
| 16-port, keyed shielded RJ45 connectors (1Base5 10Base5 10Base2 10BaseT 100BaseT 100BaseVG) | 121.72 | 133.89 | 0.33 | 4.55 | nr | **138.44** |
| 24-port, unkeyed RJ45 connectors (1Base5 10Base5 10Base2 10BaseT 100BaseT100BaseVG) | 116.85 | 128.53 | 0.57 | 7.74 | nr | **136.27** |
| 48-port, unkeyed unshielded RJ45 connectors (1Base5 10Base5 10Base2 10BaseT 100BaseT 100BaseVG) | 180.14 | 198.15 | 1.13 | 15.50 | nr | **213.65** |
| Data outlet face plate; to suit BS 4662 wall, pattress boxes or service trunking; fitted with keyed RJ45 shuttered data outlet; Category 5 twisted pair system | | | | | | |
| single gang; 1 outlet; shielded | 4.83 | 5.31 | 0.45 | 6.15 | nr | **11.46** |
| single gang; 2 outlet; shielded | 8.75 | 9.63 | 0.45 | 6.15 | nr | **15.78** |
| Data outlet face plate; to suit BS 4662 wall, pattress boxes or service trunking; fitted with keyed RJ45 shuttered data outlet; twisted pair system | | | | | | |
| single gang; 1 outlet; shielded | 6.87 | 7.56 | 0.45 | 6.15 | nr | **13.71** |
| single gang; 2 outlet; shielded | 10.13 | 11.14 | 0.45 | 6.15 | nr | **17.29** |

## W: COMMUNICATIONS/SECURITY/CONTROL SYSTEMS

| Item | Net Price £ | Material £ | Labour hours | Labour £ | Unit | Total Rate £ |
|---|---|---|---|---|---|---|
| **W30 : DATA TRANSMISSION : DATA COMMUNICATIONS (contd)** | | | | | | |
| **Local area networking; IEEE Ethernet 802.3 Systems (contd)** | | | | | | |
| **Accessories for Ethernet Systems; Fixed to backgrounds(contd)** | | | | | | |
| Data outlet face plate; to suit BS 4662 wall, pattress boxes or service trunking; fitted with unkeyed RJ45 in shuttered modules; including blanking modules where required; Category 5 twisted pair system | | | | | | |
| single gang; 1 outlet; unshielded | 6.61 | 7.75 | 0.45 | 6.15 | nr | 13.90 |
| single gang; 2 outlet; unshielded | 6.76 | 13.16 | 0.45 | 6.15 | nr | 19.31 |
| double gang; 1 outlet; unshielded | 7.56 | 9.20 | 0.45 | 6.15 | nr | 15.35 |
| double gang; 2 outlet; unshielded | 7.20 | 14.12 | 0.45 | 6.15 | nr | 20.27 |
| double gang; 3 outlet; unshielded | 7.12 | 19.69 | 0.45 | 6.15 | nr | 25.84 |
| double gang; 4 outlet; unshielded | 6.79 | 29.88 | 0.45 | 6.15 | nr | 36.03 |
| Data outlet face plate; modular; to suit BS 4662 wall, pattress boxes or service trunking; fitted with unkeyed RJ45 in shuttered modules; including blanking modules where required; twisted pair system | | | | | | |
| single gang; 1 outlet; unshielded | 5.79 | 6.83 | 0.45 | 6.15 | nr | 12.98 |
| single gang; 2 outlet; unshielded | 5.38 | 10.91 | 0.45 | 6.15 | nr | 17.06 |
| double gang; 1 outlet; unshielded | 6.71 | 8.35 | 0.45 | 6.15 | nr | 14.50 |
| double gang; 2 outlet; unshielded | 6.30 | 12.43 | 0.45 | 6.15 | nr | 18.58 |
| double gang; 3 outlet; unshielded | 6.24 | 17.30 | 0.45 | 6.15 | nr | 23.45 |
| double gang; 4 outlet; unshielded | 5.83 | 21.36 | 0.45 | 6.15 | nr | 27.51 |
| Data outlet face plate; modular; to suit BS 4662 wall, pattress boxes or service trunking; fitted with 25 pin D-type data outlet; twisted pair system | | | | | | |
| single gang; 1 outlet; shielded | 2.97 | 3.72 | 0.45 | 6.15 | nr | 9.87 |
| single gang; 2 outlet; shielded | 2.56 | 3.76 | 0.45 | 6.15 | nr | 9.91 |
| double gang; 1 outlet; shielded | 3.42 | 4.20 | 0.45 | 6.15 | nr | 10.35 |
| double gang; 2 outlet; shielded | 3.48 | 5.76 | 0.45 | 6.15 | nr | 11.91 |
| double gang; 3 outlet; shielded | 3.42 | 7.07 | 0.45 | 6.15 | nr | 13.22 |
| double gang; 4 outlet; shielded | 3.01 | 7.59 | 0.45 | 6.15 | nr | 13.74 |
| Data outlet face plate; to suit BS 4662 wall, pattress boxes or service trunking; fitted with 25 pin D-type data outlet; twisted pair system | | | | | | |
| single gang; 1 outlet; shielded | 7.43 | 8.17 | 0.45 | 6.15 | nr | 14.32 |
| single gang; 2 outlet; shielded | 13.27 | 14.60 | 0.45 | 6.15 | nr | 20.75 |
| Data outlet face plate; to suit BS 4662 wall, pattress boxes or service trunking; fitted with 25 pin D-type data outlet; coaxial cable system | | | | | | |
| single gang; 1 outlet; unshielded | 6.25 | 6.88 | 0.45 | 6.15 | nr | 13.03 |
| single gang; 2 outlet; unshielded | 9.47 | 10.42 | 0.45 | 6.15 | nr | 16.57 |

## W:COMMUNICATIONS/SECURITY/CONTROL SYSTEMS

| Item | Net Price £ | Material £ | Labour hours | Labour £ | Unit | Total Rate £ |
|---|---|---|---|---|---|---|
| Data outlet face plate; modular; to suit BS 4662 wall, pattress boxes or service trunking; fitted with BNC data outlets; including blanking modules where required; nominal impedance 50 Ohm; coaxial cable system | | | | | | |
| single gang; 1 outlet | 3.15 | 3.92 | 0.45 | 6.15 | nr | 10.07 |
| single gang; 2 outlet | 2.74 | 5.08 | 0.45 | 6.15 | nr | 11.23 |
| double gang; 1 outlet | 3.59 | 4.40 | 0.45 | 6.15 | nr | 10.55 |
| double gang; 2 outlet | 3.65 | 6.60 | 0.45 | 6.15 | nr | 12.75 |
| double gang; 3 outlet | 3.59 | 8.54 | 0.45 | 6.15 | nr | 14.69 |
| double gang; 4 outlet | 3.18 | 9.70 | 0.45 | 6.15 | nr | 15.85 |
| Data outlet face plate; modular; to suit BS 4662 wall, pattress boxes or service trunking; fitted with make-before-break safety tap data outlets; including blanking modules where required; 10Base2 coaxial cable system | | | | | | |
| single gang; 1 outlet | 1.73 | 1.91 | 0.45 | 6.15 | nr | 8.06 |
| double gang; 1 outlet | 2.59 | 3.30 | 0.45 | 6.15 | nr | 9.45 |
| double gang; 2 outlet | 2.18 | 3.37 | 0.45 | 6.15 | nr | 9.52 |
| Data outlet face plate; to suit BS 4662 wall, pattress boxes or service trunking; fitted with make-before-break safety tap data outlets; 10Base2 coaxial cable system | | | | | | |
| single gang; 2 outlet | 28.83 | 31.71 | 0.45 | 6.15 | nr | 37.86 |
| Data outlet box; surface mounted; fitted with hinged face plate | | | | | | |
| make-before-break safety tap BNC push-connector; surface mounted; (10Base2) | 19.91 | 21.90 | 0.45 | 6.15 | nr | 28.05 |
| make-before-break safety tap BNC push-connector; flush mounted; (10Base2) | 19.91 | 21.90 | 0.45 | 6.15 | nr | 28.05 |
| **Local area networking; IEEE Token Ring Systems** | | | | | | |
| **Data Transmission Cables; PVC sheathed; fixed to backgrounds or laid in trunking.** | | | | | | |
| Data cable; unshielded twisted pair; solid copper conductors; PVC insulation; nominal impedance 100 Ohm; Level 2 media/Data grade/IBM type 3 | | | | | | |
| 4 pair 24AWG; PVC sheathed; nominal outside diameter 4.8mm. (1Base5 10BaseT 100BaseVG TR4/300 TR 16/100) | 0.12 | 0.13 | 0.08 | 1.09 | m | 1.22 |

## Material Costs / Measured Work Prices - Electrical Installations

### W: COMMUNICATIONS/SECURITY/CONTROL SYSTEMS

| Item | Net Price £ | Material £ | Labour hours | Labour £ | Unit | Total Rate £ |
|---|---|---|---|---|---|---|
| **W30 : DATA TRANSMISSION : DATA COMMUNICATIONS (contd)** | | | | | | |
| **Local area networking; IEEE Token Ring Systems (contd)** | | | | | | |
| **Data Transmission Cables; PVC sheathed; fixed to backgrounds or laid in trunking.** | | | | | | |
| Data cable; unshielded twisted pair; solid copper conductors; PVC insulation; nomimal impedance 105 Ohm; Level 2 media/Data grade/IBM type 3 | | | | | | |
| 4 pair 24AWG; PVC sheathed; nominal outside diameter 4.7mm; nominal attenuation 2.1dB/100m at 1 Mhz. (1Base5 10BaseT 100BaseVG TR4/300 TR 16/100) | 0.18 | 0.19 | 0.08 | 1.09 | m | **1.28** |
| Data cable; unshielded twisted pair; solid copper conductors; PVC insulation; nomimal impedance 100 Ohm; Level 3 media/High Speed/Data grade/IBM type 3 | | | | | | |
| 4 pair 24AWG; PVC sheathed; nominal outside diameter 4.6mm; nominal attenuation 10dB/100m at 10 Mhz. (1Base5 10BaseT 100BaseVG TR4/300 TR 16/100) | 0.14 | 0.15 | 0.08 | 1.09 | m | **1.24** |
| 2 pair 24AWG; PVC sheathed; nominal outside diameter 3.9mm; nominal attenuation 10dB/100m at 10 Mhz. (1Base5 10BaseT 100BaseVG TR4/300 TR 16/100) | 0.14 | 0.15 | 0.06 | 0.82 | m | **0.97** |
| 4 pair 24AWG; PVC sheathed; nominal outside diameter 4.9mm; nominal attenuation 10dB/100m at 10 Mhz. (1Base5 10BaseT 100BaseVG TR4/300 TR 16/100) | 0.18 | 0.20 | 0.08 | 1.09 | m | **1.29** |
| 12 pair 24AWG; PVC sheathed; nominal outside diameter 7.6mm; nominal attenuation 10dB/100m at 10 Mhz. (1Base5 10BaseT 100BaseVG TR4/300 TR 16/100) | 0.59 | 0.65 | 0.08 | 1.09 | m | 1.74 |
| 25 pair 24AWG; PVC sheathed; nominal outside diameter 10.4mm; nominal attenuation 10dB/100m at 10 Mhz. (1Base5 10BaseT 100BaseVG TR4/300 TR 16/100) | 1.11 | 1.22 | 0.10 | 1.37 | m | 2.59 |
| Data cable; unshielded twisted pair; solid copper conductors; polyolefin insulation; nomimal impedance100 Ohm; Category 4 media/Extended Distance High-speed Data-grade | | | | | | |
| 4 pair 24AWG; PVC sheathed; nominal outside diameter 4.93mm; nominal attenuation 7.2dB/100m at 16 Mhz. (1Base5 10BaseT 100BaseVG TR4/300 TR 16/100) | 0.33 | 0.36 | 0.08 | 1.09 | m | **1.45** |

*Material Costs / Measured Work Prices - Electrical Installations* 373

## W:COMMUNICATIONS/SECURITY/CONTROL SYSTEMS

| Item | Net Price £ | Material £ | Labour hours | Labour £ | Unit | Total Rate £ |
|---|---|---|---|---|---|---|
| Data cable; shielded twisted pair; solid copper conductors; polyolefin insulation; nomimal impedance100 Ohm; Category 4 media/Extended Distance High-speed Data-grade | | | | | | |
| 4 pair 24AWG; PVC sheathed; nominal outside diameter 5.25mm; nominal attenuation 9.8dB/100m at 16 Mhz. (1Base5 10BaseT 100BaseT 100BaseVG TR4/300 TR 16/100) | 0.40 | 0.44 | 0.08 | 1.09 | m | **1.53** |
| Data cable; unshielded twisted pair; solid copper conductors; polythene insulation; nomimal impedance 100 Ohm; Category 5 media/Low loss Extended-Distance High-speed Data-grade | | | | | | |
| 4 pair 24AWG; PVC sheathed; nominal outside diameter 4.8mm; (1Base5 10BaseT 100BaseVG TR4/300 TR 16/300) | 0.45 | 0.50 | 0.08 | 1.09 | m | **1.59** |
| 4 pair 24AWG; PVC sheathed; nominal outside diameter 5.51mm; nominal attenuation 8.2dB/100m at 16Mhz (1Base5 10BaseT 100BaseVG TR4/300 TR 16/300) | 0.21 | 0.23 | 0.08 | 1.09 | m | **1.32** |
| 4 pair 24AWG; PVC sheathed; nominal outside diameter 5.2mm; nominal attenuation 19.1dB/100m at 100Mhz (1Base5 10BaseT 100BaseT 100BaseVG TR4/300 TR 16/300) | 0.26 | 0.29 | 0.08 | 1.09 | m | **1.38** |
| 4 pair 24AWG; PVC sheathed; nominal outside diameter 4.8mm; nominal attenuation 21.9dB/100m at 100Mhz (1Base5 10BaseT 100BaseT 100BaseVG TR4/300 TR 16/300) | 0.19 | 0.20 | 0.08 | 1.09 | m | **1.29** |
| 4 pair 24AWG; PVC sheathed; nominal attenuation 21.6dB/100m at 100Mhz (1Base5 10BaseT 100BaseT 100BaseVG TR4/300 TR16/300) | 0.25 | 0.27 | 0.08 | 1.09 | m | 1.36 |
| Data cable; shielded twisted pairs; solid copper conductors; polethene insulation; aluminium foil shield nomimal impedance 100 Ohm; Category 5 media/Low loss Extended-Distance High-speed Data-grade | | | | | | |
| 4 pair 24AWG; PVC sheathed; nominal outside diameter 5.3mm; nominal attenuation 21.98dB/100m at 100Mhz (1Base5 10BaseT 100BaseT 100BaseVG TR4/300 TR 16/300) | 0.32 | 0.35 | 0.08 | 1.09 | m | **1.44** |
| 4 pair 24AWG; PVC sheathed; nominal outside diameter 6.1mm; nominal attenuation 8.2dB/100m at 16Mhz (1Base5 10BaseT 100BaseT 100BaseVG TR4/300 TR 16/300) | 0.33 | 0.36 | 0.08 | 1.09 | m | **1.45** |
| 4 pair 24AWG; PVC sheathed; nominal attenuation 21.6dB/100m at 100Mhz (1Base5 10BaseT 100BaseT 100BaseVG TR4/300 TR16/300) | 0.28 | 0.31 | 0.08 | 1.09 | m | **1.40** |

## W: COMMUNICATIONS/SECURITY/CONTROL SYSTEMS

| Item | Net Price £ | Material £ | Labour hours | Labour £ | Unit | Total Rate £ |
|---|---|---|---|---|---|---|
| **W30 : DATA TRANSMISSION : DATA COMMUNICATIONS (contd)** | | | | | | |
| **Local area networking; IEEE Token Ring Systems (contd)** | | | | | | |
| **Data Transmission Cables; PVC sheathed; fixed to backgrounds or laid in trunking. (Contd)** | | | | | | |
| Data cable; shielded twisted pairs; solid copper conductors; polythene insulation; aluminium foil shield; overall copper braid screen; nomimal impedance 150 Ohm; IBM Type 9 Indoor | | | | | | |
| 2 pair 26AWG; PVC sheathed; nominal outside diameter 6.6mm;nominal attenuation 6.6dB/100m at 16Mhz (TR4/100TR16/100) | 0.44 | 0.48 | 0.06 | 0.82 | m | **1.30** |
| Data cable; shielded twisted pairs; solid copper conductors; polythene insulation; double layer aluminium foil shield; copper braid screen; nomimal impedance 150 Ohm; IBM Type 1A Indoor | | | | | | |
| 2 pair 22AWG; PVC sheathed; nominal outside diameter 10.9mm;nominal attenuation 4.0dB/100m at 16Mhz (TR4/600 TR16/300) | 0.38 | 0.42 | 0.06 | 0.82 | m | **1.24** |
| 2 pair 22AWG; flame resistant PVC sheathed; nominal outside diameter 10.9mm ; nominal attenuation 4.0dB/100m at 16 Mhz (TR4/600 TR16/300) | 0.47 | 0.52 | 0.06 | 0.82 | m | **1.34** |
| 2 pair 22AWG; flame resistant PVC sheathed; nominal outside diamensions 7.87 x 11.55mm nominal attenuation 4.0dB/100m at 16 Mhz (TR4/600 TR16/300) | 0.69 | 0.76 | 0.06 | 0.82 | m | **1.58** |
| Drop cable; shielded twisted pair; solid copper conductors; polythene insulation; double layer aluminium foil shield; overall copper braid screen; nominal impedance 150 Ohm; IBM Type 6 fire resistant/IBM type 6A | | | | | | |
| 2 pair 26AWG; PVC sheathed; nominal outside diameter 8.8mm; (TR4 TR16) | 3.10 | 3.41 | 0.06 | 0.82 | m | **4.23** |
| 2 pair 26AWG; PVC sheathed; nominal outside diameter 9.41mm; (TR4 TR16) | 0.79 | 0.86 | 0.06 | 0.82 | m | **1.68** |
| 2 pair 26AWG; fire resistant PVC sheathed; nominal outside diameter 8.8mm; (TR4 TR16) | 0.54 | 0.60 | 0.06 | 0.82 | m | **1.42** |

*Material Costs / Measured Work Prices - Electrical Installations* 375

## W:COMMUNICATIONS/SECURITY/CONTROL SYSTEMS

| Item | Net Price £ | Material £ | Labour hours | Labour £ | Unit | Total Rate £ |
|---|---|---|---|---|---|---|
| Patch cable; shielded twisted-pair; solid copper conductors; polythene insulation; aluminium foil shield; nominal impedance 100 Ohm; Category 5 media/Low-loss Extended- distance High speed Data grade | | | | | | |
| 4 pair 26AWG; PVC sheathed; nominal outside diameter 4.8mm (1Base5 10BaseT 100BaseT 100BaseVG) | 0.37 | 0.41 | 0.08 | 1.09 | m | **1.50** |
| Patch cable; unshielded twisted-pair; solid copper conductors; polythene insulation; nominal impedance 100 Ohm; Category 5 media/Low-loss Extended-distance High-speed Data grade | | | | | | |
| 4 pair 24AWG; PVC sheathed; nominal outside diameter 5.25mm (1Base5 10BaseT 100BaseT 100BaseVG) | 0.32 | 0.36 | 0.08 | 1.09 | m | **1.45** |
| 4 pair 26AWG; PVC sheathed; nominal outside diameter 4.5mm (1Base5 10BaseT 100BaseT 100BaseVG) | 0.30 | 0.33 | 0.08 | 1.09 | m | **1.42** |
| Outside cable; shielded twisted-pair; solid copper conductors; polythene insulation; double layer corrugated metallic shield; nominal impedance 150 Ohm; IBM Type 1 Outdoor | | | | | | |
| 2 pair 22AWG; PVC sheathed; nominal outside diameter 14.9mm (TR4/600 TR16/300) | 3.10 | 3.41 | 0.06 | 0.82 | m | **4.23** |
| Connectors; unshielded; Category 5 twisted pairs | | | | | | |
| RJ45 jack; 8-way; 24-28AWG | 0.30 | 0.33 | 0.17 | 2.27 | nr | **2.60** |
| RJ45 identification boot; PVC coloured | 0.18 | 0.20 | 0.02 | 0.22 | nr | **0.42** |
| Connectors; shielded; Category 5 twisted pairs | | | | | | |
| RJ45 jack; 8-way; 24-28AWG | 0.30 | 0.33 | 0.17 | 2.27 | nr | **2.60** |
| RJ45 identification boot; PVC coloured | 0.18 | 0.20 | 0.02 | 0.22 | nr | **0.42** |
| IBM Universal Data Connector | 2.29 | 2.52 | 0.17 | 2.27 | nr | **4.79** |
| Multiple/multi-station access units (MAU/MSAU); mains powered; electronic relays; status indicators | | | | | | |
| 8-lobe; Ring-in, Ring-out; IBM data connectors (TR4 TR16) | 278.92 | 306.81 | 1.50 | 20.52 | nr | **327.33** |
| 8-lobe; Ring-in, Ring-out; shielded RJ45 connectors (TR4 TR16) | 291.15 | 320.26 | 1.50 | 20.52 | nr | **340.78** |

## W: COMMUNICATIONS/SECURITY/CONTROL SYSTEMS

| Item | Net Price £ | Material £ | Labour hours | Labour £ | Unit | Total Rate £ |
|---|---|---|---|---|---|---|
| **W30 : DATA TRANSMISSION : DATA COMMUNICATIONS (contd)** | | | | | | |
| **Local area networking; IEEE Token Ring Systems (contd)** | | | | | | |
| **Accessories for Token Ring Systems; fixed to outlet boxes or trunking (contd)** | | | | | | |
| Multiple/multi-station access units (MAU/MSAU); mains powered; power failure protected; electronic relays; self-monitoring and diagostics; status indicators | | | | | | |
| 8-lobe; Ring-in, Ring-out; IBM data connectors (TR4 TR16) | 631.50 | 694.65 | 1.50 | 20.52 | nr | **715.17** |
| Patch panels; passive multi-port connectors; mechanical operation | | | | | | |
| 16-port, unkeyed unshielded RJ45 connectors (TR4 TR16) | 116.79 | 128.47 | 2.83 | 38.72 | nr | **167.19** |
| 16-port, unkeyed shielded RJ45 connectors (TR4 TR16) | 134.96 | 148.46 | 2.83 | 38.72 | nr | **187.18** |
| 48-port, unkeyed unshielded RJ45 connectors (TR4 TR16) | 267.78 | 294.56 | 8.20 | 112.03 | nr | **406.59** |
| 48-port, keyed unshielded RJ45 connectors (TR4 TR16) | 335.26 | 368.79 | 8.20 | 112.03 | nr | **480.82** |
| Patch panels; passive multi-port connectors; insulation-displacment operation | | | | | | |
| 16-port, unkeyed unshielded RJ45 connectors (TR4 TR16) | 78.87 | 86.76 | 2.83 | 38.72 | nr | **125.48** |
| 16-port, keyed unshielded RJ45 connectors (TR4 TR16) | 87.63 | 96.39 | 2.83 | 38.72 | nr | **135.11** |
| 16-port, keyed shielded RJ45 connectors (TR4 TR16) | 121.72 | 133.89 | 2.83 | 38.72 | nr | **172.61** |
| 24-port, unkeyed unshielded RJ45 connectors (TR4 TR16) | 116.85 | 128.53 | 4.17 | 56.95 | nr | **185.48** |
| 48-port, keyed unshielded RJ45 connectors (TR4 TR16) | 180.14 | 198.15 | 8.20 | 112.03 | nr | **310.18** |
| Data outlet face plate; to suit BS 4662 wall, pattress boxes or service trunking; fitted with keyed RJ45 shuttered data outlet; Category 5 twisted pair system | | | | | | |
| single gang; 1 outlet; shielded | 5.35 | 5.88 | 0.45 | 6.15 | nr | **12.03** |
| single gang; 2 outlet; shielded | 8.75 | 9.63 | 0.45 | 6.15 | nr | **15.78** |
| Data outlet face plate; to suit BS 4662 wall, pattress boxes or service trunking; fitted with keyed RJ45 shuttered data outlet; twisted pair system | | | | | | |
| single gang; 1 outlet; unshielded | 6.87 | 7.56 | 0.45 | 6.15 | nr | **13.71** |
| single gang; 2 outlet; unshielded | 10.13 | 11.14 | 0.45 | 6.15 | nr | **17.29** |

*Material Costs / Measured Work Prices - Electrical Installations* 377

## W:COMMUNICATIONS/SECURITY/CONTROL SYSTEMS

| Item | Net Price £ | Material £ | Labour hours | Labour £ | Unit | Total Rate £ |
|---|---|---|---|---|---|---|
| Data outlet face plate; to suit BS 4662 wall, pattress boxes or service trunking; fitted with unkeyed RJ45 in shuttered modules; including blanking modules where required; Category 5 twisted pair system | | | | | | |
| single gang; 1 outlet; unshielded | 6.61 | 7.75 | 0.45 | 6.15 | nr | **13.90** |
| single gang; 2 outlet; unshielded | 6.76 | 13.16 | 0.45 | 6.15 | nr | **19.31** |
| double gang; 1 outlet; unshielded | 7.56 | 9.20 | 0.45 | 6.15 | nr | **15.35** |
| double gang; 2 outlet; unshielded | 7.20 | 14.12 | 0.45 | 6.15 | nr | **20.27** |
| double gang; 3 outlet; unshielded | 7.12 | 19.69 | 0.45 | 6.15 | nr | **25.84** |
| double gang; 4 outlet; unshielded | 6.76 | 24.60 | 0.45 | 6.15 | nr | **30.75** |
| Data outlet face plate; modular; to suit BS 4662 wall, pattress boxes or service trunking; fitted with keyed RJ45 in shuttered modules; including blanking modules where required; twisted pair system | | | | | | |
| single gang; 1 outlet; unshielded | 5.79 | 6.83 | 0.45 | 6.15 | nr | **12.98** |
| single gang; 2 outlet; unshielded | 5.38 | 10.91 | 0.45 | 6.15 | nr | **17.06** |
| double gang; 1 outlet; unshielded | 6.71 | 8.35 | 0.45 | 6.15 | nr | **14.50** |
| double gang; 2 outlet; unshielded | 6.30 | 12.43 | 0.45 | 6.15 | nr | **18.58** |
| double gang; 3 outlet; unshielded | 6.24 | 17.30 | 0.45 | 6.15 | nr | **23.45** |
| double gang; 4 outlet; unshielded | 5.83 | 21.36 | 0.45 | 6.15 | nr | **27.51** |
| Data outlet face plate; to suit BS 4662 wall, pattress boxes or service trunking; fitted with IBM Universal Data connector; twisted pair system | | | | | | |
| single gang; 1 outlet; unshielded | 7.35 | 8.09 | 0.45 | 6.15 | nr | **14.24** |
| Data outlet face plate; modular; to suit BS 4662 wall, pattress boxes or service trunking; fitted with IBM Universal Data Connector and hinged fascia plate; including blanking modules where required; twisted pair systems | | | | | | |
| single gang; 1 outlet; unshielded | 4.02 | 4.43 | 0.45 | 6.15 | nr | **10.58** |
| double gang; 1 outlet; unshielded | 4.88 | 5.82 | 0.45 | 6.15 | nr | **11.97** |
| double gang; 2 outlet; unshielded | 4.47 | 8.40 | 0.45 | 6.15 | nr | **14.55** |
| Data outlet face plate; to suit BS 4662 wall, pattress boxes or service trunking; fitted with 25 pin D-type data outlet; twisted pair systems | | | | | | |
| single gang; 1 outlet; shielded | 2.56 | 3.72 | 0.45 | 6.15 | nr | **9.87** |
| single gang; 2 outlet; shielded | 2.56 | 3.76 | 0.45 | 6.15 | nr | **9.91** |
| double gang; 1 outlet; shielded | 3.42 | 4.20 | 0.45 | 6.15 | nr | **10.35** |
| double gang; 2 outlet; shielded | 3.48 | 5.76 | 0.45 | 6.15 | nr | **11.91** |
| double gang; 3 outlet; shielded | 3.42 | 7.07 | 0.45 | 6.15 | nr | **13.22** |
| double gang; 4 outlet; shielded | 3.01 | 7.59 | 0.45 | 6.15 | nr | **13.74** |
| Data outlet face plate; to suit BS 4662 wall, pattress boxes or service trunking; fitted with 25 pin D-type data outlet; twisted pair systems | | | | | | |
| single gang; 1 outlet; unshielded | 7.43 | 8.17 | 0.45 | 6.15 | nr | **14.32** |
| single gang; 2 outlet; unshielded | 13.27 | 14.60 | 0.45 | 6.15 | nr | **20.75** |

## W: COMMUNICATIONS/SECURITY/CONTROL SYSTEMS

| Item | Net Price £ | Material £ | Labour hours | Labour £ | Unit | Total Rate £ |
|---|---|---|---|---|---|---|
| **W50 : FIRE DETECTION AND ALARM : FIRE DETECTION SYSTEMS** | | | | | | |
| **Fire Detection Equipment; fixed to backgrounds. (Including supports and fixings.)** | | | | | | |
| Zone control panel - small installation, complete with Batteries and charger; mild steel case; flush or surface mounting | | | | | | |
| 1 zone unit | 131.20 | 144.32 | 3.00 | 41.05 | nr | 185.37 |
| 2 zone unit | 181.43 | 199.57 | 3.51 | 47.96 | nr | 247.53 |
| 4 zone unit | 308.52 | 339.37 | 4.00 | 54.67 | nr | 394.04 |
| 8 zone unit | 453.05 | 498.36 | 5.00 | 68.34 | nr | 566.70 |
| 8 zone repeat indicator | 222.43 | 244.67 | 4.00 | 54.67 | nr | 299.34 |
| 5 zone unit complete with batteries and charger and 5 alarm sectors | 1079.33 | 1187.26 | 4.00 | 54.67 | nr | 1241.93 |
| 10 zone unit complete with batteries and charger and 10 alarm sectors | 1366.33 | 1502.96 | 5.00 | 68.34 | nr | 1571.30 |
| Multi-zone addressable Fire Alarm Pane ; BS 5839 Part 4; single loop arangement; incorporating 16 detection circuits and 2 master alarm circuits; integral battery standby providing 72 hour back-up; mild steel case; surface fixed | | | | | | |
| 16 zone unit | 523.21 | 575.53 | 7.04 | 96.25 | nr | 671.78 |
| Repeat indicator with LCD display and integral batteries and charger | 441.26 | 485.39 | 6.02 | 82.34 | nr | 567.73 |
| Zone module including end-of-line device | 56.38 | 62.02 | 5.00 | 68.34 | nr | 130.36 |
| Power supply/charger; 24 volt; free standing | | | | | | |
| 1.5 Amp Lead Acid | 171.18 | 188.30 | 6.02 | 82.34 | nr | 270.64 |
| 6.0 Amp Lead Acid | 451.00 | 496.10 | 6.02 | 82.34 | nr | 578.44 |
| Battery Packs; 24 volt lead acid; sealed | | | | | | |
| 6.0 Amp/hr | 68.93 | 75.82 | 8.00 | 109.34 | nr | 185.16 |
| 12.0 Amp/hr | 138.38 | 152.22 | 8.00 | 109.34 | nr | 261.56 |
| 24.0 Amp/hr | 209.10 | 230.01 | 8.00 | 109.34 | nr | 339.35 |
| Accessories | | | | | | |
| Manual call point units: surface mounted | | | | | | |
| Plastic covered push to break glass unit | 9.39 | 10.33 | 0.50 | 6.83 | nr | 17.16 |
| Plastic covered push to break glass unit; Waterproof | 14.64 | 16.10 | 0.80 | 10.93 | nr | 27.03 |
| Manual call point units: flush mounted | | | | | | |
| Plastic covered push to break glass unit | 8.81 | 9.69 | 0.56 | 7.66 | nr | 17.35 |
| Plastic covered push to break glass unit; Waterproof | 14.06 | 15.47 | 0.86 | 11.76 | nr | 27.23 |

## W:COMMUNICATIONS/SECURITY/CONTROL SYSTEMS

| Item | Net Price £ | Material £ | Labour hours | Labour £ | Unit | Total Rate £ |
|---|---|---|---|---|---|---|
| Addressable Manual Call Point | | | | | | |
| Plastic covered push to break glass unit | 39.98 | 43.98 | 1.00 | 13.67 | nr | **57.65** |
| Plastic covered push to break glass unit; Weather resistant | 47.71 | 52.48 | 1.25 | 17.09 | nr | **69.57** |
| Detectors | | | | | | |
| Smoke, ionisation type with mounting base | 42.79 | 47.07 | 0.75 | 10.25 | nr | **57.32** |
| Smoke, optical type with mounting base | 39.21 | 43.13 | 0.75 | 10.25 | nr | **53.38** |
| Fixed temperature heat detector with mounting base (60 degrees) | 18.04 | 19.84 | 0.75 | 10.25 | nr | **30.09** |
| Rate of Rise heat detector with mounting base (90 degrees) | 18.04 | 19.84 | 0.75 | 10.25 | nr | **30.09** |
| Addressable Duct Detector including optical smoke detector and addressable base | 214.22 | 235.64 | 2.00 | 27.34 | nr | **262.98** |
| Sundries for detection system | | | | | | |
| Xenon flasher, 24 volt, conduit box | 34.85 | 38.34 | 0.50 | 6.83 | nr | **45.17** |
| 6" bell, conduit box | 19.88 | 21.87 | 0.75 | 10.25 | nr | **32.12** |
| 6" bell, conduit box; weather resistant | 31.47 | 34.62 | 1.00 | 13.67 | nr | **48.29** |
| Siren; 24V polarised | 26.55 | 43.85 | 1.00 | 13.67 | nr | **57.52** |
| Siren; 240V | 72.97 | 80.27 | 1.25 | 17.09 | nr | **97.36** |
| Magnetic Door Holder; 240V ; surface fixed | 50.53 | 55.58 | 1.50 | 20.52 | nr | **76.10** |

## W: COMMUNICATIONS/SECURITY/CONTROL SYSTEMS

| Item | Net Price £ | Material £ | Labour hours | Labour £ | Unit | Total Rate £ |
|---|---|---|---|---|---|---|
| **W52 : LIGHTNING PROTECTION : GENERAL** | | | | | | |
| **Lightning Protection Equipment; fixed to backgrounds. (Including supports and fixings.)** | | | | | | |
| Multiple point air terminal fixed to structure; copper | | | | | | |
| 15 mm | 37.85 | 41.63 | 1.00 | 13.67 | nr | **55.30** |
| Elevation rod, fixed to structure | | | | | | |
| 500 x 10 mm | 7.10 | 7.81 | 0.75 | 10.25 | nr | **18.06** |
| 1000 x 10 mm | 10.78 | 11.86 | 0.90 | 12.30 | nr | **24.16** |
| Tapered rod, fixed to structure | | | | | | |
| 500 x 15 mm | 55.28 | 60.81 | 0.75 | 10.25 | nr | **71.06** |
| 1000 x 15 mm | 79.30 | 87.24 | 0.90 | 12.30 | nr | **99.54** |
| Down conductor, PVC covered copper tape | | | | | | |
| 12.5 x 1.5 mm | 1.65 | 1.90 | 0.20 | 2.73 | m | **4.63** |
| 25.0 x 3.0 mm | 5.42 | 6.23 | 0.20 | 2.73 | m | **8.96** |
| 25.0 x 6.0 mm | 9.56 | 10.99 | 0.30 | 4.10 | m | **15.09** |
| Down conductor, bare copper tape | | | | | | |
| 20.0 x 3 mm | 2.96 | 3.41 | 0.20 | 2.73 | m | **6.14** |
| 25.0 x 3 mm | 3.70 | 4.07 | 0.20 | 2.73 | m | **6.80** |
| 31.0 x 3 mm | 4.67 | 5.37 | 0.30 | 4.10 | m | **9.47** |
| 38.0 x 3 mm | 5.49 | 6.31 | 0.30 | 4.10 | m | **10.41** |
| Down conductor, PVC covered aluminium tape | | | | | | |
| 20.0 x 3 mm | 3.35 | 3.87 | 0.20 | 2.73 | m | **6.60** |
| 25.0 x 3 mm | 3.38 | 3.90 | 0.20 | 2.73 | m | **6.63** |
| Down conductor, bare aluminium tape | | | | | | |
| 12.5 x 1.5 mm | 0.69 | 0.80 | 0.20 | 2.73 | m | **3.53** |
| 20.0 x 3.0 mm | 1.93 | 2.23 | 0.20 | 2.73 | m | **4.96** |
| 25.0 x 3.0 mm | 2.07 | 2.39 | 0.20 | 2.73 | m | **5.12** |
| 25.0 x 6.0 mm | 4.15 | 4.79 | 0.20 | 2.73 | m | **7.52** |
| Rod to tape gunmetal coupling | | | | | | |
| 15 mm | 7.68 | 8.45 | 0.20 | 2.73 | nr | **11.18** |
| Test clamp | | | | | | |
| Oblong | 4.84 | 5.32 | 0.15 | 2.05 | nr | **7.37** |
| Screwdown | 11.97 | 13.17 | 0.18 | 2.46 | nr | **15.63** |
| Bolted | 13.70 | 15.07 | 0.25 | 3.42 | nr | **18.49** |

*Material Costs / Measured Work Prices - Electrical Installations*

## W:COMMUNICATIONS/SECURITY/CONTROL SYSTEMS

| Item | Net Price £ | Material £ | Labour hours | Labour £ | Unit | Total Rate £ |
|---|---|---|---|---|---|---|
| Copper lattice earth plate; laid in ground and connected | | | | | | |
| 600 x 600 x 3 mm | 40.14 | 44.15 | 0.33 | 4.51 | nr | **48.66** |
| 900 x 900 x 3 mm | 71.56 | 78.72 | 0.33 | 4.51 | nr | **83.23** |
| Copper earth electrodes; driven into ground and connected | | | | | | |
| 1220 x 15 mm | 11.28 | 12.41 | 1.20 | 16.41 | nr | **28.82** |
| 1220 x 19 mm | 20.09 | 22.09 | 1.30 | 17.77 | nr | **39.86** |
| Steel cored copper bonded earth electrodes driven into ground and connected | | | | | | |
| 1200 x 16 mm | 5.64 | 6.20 | 1.70 | 23.25 | nr | **29.45** |
| 2400 x 16 mm | 11.12 | 12.23 | 1.80 | 24.63 | nr | **36.86** |
| 3000 x 16 mm | 13.81 | 15.19 | 1.90 | 25.98 | nr | **41.17** |
| Driving stud, coupling and spike fitted to earth rod | | | | | | |
| 16 mm | 0.69 | 0.76 | 0.20 | 2.73 | nr | **3.49** |
| 20 mm | 1.49 | 1.64 | 0.20 | 2.73 | nr | **4.37** |

## Y:MECHANICAL AND ELECTRICAL SERVICES

| Item | Net Price £ | Material £ | Labour hours | Labour £ | Unit | Total Rate £ |
|---|---|---|---|---|---|---|
| **Y24 : TRACE HEATING** | | | | | | |
| Trace Heating; for freeze protection or temperature maintenance of pipework; to BS 6351; including fixing to parent structures by plastic pull ties. | | | | | | |
| Straight Laid | | | | | | |
| 15mm | 5.22 | 6.03 | 0.33 | 4.52 | m | **10.55** |
| 25mm | 5.22 | 6.03 | 0.33 | 4.52 | m | **10.55** |
| 28mm | 5.22 | 6.03 | 0.33 | 4.52 | m | **10.55** |
| 32mm | 5.22 | 6.03 | 0.35 | 4.82 | m | **10.85** |
| 35mm | 5.22 | 6.03 | 0.36 | 4.92 | m | **10.95** |
| 50mm | 5.22 | 6.03 | 0.38 | 5.23 | m | **11.26** |
| 100mm | 5.94 | 6.86 | 0.43 | 5.93 | m | **12.79** |
| 150mm | 5.94 | 6.86 | 0.43 | 5.93 | m | **12.79** |
| Helically Wound | | | | | | |
| 15mm | 6.01 | 6.94 | 0.38 | 5.21 | m | **12.15** |
| 25mm | 6.01 | 6.94 | 0.38 | 5.21 | m | **12.15** |
| 28mm | 6.01 | 6.94 | 0.38 | 5.21 | m | **12.15** |
| 32mm | 6.01 | 6.94 | 0.41 | 5.55 | m | **12.49** |
| 35mm | 6.01 | 6.94 | 0.41 | 5.67 | m | **12.61** |
| 50mm | 6.01 | 6.94 | 0.44 | 6.01 | m | **12.95** |
| 100m | 6.83 | 7.89 | 0.50 | 6.82 | m | **14.71** |
| 150mm | 6.83 | 7.89 | 0.50 | 6.82 | m | **14.71** |
| Accessories for Trace Heating; Weatherproof; polycarbonate enclosure to IP Standards; fully installed. | | | | | | |
| Connection Junction Box | | | | | | |
| 100*100*75mm | 23.75 | 26.13 | 1.84 | 25.12 | nr | **51.25** |
| Single Air Thermostat | | | | | | |
| 150*150*75mm | 52.25 | 57.48 | 1.84 | 25.12 | nr | **82.60** |
| Single Capillary Thermostat | | | | | | |
| 150*150*75mm | 57.00 | 62.70 | 1.84 | 25.12 | nr | **87.82** |
| Twin Capillary Thermostat | | | | | | |
| 150*150*75mm | 71.25 | 78.38 | 1.84 | 25.12 | nr | **103.50** |

## Y:MECHANICAL AND ELECTRICAL SERVICES

| Item | Net Price £ | Material £ | Labour hours | Labour £ | Unit | Total Rate £ |
|---|---|---|---|---|---|---|
| **Y60 : CONDUIT AND CABLE TRUNKING : METAL AND PLASTICS CONDUIT** | | | | | | |
| Heavy Gauged, Screwed Welded Steel; surface fixed on saddles to backgrounds, with standard pattern boxes and fittings including all fixings and supports. (Forming holes, conduit entry, draw wires etc. and components for earth continuity are included.) | | | | | | |
| Black Enamelled | | | | | | |
| 20 mm dia. | 0.73 | 0.85 | 0.50 | 6.83 | m | 7.68 |
| 25 mm dia. | 1.00 | 1.16 | 0.55 | 7.52 | m | 8.68 |
| 32 mm dia. | 1.60 | 1.85 | 0.60 | 8.20 | m | 10.05 |
| 38 mm dia. | 2.07 | 2.40 | 0.70 | 9.57 | m | 11.97 |
| 50 mm dia. | 3.35 | 3.87 | 1.15 | 15.73 | m | 19.60 |
| Galvanised | | | | | | |
| 20 mm dia. | 0.90 | 1.04 | 0.50 | 6.83 | m | 7.87 |
| 25 mm dia. | 1.23 | 1.42 | 0.55 | 7.52 | m | 8.94 |
| 32 mm dia. | 2.00 | 2.31 | 0.60 | 8.20 | m | 10.51 |
| 38 mm dia. | 2.55 | 2.94 | 0.70 | 9.57 | m | 12.51 |
| 50 mm dia. | 3.77 | 4.36 | 1.15 | 15.73 | m | 20.09 |
| Heavy Duty; Galvanised Steel core; IP67 Standards; grey or black PVC covering; surface fixed to backgrounds with standard connectors and components for earth continuity. | | | | | | |
| Temperature Range -10C to +70C | | | | | | |
| 16mm | 2.98 | 3.28 | 0.40 | 5.47 | nr | 8.75 |
| 20mm | 3.42 | 3.76 | 0.40 | 5.47 | nr | 9.23 |
| 25mm | 4.54 | 4.99 | 0.40 | 5.47 | nr | 10.46 |
| 32mm | 6.78 | 7.46 | 0.40 | 5.47 | nr | 12.93 |
| 40mm | 8.40 | 9.24 | 0.40 | 5.47 | nr | 14.71 |
| Temperature Range -25C to +105C | | | | | | |
| 10mm | 2.89 | 3.18 | 0.40 | 5.47 | nr | 8.65 |
| 12mm | 2.89 | 3.18 | 0.40 | 5.47 | nr | 8.65 |
| 16mm | 3.54 | 3.89 | 0.40 | 5.47 | nr | 9.36 |
| 20mm | 4.16 | 4.58 | 0.40 | 5.47 | nr | 10.05 |
| 25mm | 5.60 | 6.16 | 0.40 | 5.47 | nr | 11.63 |
| 32mm | 8.18 | 9.00 | 0.40 | 5.47 | nr | 14.47 |
| 40mm | 9.98 | 10.98 | 0.40 | 5.47 | nr | 16.45 |
| 50mm | 14.38 | 15.82 | 0.40 | 5.47 | nr | 21.29 |
| 63mm | 17.18 | 18.90 | 0.40 | 5.47 | nr | 24.37 |

## Y:MECHANICAL AND ELECTRICAL SERVICES

| Item | Net Price £ | Material £ | Labour hours | Labour £ | Unit | Total Rate £ |
|---|---|---|---|---|---|---|
| **Y60 : CONDUIT AND CABLE TRUNKING : METAL AND PLASTICS CONDUIT (contd)** | | | | | | |
| Heavy Duty; Galvanised Steel Core; IP67 Standards; Copper packed; grey or black PVC covering; surface fixed to backgrounds; jointed with standard connectors and components for earth continuity | | | | | | |
| Temperature Range -10C to +60C | | | | | | |
| 16mm | 5.06 | 5.57 | 0.45 | 6.15 | nr | **11.72** |
| 20mm | 6.14 | 6.75 | 0.45 | 6.15 | nr | **12.90** |
| 25mm | 8.17 | 8.99 | 0.45 | 6.15 | nr | **15.14** |
| 32mm | 11.36 | 12.50 | 0.45 | 6.15 | nr | **18.65** |
| 40mm | 14.04 | 15.44 | 0.45 | 6.15 | nr | **21.59** |
| 50mm | 21.24 | 23.36 | 0.45 | 6.15 | nr | **29.51** |
| 63mm | 26.80 | 29.48 | 0.45 | 6.15 | nr | **35.63** |
| Heavy Duty; Galvanised Steel Core Fittings; to IP67 standards; grey or black; including surface fixing. | | | | | | |
| Male Connector | | | | | | |
| 10mm | 2.10 | 2.31 | 0.08 | 1.13 | nr | **3.44** |
| 12mm | 2.10 | 2.31 | 0.08 | 1.13 | nr | **3.44** |
| 16mm | 2.51 | 2.76 | 0.08 | 1.13 | nr | **3.89** |
| 20mm | 2.63 | 2.89 | 0.08 | 1.13 | nr | **4.02** |
| 25mm | 3.76 | 4.14 | 0.08 | 1.13 | nr | **5.27** |
| 32mm | 6.13 | 6.74 | 0.08 | 1.13 | nr | **7.87** |
| 40mm | 10.18 | 11.19 | 0.08 | 1.13 | nr | **12.32** |
| 50mm | 15.37 | 16.91 | 0.08 | 1.13 | nr | **18.04** |
| 63mm | 28.32 | 31.15 | 0.08 | 1.13 | nr | **32.28** |
| Female Connector | | | | | | |
| 16mm | 2.37 | 2.60 | 0.10 | 1.37 | nr | **3.97** |
| 20mm | 2.48 | 2.73 | 0.10 | 1.37 | nr | **4.10** |
| 25mm | 3.55 | 3.91 | 0.10 | 1.37 | nr | **5.28** |
| 32mm | 5.78 | 6.36 | 0.10 | 1.37 | nr | **7.73** |
| 40mm | 9.33 | 10.26 | 0.10 | 1.37 | nr | **11.63** |
| 50mm | 13.89 | 15.28 | 0.10 | 1.37 | nr | **16.65** |
| 63mm | 26.72 | 29.39 | 0.10 | 1.37 | nr | **30.76** |
| Elbows | | | | | | |
| 16mm | 4.47 | 4.92 | 0.11 | 1.50 | nr | **6.42** |
| 20mm | 4.54 | 5.00 | 0.12 | 1.64 | nr | **6.64** |
| 25mm | 6.82 | 7.51 | 0.12 | 1.64 | nr | **9.15** |
| 32mm | 12.11 | 13.32 | 0.14 | 1.91 | nr | **15.23** |
| 40mm | 21.48 | 23.63 | 0.14 | 1.91 | nr | **25.54** |
| 50mm | 25.09 | 27.60 | 0.19 | 2.60 | nr | **30.20** |
| 63mm | 37.41 | 41.15 | 0.21 | 2.87 | nr | **44.02** |

*Material Costs / Measured Work Prices - Electrical Installations* 385

## Y:MECHANICAL AND ELECTRICAL SERVICES

| Item | Net Price £ | Material £ | Labour hours | Labour £ | Unit | Total Rate £ |
|---|---|---|---|---|---|---|
| **Medium Duty; Nylon to IP66 Standards; Grey or Black; Surface Fixed to backgrounds with standard components for earth continuity** | | | | | | |
| Temperature Range -40C to +100C | | | | | | |
| 10mm | 2.34 | 2.83 | 0.40 | 5.47 | nr | 8.30 |
| 12mm | 2.58 | 3.12 | 0.40 | 5.47 | nr | 8.59 |
| 16mm | 2.95 | 3.57 | 0.40 | 5.47 | nr | 9.04 |
| 20mm | 3.07 | 3.71 | 0.40 | 5.47 | nr | 9.18 |
| 25mm | 3.57 | 4.32 | 0.40 | 5.47 | nr | 9.79 |
| 32mm | 6.02 | 7.28 | 0.40 | 5.47 | nr | 12.75 |
| 40mm | 8.13 | 9.84 | 0.45 | 6.15 | nr | 15.99 |
| 50mm | 10.91 | 13.20 | 0.45 | 6.15 | nr | 19.35 |
| Temperature Range -40C to +120C | | | | | | |
| 12mm | 1.61 | 1.82 | 0.40 | 5.47 | nr | 7.29 |
| 16mm | 1.86 | 2.10 | 0.40 | 5.47 | nr | 7.57 |
| 20mm | 2.18 | 2.46 | 0.40 | 5.47 | nr | 7.93 |
| 25mm | 2.72 | 3.07 | 0.40 | 5.47 | nr | 8.54 |
| 32mm | 3.88 | 4.37 | 0.40 | 5.47 | nr | 9.84 |
| **Medium Duty; Nylon Fittings to IP66 Standards; Grey or black including surface fixing** | | | | | | |
| Male Connector | | | | | | |
| 10mm | 1.00 | 1.10 | 0.08 | 1.13 | nr | 2.23 |
| 12mm | 0.99 | 1.09 | 0.08 | 1.13 | nr | 2.22 |
| 16mm | 1.03 | 1.13 | 0.08 | 1.13 | nr | 2.26 |
| 20mm | 1.28 | 1.41 | 0.08 | 1.13 | nr | 2.54 |
| 25mm | 1.91 | 2.10 | 0.08 | 1.13 | nr | 3.23 |
| 32mm | 2.68 | 2.95 | 0.08 | 1.13 | nr | 4.08 |
| 40mm | 5.60 | 6.16 | 0.08 | 1.13 | nr | 7.29 |
| 50mm | 8.65 | 9.52 | 0.08 | 1.13 | nr | 10.65 |
| Female Connector | | | | | | |
| 10mm | 2.80 | 3.08 | 0.09 | 1.23 | nr | 4.31 |
| 12mm | 1.80 | 1.98 | 0.09 | 1.23 | nr | 3.21 |
| 16mm | 1.95 | 2.15 | 0.09 | 1.23 | nr | 3.38 |
| 20mm | 2.05 | 2.25 | 0.09 | 1.23 | nr | 3.48 |
| 25mm | 3.80 | 4.18 | 0.09 | 1.23 | nr | 5.41 |
| 32mm | 7.91 | 8.70 | 0.09 | 1.23 | nr | 9.93 |
| Elbows | | | | | | |
| 10mm | 1.62 | 1.78 | 0.12 | 1.64 | nr | 3.42 |
| 12mm | 1.66 | 1.83 | 0.12 | 1.64 | nr | 3.47 |
| 16mm | 1.67 | 1.84 | 0.12 | 1.64 | nr | 3.48 |
| 20mm | 1.94 | 2.13 | 0.12 | 1.64 | nr | 3.77 |
| 25mm | 3.32 | 3.65 | 0.12 | 1.64 | nr | 5.29 |
| 32mm | 4.51 | 4.96 | 0.12 | 1.64 | nr | 6.60 |
| 40mm | 9.18 | 10.10 | 0.12 | 1.64 | nr | 11.74 |
| 50mm | 11.69 | 12.86 | 0.12 | 1.64 | nr | 14.50 |

## Y:MECHANICAL AND ELECTRICAL SERVICES

| Item | Net Price £ | Material £ | Labour hours | Labour £ | Unit | Total Rate £ |
|---|---|---|---|---|---|---|
| **Y60 : CONDUIT AND CABLE TRUNKING : METAL AND PLASTICS CONDUIT (contd)** | | | | | | |
| **High Impact Unscrewed PVC; surface fixed on saddles to backgrounds; with standard pattern boxes and fittings; including all fixings and supports.** | | | | | | |
| Light Gauge | | | | | | |
| 16 mm dia. | 0.86 | 0.99 | 0.25 | 3.42 | m | 4.41 |
| 20 mm dia. | 1.07 | 1.24 | 0.25 | 3.42 | m | 4.66 |
| 25 mm dia. | 1.80 | 2.08 | 0.30 | 4.10 | m | 6.18 |
| 32 mm dia. | 2.42 | 2.80 | 0.30 | 4.10 | m | 6.90 |
| 38 mm dia. | 3.08 | 3.56 | 0.35 | 4.78 | m | 8.34 |
| 50 mm dia. | 5.06 | 5.84 | 0.35 | 4.78 | m | 10.62 |
| Heavy Gauge | | | | | | |
| 16 mm dia. | 1.21 | 1.40 | 0.25 | 3.42 | m | 4.82 |
| 20 mm dia. | 1.47 | 1.70 | 0.25 | 3.42 | m | 5.12 |
| 25 mm dia. | 1.97 | 2.28 | 0.30 | 4.10 | m | 6.38 |
| 32 mm dia. | 3.16 | 3.65 | 0.30 | 4.10 | m | 7.75 |
| 38 mm dia. | 4.16 | 4.80 | 0.35 | 4.78 | m | 9.58 |
| 50 mm dia. | 6.58 | 7.60 | 0.35 | 4.78 | m | 12.38 |
| **Flexible Conduits; including adaptors and locknuts. (For connections to equipment.)** | | | | | | |
| Metallic, PVC covered conduit; not exceeding 1m long; including zinc plated mild steel adaptors, lock nuts and earth conductor | | | | | | |
| 16 mm dia. | 3.69 | 4.26 | 0.50 | 6.83 | nr | 11.09 |
| 20 mm dia. | 4.31 | 4.98 | 0.50 | 6.83 | nr | 11.81 |
| 25 mm dia. | 5.70 | 6.58 | 0.50 | 6.83 | nr | 13.41 |
| 32 mm dia. | 8.36 | 9.66 | 0.60 | 8.20 | nr | 17.86 |
| 38 mm dia. | 10.41 | 12.02 | 0.60 | 8.20 | nr | 20.22 |
| 50 mm dia. | 16.89 | 13.66 | 1.00 | 13.67 | nr | 27.33 |
| PVC Conduit; not exceeding 1m long; including nylon adaptors, lock nuts | | | | | | |
| 16 mm dia. | 1.32 | 1.52 | 0.50 | 6.83 | nr | 8.35 |
| 20 mm dia. | 1.64 | 1.89 | 0.50 | 6.83 | nr | 8.72 |
| 25 mm dia. | 2.37 | 2.74 | 0.50 | 6.83 | nr | 9.57 |
| 32 mm.dia. | 3.69 | 4.26 | 0.60 | 8.20 | nr | 12.46 |
| **Grey Cast Iron Adaptable Boxes; with heavy covers, fixed to backgrounds; including all supports and fixings. (Cutting and connecting conduit to boxesis included.)** | | | | | | |
| Square Pattern - Black | | | | | | |
| 150 x 150 x 50 mm | 19.18 | 21.10 | 0.92 | 12.57 | nr | 33.67 |
| 150 x 150 x 75 mm | 24.01 | 26.41 | 0.92 | 12.57 | nr | 38.98 |
| 150 x 150 x 100 mm | 31.74 | 34.91 | 0.92 | 12.57 | nr | 47.48 |
| 225 x 225 x 75 mm | 47.19 | 51.91 | 1.00 | 13.67 | nr | 65.58 |
| 300 x 300 x 100 mm | 105.35 | 115.89 | 1.05 | 14.36 | nr | 130.25 |

## Y:MECHANICAL AND ELECTRICAL SERVICES

| Item | Net Price £ | Material £ | Labour hours | Labour £ | Unit | Total Rate £ |
|---|---|---|---|---|---|---|
| **Square Pattern - Galvanised** | | | | | | |
| 150 x 150 x 50 mm | 26.86 | 29.55 | 0.92 | 12.57 | nr | **42.12** |
| 150 x 150 x 75 mm | 30.98 | 34.08 | 0.92 | 12.57 | nr | **46.65** |
| 150 x 150 x 100 mm | 41.15 | 45.27 | 0.92 | 12.57 | nr | **57.84** |
| 225 x 225 x 75 mm | 64.96 | 71.46 | 1.00 | 13.67 | nr | **85.13** |
| 300 x 300 x 100 mm | 134.34 | 147.77 | 1.05 | 14.36 | nr | **162.13** |
| **Rectangular Pattern - Black** | | | | | | |
| 150 x 75 x 50 mm | 12.66 | 13.93 | 0.92 | 12.57 | nr | **26.50** |
| 150 x 100 x 50 mm | 19.59 | 21.55 | 0.92 | 12.57 | nr | **34.12** |
| 150 x 100 x 75 mm | 21.22 | 23.34 | 0.92 | 12.57 | nr | **35.91** |
| 225 x 150 x 75 mm | 36.99 | 40.69 | 0.98 | 13.40 | nr | **54.09** |
| 225 x 150 x 100 mm | 49.96 | 54.96 | 1.00 | 13.67 | nr | **68.63** |
| 300 x 150 x 75 mm | 68.73 | 75.60 | 1.05 | 14.36 | nr | **89.96** |
| **Rectangular Pattern - Galvanised** | | | | | | |
| 150 x 75 x 50 mm | 17.24 | 18.96 | 0.92 | 12.57 | nr | **31.53** |
| 150 x 100 x 50 mm | 27.15 | 29.86 | 0.92 | 12.57 | nr | **42.43** |
| 150 x 100 x 75 mm | 28.13 | 30.94 | 0.92 | 12.57 | nr | **43.51** |
| 225 x 150 x 75 mm | 51.53 | 56.68 | 0.98 | 13.40 | nr | **70.08** |
| 225 x 150 x 100 mm | 67.34 | 74.07 | 1.00 | 13.67 | nr | **87.74** |
| 300 x 150 x 75 mm | 89.21 | 98.13 | 1.05 | 14.36 | nr | **112.49** |
| **Sheet Steel Adaptable Boxes; with plain or knockout sides; fixed to backgrounds; including supports and fixings. (Cutting and connecting conduit to boxes is included.)** | | | | | | |
| **Square Pattern - Black** | | | | | | |
| 75 x 75 x 37 mm | 4.97 | 0.55 | 0.88 | 12.03 | nr | **12.58** |
| 75 x 75 x 50 mm | 5.64 | 0.62 | 0.88 | 12.03 | nr | **12.65** |
| 75 x 75 x 75 mm | 6.77 | 1.86 | 0.88 | 12.03 | nr | **13.89** |
| 100 x 100 x 50 mm | 5.99 | 0.66 | 0.90 | 12.30 | nr | **12.96** |
| 150 x 150 x 50 mm | 8.01 | 1.76 | 0.90 | 12.30 | nr | **14.06** |
| 150 x 150 x 75 mm | 8.89 | 2.44 | 0.92 | 12.57 | nr | **15.01** |
| 150 x 150 x 100 mm | 11.08 | 3.05 | 0.92 | 12.57 | nr | **15.62** |
| 225 x 225 x 50 mm | 13.93 | 7.66 | 0.98 | 13.40 | nr | **21.06** |
| 225 x 225 x 100 mm | 17.11 | 9.41 | 1.00 | 13.67 | nr | **23.08** |
| 300 x 300 x 100 mm | 27.65 | 30.41 | 1.05 | 14.36 | nr | **44.77** |
| **Square Pattern - Galvanised** | | | | | | |
| 75 x 75 x 37 mm | 5.37 | 0.59 | 0.88 | 12.03 | nr | **12.62** |
| 75 x 75 x 50 mm | 5.75 | 0.63 | 0.88 | 12.03 | nr | **12.66** |
| 75 x 75 x 75 mm | 7.18 | 1.97 | 0.90 | 12.30 | nr | **14.27** |
| 100 x 100 x 50 mm | 6.40 | 0.70 | 0.90 | 12.30 | nr | **13.00** |
| 150 x 150 x 50 mm | 8.35 | 1.65 | 1.00 | 13.67 | nr | **15.32** |
| 150 x 150 x 75 mm | 10.04 | 2.76 | 0.92 | 12.57 | nr | **15.33** |
| 150 x 150 x 100 mm | 12.71 | 3.50 | 0.92 | 12.57 | nr | **16.07** |
| 225 x 225 x 50 mm | 17.24 | 9.48 | 0.98 | 13.40 | nr | **22.88** |
| 225 x 225 x 100 mm | 20.70 | 11.38 | 1.00 | 13.67 | nr | **25.05** |
| 300 x 300 x 100 mm | 34.05 | 37.45 | 1.05 | 14.36 | nr | **51.81** |

*Material Costs / Measured Work Prices - Electrical Installations*

## Y:MECHANICAL AND ELECTRICAL SERVICES

| Item | Net Price £ | Material £ | Labour hours | Labour £ | Unit | Total Rate £ |
|---|---|---|---|---|---|---|
| **Y60 : CONDUIT AND CABLE TRUNKING : METAL AND PLASTICS CONDUIT (contd)** | | | | | | |
| Sheet Steel Adaptable Boxes; with plain or knockout sides; fixed to backgrounds; including supports and fixings. (Cutting and connecting conduit to boxes is included.) | | | | | | |
| Rectangular Pattern - Black | | | | | | |
| 100 x 75 x 50 mm | 5.75 | 0.63 | 0.88 | 12.03 | nr | 12.66 |
| 150 x 75 x 50 mm | 6.11 | 1.34 | 0.88 | 12.03 | nr | 13.37 |
| 150 x 75 x 75 mm | 7.31 | 4.02 | 0.90 | 12.30 | nr | 16.32 |
| 150 x 100 x 75 mm | 8.01 | 2.20 | 0.90 | 12.30 | nr | 14.50 |
| 225 x 75 x 50 mm | 7.74 | 4.26 | 0.90 | 12.30 | nr | 16.56 |
| 225 x 150 x 75 mm | 12.60 | 6.93 | 0.92 | 12.57 | nr | 19.50 |
| 225 x 150 x 100 mm | 13.87 | 15.26 | 0.92 | 12.57 | nr | 27.83 |
| 300 x 150 x 50 mm | 14.42 | 15.86 | 0.98 | 13.40 | nr | 29.26 |
| 300 x 150 x 75 mm | 15.39 | 16.93 | 1.00 | 13.67 | nr | 30.60 |
| 300 x 150 x 100 mm | 17.63 | 19.39 | 1.05 | 14.36 | nr | 33.75 |
| Rectangular Pattern - Galvanised | | | | | | |
| 100 x 75 x 50 mm | 7.45 | 0.82 | 0.88 | 12.03 | nr | 12.85 |
| 150 x 75 x 50 mm | 6.66 | 1.47 | 0.88 | 12.03 | nr | 13.50 |
| 150 x 75 x 75 mm | 8.25 | 4.54 | 0.90 | 12.30 | nr | 16.84 |
| 150 x 100 x 75 mm | 8.14 | 2.24 | 0.90 | 12.30 | nr | 14.54 |
| 225 x 75 x 50 mm | 8.24 | 4.53 | 0.90 | 12.30 | nr | 16.83 |
| 225 x 150 x 75 mm | 14.97 | 8.23 | 0.92 | 12.57 | nr | 20.80 |
| 225 x 150 x 100 mm | 16.42 | 18.06 | 0.92 | 12.57 | nr | 30.63 |
| 300 x 150 x 50 mm | 17.99 | 19.79 | 0.98 | 13.40 | nr | 33.19 |
| 300 x 150 x 75 mm | 18.93 | 20.82 | 1.00 | 13.67 | nr | 34.49 |
| 300 x 150 x 100 mm | 21.34 | 23.47 | 1.05 | 14.36 | nr | 37.83 |
| PVC Adaptable Boxes; fixed to backgrounds; including all supports and fixings. (Cutting and connecting conduit to boxes is included.) | | | | | | |
| Square Pattern | | | | | | |
| 75 x 75 x 53 mm | 3.40 | 3.74 | 0.88 | 12.03 | nr | 15.77 |
| 100 x 100 x 75 mm | 5.60 | 6.16 | 0.90 | 12.30 | nr | 18.46 |
| 150 x 150 x 75 mm | 7.18 | 7.90 | 0.92 | 12.57 | nr | 20.47 |

# Keep your figures up to date, free of charge

This section, and most of the other information in this Price Book, is brought up to date every three months, until the next annual edition, in the *Price Book Update*.

The *Update* is available free to all Price Book purchasers.

To ensure you receive your copy, simply complete the reply card from the centre of the book and return it to us.

## Y:MECHANICAL AND ELECTRICAL SERVICES

| Item | Net Price £ | Material £ | Labour hours | Labour £ | Unit | Total Rate £ |
|---|---|---|---|---|---|---|
| **Terminal Strips; (To be fixed in metal or polythene adaptable boxes.)** | | | | | | |
| 20 Amp High Density Polythene | | | | | | |
| 2 way | 0.56 | 0.62 | 0.25 | 3.42 | nr | 4.04 |
| 3 way | 0.71 | 0.78 | 0.25 | 3.42 | nr | 4.20 |
| 4 way | 0.87 | 0.96 | 0.25 | 3.42 | nr | 4.38 |
| 5 way | 1.02 | 1.12 | 0.25 | 3.42 | nr | 4.54 |
| 6 way | 1.16 | 1.28 | 0.25 | 3.42 | nr | 4.70 |
| 7 way | 1.32 | 1.45 | 0.25 | 3.42 | nr | 4.87 |
| 8 way | 1.47 | 1.61 | 0.30 | 4.10 | nr | 5.71 |
| 9 way | 1.61 | 1.78 | 0.30 | 4.10 | nr | 5.88 |
| 10 way | 1.76 | 1.94 | 0.35 | 4.78 | nr | 6.72 |
| 11 way | 1.91 | 2.10 | 0.35 | 4.78 | nr | 6.88 |
| 12 way | 2.07 | 2.27 | 0.35 | 4.78 | nr | 7.05 |
| 13 way | 2.22 | 2.44 | 0.35 | 4.78 | nr | 7.22 |
| 14 way | 2.36 | 2.60 | 0.35 | 4.78 | nr | 7.38 |
| 15 way | 2.51 | 2.76 | 0.40 | 5.47 | nr | 8.23 |
| 16 way | 2.66 | 2.93 | 0.40 | 5.47 | nr | 8.40 |
| 18 way | 2.53 | 2.78 | 0.40 | 5.47 | nr | 8.25 |
| **Galvanised Steel Trunking; fixed to backgrounds; jointed with standard connectors (including plates for air gap between trunking and background; earth continuity straps included).** | | | | | | |
| Single Compartment | | | | | | |
| 50 x 50 mm | 5.46 | 6.31 | 0.42 | 5.74 | m | 12.05 |
| 75 x 50 mm | 7.36 | 8.50 | 0.50 | 6.83 | m | 15.33 |
| 75 x 75 mm | 8.18 | 9.45 | 0.54 | 7.38 | m | 16.83 |
| 100 x 50 mm | 8.41 | 9.71 | 0.54 | 7.38 | m | 17.09 |
| 100 x 75 mm | 9.70 | 11.20 | 0.67 | 9.16 | m | 20.36 |
| 100 x 100 mm | 10.38 | 11.99 | 0.75 | 10.25 | m | 22.24 |
| 150 x 50 mm | 12.42 | 12.63 | 1.00 | 13.67 | m | 26.30 |
| 150 x 100 mm | 15.25 | 17.61 | 0.96 | 13.13 | m | 30.74 |
| 150 x 150 mm | 18.03 | 20.82 | 1.08 | 14.78 | m | 35.60 |
| 225 x 75 mm | 19.51 | 22.53 | 1.12 | 15.32 | m | 37.85 |
| 225 x 150 mm | 23.17 | 30.00 | 1.00 | 13.67 | m | 43.67 |
| 225 x 225 mm | 29.24 | 33.77 | 1.20 | 16.41 | m | 50.18 |
| 300 x 75 mm | 23.17 | 26.76 | 1.20 | 16.41 | m | 43.17 |
| 300 x 100 mm | 27.68 | 31.97 | 1.25 | 17.09 | m | 49.06 |
| 300 x 150 mm | 29.24 | 33.77 | 1.25 | 17.09 | m | 50.86 |
| 300 x 225 mm | 32.35 | 37.36 | 1.35 | 18.47 | m | 55.83 |
| 300 x 300 mm | 35.35 | 40.83 | 1.40 | 19.14 | m | 59.97 |

## Y:MECHANICAL AND ELECTRICAL SERVICES

| Item | Net Price £ | Material £ | Labour hours | Labour £ | Unit | Total Rate £ |
|---|---|---|---|---|---|---|
| **Y60 : CONDUIT AND CABLE TRUNKING : METAL AND PLASTICS CONDUIT (contd)** | | | | | | |
| **Galvanised Steel Trunking; fixed to backgrounds; jointed with standard connectors (including plates for air gap between trunking and background; earth continuity straps included). (contd)** | | | | | | |
| Double Compartment | | | | | | |
| 50 x 50 mm | 8.00 | 9.24 | 0.45 | 6.15 | m | 15.39 |
| 75 x 50 mm | 8.88 | 10.26 | 0.55 | 7.52 | m | 17.78 |
| 75 x 75 mm | 9.84 | 11.37 | 0.60 | 8.20 | m | 19.57 |
| 100 x 50 mm | 9.94 | 11.48 | 0.60 | 8.20 | m | 19.68 |
| 100 x 75 mm | 10.89 | 12.58 | 0.75 | 10.25 | m | 22.83 |
| 100 x 100 mm | 12.02 | 13.88 | 0.83 | 11.35 | m | 25.23 |
| 150 x 50 mm | 13.94 | 16.10 | 0.83 | 11.35 | m | 27.45 |
| 150 x 100 mm | 17.41 | 20.11 | 1.05 | 14.36 | m | 34.47 |
| 150 x 150 mm | 21.22 | 24.51 | 1.20 | 16.41 | m | 40.92 |
| Triple Compartment | | | | | | |
| 75 x 50 mm | 10.72 | 12.38 | 0.65 | 8.89 | m | 21.27 |
| 75 x 75 mm | 11.78 | 13.61 | 0.70 | 9.57 | m | 23.18 |
| 100 x 50 mm | 11.49 | 13.27 | 0.70 | 9.57 | m | 22.84 |
| 100 x 75 mm | 12.81 | 14.80 | 0.85 | 11.62 | m | 26.42 |
| 100 x 100 mm | 14.38 | 16.61 | 0.93 | 12.71 | m | 29.32 |
| 150 x 50 mm | 15.51 | 17.91 | 0.93 | 12.71 | m | 30.62 |
| 150 x 100 mm | 19.75 | 22.81 | 0.87 | 11.89 | m | 34.70 |
| 150 x 150 mm | 26.05 | 30.09 | 1.30 | 17.77 | m | 47.86 |
| Four Compartment | | | | | | |
| 100 x 50 mm | 13.02 | 15.04 | 0.75 | 10.25 | m | 25.29 |
| 100 x 75 mm | 14.77 | 17.06 | 0.90 | 12.30 | m | 29.36 |
| 100 x 100 mm | 16.73 | 19.32 | 0.98 | 13.40 | m | 32.72 |
| 150 x 50 mm | 17.03 | 19.67 | 0.98 | 13.40 | m | 33.07 |
| 150 x 100 mm | 22.10 | 25.53 | 1.20 | 16.41 | m | 41.94 |
| 150 x 150 mm | 30.82 | 35.60 | 1.35 | 18.47 | m | 54.07 |
| **Galvanised Steel Trunking Fittings; (Cutting and jointing trunking to fittings is included.)** | | | | | | |
| Additional Connector or Stop End | | | | | | |
| 50 x 50 mm | 1.12 | 1.23 | 0.16 | 2.19 | nr | 3.42 |
| 75 x 50 mm | 1.20 | 1.32 | 0.16 | 2.19 | nr | 3.51 |
| 75 x 75 mm | 1.29 | 1.42 | 0.16 | 2.19 | nr | 3.61 |
| 100 x 50 mm | 1.35 | 1.49 | 0.20 | 2.73 | nr | 4.22 |
| 100 x 75 mm | 1.46 | 1.61 | 0.20 | 2.73 | nr | 4.34 |
| 100 x 100 mm | 1.53 | 1.68 | 0.20 | 2.73 | nr | 4.41 |
| 150 x 50 mm | 1.68 | 1.85 | 0.20 | 2.73 | nr | 4.58 |
| 150 x 100 mm | 1.81 | 1.99 | 0.20 | 2.73 | nr | 4.72 |
| 150 x 150 mm | 2.49 | 2.74 | 0.20 | 2.73 | nr | 5.47 |
| 225 x 75 mm | 2.68 | 2.95 | 0.25 | 3.42 | nr | 6.37 |
| 225 x 150 mm | 3.88 | 4.27 | 0.25 | 3.42 | nr | 7.69 |
| 225 x 225 mm | 4.67 | 5.14 | 0.25 | 3.42 | nr | 8.56 |
| 300 x 75 mm | 3.88 | 4.27 | 0.30 | 4.10 | nr | 8.37 |
| 300 x 100 mm | 4.67 | 5.14 | 0.30 | 4.10 | nr | 9.24 |
| 300 x 150 mm | 4.83 | 5.31 | 0.30 | 4.10 | nr | 9.41 |

## Y:MECHANICAL AND ELECTRICAL SERVICES

| Item | Net Price £ | Material £ | Labour hours | Labour £ | Unit | Total Rate £ |
|---|---|---|---|---|---|---|
| Additional Connector or Stop End (contd) | | | | | | |
| 300 x 225 mm | 5.07 | 5.58 | 0.30 | 4.10 | nr | **9.68** |
| 300 x 300 mm | 5.28 | 5.81 | 0.30 | 4.10 | nr | **9.91** |
| Flanged Connector or Stop End | | | | | | |
| 50 x 50 mm | 1.12 | 1.23 | 0.16 | 2.19 | nr | **3.42** |
| 75 x 50 mm | 1.20 | 1.32 | 0.16 | 2.19 | nr | **3.51** |
| 75 x 75 mm | 1.29 | 1.42 | 0.16 | 2.19 | nr | **3.61** |
| 100 x 50 mm | 1.35 | 1.49 | 0.20 | 2.73 | nr | **4.22** |
| 100 x 75 mm | 1.46 | 1.61 | 0.20 | 2.73 | nr | **4.34** |
| 100 x 100 mm | 1.53 | 1.68 | 0.20 | 2.73 | nr | **4.41** |
| 150 x 50 mm | 1.68 | 1.85 | 0.20 | 2.73 | nr | **4.58** |
| 150 x 100 mm | 1.81 | 1.99 | 0.20 | 2.73 | nr | **4.72** |
| 150 x 150 mm | 2.49 | 2.74 | 0.20 | 2.73 | nr | **5.47** |
| 225 x 75 mm | 2.68 | 2.95 | 0.25 | 3.42 | nr | **6.37** |
| 225 x 150 mm | 3.88 | 4.27 | 0.25 | 3.42 | nr | **7.69** |
| 225 x 225 mm | 4.67 | 5.14 | 0.25 | 3.42 | nr | **8.56** |
| 300 x 75 mm | 3.88 | 4.27 | 0.30 | 4.10 | nr | **8.37** |
| 300 x 100 mm | 4.67 | 5.14 | 0.30 | 4.10 | nr | **9.24** |
| 300 x 150 mm | 4.83 | 5.31 | 0.30 | 4.10 | nr | **9.41** |
| 300 x 225 mm | 5.07 | 5.58 | 0.30 | 4.10 | nr | **9.68** |
| 300 x 300 mm | 5.28 | 5.81 | 0.30 | 4.10 | nr | **9.91** |
| Bends 90 Degree; Single Compartment | | | | | | |
| 50 x 50 mm | 6.87 | 7.56 | 0.50 | 6.83 | nr | **14.39** |
| 75 x 50 mm | 7.97 | 8.77 | 0.55 | 7.52 | nr | **16.29** |
| 75 x 75 mm | 8.63 | 9.49 | 0.60 | 8.20 | nr | **17.69** |
| 100 x 50 mm | 8.88 | 9.77 | 0.62 | 8.48 | nr | **18.25** |
| 100 x 75 mm | 9.13 | 10.04 | 0.67 | 9.16 | nr | **19.20** |
| 100 x 100 mm | 9.50 | 10.45 | 0.70 | 9.57 | nr | **20.02** |
| 150 x 50 mm | 10.18 | 11.20 | 0.75 | 10.25 | nr | **21.45** |
| 150 x 100 mm | 13.75 | 15.13 | 1.25 | 17.09 | nr | **32.22** |
| 150 x 150 mm | 15.31 | 16.84 | 1.18 | 16.08 | nr | **32.92** |
| 225 x 75 mm | 22.38 | 24.62 | 0.85 | 11.62 | nr | **36.24** |
| 225 x 150 mm | 29.34 | 32.27 | 0.95 | 12.99 | nr | **45.26** |
| 225 x 225 mm | 34.22 | 37.64 | 0.95 | 12.99 | nr | **50.63** |
| 300 x 75 mm | 31.57 | 34.73 | 0.95 | 13.02 | nr | **47.75** |
| 300 x 100 mm | 32.24 | 35.46 | 1.05 | 14.36 | nr | **49.82** |
| 300 x 150 mm | 34.19 | 37.61 | 1.15 | 15.73 | nr | **53.34** |
| 300 x 225 mm | 39.70 | 43.67 | 1.15 | 15.73 | nr | **59.40** |
| 300 x 300 mm | 49.84 | 54.82 | 1.25 | 17.09 | nr | **71.90** |
| 90 Degree; Double Compartment | | | | | | |
| 50 x 50 mm | 8.53 | 9.38 | 0.50 | 6.83 | nr | **16.21** |
| 75 x 50 mm | 9.66 | 10.63 | 0.55 | 7.52 | nr | **18.15** |
| 75 x 75 mm | 10.54 | 11.59 | 0.60 | 8.20 | nr | **19.79** |
| 100 x 50 mm | 10.69 | 11.76 | 0.62 | 8.48 | nr | **20.24** |
| 100 x 75 mm | 11.01 | 12.11 | 0.67 | 9.16 | nr | **21.27** |
| 100 x 100 mm | 11.61 | 12.77 | 0.70 | 9.57 | nr | **22.34** |
| 150 x 50 mm | 12.28 | 13.51 | 0.75 | 10.25 | nr | **23.76** |
| 150 x 100 mm | 15.92 | 17.51 | 0.80 | 10.93 | nr | **28.44** |
| 150 x 150 mm | 17.74 | 19.51 | 0.85 | 11.62 | nr | **31.13** |

## Y:MECHANICAL AND ELECTRICAL SERVICES

| Item | Net Price £ | Material £ | Labour hours | Labour £ | Unit | Total Rate £ |
|---|---|---|---|---|---|---|
| **Y60 : CONDUIT AND CABLE TRUNKING : METAL AND PLASTICS CONDUIT (contd)** | | | | | | |
| **Galvanised Steel Trunking Fittings; (Cutting and jointing trunking to fittings is included.) (Contd)** | | | | | | |
| Bends 90 Degree; Triple Compartment | | | | | | |
| 75 x 50 mm | 11.20 | 12.32 | 0.55 | 7.52 | nr | **19.84** |
| 75 x 75 mm | 12.16 | 13.38 | 0.60 | 8.20 | nr | **21.58** |
| 100 x 50 mm | 12.35 | 13.59 | 0.62 | 8.48 | nr | **22.07** |
| 100 x 75 mm | 12.63 | 13.89 | 0.67 | 9.16 | nr | **23.05** |
| 100 x 100 mm | 13.39 | 14.73 | 0.70 | 9.57 | nr | **24.30** |
| 150 x 50 mm | 13.77 | 15.15 | 0.75 | 10.25 | nr | **25.40** |
| 150 x 100 mm | 17.75 | 19.52 | 0.80 | 10.93 | nr | **30.45** |
| 150 x 150 mm | 19.70 | 21.67 | 0.85 | 11.62 | nr | **33.29** |
| Bends 90 Degrees; Four Compartments | | | | | | |
| 100 x 50 mm | 13.63 | 14.99 | 0.62 | 8.48 | nr | **23.47** |
| 100 x 75 mm | 14.23 | 15.65 | 0.67 | 9.16 | nr | **24.81** |
| 100 x 100 mm | 15.19 | 16.71 | 0.70 | 9.57 | nr | **26.28** |
| 150 x 50 mm | 15.27 | 16.80 | 0.75 | 10.25 | nr | **27.05** |
| 150 x 100 mm | 19.50 | 21.45 | 0.80 | 10.93 | nr | **32.38** |
| 150 x 150 mm | 21.63 | 23.79 | 0.85 | 11.62 | nr | **35.41** |
| Tees; Single Compartment | | | | | | |
| 50 x 50 mm | 7.94 | 8.73 | 0.71 | 9.71 | nr | **18.44** |
| 75 x 50 mm | 9.19 | 10.11 | 0.71 | 9.71 | nr | **19.82** |
| 75 x 75 mm | 9.50 | 10.45 | 0.75 | 10.25 | nr | **20.70** |
| 100 x 50 mm | 10.71 | 11.78 | 0.75 | 10.25 | nr | **22.03** |
| 100 x 75 mm | 11.10 | 12.21 | 0.85 | 11.62 | nr | **23.83** |
| 100 x 100 mm | 11.43 | 12.57 | 0.85 | 11.62 | nr | **24.19** |
| 150 x 50 mm | 11.76 | 12.94 | 0.95 | 12.99 | nr | **25.93** |
| 150 x 100 mm | 16.52 | 18.17 | 0.95 | 12.99 | nr | **31.16** |
| 150 x 150 mm | 18.07 | 19.88 | 1.05 | 14.36 | nr | **34.24** |
| 225 x 75 mm | 30.26 | 33.29 | 1.05 | 14.36 | nr | **47.65** |
| 225 x 150 mm | 41.43 | 45.57 | 1.15 | 15.73 | nr | **61.30** |
| 225 x 225 mm | 48.24 | 53.06 | 1.15 | 15.73 | nr | **68.79** |
| 300 x 75 mm | 44.45 | 48.90 | 1.25 | 17.09 | nr | **65.99** |
| 300 x 100 mm | 46.58 | 51.24 | 1.25 | 17.09 | nr | **68.33** |
| 300 x 150 mm | 48.49 | 53.34 | 1.35 | 18.47 | nr | **71.81** |
| 300 x 225 mm | 55.93 | 61.52 | 1.40 | 19.14 | nr | **80.66** |
| 300 x 300mm | 66.99 | 73.69 | 1.50 | 20.52 | nr | **94.21** |
| Tees; Double Compartment | | | | | | |
| 50 x 50 mm | 10.99 | 12.09 | 0.71 | 9.71 | nr | **21.80** |
| 75 x 50 mm | 12.27 | 13.50 | 0.71 | 9.71 | nr | **23.21** |
| 75 x 75 mm | 12.62 | 13.88 | 0.75 | 10.25 | nr | **24.13** |
| 100 x 50 mm | 13.61 | 14.97 | 0.75 | 10.25 | nr | **25.22** |
| 100 x 75 mm | 14.23 | 15.65 | 0.85 | 11.62 | nr | **27.27** |
| 100 x 100 mm | 14.61 | 16.07 | 0.85 | 11.62 | nr | **27.69** |
| 150 x 50 mm | 14.89 | 16.38 | 0.95 | 12.99 | nr | **29.37** |
| 150 x 100 mm | 19.81 | 21.79 | 0.95 | 12.99 | nr | **34.78** |
| 150 x 150 mm | 21.39 | 23.53 | 1.05 | 14.36 | nr | **37.89** |

## Y:MECHANICAL AND ELECTRICAL SERVICES

| Item | Net Price £ | Material £ | Labour hours | Labour £ | Unit | Total Rate £ |
|---|---|---|---|---|---|---|
| **Tees; Triple Compartment** | | | | | | |
| 75 x 50 mm | 15.11 | 16.62 | 0.71 | 9.71 | nr | 26.33 |
| 75 x 75 mm | 15.41 | 16.95 | 0.75 | 10.25 | nr | 27.20 |
| 100 x 50 mm | 16.42 | 18.06 | 0.75 | 10.25 | nr | 28.31 |
| 100 x 75 mm | 17.08 | 18.79 | 0.85 | 11.62 | nr | 30.41 |
| 100 x 100 mm | 17.40 | 19.14 | 0.85 | 11.62 | nr | 30.76 |
| 150 x 50 mm | 17.74 | 19.51 | 0.95 | 12.99 | nr | 32.50 |
| 150 x 100 mm | 22.61 | 24.87 | 0.95 | 12.99 | nr | 37.86 |
| 150 x 150 mm | 24.24 | 26.66 | 1.05 | 14.36 | nr | 41.02 |
| **Tees; Four Compartment** | | | | | | |
| 100 x 50 mm | 19.22 | 21.14 | 0.75 | 10.25 | nr | 31.39 |
| 100 x 75 mm | 19.83 | 21.81 | 0.85 | 11.62 | nr | 33.43 |
| 100 x 100 mm | 20.18 | 22.20 | 0.85 | 11.62 | nr | 33.82 |
| 150 x 50 mm | 20.49 | 22.54 | 0.95 | 12.99 | nr | 35.53 |
| 150 x 100 mm | 25.37 | 27.91 | 0.95 | 12.99 | nr | 40.90 |
| 150 x 150 mm | 27.03 | 29.73 | 1.05 | 14.36 | nr | 44.09 |
| **Crossovers; Single Compartment** | | | | | | |
| 50 x 50 mm | 10.01 | 11.01 | 0.81 | 11.08 | nr | 22.09 |
| 75 x 50 mm | 13.12 | 14.43 | 0.81 | 11.08 | nr | 25.51 |
| 75 x 75 mm | 15.18 | 16.70 | 0.85 | 11.62 | nr | 28.32 |
| 100 x 50 mm | 16.13 | 17.74 | 0.85 | 11.62 | nr | 29.36 |
| 100 x 75 mm | 17.24 | 18.96 | 0.95 | 12.99 | nr | 31.95 |
| 100 x 100 mm | 18.33 | 20.16 | 0.95 | 12.99 | nr | 33.15 |
| 150 x 50 mm | 19.19 | 21.11 | 1.05 | 14.36 | nr | 35.47 |
| 150 x 100 mm | 22.42 | 24.66 | 1.05 | 14.36 | nr | 39.02 |
| 150 x 150 mm | 25.59 | 28.15 | 1.15 | 15.73 | nr | 43.88 |
| 225 x 75 mm | 41.46 | 45.61 | 1.15 | 15.73 | nr | 61.34 |
| 225 x 150 mm | 55.13 | 60.64 | 1.25 | 17.09 | nr | 77.73 |
| 225 x 225 mm | 63.23 | 69.55 | 1.25 | 17.09 | nr | 86.64 |
| 300 x 75 mm | 59.42 | 65.36 | 1.30 | 17.77 | nr | 83.13 |
| 300 x 100 mm | 61.24 | 67.36 | 1.30 | 17.77 | nr | 85.13 |
| 300 x 150 mm | 63.23 | 69.55 | 1.35 | 18.47 | nr | 88.02 |
| 300 x 225 mm | 72.04 | 79.24 | 1.35 | 18.47 | nr | 97.71 |
| 300 x 300mm | 85.84 | 94.42 | 1.50 | 20.52 | nr | 114.94 |
| **Crossovers; Double Compartment** | | | | | | |
| 50 x 50 mm | 13.50 | 14.85 | 0.81 | 11.08 | nr | 25.93 |
| 75 x 50 mm | 16.72 | 18.39 | 0.81 | 11.08 | nr | 29.47 |
| 75 x 75 mm | 18.78 | 20.66 | 0.85 | 11.62 | nr | 32.28 |
| 100 x 50 mm | 20.25 | 22.27 | 0.85 | 11.62 | nr | 33.89 |
| 100 x 75 mm | 20.92 | 23.01 | 0.95 | 12.99 | nr | 36.00 |
| 100 x 100 mm | 21.97 | 24.17 | 0.95 | 12.99 | nr | 37.16 |
| 150 x 50 mm | 22.89 | 25.18 | 0.95 | 12.99 | nr | 38.17 |
| 150 x 100 mm | 26.25 | 28.88 | 1.05 | 14.36 | nr | 43.24 |
| 150 x 150 mm | 27.64 | 30.40 | 1.15 | 15.73 | nr | 46.13 |
| **Crossovers; Triple Compartment** | | | | | | |
| 75 x 50 mm | 19.87 | 21.86 | 0.81 | 11.08 | nr | 32.94 |
| 75 x 75 mm | 21.97 | 24.17 | 0.85 | 11.62 | nr | 35.79 |
| 100 x 50 mm | 23.43 | 25.77 | 0.85 | 11.62 | nr | 37.39 |
| 100 x 75 mm | 24.10 | 26.51 | 0.95 | 12.99 | nr | 39.50 |
| 100 x 100 mm | 25.18 | 27.70 | 0.95 | 12.99 | nr | 40.69 |
| 150 x 50 mm | 26.04 | 28.64 | 1.05 | 14.36 | nr | 43.00 |

Material Costs / Measured Work Prices - Electrical Installations

## Y:MECHANICAL AND ELECTRICAL SERVICES

| Item | Net Price £ | Material £ | Labour hours | Labour £ | Unit | Total Rate £ |
|---|---|---|---|---|---|---|
| **Y60 : CONDUIT AND CABLE TRUNKING : METAL AND PLASTICS CONDUIT (contd)** | | | | | | |
| **Galvanised Steel Trunking Fittings; (Cutting and jointing trunking to fittings is included.) (Contd)** | | | | | | |
| Crossovers; Triple Compartment (contd) | | | | | | |
| 150 x 100 mm | 29.46 | 32.41 | 1.05 | 14.36 | nr | **46.77** |
| 150 x 150 mm | 30.87 | 33.96 | 1.15 | 15.73 | nr | **49.69** |
| Crossovers; Four Compartments | | | | | | |
| 100 x 50 mm | 26.59 | 29.25 | 0.85 | 11.62 | nr | **40.87** |
| 100 x 75 mm | 27.31 | 30.04 | 0.95 | 12.99 | nr | **43.03** |
| 100 x 100 mm | 28.39 | 31.23 | 0.95 | 12.99 | nr | **44.22** |
| 150 x 50 mm | 29.25 | 32.17 | 1.05 | 14.36 | nr | **46.53** |
| 150 x 100 mm | 32.64 | 35.90 | 1.05 | 14.36 | nr | **50.26** |
| 150 x 150 mm | 34.02 | 37.42 | 1.15 | 15.73 | nr | **53.15** |
| **Anodised Aluminium Lightweight Trunking; fixed to brackets; standard coupling joints; earth continuity straps** | | | | | | |
| Suspended Surface Trunking | | | | | | |
| 66 x 32mm | 5.46 | 6.31 | 0.45 | 6.15 | m | **12.46** |
| Suspended Recessed Trunking | | | | | | |
| 90 x 65mm | 8.95 | 10.34 | 0.45 | 6.15 | m | **16.49** |
| **Anodised Aluminium Lightweight Trunking Fittings; (Cutting and jointing trunking to fittings is included.)** | | | | | | |
| Extra on 66 x 32mm trunking for: | | | | | | |
| Bend | 10.89 | 11.98 | 0.50 | 6.83 | nr | **18.81** |
| Tee | 10.53 | 11.58 | 0.55 | 7.52 | nr | **19.10** |
| Cross | 15.51 | 17.06 | 0.60 | 8.20 | nr | **25.26** |
| End cap | 16.08 | 17.69 | 0.15 | 2.05 | nr | **19.74** |
| PVC cover (per 1.8m length) | 3.36 | 3.70 | 0.12 | 1.64 | nr | **5.34** |
| T bolt fitting suspension unit | 2.01 | 2.21 | 0.16 | 2.19 | nr | **4.40** |
| Extra on 90 x 65mm trunking for: | | | | | | |
| Bend | 17.43 | 19.17 | 0.50 | 6.83 | nr | **26.00** |
| Tee | 21.86 | 24.05 | 0.55 | 7.52 | nr | **31.57** |
| Cross | 23.99 | 26.39 | 0.60 | 8.20 | nr | **34.59** |
| End cap | 2.68 | 2.95 | 0.15 | 2.05 | nr | **5.00** |
| PVC cover (per 1.8m length) | 3.36 | 3.70 | 0.12 | 1.64 | nr | **5.34** |
| T bolt fitting suspension unit | 3.62 | 3.98 | 0.16 | 2.19 | nr | **6.17** |

## Y:MECHANICAL AND ELECTRICAL SERVICES

| Item | Net Price £ | Material £ | Labour hours | Labour £ | Unit | Total Rate £ |
|---|---|---|---|---|---|---|
| **Galvanised Steel Flush Floor Trunking; fixed to backgrounds; supports and fixings; standard coupling joints; (including plates for air gap between trunking and background; earth continuity straps included.)** | | | | | | |
| Triple Compartment | | | | | | |
| 350 x 60mm | 35.17 | 45.89 | 1.80 | 24.63 | m | 70.52 |
| Four Compartment | | | | | | |
| 350 x 60mm | 36.67 | 47.84 | 1.80 | 24.63 | m | 72.47 |
| **Galvanised Steel Flush Floor Trunking; Fittings (Cutting and jointing trunking to fittings is included)** | | | | | | |
| Stop End; Triple Compartment | | | | | | |
| 350 x 60mm | 1.88 | 2.34 | 0.55 | 7.52 | nr | 9.86 |
| Stop End; Four Compartment | | | | | | |
| 350 x 60mm | 1.88 | 2.34 | 0.55 | 7.52 | nr | 9.86 |
| Rising Bend; Standard; Triple Compartment | | | | | | |
| 350 x 60mm | 30.76 | 38.21 | 1.80 | 24.63 | nr | 62.84 |
| Rising Bend; Standard; Four Compartment | | | | | | |
| 350 x 60mm | 32.26 | 40.08 | 1.80 | 24.63 | nr | 64.71 |
| Rising Bend; Skirting; Triple Compartment | | | | | | |
| 350 x 60mm | 31.41 | 39.02 | 1.80 | 24.63 | nr | 63.65 |
| Rising Bend; Skirting; Four Compartment | | | | | | |
| 350 x 60mm | 32.91 | 40.89 | 1.80 | 24.63 | nr | 65.52 |
| Junction Box; Triple Compartment | | | | | | |
| 350 x 60mm | 37.08 | 46.07 | 1.25 | 17.09 | nr | 63.16 |
| Junction Box; Four Compartment | | | | | | |
| 350 x 60mm | 38.58 | 47.93 | 1.25 | 17.09 | nr | 65.02 |
| Body Coupler (pair) | | | | | | |
| 3 and 4 Compartment | 0.72 | 0.89 | 0.16 | 2.19 | nr | 3.08 |
| Service Outlet Module comprising flat lid with flanged carpet trim; twin 13 A outlet and drilled plate for mounting 2 telephone outlets; One blank plate; Triple compartment | | | | | | |
| 3 Compartment | 49.64 | 54.60 | 0.25 | 3.42 | nr | 58.02 |

## Y:MECHANICAL AND ELECTRICAL SERVICES

| Item | Net Price £ | Material £ | Labour hours | Labour £ | Unit | Total Rate £ |
|---|---|---|---|---|---|---|
| **Y60 : CONDUIT AND CABLE TRUNKING : METAL AND PLASTICS CONDUIT (contd)** | | | | | | |
| **Galvanised Steel Flush Floor Trunking; Fittings (Cutting and jointing trunking to fittings is included) (contd)** | | | | | | |
| Service Outlet Module comprising flat lid with flanged carpet trim; twin 13 A outlet and drilled plate for mounting 2 telephone outlets; Two blank plates; Four compartments | | | | | | |
| 4 Compartment | 55.89 | 61.48 | 0.25 | 3.42 | nr | **64.90** |
| **Single Compartment PVC Trunking; grey finish; clip on lid; fixed to backgrounds; including supports and fixings (standard coupling joints)** | | | | | | |
| Single Compartment | | | | | | |
| 50 x 50mm | 9.50 | 10.97 | 0.25 | 3.42 | m | **14.39** |
| 75 x 50mm | 10.92 | 12.61 | 0.25 | 3.42 | m | **16.03** |
| 75 x 75mm | 13.32 | 15.38 | 0.25 | 3.42 | m | **18.80** |
| 100 x 50mm | 15.60 | 18.02 | 0.30 | 4.10 | m | **22.12** |
| 100 x 75mm | 17.32 | 20.00 | 0.35 | 4.78 | m | **24.78** |
| 100 x 100mm | 19.31 | 22.30 | 0.35 | 4.78 | m | **27.08** |
| 150 x 75mm | 33.86 | 39.11 | 0.40 | 5.47 | m | **44.58** |
| 150 x 100mm | 41.64 | 48.09 | 0.40 | 5.47 | m | **53.56** |
| 150 x 50mm | 42.56 | 49.16 | 0.45 | 6.15 | m | **55.31** |
| **Single Compartment PVC Trunking Fittings; (cutting and jointing trunking to fittings is included)** | | | | | | |
| Crossover | | | | | | |
| 50 x 50mm | 25.10 | 27.61 | 0.18 | 2.46 | nr | **30.07** |
| 75 x 50mm | 27.52 | 30.27 | 0.18 | 2.46 | nr | **32.73** |
| 75 x 75mm | 29.90 | 32.89 | 0.18 | 2.46 | nr | **35.35** |
| 100 x 50mm | 46.47 | 51.12 | 0.20 | 2.73 | nr | **53.85** |
| 100 x 75mm | 50.23 | 55.25 | 0.20 | 2.73 | nr | **57.98** |
| 100 x 100mm | 51.73 | 56.90 | 0.25 | 3.42 | nr | **60.32** |
| 150 x 75mm | 60.29 | 66.32 | 0.25 | 3.42 | nr | **69.74** |
| 150 x 100mm | 75.63 | 83.19 | 0.25 | 3.42 | nr | **86.61** |
| 150 x 150mm | 95.46 | 105.01 | 0.25 | 3.42 | nr | **108.43** |
| Stop End | | | | | | |
| 50 x 50mm | 1.19 | 1.31 | 0.05 | 0.68 | nr | **1.99** |
| 75 x 50mm | 1.66 | 1.83 | 0.05 | 0.68 | nr | **2.51** |
| 75 x 75mm | 2.20 | 2.42 | 0.05 | 0.68 | nr | **3.10** |
| 100 x 50mm | 2.99 | 3.29 | 0.05 | 0.68 | nr | **3.97** |
| 100 x 75mm | 4.65 | 5.12 | 0.05 | 0.68 | nr | **5.80** |
| 100 x 100mm | 4.65 | 5.12 | 0.08 | 1.09 | nr | **6.21** |
| 150 x 75mm | 15.09 | 16.60 | 0.08 | 1.09 | nr | **17.69** |
| 150 x 100mm | 15.09 | 16.60 | 0.08 | 1.09 | nr | **17.69** |
| 150 x 150mm | 20.97 | 23.07 | 0.08 | 1.09 | nr | **24.16** |

## Y:MECHANICAL AND ELECTRICAL SERVICES

| Item | Net Price £ | Material £ | Labour hours | Labour £ | Unit | Total Rate £ |
|---|---|---|---|---|---|---|
| **Flanged Coupling** | | | | | | |
| 50 x 50mm | 6.99 | 7.69 | 0.20 | 2.73 | nr | **10.42** |
| 75 x 50mm | 7.97 | 8.77 | 0.20 | 2.73 | nr | **11.50** |
| 75 x 75mm | 9.59 | 10.55 | 0.20 | 2.73 | nr | **13.28** |
| 100 x 50mm | 10.45 | 11.49 | 0.30 | 4.10 | nr | **15.59** |
| 100 x 75mm | 12.94 | 14.23 | 0.30 | 4.10 | nr | **18.33** |
| 100 x 100mm | 12.94 | 14.23 | 0.30 | 4.10 | nr | **18.33** |
| 150 x 75mm | 15.09 | 16.60 | 0.40 | 5.47 | nr | **22.07** |
| 150 x 100mm | 18.76 | 20.54 | 0.40 | 5.47 | nr | **26.01** |
| 150 x 150mm | 23.76 | 26.14 | 0.40 | 5.47 | nr | **31.61** |
| **Internal Coupling** | | | | | | |
| 50 x 50mm | 2.53 | 2.78 | 0.06 | 0.82 | nr | **3.60** |
| 75 x 50mm | 2.83 | 3.11 | 0.06 | 0.82 | nr | **3.93** |
| 75 x 75mm | 2.83 | 3.11 | 0.06 | 0.82 | nr | **3.93** |
| 100 x 50mm | 3.76 | 4.14 | 0.08 | 1.09 | nr | **5.23** |
| 100 x 75mm | 4.72 | 5.19 | 0.08 | 1.09 | nr | **6.28** |
| 100 x 100mm | 4.72 | 5.19 | 0.08 | 1.09 | nr | **6.28** |
| **External Coupling** | | | | | | |
| 50 x 50mm | 3.14 | 3.45 | 0.06 | 0.82 | nr | **4.27** |
| 75 x 50mm | 3.91 | 4.30 | 0.06 | 0.82 | nr | **5.12** |
| 75 x 75mm | 4.58 | 5.04 | 0.06 | 0.82 | nr | **5.86** |
| 100 x 50mm | 7.23 | 7.95 | 0.08 | 1.09 | nr | **9.04** |
| 100 x 75mm | 9.08 | 9.99 | 0.08 | 1.09 | nr | **11.08** |
| 100 x 100mm | 9.08 | 9.99 | 0.08 | 1.09 | nr | **11.08** |
| 150 x 75mm | 13.32 | 14.65 | 0.10 | 1.37 | nr | **16.02** |
| 150 x 100mm | 16.45 | 18.09 | 0.10 | 1.37 | nr | **19.46** |
| 150 x 150mm | 20.39 | 22.43 | 0.10 | 1.37 | nr | **23.80** |
| **Angle, Flat Cover** | | | | | | |
| 50 x 50mm | 5.63 | 6.19 | 0.10 | 1.37 | nr | **7.56** |
| 75 x 50mm | 7.61 | 8.37 | 0.10 | 1.37 | nr | **9.74** |
| 75 x 75mm | 9.36 | 10.30 | 0.10 | 1.37 | nr | **11.67** |
| 100 x 50mm | 17.35 | 19.09 | 0.10 | 1.37 | nr | **20.46** |
| 100 x 75mm | 24.89 | 27.38 | 0.15 | 2.05 | nr | **29.43** |
| 100 x 100mm | 33.98 | 37.38 | 0.15 | 2.05 | nr | **39.43** |
| 150 x 75mm | 41.02 | 45.12 | 0.15 | 2.05 | nr | **47.17** |
| 150 x 100mm | 46.77 | 51.45 | 0.20 | 2.73 | nr | **54.18** |
| 150 x 150mm | 71.43 | 78.57 | 0.20 | 2.73 | nr | **81.30** |
| **Angle, Internal or External Cover** | | | | | | |
| 50 x 50mm | 5.63 | 6.19 | 0.10 | 1.37 | nr | **7.56** |
| 75 x 50mm | 7.61 | 8.37 | 0.10 | 1.37 | nr | **9.74** |
| 75 x 75mm | 16.81 | 18.49 | 0.10 | 1.37 | nr | **19.86** |
| 100 x 50mm | 18.96 | 20.86 | 0.10 | 1.37 | nr | **22.23** |
| 100 x 75mm | 24.89 | 27.38 | 0.15 | 2.05 | nr | **29.43** |
| 100 x 100mm | 33.98 | 37.38 | 0.15 | 2.05 | nr | **39.43** |
| 150 x 75mm | 39.34 | 43.27 | 0.15 | 2.05 | nr | **45.32** |
| 150 x 100mm | 46.77 | 51.45 | 0.20 | 2.73 | nr | **54.18** |
| 150 x 150mm | 71.43 | 78.57 | 0.20 | 2.73 | nr | **81.30** |

## Material Costs / Measured Work Prices - Electrical Installations
### Y:MECHANICAL AND ELECTRICAL SERVICES

| Item | Net Price £ | Material £ | Labour hours | Labour £ | Unit | Total Rate £ |
|---|---|---|---|---|---|---|
| **Y60 : CONDUIT AND CABLE TRUNKING : METAL AND PLASTICS CONDUIT (contd)** | | | | | | |
| **Single Compartment PVC Trunking Fittings; (cutting and jointing trunking to fittings is included) (contd)** | | | | | | |
| Tee, Flat Cover | | | | | | |
| 50 x 50mm | 8.81 | 9.69 | 0.15 | 2.05 | nr | 11.74 |
| 75 x 50mm | 11.20 | 12.32 | 0.15 | 2.05 | nr | 14.37 |
| 75 x 75mm | 12.66 | 13.93 | 0.15 | 2.05 | nr | 15.98 |
| 100 x 50mm | 20.52 | 22.57 | 0.20 | 2.73 | nr | 25.30 |
| 100 x 75mm | 29.41 | 32.35 | 0.20 | 2.73 | nr | 35.08 |
| 100 x 100mm | 29.41 | 32.35 | 0.20 | 2.73 | nr | 35.08 |
| 150 x 75mm | 50.73 | 55.80 | 0.25 | 3.42 | nr | 59.22 |
| 150 x 100mm | 65.10 | 71.61 | 0.25 | 3.42 | nr | 75.03 |
| 150 x 150mm | 81.89 | 90.08 | 0.25 | 3.42 | nr | 93.50 |
| Tee, Internal or External Cover | | | | | | |
| 50 x 50mm | 18.24 | 20.06 | 0.15 | 2.05 | nr | 22.11 |
| 75 x 50mm | 19.89 | 21.88 | 0.15 | 2.05 | nr | 23.93 |
| 75 x 75mm | 23.71 | 26.08 | 0.15 | 2.05 | nr | 28.13 |
| 100 x 50mm | 28.27 | 31.10 | 0.20 | 2.73 | nr | 33.83 |
| 100 x 75mm | 37.81 | 41.59 | 0.20 | 2.73 | nr | 44.32 |
| 100 x 100mm | 37.81 | 41.59 | 0.20 | 2.73 | nr | 44.32 |
| 150 x 75mm | 54.19 | 59.61 | 0.25 | 3.42 | nr | 63.03 |
| 150 x 100mm | 69.53 | 76.48 | 0.25 | 3.42 | nr | 79.90 |
| 150 x 150mm | 81.09 | 89.20 | 0.25 | 3.42 | nr | 92.62 |
| Division Strip (1.8m long) | | | | | | |
| 50mm | 6.92 | 7.61 | 0.05 | 0.68 | nr | 8.29 |
| 75mm | 9.18 | 10.10 | 0.05 | 0.68 | nr | 10.78 |
| 100mm | 11.67 | 12.84 | 0.07 | 0.96 | nr | 13.80 |
| **PVC Minature Trunking; white finish; fixed to backgrounds; including supports and fixing; standard coupling joints** | | | | | | |
| Single Compartment | | | | | | |
| 16 x 16mm | 2.07 | 2.39 | 0.20 | 2.73 | m | 5.12 |
| 25 x 16mm | 2.60 | 3.00 | 0.20 | 2.73 | m | 5.73 |
| 38 x 16mm | 3.27 | 3.77 | 0.22 | 3.01 | m | 6.78 |
| 38 x 25mm | 3.94 | 4.55 | 0.25 | 3.42 | m | 7.97 |
| Compartmented | | | | | | |
| 38 x 16mm | 3.82 | 4.42 | 0.22 | 3.01 | m | 7.43 |
| 38 x 25mm | 4.63 | 5.34 | 0.25 | 3.42 | m | 8.76 |

## Y:MECHANICAL AND ELECTRICAL SERVICES

| Item | Net Price £ | Material £ | Labour hours | Labour £ | Unit | Total Rate £ |
|---|---|---|---|---|---|---|
| **PVC Miniature Trunking Fittings; single compartment; white finish; (cutting and jointing trunking to fittings is included)** | | | | | | |
| Coupling | | | | | | |
| 16 x 16mm | 0.66 | 0.73 | 0.10 | 1.37 | nr | **2.10** |
| 25 x 16mm | 0.66 | 0.73 | 0.10 | 1.37 | nr | **2.10** |
| 38 x 16mm | 0.66 | 0.73 | 0.15 | 2.05 | nr | **2.78** |
| 38 x 25mm | 1.62 | 1.78 | 0.20 | 2.73 | nr | **4.51** |
| Stop End | | | | | | |
| 16 x 16mm | 0.66 | 0.72 | 0.10 | 1.37 | nr | **2.09** |
| 25 x 16mm | 0.66 | 0.72 | 0.10 | 1.37 | nr | **2.09** |
| 38 x 16mm | 0.66 | 0.72 | 0.15 | 2.05 | nr | **2.77** |
| 38 x 25mm | 0.81 | 0.89 | 0.20 | 2.73 | nr | **3.62** |
| Bend; Flat, Internal or External | | | | | | |
| 16 x 16mm | 0.66 | 0.73 | 0.15 | 2.05 | nr | **2.78** |
| 25 x 16mm | 0.66 | 0.73 | 0.15 | 2.05 | nr | **2.78** |
| 38 x 16mm | 0.66 | 0.73 | 0.20 | 2.73 | nr | **3.46** |
| 38 x 25mm | 1.62 | 1.78 | 0.25 | 3.42 | nr | **5.20** |
| Tee | | | | | | |
| 16 x 16mm | 1.12 | 1.23 | 0.20 | 2.73 | nr | **3.96** |
| 25 x 16mm | 1.12 | 1.23 | 0.20 | 2.73 | nr | **3.96** |
| 38 x 16mm | 1.12 | 1.23 | 0.25 | 3.42 | nr | **4.65** |
| 38 x 25mm | 1.12 | 1.23 | 0.30 | 4.10 | nr | **5.33** |
| **PVC Bench Trunking; White or Grey Finish; fixed to backgrounds; including supports and fixings; standard coupling joints** | | | | | | |
| Trunking | | | | | | |
| 90 x 90mm | 30.14 | 34.81 | 0.35 | 4.78 | m | **39.59** |
| **PVC Bench Trunking Fittings; White or Grey Finish. (Cutting and jointing trunking to fittings is included.)** | | | | | | |
| Stop End | | | | | | |
| 90 x 90mm | 6.70 | 7.37 | 0.05 | 0.68 | nr | **8.05** |
| Coupling | | | | | | |
| 90 x 90mm | 11.48 | 12.63 | 0.10 | 1.37 | nr | **14.00** |
| Internal or External Bend | | | | | | |
| 90 x 90mm | 33.68 | 37.05 | 0.15 | 2.05 | nr | **39.10** |
| Socket Plate - 1 Gang | | | | | | |
| 90 x 90mm - 1 gang | 14.52 | 15.97 | 0.10 | 1.37 | nr | **17.34** |
| 90 x 90mm - 2 gang | 17.30 | 19.03 | 0.10 | 1.37 | nr | **20.40** |

## Y:MECHANICAL AND ELECTRICAL SERVICES

| Item | Net Price £ | Material £ | Labour hours | Labour £ | Unit | Total Rate £ |
|---|---|---|---|---|---|---|
| **Y60 : CONDUIT AND CABLE TRUNKING : METAL AND PLASTICS CONDUIT (contd)** | | | | | | |
| **PVC Underfloor Trunking; Single Compartment; fitted in floor screed; standard coupling joints** | | | | | | |
| Trunking | | | | | | |
| 60 x 25mm | 1.62 | 1.87 | 0.25 | 3.42 | m | **5.29** |
| 90 x 35mm | 2.57 | 2.97 | 0.30 | 4.10 | m | **7.07** |
| **PVC Underfloor Trunking Fittings; SingleCompartment; fitted in floor screed; (Cutting and jointing trunking to fittings is included.)** | | | | | | |
| Jointing Sleeve | | | | | | |
| 60 x 25mm | 0.22 | 0.24 | 0.08 | 1.09 | nr | **1.33** |
| 90 x 35mm | 0.43 | 0.47 | 0.10 | 1.37 | nr | **1.84** |
| Duct Connector | | | | | | |
| 90 x 35mm | 0.36 | 0.40 | 0.18 | 2.46 | nr | **2.86** |
| Socket Reducer | | | | | | |
| 90 x 35mm | 0.65 | 0.71 | 0.10 | 1.37 | nr | **2.08** |
| Vertical Access Box; 2 compartment | | | | | | |
| Shallow | 33.76 | 37.14 | 0.33 | 4.51 | nr | **41.65** |
| Duct Bend; Vertical | | | | | | |
| 60 x 25mm | 7.32 | 8.05 | 0.30 | 4.10 | nr | **12.15** |
| 90 x 35mm | 8.27 | 9.10 | 0.40 | 5.47 | nr | **14.57** |
| Duct Bend; Horizontal | | | | | | |
| 60 x 25mm | 8.63 | 9.49 | 0.35 | 4.78 | nr | **14.27** |
| 90 x 35mm | 6.84 | 7.52 | 0.45 | 6.15 | nr | **13.67** |
| **Zinc Coated Steel Underfloor Ducting; fixed to backgrounds; standard coupling joints; earth continuity straps; (Including supports and fixing, packing shims where required)** | | | | | | |
| Double Compartment | | | | | | |
| 150 x 25mm | 9.75 | 11.27 | 0.60 | 8.20 | m | **19.47** |
| Triple Compartment | | | | | | |
| 225 x 25mm | 16.01 | 18.49 | 1.00 | 13.67 | m | **32.16** |

## Y:MECHANICAL AND ELECTRICAL SERVICES

| Item | Net Price £ | Material £ | Labour hours | Labour £ | Unit | Total Rate £ |
|---|---|---|---|---|---|---|
| **Zinc Coated Steel Underfloor Ducting Fittings; (cutting and jointing to fittings is included.)** | | | | | | |
| Stop End; Double Compartment | | | | | | |
| 150 x 25mm | 1.46 | 1.61 | 0.20 | 2.73 | nr | **4.34** |
| Stop End; Triple Compartment | | | | | | |
| 225 x 25mm | 1.93 | 2.12 | 0.30 | 4.10 | nr | **6.22** |
| Rising Bend; Double Compartment; Standard Trunking | | | | | | |
| 150 x 25mm | 13.74 | 15.11 | 0.70 | 9.57 | nr | **24.68** |
| Rising Bend; Triple Compartment; Standard Trunking | | | | | | |
| 225 x 25mm | 18.73 | 20.60 | 0.90 | 12.30 | nr | **32.90** |
| Rising Bend; Double Compartment; To Skirting | | | | | | |
| 150 x 25 | 18.50 | 20.35 | 0.90 | 12.30 | nr | **32.65** |
| Rising Bend; Triple Compartment; To Skirting | | | | | | |
| 225 x 25 | 23.83 | 26.21 | 0.90 | 12.30 | nr | **38.51** |
| Horizontal Bend; Double Compartment | | | | | | |
| 150 x 25mm | 20.10 | 22.11 | 0.55 | 7.52 | nr | **29.63** |
| Horizontal Bend; Triple Compartment | | | | | | |
| 225 x 25mm | 24.61 | 27.07 | 0.75 | 10.25 | nr | **37.32** |
| Junction or Service Outlet Boxes; Terminal; Double Compartment | | | | | | |
| 150mm | 32.57 | 35.83 | 0.85 | 11.62 | nr | **47.45** |
| Junction or Service Outlet Boxes; Terminal; Triple Compartment | | | | | | |
| 225mm | 36.59 | 40.25 | 1.25 | 17.09 | nr | **57.34** |
| Junction or Service Outlet Boxes; Through or Angle; Double Compartment | | | | | | |
| 150mm | 32.57 | 35.83 | 0.85 | 11.62 | nr | **47.45** |
| Junction or Service Outlet Boxes; Through or Angle; Triple Compartment | | | | | | |
| 225mm | 36.59 | 40.25 | 1.25 | 17.09 | nr | **57.34** |

## Y:MECHANICAL AND ELECTRICAL SERVICES

| Item | Net Price £ | Material £ | Labour hours | Labour £ | Unit | Total Rate £ |
|---|---|---|---|---|---|---|
| **Y60 : CONDUIT AND CABLE TRUNKING : METAL AND PLASTICS CONDUIT (contd)** | | | | | | |
| **Zinc Coated Steel Underfloor Ducting Fittings; (cutting and jointing to fittings is included.) (Contd)** | | | | | | |
| Junction or Service Outlet Boxes; Tee; Double Compartment | | | | | | |
| 150mm | 32.57 | 35.83 | 0.85 | 11.62 | nr | 47.45 |
| Junction or Service Outlet Boxes; Tee; Triple Compartment | | | | | | |
| 225mm | 36.59 | 40.25 | 1.25 | 17.09 | nr | 57.34 |
| Junction or Service Outlet Boxes; Cross;Double Compartment | | | | | | |
| up to 150mm | 32.57 | 35.83 | 0.85 | 11.62 | nr | 47.45 |
| Junction or Service Outlet Boxes; Cross; Triple Compartment | | | | | | |
| 225mm | 36.59 | 40.25 | 1.25 | 17.09 | nr | 57.34 |
| Plates for Junction / Inspection Boxes; Double and Triple Compartment | | | | | | |
| Blank Plate | 2.56 | 2.82 | 1.50 | 20.50 | nr | 23.32 |
| Conduit Entry Plate | 2.56 | 2.82 | 1.50 | 20.50 | nr | 23.32 |
| Trunking Entry Plate | 2.56 | 2.82 | 1.50 | 20.50 | nr | 23.32 |
| Service outlet box comprising flat lid with flanged carpet trim; twin 13 A outlet and drilled plate for mounting 2 telephone outlets and terminal blocks; terminal outlet box; double compartment | | | | | | |
| 150 x 25mm trunking | 50.88 | 55.97 | 2.00 | 27.34 | nr | 83.31 |
| Service outlet box comprising flat lid with flanged carpet trim; twin 13 A outlet and drilled plate for mounting 2 telephone outlets and terminal blocks; terminal outlet box; triple compartment | | | | | | |
| 225 x 25mm trunking | 50.88 | 55.97 | 2.50 | 34.17 | nr | 90.14 |
| **PVC Skirting/Dado Modular Trunking; White. (Cutting and jointing trunking to fittings and backplates for fixing to walls is included)** | | | | | | |
| Main carrier/backplate | | | | | | |
| 50 x 170mm | 12.62 | 14.29 | 3.52 | 48.07 | m | 62.36 |
| Extension carrier/backplate | | | | | | |
| 50 x 42mm | 7.26 | 8.22 | 8.33 | 113.85 | m | 19.16 |

Material Costs / Measured Work Prices - Electrical Installations

## Y:MECHANICAL AND ELECTRICAL SERVICES

| Item | Net Price £ | Material £ | Labour hours | Labour £ | Unit | Total Rate £ |
|---|---|---|---|---|---|---|
| Carrier/backplate; Including cover seal | 1.57 | 1.78 | 0.80 | 4.78 | m | 12.71 |
| Chamfered covers for fixing to packplates | | | | | | |
| 50 x 42mm | 2.98 | 3.38 | 0.35 | 4.78 | m | 8.16 |
| Square covers for fixing to packplates | | | | | | |
| 50 x 42 mm | 2.98 | 3.38 | 0.35 | 4.78 | m | 8.16 |
| Plain covers for fixing to backplates | | | | | | |
| 85 mm | 3.65 | 4.02 | 0.35 | 4.78 | m | 8.80 |
| Retainers-clip to backplates to hold cables | | | | | | |
| For chamfered covers | 0.55 | 0.60 | 0.06 | 0.82 | m | 1.43 |
| For square-recessed covers | 0.55 | 0.60 | 0.06 | 0.82 | m | 1.43 |
| For plain covers | 0.65 | 0.71 | 0.06 | 0.82 | m | 1.54 |
| Prepackaged corner assemblies | | | | | | |
| Internal ; for 170 x 50 Assy | 5.14 | 5.65 | 0.50 | 6.83 | m | 12.49 |
| Internal; for 215 x 50 Assy | 7.63 | 8.39 | 0.50 | 6.83 | m | 15.23 |
| Internal; for 254 x 50 Assy | 7.63 | 8.39 | 0.50 | 6.83 | m | 15.23 |
| External; for 170 x 50 Assy | 5.14 | 5.65 | 0.60 | 8.20 | m | 13.85 |
| External ; for 215 x 50 Assy | 6.40 | 7.04 | 0.60 | 8.20 | m | 15.24 |
| External ; for 254 x 50 Assy | 7.63 | 8.39 | 0.60 | 8.20 | m | 16.59 |
| Clip on end caps | | | | | | |
| 170 x 50 Assy | 3.03 | 1.67 | 0.12 | 1.64 | m | 3.31 |
| 215 x 50 Assy | 2.78 | 1.53 | 0.12 | 1.64 | m | 3.17 |
| 254 x 50 Assy | 2.98 | 1.64 | 0.12 | 1.64 | m | 3.28 |
| Outlet box | | | | | | |
| 1Gang; in horizontal trunking; clip in | 2.14 | 2.14 | 0.55 | 7.52 | m | 9.66 |
| 2 Gang; in horizontal trunking; clip in | 2.62 | 2.62 | 0.55 | 7.52 | m | 10.14 |
| 1 Gang; in vertical trunking; clip in | 2.67 | 2.67 | 0.55 | 7.52 | m | 10.19 |

ReSPONse

SPON's price data now available in an electronic estimating system call us on 0171-522-9966 for details

## Y:MECHANICAL AND ELECTRICAL SERVICES

| Item | Net Price £ | Material £ | Labour hours | Labour £ | Unit | Total Rate £ |
|---|---|---|---|---|---|---|
| **Y61 : HV/LV CABLES AND WIRING : CABLE** | | | | | | |
| Cable; Lead Covered, Paper Insulated; Textile bedded; SWA; PVC sheathed copper; clipped to backgrounds; BS 6350 and BS 6480/69 | | | | | | |
| 6350/11000 Volt Grade, including fixings and connecting tails; Single Core | | | | | | |
| 120 mm2 | 11.38 | 13.14 | 0.40 | 5.47 | m | 18.61 |
| 150 mm2 | 11.85 | 13.69 | 0.42 | 5.74 | m | 19.43 |
| 185 mm2 | 13.27 | 15.33 | 0.47 | 6.43 | m | 21.76 |
| 240 mm2 | 16.12 | 18.62 | 0.53 | 7.25 | m | 25.87 |
| 300 mm2 | 18.01 | 20.80 | 0.60 | 8.20 | m | 29.00 |
| 6350/11000 Volt Grade, including fixings and connecting tails; Triple Core | | | | | | |
| 95 mm2 | 22.75 | 26.28 | 0.45 | 6.15 | m | 32.43 |
| 120 mm2 | 25.60 | 29.57 | 0.48 | 6.56 | m | 36.13 |
| 150 mm2 | 29.39 | 33.95 | 0.52 | 7.11 | m | 41.06 |
| 185 mm2 | 33.18 | 38.32 | 0.53 | 7.25 | m | 45.57 |
| 240 mm2 | 39.82 | 45.99 | 0.60 | 8.20 | m | 54.19 |
| 300 mm2 | 47.41 | 54.76 | 0.70 | 9.57 | m | 64.33 |
| Cable; Lead Covered, Paper Insulated; Textile bedded; SWA; PVC sheathed copper; laid in trench with marker tape; BS 6350 and BS 6480/69 (Cable tiles measured elsewhere). | | | | | | |
| Single Core | | | | | | |
| 120 mm2 | 11.38 | 13.14 | 0.18 | 2.46 | m | 15.60 |
| 150 mm2 | 11.85 | 13.69 | 0.18 | 2.46 | m | 16.15 |
| 185 mm2 | 13.27 | 15.33 | 0.18 | 2.46 | m | 17.79 |
| 240 mm2 | 16.12 | 18.62 | 0.27 | 3.69 | m | 22.31 |
| 300 mm2 | 18.01 | 20.80 | 0.30 | 4.10 | m | 24.90 |
| Triple Core | | | | | | |
| 95 mm2 | 22.75 | 26.28 | 0.20 | 2.73 | m | 29.01 |
| 120 mm2 | 25.60 | 29.57 | 0.20 | 2.73 | m | 32.30 |
| 150 mm2 | 29.39 | 33.95 | 0.20 | 2.73 | m | 36.68 |
| 185 mm2 | 33.18 | 38.32 | 0.20 | 2.73 | m | 41.05 |
| 240 mm2 | 39.82 | 45.99 | 0.30 | 4.10 | m | 50.09 |
| 300 mm2 | 47.41 | 54.76 | 0.40 | 5.47 | m | 60.23 |
| Cable Termination; for lead covered, paper insulated; textile bedded; SWA; PVC sheathed copper cable. | | | | | | |
| Bolt on Cable Gland; Two-bolt fixing wiping gland for metal sheathed single wire or steel tape armoured or unarmoured cables. | | | | | | |
| Single Core | | | | | | |
| 120 mm2 | 2.63 | 2.90 | 1.60 | 21.87 | nr | 24.77 |
| 150 mm2 | 2.63 | 2.90 | 1.60 | 21.87 | nr | 24.77 |
| 185 mm2 | 2.63 | 2.90 | 1.60 | 21.87 | nr | 24.77 |
| 240 mm2 | 2.63 | 2.90 | 1.60 | 21.87 | nr | 24.77 |

*Material Costs / Measured Work Prices - Electrical Installations* 405

## Y:MECHANICAL AND ELECTRICAL SERVICES

| Item | Net Price £ | Material £ | Labour hours | Labour £ | Unit | Total Rate £ |
|---|---|---|---|---|---|---|
| Bolt on Cable Gland; Two-bolt fixing wiping gland for metal sheathed single wire or steel tape armoured or unarmoured cables. | | | | | | |
| Triple Core | | | | | | |
| 95 mm2 | 2.82 | 3.10 | 1.70 | 23.25 | nr | 26.35 |
| 120 mm2 | 3.07 | 3.37 | 1.70 | 23.25 | nr | 26.62 |
| 185 mm2 | 3.07 | 3.37 | 1.70 | 23.25 | nr | 26.62 |
| 240 mm2 | 3.20 | 3.52 | 1.70 | 23.25 | nr | 26.77 |
| Cable; XLPE Insulated; SWA; PVC sheathed copper; BS 5467; clipped to backgrounds;(Supports, fixings and connecting tails are excluded). | | | | | | |
| 600/1000 Volt Grade; Two Core | | | | | | |
| 25 mm2 | 2.27 | 2.62 | 0.30 | 4.10 | m | 6.72 |
| 35 mm2 | 2.73 | 3.15 | 0.32 | 4.37 | m | 7.52 |
| 50 mm2 | 3.45 | 3.98 | 0.33 | 4.51 | m | 8.49 |
| 70 mm2 | 4.22 | 4.87 | 0.35 | 4.78 | m | 9.65 |
| 90 mm2 | 5.35 | 6.18 | 0.37 | 5.06 | m | 11.24 |
| 120 mm2 | 6.43 | 7.43 | 0.38 | 5.20 | m | 12.63 |
| 150 mm2 | 7.70 | 8.89 | 0.40 | 5.47 | m | 14.36 |
| 185 mm2 | 9.99 | 11.54 | 0.43 | 5.88 | m | 17.42 |
| 240 mm2 | 12.63 | 14.59 | 0.46 | 6.29 | m | 20.88 |
| 300 mm2 | 15.35 | 17.73 | 0.48 | 6.56 | m | 24.29 |
| 600/1000 Volt Grade; Three Core | | | | | | |
| 25 mm2 | 2.85 | 3.29 | 0.35 | 4.78 | m | 8.07 |
| 35 mm2 | 3.46 | 4.00 | 0.37 | 5.06 | m | 9.06 |
| 50 mm2 | 4.55 | 5.26 | 0.38 | 5.20 | m | 10.46 |
| 70 mm2 | 5.65 | 6.53 | 0.39 | 5.33 | m | 11.86 |
| 90 mm2 | 7.36 | 8.50 | 0.40 | 5.47 | m | 13.97 |
| 120 mm2 | 8.87 | 10.24 | 0.44 | 6.02 | m | 16.26 |
| 150 mm2 | 11.53 | 13.32 | 0.44 | 6.02 | m | 19.34 |
| 185 mm2 | 13.70 | 15.82 | 0.47 | 6.43 | m | 22.25 |
| 240 mm2 | 17.92 | 20.70 | 0.49 | 6.70 | m | 27.40 |
| 300 mm2 | 21.72 | 25.09 | 0.52 | 7.11 | m | 32.20 |
| 400 mm2 | 27.62 | 31.90 | 0.60 | 8.20 | m | 40.10 |
| 600/1000 Volt Grade; Four Core | | | | | | |
| 25 mm2 | 3.32 | 3.83 | 0.37 | 5.06 | m | 8.89 |
| 35 mm2 | 4.20 | 4.85 | 0.39 | 5.33 | m | 10.18 |
| 50 mm2 | 5.66 | 6.54 | 0.40 | 5.47 | m | 12.01 |
| 70 mm2 | 7.32 | 8.45 | 0.44 | 6.02 | m | 14.47 |
| 90 mm2 | 9.33 | 10.78 | 0.46 | 6.29 | m | 17.07 |
| 120 mm2 | 12.21 | 14.10 | 0.46 | 6.29 | m | 20.39 |
| 150 mm2 | 14.41 | 16.64 | 0.49 | 6.70 | m | 23.34 |
| 185 mm2 | 17.35 | 20.04 | 0.51 | 6.97 | m | 27.01 |
| 240 mm2 | 23.11 | 26.69 | 0.53 | 7.25 | m | 33.94 |
| 300 mm2 | 28.10 | 32.46 | 0.57 | 7.79 | m | 40.25 |
| 400 mm2 | 37.78 | 43.64 | 0.63 | 8.61 | m | 52.25 |

## Material Costs / Measured Work Prices - Electrical Installations

### Y:MECHANICAL AND ELECTRICAL SERVICES

| Item | Net Price £ | Material £ | Labour hours | Labour £ | Unit | Total Rate £ |
|---|---|---|---|---|---|---|
| **Y61 : HV/LV CABLES AND WIRING : CABLE (contd)** | | | | | | |
| Cable; XLPE insulated; SWA; PVC sheathed copper; BS 5467; laid in trench including marker tape. (Cable tiles measured elsewhere.) | | | | | | |
| 600/1000 Volt Grade; Two Core | | | | | | |
| 25 mm2 | 2.27 | 2.62 | 0.14 | 1.91 | m | **4.53** |
| 35 mm2 | 2.73 | 3.15 | 0.14 | 1.91 | m | **5.06** |
| 50 mm2 | 3.45 | 3.98 | 0.16 | 2.19 | m | **6.17** |
| 70 mm2 | 4.22 | 4.87 | 0.16 | 2.19 | m | **7.06** |
| 90 mm2 | 5.35 | 6.18 | 0.18 | 2.46 | m | **8.64** |
| 120 mm2 | 6.43 | 7.43 | 0.18 | 2.46 | m | **9.89** |
| 150 mm2 | 7.70 | 8.89 | 0.20 | 2.73 | m | **11.62** |
| 185 mm2 | 9.99 | 11.54 | 0.20 | 2.73 | m | **14.27** |
| 240 mm2 | 12.63 | 14.59 | 0.22 | 3.01 | m | **17.60** |
| 300 mm2 | 15.35 | 17.73 | 0.22 | 3.01 | m | **20.74** |
| 600/1000 Volt Grade; Three Core | | | | | | |
| 25 mm2 | 2.85 | 3.29 | 0.15 | 2.05 | m | **5.34** |
| 35 mm2 | 3.46 | 4.00 | 0.15 | 2.05 | m | **6.05** |
| 50 mm2 | 4.55 | 5.26 | 0.18 | 2.46 | m | **7.72** |
| 70 mm2 | 5.65 | 6.53 | 0.18 | 2.46 | m | **8.99** |
| 90 mm2 | 7.36 | 8.50 | 0.20 | 2.73 | m | **11.23** |
| 120 mm2 | 8.87 | 10.24 | 0.20 | 2.73 | m | **12.97** |
| 150 mm2 | 11.53 | 13.32 | 0.22 | 3.01 | m | **16.33** |
| 185 mm2 | 13.70 | 15.82 | 0.22 | 3.01 | m | **18.83** |
| 240 mm2 | 17.92 | 20.70 | 0.24 | 3.28 | m | **23.98** |
| 300 mm2 | 21.72 | 25.09 | 0.24 | 3.28 | m | **28.37** |
| 400 mm2 | 27.62 | 31.90 | 0.26 | 3.55 | m | **35.45** |
| 600/1000 Volt Grade; Four Core | | | | | | |
| 25 mm2 | 3.32 | 3.83 | 0.18 | 2.46 | m | **6.29** |
| 35 mm2 | 4.20 | 4.85 | 0.18 | 2.46 | m | **7.31** |
| 50 mm2 | 5.66 | 6.54 | 0.21 | 2.87 | m | **9.41** |
| 70 mm2 | 7.32 | 8.45 | 0.21 | 2.87 | m | **11.32** |
| 90 mm2 | 9.33 | 10.78 | 0.23 | 3.14 | m | **13.92** |
| 120 mm2 | 12.21 | 14.10 | 0.23 | 3.14 | m | **17.24** |
| 150 mm2 | 14.41 | 16.64 | 0.25 | 3.42 | m | **20.06** |
| 185 mm2 | 17.35 | 20.04 | 0.25 | 3.42 | m | **23.46** |
| 240 mm2 | 23.11 | 26.69 | 0.27 | 3.69 | m | **30.38** |
| 300 mm2 | 28.10 | 32.46 | 0.27 | 3.69 | m | **36.15** |
| 400 mm2 | 37.78 | 43.64 | 0.29 | 3.96 | m | **47.60** |

*Material Costs / Measured Work Prices - Electrical Installations*

## Y:MECHANICAL AND ELECTRICAL SERVICES

| Item | Net Price £ | Material £ | Labour hours | Labour £ | Unit | Total Rate £ |
|---|---|---|---|---|---|---|
| **Cable Termination; XLPE or PVC insulated armoured cables (including drilling and cutting mild steel gland plate).** | | | | | | |
| Brass weatherproof gland with inner and outer seal; PVC shroud, brass locknut, brass earthing ring (Type E1W); Two Core | | | | | | |
| 25 mm2 | 10.49 | 11.54 | 0.98 | 13.40 | nr | **24.94** |
| 35 mm2 | 14.43 | 15.86 | 1.05 | 14.36 | nr | **30.22** |
| 50 mm2 | 10.49 | 11.54 | 1.10 | 15.04 | nr | **26.58** |
| 70 mm2 | 14.43 | 15.86 | 1.12 | 15.32 | nr | **31.18** |
| 90 mm2 | 14.43 | 15.86 | 1.15 | 15.73 | nr | **31.59** |
| 120 mm2 | 26.31 | 28.94 | 1.20 | 16.41 | nr | **45.35** |
| 150 mm2 | 26.31 | 28.94 | 1.28 | 17.50 | nr | **46.44** |
| 185 mm2 | 37.07 | 40.78 | 1.72 | 23.53 | nr | **64.31** |
| 240 mm2 | 38.47 | 42.32 | 1.83 | 25.03 | nr | **67.35** |
| 300 mm2 | 64.57 | 71.02 | 1.93 | 26.39 | nr | **97.41** |
| Brass weatherproof gland with inner and outer seal; PVC shroud, brass locknut, brass earthing ring, (type E1W); Three Core | | | | | | |
| 25 mm2 | 14.43 | 15.86 | 1.05 | 14.36 | nr | **30.22** |
| 35 mm2 | 14.43 | 15.86 | 1.15 | 15.73 | nr | **31.59** |
| 50 mm2 | 14.43 | 15.86 | 1.20 | 16.41 | nr | **32.27** |
| 70 mm2 | 14.43 | 15.86 | 1.25 | 17.09 | nr | **32.95** |
| 90 mm2 | 26.31 | 28.94 | 1.28 | 17.50 | nr | **46.44** |
| 120 mm2 | 37.07 | 40.78 | 1.37 | 18.75 | nr | **59.53** |
| 150 mm2 | 37.07 | 40.78 | 1.80 | 24.63 | nr | **65.41** |
| 185 mm2 | 38.47 | 42.32 | 1.80 | 24.63 | nr | **66.95** |
| 240 mm2 | 64.57 | 71.02 | 2.00 | 27.34 | nr | **98.36** |
| 300 mm2 | 65.93 | 72.51 | 2.17 | 29.71 | nr | **102.22** |
| 400 mm2 | 83.85 | 92.24 | 2.83 | 38.72 | nr | **130.96** |
| Brass weatherproof gland with inner and outer seal; PVC shroud, brass locknut, brass earthing ring, (Type E1W); Four Core | | | | | | |
| 25 mm2 | 14.43 | 15.86 | 1.15 | 15.73 | nr | **31.59** |
| 35 mm2 | 14.43 | 15.86 | 1.30 | 17.77 | nr | **33.63** |
| 50 mm2 | 14.43 | 15.86 | 1.34 | 18.30 | nr | **34.16** |
| 70 mm2 | 26.31 | 28.94 | 1.45 | 19.84 | nr | **48.78** |
| 90 mm2 | 37.07 | 40.78 | 1.48 | 20.25 | nr | **61.03** |
| 120 mm2 | 38.47 | 42.32 | 1.62 | 22.15 | nr | **64.47** |
| 150 mm2 | 38.47 | 42.32 | 1.70 | 23.25 | nr | **65.57** |
| 185 mm2 | 64.57 | 71.02 | 1.87 | 25.60 | nr | **96.62** |
| 240 mm2 | 65.93 | 72.51 | 2.40 | 32.86 | nr | **105.37** |
| 300 mm2 | 83.85 | 92.24 | 2.50 | 34.17 | nr | **126.41** |
| 400 mm2 | 108.79 | 119.67 | 3.00 | 41.05 | nr | **160.72** |

## Y:MECHANICAL AND ELECTRICAL SERVICES

| Item | Net Price £ | Material £ | Labour hours | Labour £ | Unit | Total Rate £ |
|---|---|---|---|---|---|---|
| **Y61 : HV/LV CABLES AND WIRING : CABLE (contd)** | | | | | | |
| Cable; XLPE Ins; PVC Sheathed; SWA; PVC Insulated; copper; BS 5467 : 1989; fixed with clips to backgrounds. (Supports and fixings, connecting tails are included.) | | | | | | |
| 600/1000 Volt Grade (Multi Strand); Two Core | | | | | | |
| 1.5 mm2 | 1.47 | 1.70 | 0.09 | 1.23 | m | **2.93** |
| 2.5 mm2 | 1.82 | 2.11 | 0.10 | 1.37 | m | **3.48** |
| 4.0 mm2 | 3.07 | 3.55 | 0.11 | 1.50 | m | **5.05** |
| 6.0 mm2 | 3.73 | 4.31 | 0.12 | 1.64 | m | **5.95** |
| 10.0 mm2 | 6.41 | 7.40 | 0.13 | 1.78 | m | **9.18** |
| 16.0 mm2 | 7.76 | 8.97 | 0.14 | 1.91 | m | **10.88** |
| 600/1000 Volt Grade (Multi Stand); Three Core | | | | | | |
| 1.5 mm2 | 1.78 | 2.06 | 0.10 | 1.37 | m | **3.43** |
| 2.5 mm2 | 2.18 | 2.52 | 0.11 | 1.50 | m | **4.02** |
| 4.0 mm2 | 3.80 | 4.39 | 0.12 | 1.64 | m | **6.03** |
| 6.0 mm2 | 4.83 | 5.58 | 0.13 | 1.78 | m | **7.36** |
| 10.0 mm2 | 8.73 | 10.08 | 0.14 | 1.91 | m | **11.99** |
| 16.0 mm2 | 11.17 | 12.90 | 0.15 | 2.05 | m | **14.95** |
| 600/1000 Volt Grade (Multi Strand); Four Core | | | | | | |
| 1.5 mm2 | 1.98 | 2.28 | 0.11 | 1.50 | m | **3.78** |
| 2.5 mm2 | 2.52 | 2.91 | 0.12 | 1.64 | m | **4.55** |
| 4.0 mm2 | 4.54 | 5.24 | 0.13 | 1.78 | m | **7.02** |
| 6.0 mm2 | 6.56 | 7.57 | 0.14 | 1.91 | m | **9.48** |
| 10.0 mm2 | 10.20 | 11.78 | 0.15 | 2.05 | m | **13.83** |
| 16.0 mm2 | 12.44 | 14.37 | 0.16 | 2.19 | m | **16.56** |
| 600/1000 Volt Grade (Multi Strand); Seven Core | | | | | | |
| 1.5 mm2 | 3.66 | 4.23 | 0.14 | 1.91 | m | **6.14** |
| 2.5 mm2 | 4.98 | 5.75 | 0.15 | 2.05 | m | **7.80** |
| 4.0 mm2 | 7.30 | 8.43 | 0.16 | 2.19 | m | **10.62** |
| 600/1000 Volt Grade (Multi Strand); Twelve Core | | | | | | |
| 1.5 mm2 | 5.94 | 6.86 | 0.17 | 2.32 | m | **9.18** |
| 2.5 mm2 | 7.97 | 9.20 | 0.18 | 2.46 | m | **11.66** |
| 600/1000 Volt Grade (Multi Strand); Nineteen Core | | | | | | |
| 1.5 mm2 | 8.98 | 10.37 | 0.22 | 3.01 | m | **13.38** |
| 2.5 mm2 | 12.49 | 14.42 | 0.23 | 3.14 | m | **17.56** |
| 600/1000 Volt Grade (Multi Strand); Twenty Seven Core | | | | | | |
| 1.5 mm2 | 13.18 | 15.22 | 0.25 | 3.42 | m | **18.64** |
| 2.5 mm2 | 16.90 | 19.52 | 0.26 | 3.55 | m | **23.07** |

## Y:MECHANICAL AND ELECTRICAL SERVICES

| Item | Net Price £ | Material £ | Labour hours | Labour £ | Unit | Total Rate £ |
|---|---|---|---|---|---|---|
| 600/1000 Volt Grade (Multi Strand); Thirty Seven Core | | | | | | |
| 1.5 mm2 | 16.94 | 19.57 | 0.29 | 3.96 | m | **23.53** |
| 2.5 mm2 | 22.40 | 25.87 | 0.30 | 4.10 | m | **29.97** |
| Cable; XLPE; PVC Sheathed; SWA; PVC Insulated; copper; BS 5467 : 1989; laid in trench; including marker tape. (Cable tiles measured eleswhere). | | | | | | |
| 600/1000 Volt Grade (Multi Strand); Two Core | | | | | | |
| 1.5 mm2 | 1.47 | 1.70 | 0.03 | 0.41 | m | **2.11** |
| 2.5 mm2 | 1.82 | 2.11 | 0.03 | 0.41 | m | **2.52** |
| 4 mm2 | 3.07 | 3.55 | 0.05 | 0.68 | m | **4.23** |
| 6 mm2 | 3.73 | 4.31 | 0.05 | 0.68 | m | **4.99** |
| 10 mm2 | 6.41 | 7.40 | 0.06 | 0.82 | m | **8.22** |
| 16 mm2 | 7.76 | 8.97 | 0.06 | 0.82 | m | **9.79** |
| 600/1000 Volt Grade (Multi Strand); Three Core | | | | | | |
| 1.5 mm2 | 1.78 | 2.06 | 0.05 | 0.68 | m | **2.74** |
| 2.5 mm2 | 2.18 | 2.52 | 0.05 | 0.68 | m | **3.20** |
| 4 mm2 | 3.80 | 4.39 | 0.07 | 0.96 | m | **5.35** |
| 6 mm2 | 4.83 | 5.58 | 0.07 | 0.96 | m | **6.54** |
| 10 mm2 | 8.73 | 10.08 | 0.07 | 0.96 | m | **11.04** |
| 16 mm2 | 11.17 | 12.90 | 0.07 | 0.96 | m | **13.86** |
| 600/1000 Volt Grade (Multi Strand); FourCore | | | | | | |
| 1.5 mm2 | 1.98 | 2.28 | 0.07 | 0.96 | m | **3.24** |
| 2.5 mm2 | 2.52 | 2.91 | 0.07 | 0.96 | m | **3.87** |
| 4 mm2 | 4.54 | 5.24 | 0.07 | 0.96 | m | **6.20** |
| 6 mm2 | 6.56 | 7.57 | 0.07 | 0.96 | m | **8.53** |
| 10 mm2 | 10.20 | 11.78 | 0.08 | 1.09 | m | **12.87** |
| 16 mm2 | 12.44 | 14.37 | 0.08 | 1.09 | m | **15.46** |
| 600/1000 Volt Grade (Multi Strand); Seven Core | | | | | | |
| 1.5 mm2 | 3.66 | 4.23 | 0.08 | 1.09 | m | **5.32** |
| 2.5 mm2 | 4.98 | 5.75 | 0.08 | 1.09 | m | **6.84** |
| 4 mm2 | 7.30 | 8.43 | 0.08 | 1.09 | m | **9.52** |
| 600/1000 Volt Grade (Multi Strand); Twelve Core | | | | | | |
| 1.5 mm2 | 5.94 | 6.86 | 0.09 | 1.23 | m | **8.09** |
| 2.5 mm2 | 7.97 | 9.20 | 0.09 | 1.23 | m | **10.43** |
| 600/1000 Volt Grade (Multi Strand); Nineteen Core | | | | | | |
| 1.5 mm2 | 8.98 | 10.37 | 0.11 | 1.50 | m | **11.87** |
| 2.5 mm2 | 12.49 | 14.42 | 0.11 | 1.50 | m | **15.92** |

## Y:MECHANICAL AND ELECTRICAL SERVICES

| Item | Net Price £ | Material £ | Labour hours | Labour £ | Unit | Total Rate £ |
|---|---|---|---|---|---|---|
| **Y61 : HV/LV CABLES AND WIRING : CABLE (contd)** | | | | | | |
| Cable; XLPE; PVC Sheathed; SWA; PVC Insulated; copper; BS 5467 : 1989; laid in trench; including marker tape. (Cable tiles measured eleswhere). (contd) | | | | | | |
| 600/1000 Volt Grade (Multi Strand); Twenty Seven Core | | | | | | |
| 1.5 mm2 | 13.18 | 15.22 | 0.12 | 1.64 | m | 16.86 |
| 2.5 mm2 | 16.90 | 19.52 | 0.12 | 1.64 | m | 21.16 |
| 600/1000 Volt Grade (Multi Strand); Thirty Seven Core | | | | | | |
| 1.5 mm2 | 16.94 | 19.57 | 0.13 | 1.78 | m | 21.35 |
| 2.5 mm2 | 22.40 | 25.87 | 0.13 | 1.78 | m | 27.65 |
| **Cable Terminations; for PVC Insulated SWA PVC sheathed copper (including drilling and cutting mild steel gland plate)** | | | | | | |
| Brass weatherproof gland with outer seal; brass locknut and earthing ring; PVC shroud (Type E1W); Two Core | | | | | | |
| 1.5 mm2 | 5.81 | 6.39 | 0.48 | 6.56 | nr | 12.95 |
| 2.5 mm2 | 5.96 | 6.56 | 0.48 | 6.56 | nr | 13.12 |
| 4 mm2 | 5.96 | 6.56 | 0.48 | 6.56 | nr | 13.12 |
| 6 mm2 | 7.00 | 7.70 | 0.50 | 6.83 | nr | 14.53 |
| 10 mm2 | 7.00 | 7.70 | 0.65 | 8.89 | nr | 16.59 |
| 16 mm2 | 10.49 | 11.54 | 0.77 | 10.53 | nr | 22.07 |
| Brass weatherproof gland with outer seal; brass locknut and earthing ring; PVC shroud (Type E1W); Three Core | | | | | | |
| 1.5 mm2 | 5.81 | 6.39 | 0.52 | 7.11 | nr | 13.50 |
| 2.5 mm2 | 5.96 | 6.56 | 0.52 | 7.11 | nr | 13.67 |
| 4 mm2 | 5.96 | 6.56 | 0.52 | 7.11 | nr | 13.67 |
| 6 mm2 | 7.00 | 7.70 | 0.53 | 7.25 | nr | 14.95 |
| 10 mm2 | 6.65 | 7.70 | 0.72 | 9.85 | nr | 17.55 |
| 16 mm2 | 10.49 | 11.54 | 0.83 | 11.35 | nr | 22.89 |
| Brass weatherproof gland with outer seal; brass locknut and earthing ring; PVC shroud (Type E1W); Four Core | | | | | | |
| 1.5 mm2 | 5.96 | 6.56 | 0.57 | 7.79 | nr | 14.35 |
| 2.5 mm2 | 5.96 | 6.56 | 0.57 | 7.79 | nr | 14.35 |
| 4 mm2 | 7.00 | 7.70 | 0.57 | 7.79 | nr | 15.49 |
| 6 mm2 | 7.00 | 7.70 | 0.58 | 7.93 | nr | 15.63 |
| 10 mm2 | 10.49 | 11.54 | 0.82 | 11.21 | nr | 22.75 |
| 16 mm2 | 10.49 | 11.54 | 0.93 | 12.71 | nr | 24.25 |

## Y:MECHANICAL AND ELECTRICAL SERVICES

| Item | Net Price £ | Material £ | Labour hours | Labour £ | Unit | Total Rate £ |
|---|---|---|---|---|---|---|
| Brass weatherproof gland with outer seal; brass locknut and earthing ring; PVC shroud (Type E1W); Five Core | | | | | | |
| 1.5 mm2 | 5.96 | 6.56 | 0.63 | 8.61 | nr | 15.17 |
| 2.5 mm2 | 5.96 | 6.56 | 0.63 | 8.61 | nr | 15.17 |
| Brass weatherproof gland with outer seal; brass locknut and earthing ring; PVC shroud (Type E1W); Seven Core | | | | | | |
| 1.5 mm2 | 7.00 | 7.70 | 0.73 | 9.98 | nr | 17.68 |
| 2.5 mm2 | 7.00 | 7.70 | 0.73 | 9.98 | nr | 17.68 |
| 4 mm2 | 7.00 | 7.70 | 0.76 | 10.39 | nr | 18.09 |
| Brass weatherproof gland with outer seal; brass locknut and earthing ring; PVC shroud (Type E1W); Ten Core | | | | | | |
| 1.5 mm2 | 7.00 | 7.70 | 0.85 | 11.62 | nr | 19.32 |
| Brass weatherproof gland with outer seal; brass locknut and earthing ring; PVC shroud (Type E1W); Twelve Core | | | | | | |
| 1.5 mm2 | 10.25 | 11.26 | 0.85 | 11.62 | nr | 22.88 |
| 2.5 mm2 | 10.49 | 11.54 | 0.85 | 11.62 | nr | 23.16 |
| Brass weatherproof gland with outer seal; brass locknut and earthing ring; PVC shroud (Type EIW); Nineteen Core | | | | | | |
| 1.5 mm2 | 10.49 | 11.54 | 1.05 | 14.36 | nr | 25.90 |
| 2.5 mm2 | 10.49 | 11.54 | 1.05 | 14.36 | nr | 25.90 |
| Brass weatherproof gland with outer seal; brass locknut and earthing ring; PVC shroud (Type E1W); Twenty Seven Core | | | | | | |
| 1.5 mm2 | 14.43 | 15.86 | 1.28 | 17.50 | nr | 33.36 |
| 2.5 mm2 | 14.43 | 15.86 | 1.28 | 17.50 | nr | 33.36 |
| Brass weatherproof gland with outer seal; brass locknut and earthing ring; PVC shroud (Type CW); Thirty Seven Core | | | | | | |
| 1.5 mm2 | 14.43 | 15.86 | 1.55 | 21.19 | nr | 37.05 |
| 2.5 mm2 | 26.31 | 28.94 | 1.55 | 21.19 | nr | 50.13 |
| Brass weatherproof gland with outer seal; brass locknut and earthing ring; PVC shroud (Type E1W); Forty Eight Core | | | | | | |
| 1.5 mm2 | 13.82 | 15.19 | 2.13 | 29.14 | nr | 44.33 |
| 2.5 mm2 | 25.22 | 27.74 | 2.13 | 29.14 | nr | 56.88 |

## Y:MECHANICAL AND ELECTRICAL SERVICES

| Item | Net Price £ | Material £ | Labour hours | Labour £ | Unit | Total Rate £ |
|---|---|---|---|---|---|---|
| **Y61 : HV/LV CABLES AND WIRING : CABLE (contd)** | | | | | | |
| Cable, Mineral Insulated; copper sheathed with copper conductors; fixed with clips to backgrounds. BASEC approval to BS 6207 Part 1 1995; complies with BS 6387 Category CWZ | | | | | | |
| Light duty 500 Volt grade; Bare | | | | | | |
| 2L 1.0 | 1.13 | 1.30 | 0.16 | 2.23 | m | **3.53** |
| 2L 1.5 | 1.35 | 1.55 | 0.16 | 2.24 | m | **3.79** |
| 2L 2.5 | 1.75 | 2.02 | 0.17 | 2.32 | m | **4.34** |
| 2L 4.0 | 2.63 | 3.03 | 0.17 | 2.39 | m | **5.42** |
| 3L 1.0 | 1.38 | 1.60 | 0.17 | 2.26 | m | **3.86** |
| Light duty 500 Volt grade; Bare (contd) | | | | | | |
| 3L 1.5 | 1.76 | 2.03 | 0.17 | 2.31 | m | **4.34** |
| 3L 2.5 | 2.76 | 3.19 | 0.17 | 2.36 | m | **5.55** |
| 4L 1.0 | 1.66 | 1.91 | 0.17 | 2.30 | m | **4.21** |
| 4L 1.5 | 2.12 | 2.45 | 0.17 | 2.35 | m | **4.80** |
| 4L 2.5 | 3.35 | 3.87 | 0.18 | 2.45 | m | **6.32** |
| 7L 1.0 | 2.49 | 2.88 | 0.17 | 2.38 | m | **5.26** |
| 7L 1.5 | 3.08 | 3.56 | 0.18 | 2.47 | m | **6.03** |
| 7L 2.5 | 4.05 | 4.68 | 0.16 | 2.21 | m | **6.89** |
| Light duty 500 Volt grade; LSF sheathed | | | | | | |
| 2L 1.0 | 1.32 | 1.52 | 0.16 | 2.23 | m | **3.75** |
| 2L 1.5 | 1.47 | 1.70 | 0.16 | 2.24 | m | **3.94** |
| 2L 2.5 | 1.87 | 2.16 | 0.17 | 2.32 | m | **4.48** |
| 2L 4.0 | 2.78 | 3.22 | 0.17 | 2.39 | m | **5.61** |
| 3L 1.0 | 1.60 | 1.84 | 0.17 | 2.26 | m | **4.10** |
| 3L 1.5 | 1.95 | 2.25 | 0.17 | 2.30 | m | **4.55** |
| 3L 2.5 | 2.92 | 3.37 | 0.17 | 2.36 | m | **5.73** |
| 4L 1.0 | 1.88 | 2.17 | 0.17 | 2.30 | m | **4.47** |
| 4L 1.5 | 2.37 | 2.74 | 0.17 | 2.35 | m | **5.09** |
| 4L 2.5 | 3.49 | 4.03 | 0.18 | 2.45 | m | **6.48** |
| 7L 1.0 | 2.85 | 3.29 | 0.17 | 2.38 | m | **5.67** |
| 7L 1.5 | 3.52 | 4.07 | 0.18 | 2.47 | m | **6.54** |
| 7L 2.5 | 4.52 | 5.22 | 0.16 | 2.21 | m | **7.43** |
| Cable, Mineral Insulated; copper sheathed with copper conductors; fixed with clips to backgrounds; BASEC approval to BS 6207 Part 1 1995; complies with BS 6387 Category CWZ | | | | | | |
| Heavy duty 750 Volt grade; Bare | | | | | | |
| 1H 10 | 2.55 | 2.95 | 0.18 | 2.49 | m | **5.44** |
| 1H 16 | 3.50 | 4.04 | 0.19 | 2.60 | m | **6.64** |
| 1H 25 | 4.89 | 5.64 | 0.17 | 2.36 | m | **8.00** |
| 1H 35 | 6.52 | 7.53 | 0.19 | 2.60 | m | **10.13** |
| 1H 50 | 8.28 | 9.56 | 0.22 | 2.98 | m | **12.54** |
| 1H 70 | 10.75 | 12.41 | 0.26 | 3.54 | m | **15.95** |
| 1H 95 | 14.12 | 16.31 | 0.28 | 3.85 | m | **20.16** |
| 1H 120 | 17.26 | 19.94 | 0.33 | 4.47 | m | **24.41** |
| 1H 150 | 21.40 | 24.71 | 0.39 | 5.39 | m | **30.10** |

## Y:MECHANICAL AND ELECTRICAL SERVICES

| Item | Net Price £ | Material £ | Labour hours | Labour £ | Unit | Total Rate £ |
|---|---|---|---|---|---|---|
| Heavy duty 750 Volt grade; Bare (contd) | | | | | | |
| 1H 185 | 26.05 | 30.08 | 0.48 | 6.55 | m | 36.63 |
| 1H 240 | 33.82 | 39.07 | 0.68 | 9.28 | m | 48.35 |
| 2H 1.5 | 2.17 | 2.51 | 0.18 | 2.42 | m | 4.93 |
| 2H 2.5 | 2.67 | 3.08 | 0.18 | 2.53 | m | 5.61 |
| 2H 4 | 3.37 | 3.89 | 0.16 | 2.23 | m | 6.12 |
| 2H 6 | 4.48 | 5.18 | 0.18 | 2.43 | m | 7.61 |
| 2H 10 | 5.80 | 6.70 | 0.21 | 2.84 | m | 9.54 |
| 2H 16 | 8.36 | 9.65 | 0.25 | 3.44 | m | 13.09 |
| 2H 25 | 11.73 | 13.55 | 0.29 | 4.02 | m | 17.57 |
| 3H 1.5 | 2.40 | 2.78 | 0.18 | 2.46 | m | 5.24 |
| 3H 2.5 | 3.03 | 3.50 | 0.16 | 2.15 | m | 5.65 |
| Heavy duty 750 Volt grade; Bare | | | | | | |
| 3H 4 | 3.84 | 4.43 | 0.17 | 2.35 | m | 6.78 |
| 3H 6 | 7.16 | 8.27 | 0.19 | 2.57 | m | 10.84 |
| 3H 10 | 7.16 | 8.27 | 0.23 | 3.10 | m | 11.37 |
| 3H 16 | 10.05 | 11.61 | 0.25 | 3.42 | m | 15.03 |
| 3H 25 | 15.43 | 17.82 | 0.33 | 4.51 | m | 22.33 |
| 4H 1.5 | 2.99 | 3.46 | 0.15 | 2.12 | m | 5.58 |
| 4H 2.5 | 3.76 | 4.34 | 0.17 | 2.28 | m | 6.62 |
| 4H 4 | 4.70 | 5.43 | 0.19 | 2.54 | m | 7.97 |
| 4H 6 | 6.27 | 7.24 | 0.21 | 2.84 | m | 10.08 |
| 4H 10 | 8.90 | 10.28 | 0.26 | 3.49 | m | 13.77 |
| 4H 16 | 12.98 | 14.99 | 0.30 | 4.10 | m | 19.09 |
| 4H 25 | 18.86 | 21.78 | 0.40 | 5.43 | m | 27.21 |
| 7H 1.5 | 4.13 | 4.77 | 0.18 | 2.42 | m | 7.19 |
| 7H 2.5 | 5.62 | 6.49 | 0.20 | 2.71 | m | 9.20 |
| 12H 1.5 | 7.38 | 8.52 | 0.24 | 3.24 | m | 11.76 |
| 12H 2.5 | 9.80 | 11.32 | 0.25 | 3.42 | m | 14.74 |
| 19H 1.5 | 15.12 | 17.47 | 0.28 | 3.81 | m | 21.28 |
| Heavy duty 750 Volt grade; LSF Sheathed | | | | | | |
| 1H 10 | 2.78 | 3.21 | 0.18 | 2.49 | m | 5.70 |
| 1H 16 | 3.80 | 4.39 | 0.19 | 2.60 | m | 6.99 |
| 1H 25 | 5.25 | 6.07 | 0.17 | 2.36 | m | 8.43 |
| 1H 35 | 6.89 | 7.96 | 0.19 | 2.60 | m | 10.56 |
| 1H 50 | 8.70 | 10.04 | 0.22 | 2.98 | m | 13.02 |
| 1H 70 | 11.30 | 13.05 | 0.26 | 3.54 | m | 16.59 |
| 1H 95 | 14.83 | 17.13 | 0.28 | 3.85 | m | 20.98 |
| 1H 120 | 18.09 | 20.89 | 0.33 | 4.47 | m | 25.36 |
| 1H 150 | 22.29 | 25.75 | 0.39 | 5.39 | m | 31.14 |
| 1H 185 | 27.36 | 31.60 | 0.48 | 6.55 | m | 38.15 |
| 1H 240 | 35.27 | 40.74 | 0.68 | 9.28 | m | 50.02 |
| 2H 1.5 | 2.45 | 2.83 | 0.18 | 2.42 | m | 5.25 |
| 2H 2.5 | 2.91 | 3.36 | 0.18 | 2.53 | m | 5.89 |
| 2H 4 | 3.67 | 4.23 | 0.16 | 2.23 | m | 6.46 |
| 2H 6 | 4.86 | 5.61 | 0.18 | 2.43 | m | 8.04 |
| 2H 10 | 6.27 | 7.24 | 0.21 | 2.84 | m | 10.08 |
| 2H 16 | 8.85 | 10.23 | 0.25 | 3.44 | m | 13.67 |
| 2H 25 | 11.73 | 13.55 | 0.29 | 4.02 | m | 17.57 |
| 3H 1.5 | 2.70 | 3.12 | 0.18 | 2.46 | m | 5.58 |
| 3H 2.5 | 3.32 | 3.84 | 0.16 | 2.15 | m | 5.99 |
| 3H 4 | 4.23 | 4.89 | 0.17 | 2.35 | m | 7.24 |
| 3H 6 | 5.32 | 6.14 | 0.19 | 2.57 | m | 8.71 |
| 3H 10 | 7.64 | 8.83 | 0.23 | 3.10 | m | 11.93 |
| 3H 16 | 10.79 | 12.46 | 0.25 | 3.42 | m | 15.88 |
| 3H 25 | 16.27 | 18.79 | 0.33 | 4.51 | m | 23.30 |

## Y:MECHANICAL AND ELECTRICAL SERVICES

| Item | Net Price £ | Material £ | Labour hours | Labour £ | Unit | Total Rate £ |
|---|---|---|---|---|---|---|
| **Y61 : HV/LV CABLES AND WIRING : CABLE (contd)** | | | | | | |
| Cable, Mineral Insulated; copper sheathed with copper conductors; fixed with clips to backgrounds; BASEC approval to BS 6207 Part 1 1995; complies with BS 6387 Category CWZ (contd) | | | | | | |
| Heavy duty 750 Volt grade; LSF Sheathed (contd) | | | | | | |
| 4H 1.5 | 3.26 | 3.77 | 0.15 | 2.12 | m | 5.89 |
| 4H 2.5 | 4.10 | 4.73 | 0.17 | 2.28 | m | 7.01 |
| 4H 4 | 5.08 | 5.87 | 0.19 | 2.54 | m | 8.41 |
| 4H 6 | 6.68 | 7.72 | 0.21 | 2.84 | m | 10.56 |
| 4H 10 | 9.43 | 10.89 | 0.26 | 3.49 | m | 14.38 |
| 4H 16 | 13.75 | 15.88 | 0.30 | 4.10 | m | 19.98 |
| 4H 25 | 20.04 | 23.15 | 0.40 | 5.43 | m | 28.58 |
| 7H 1.5 | 4.52 | 5.22 | 0.18 | 2.42 | m | 7.64 |
| 7H 2.5 | 6.06 | 7.00 | 0.20 | 2.71 | m | 9.71 |
| 12H 1.5 | 7.86 | 9.08 | 0.24 | 3.24 | m | 12.32 |
| 12H 2.5 | 10.48 | 12.10 | 0.25 | 3.42 | m | 15.52 |
| 19H 1.5 | 15.94 | 18.41 | 0.28 | 3.81 | m | 22.22 |
| Cable terminations for M.I. Cable; Polymeric one piece moulding; containing grey sealing compound; testing; phase marking and connection | | | | | | |
| Light Duty 500 Volt grade; Brass gland; polymeric one moulding containing grey sealing compound; coloured conductor sleeving; Earth tag; plastic gland shroud | | | | | | |
| 2L 1.5 | 2.33 | 2.56 | 0.15 | 2.05 | m | 4.61 |
| 2L 2.5 | 2.34 | 2.57 | 0.15 | 2.05 | m | 4.62 |
| 3L 1.5 | 2.33 | 2.56 | 0.15 | 2.05 | m | 4.61 |
| 4L 1.5 | 2.33 | 2.56 | 0.15 | 2.05 | m | 4.61 |
| Cable Terminations; for MI copper sheathed cable. Certified for installation in potentially explosive atmospheres; testing; phase marking and connection; BS 6207 Part 2 1995 | | | | | | |
| Light duty 500 Volt grade; brass gland; brass pot with earth tail; pot closure; sealing compound; conductor sleeving; plastic gland shroud; identification markers | | | | | | |
| 2L 1.0 | 2.08 | 2.29 | 0.30 | 4.10 | nr | 6.39 |
| 2L 1.5 | 2.08 | 2.29 | 0.30 | 4.10 | nr | 6.39 |
| 2L 2.5 | 2.08 | 2.29 | 0.30 | 4.10 | nr | 6.39 |
| 2L 4.0 | 2.08 | 2.29 | 0.30 | 4.10 | nr | 6.39 |
| 3L 1.0 | 2.08 | 2.29 | 0.30 | 4.10 | nr | 6.39 |
| 3L 1.5 | 2.08 | 2.29 | 0.30 | 4.10 | nr | 6.39 |
| 3L 2.5 | 2.08 | 2.29 | 0.30 | 4.10 | nr | 6.39 |
| 4L 1.0 | 2.08 | 2.29 | 0.33 | 4.51 | nr | 6.80 |
| 4L 1.5 | 2.08 | 2.29 | 0.33 | 4.51 | nr | 6.80 |
| 4L 2.5 | 2.08 | 2.29 | 0.33 | 4.51 | nr | 6.80 |

## Y:MECHANICAL AND ELECTRICAL SERVICES

| Item | Net Price £ | Material £ | Labour hours | Labour £ | Unit | Total Rate £ |
|---|---|---|---|---|---|---|
| 7L 1.0 | 5.58 | 6.14 | 0.53 | 7.25 | nr | 13.39 |
| 7L 1.5 | 5.58 | 6.14 | 0.53 | 7.25 | nr | 13.39 |
| 7L 2.5 | 5.58 | 6.14 | 0.53 | 7.25 | nr | 13.39 |
| Heavy duty 750 Volt grade; brass gland; brass pot with earth tail; pot closure; sealing compound; conductor sleeving; plastic gland shroud; identification markers | | | | | | |
| 1H 10 | 5.32 | 5.85 | 0.22 | 3.01 | nr | 8.86 |
| 1H 16 | 5.32 | 5.85 | 0.22 | 3.01 | nr | 8.86 |
| 1H 25 | 8.84 | 9.73 | 0.30 | 4.10 | nr | 13.83 |
| 1H 35 | 8.84 | 9.73 | 0.30 | 4.10 | nr | 13.83 |
| 1H 50 | 16.49 | 18.14 | 0.30 | 4.10 | nr | 22.24 |
| 1H 70 | 3.69 | 4.06 | 0.38 | 5.20 | nr | 9.26 |
| 1H 95 | 3.69 | 4.06 | 0.38 | 5.20 | nr | 9.26 |
| 1H 120 | 10.09 | 11.09 | 0.55 | 7.52 | nr | 18.61 |
| 1H 150 | 7.37 | 8.11 | 0.55 | 7.52 | nr | 15.63 |
| 1H 185 | 7.92 | 8.72 | 0.75 | 10.25 | nr | 18.97 |
| 1H 240 | 16.47 | 18.12 | 0.75 | 10.25 | nr | 28.37 |
| 2H 1.5 | 4.45 | 4.89 | 0.30 | 4.10 | nr | 8.99 |
| 2H 2.5 | 2.65 | 2.92 | 0.30 | 4.10 | nr | 7.02 |
| 2H 4 | 5.58 | 6.14 | 0.30 | 4.10 | nr | 10.24 |
| 2H 6 | 5.58 | 6.14 | 0.38 | 5.20 | nr | 11.34 |
| 2H 10 | 9.31 | 10.22 | 0.38 | 5.20 | nr | 15.42 |
| 2H 16 | 18.39 | 20.23 | 0.41 | 5.60 | nr | 25.83 |
| 2H 25 | 19.38 | 21.32 | 0.41 | 5.60 | nr | 26.92 |
| 3H 1.5 | 4.45 | 4.89 | 0.30 | 4.10 | nr | 8.99 |
| 3H 2.5 | 5.32 | 5.86 | 0.30 | 4.10 | nr | 9.96 |
| 3H 4 | 5.58 | 6.14 | 0.42 | 5.74 | nr | 11.88 |
| 3H 6 | 5.58 | 6.14 | 0.42 | 5.74 | nr | 11.88 |
| 3H 10 | 9.86 | 10.84 | 0.42 | 5.74 | nr | 16.58 |
| 3H 16 | 18.39 | 20.23 | 0.45 | 6.15 | nr | 26.38 |
| 3H 25 | 19.38 | 21.32 | 0.45 | 6.15 | nr | 27.47 |
| 4H 1.5 | 4.45 | 4.89 | 0.40 | 5.47 | nr | 10.36 |
| 4H 2.5 | 5.58 | 6.14 | 0.40 | 5.47 | nr | 11.61 |
| 4H 4 | 5.58 | 6.14 | 0.38 | 5.20 | nr | 11.34 |
| 4H 6 | 9.86 | 10.84 | 0.43 | 5.88 | nr | 16.72 |
| 4H 10 | 9.86 | 10.84 | 0.43 | 5.88 | nr | 16.72 |
| 4H 16 | 19.38 | 21.32 | 0.46 | 6.29 | nr | 27.61 |
| 4H 25 | 19.38 | 21.32 | 0.46 | 6.29 | nr | 27.61 |
| 7H 1.5 | 5.58 | 6.14 | 0.53 | 7.25 | nr | 13.39 |
| 7H 2.5 | 5.58 | 6.14 | 0.53 | 7.25 | nr | 13.39 |
| 12H 1.5 | 9.90 | 10.88 | 0.58 | 7.93 | nr | 18.81 |
| 12H 2.5 | 9.90 | 10.88 | 0.58 | 7.93 | nr | 18.81 |
| 19H 2.5 | 19.05 | 20.96 | 0.75 | 10.25 | nr | 31.21 |
| PVC Insulated PVC Sheathed Copper Cable BS 6004 Tables 4 and 5; clipped to backgrounds; (Supports and fixings included.) | | | | | | |
| 300/500 Volt Grade; Single Core | | | | | | |
| 1.0 mm2 | 0.38 | 0.44 | 0.03 | 0.41 | m | 0.85 |
| 1.5 mm2 | 0.50 | 0.58 | 0.03 | 0.41 | m | 0.99 |

## Y:MECHANICAL AND ELECTRICAL SERVICES

| Item | Net Price £ | Material £ | Labour hours | Labour £ | Unit | Total Rate £ |
|---|---|---|---|---|---|---|
| **Y61 : HV/LV CABLES AND WIRING : CABLE (contd)** | | | | | | |
| **PVC Insulated PVC Sheathed Copper Cable BS 6004 Tables 4 and 5; clipped to backgrounds; (Supports and fixings included.) (Contd)** | | | | | | |
| 300/500 Volt Grade; Single Core (contd) | | | | | | |
| 2.5 mm2 | 0.84 | 0.97 | 0.03 | 0.41 | m | 1.38 |
| 4.0 mm2 | 1.36 | 1.57 | 0.04 | 0.55 | m | 2.12 |
| 6.0 mm2 | 1.81 | 2.09 | 0.04 | 0.55 | m | 2.64 |
| 10.0 mm2 | 2.88 | 3.33 | 0.05 | 0.68 | m | 4.01 |
| 16.0 mm2 | 3.75 | 4.33 | 0.05 | 0.68 | m | 5.01 |
| 25.0 mm2 | 7.18 | 8.30 | 0.06 | 0.82 | m | 9.12 |
| 35.0 mm2 | 11.03 | 12.74 | 0.06 | 0.82 | m | 13.56 |
| 300/500 Volt Grade; Twin Core | | | | | | |
| 1.0 mm2 | 0.67 | 0.78 | 0.04 | 0.55 | m | 1.33 |
| 1.5 mm2 | 0.86 | 0.99 | 0.04 | 0.55 | m | 1.54 |
| 2.5 mm2 | 1.22 | 1.40 | 0.04 | 0.55 | m | 1.95 |
| 4.0 mm2 | 1.80 | 2.08 | 0.05 | 0.68 | m | 2.76 |
| 6.0 mm2 | 2.46 | 2.84 | 0.05 | 0.68 | m | 3.52 |
| 10.0 mm2 | 4.00 | 4.62 | 0.05 | 0.68 | m | 5.30 |
| 16.0 mm2 | 6.22 | 7.18 | 0.06 | 0.82 | m | 8.00 |
| 300/500 Volt Grade; Twin Core with EarthWire | | | | | | |
| 1.0 mm2 | 0.58 | 0.67 | 0.05 | 0.68 | m | 1.35 |
| 1.5 mm2 | 0.72 | 0.83 | 0.05 | 0.68 | m | 1.51 |
| 2.5 mm2 | 0.99 | 1.14 | 0.05 | 0.68 | m | 1.82 |
| 4.0 mm2 | 3.06 | 3.54 | 0.05 | 0.68 | m | 4.22 |
| 6.0 mm2 | 3.64 | 4.20 | 0.05 | 0.68 | m | 4.88 |
| 10.0 mm2 | 5.87 | 6.78 | 0.06 | 0.82 | m | 7.60 |
| 16.0 mm2 | 9.36 | 10.81 | 0.06 | 0.82 | m | 11.63 |
| 300/500 Volt Grade; Three Core | | | | | | |
| 1.0 mm2 | 1.15 | 1.33 | 0.05 | 0.68 | m | 2.01 |
| 1.5 mm2 | 1.60 | 1.85 | 0.05 | 0.68 | m | 2.53 |
| 2.5 mm2 | 2.61 | 3.02 | 0.05 | 0.68 | m | 3.70 |
| 4.0 mm2 | 2.99 | 3.45 | 0.06 | 0.82 | m | 4.27 |
| 6.0 mm2 | 4.06 | 4.69 | 0.06 | 0.82 | m | 5.51 |
| 10.0 mm2 | 6.27 | 7.24 | 0.06 | 0.82 | m | 8.06 |
| 16.0 mm2 | 9.89 | 11.42 | 0.07 | 0.96 | m | 12.38 |
| 300/500 Volt Grade; Three Core with Earth Wire | | | | | | |
| 1.0 mm2 | 1.39 | 1.60 | 0.06 | 0.82 | m | 2.42 |
| 1.5 mm2 | 2.19 | 2.53 | 0.06 | 0.82 | m | 3.35 |

## Y:MECHANICAL AND ELECTRICAL SERVICES

| Item | Net Price £ | Material £ | Labour hours | Labour £ | Unit | Total Rate £ |
|---|---|---|---|---|---|---|
| **Single Core, PVC Insulated Cable; non-sheathed copper; BS 6004 Table 1a** | | | | | | |
| 450/700 Volt Grade; Single Strand - laid in trunking | | | | | | |
| 1.0 mm2 | 0.23 | 0.27 | 0.03 | 0.41 | m | **0.68** |
| 1.5 mm2 | 0.38 | 0.44 | 0.03 | 0.41 | m | **0.85** |
| 2.5 mm2 | 0.55 | 0.63 | 0.03 | 0.41 | m | **1.04** |
| 450/700 Volt Grade; Multi Strand - laid in trunking | | | | | | |
| 1.5 mm2 | 0.30 | 0.35 | 0.03 | 0.41 | m | **0.76** |
| 2.5 mm2 | 0.44 | 0.51 | 0.03 | 0.41 | m | **0.92** |
| 4.0 mm2 | 0.90 | 1.04 | 0.03 | 0.41 | m | **1.45** |
| 6.0 mm2 | 1.30 | 1.51 | 0.04 | 0.55 | m | **2.06** |
| 10.0 mm2 | 2.44 | 2.82 | 0.04 | 0.55 | m | **3.37** |
| 16.0 mm2 | 3.77 | 4.35 | 0.04 | 0.55 | m | **4.90** |
| 25.0 mm2 | 3.92 | 4.52 | 0.05 | 0.68 | m | **5.20** |
| 35.0 mm2 | 4.85 | 5.61 | 0.05 | 0.68 | m | **6.29** |
| 50.0 mm2 | 7.06 | 8.15 | 0.05 | 0.68 | m | **8.83** |
| 70.0 mm2 | 8.30 | 9.58 | 0.05 | 0.68 | m | **10.26** |
| 95.0 mm2 | 13.22 | 15.26 | 0.06 | 0.82 | m | **16.08** |
| 120.0 mm2 | 17.95 | 20.73 | 0.07 | 0.96 | m | **21.69** |
| 150.0 mm2 | 21.95 | 25.35 | 0.08 | 1.09 | m | **26.44** |
| 450/700 Volt Grade; Single Strand - drawn into conduit | | | | | | |
| 1.0 mm2 | 0.22 | 0.25 | 0.04 | 0.55 | m | **0.80** |
| 1.5 mm2 | 0.35 | 0.40 | 0.04 | 0.55 | m | **0.95** |
| 2.5 mm2 | 0.51 | 0.59 | 0.04 | 0.55 | m | **1.14** |
| 450/700 Volt Grade; Multi Strand - drawn into conduit | | | | | | |
| 1.5 mm2 | 0.35 | 0.40 | 0.03 | 0.41 | m | **0.81** |
| 2.5 mm2 | 0.51 | 0.58 | 0.03 | 0.41 | m | **0.99** |
| 4.0 mm2 | 0.83 | 0.96 | 0.04 | 0.55 | m | **1.51** |
| 6.0 mm2 | 1.21 | 1.40 | 0.04 | 0.55 | m | **1.95** |
| 10.0 mm2 | 2.27 | 2.62 | 0.05 | 0.68 | m | **3.30** |
| 16.0 mm2 | 3.50 | 4.04 | 0.05 | 0.68 | m | **4.72** |
| 25.0 mm2 | 3.95 | 4.57 | 0.05 | 0.68 | m | **5.25** |
| 35.0 mm2 | 4.92 | 5.68 | 0.07 | 0.96 | m | **6.64** |

## Y:MECHANICAL AND ELECTRICAL SERVICES

| Item | Net Price £ | Material £ | Labour hours | Labour £ | Unit | Total Rate £ |
|---|---|---|---|---|---|---|
| **Y61 : HV/LV CABLES AND WIRING : CABLE (contd)** | | | | | | |
| H.O.F.R. (Heat, Oil, Fire Retardent) instrument/Control cable; silicon rubber insulated copper conductor; aluminium/ hard grade L.S.O.H. laminate sheath; clipped to backgrounds; (Including supports fixings and connecting tails) BS 6387; IEC 331 Part 1. | | | | | | |
| 250/440 Volt Grade; with drain wire; Two Core | | | | | | |
| 1.0 mm2 | 1.24 | 1.43 | 0.07 | 0.96 | m | 2.39 |
| 1.5 mm2 | 1.50 | 1.73 | 0.07 | 0.96 | m | 2.69 |
| 2.5 mm2 | 1.92 | 2.22 | 0.07 | 0.96 | m | 3.18 |
| 4.0 mm2 | 3.03 | 3.50 | 0.07 | 0.96 | m | 4.46 |
| 250/440 Volt Grade; with drain wire; Three Core | | | | | | |
| 1.0 mm2 | 1.57 | 1.82 | 0.07 | 0.96 | m | 2.78 |
| 1.5 mm2 | 1.96 | 2.27 | 0.07 | 0.96 | m | 3.23 |
| 2.5 mm2 | 2.47 | 2.86 | 0.08 | 1.09 | m | 3.95 |
| 4.0 mm2 | 4.08 | 4.71 | 0.08 | 1.09 | m | 5.80 |
| 250/440 Volt Grade; with drain wire; Four Core | | | | | | |
| 1.0 mm2 | 1.91 | 2.21 | 0.08 | 1.09 | m | 3.30 |
| 1.5 mm2 | 2.39 | 2.76 | 0.08 | 1.09 | m | 3.85 |
| 2.5 mm2 | 3.37 | 3.90 | 0.09 | 1.23 | m | 5.13 |
| 4.0 mm2 | 5.30 | 6.12 | 0.09 | 1.23 | m | 7.35 |
| 250/440 Volt Grade; with drain wire; Seven Core | | | | | | |
| 1.0 mm2 | 2.81 | 3.25 | 0.09 | 1.23 | m | 4.48 |
| 1.5 mm2 | 3.47 | 4.01 | 0.09 | 1.23 | m | 5.24 |
| 2.5 mm2 | 4.55 | 5.26 | 0.09 | 1.23 | m | 6.49 |
| 250/440 Volt Grade; with drain wire; Twelve Core | | | | | | |
| 1.0 mm2 | 8.37 | 9.67 | 0.10 | 1.37 | m | 11.04 |
| 1.5 mm2 | 10.99 | 12.69 | 0.10 | 1.37 | m | 14.06 |
| 250/440 Volt Grade; with drain wire; Nineteen Core | | | | | | |
| 1.5 mm2 | 19.78 | 22.85 | 0.11 | 1.50 | m | 24.35 |
| 2.5 mm2 | 15.23 | 17.59 | 0.13 | 1.78 | m | 19.37 |
| **Cable Terminations for H.O.F.R. Cable (Including drilling and cutting mild steel gland plate)** | | | | | | |
| 250 Volt grade; with drain wire; gland; nylon compression seal; locknut. Two Core | | | | | | |
| 1.0 mm2 | 0.33 | 0.36 | 0.13 | 1.78 | m | 2.14 |
| 1.5 mm2 | 0.33 | 0.36 | 0.13 | 1.78 | m | 2.14 |
| 2.5 mm2 | 0.44 | 0.49 | 0.15 | 2.05 | m | 2.54 |
| 4.0 mm2 | 0.76 | 0.84 | 0.15 | 2.05 | m | 2.89 |

## Y:MECHANICAL AND ELECTRICAL SERVICES

| Item | Net Price £ | Material £ | Labour hours | Labour £ | Unit | Total Rate £ |
|---|---|---|---|---|---|---|
| 250 Volt grade; with drain wire; gland; nylon compression seal; locknut; shroud;i dentification ferrules; Three Core | | | | | | |
| 1.0 mm2 | 0.33 | 0.36 | 0.16 | 2.19 | m | 2.55 |
| 1.5 mm2 | 0.44 | 0.49 | 0.16 | 2.19 | m | 2.68 |
| 2.5 mm2 | 0.44 | 0.49 | 0.18 | 2.46 | m | 2.95 |
| 4.0 mm2 | 0.76 | 0.84 | 0.18 | 2.46 | m | 3.30 |
| 250 Volt grade; with drain wire; gland; nylon compression seal; locknut; shroud; identification ferrules; Four Core | | | | | | |
| 1.0 mm2 | 0.44 | 0.49 | 0.21 | 2.87 | m | 3.36 |
| 1.5 mm2 | 0.44 | 0.49 | 0.23 | 3.14 | m | 3.63 |
| 2.5 mm2 | 0.76 | 0.84 | 0.27 | 3.69 | m | 4.53 |
| 4.0 mm2 | 0.76 | 0.84 | 0.30 | 4.10 | m | 4.94 |
| 250 Volt grade; with drain wire; gland; nylon compression seal; locknut; shroud; identification ferrules; Seven Core | | | | | | |
| 1.0 mm2 | 0.44 | 0.49 | 0.35 | 4.78 | m | 5.27 |
| 1.5 mm2 | 0.76 | 0.84 | 0.35 | 4.78 | m | 5.62 |
| 2.5 mm2 | 0.76 | 0.84 | 0.40 | 5.47 | m | 6.31 |
| 250 Volt grade; with drain wire; gland; nylon compression seal; locknut; shroud; identification ferrules; Twelve Core | | | | | | |
| 1.5 mm2 | 1.05 | 1.15 | 0.55 | 7.52 | m | 8.67 |
| 2.5 mm2 | 0.76 | 0.84 | 0.60 | 8.20 | m | 9.04 |
| 250 Volt grade; with drain wire; gland; nylon compression seal; locknut; shroud; identification ferrules; Nineteen Core | | | | | | |
| 1.5 mm2 | 1.05 | 1.15 | 0.80 | 10.93 | m | 12.08 |
| 2.5 mm2 | 2.15 | 2.37 | 0.85 | 11.62 | m | 13.99 |
| 250/440 Volt grade; with drain wire; gland; brass A1/A2 weatherproof rated I.P.66 PVC Shroud; brass locknut, identification ferrules; Two Core | | | | | | |
| 1.0 mm2 | 3.72 | 4.09 | 0.13 | 1.78 | nr | 5.87 |
| 1.5 mm2 | 3.72 | 4.09 | 0.13 | 1.78 | nr | 5.87 |
| 250/440 Volt grade; with drain wire; gland; brass A1/A2 weatherproof rated I.P.66 PVC Shroud; brass locknut, identification ferrules; Three Core | | | | | | |
| 1.0 mm2 | 3.72 | 4.09 | 0.16 | 2.19 | nr | 6.28 |
| 1.5 mm2 | 3.72 | 4.09 | 0.16 | 2.19 | nr | 6.28 |

## Y:MECHANICAL AND ELECTRICAL SERVICES

| Item | Net Price £ | Material £ | Labour hours | Labour £ | Unit | Total Rate £ |
|---|---|---|---|---|---|---|
| **Y61 : HV/LV CABLES AND WIRING : CABLE (contd)** | | | | | | |
| **Cable Terminations for H.O.F.R. Cable (Including drilling and cutting mild steel gland plate) (contd)** | | | | | | |
| 250/440 Volt grade; with drain wire; gland; brass A1/A2 weatherporrf rated I.P.66 PVC Shroud; brass locknut, identification ferrules; | | | | | | |
| Four Core | | | | | | |
| 1.0 mm2 | 3.79 | 4.16 | 0.23 | 3.14 | nr | **7.30** |
| 1.5 mm2 | 3.76 | 4.14 | 0.23 | 3.14 | nr | **7.28** |
| 250/440 Volt grade; with drain wire; gland; brass A1/A2 weatherproof rated I.P.66 PVC Shroud; brass locknut, identification ferrules; | | | | | | |
| Seven Core | | | | | | |
| 1.0 mm2 | 3.79 | 4.17 | 0.35 | 4.78 | nr | **8.95** |
| 1.5 mm2 | 3.79 | 4.17 | 0.35 | 4.78 | nr | **8.95** |
| 250/440 Volt grade; with drain wire; gland; brass A1/A2 weatherproof rated I.P.66 PVC Shroud; brass locknut, identification ferrules; | | | | | | |
| Twelve Core | | | | | | |
| 1.5 mm2 | 4.18 | 4.60 | 0.53 | 7.25 | nr | **11.85** |
| 250/440 Volt grade; with drain wire; gland; brass A1/A2 weatherproof rated I.P.66 PVC Shroud; brass locknut, identification ferrules; | | | | | | |
| Nineteen Core | | | | | | |
| 1.5 mm2 | 6.73 | 7.41 | 0.83 | 11.35 | nr | **18.76** |

**Material Costs / Measured Work Prices - Electrical Installations**

## Y:MECHANICAL AND ELECTRICAL SERVICES

| Item | Net Price £ | Material £ | Labour hours | Labour £ | Unit | Total Rate £ |
|---|---|---|---|---|---|---|
| **Y61 : HV/LV CABLES AND WIRING : FLEXIBLECORDS** | | | | | | |
| PVC Insulated PVC Sheathed, Flexible cord; heat resistant; copper conductor; BS 6141 (loose laid). | | | | | | |
| 300/500 Volt Grade; Two Core | | | | | | |
| 0.50 mm2 | 0.80 | 0.92 | 0.07 | 0.96 | m | **1.88** |
| 0.75 mm2 | 0.97 | 1.12 | 0.07 | 0.96 | m | **2.08** |
| 300/500 Volt Grade; Three Core | | | | | | |
| 0.50 mm2 | 1.08 | 1.25 | 0.07 | 0.96 | m | **2.21** |
| 0.75 mm2 | 1.15 | 1.33 | 0.07 | 0.96 | m | **2.29** |
| 1.00 mm2 | 1.56 | 1.80 | 0.07 | 0.96 | m | **2.76** |
| 1.50 mm2 | 2.18 | 2.52 | 0.08 | 1.09 | m | **3.61** |
| 2.50 mm2 | 3.27 | 3.78 | 0.08 | 1.09 | m | **4.87** |
| VR Insulated, Tough Rubber Sheathed Cable; copper conductor; BS 6500 Table 6 (loose laid). | | | | | | |
| 300/500 Volt Grade; Two Core | | | | | | |
| 0.50 mm2 | 1.30 | 1.50 | 0.07 | 0.96 | m | **2.46** |
| 0.75 mm2 | 0.91 | 1.05 | 0.07 | 0.96 | m | **2.01** |
| 1.00 mm2 | 1.12 | 1.30 | 0.07 | 0.96 | m | **2.26** |
| 1.50 mm2 | 1.56 | 1.80 | 0.07 | 0.96 | m | **2.76** |
| 2.50 mm2 | 2.29 | 2.64 | 0.07 | 0.96 | m | **3.60** |
| 300/500 Volt Grade; Three Core | | | | | | |
| 0.50 mm2 | 1.60 | 1.85 | 0.07 | 0.96 | m | **2.81** |
| 0.75 mm2 | 1.16 | 1.34 | 0.07 | 0.96 | m | **2.30** |
| 1.00 mm2 | 1.36 | 1.57 | 0.07 | 0.96 | m | **2.53** |
| 1.50 mm2 | 1.88 | 2.17 | 0.07 | 0.96 | m | **3.13** |
| 2.50 mm2 | 2.78 | 3.21 | 0.07 | 0.96 | m | **4.17** |
| 300/500 Volt Grade; Four Core | | | | | | |
| 0.75 mm2 | 1.69 | 1.95 | 0.08 | 1.09 | m | **3.04** |
| 1.00 mm2 | 2.00 | 2.31 | 0.08 | 1.09 | m | **3.40** |
| 1.50 mm2 | 2.51 | 2.90 | 0.08 | 1.09 | m | **3.99** |
| 2.50 mm2 | 3.74 | 4.32 | 0.08 | 1.09 | m | **5.41** |
| Rubber Insulated H.O.F.R. Sheathed, Circular flexible Cords; copper conductor; BS 6500 Table 9 (loose laid). | | | | | | |
| 300/500 Volt Grade; Two Core | | | | | | |
| 0.50 mm2 | 1.94 | 2.24 | 0.07 | 0.96 | m | **3.20** |
| 0.75 mm2 | 1.61 | 1.86 | 0.07 | 0.96 | m | **2.82** |
| 1.00 mm2 | 2.27 | 2.62 | 0.07 | 0.96 | m | **3.58** |
| 1.50 mm2 | 2.53 | 2.92 | 0.07 | 0.96 | m | **3.88** |
| 2.50 mm2 | 3.89 | 4.49 | 0.08 | 1.09 | m | **5.58** |
| 4.00 mm2 | 5.29 | 6.11 | 0.09 | 1.23 | m | **7.34** |

## Y:MECHANICAL AND ELECTRICAL SERVICES

| Item | Net Price £ | Material £ | Labour hours | Labour £ | Unit | Total Rate £ |
|---|---|---|---|---|---|---|
| **Y61 : HV/LV CABLES AND WIRING : FLEXIBLECORDS (contd)** | | | | | | |
| Rubber Insulated H.O.F.R. Sheathed, Circular flexible Cords; copper conductor; BS 6500 Table 9 (loose laid). (contd) | | | | | | |
| 300/500 Volt Grade; Three Core | | | | | | |
| 0.50 mm2 | 1.70 | 1.97 | 0.07 | 0.96 | m | 2.93 |
| 0.75 mm2 | 1.87 | 2.16 | 0.07 | 0.96 | m | 3.12 |
| 1.00 mm2 | 2.21 | 2.56 | 0.07 | 0.96 | m | 3.52 |
| 1.50 mm2 | 2.73 | 3.15 | 0.08 | 1.09 | m | 4.24 |
| 2.50 mm2 | 3.91 | 4.52 | 0.08 | 1.09 | m | 5.61 |
| 4.00 mm2 | 6.88 | 7.94 | 0.10 | 1.37 | m | 9.31 |
| 300/500 Volt Grade; Four Core | | | | | | |
| 0.75 mm2 | 3.31 | 3.82 | 0.07 | 0.96 | m | 4.78 |
| 1.00 mm2 | 3.85 | 4.44 | 0.07 | 0.96 | m | 5.40 |
| 1.50 mm2 | 4.89 | 5.64 | 0.08 | 1.09 | m | 6.73 |
| 2.50 mm2 | 6.27 | 7.25 | 0.08 | 1.09 | m | 8.34 |
| 4.00 mm2 | 15.25 | 17.61 | 0.10 | 1.37 | m | 18.98 |
| PVC Insulated PVC Sheathed, Light flexible Cords; copper conductor; BS 6500 Table 15 (loose laid). | | | | | | |
| 300/300 Volt Grade; Two Core | | | | | | |
| 0.50 mm2 | 0.33 | 0.39 | 0.07 | 0.96 | m | 1.35 |
| 0.75 mm2 | 0.52 | 0.60 | 0.07 | 0.96 | m | 1.56 |
| 300/500 Volt Grade; Three Core | | | | | | |
| 0.50 mm2 | 0.45 | 0.52 | 0.07 | 0.96 | m | 1.48 |
| 0.75 mm2 | 0.71 | 0.82 | 0.07 | 0.96 | m | 1.78 |
| 300/500 Volt Grade; Parallel Twin | | | | | | |
| 0.50 mm2 | 0.43 | 0.50 | 0.07 | 0.96 | m | 1.46 |
| 0.75 mm2 | 0.53 | 0.61 | 0.07 | 0.96 | m | 1.57 |
| PVC Insulated, Non sheathed, Parallel Twin Flexible Cord; figure 8 type; copper conductor; BS 6500 Table 15 (loose laid). | | | | | | |
| 300/300 Volt Grade; Two Core | | | | | | |
| 0.50 mm2 | 0.12 | 0.14 | 0.07 | 0.96 | m | 1.10 |
| 0.75 mm2 | 0.20 | 0.23 | 0.07 | 0.96 | m | 1.19 |
| PVC insulated, PVC Sheathed Circular flexible cord copper conductor; BS 6500 Table 16 (loose laid). | | | | | | |
| 300/500 Volt Grade; Two Core | | | | | | |
| 0.75 mm2 | 0.53 | 0.61 | 0.07 | 0.96 | m | 1.57 |
| 1.00 mm2 | 0.68 | 0.79 | 0.07 | 0.96 | m | 1.75 |
| 1.50 mm2 | 0.97 | 1.12 | 0.07 | 0.96 | m | 2.08 |

# Heating and Water Services Design in Buildings

**Keith J. Moss**, Consultant Engineer and Visiting Lecturer to City of Bath College and the University of Bath, UK

This book gives comprehensive coverage of the design of heating and water services in buildings. Each chapter starts with the information needed to understand the specific area, which is then reinforced by many examples and case studies with worked solutions. Mathematics and fluids are introduced as core skills where they are required as part of the design solution.

- **provides a thorough understanding of design procedures**

- **does not require a detailed knowledge of mathematical theory**

### Contents
Heat requirements of heated buildings in temperate climates. Low-temperature hot water heating systems. Pump and system. High temperature hot water systems. Steam systems. Plant connections and controls. The application of probability and demand units in design. Hot and cold water supply systems utilising the static head. Hot and cold water supply systems using booster pumps. Loose ends. Sources of information. Index.

*E & F N Spon*
*246x189: 264pp :98 line illus: May 1996*
*Paperback: 0-419-20110-6: £19.99*

# Marketing for Architects and Engineers
*A new approach*
B. Richardson, Director, Northern Architecture Centre Ltd,
Newcastle-upon-Tyne, UK
**Foreword by Francis Duffy**

Professional services marketing is a relatively new form of marketing that has been recogonised only since the late 1980s. Most of the attempts to write about marketing for professional services have been a regurgitation of the traditional marketing approach that has evolved since the 1960s and have concentrated on minor differences and adjustments. In many ways, what is needed is a fresh approach which takes into account the complex political, social, economic, legislative and cultural backdrop and provides a way for design professionals, such as architects and engineers, to look to the future. This book does just that.

- offers a way for architects and engineers to take charge of their own future

- provides a comprehensive, forward looking marketing discipline for design professionals

- introduces the new concepts of synthesis marketing and strategic mapping

**Contents:**

Introduction. Markets and marketing. Scenario planning. Synthesis marketing. Strategic mapping. A synthesis marketing programme. Architecture centres. Bibliography and references. Index.

**For further information and to order please contact**
The Marketing Dept., E & FN Spon, 2-6 Boundary Row, London SE1 8HN
Tel: 0171 865 0066  Fax: 0171 522 9621

**E & F N Spon**
**234x156: 152pp :25 line illus: November 1996**
**Paperback:0-419-20290-0: £17.50**

## Y:MECHANICAL AND ELECTRICAL SERVICES

| Item | Net Price £ | Material £ | Labour hours | Labour £ | Unit | Total Rate £ |
|---|---|---|---|---|---|---|
| 2.50 mm2 | 2.06 | 2.39 | 0.07 | 0.96 | m | 3.35 |
| **PVC insulated, PVC Sheathed Circular flexible cord copper conductor; BS 6500 Table 16 (loose laid).** | | | | | | |
| 300/500 Volt Grade; Three Core | | | | | | |
| 0.75 mm2 | 0.51 | 0.59 | 0.07 | 0.96 | m | **1.55** |
| 1.00 mm2 | 0.62 | 0.72 | 0.07 | 0.96 | m | **1.68** |
| 1.50 mm2 | 0.87 | 1.00 | 0.07 | 0.96 | m | **1.96** |
| 2.50 mm2 | 1.77 | 2.05 | 0.07 | 0.96 | m | **3.01** |
| 300/500 Volt Grade; Four Core | | | | | | |
| 0.75 mm2 | 1.43 | 1.65 | 0.07 | 0.96 | m | **2.61** |
| 1.00 mm2 | 1.67 | 1.93 | 0.07 | 0.96 | m | **2.89** |
| 1.50 mm2 | 2.38 | 2.75 | 0.07 | 0.96 | m | **3.71** |
| 2.50 mm2 | 3.68 | 4.25 | 0.07 | 0.96 | m | **5.21** |
| 300/500 Volt Grade; Five Core | | | | | | |
| 1.00 mm2 | 3.15 | 3.64 | 0.08 | 1.09 | m | **4.73** |
| 1.50 mm2 | 4.41 | 5.10 | 0.08 | 1.09 | m | **6.19** |
| 2.50 mm2 | 6.95 | 8.02 | 0.08 | 1.09 | m | **9.11** |
| 300/500 Volt Grade; Parallel Twin | | | | | | |
| 0.75 mm2 | 0.56 | 0.64 | 0.07 | 0.96 | m | **1.60** |
| **PVC Insulated Flexible Cord; copper conductor; BS 6500 Table 19 and BS 6004 Table 1c (loose laid).** | | | | | | |
| 300/500 Volt Grade; Single Core | | | | | | |
| 0.50 mm2 | 0.04 | 0.05 | 0.07 | 0.96 | m | **1.01** |
| 0.75 mm2 | 0.05 | 0.05 | 0.07 | 0.96 | m | **1.01** |
| 1.00 mm2 | 0.05 | 0.06 | 0.07 | 0.96 | m | **1.02** |
| 1.50 mm2 | 0.08 | 0.09 | 0.08 | 1.09 | m | **1.18** |
| 2.50 mm2 | 0.15 | 0.17 | 0.08 | 1.09 | m | **1.26** |
| **V.R. Insulated, Tough Rubber Sheathed Flexible cord cooper conductor; screened; PVC sheathed; BS 6500 Table 7 (loose laid).** | | | | | | |
| 300/500 Volt Grade; Three Core | | | | | | |
| 0.75 mm2 | 8.37 | 9.67 | 0.07 | 0.96 | m | **10.63** |
| 1.00 mm2 | 11.62 | 13.42 | 0.07 | 0.96 | m | **14.38** |
| 1.50 mm2 | 10.22 | 11.80 | 0.08 | 1.09 | m | **12.89** |
| 2.50 mm2 | 13.61 | 15.72 | 0.08 | 1.09 | m | **16.81** |
| 300/600 Volt Grade; Four Core | | | | | | |
| 0.75 mm2 | 12.44 | 14.37 | 0.07 | 0.96 | m | **15.33** |
| 1.00 mm2 | 13.16 | 15.20 | 0.08 | 1.09 | m | **16.29** |
| 1.50 mm2 | 11.54 | 13.33 | 0.08 | 1.09 | m | **14.42** |
| 2.50 mm2 | 14.41 | 16.64 | 0.08 | 1.09 | m | **17.73** |

## Y:MECHANICAL AND ELECTRICAL SERVICES

| Item | Net Price £ | Material £ | Labour hours | Labour £ | Unit | Total Rate £ |
|---|---|---|---|---|---|---|
| **Y62 : BUSBAR TRUNKING : BUSBARS** | | | | | | |
| Rising Main Busbar System; fixed to backgrounds including supports, fixings and connections/jointing to equipment. (Provision and fixing of plates, discs etc, for identification is included.) | | | | | | |
| 380 Amp TP&N | | | | | | |
| Copper busbar enclosed in metal trunking with insulated supports, earth continuity bar fixed to outside of trunking, single hole joints and hangers. | 59.40 | 65.34 | 6.02 | 82.34 | m | 147.68 |
| Extra for trunking fittings | | | | | | |
| End feed unit | 136.32 | 149.95 | 4.00 | 54.67 | nr | 204.62 |
| Supplement for elbow | 107.11 | 117.82 | 1.00 | 13.67 | nr | 131.49 |
| End cap | 14.61 | 16.07 | 1.00 | 13.67 | nr | 29.74 |
| Fire Resistant Barriers | 37.00 | 40.70 | 1.00 | 13.67 | nr | 54.37 |
| Direct Unswitched Tap-Off unit excluding HRC fuses | | | | | | |
| 32 Amp | 61.35 | 67.48 | 3.00 | 41.05 | nr | 108.53 |
| 63 Amp | 67.19 | 73.91 | 3.00 | 41.05 | nr | 114.96 |
| 100 Amp | 128.53 | 141.38 | 3.00 | 41.05 | nr | 182.43 |
| 125 Amp | 160.67 | 176.74 | 3.00 | 41.05 | nr | 217.79 |
| 250 Amp | 391.45 | 430.60 | 3.00 | 41.05 | nr | 471.65 |
| Direct switch fuse Tap-Off excluding HRC fuses | | | | | | |
| 32 Amp | 90.56 | 99.62 | 3.00 | 41.05 | nr | 140.67 |
| 63 Amp | 111.98 | 123.18 | 3.00 | 41.05 | nr | 164.23 |
| 100 Amp | 154.83 | 170.31 | 3.00 | 41.05 | nr | 211.36 |
| Direct fuse switch excluding HRC fuses | | | | | | |
| 200 Amp | 418.71 | 460.58 | 3.00 | 41.05 | nr | 501.63 |
| 315 Amp | 655.33 | 720.86 | 3.00 | 41.05 | nr | 761.91 |
| 400 Amp | 741.02 | 815.12 | 3.00 | 41.05 | nr | 856.17 |
| 440 Amp TP&N | | | | | | |
| Copper busbar enclosed in metal trunking with insulated supports, earth continuity bar fixed to outside of trunking, single hole joints and hangers | 70.11 | 77.12 | 6.02 | 82.34 | m | 159.46 |
| Extra for trunking fittings | | | | | | |
| End feed unit | 166.51 | 183.16 | 4.00 | 54.67 | nr | 237.83 |
| Centre feed unit | 130.48 | 143.53 | 6.02 | 82.34 | nr | 225.87 |
| End cap | 15.58 | 17.14 | 1.00 | 13.67 | nr | 30.81 |
| Fire Resistant Barrier | 37.00 | 40.70 | 1.00 | 13.67 | nr | 54.37 |

## Y:MECHANICAL AND ELECTRICAL SERVICES

| Item | Net Price £ | Material £ | Labour hours | Labour £ | Unit | Total Rate £ |
|---|---|---|---|---|---|---|
| Direct unswitched tap-off unit excluding HRC fuses | | | | | | |
| 32 Amp | 61.35 | 67.48 | 3.00 | 41.05 | nr | 108.53 |
| 63 Amp | 67.19 | 73.91 | 3.00 | 41.05 | nr | 114.96 |
| 100 Amp | 128.53 | 141.38 | 3.00 | 41.05 | nr | 182.43 |
| 125 Amp | 160.67 | 176.74 | 3.00 | 41.05 | nr | 217.79 |
| 250 Amp | 391.45 | 430.60 | 3.00 | 41.05 | nr | 471.65 |
| Direct switch fuse excluding HRC fuses | | | | | | |
| 32 Amp | 90.56 | 99.62 | 3.00 | 41.05 | nr | 140.67 |
| 60 Amp | 111.98 | 123.18 | 3.00 | 41.05 | nr | 164.23 |
| 100 Amp | 154.83 | 170.31 | 3.00 | 41.05 | nr | 211.36 |
| Direct fuse switch excluding HRC fuses | | | | | | |
| 200 Amp | 418.71 | 460.58 | 3.00 | 41.05 | nr | 501.63 |
| 315 Amp | 655.33 | 720.86 | 3.00 | 41.05 | nr | 761.91 |
| 400 Amp | 741.02 | 815.12 | 3.00 | 41.05 | nr | 856.17 |
| 750 Amp TP&N | | | | | | |
| Copper busbar enclosed in metal trunking with insulated supports, earth continuity bar fixed to outside of trunking, single hole joints and hangers | 107.11 | 117.82 | 6.02 | 82.34 | m | 200.16 |
| Extra for trunking fittings | | | | | | |
| End feed unit | 237.59 | 261.35 | 4.00 | 54.67 | nr | 316.02 |
| Centre feed unit | 197.67 | 217.44 | 6.02 | 82.34 | nr | 299.78 |
| End cap | 19.48 | 21.43 | 1.00 | 13.67 | nr | 35.10 |
| Fire Resistant Barrier | 48.69 | 53.56 | 1.00 | 13.67 | nr | 67.23 |
| Direct tap-off unit including HRC fuses | | | | | | |
| 32 Amp | 61.35 | 67.48 | 3.00 | 41.05 | nr | 108.53 |
| 63 Amp | 67.19 | 73.91 | 3.00 | 41.05 | nr | 114.96 |
| 100 Amp | 128.53 | 141.38 | 3.00 | 41.05 | nr | 182.43 |
| 125 Amp | 160.67 | 176.74 | 3.00 | 41.05 | nr | 217.79 |
| 250 Amp | 391.45 | 430.60 | 3.00 | 41.05 | nr | 471.65 |
| Direct switch fused including HRC fuses | | | | | | |
| 32 Amp | 90.56 | 99.62 | 3.00 | 41.05 | nr | 140.67 |
| 63 Amp | 111.98 | 123.18 | 3.00 | 41.05 | nr | 164.23 |
| 100 Amp | 154.83 | 170.31 | 3.00 | 41.05 | nr | 211.36 |
| Direct fuse switch including HRC fuses | | | | | | |
| 200 Amp | 418.71 | 460.58 | 3.00 | 41.05 | nr | 501.63 |
| 315 Amp | 655.33 | 720.86 | 3.00 | 41.05 | nr | 761.91 |
| 400 Amp | 741.02 | 815.12 | 3.00 | 41.05 | nr | 856.17 |
| 1070 Amp TP&N | | | | | | |
| Copper busbar enclosed in metal trunking with insulated supports, earth continuity bar fixed to outside of trunking, single hole joints and hangers | 152.88 | 168.17 | 8.00 | 109.34 | m | 277.51 |

## Y:MECHANICAL AND ELECTRICAL SERVICES

| Item | Net Price £ | Material £ | Labour hours | Labour £ | Unit | Total Rate £ |
|---|---|---|---|---|---|---|
| **Y62 : BUSBAR TRUNKING : BUSBARS (contd)** | | | | | | |
| **Rising Main Busbar System; fixed to backgrounds including supports, fixings and connections/jointing to equipment. (Provision and fixing of plates, discs etc, for identification is included.) (Contd)** | | | | | | |
| Extra for trunking fittings | | | | | | |
| End feed unit | 345.68 | 380.25 | 4.00 | 54.67 | nr | **434.92** |
| Centre feed unit | 289.20 | 318.12 | 6.02 | 82.34 | nr | **400.46** |
| End cap | 24.34 | 26.77 | 1.00 | 13.67 | nr | **40.44** |
| Fire Resistant Barrier | 58.42 | 64.26 | 1.00 | 13.67 | nr | **77.93** |
| Direct tap-off unit excluding HRC fuses | | | | | | |
| 32 Amp | 61.35 | 67.48 | 3.00 | 41.05 | nr | **108.53** |
| 100 Amp | 128.53 | 141.38 | 3.00 | 41.05 | nr | **182.43** |
| 125 Amp | 160.67 | 176.74 | 3.00 | 41.05 | nr | **217.79** |
| 250 Amp | 391.45 | 430.60 | 3.00 | 41.05 | nr | **471.65** |
| Direct fused switch excluding HRC fuses | | | | | | |
| 32 Amp | 90.56 | 99.62 | 3.00 | 41.05 | nr | **140.67** |
| 63 Amp | 111.98 | 123.18 | 3.00 | 41.05 | nr | **164.23** |
| 100 Amp | 154.83 | 170.31 | 3.00 | 41.05 | nr | **211.36** |
| Direct fuse switch excluding HRC fuses | | | | | | |
| 200 Amp | 418.71 | 460.58 | 3.00 | 41.05 | nr | **501.63** |
| 315 Amp | 655.33 | 720.86 | 3.00 | 41.05 | nr | **761.91** |
| 400 Amp | 741.02 | 815.12 | 3.00 | 41.05 | nr | **856.17** |
| **Rising Main Busbar System; fixed to backgrounds including supports, fixings and connections/jointing to equipment. (Provision and fixing of plates, discs etc, for identification is included.)** | | | | | | |
| Sheet steel case enclosing 4 pole 550 Volt copper bars, detachable metal end plates; 600mm long | | | | | | |
| 200 Amp | 217.50 | 239.25 | 3.51 | 47.96 | nr | **287.21** |
| 300 Amp | 217.27 | 239.00 | 4.00 | 54.67 | nr | **293.67** |
| 500 Amp | 317.92 | 349.71 | 5.00 | 68.34 | nr | **418.05** |
| Sheet steel case enclosing 4 pole 550 Volt copper bars, detachable metal end plates; 900mm long | | | | | | |
| 200 Amp | 313.07 | 344.38 | 3.76 | 51.38 | nr | **395.76** |
| 300 Amp | 369.25 | 406.18 | 4.26 | 58.16 | nr | **464.34** |
| 500 Amp | 427.53 | 470.28 | 5.26 | 71.94 | nr | **542.22** |

## Y:MECHANICAL AND ELECTRICAL SERVICES

| Item | Net Price £ | Material £ | Labour hours | Labour £ | Unit | Total Rate £ |
|---|---|---|---|---|---|---|
| Sheet steel case enclosing 4 pole 550 Volt copper bars, detachable metal end plates; 1350mm long | | | | | | |
| 200 Amp | 427.93 | 470.72 | 4.00 | 54.67 | nr | **525.39** |
| 300 Amp | 503.69 | 554.06 | 4.50 | 61.57 | nr | **615.63** |
| 500 Amp | 563.95 | 620.35 | 5.52 | 75.51 | nr | **695.86** |
| **Pre-Wired Busbar, Plug-In Trunking for Lighting; Galvanised Sheet Steel Housing (PE); Tin-Plated Copper Conducters with Tap-Off units at 1m intervals.** | | | | | | |
| Straight Lengths - 25 Amp | | | | | | |
| 2 Pole & PE | 8.72 | 9.60 | 0.05 | 0.68 | m | **10.28** |
| 4 Pole & PE | 10.01 | 11.01 | 0.05 | 0.68 | m | **11.69** |
| Straight Lengths - 40 Amp | | | | | | |
| 2 Pole & PE | 9.94 | 10.94 | 0.05 | 0.68 | m | **11.62** |
| 4 Pole & PE | 12.77 | 14.05 | 0.05 | 0.68 | m | **14.73** |
| **Components for Pre-Wired Busbars, Plug-In Trunking for Lighting.** | | | | | | |
| Plug-In Tap-Off Units | | | | | | |
| 10 Amp with phase selection, 2P & PE; 2m of cable | 6.78 | 7.46 | 0.03 | 0.46 | | **7.92** |
| 10 Amp 4 Pole & PE; 3m of cable | 10.03 | 10.48 | 0.03 | 0.46 | | **10.94** |
| 16 Amp 4 Pole & PE; 3m of cable | 9.68 | 10.10 | 0.03 | 0.46 | | **10.58** |
| 16 Amp with phase selection, 2P & PE; no cable | 8.41 | 8.78 | 0.03 | 0.46 | | **9.24** |
| Trunking Components | | | | | | |
| End Feed Unit & Cover; 4P & PE | 16.68 | 18.35 | 0.17 | 2.32 | nr | **20.67** |
| Centre Feed Unit | 92.17 | 101.39 | 0.17 | 2.32 | nr | **103.71** |
| Right hand, Intermediate Terminal Box Feed unit | 17.40 | 19.14 | 0.17 | 2.32 | nr | **21.46** |
| End Cover (for R/hand feed) | 3.15 | 3.46 | 0.03 | 0.46 | nr | **3.92** |
| Flexible Elbow Unit | 43.67 | 48.04 | 0.05 | 0.68 | nr | **48.72** |
| Fixing Bracket - Universal | 1.02 | 1.12 | 0.10 | 1.37 | nr | **2.49** |
| Suspension Bracket - Flat | 0.71 | 0.78 | 0.10 | 1.37 | nr | **2.15** |

# Keep your figures up to date, free of charge

This section, and most of the other information in this Price Book, is brought up to date every three months, until the next annual edition, in the *Price Book Update*.

The *Update* is available free to all Price Book purchasers.

To ensure you receive your copy, simply complete the reply card from the centre of the book and return it to us.

## Y:MECHANICAL AND ELECTRICAL SERVICES

| Item | Net Price £ | Material £ | Labour hours | Labour £ | Unit | Total Rate £ |
|---|---|---|---|---|---|---|
| **Y63 : SUPPORT COMPONENTS - CABLES : CABLE TRAY** | | | | | | |
| Galvanised Steel Cable Tray to BS 729; including standard coupling joints, fixings and earth continuity straps. (Supports and hangers are excluded.) | | | | | | |
| Tray | | | | | | |
| 75 mm | 4.99 | 5.49 | 0.22 | 3.01 | m | 8.50 |
| 100 mm | 6.10 | 6.71 | 0.32 | 4.37 | m | 11.08 |
| 150 mm | 7.33 | 8.06 | 0.32 | 4.37 | m | 12.43 |
| 225 mm | 9.47 | 10.42 | 0.38 | 5.20 | m | 15.62 |
| 300 mm | 14.04 | 15.44 | 0.50 | 6.83 | m | 22.27 |
| 450 mm | 21.53 | 23.68 | 0.57 | 7.79 | m | 31.47 |
| 600 mm | 27.17 | 29.88 | 0.80 | 10.93 | m | 40.81 |
| Tray with return flange | | | | | | |
| 75 mm | 9.68 | 10.65 | 0.35 | 4.78 | m | 15.43 |
| 100 mm | 10.14 | 11.15 | 0.35 | 4.78 | m | 15.93 |
| 150 mm | 12.30 | 13.53 | 0.35 | 4.78 | m | 18.31 |
| 225 mm | 15.65 | 17.21 | 0.40 | 5.47 | m | 22.68 |
| 300 mm | 17.94 | 19.74 | 0.55 | 7.52 | m | 27.26 |
| 450 mm | 28.04 | 30.84 | 0.60 | 8.20 | m | 39.04 |
| 600 mm | 35.25 | 38.77 | 0.85 | 11.62 | m | 50.39 |
| 750 mm | 43.92 | 48.32 | 0.90 | 12.30 | m | 60.62 |
| 900 mm | 51.57 | 56.72 | 1.00 | 13.67 | m | 70.39 |
| Tray with return flange with epoxy bonding coat | | | | | | |
| 75 mm | 18.91 | 20.80 | 0.37 | 5.06 | m | 25.86 |
| 100 mm | 21.09 | 23.20 | 0.37 | 5.06 | m | 28.26 |
| 150 mm | 25.31 | 27.84 | 0.38 | 5.20 | m | 33.04 |
| 225 mm | 31.09 | 34.20 | 0.42 | 5.74 | m | 39.94 |
| 300 mm | 37.42 | 41.17 | 0.57 | 7.79 | m | 48.96 |
| 450 mm | 58.21 | 64.03 | 0.62 | 8.48 | m | 72.51 |
| 600 mm | 71.77 | 78.95 | 0.88 | 12.03 | m | 90.98 |
| 750 mm | 89.85 | 98.83 | 0.93 | 12.71 | m | 111.54 |
| 900 mm | 108.02 | 118.83 | 1.03 | 14.09 | m | 132.92 |
| Galvanised Steel Tray and Fittings to BS 729; (Cutting and jointing tray to fittings is included) | | | | | | |
| Bend | | | | | | |
| 75 mm | 3.93 | 4.32 | 0.20 | 2.73 | nr | 7.05 |
| 100 mm | 4.13 | 4.54 | 0.22 | 3.01 | nr | 7.55 |
| 150 mm | 5.28 | 5.81 | 0.22 | 3.01 | nr | 8.82 |
| 225 mm | 7.25 | 7.97 | 0.28 | 3.83 | nr | 11.80 |
| 300 mm | 10.35 | 11.38 | 0.32 | 4.37 | nr | 15.75 |
| 450 mm | 17.82 | 19.60 | 0.53 | 7.25 | nr | 26.85 |
| 600 mm | 26.14 | 28.75 | 0.67 | 9.16 | nr | 37.91 |

## Y:MECHANICAL AND ELECTRICAL SERVICES

| Item | Net Price £ | Material £ | Labour hours | Labour £ | Unit | Total Rate £ |
|---|---|---|---|---|---|---|
| Bend; with return flange | | | | | | |
| 75 mm | 24.36 | 26.80 | 0.22 | 3.01 | nr | 29.81 |
| 100 mm | 27.69 | 30.46 | 0.22 | 3.01 | nr | 33.47 |
| 150 mm | 29.46 | 32.41 | 0.25 | 3.42 | nr | 35.83 |
| 225 mm | 35.04 | 38.54 | 0.33 | 4.51 | nr | 43.05 |
| 300 mm | 42.54 | 46.79 | 0.35 | 4.78 | nr | 51.57 |
| 450 mm | 65.95 | 72.55 | 0.55 | 7.52 | nr | 80.07 |
| 600 mm | 92.67 | 101.94 | 0.70 | 9.57 | nr | 111.51 |
| 750 mm | 99.48 | 109.43 | 1.00 | 13.67 | nr | 123.10 |
| 900 mm | 150.90 | 165.99 | 1.00 | 13.67 | nr | 179.66 |
| Bend; with return flange with epoxy bonding coat | | | | | | |
| 75 mm | 48.15 | 52.97 | 0.24 | 3.28 | nr | 56.25 |
| 100 mm | 49.26 | 54.19 | 0.24 | 3.28 | nr | 57.47 |
| 150 mm | 54.98 | 60.48 | 0.27 | 3.69 | nr | 64.17 |
| 225 mm | 70.20 | 77.22 | 0.32 | 4.37 | nr | 81.59 |
| 300 mm | 83.21 | 91.53 | 0.37 | 5.06 | nr | 96.59 |
| 450 mm | 136.90 | 150.59 | 0.57 | 7.79 | nr | 158.38 |
| 600 mm | 181.81 | 200.00 | 0.70 | 9.57 | nr | 209.57 |
| 750 mm | 227.30 | 250.03 | 0.85 | 11.62 | nr | 261.65 |
| 900 mm | 308.30 | 339.13 | 1.02 | 13.95 | nr | 353.08 |
| Equal Tee | | | | | | |
| 50 mm | 5.35 | 5.88 | 0.30 | 4.10 | nr | 9.98 |
| 75 mm | 5.50 | 6.05 | 0.30 | 4.10 | nr | 10.15 |
| 100 mm | 5.76 | 6.34 | 0.32 | 4.37 | nr | 10.71 |
| 150 mm | 6.66 | 7.33 | 0.32 | 4.37 | nr | 11.70 |
| 225 mm | 10.46 | 4.60 | 1.00 | 13.67 | nr | 18.27 |
| 300 mm | 16.05 | 17.66 | 0.50 | 6.83 | nr | 24.49 |
| 450 mm | 27.21 | 29.93 | 0.77 | 10.53 | nr | 40.46 |
| 600 mm | 40.41 | 44.45 | 1.10 | 15.04 | nr | 59.49 |
| Equal Tee; with return flange | | | | | | |
| 75 mm | 28.51 | 31.36 | 0.32 | 4.37 | nr | 35.73 |
| 100 mm | 32.68 | 35.95 | 0.32 | 4.37 | nr | 40.32 |
| 150 mm | 34.82 | 38.30 | 0.35 | 4.78 | nr | 43.08 |
| 225 mm | 41.11 | 45.22 | 0.42 | 5.74 | nr | 50.96 |
| 300 mm | 46.10 | 50.71 | 0.53 | 7.25 | nr | 57.96 |
| 450 mm | 77.47 | 85.22 | 0.80 | 10.93 | nr | 96.15 |
| 600 mm | 108.13 | 118.94 | 1.13 | 15.44 | nr | 134.38 |
| 750 mm | 135.45 | 149.00 | 1.19 | 16.27 | nr | 165.27 |
| 900 mm | 159.21 | 175.13 | 1.50 | 20.52 | nr | 195.65 |
| Equal Tee; with return flange with epoxy bonding coat | | | | | | |
| 75 mm | 28.51 | 31.36 | 0.34 | 4.65 | nr | 36.01 |
| 100 mm | 64.76 | 71.24 | 0.34 | 4.65 | nr | 75.89 |
| 150 mm | 71.86 | 79.05 | 0.37 | 5.05 | nr | 84.10 |
| 225 mm | 93.26 | 102.59 | 0.47 | 6.43 | nr | 109.02 |
| 300 mm | 107.38 | 118.12 | 0.55 | 7.52 | nr | 125.64 |
| 450 mm | 174.08 | 191.49 | 0.88 | 12.03 | nr | 203.52 |
| 600 mm | 245.29 | 269.82 | 1.15 | 15.73 | nr | 285.55 |
| 750 mm | 306.64 | 337.30 | 1.21 | 16.55 | nr | 353.85 |
| 900 mm | 369.83 | 406.81 | 1.52 | 20.80 | nr | 427.61 |

Material Costs / Measured Work Prices - Electrical Installations

Y:MECHANICAL AND ELECTRICAL SERVICES

| Item | Net Price £ | Material £ | Labour hours | Labour £ | Unit | Total Rate £ |
|---|---|---|---|---|---|---|
| **Y63 : SUPPORT COMPONENTS - CABLES : CABLE TRAY (contd)** | | | | | | |
| **Galvanised Steel Tray and Fittings to BS 729; (Cutting and jointing tray to fitings is included) (contd)** | | | | | | |
| Reducer | | | | | | |
| 75 mm | 7.25 | 7.97 | 0.15 | 2.05 | nr | **10.02** |
| 100 mm | 7.42 | 8.16 | 0.15 | 2.05 | nr | **10.21** |
| 150 mm | 10.56 | 11.62 | 0.15 | 2.05 | nr | **13.67** |
| 225 mm | 14.07 | 15.48 | 0.23 | 3.14 | nr | **18.62** |
| 300 mm | 17.90 | 19.69 | 0.23 | 3.14 | nr | **22.83** |
| 450 mm | 28.90 | 31.79 | 0.37 | 5.06 | nr | **36.85** |
| 600 mm | 37.10 | 40.81 | 0.37 | 5.05 | nr | **45.86** |
| Reducer; with return flange | | | | | | |
| 100 mm | 17.70 | 19.47 | 0.17 | 2.32 | nr | **21.79** |
| 150 mm | 19.60 | 21.56 | 0.17 | 2.32 | nr | **23.88** |
| 225 mm | 22.69 | 24.96 | 0.25 | 3.42 | nr | **28.38** |
| 300 mm | 25.54 | 28.09 | 0.25 | 3.42 | nr | **31.51** |
| 450 mm | 40.40 | 44.44 | 0.40 | 5.47 | nr | **49.91** |
| 600 mm | 54.65 | 60.12 | 0.40 | 5.47 | nr | **65.59** |
| 750 mm | 77.23 | 84.95 | 0.61 | 8.34 | nr | **93.29** |
| 900 mm | 96.48 | 106.13 | 0.61 | 8.34 | nr | **114.47** |
| Reducer; with return flange with epoxy bonding coat | | | | | | |
| 100 mm | 41.24 | 45.36 | 0.19 | 2.60 | nr | **47.96** |
| 150 mm | 44.10 | 48.51 | 0.19 | 2.60 | nr | **51.11** |
| 225 mm | 53.60 | 58.96 | 0.27 | 3.69 | nr | **62.65** |
| 300 mm | 65.04 | 71.54 | 0.27 | 3.69 | nr | **75.23** |
| 450 mm | 97.32 | 107.05 | 0.42 | 5.74 | nr | **112.79** |
| 600 mm | 101.66 | 111.83 | 0.42 | 5.74 | nr | **117.57** |
| 750 mm | 106.83 | 117.51 | 0.63 | 8.61 | nr | **126.12** |
| 900 mm | 111.90 | 123.09 | 0.63 | 8.61 | nr | **131.70** |
| **Rigid PVC Cable Tray; including standard coupling joints and fixings. (Excluding supports or hangers.)** | | | | | | |
| Light Duty Tray | | | | | | |
| 50 mm | 3.52 | 4.07 | 0.40 | 5.47 | m | **9.54** |
| 75 mm | 3.90 | 4.51 | 0.40 | 5.47 | m | **9.98** |
| 100 mm | 4.36 | 5.03 | 0.40 | 5.47 | m | **10.50** |
| 150 mm | 5.33 | 6.15 | 0.45 | 6.15 | m | **12.30** |
| 200 mm | 9.51 | 10.98 | 0.45 | 6.15 | m | **17.13** |
| 300 mm | 11.60 | 13.40 | 0.45 | 6.15 | m | **19.55** |

*Material Costs / Measured Work Prices - Electrical Installations* 431

## Y:MECHANICAL AND ELECTRICAL SERVICES

| Item | Net Price £ | Material £ | Labour hours | Labour £ | Unit | Total Rate £ |
|---|---|---|---|---|---|---|
| **Medium Duty Tray** | | | | | | |
| 50 mm | 3.57 | 4.12 | 0.40 | 5.47 | m | 9.59 |
| 75 mm | 4.12 | 4.76 | 0.40 | 5.47 | m | 10.23 |
| 100 mm | 4.48 | 5.17 | 0.40 | 5.47 | m | 10.64 |
| 150 mm | 5.43 | 6.28 | 0.40 | 5.47 | m | 11.75 |
| **Medium Duty Cover** | | | | | | |
| 50 mm | 1.77 | 2.04 | 0.10 | 1.37 | m | 3.41 |
| 75 mm | 2.12 | 2.44 | 0.10 | 1.37 | m | 3.81 |
| 100 mm | 3.68 | 4.25 | 0.10 | 1.37 | m | 5.62 |
| 150 mm | 4.23 | 4.88 | 0.10 | 1.37 | m | 6.25 |
| **Heavy Duty Tray** | | | | | | |
| 100 mm | 5.85 | 6.76 | 0.40 | 5.47 | m | 12.23 |
| 150 mm | 6.04 | 6.97 | 0.40 | 5.47 | m | 12.44 |
| 200 mm | 9.74 | 11.25 | 0.45 | 6.15 | m | 17.40 |
| 300 mm | 10.68 | 12.34 | 0.45 | 6.15 | m | 18.49 |
| 400 mm | 17.03 | 19.67 | 0.55 | 7.52 | m | 27.19 |
| 600 mm | 40.33 | 46.58 | 0.55 | 7.52 | m | 54.10 |
| **Heavy Duty Cover** | | | | | | |
| 100 mm | 3.90 | 4.50 | 0.10 | 1.37 | m | 5.87 |
| 150 mm | 4.40 | 5.08 | 0.10 | 1.37 | m | 6.45 |
| 200 mm | 7.85 | 9.07 | 0.10 | 1.37 | m | 10.44 |
| 300 mm | 10.68 | 12.34 | 0.10 | 1.37 | m | 13.71 |
| 400 mm | 17.03 | 19.67 | 0.15 | 2.05 | m | 21.72 |
| 600 mm | 24.99 | 28.86 | 0.15 | 2.05 | m | 30.91 |
| **Rigid PVC Tray and Fittings; (Cutting and jointing to fittings included.)** | | | | | | |
| **Bend** | | | | | | |
| 50 mm | 10.31 | 11.34 | 0.40 | 5.47 | nr | 16.81 |
| 100 mm | 13.05 | 14.36 | 0.40 | 5.47 | nr | 19.83 |
| 150 mm | 15.71 | 17.28 | 0.40 | 5.47 | nr | 22.75 |
| 200 mm | 26.82 | 29.50 | 0.45 | 6.15 | nr | 35.65 |
| 300 mm | 36.01 | 39.61 | 0.45 | 6.15 | nr | 45.76 |
| 400 mm | 73.43 | 80.77 | 0.55 | 7.52 | nr | 88.29 |
| 600 mm | 100.53 | 110.58 | 0.55 | 7.52 | nr | 118.10 |
| **Bend Cover** | | | | | | |
| 50 mm | 7.89 | 8.68 | 0.10 | 1.37 | nr | 10.05 |
| 100 mm | 7.89 | 8.68 | 0.10 | 1.37 | nr | 10.05 |
| 150 mm | 11.01 | 12.11 | 0.10 | 1.37 | nr | 13.48 |
| 200 mm | 12.58 | 13.84 | 0.10 | 1.37 | nr | 15.21 |
| 300 mm | 21.22 | 23.34 | 0.10 | 1.37 | nr | 24.71 |
| 400 mm | 37.19 | 40.91 | 0.15 | 2.05 | nr | 42.96 |
| 600 mm | 47.32 | 52.05 | 0.15 | 2.05 | nr | 54.10 |

*Material Costs / Measured Work Prices - Electrical Installations*

## Y:MECHANICAL AND ELECTRICAL SERVICES

| Item | Net Price £ | Material £ | Labour hours | Labour £ | Unit | Total Rate £ |
|---|---|---|---|---|---|---|
| **Y63 : SUPPORT COMPONENTS - CABLES : CABLE TRAY(contd)** | | | | | | |
| **Rigid PVC Tray and Fittings; (Cutting and jointing to fittings included.) (Contd)** | | | | | | |
| Reducer | | | | | | |
| 100 mm | 28.32 | 31.15 | 0.40 | 5.47 | nr | 36.62 |
| 150 mm | 42.02 | 46.22 | 0.40 | 5.47 | nr | 51.69 |
| 200 mm | 47.42 | 52.16 | 0.45 | 6.15 | nr | 58.31 |
| 300 mm | 61.93 | 68.12 | 0.45 | 6.15 | nr | 74.27 |
| 400 mm | 76.50 | 84.15 | 0.55 | 7.52 | nr | 91.67 |
| 600 mm | 76.50 | 84.15 | 0.55 | 7.52 | nr | 91.67 |
| **GRP Cable Tray including standard coupling joints and fixings; (Supports and hangers excluded).** | | | | | | |
| Tray | | | | | | |
| 100 mm | 9.47 | 10.94 | 0.40 | 5.47 | m | 16.41 |
| 200 mm | 12.69 | 14.65 | 0.45 | 6.15 | m | 20.80 |
| 400 mm | 28.50 | 32.92 | 0.55 | 7.52 | m | 40.44 |
| Cover | | | | | | |
| 100 mm | 7.17 | 8.28 | 0.10 | 1.37 | m | 9.65 |
| 200 mm | 10.73 | 12.39 | 0.10 | 1.37 | m | 13.76 |
| 400 mm | 16.45 | 19.00 | 0.15 | 2.05 | m | 21.05 |
| **GRP Tray and Fittings; (Cutting and jointing to fittings included).** | | | | | | |
| Bend | | | | | | |
| 100 mm | 25.75 | 28.32 | 0.40 | 5.47 | nr | 33.79 |
| 200 mm | 35.00 | 38.50 | 0.45 | 6.15 | nr | 44.65 |
| 400 mm | 64.45 | 70.89 | 0.15 | 2.05 | nr | 72.94 |
| Bend Cover | | | | | | |
| 100 mm | 11.26 | 12.39 | 0.10 | 1.37 | nr | 13.76 |
| 200 mm | 13.19 | 14.51 | 0.10 | 1.37 | nr | 15.88 |
| 400 mm | 16.10 | 17.71 | 0.15 | 2.05 | nr | 19.76 |
| Tee | | | | | | |
| 100 mm | 30.04 | 33.04 | 0.40 | 5.47 | nr | 38.51 |
| 200 mm | 33.74 | 37.11 | 0.45 | 6.15 | nr | 43.26 |
| 400 mm | 50.74 | 55.81 | 0.55 | 7.52 | nr | 63.33 |
| Tee Cover | | | | | | |
| 100 mm | 16.28 | 17.91 | 0.40 | 5.47 | nr | 23.38 |
| 200 mm | 17.41 | 19.15 | 0.45 | 6.15 | nr | 25.30 |
| 400 mm | 19.29 | 21.22 | 0.55 | 7.52 | nr | 28.74 |

## Y:MECHANICAL AND ELECTRICAL SERVICES

| Item | Net Price £ | Material £ | Labour hours | Labour £ | Unit | Total Rate £ |
|---|---|---|---|---|---|---|
| Reducer | | | | | | |
| 200 mm | 30.04 | 33.04 | 0.10 | 1.37 | nr | **34.41** |
| 400 mm | 77.71 | 85.48 | 0.15 | 2.05 | nr | **87.53** |
| Reducer Cover | | | | | | |
| 200 mm | 15.83 | 17.41 | 0.40 | 5.47 | nr | **22.88** |
| 400 mm | 16.30 | 17.93 | 0.45 | 6.15 | nr | **24.08** |

## Y:MECHANICAL AND ELECTRICAL SERVICES

| Item | Net Price £ | Material £ | Labour hours | Labour £ | Unit | Total Rate £ |
|---|---|---|---|---|---|---|
| **Y63 : SUPPORT COMPONENTS - CABLES : CABLE RACK** | | | | | | |
| **Galvanised Steel Ladder Rack; fixed to backgrounds; including supports, fixings and brackets; earth continuity straps.** | | | | | | |
| Straight | | | | | | |
| 150 mm wide ladder | 11.08 | 12.79 | 0.67 | 9.16 | m | **21.95** |
| 200 mm wide ladder | 11.31 | 13.06 | 0.67 | 9.16 | m | **22.22** |
| 300 mm wide ladder | 11.74 | 13.57 | 0.87 | 11.90 | m | **25.47** |
| 400 mm wide ladder | 13.68 | 15.79 | 1.08 | 14.78 | m | **30.57** |
| 500 mm wide ladder | 15.59 | 18.00 | 1.30 | 17.77 | m | **35.77** |
| 600 mm wide ladder | 16.02 | 18.50 | 1.58 | 21.63 | m | **40.13** |
| 800 mm wide ladder | 18.53 | 21.40 | 1.75 | 23.94 | m | **45.34** |
| 1000 mm wide ladder | 19.31 | 22.31 | 1.90 | 25.98 | m | **48.29** |
| **Galvanised Steel Ladder Rack Fittings; (Cutting and jointing racking to fittings is included.)** | | | | | | |
| Internal Radius Bend | | | | | | |
| 150 mm wide ladder | 26.60 | 29.26 | 0.27 | 3.69 | nr | **32.95** |
| 200 mm wide ladder | 27.14 | 29.85 | 0.27 | 3.69 | nr | **33.54** |
| 300 mm wide ladder | 30.75 | 33.83 | 0.50 | 6.83 | nr | **40.66** |
| 400 mm wide ladder | 35.97 | 39.57 | 0.83 | 11.35 | nr | **50.92** |
| 500 mm wide ladder | 38.20 | 42.02 | 0.92 | 12.57 | nr | **54.59** |
| 600 mm wide ladder | 40.22 | 44.24 | 1.00 | 13.67 | nr | **57.91** |
| 800 mm wide ladder | 50.33 | 55.36 | 1.02 | 13.95 | nr | **69.31** |
| 1000 mm wide ladder | 73.74 | 81.11 | 1.03 | 14.09 | nr | **95.20** |
| External Radius Bend | | | | | | |
| 200 mm wide ladder | 42.45 | 46.70 | 0.27 | 3.69 | nr | **50.39** |
| 300 mm wide ladder | 48.00 | 52.80 | 0.33 | 4.55 | nr | **57.35** |
| 400 mm wide ladder | 51.61 | 56.77 | 0.83 | 11.35 | nr | **68.12** |
| 500 mm wide ladder | 61.30 | 67.43 | 0.58 | 7.93 | nr | **75.36** |
| 600 mm wide ladder | 67.15 | 73.86 | 0.83 | 11.35 | nr | **85.21** |
| 800 mm wide ladder | 101.10 | 111.21 | 1.00 | 13.67 | nr | **124.88** |
| T-Junction | | | | | | |
| 200 mm wide ladder | 44.16 | 48.58 | 0.37 | 5.06 | nr | **53.64** |
| 300 mm wide ladder | 47.46 | 52.21 | 0.57 | 7.79 | nr | **60.00** |
| 400 mm wide ladder | 50.87 | 55.96 | 1.17 | 16.01 | nr | **71.97** |
| 500 mm wide ladder | 53.85 | 59.23 | 1.17 | 16.01 | nr | **75.24** |
| 600 mm wide ladder | 56.29 | 61.92 | 1.17 | 16.01 | nr | **77.93** |
| 800 mm wide ladder | 96.62 | 106.28 | 1.25 | 17.09 | nr | **123.37** |
| 1000 mm wide ladder | 106.42 | 117.06 | 1.33 | 18.20 | nr | **135.26** |
| X-Junction | | | | | | |
| 200 mm wide ladder | 58.00 | 63.80 | 0.50 | 6.83 | nr | **70.63** |
| 300 mm wide ladder | 63.00 | 69.30 | 0.67 | 9.16 | nr | **78.46** |
| 400 mm wide ladder | 68.64 | 75.50 | 1.17 | 16.01 | nr | **91.51** |
| 500 mm wide ladder | 77.26 | 84.99 | 1.27 | 17.37 | nr | **102.36** |
| 600 mm wide ladder | 82.68 | 90.95 | 1.37 | 18.75 | nr | **109.70** |
| 800 mm wide ladder | 94.29 | 103.72 | 1.44 | 19.69 | nr | **123.41** |
| 1000 mm wide ladder | 106.42 | 117.06 | 1.50 | 20.52 | nr | **137.58** |

## Y:MECHANICAL AND ELECTRICAL SERVICES

| Item | Net Price £ | Material £ | Labour hours | Labour £ | Unit | Total Rate £ |
|---|---|---|---|---|---|---|
| Riser | | | | | | |
| 150 mm wide ladder | 37.25 | 40.98 | 0.33 | 4.55 | nr | **45.53** |
| 200 mm wide ladder | 40.76 | 44.84 | 0.50 | 6.83 | nr | **51.67** |
| 300 mm wide ladder | 44.91 | 49.40 | 0.67 | 9.14 | nr | **58.54** |
| 400 mm wide ladder | 48.20 | 53.02 | 0.83 | 11.35 | nr | **64.37** |
| 500 mm wide ladder | 49.27 | 54.20 | 0.92 | 12.57 | nr | **66.77** |
| 600 mm wide ladder | 50.02 | 55.02 | 1.03 | 14.09 | nr | **69.11** |
| 800 mm wide ladder | 53.21 | 58.53 | 1.10 | 15.04 | nr | **73.57** |
| 1000 mm wide ladder | 65.98 | 72.58 | 1.17 | 16.01 | nr | **88.59** |
| End Connector | | | | | | |
| Any ladder width | 2.66 | 2.93 | 0.27 | 3.69 | nr | **6.62** |
| Universal Coupling | | | | | | |
| Any ladder width | 3.51 | 3.86 | 0.27 | 3.69 | nr | **7.55** |

## Y:MECHANICAL AND ELECTRICAL SERVICES

| Item | Net Price £ | Material £ | Labour hours | Labour £ | Unit | Total Rate £ |
|---|---|---|---|---|---|---|
| **Y70 : HV SWITCHGEAR : CIRCUIT BREAKERS** | | | | | | |
| **H.V. Circuit Breakers; installed on prepared foundations including all supports, fixings and inter panel connections (Excluding main and multi core cabling and heat shrink cable termination kits.)** | | | | | | |
| Three phase 11kV, 630 Amp circuit breaker panels; hand charged spring closing operation; prospective fault level up to 20 kA; feeders include ammeter with phase selector switch, 3 pole IDMTL, overcurrent and earth fault relays with necessary current relays with necessary current transformers; incomers include 3 phase voltage transformer, voltmeter and phase selector switch; vacuum insulation | | | | | | |
| Single panel with incoming and outgoing cable box | 16000.00 | 17600.00 | 30.30 | 414.18 | nr | 18014.18 |
| Three panel with one incomer and two feeders | 45000.00 | 49500.00 | 50.00 | 683.40 | nr | 50183.40 |
| Five panel with two incoming, two feeders and a bus section | 77000.00 | 84700.00 | 62.50 | 854.25 | nr | 85554.25 |
| HV Circuit Breaker; 3 way Ring Main Unit | 45000.00 | 4950.00 | 50.00 | 683.40 | nr | 5633.40 |

*Material Costs / Measured Work Prices - Electrical Installations*

## Y:MECHANICAL AND ELECTRICAL SERVICES

| Item | Net Price £ | Material £ | Labour hours | Labour £ | Unit | Total Rate £ |
|---|---|---|---|---|---|---|
| **Y71 : LV SWITCHGEAR AND DISTRIBUTION BOARDS : SWITCHES** | | | | | | |
| Switches; in sheet steel case; surface fixed to backgrounds including supports, fixings, connections/jointing to equipment. (Provision and fixing of plates, discs, for identification is included.) NOTE: Prices are exclusive of fuse links. | | | | | | |
| 20 Amp/500V; Isolating Switches; Sheet Steel Case | | | | | | |
| D.P. | 44.66 | 49.13 | 1.00 | 13.67 | nr | **62.80** |
| T.P.&N. | 50.89 | 55.98 | 1.25 | 17.09 | nr | **73.07** |
| 20 Amp/500V; Switch Fuses; Sheet Steel Case | | | | | | |
| S.P.&N. | 50.47 | 55.52 | 1.00 | 13.67 | nr | **69.19** |
| D.P. | 53.90 | 59.29 | 1.00 | 13.67 | nr | **72.96** |
| T.P.&N. | 63.33 | 69.66 | 1.75 | 23.94 | nr | **93.60** |
| 32 Amp/500V; Isolating Switches; Sheet Steel Case | | | | | | |
| D.P. | 54.55 | 60.01 | 1.00 | 13.67 | nr | **73.68** |
| T.P.&N. | 62.74 | 69.01 | 1.25 | 17.09 | nr | **86.10** |
| 32 Amp/500V; Switch Fuses; Sheet Steel Case | | | | | | |
| S.P.&N. | 62.16 | 68.38 | 1.25 | 17.09 | nr | **85.47** |
| D.P. | 66.34 | 72.97 | 1.25 | 17.09 | nr | **90.06** |
| T.P.&N | 81.42 | 89.56 | 2.00 | 27.34 | nr | **116.90** |
| 63 Amp/500V; Isolating Switches; Sheet Steel Case | | | | | | |
| D.P. | 79.59 | 87.55 | 1.50 | 20.52 | nr | **108.07** |
| T.P.&N | 102.26 | 112.49 | 1.75 | 23.94 | nr | **136.43** |
| 63 Amp/500V; Switch Fuses; Sheet Steel Case | | | | | | |
| S.P.&N | 98.52 | 108.37 | 0.67 | 9.11 | nr | **117.48** |
| D.P. | 108.35 | 119.19 | 1.50 | 20.52 | nr | **139.71** |
| T.P.&N | 139.92 | 153.91 | 2.25 | 30.78 | nr | **184.69** |
| 100 Amp; Isolating Switches; Sheet SteelCase | | | | | | |
| D.P. | 124.35 | 136.78 | 1.50 | 20.52 | nr | **157.30** |
| T.P.&N. | 159.25 | 175.18 | 1.75 | 23.94 | nr | **199.12** |
| 100 Amp; Switch Fuses; Sheet Steel Case | | | | | | |
| S.P.&N | 155.63 | 171.19 | 1.75 | 23.94 | nr | **195.13** |
| D.P. | 181.60 | 199.76 | 1.75 | 23.94 | nr | **223.70** |
| T.P.&N | 236.53 | 260.18 | 2.50 | 34.17 | nr | **294.35** |

## Y:MECHANICAL AND ELECTRICAL SERVICES

| Item | Net Price £ | Material £ | Labour hours | Labour £ | Unit | Total Rate £ |
|---|---|---|---|---|---|---|
| **Y71 : LV SWITCHGEAR AND DISTRIBUTION BOARDS : SWITCHES (contd)** | | | | | | |
| Switches; in sheet steel case; surface fixed to backgrounds including supports,fixings, connections/jointing to equipment. (Provision and fixing of plates, discs, for identification is included.) NOTE: Prices are exclusive of fuse links. (Contd) | | | | | | |
| 63 Amp; Fuse Switches; Sheet Steel Case H.R.C. | | | | | | |
| S.P.&N. | 98.52 | 108.37 | 1.50 | 20.52 | nr | 128.89 |
| D.P. | 107.04 | 117.74 | 1.50 | 20.52 | nr | 138.26 |
| T.P.&N. | 134.53 | 147.98 | 1.75 | 23.94 | nr | 171.92 |
| 100 Amp; Fuse Switches; Sheet Steel Case H.R.C. | | | | | | |
| S.P.&N. | 155.60 | 171.16 | 1.75 | 23.94 | nr | 195.10 |
| D.P. | 178.52 | 196.37 | 1.75 | 23.94 | nr | 220.31 |
| T.P.&N. | 234.55 | 258.00 | 2.25 | 30.78 | nr | 288.78 |
| T.P.&S.N | 284.40 | 312.84 | 2.50 | 34.17 | nr | 347.01 |
| 200 Amp; Fuse Switches; Sheet Steel Case H.R.C. | | | | | | |
| S.P.& N. | 309.99 | 340.99 | 2.25 | 30.78 | nr | 371.77 |
| D.P. | 330.11 | 363.12 | 2.25 | 30.78 | nr | 393.90 |
| T.P.&N. | 360.34 | 396.37 | 2.50 | 34.17 | nr | 430.54 |
| T.P.&S.N. | 421.90 | 464.09 | 2.75 | 37.65 | nr | 501.74 |
| 315 Amp; Fuse Switches; Sheet Steel Case; H.R.C. | | | | | | |
| T.P.&N. | 644.60 | 709.06 | 2.75 | 37.65 | nr | 746.71 |
| T.P.&S.N. | 688.68 | 757.55 | 3.26 | 44.52 | nr | 802.07 |
| 400 Amp; Fuse Switches; Sheet Steel Case; H.R.C. | | | | | | |
| T.P.&N. | 704.35 | 774.78 | 3.00 | 41.05 | nr | 815.83 |
| T.P.&S.N. | 818.71 | 900.58 | 4.50 | 61.57 | nr | 962.15 |
| 630 Amp; Fuse Switches; Sheet Steel Case; H.R.C. | | | | | | |
| T.P.&N. | 1362.66 | 1498.93 | 3.76 | 51.38 | nr | 1550.31 |
| T.P.&S.N. | 1585.40 | 1743.94 | 4.50 | 61.57 | nr | 1805.51 |
| 800 Amp; Fuse Switches; Sheet Steel Case; H.R.C. | | | | | | |
| T.P.&N. | 1530.36 | 1683.40 | 4.50 | 61.57 | nr | 1744.97 |
| T.P.&S.N. | 1757.63 | 1933.39 | 4.76 | 65.09 | nr | 1998.48 |

*Material Costs / Measured Work Prices - Electrical Installations*

## Y:MECHANICAL AND ELECTRICAL SERVICES

| Item | Net Price £ | Material £ | Labour hours | Labour £ | Unit | Total Rate £ |
|---|---|---|---|---|---|---|
| **Y71 : LV SWITCHGEAR AND DISTRIBUTION BOARDS : CIRCUIT BREAKERS** | | | | | | |
| **L.V. Circuit Breakers; Non-automatic air circuit breakers for purpose made switchboard.** | | | | | | |
| Up to 65kA at 415/500V; 1 second; Fixed Manual; Triple Pole | | | | | | |
| 800 Amp | 1325.00 | 1457.50 | 2.00 | 27.34 | nr | 1484.84 |
| 1000 Amp | 1334.00 | 1467.40 | 2.00 | 27.34 | nr | 1494.74 |
| 1250 Amp | 1356.00 | 1491.60 | 2.00 | 27.34 | nr | 1518.94 |
| 1600 Amp | 1536.00 | 1689.60 | 2.00 | 27.34 | nr | 1716.94 |
| 2000 Amp | 1932.00 | 2125.20 | 3.00 | 41.05 | nr | 2166.25 |
| 2500 Amp | 2548.00 | 2802.80 | 3.00 | 41.05 | nr | 2843.85 |
| 1 Second; Fixed Manual; Four Pole | | | | | | |
| 800 Amp | 1528.00 | 1680.80 | 3.00 | 41.05 | nr | 1721.85 |
| 1000 Amp | 1562.00 | 1718.20 | 3.00 | 41.05 | nr | 1759.25 |
| 1250 Amp | 1578.00 | 1735.80 | 3.00 | 41.05 | nr | 1776.85 |
| 1600 Amp | 1800.00 | 1980.00 | 3.00 | 41.05 | nr | 2021.05 |
| 2000 Amp | 2282.00 | 2510.20 | 4.00 | 54.67 | nr | 2564.87 |
| 2500 Amp | 3040.00 | 3344.00 | 4.00 | 54.67 | nr | 3398.67 |
| 1 Second; Drawout Manual; Triple Pole | | | | | | |
| 800 Amp | 1855.00 | 2040.50 | 3.00 | 41.05 | nr | 2081.55 |
| 1000 Amp | 1866.00 | 2052.60 | 3.00 | 41.05 | nr | 2093.65 |
| 1250 Amp | 1906.00 | 2096.60 | 3.00 | 41.05 | nr | 2137.65 |
| 1600 Amp | 2216.00 | 2437.60 | 3.00 | 41.05 | nr | 2478.65 |
| 2000 Amp | 3352.00 | 3687.20 | 4.00 | 54.67 | nr | 3741.87 |
| 2500 Amp | 3582.00 | 3940.20 | 4.00 | 54.67 | nr | 3994.87 |
| 1 Second; Drawout Manual; Four Pole | | | | | | |
| 800 Amp | 2260.00 | 2486.00 | 4.00 | 54.67 | nr | 2540.67 |
| 1000 Amp | 2278.00 | 2505.80 | 4.00 | 54.67 | nr | 2560.47 |
| 1250 Amp | 2320.00 | 2552.00 | 4.00 | 54.67 | nr | 2606.67 |
| 1600 Amp | 2710.00 | 2981.00 | 4.00 | 54.67 | nr | 3035.67 |
| 2000 Amp | 3582.00 | 3940.20 | 5.00 | 68.34 | nr | 4008.54 |
| 2500 Amp | 4398.00 | 4837.80 | 5.00 | 68.34 | nr | 4906.14 |
| Up to 80kA at 415/500V; 3 Second (at 50kA); Fixed Manual; Triple Pole | | | | | | |
| 800 Amp | 1680.00 | 1848.00 | 5.00 | 68.34 | nr | 1916.34 |
| 1000 Amp | 1754.00 | 1929.40 | 5.00 | 68.34 | nr | 1997.74 |
| 1250 Amp | 1819.00 | 2000.90 | 5.00 | 68.34 | nr | 2069.24 |
| 1600 Amp | 1875.00 | 2062.50 | 5.00 | 68.34 | nr | 2130.84 |
| 2000 Amp | 2705.00 | 2975.50 | 6.02 | 82.34 | nr | 3057.84 |
| 2500 Amp | 3128.00 | 3440.80 | 6.02 | 82.34 | nr | 3523.14 |
| 3200 Amp | 4020.00 | 4422.00 | 7.04 | 96.25 | nr | 4518.25 |
| 4000 Amp | 6160.00 | 6776.00 | 8.00 | 109.34 | nr | 6885.34 |

## Y:MECHANICAL AND ELECTRICAL SERVICES

| Item | Net Price £ | Material £ | Labour hours | Labour £ | Unit | Total Rate £ |
|---|---|---|---|---|---|---|
| **Y71 : LV SWITCHGEAR AND DISTRIBUTION BOARDS : CIRCUIT BREAKERS(contd)** | | | | | | |
| L.V. Circuit Breakers; Non-automatic air circuit breakers for purpose made switchboard. (Contd) | | | | | | |
| 3 Second; Fixed Manual; Four Pole | | | | | | |
| 800 Amp | 2020.00 | 2222.00 | 6.02 | 82.34 | nr | 2304.34 |
| 1000 Amp | 2036.00 | 2239.60 | 6.02 | 82.34 | nr | 2321.94 |
| 1250 Amp | 2438.00 | 2681.80 | 6.02 | 82.34 | nr | 2764.14 |
| 1600 Amp | 2635.00 | 2898.50 | 6.02 | 82.34 | nr | 2980.84 |
| 2000 Amp | 3430.00 | 3773.00 | 7.04 | 96.25 | nr | 3869.25 |
| 2500 Amp | 4010.00 | 4411.00 | 7.04 | 96.25 | nr | 4507.25 |
| 3200 Amp | 4922.00 | 5414.20 | 7.04 | 96.25 | nr | 5510.45 |
| 4000 Amp | 7440.00 | 8184.00 | 8.00 | 109.34 | nr | 8293.34 |
| 3 Second; Drawout Manual; Triple Pole | | | | | | |
| 800 Amp | 2075.00 | 2282.50 | 2.00 | 27.34 | nr | 2309.84 |
| 1000 Amp | 2165.00 | 2381.50 | 2.00 | 27.34 | nr | 2408.84 |
| 1250 Amp | 2334.00 | 2567.40 | 2.00 | 27.34 | nr | 2594.74 |
| 1600 Amp | 2726.00 | 2998.60 | 2.00 | 27.34 | nr | 3025.94 |
| 2000 Amp | 3430.00 | 3773.00 | 3.00 | 41.05 | nr | 3814.05 |
| 2500 Amp | 4480.00 | 4928.00 | 3.00 | 41.05 | nr | 4969.05 |
| 3200 Amp | 5584.00 | 6142.40 | 3.00 | 41.05 | nr | 6183.45 |
| 4000 Amp | 8346.00 | 9180.60 | 4.00 | 54.67 | nr | 9235.27 |
| 3 Second; Drawout Manual; Four Pole | | | | | | |
| 800 Amp | 2784.00 | 3062.40 | 3.00 | 41.05 | nr | 3103.45 |
| 1000 Amp | 2819.00 | 3100.90 | 3.00 | 41.05 | nr | 3141.95 |
| 1250 Amp | 2915.00 | 3206.50 | 3.00 | 41.05 | nr | 3247.55 |
| 1600 Amp | 3586.00 | 3944.60 | 3.00 | 41.05 | nr | 3985.65 |
| 2000 Amp | 4142.00 | 4556.20 | 4.00 | 54.67 | nr | 4610.87 |
| 2500 Amp | 5128.00 | 5640.80 | 4.00 | 54.67 | nr | 5695.47 |
| 3200 Amp | 6838.00 | 7521.80 | 4.00 | 54.67 | nr | 7576.47 |
| 4000 Amp | 10298.00 | 11327.80 | 5.00 | 68.34 | nr | 11396.14 |
| Extra For | | | | | | |
| Closing Coil | 49.00 | 53.90 | 2.00 | 27.34 | nr | 81.24 |
| Shunt Trip Coil | 49.00 | 53.90 | 2.00 | 27.34 | nr | 81.24 |
| Motorised Charging Unit | 320.00 | 352.00 | 2.00 | 27.34 | nr | 379.34 |
| Under voltage release device - instant | 67.00 | 73.70 | 2.00 | 27.34 | nr | 101.04 |
| Under voltage release device - time delay | 142.00 | 156.20 | 2.00 | 27.34 | nr | 183.54 |
| Carriage Position Switch | 58.00 | 63.80 | 3.00 | 41.05 | nr | 104.85 |
| Operations Counter | 64.00 | 70.40 | 3.00 | 41.05 | nr | 111.45 |
| Breaker Insertion Interlock | 23.00 | 25.30 | 3.00 | 41.05 | nr | 66.35 |
| IP54 Door Panel | 93.00 | 102.30 | 0.50 | 6.83 | nr | 109.13 |
| Earthing Device | | | | | | |
| 3 Pole up to 65kA | 353.00 | 388.30 | 2.00 | 27.34 | nr | 415.64 |
| 4 Pole up to 65kA | 471.00 | 518.10 | 2.00 | 27.34 | nr | 545.44 |
| 3 Pole up to 80kA | 424.00 | 466.40 | 3.00 | 41.05 | nr | 507.45 |
| 4 Pole up to 80kA | 565.00 | 621.50 | 4.00 | 54.67 | nr | 676.17 |

## Y:MECHANICAL AND ELECTRICAL SERVICES

| Item | Net Price £ | Material £ | Labour hours | Labour £ | Unit | Total Rate £ |
|---|---|---|---|---|---|---|
| **Protection Units** | | | | | | |
| Overload Protection - Adjustable instantaneous | 165.00 | 181.50 | 2.00 | 27.34 | nr | 208.84 |
| Overload Protection - Inverse overload and short circuit | 260.00 | 286.00 | 2.00 | 27.34 | nr | 313.34 |
| Unrestricted Earth Fault Protection | 200.00 | 220.00 | 2.00 | 27.34 | nr | 247.34 |
| Load Monitoring - Pre-trip alarm | 97.00 | 106.70 | 2.00 | 27.34 | nr | 134.04 |
| Load Monitoring - Load shedding warning | 123.00 | 135.30 | 2.00 | 27.34 | nr | 162.64 |
| Communication - Remote signalling | 220.00 | 242.00 | 2.00 | 27.34 | nr | 269.34 |
| Communication - Data reception / transmission | 236.00 | 259.60 | 2.00 | 27.34 | nr | 286.94 |
| Communication - Electronic Interlocks / interface | 281.00 | 309.10 | 2.00 | 27.34 | nr | 336.44 |
| **Moulded Case Circuit Breakers; in sheet steel enclosure with thermal/magnetic trips; including supports connections jointing to equipment; BS 4572/IEC947; (provision and fixing of plates, discs etc, for identification is included.)** | | | | | | |
| 25 -200 Amp at 25kV Icu range in nominal ratings indicated; T.P.& S.N. | | | | | | |
| 25-63 Amp | 136.99 | 150.69 | 2.75 | 37.65 | nr | 188.34 |
| 80 - 100 Amp | 146.78 | 161.46 | 2.75 | 37.65 | nr | 199.11 |
| 125 Amp | 195.70 | 215.27 | 2.75 | 37.65 | nr | 252.92 |
| 160 Amp | 210.38 | 231.42 | 3.00 | 41.05 | nr | 272.47 |
| 200 Amp | 244.63 | 269.09 | 3.51 | 47.96 | nr | 317.05 |
| 250 Amp | 293.55 | 322.90 | 3.51 | 47.96 | nr | 370.86 |
| 250-800 Amp at 32kV Icu range in nominal ratings indicated; T.P.& S.N. | | | | | | |
| 250 Amp | 308.23 | 339.05 | 3.76 | 51.38 | nr | 390.43 |
| 315 Amp | 313.12 | 344.43 | 3.76 | 51.38 | nr | 395.81 |
| 400 Amp | 318.01 | 349.81 | 3.76 | 51.38 | nr | 401.19 |
| 250-800 Amp at 50kV Icu range in nominal ratings indicated; T.P.& S.N. | | | | | | |
| 400 Amp | 391.40 | 430.54 | 4.00 | 54.67 | nr | 485.21 |
| 630 Amp | 636.02 | 699.62 | 4.50 | 61.57 | nr | 761.19 |
| 800 Amp | 782.80 | 861.08 | 4.76 | 65.09 | nr | 926.17 |
| **Miniature Circuit Breakers for Distribution Boards; BS EN 60 898; DIN rail mounting; including connecting to circuit** | | | | | | |
| Extra for miniature circuit breakers installation and connecting of wiring; single pole | | | | | | |
| 6 Amp | 5.56 | 6.12 | 0.10 | 1.37 | nr | 7.49 |
| 10 - 40 Amp | 5.28 | 5.81 | 0.10 | 1.37 | nr | 7.18 |
| 50 - 60 Amp | 5.56 | 6.12 | 0.15 | 2.05 | nr | 8.17 |

Material Costs / Measured Work Prices - Electrical Installations

## Y:MECHANICAL AND ELECTRICAL SERVICES

| Item | Net Price £ | Material £ | Labour hours | Labour £ | Unit | Total Rate £ |
|---|---|---|---|---|---|---|
| **Y71 : LV SWITCHGEAR AND DISTRIBUTION BOARDS : CIRCUIT BREAKERS (contd)** | | | | | | |
| **Miniature Circuit Breakers for Distribution Boards; BS EN 60 898; DIN rail mounting; including connecting to circuit. (Contd)** | | | | | | |
| Extra for miniature circuit breakers installation and connecting of wiring; double pole | | | | | | |
| 6 Amp | 16.40 | 18.04 | 0.20 | 2.73 | nr | **20.77** |
| 10 - 40 Amp | 15.40 | 16.94 | 0.20 | 2.73 | nr | **19.67** |
| 50 - 60 Amp | 16.40 | 18.04 | 0.30 | 4.10 | nr | **22.14** |
| Extra for miniature circuit breakers installation and connecting of wiring; triple pole | | | | | | |
| 6 Amp | 23.80 | 26.18 | 0.30 | 4.10 | nr | **30.28** |
| 10 - 40 Amp | 22.80 | 25.08 | 0.45 | 6.15 | nr | **31.23** |
| 50 - 60 Amp | 23.80 | 26.18 | 0.45 | 6.15 | nr | **32.33** |

## Y:MECHANICAL AND ELECTRICAL SERVICES

| Item | Net Price £ | Material £ | Labour hours | Labour £ | Unit | Total Rate £ |
|---|---|---|---|---|---|---|
| **Y71 : LV SWITCHGEAR AND DISTRIBUTION BOARDS : DISTRIBUTION BOARDS** | | | | | | |
| Distribution Boards; sheet steel case; fully shrouded; fixed to backgrounds; including supports, fixings/connections and equipment. (Provision and fixing of plates, discs etc, for identification is included.) NOTE: prices exclude fuse links. | | | | | | |
| 500V suitable for 2-20 Amp H.R.C. fuses; S.P.&N. | | | | | | |
| 4 way | 81.91 | 90.10 | 1.00 | 13.67 | nr | 103.77 |
| 6 way | 98.89 | 108.78 | 1.25 | 17.09 | nr | 125.87 |
| 8 way | 115.90 | 127.49 | 1.50 | 20.49 | nr | 147.98 |
| 12 way | 147.96 | 162.76 | 2.00 | 27.34 | nr | 190.10 |
| 500V suitable for 2-20 Amp H.R.C. fuses; T.P.&N. | | | | | | |
| 4 way | 144.77 | 159.25 | 1.50 | 20.49 | nr | 179.74 |
| 6 way | 180.88 | 198.97 | 1.90 | 25.98 | nr | 224.95 |
| 8 way | 231.70 | 254.86 | 2.30 | 31.49 | nr | 286.35 |
| 12 way | 295.61 | 325.17 | 3.40 | 46.49 | nr | 371.66 |
| 500V suitable for 2-30 Amp H.R.C. fuses; S.P.&N. | | | | | | |
| 4 way | 97.50 | 107.25 | 1.50 | 20.49 | nr | 127.74 |
| 6 way | 128.56 | 141.42 | 1.40 | 19.14 | nr | 160.56 |
| 8 way | 151.42 | 166.56 | 1.65 | 22.55 | nr | 189.11 |
| 12 way | 186.67 | 205.34 | 2.15 | 29.39 | nr | 234.73 |
| 500V suitable for 2-30 Amp H.R.C. fuses; T.P.&N. | | | | | | |
| 2 way | 118.22 | 130.04 | 1.00 | 13.67 | nr | 143.71 |
| 4 way | 171.08 | 188.19 | 1.90 | 25.98 | nr | 214.17 |
| 6 way | 227.77 | 250.55 | 2.40 | 32.86 | nr | 283.41 |
| 8 way | 227.77 | 250.55 | 2.80 | 38.29 | nr | 288.84 |
| 10 way | 330.40 | 363.44 | 3.21 | 43.81 | nr | 407.25 |
| 12 way | 380.56 | 418.62 | 3.80 | 51.97 | nr | 470.59 |
| 500V suitable for 35-63 Amp H.R.C. fuses; T.P.&N. | | | | | | |
| 2 way | 273.48 | 300.83 | 1.60 | 21.87 | nr | 322.70 |
| 4 way | 367.29 | 404.02 | 2.50 | 34.17 | nr | 438.19 |
| 6 way | 465.87 | 512.46 | 3.00 | 41.05 | nr | 553.51 |
| 8 way | 556.11 | 611.72 | 3.51 | 47.96 | nr | 659.68 |
| 10 way | 667.89 | 734.68 | 3.75 | 51.19 | nr | 785.87 |
| 500V suitable for 35-100 Amp H.R.C. fuses; T.P.&N. | | | | | | |
| 2 way | 365.60 | 402.16 | 1.90 | 25.98 | nr | 428.14 |
| 3 way | 433.87 | 477.26 | 2.40 | 32.86 | nr | 510.12 |
| 4 way | 556.88 | 612.57 | 2.75 | 37.65 | nr | 650.22 |
| 6 way | 718.17 | 789.99 | 3.26 | 44.52 | nr | 834.51 |
| 8 way | 863.52 | 949.87 | 3.51 | 47.96 | nr | 997.83 |

## Y:MECHANICAL AND ELECTRICAL SERVICES

| Item | Net Price £ | Material £ | Labour hours | Labour £ | Unit | Total Rate £ |
|---|---|---|---|---|---|---|
| **Y71 : LV SWITCHGEAR AND DISTRIBUTION BOARDS : DISTRIBUTION BOARDS (contd)** | | | | | | |
| **Distribution Boards; Sheet steel case; fully shrouded; fixed to backgrounds; including supports, fixings/connections to equipment. (Provision and fixings of plates, discs etc, for identification is included.) NOTE: prices include MCB's and doors. (Contd)** | | | | | | |
| 500V suitable for 6-63 Amp M.C.B's; S.P.&N. | | | | | | |
|    12 way | 116.73 | 128.40 | 3.00 | 41.05 | nr | **169.45** |
|    16 way | 146.43 | 161.14 | 4.00 | 54.67 | nr | **215.81** |
| 500V suitable for 6-63 Amp M.C.B's; T.P.&N | | | | | | |
|    4 way | 112.59 | 123.86 | 3.00 | 41.05 | nr | **164.91** |
|    8 way | 236.25 | 259.88 | 3.00 | 41.05 | nr | **300.93** |
|    12 way | 317.43 | 349.17 | 4.00 | 54.67 | nr | **403.84** |

## Y:MECHANICAL AND ELECTRICAL SERVICES

| Item | Net Price £ | Material £ | Labour hours | Labour £ | Unit | Total Rate £ |
|---|---|---|---|---|---|---|
| **Y71 : LV SWITCHGEAR AND DISTRIBUTION BOARDS : CONSUMER UNITS** | | | | | | |
| Consumer units; sheet steel case; fully shrouded; fixed to backgrounds; including supports, fixings/connections to equipment. (Provision and fixings of plates, discs etc, for identification is included.) NOTE: prices include MCB's and doors. | | | | | | |
| 500V suitable for 6-40 Amp M.C.B's; S.P.&N. | | | | | | |
| 7 way | 66.57 | 73.22 | 3.00 | 41.05 | nr | 112.88 |
| 10 way | 88.03 | 96.84 | 4.00 | 54.67 | nr | 151.51 |
| 13 way | 111.68 | 122.86 | 4.00 | 54.67 | nr | 177.53 |
| Consumer units; fixed to backgrounds; including supports, fixings/connections jointing to equipment. (Provision and fixings of plates, discs etc, for identification is included.) | | | | | | |
| Switched and insulated; moulded plastic case, 60 Amp 240 Volt D.P./A.C.; fitted rewireable fuses | | | | | | |
| 2 way | 14.05 | 17.56 | 1.25 | 17.09 | nr | 34.65 |
| 3 way | 17.74 | 23.72 | 1.50 | 20.52 | nr | 44.24 |
| 4 way | 21.01 | 29.41 | 1.50 | 20.52 | nr | 49.93 |
| Switched and insulated; moulded plastic case, 60 Amp 240 Volt D.P./A.C.; fitted cartridge fuses | | | | | | |
| 2 way | 14.52 | 18.59 | 1.25 | 17.09 | nr | 35.68 |
| 3 way | 18.21 | 25.27 | 1.50 | 20.52 | nr | 45.79 |
| 4 way | 21.48 | 28.87 | 1.50 | 20.52 | nr | 49.39 |
| Switched and insulated; moulded plastic case, 60 Amp 240 Volt D.P./A.C.; fitted MCB's | | | | | | |
| 2 way | 18.64 | 27.63 | 1.25 | 17.09 | nr | 44.72 |
| 3 way | 22.33 | 38.85 | 1.50 | 20.52 | nr | 59.37 |
| 4 way | 25.59 | 49.57 | 1.50 | 20.52 | nr | 70.09 |
| Switched and insulated; moulded plastic case, 100 Amp 240 Volt D.P./A.C.; fitted rewireable fuses | | | | | | |
| 6 way | 25.40 | 38.50 | 1.83 | 25.03 | nr | 63.53 |
| 8 way | 31.17 | 48.99 | 2.00 | 27.34 | nr | 76.33 |
| Switched and insulated; moulded plastic case, 100 Amp 240 Volt D.P./A.C.; fitted cartridge fuses | | | | | | |
| 6 way | 25.87 | 41.61 | 1.83 | 25.03 | nr | 66.64 |
| 8 way | 31.64 | 53.13 | 2.00 | 27.34 | nr | 80.47 |

## Y:MECHANICAL AND ELECTRICAL SERVICES

| Item | Net Price £ | Material £ | Labour hours | Labour £ | Unit | Total Rate £ |
|---|---|---|---|---|---|---|
| **Y71 : LV SWITCHGEAR AND DISTRIBUTION BOARDS : CONSUMER UNITS (contd)** | | | | | | |
| Consumer units; fixed to backgrounds; including supports, fixings/connections jointing to equipment. (Provision and fixings of plates, discs etc, for identification is included.) (Contd) | | | | | | |
| Switched and insulated; moulded plastic case, 100 Amp 240 Volt D.P./A.C.; fitted MCBs | | | | | | |
| 6 way | 29.99 | 68.84 | 1.83 | 25.03 | nr | **93.87** |
| 8 way | 35.75 | 89.30 | 2.00 | 27.34 | nr | **116.64** |
| Switched and insulated; enamelled steel case; 60 Amp 240 Volt D.P./A.C.; fitted rewireable fuses | | | | | | |
| 2 way | 17.08 | 20.89 | 1.25 | 17.09 | nr | **37.98** |
| 3 way | 22.99 | 29.50 | 1.50 | 20.52 | nr | **50.02** |
| 4 way | 25.18 | 34.00 | 1.50 | 20.52 | nr | **54.52** |
| Switched and insulated; enamelled steel case; 60 Amp 240 Volt D.P./A.C.; fitted cartridge fuses | | | | | | |
| 2 way | 17.55 | 21.92 | 1.25 | 17.09 | nr | **39.01** |
| 3 way | 23.46 | 31.05 | 1.50 | 20.52 | nr | **51.57** |
| 4 way | 25.65 | 36.07 | 1.50 | 20.52 | nr | **56.59** |
| Switched and insulated; enamelled steel case; 60 Amp 240 Volt D.P./A.C.; fitted MCB's | | | | | | |
| 2 way | 21.66 | 30.97 | 1.25 | 17.09 | nr | **48.06** |
| 3 way | 27.58 | 44.63 | 1.50 | 20.52 | nr | **65.15** |
| 4 way | 29.77 | 54.15 | 1.50 | 20.52 | nr | **74.67** |
| Switched and insulated; enamelled steel case; 100 Amp 240 Volt D.P./A.C.; fitted rewireable fuses | | | | | | |
| 6 way | 29.96 | 43.51 | 1.83 | 25.03 | nr | **68.54** |
| 8 way | 1.91 | 53.05 | 1.83 | 25.03 | nr | **78.08** |
| 12 way | 65.39 | 95.14 | 1.83 | 25.03 | nr | **120.17** |
| Switched and insulated; enamelled steel case; 100 Amp 240 Volt D.P./A.C.; fitted cartridge fuses | | | | | | |
| 6 way | 30.43 | 46.63 | 1.83 | 25.03 | nr | **71.66** |
| 8 way | 35.33 | 57.19 | 2.00 | 27.34 | nr | **84.53** |
| 12 way | 65.85 | 101.37 | 3.00 | 41.05 | nr | **142.42** |
| Switched and insulated; enamelled steel case; 100 Amp 240 Volt D.P./A.C.; fitted MCB's | | | | | | |
| 6 way | 34.55 | 73.86 | 1.83 | 25.03 | nr | **98.89** |
| 8 way | 39.44 | 93.36 | 2.00 | 27.34 | nr | **120.70** |
| 12 way | 69.97 | 155.84 | 3.00 | 41.05 | nr | **196.89** |

*Material Costs / Measured Work Prices - Electrical Installations*

## Y:MECHANICAL AND ELECTRICAL SERVICES

| Item | Net Price £ | Material £ | Labour hours | Labour £ | Unit | Total Rate £ |
|---|---|---|---|---|---|---|
| Switched and insulated; moulded plastic case; D.P. RCCB control on MCB ways; 240 Volt A.C.; 80 Amp; 30 mA | | | | | | |
| 3 way | 69.47 | 76.42 | 1.20 | 16.41 | nr | 92.83 |
| 5 way | 73.07 | 80.38 | 1.20 | 16.41 | nr | 96.79 |
| 7 way | 76.95 | 84.64 | 1.50 | 20.52 | nr | 105.16 |
| 11 way | 84.09 | 92.50 | 1.90 | 25.98 | nr | 118.48 |
| 17 way | 107.95 | 118.75 | 2.10 | 28.71 | nr | 147.46 |
| Switched and insulated; moulded plastic case; D.P. RCCB control for all MCB ways; 240 Volt A.C.; 80 Amp; 100mA | | | | | | |
| 3 way | 67.30 | 74.03 | 1.20 | 16.41 | nr | 90.44 |
| 5 way | 70.90 | 77.99 | 1.40 | 19.14 | nr | 97.13 |
| 7 way | 74.77 | 82.25 | 1.50 | 20.52 | nr | 102.77 |
| 9 way | 81.92 | 90.11 | 1.90 | 25.98 | nr | 116.09 |
| 11 way | 105.78 | 116.36 | 2.10 | 28.71 | nr | 145.07 |
| Switched and insulated; enamelled steel case; D.P. RCCB control for MCB ways; 240 Volt A.C.; 80 Amps; 30mA | | | | | | |
| 5 way | 75.26 | 82.79 | 1.40 | 19.14 | nr | 101.93 |
| 7 way | 79.66 | 87.63 | 1.75 | 23.94 | nr | 111.57 |
| 11 way | 87.68 | 96.45 | 2.10 | 28.71 | nr | 125.16 |
| Switched and insulated; enamelled steel case; D.P. RCCB control foe MCB ways; 240 Volt A.C.; 80Amp; 100mA | | | | | | |
| 5 way | 73.08 | 80.39 | 1.40 | 19.14 | nr | 99.53 |
| 7 way | 77.49 | 85.24 | 1.50 | 20.52 | nr | 105.76 |
| 11 way | 85.51 | 94.06 | 1.90 | 25.98 | nr | 120.04 |
| Switched and insulated; moulded plastic case; split load range; 100 Amp D.P. main switch with 63 Amp D.P. RCCB on selected ways; 240 Volt AC. | | | | | | |
| 9 MCB way; 6 RCCB Way; 100mA | 92.96 | 102.26 | 1.75 | 23.94 | nr | 126.20 |
| 9 MCB way; 6 RCCB Way; 30mA | 97.12 | 106.83 | 1.75 | 23.94 | nr | 130.77 |
| 9 MCB way; 5 RCCB Way; 100mA | 92.96 | 102.26 | 1.75 | 23.94 | nr | 126.20 |
| 9 MCB way; 5 RCCB Way; 30mA | 97.12 | 106.83 | 1.75 | 23.94 | nr | 130.77 |
| 9 MCB way; 4 RCCB Way; 100mA | 92.96 | 102.26 | 1.75 | 23.94 | nr | 126.20 |
| 9 MCB way; 4 RCCB Way; 30mA | 97.12 | 106.83 | 1.75 | 23.94 | nr | 130.77 |
| 9 MCB way; 3 RCCB Way; 100mA | 92.96 | 102.26 | 1.75 | 23.94 | nr | 126.20 |
| 9 MCB way; 3 RCCB Way; 30mA | 97.12 | 106.83 | 1.75 | 23.94 | nr | 130.77 |
| Switched and insulated; enamelled steel case; split load range; 100 Amp D.P. main switch with 63 Amp D.P. RCCB on selected ways; 240 Volt AC; 30mA | | | | | | |
| 5 MCCB way; 3 RCCB ways; 100mA | 88.98 | 97.88 | 1.60 | 21.87 | nr | 119.75 |
| 5 MCCB way; 3 RCCB ways; 30mA | 93.14 | 102.45 | 1.60 | 21.87 | nr | 124.32 |
| 5 MCCB way; 2 RCCB ways; 100mA | 88.98 | 97.88 | 1.60 | 21.87 | nr | 119.75 |
| 5 MCCB way; 2 RCCB ways; 30mA | 93.14 | 102.45 | 1.60 | 21.87 | nr | 124.32 |

## Y:MECHANICAL AND ELECTRICAL SERVICES

| Item | Net Price £ | Material £ | Labour hours | Labour £ | Unit | Total Rate £ |
|---|---|---|---|---|---|---|
| **Y71 : LV SWITCHGEAR AND DISTRIBUTION BOARDS : CONSUMER UNITS (contd)** | | | | | | |
| Consumer units; fixed to backgrounds; including supports, fixings/connections jointing to equipment. (Provision and fixings of plates, discs etc, for identification is included.) (Contd) | | | | | | |
| Switched and insulated; enamelled steel case; split load range; 100 Amp D.P. main switch with 63 Amp D.P. RCCB on selected ways; 240 Volt AC; 30mA (contd) | | | | | | |
| 9 MCCB way; 6 RCCB ways; 100mA | 96.54 | 106.19 | 1.60 | 21.87 | nr | **128.06** |
| 9 MCCB way; 6 RCCB ways; 30mA | 100.71 | 110.78 | 1.75 | 23.94 | nr | **134.72** |
| 9 MCCB way; 5 RCCB ways; 100mA | 96.54 | 106.19 | 1.75 | 23.94 | nr | **130.13** |
| 9 MCCB way; 5 RCCB ways; 30mA | 100.71 | 110.78 | 1.75 | 23.94 | nr | **134.72** |
| 9 MCCB way; 4 RCCB ways; 100mA | 96.54 | 106.19 | 1.75 | 23.94 | nr | **130.13** |
| 9 MCCB way; 4 RCCB ways; 30mA | 100.71 | 110.78 | 1.75 | 23.94 | nr | **134.72** |
| 9 MCCB way; 3 RCCB ways; 100mA | 96.54 | 106.19 | 1.75 | 23.94 | nr | **130.13** |
| 9 MCCB way; 3 RCCB ways; 30mA | 100.71 | 110.78 | 1.75 | 23.94 | nr | **134.72** |

## Y:MECHANICAL AND ELECTRICAL SERVICES

| Item | Net Price £ | Material £ | Labour hours | Labour £ | Unit | Total Rate £ |
|---|---|---|---|---|---|---|
| **Y71 : LV SWITCHGEAR AND DISTRIBUTION BOARDS : VOLTAGE STABLISERS/LINE CONDITIONERS** | | | | | | |
| Voltage Stabilisers; solid state circuitry; sheet steel enclosure; maintains output +/- 3% over zero to full load with input voltage swings of +/- 15%; positioned on site, final connections and commissioning. (Include delivery up to 75 miles.) | | | | | | |
| Electronic; Single Phase; 220/230/240 Volt AC, 50 Hz | | | | | | |
| 0.3 kVA | 333.90 | 367.29 | 2.00 | 27.34 | nr | 394.63 |
| 0.6 kVA | 350.70 | 385.77 | 2.00 | 27.34 | nr | 413.11 |
| 1 kVA | 382.20 | 420.42 | 2.00 | 27.34 | nr | 447.76 |
| 2 kVA | 497.70 | 547.47 | 2.00 | 27.34 | nr | 574.81 |
| 3 kVA | 585.90 | 644.49 | 3.00 | 41.05 | nr | 685.54 |
| 4.5 kVA | 814.80 | 896.28 | 4.00 | 54.67 | nr | 950.95 |
| 6 kVA | 1090.95 | 1200.05 | 4.00 | 54.67 | nr | 1254.72 |
| 9 kVA | 1375.50 | 1513.05 | 6.02 | 82.34 | nr | 1595.39 |
| 12 kVA | 1535.10 | 1688.61 | 7.04 | 96.25 | nr | 1784.86 |
| 15 kVA | 1738.80 | 1912.68 | 7.52 | 102.77 | nr | 2015.45 |
| 20 kVA | 2032.80 | 2236.08 | 8.00 | 109.34 | nr | 2345.42 |
| 25 kVA | 2361.45 | 2597.59 | 8.00 | 109.34 | nr | 2706.93 |
| 30 kVA | 2730.00 | 3003.00 | 10.00 | 136.68 | nr | 3139.68 |
| 40 kVA | 3106.95 | 3417.64 | 12.05 | 164.67 | nr | 3582.31 |
| 50 kVA | 3771.60 | 4148.76 | 14.08 | 192.51 | nr | 4341.27 |
| Three phase, 380/400/415 Volt AC, 50 Hz | | | | | | |
| 3.0 kVA | 987.00 | 1085.70 | 4.00 | 54.67 | nr | 1140.37 |
| 6.0 kVA | 1207.50 | 1328.25 | 4.00 | 54.67 | nr | 1382.92 |
| 9.0 kVA | 1522.50 | 1674.75 | 6.02 | 82.34 | nr | 1757.09 |
| 13.0 kVA | 2056.95 | 2262.64 | 7.04 | 96.25 | nr | 2358.89 |
| 18.0 kVA | 2644.95 | 2909.45 | 8.00 | 109.34 | nr | 3018.79 |
| 27.0 kVA | 3162.60 | 3478.86 | 8.00 | 109.34 | nr | 3588.20 |
| 36.0 kVA | 3541.65 | 3895.82 | 10.00 | 136.68 | nr | 4032.50 |
| 50.0 kVA | 4440.45 | 4884.49 | 14.08 | 192.51 | nr | 5077.00 |
| 60.0 kVA | 7295.40 | 8024.94 | 16.13 | 220.45 | nr | 8245.39 |
| 75.0 kVA | 8415.75 | 9257.33 | 16.13 | 220.45 | nr | 9477.78 |
| 100.0 kVA | 7951.65 | 8746.82 | 16.13 | 220.45 | nr | 8967.27 |
| 125.0 kVA | 8523.90 | 9376.29 | 18.18 | 248.51 | nr | 9624.80 |
| 150.0 kVA | 8941.80 | 9835.98 | 24.39 | 333.37 | nr | 10169.35 |
| Line conditioners; solid state circuitry; sheet steel enclosure; maintains output +/- 2% over zero to full load with input voltage swings of +/- 15%; discrimination against voltage spikes and r.f. interference; positioned on site, final connections and commissioning. (Includes delivery up to 75 miles.) | | | | | | |
| Single phase, 240 Volt AC, 50 Hz | | | | | | |
| 4.5 kVA | 1152.90 | 1268.19 | 4.00 | 54.67 | nr | 1322.86 |
| 6.0 kVA | 1503.60 | 1653.96 | 4.00 | 54.67 | nr | 1708.63 |
| 9.0 kVA | 1776.60 | 1954.26 | 6.02 | 82.34 | nr | 2036.60 |
| 12.0 kVA | 1932.00 | 2125.20 | 7.04 | 96.25 | nr | 2221.45 |

## Y:MECHANICAL AND ELECTRICAL SERVICES

| Item | Net Price £ | Material £ | Labour hours | Labour £ | Unit | Total Rate £ |
|---|---|---|---|---|---|---|
| **Y71 : LV SWITCHGEAR AND DISTRIBUTION BOARDS : VOLTAGE STABLISERS/LINE CONDITIONERS (contd)** | | | | | | |
| Line conditioners; solid state circuitry; sheet steel enclosure; maintains output +/- 2% over zero to full load with input voltage swings of +/- 15%; discrimination against voltage spikes and r.f. interference; positioned on site, final connections and commissioning. (Includes delivery up to 75 miles.) (Contd) | | | | | | |
| Three phase, 415 Volt AC, 50Hz | | | | | | |
| 18.0 kVA | 3354.75 | 3690.22 | 8.00 | 109.34 | nr | **3799.56** |
| 27.0 kVA | 4336.50 | 4770.15 | 8.00 | 109.34 | nr | **4879.49** |
| 60.0 kVA | 7295.40 | 8024.94 | 16.13 | 220.45 | nr | **8245.39** |
| 100.0 kVA | 10506.30 | 11556.93 | 16.13 | 220.45 | nr | **11777.38** |
| 150.0 kVA | 12348.00 | 13582.80 | 24.39 | 333.37 | nr | **13916.17** |

## Y:MECHANICAL AND ELECTRICAL SERVICES

| Item | Net Price £ | Material £ | Labour hours | Labour £ | Unit | Total Rate £ |
|---|---|---|---|---|---|---|
| **Y71 : LV SWITCHGEAR AND DISTRIBUTION BOARDS : FUSIBLE BOARDS** | | | | | | |
| **Fusible Links for Distribution Boards** | | | | | | |
| Extra for fuse links; Q1 fusing factor; BS 88 Part 2; fitting and connecting | | | | | | |
| 6 Amp | 4.25 | 4.67 | 0.10 | 1.37 | nr | 6.04 |
| 10 Amp | 4.25 | 4.67 | 0.10 | 1.37 | nr | 6.04 |
| 16 Amp | 4.25 | 4.67 | 0.10 | 1.37 | nr | 6.04 |
| 20 Amp | 4.25 | 4.67 | 0.10 | 1.37 | nr | 6.04 |
| 32 Amp | 4.60 | 5.06 | 0.10 | 1.37 | nr | 6.43 |
| 50 Amp | 4.85 | 5.33 | 0.10 | 1.37 | nr | 6.70 |
| 63 Amp | 4.85 | 5.33 | 0.10 | 1.37 | nr | 6.70 |
| 80 Amp | 7.96 | 8.76 | 0.10 | 1.37 | nr | 10.13 |
| 100 Amp | 7.96 | 8.76 | 0.10 | 1.37 | nr | 10.13 |
| 125 Amp | 13.25 | 14.57 | 0.10 | 1.37 | nr | 15.94 |
| 160 Amp | 14.31 | 15.74 | 0.10 | 1.37 | nr | 17.11 |
| 200 Amp | 15.12 | 16.63 | 0.10 | 1.37 | nr | 18.00 |
| 250 Amp | 19.49 | 21.44 | 0.10 | 1.37 | nr | 22.81 |
| 315 Amp | 23.58 | 25.94 | 0.10 | 1.37 | nr | 27.31 |
| 355 Amp | 29.43 | 32.37 | 0.10 | 1.37 | nr | 33.74 |
| 400 Amp | 32.31 | 35.54 | 0.10 | 1.37 | nr | 36.91 |
| 560 Amp | 50.29 | 55.32 | 0.10 | 1.37 | nr | 56.69 |
| 630 Amp | 59.74 | 65.71 | 0.10 | 1.37 | nr | 67.08 |
| 710 Amp | 75.53 | 83.08 | 0.10 | 1.37 | nr | 84.45 |
| 750 Amp | 82.23 | 90.45 | 0.10 | 1.37 | nr | 91.82 |
| 800 Amp | 96.36 | 106.00 | 0.10 | 1.37 | nr | 107.37 |
| Extra for fuse links; extended motor range, fitting and connecting motor circuit | | | | | | |
| 20 M25 (3HP) | 2.03 | 2.23 | 0.10 | 1.37 | nr | 3.60 |
| 20 M32 (4HP) | 2.03 | 2.23 | 0.10 | 1.37 | nr | 3.60 |
| 32 M35 (5.5HP) | 2.77 | 3.05 | 0.10 | 1.37 | nr | 4.42 |
| 32 M50 (7.5HP) | 2.94 | 3.23 | 0.10 | 1.37 | nr | 4.60 |
| 32 M63 (10HP) | 3.23 | 3.55 | 0.10 | 1.37 | nr | 4.92 |
| 63 M80 (15HP) | 6.09 | 6.70 | 0.10 | 1.37 | nr | 8.07 |
| 63 M100 (20HP) | 6.33 | 6.96 | 0.10 | 1.37 | nr | 8.33 |
| 100 M125 (25HP) | 9.69 | 10.66 | 0.10 | 1.37 | nr | 12.03 |
| 100 M160 (30HP) | 9.83 | 10.81 | 0.10 | 1.37 | nr | 12.18 |
| 100 M200 (50HP) | 11.05 | 12.15 | 0.10 | 1.37 | nr | 13.52 |
| 200 M250 (60HP) | 16.99 | 18.69 | 0.10 | 1.37 | nr | 20.06 |
| 200 M315 (70HP) | 19.52 | 21.47 | 0.10 | 1.37 | nr | 22.84 |
| 315 M355 (100HP) | 29.37 | 32.31 | 0.10 | 1.37 | nr | 33.68 |
| 400 M450 (150HP) | 40.93 | 45.02 | 0.10 | 1.37 | nr | 46.39 |

Material Costs / Measured Work Prices - Electrical Installations

Y:MECHANICAL AND ELECTRICAL SERVICES

| Item | Net Price £ | Material £ | Labour hours | Labour £ | Unit | Total Rate £ |
|---|---|---|---|---|---|---|
| **Y72 : CONTACTORS AND STARTERS : CONTACTORS/PUSH BUTTONS** | | | | | | |
| Contactor Relays; pressed steel enclosure; fixed to backgrounds including supports, fixings, connections/jointing to equipment. (Provision and fixing of plates, discs, etc, for identification is included.) | | | | | | |
| Relays | | | | | | |
| 6 Amp, 415/240 Volt, 4 pole N/O | 25.99 | 28.59 | 0.50 | 6.83 | nr | **35.42** |
| 6 Amp, 415/240 Volt, 8 pole N/O | 31.69 | 34.86 | 0.75 | 10.25 | nr | **45.11** |
| Push Button Stations; Heavy gauge pressed steel enclosure; polycarbonate cover; IP65; fixed to backgrounds including supports, fixings,connections/joinging to equipment. (plates, discs, etc, for identification is included) | | | | | | |
| Standard Units | | | | | | |
| One button (start or stop) | 9.90 | 10.89 | 0.33 | 4.51 | nr | **15.40** |
| Two button (start or stop) | 15.60 | 17.16 | 0.42 | 5.74 | nr | **22.90** |
| Three button (forward-reverse-stop) | 21.58 | 23.74 | 0.50 | 6.83 | nr | **30.57** |

## Y:MECHANICAL AND ELECTRICAL SERVICES

| Item | Net Price £ | Material £ | Labour hours | Labour £ | Unit | Total Rate £ |
|---|---|---|---|---|---|---|
| **Y73 : LUMINAIRES AND LAMPS : LUMINAIRES** | | | | | | |
| **Fluorescent Luminaires; surface fixed to backgrounds. (Including supports and fixings.)** | | | | | | |
| Fluorescent light fittings; batten type; tubes fitted; surface mounted | | | | | | |
| 600 mm Single - 18 W | 12.34 | 13.57 | 0.50 | 6.83 | nr | **20.40** |
| 600 mm Twin - 18 W | 21.85 | 24.04 | 0.50 | 6.83 | nr | **30.87** |
| 1200 mm Single - 36 W | 16.32 | 17.95 | 0.65 | 8.89 | nr | **26.84** |
| 1200 mm Twin - 36 W | 31.29 | 34.42 | 0.65 | 8.89 | nr | **43.31** |
| 1500 mm Single - 58 W | 18.46 | 20.31 | 0.75 | 10.25 | nr | **30.56** |
| 1500 mm Twin - 58 W | 37.17 | 40.89 | 0.75 | 10.25 | nr | **51.14** |
| 1800 mm Single - 70 W | 22.30 | 24.53 | 1.00 | 13.67 | nr | **38.20** |
| 1800 mm Twin - 70 W | 40.75 | 44.83 | 1.00 | 13.67 | nr | **58.50** |
| 2400 mm Single - 100 W | 30.50 | 33.55 | 1.25 | 17.09 | nr | **50.64** |
| 2400 mm Twin - 100 W | 53.39 | 58.73 | 1.25 | 17.09 | nr | **75.82** |
| Surface mounted, reeded diffuser; tubes fitted | | | | | | |
| 600 mm Twin - 18 W | 8.47 | 9.32 | 0.50 | 6.83 | nr | **16.15** |
| 1200 mm Single - 36 W | 9.23 | 10.15 | 0.65 | 8.89 | nr | **19.04** |
| 1200 mm Twin - 36 W | 18.93 | 20.82 | 0.65 | 8.89 | nr | **29.71** |
| 1500 mm Single - 58 W | 10.41 | 11.45 | 0.75 | 10.25 | nr | **21.70** |
| 1500 mm Twin - 58 W | 23.54 | 25.89 | 0.75 | 10.25 | nr | **36.14** |
| 1800 mm Single - 70 W | 11.96 | 13.16 | 1.00 | 13.67 | nr | **26.83** |
| 1800 mm Twin - 70 W | 27.34 | 30.07 | 1.00 | 13.67 | nr | **43.74** |
| 2400 mm Single - 100 W | 17.66 | 19.43 | 1.25 | 17.09 | nr | **36.52** |
| 2400 mm Twin - 100 W | 36.14 | 39.75 | 1.25 | 17.09 | nr | **56.84** |
| Modular type; recessed mounting; opal diffuser; switchstart; tubes fitted | | | | | | |
| 300 x 1200 mm - 36 W, 1 lamp | 101.35 | 111.48 | 1.50 | 20.52 | nr | **132.00** |
| 300 x 1200 mm - 36 W, 2 lamp | 103.25 | 113.58 | 1.50 | 20.52 | nr | **134.10** |
| 600 x 600 mm - 40 W, 2 lamp | 96.79 | 106.48 | 1.50 | 20.52 | nr | **127.00** |
| 600 x 600 mm - 40 W, 3 lamp | 110.96 | 122.07 | 1.50 | 20.52 | nr | **142.59** |
| 600 x 1200 mm - 36 W, 3 lamp | 162.50 | 178.75 | 1.75 | 23.94 | nr | **202.69** |
| 600 x 1200 mm - 36 W, 4 lamp | 167.26 | 183.99 | 1.75 | 23.94 | nr | **207.93** |
| 600 x 1800 mm - 70 W, 2 lamp | 113.06 | 124.37 | 2.00 | 27.34 | nr | **151.71** |
| 600 x 1800 mm - 70 W, 3 lamp | 129.54 | 142.48 | 2.15 | 29.39 | nr | **171.87** |
| Corrosion resistant GRP body; gasket sealed; acrylic diffuser; tubes fitted | | | | | | |
| 600 mm Single - 18 W | 24.93 | 27.42 | 0.50 | 6.83 | nr | **34.25** |
| 600 mm Twin - 18 W | 34.57 | 38.03 | 0.50 | 6.83 | nr | **44.86** |
| 1200 mm Single - 36 W | 31.14 | 34.25 | 0.65 | 8.89 | nr | **43.14** |
| 1200 mm Twin - 36 W | 41.89 | 46.08 | 0.65 | 8.89 | nr | **54.97** |
| 1500 mm Single - 58 W | 32.67 | 35.94 | 0.75 | 10.25 | nr | **46.19** |
| 1500 mm Twin - 58 W | 45.14 | 49.65 | 0.75 | 10.25 | nr | **59.90** |
| 1800 mm Single - 70 W | 38.21 | 42.03 | 1.00 | 13.67 | nr | **55.70** |
| 1800 mm Twin - 70 W | 55.19 | 60.71 | 1.00 | 13.67 | nr | **74.38** |

*Material Costs / Measured Work Prices - Electrical Installations*

## Y:MECHANICAL AND ELECTRICAL SERVICES

| Item | Net Price £ | Material £ | Labour hours | Labour £ | Unit | Total Rate £ |
|---|---|---|---|---|---|---|
| **Y73 : LUMINAIRES AND LAMPS : LUMINAIRES (contd)** | | | | | | |
| **Fluorescent Luminaires; surface fixed to backgrounds. (Including supports and fixings (contd)** | | | | | | |
| Flameproof to IIA/IIB,I.P. 64; Aluminium Body; BS 229 and 899; Tubes fitted | | | | | | |
| 600 mm Single - 18 W | 298.46 | 328.31 | 1.00 | 13.67 | nr | **341.98** |
| 600 mm Twin - 18 W | 371.82 | 409.00 | 1.00 | 13.67 | nr | **422.67** |
| 1200 mm Single - 36 W | 326.94 | 359.63 | 1.25 | 17.09 | nr | **376.72** |
| 1200 mm Twin - 36 W | 404.20 | 444.62 | 1.00 | 13.67 | nr | **458.29** |
| 1500 mm Single - 58 W | 350.01 | 385.01 | 1.50 | 20.52 | nr | **405.53** |
| 1500 mm Twin - 58 W | 422.64 | 464.90 | 1.50 | 20.52 | nr | **485.42** |
| 1800 mm Single - 70 W | 383.69 | 422.06 | 1.75 | 23.94 | nr | **446.00** |
| 1800 mm Twin - 70 W | 442.65 | 486.92 | 1.75 | 23.98 | nr | **510.90** |
| **Emergency Lighting Fittings; surface fixed to backgrounds. (Including supports and fixings.)** | | | | | | |
| Self contained non-maintained 3 hour duration; aluminium body; glass diffuser; vandel resistant fixings; IP65; | | | | | | |
| 4 W | 63.25 | 69.58 | 1.00 | 13.67 | nr | **83.25** |
| Self contained non-maintained 3 hour duration; polycarbonate body; prismatic diffuser; vandel resistant fixings; IP65; | | | | | | |
| 4 W | 53.34 | 58.67 | 1.00 | 13.67 | nr | **72.34** |

## Y:MECHANICAL AND ELECTRICAL SERVICES

| Item | Net Price £ | Material £ | Labour hours | Labour £ | Unit | Total Rate £ |
|---|---|---|---|---|---|---|
| **Y74 : ACCESSORIES FOR ELECTRICAL SERVICES : SWITCHES** | | | | | | |
| **Local Switches; fixed to backgrounds (Including fixings.)** | | | | | | |
| 6 Amp metal clad surface mounted switch, gridswitch; one way | | | | | | |
| 1 Gang | 2.87 | 3.17 | 0.50 | 6.83 | nr | **10.00** |
| 2 Gang | 2.89 | 3.18 | 0.70 | 9.57 | nr | **12.75** |
| 3 Gang | 4.97 | 5.47 | 1.00 | 13.67 | nr | **19.14** |
| 4 Gang | 4.97 | 5.47 | 1.17 | 16.01 | nr | **21.48** |
| 6 Gang | 8.59 | 9.46 | 1.35 | 18.47 | nr | **27.93** |
| 8 Gang | 8.59 | 9.46 | 1.50 | 20.52 | nr | **29.98** |
| 12 Gang | 12.44 | 13.67 | 1.75 | 23.94 | nr | **37.61** |
| Extra for; Blank inset | | | | | | |
| 6 Amp - Two way switch | 1.93 | 2.12 | 0.03 | 0.41 | nr | **2.53** |
| 20 Amp - Two way switch | 2.49 | 2.74 | 0.04 | 0.55 | nr | **3.29** |
| 20 Amp - Intermediate | 4.48 | 4.93 | 0.09 | 1.23 | nr | **6.16** |
| 20 Amp - One way SP switch | 2.82 | 3.10 | 0.09 | 1.23 | nr | **4.33** |
| Steel blank plate; 1 Gang | 0.98 | 1.08 | 0.03 | 0.41 | nr | **1.49** |
| Steel blank plate; 2 Gang | 1.44 | 1.58 | 0.03 | 0.41 | nr | **1.99** |
| 6 Amp modular type switch; galvanised steel box, bronze or satin chrome coverplate; metalclad switches; flush mounting; one way | | | | | | |
| 1 Gang | 12.41 | 13.65 | 0.50 | 6.83 | nr | **20.48** |
| 2 Gang | 11.26 | 12.39 | 0.70 | 9.57 | nr | **21.96** |
| 3 Gang | 15.19 | 16.70 | 1.00 | 13.67 | nr | **30.37** |
| 4 Gang | 15.46 | 17.00 | 1.17 | 16.01 | nr | **33.01** |
| 6 Gang | 21.03 | 23.13 | 1.50 | 20.52 | nr | **43.65** |
| 8 Gang | 21.04 | 23.14 | 2.20 | 30.11 | nr | **53.25** |
| 9 Gang | 24.65 | 27.11 | 2.50 | 34.17 | nr | **61.28** |
| 12 Gang | 24.65 | 27.11 | 3.00 | 41.05 | nr | **68.16** |
| 6 Amp modular type swtich; galvanised steel box; bronze or satin chrome coverplate; flush mounting; two way | | | | | | |
| 1 Gang | 12.97 | 14.27 | 0.50 | 6.83 | nr | **21.10** |
| 2 Gang | 11.81 | 12.99 | 0.70 | 9.57 | nr | **22.56** |
| 3 Gang | 15.73 | 17.30 | 1.00 | 13.67 | nr | **30.97** |
| 4 Gang | 16.01 | 17.61 | 1.17 | 16.01 | nr | **33.62** |
| 6 Gang | 16.01 | 17.61 | 1.50 | 20.52 | nr | **38.13** |
| 8 Gang | 21.60 | 23.76 | 2.20 | 30.11 | nr | **53.87** |
| 9 Gang | 25.20 | 27.73 | 2.50 | 34.17 | nr | **61.90** |
| 12 Gang | 25.20 | 27.73 | 3.00 | 41.05 | nr | **68.78** |

## Material Costs / Measured Work Prices - Electrical Installations

### Y:MECHANICAL AND ELECTRICAL SERVICES

| Item | Net Price £ | Material £ | Labour hours | Labour £ | Unit | Total Rate £ |
|---|---|---|---|---|---|---|
| **Y74 : ACCESSORIES FOR ELECTRICAL SERVICES : SWITCHES (contd)** | | | | | | |
| **Local Switches; fixed to backgrounds (Including fixings.) (Contd)** | | | | | | |
| Plate switches; 10 Amp flush mounted, white plastic fronted; 16mm metal box; fitted brass earth terminal | | | | | | |
| 1 Gang 1 Way, Single Pole | 2.24 | 2.46 | 0.25 | 3.42 | nr | 5.88 |
| 1 Gang 2 Way, Single Pole | 2.49 | 2.74 | 0.33 | 4.51 | nr | 7.25 |
| 2 Gang 2 Way, Single Pole | 3.85 | 4.24 | 0.49 | 6.70 | nr | 10.94 |
| 3 Gang 2 Way, Single Pole | 5.72 | 6.29 | 0.60 | 8.20 | nr | 14.49 |
| 1 Gang Intermediate | 5.46 | 6.01 | 0.50 | 6.83 | nr | 12.84 |
| 1 Gang 1 Way, Double Pole | 4.90 | 5.39 | 0.33 | 4.51 | nr | 9.90 |
| 1 Gang Single Pole with bell symbol | 4.01 | 4.41 | 0.25 | 3.42 | nr | 7.83 |
| 1 Gang Single Pole marked "PRESS" | 3.71 | 4.08 | 0.25 | 3.42 | nr | 7.50 |
| Time delay switch, suppressed | 31.19 | 34.31 | 0.44 | 6.02 | nr | 40.33 |
| Plate switches; 6 Amp flush mounted white plastic fronted; 25mm metal box; fitted brass earth terminal | | | | | | |
| 4 Gang 2 Way, Single Pole | 9.63 | 10.58 | 0.25 | 3.42 | nr | 14.00 |
| 6 Gang 2 Way, Single Way | 18.10 | 19.91 | 0.25 | 3.42 | nr | 23.33 |
| Architrave plate switches; 6 Amp flush mounted, white plastic fronted; 27mm metal box; brass earth terminal | | | | | | |
| 1 Gang 2 Way, Single Pole | 2.65 | 2.92 | 0.25 | 3.42 | nr | 6.34 |
| 2 Gang 2 Way, Single Pole | 5.49 | 6.04 | 0.25 | 3.42 | nr | 9.46 |
| Ceiling switches, white moulded plastic, pull cord; standard unit | | | | | | |
| 6 Amp, 1 Way, Single Pole | 3.55 | 3.90 | 0.25 | 3.42 | nr | 7.32 |
| 6 Amp, 2 Way, Single Pole | 4.06 | 4.47 | 0.30 | 4.10 | nr | 8.57 |
| 16 Amp, 1 Way, Double Pole | 5.45 | 6.00 | 0.35 | 4.78 | nr | 10.78 |
| 45 Amp, 1 Way, Double Pole with neon indicator | 10.50 | 11.55 | 0.40 | 5.47 | nr | 17.02 |
| 10 Amp splash proof moulded switch with plain, threaded or PVC entry | | | | | | |
| 1 Gang,2 Way; (2630) | 10.81 | 11.89 | 0.35 | 4.78 | nr | 16.67 |
| 2 Gang, 1 Way;(2631) | 12.91 | 14.20 | 0.40 | 5.47 | nr | 19.67 |
| 2 Gang, 2 Way;(2632) | 19.48 | 21.43 | 0.40 | 5.47 | nr | 26.90 |
| 6 Amp Watertight switch; metalclad; BS 3676; ingress protected to IP65 surface mounted | | | | | | |
| 1 Gang, 2 Way; terminal entry | 16.32 | 19.30 | 0.38 | 5.20 | nr | 24.50 |
| 1 Gang, 2 Way; through entry | 17.44 | 20.62 | 0.41 | 5.60 | nr | 26.22 |
| 2 Gang, 2 Way; terminal entry | 26.62 | 31.48 | 0.54 | 7.38 | nr | 38.86 |
| 2 Gang, 2 Way; through entry | 27.73 | 32.79 | 0.57 | 7.79 | nr | 40.58 |
| 2 Way replacement switch | 2.98 | 3.52 | 0.06 | 0.82 | nr | 4.34 |

## Y:MECHANICAL AND ELECTRICAL SERVICES

| Item | Net Price £ | Material £ | Labour hours | Labour £ | Unit | Total Rate £ |
|---|---|---|---|---|---|---|
| 15 Amp Watertight switch; metalclad; BS 3676; ingress protected to IP65; surface mounted | | | | | | |
| 1 Gang 2 Way, terminal entry | 17.25 | 20.40 | 0.40 | 5.47 | nr | **25.87** |
| 1 Gang 2 Way, through entry | 18.47 | 21.84 | 0.43 | 5.88 | nr | **27.72** |
| 2 Gang 2 Way, terminal entry | 28.67 | 33.90 | 0.56 | 7.66 | nr | **41.56** |
| 2 Gang 2 Way, through entry | 29.61 | 35.01 | 0.59 | 8.07 | nr | **43.08** |
| Intermediate interior only | 5.18 | 6.13 | 0.08 | 1.09 | nr | **7.22** |
| 2 way interior only | 3.84 | 4.54 | 0.08 | 1.09 | nr | **5.63** |
| Double pole interior only | 4.64 | 5.49 | 0.08 | 1.09 | nr | **6.58** |
| **Electrical Accessories; fixed to backgrounds (Including fixings.)** | | | | | | |
| Dimmer switches; Rotary action; for individual lights; moulded plastic case; metal backbox; flush mounted | | | | | | |
| 1 Gang, 1 Way; 250 Watt | 9.39 | 10.33 | 0.35 | 4.78 | nr | **15.11** |
| 1 Gang, 1 Way; 400 Watt | 14.39 | 15.83 | 0.35 | 4.78 | nr | **20.61** |
| Dimmer switches; Push on/off action; for individual lights; moulded plastic case; metal backbox; flush mounted | | | | | | |
| 1 Gang, 2 Way; 250 Watt | 14.39 | 15.83 | 0.38 | 5.20 | nr | **21.03** |
| 3 Gang, 2 Way; 250 Watt | 50.24 | 55.26 | 0.53 | 7.25 | nr | **62.51** |
| 4 Gang, 2 Way; 250 Watt | 69.20 | 76.13 | 0.62 | 8.48 | nr | **84.61** |
| Dimmer switches; Rotary action; metal clad; metal backbox; BS 5518 and BS 800;flush mounted | | | | | | |
| 1 Gang, 1 Way; 400 Watt | 28.29 | 31.12 | 0.35 | 4.78 | nr | **35.90** |

## Y:MECHANICAL AND ELECTRICAL SERVICES

| Item | Net Price £ | Material £ | Labour hours | Labour £ | Unit | Total Rate £ |
|---|---|---|---|---|---|---|
| **Y74 : ACCESSORIES FOR ELECTRICAL SERVICES : SOCKET OUTLETS** | | | | | | |
| **Electrical Accessories; fixed to backgrounds. (Including fixings.)** | | | | | | |
| Socket outlet: unswitched; 13 Amp metal clad; BS 1363; galvanised steel box and coverplate with white plastic inserts; fixed surface mounted | | | | | | |
| 1 Gang | 4.28 | 6.00 | 0.45 | 6.15 | nr | **12.15** |
| 2 Gang | 9.88 | 10.87 | 0.45 | 6.15 | nr | **17.02** |
| Socket outlet: switched; 13 Amp metal clad; BS 1363; galvanised steel box and coverplate with white plastic inserts; fixed surface mounted | | | | | | |
| 1 Gang | 7.23 | 7.95 | 0.45 | 6.15 | nr | **14.10** |
| 2 Gang | 13.26 | 14.59 | 0.45 | 6.15 | nr | **20.74** |
| Socket outlet: switched with neon indicator; 13 Amp metal clad; BS 1363; galvanised steel box and coverplate with white plastic inserts; fixed surface mounted | | | | | | |
| 1 Gang | 9.62 | 10.58 | 0.45 | 6.15 | nr | **16.73** |
| 2 Gang | 15.98 | 19.49 | 0.45 | 6.15 | nr | **25.64** |
| Socket outlet: unswitched; 13 Amp; BS 1363; white moulded plastic box and coverplate; fixed surface mounted | | | | | | |
| 1 Gang | 4.27 | 4.70 | 0.45 | 6.15 | nr | **10.85** |
| 2 Gang | 8.16 | 8.98 | 0.45 | 6.15 | nr | **15.13** |
| Socket outlet; switched; 13 Amp; BS 1363; white moulded plastic box and coverplate; fixed surface mounted | | | | | | |
| 1 Gang | 4.80 | 5.28 | 0.45 | 6.15 | nr | **11.43** |
| 2 Gang | 9.25 | 10.18 | 0.45 | 6.15 | nr | **16.33** |
| Socket outlet: switched with neon indicator; 13 Amp; BS 1363; white moulded plastic box and coverplate; fixed surface mounted | | | | | | |
| 1 Gang | 7.05 | 7.75 | 0.45 | 6.15 | nr | **13.90** |
| 2 Gang | 13.27 | 14.60 | 0.45 | 6.15 | nr | **20.75** |
| Socket outlet: switched; 13 Amp; BS 1363; galvanised steel box, white moulded coverplate; flush fitted | | | | | | |
| 1 Gang | 4.22 | 4.64 | 0.45 | 6.15 | nr | **10.79** |
| 2 Gang | 7.49 | 8.24 | 0.45 | 6.15 | nr | **14.39** |

## Y:MECHANICAL AND ELECTRICAL SERVICES

| Item | Net Price £ | Material £ | Labour hours | Labour £ | Unit | Total Rate £ |
|---|---|---|---|---|---|---|
| Socket outlet: switched with neon indicator; 13 Amp; BS 1363; galvanised steel box, white moulded coverplate; flush fixed | | | | | | |
| 1 Gang | 6.46 | 7.12 | 0.45 | 6.15 | nr | **13.27** |
| 2 Gang | 11.51 | 12.66 | 0.45 | 6.15 | nr | **18.81** |
| Socket outlet: switched; 13 Amp; BS 1363; galvanised steel box, satin chrome coverplate; BS 4662; flush fixed | | | | | | |
| 1 Gang | 11.08 | 12.19 | 0.45 | 6.15 | nr | **18.34** |
| 2 Gang | 13.91 | 15.30 | 0.45 | 6.15 | nr | **21.45** |
| Socket outlet: switched with neon indicator; 13 Amp; BS 1363; steel backbox, satin chrome coverplate; BS 4662; flush fixed | | | | | | |
| 1 Gang | 13.12 | 14.43 | 0.45 | 6.15 | nr | **20.58** |
| 2 Gang | 22.02 | 26.98 | 0.45 | 6.15 | nr | **33.13** |
| Weatherproof socket outlet: 13 Amp; switched; single gang; RCD protected; water and dust protected to I.P.66; surface mounted | | | | | | |
| 40A 30mA tripping current protecting 1 socket | 65.78 | 72.36 | 0.75 | 10.25 | nr | **82.61** |
| 40A 30mA tripping current protecting 2 sockets | 80.95 | 89.05 | 0.75 | 10.25 | nr | **99.30** |
| Plug for weatherproof socket outlet: protected to I.P.66 | | | | | | |
| 13Amp plug | 11.29 | 12.41 | 0.20 | 2.73 | nr | **15.15** |

## Material Costs / Measured Work Prices - Electrical Installations

### Y:MECHANICAL AND ELECTRICAL SERVICES

| Item | Net Price £ | Material £ | Labour hours | Labour £ | Unit | Total Rate £ |
|---|---|---|---|---|---|---|
| **Y74 : ACCESSORIES FOR ELECTRICAL SERVICES : CONNECTOR UNITS** | | | | | | |
| **Electrical Accessories; fixed to backgrounds (Including fixings.)** | | | | | | |
| Connection units: Moulded pattern; BS 5733; moulded plastic box; white coverplate; knockout for flex outlet; surface mounted - standard fused | | | | | | |
|   DP Switched | 7.20 | 7.92 | 0.45 | 6.15 | nr | **14.07** |
|   Unswitched | 6.64 | 7.30 | 0.45 | 6.15 | nr | **13.45** |
|   DP Switched with neon indicator | 9.28 | 10.21 | 0.45 | 6.15 | nr | **16.36** |
| Connection units: moulded pattern; BS 5733; galvanised steel box; white coverplate; knockout for flex outlet; surface mounted | | | | | | |
|   DP Switched | 7.21 | 7.93 | 0.45 | 6.15 | nr | **14.08** |
|   Unswitched | 6.65 | 7.31 | 0.45 | 6.15 | nr | **13.46** |
|   DP Switched with neon indicator | 9.29 | 10.22 | 0.45 | 6.15 | nr | **16.37** |
| Connection units: galvanised pressed steel pattern; galvanised steel box; satin chrome or satin brass finish; white moulded plastic inserts; flush mounted - standard fused | | | | | | |
|   DP Switched | 9.67 | 10.64 | 0.45 | 6.15 | nr | **16.79** |
|   Unswitched | 8.56 | 9.43 | 0.45 | 6.15 | nr | **15.58** |
|   DP Switched with neon indicator | 10.53 | 12.87 | 0.45 | 6.15 | nr | **19.02** |
| Connection units: galvanised steel box; satin chrome or satin brass finish; white moulded plastic inserts; flex outlet; flush mounted - standard fused | | | | | | |
|   Switched | 10.93 | 12.03 | 0.45 | 6.15 | nr | **18.18** |
|   Unswitched | 10.32 | 11.35 | 0.45 | 6.15 | nr | **17.50** |
|   Switched with neon indicator | 13.04 | 14.36 | 0.45 | 6.15 | nr | **20.51** |

## Y:MECHANICAL AND ELECTRICAL SERVICES

| Item | Net Price £ | Material £ | Labour hours | Labour £ | Unit | Total Rate £ |
|---|---|---|---|---|---|---|
| **Y74 : ACCESSORIES FOR ELECTRICAL SERVICES : LAMP HOLDERS, OUTLETS AND CONTROL UNITS** | | | | | | |
| **Electrical Accessories; fixed to backgrounds. (Including fixings.)** | | | | | | |
| Telephone outlet: moulded plastic plate with box; fitted and connected; flush or surface mounted | | | | | | |
|    Single Master outlet | 5.49 | 6.04 | 0.30 | 4.10 | nr | 10.14 |
|    Single Secondary outlet | 3.84 | 4.22 | 0.30 | 4.10 | nr | 8.32 |
| Telephone outlet: bronze or satin chromeplate; with box; fitted and connected; flush or surface mounted | | | | | | |
|    Single Master outlet | 9.86 | 10.85 | 0.30 | 4.10 | nr | 14.95 |
|    Single Secondary outlet | 8.22 | 9.03 | 0.30 | 4.10 | nr | 13.13 |
| TV Co-Axial socket outlet: moulded plastic box: flush or surface mounted | | | | | | |
|    One way Direct Connection | 4.24 | 4.66 | 0.30 | 4.10 | nr | 8.76 |
|    Two way Direct Connection | 5.82 | 6.40 | 0.30 | 4.10 | nr | 10.50 |
|    One way Isolated UHF/VHF | 7.34 | 8.07 | 0.30 | 4.10 | nr | 12.17 |
|    Two way Isolated UHF/VHF | 10.01 | 11.01 | 0.30 | 4.10 | nr | 15.11 |
| Ceiling Rose: white moulded plastic; flush fixed to conduit box | | | | | | |
|    Plug in type; ceiling socket with 2 terminals, loop-in and ceiling plug with 3 terminals and cover | 3.57 | 3.93 | 0.33 | 4.51 | nr | 8.44 |
| BC Lampholder; white moulded plastic; Heat resistant PVC insulated and sheathed cable; flush fixed | | | | | | |
|    2 Core; 0.75mm2 | 1.49 | 1.64 | 0.30 | 4.10 | nr | 5.74 |
| Battern Holder: white moulded plastic; 3 terminals; BS 5042; fixed to conduit | | | | | | |
|    Straight pattern; 2 terminals with loop-in and Earth | 2.87 | 3.16 | 0.25 | 3.42 | nr | 6.58 |
|    Angled pattern ; looped in terminal | 3.57 | 3.93 | 0.25 | 3.42 | nr | 7.35 |
| Shaver Unit: self setting overload device; 200/250 voltage supply; white moulded plastic faceplate; unswitched | | | | | | |
|    Surface type with moulded plastic box | 14.22 | 15.64 | 0.42 | 5.74 | nr | 21.38 |
|    Flush type with galvanised steel box | 14.11 | 15.52 | 0.45 | 6.15 | nr | 21.67 |

## Y:MECHANICAL AND ELECTRICAL SERVICES

| Item | Net Price £ | Material £ | Labour hours | Labour £ | Unit | Total Rate £ |
|---|---|---|---|---|---|---|
| **Y74 : ACCESSORIES FOR ELECTRICAL SERVICES : LAMP HOLDERS, OUTLETS AND CONTROL UNITS (contd)** | | | | | | |
| **Electrical Accessories; fixed to backgrounds. (Including fixings.) (Contd)** | | | | | | |
| Shaver Unit: dual voltage supply unit; white moulded plastic faceplate; unswitched | | | | | | |
|   Surface type with moulded plastic box | 33.35 | 36.69 | 0.55 | 7.52 | nr | **44.21** |
|   Flush type with galvanised steel box | 32.59 | 35.85 | 0.60 | 8.20 | nr | **44.05** |
| Cooker Control Unit: BS 4177; 45 amp D.P. main switch; 13 Amp switched socket outlet; metal coverplate; plastic inserts; neon indicators | | | | | | |
|   Surface mounted with mounting box | 25.65 | 28.21 | 0.45 | 6.15 | nr | **34.36** |
|   Flush mounted with galvanised steel box | 22.89 | 25.17 | 0.45 | 6.15 | nr | **31.32** |
| Cooker Control Unit: BS 4177; 45 Amp D.P. main switch; 13 Amp switched socket outlet; moulded plastic box and coverplate; surface mounted | | | | | | |
|   Standard | 14.49 | 15.94 | 0.45 | 6.15 | nr | **22.09** |
|   With neon indicators | 19.03 | 20.93 | 0.45 | 6.15 | nr | **27.08** |

## Y:MECHANICAL AND ELECTRICAL SERVICES

| Item | Net Price £ | Material £ | Labour hours | Labour £ | Unit | Total Rate £ |
|---|---|---|---|---|---|---|
| **Y74 : ACCESSORIES FOR ELECTRICAL SERVICES : CABLE TILES** | | | | | | |
| Cable Tiles; single width; laid in trench above cables on prepared sand bed. (Costs of excavation and filling sand excluded.) | | | | | | |
| Reinforced concrete covers; concave/ convex ends | | | | | | |
| 914 * 152 * 63/38 mm | 1.67 | 1.84 | 0.10 | 1.37 | m | 3.21 |
| 914 * 229 * 63/38 mm | 2.45 | 2.69 | 0.10 | 1.37 | m | 4.06 |
| 914 * 305 * 63/38 mm | 3.28 | 3.61 | 0.10 | 1.37 | m | 4.98 |

# Keep your figures up to date, free of charge

This section, and most of the other information in this Price Book, is brought up to date every three months, until the next annual edition, in the *Price Book Update*.

The *Update* is available free to all Price Book purchasers.

To ensure you receive your copy, simply complete the reply card from the centre of the book and return it to us.

## Y:MECHANICAL AND ELECTRICAL SERVICES

| Item | Net Price £ | Material £ | Labour hours | Labour £ | Unit | Total Rate £ |
|---|---|---|---|---|---|---|
| **Y89 : SUNDRY COMMON ELECTRICAL ITEMS : JUNCTION BOXES** | | | | | | |
| **Weatherproof Junction Boxes; enclosures with rail mounted terminal blocks; side hung door to receive padlock; fixed to backgrounds, including all supports and fixings. (Suitable for cable up to 2.5mm2; including glandplates and gaskets.)** | | | | | | |
| Sheet steel with zinc spray finish enclosure | | | | | | |
| Overall Size 229 * 152; suitable to receive 3 * 20(A) glands per gland plate | 62.38 | 68.62 | 1.65 | 22.55 | nr | **91.17** |
| Overall Size 306 * 306; suitable to receive 14*20(A) glands per gland plate | 86.38 | 95.02 | 2.50 | 34.17 | nr | **129.19** |
| Overall Size 458 * 382; suitable to receive 18*20(A) glands per gland plate | 118.03 | 129.83 | 4.26 | 58.16 | nr | **187.99** |
| Overall Size 762 * 508; suitable to receive 26*20(A) glands per gland plate | 208.49 | 229.34 | 6.02 | 82.34 | nr | **311.68** |
| Overall Size 914 * 610; suitable to receive 45*20(A) glands per gland plate | 278.00 | 305.80 | 9.01 | 123.14 | nr | **428.94** |
| **Weatherproof Junction Boxes; enclosures with rail mounted terminal blocks; screwfixed lid; fixed to backgrounds, including all supports and fixings. (Suitable for cable up to 2.5mm2; including glandplates and gaskets.)** | | | | | | |
| In Fex glassfibre reinforced polycarbonate enclosure | | | | | | |
| Overall Size 190 * 190 * 130 | 19.52 | 21.47 | 1.65 | 22.55 | nr | **44.02** |
| Overall Size 190 * 190 * 180 | 23.35 | 25.68 | 1.80 | 24.58 | nr | **50.26** |
| Overall Size 280 * 190 * 130 | 25.70 | 28.27 | 2.50 | 34.17 | nr | **62.44** |
| Overall Size 280 * 190 * 180 | 28.50 | 31.35 | 2.80 | 38.29 | nr | **69.64** |
| Overall Size 380 * 190 * 130 | 30.44 | 33.48 | 4.00 | 54.67 | nr | **88.15** |
| Overall Size 380 * 190 * 180 | 34.69 | 38.16 | 3.95 | 54.02 | nr | **92.18** |
| Overall Size 380 * 280 * 130 | 39.92 | 43.91 | 5.99 | 81.84 | nr | **125.75** |
| Overall Size 380 * 280 * 180 | 45.61 | 50.17 | 7.04 | 96.25 | nr | **146.42** |
| Overall Size 560 * 280 * 130 | 57.62 | 63.38 | 9.01 | 123.14 | nr | **186.52** |
| Overall Size 560 * 380 * 180 | 80.99 | 89.09 | 10.00 | 136.68 | nr | **225.77** |

# PART FOUR
# Daywork

Heating and Ventilating Industry, *page 467*
Electrical Industry, *page 472*
Building Industry Plant Hire Costs, *page 476*

When work is carried out in connection with a contract which cannot be valued in any other way, it is usual to assess the value on a cost basis with suitable allowances to cover overheads and profit. The basis of costing is a matter for agreement between the parties concerned but definitions of prime cost for the Heating and Ventilating and Electrical Industries have been prepared and published jointly by the Royal Institution of Chartered Surveyors and the appropriate bodies of the industries concerned for the convenience of those who wish to use them together with a schedule of basic plant charges published by the Royal Institution of Chartered Surveyors. These documents are reproduced by kind permission of the publishers.

# THE CONSTRUCTION NET
### Online information sources for the construction industry

by *Alan H Bridges*
*Universities of Strathclyde (Glasgow) and Delft (the Netherlands)*

You are a busy architect/engineer/contractor who has just received an Internet connection and are thinking "so what!" With slow connections it is costing you time and money to search for information - and often the information you find is less than useful. This book provides a guide to the use of tools to improve use of the Net and a comprehensive guide to where the best information sources are to be found for the building design and construction professional.

Learn where to find design images or hard technical information; where to join discussion groups for advice on CAD software; and how to join mailing lists to automatically receive up-to-date information on your chosen special subject areas.

Using this book can help the busy professional to optimize online time by determining the key sites to visit before connection to the Internet. Topics are conveniently arranged by subject showing where to find the key "index sites" together with details of many specialist sites.

- helps the user access information faster

- quick guide to specialized information

- regular online updates

---

**For further information and to order please contact**
The Marketing Dept., E & FN Spon, 2-6 Boundary Row, London SE1 8HN
Tel: 0171 865 0066  Fax: 0171 522 9621

---

Published by E & F N Spon
246 x 189mm 256 pages
October 1996
Paperback 0-419-21780-0
£24.99

## HEATING AND VENTILATING INDUSTRY

## DEFINITION OF PRIME COST OF DAYWORK CARRIED OUT UNDER A HEATING, VENTILATING, AIR CONDITIONING, REFRIGERATION, PIPEWORK AND/OR DOMESTIC ENGINEERING CONTRACT (JULY 1980 EDITION)

This Definition of Prime Cost is published by the Royal Institution of Chartered Surveyors and the Heating and Ventilating Contractors Association for convenience, and for use by people who choose to use it. Members of the Heating and Ventilating Contractors Association are not in any way debarred from defining Prime Cost and rendering accounts for work carried out on that basis in any way they choose. Building owners are advised to reach agreement with contractors on the Definition of Prime Cost to be used prior to entering into a contract or sub-contract.

SECTION 1: APPLICATION

1.1 This Definition provides a basis for the valuation of daywork executed under such heating, ventilating, air conditioning, refrigeration, pipework and or domestic engineering contracts as provide for its use.

1.2 It is not applicable in any other circumstances, such as jobbing or other work carried out as a separate or main contract nor in the case of daywork executed after a date of practical completion.

1.3 The terms 'contract' and 'contractor' herein shall be read as 'sub-contract' and 'sub-contractor' as applicable.

SECTION 2: COMPOSITION OF TOTAL CHARGES

2.1 The Prime Cost of daywork comprises the sum of the following costs:
 (a) Labour as defined in Section 3.
 (b) Materials and goods as defined in Section 4.
 (c) Plant as defined in Section 5.

2.2 Incidental costs, overheads and profit as defined in Section 6, as provided in the contract and expressed therein as percentage adjustments, are applicable to each of 2.1 (a)-(c).

SECTION 3: LABOUR

3.1 The standard wage rates, emoluments and expenses referred to below and the standard working hours referred to in 3.2 are those laid down for the time being in the rules or decisions or agreements of the Joint Conciliation Committee of the Heating, Ventilating and Domestic Engineering Industry applicable to the works (or those of such other body as may be appropriate) and to the grade of operative concerned at the time when and the area where the daywork is executed.

3.2 Hourly base rates for labour are computed by dividing the annual prime cost of labour, based upon the standard working hours and as defined in 3.4, by the number of standard working hours per annum. See example.

3.3 The hourly rates computed in accordance with 3.2 shall be applied in respect of the time spent by operatives directly engaged on daywork, including those operating mechanical plant and transport and erecting and dismantling other plant (unless otherwise expressly provided in the contract) and handling and distributing the materials and goods used in the daywork.

3.4 The annual prime cost of labour comprises the following:
 (a) Standard weekly earnings (i.e. the standard working week as determined at the appropriate rate for the operative concerned).
 (b) Any supplemental payments.
 (c) Any guaranteed minimum payments (unless included in Section 6.1(a)-(p)).
 (d) Merit money.
 (e) Differentials or extra payments in respect of skill, responsibility, discomfort, inconvenience or risk (excluding those in respect of supervisory responsibility - see 3.5)
 (f) Payments in respect of public holidays.
 (g) Any amounts which may become payable by the contractor to or in respect of operatives arising from the rules etc. referred to in 3.1 which are not provided for in 3.4 (a)-(f) nor in Section 6.1 (a)-(p).

## *Daywork*

### HEATING AND VENTILATING INDUSTRY

    (h)    Employers contributions to the WELPLAN, the HVACR Welfare and Holiday Scheme or payments in lieu thereof.
    (i)    Employers National Insurance contributions as applicable to 3.4 (a)-(h).
    (j)    Any contribution, levy or tax imposed by Statute, payable by the contractor in his capacity as an employer.

3.5    Differentials or extra payments in respect of supervisory responsibility are excluded from the annual prime cost (see Section 6). The time of principals, staff, foremen, chargehands and the like when working manually is admissible under this Section at the rates for the appropriate grades.

### SECTION 4: MATERIALS AND GOODS

4.1    The prime cost of materials and goods obtained specifically for the daywork is the invoice cost after deducting all trade discounts and any portion of cash discounts in excess of 5%.

4.2    The prime cost of all other materials and goods used in the daywork is based upon the current market prices plus any appropriate handling charges.

4.3    The prime cost referred to in 4.1 and 4.2 includes the cost of delivery to site.

4.4    Any Value Added Tax which is treated, or is capable of being treated, as input tax (as defined by the Finance Act 1972, or any re-enactment or amendment thereof or substitution therefore) by the contractor is excluded.

### SECTION 5: PLANT

5.1    Unless otherwise stated in the contract, the prime cost of plant comprises the cost of the following:
    (a)    use or hire of mechanically-operated plant and transport for the time employed on and/or provided or retained for the daywork;
    (b)    use of non-mechanical plant (excluding non-mechanical hand tools) for the time employed on and/or provided or retained for the daywork;
    (c)    transport to and from the site and erection and dismantling where applicable.

5.2    The use of non-mechanical hand tools and of erected scaffolding, staging, trestles or the like is excluded (see Section 6), unless specifically retained for the daywork.

### SECTION 6: INCIDENTAL COSTS, OVERHEADS AND PROFIT

6.1    The percentage adjustments provided in the contract which are applicable to each of the totals of Sections 3, 4 and 5 comprise the following:
    (a)    Head office charges.
    (b)    Site staff including site supervision.
    (c)    The additional cost of overtime (other than that referred to in 6.2).
    (d)    Time lost due to inclement weather.
    (e)    The additional cost of bonuses and all other incentive payments in excess of any included in 3.4.
    (f)    Apprentices' study time.
    (g)    Fares and travelling allowances.
    (h)    Country, lodging and periodic allowances.
    (i)    Sick pay or insurances in respect thereof, other than as included in 3.4.
    (j)    Third party and employers' liability insurance.
    (k)    Liability in respect of redundancy payments to employees.
    (l)    Employer's National Insurance contributions not included in 3.4.
    (m)    Use and maintenance of non-mechanical hand tools.
    (n)    Use of erected scaffolding, staging, trestles or the like (but see 5.2).
    (o)    Use of tarpaulins, protective clothing, artificial lighting, safety and welfare facilities, storage and the like that may be available on site.
    (p)    Any variation to basic rates required by the contractor in cases where the contract provides for the use of a specified schedule of basic plant charges (to the extent that no other provision is made for such variation - see 5.1).

## HEATING AND VENTILATING INDUSTRY

- (q) In the case of a sub-contract which provides that the sub-contractor shall allow a cash discount, such provision as is necessary for the allowance of the prescribed rate of discount.
- (r) All other liabilities and obligations whatsoever not specifically referred to in this Section nor chargeable under any other Section.
- (s) Profit.

6.2 The additional cost of overtime where specifically ordered by the Architect/Supervising Officer shall only be chargeable in the terms of a prior written agreement between the parties.

## Daywork

### HEATING AND VENTILATING INDUSTRY

*Example of calculation of a typical standard hourly base rate (as defined in Section 3) for Advanced Fitter (qualified gas and arc) and Mate employed in the Heating, Ventilating, Air Conditioning, Piping and Domestic Engineering Industry under NJIC Rules based upon rates ruling at 11 November 1996.*

|  | Adv Fitter (qual gas/arc) | | Mate | |
|---|---|---|---|---|
|  | Rate/hr £ | Rate/annum £ | Rate/hr £ | Rate/annum £ |
| Standard Weekly Earnings - 46.2 wks x 38 hrs | 6.13 | 10761.28 | 4.49 | 7882.64 |
| Welding Supplement (gas and arc) - 46.2 wks x 38 hrs | 0.52 | 912.91 |  |  |
| Merit Money and other variables as applicable [See note (a)] |  | * |  | * |
|  |  | 11674.19 |  | 7882.64 |
| Employer's National Insurance Contribution | 10% | 1167.42 | 7% | 551.78 |
|  |  | 12841.61 |  | 8434.42 |
| Weekly HolidayCredit 52 weeks | 28.75 | 1495.00 | 18.46 | 959.92 |
| Welfare Premium 52 weeks | 3.85 | 200.20 | 3.85 | 200.20 |
| WELPLAN Sickness / Accident Welfare Premium - |  |  |  |  |
| 28 weeks | 127.33 | 3565.24 | 61.25 | 1715.00 |
| 24 weeks | 63.70 | 1528.80 | 30.59 | 734.16 |
| Annual Labour Cost as defined in Section 3 |  | £19630.85 |  | £12043.70 |
| Hourly base rates as defined in Section 3, Clause 3.2 | £19630.85 / 1755.6 = | **£11.18** | £12043.70 / 1755.6 = | **£6.86** |

Standard working hours per annum are calculated as follows:
  52 weeks at 38 hours                                        1976.0
Less:
  4.2 weeks holiday at 38 hours                       =        159.6
  8 days Public Holidays at (av.) 7.6 hrs per day    =         60.8        220.4
                                                                          1755.6

Notes

(a)  Where applicable, Merit Money and other variables (e.g. Daily Abnormal Conditions Money), which attract Employer's National Insurance Contribution, should be included at *. It should be noted that all labour costs incurred by the Contractor in his capacity as an Employer, other than those contained in the hourly base rate, as defined under Section 3, are to be taken into account under Section 6.

(b)  The above example is for the convenience of users only and does not form part of the Definition; all the basic costs are subject to re-examination according to the time when and where the daywork is executed.

# ELECTRICAL INDUSTRY

## DEFINITION OF PRIME COST OF DAYWORK CARRIED OUT UNDER AN ELECTRICAL CONTRACT (MARCH 1981 EDITION)

This Definition of Prime Cost is published by The Royal Institution of Chartered Surveyors and The Electrical Contractors' Associations for convenience and for use by people who choose to use it. Members of The Electrical Contractors' Association are not in any way debarred from defining Prime Cost and rendering accounts for work carried out on that basis in any way they choose. Building owners are advised to reach agreement with contractors on the Definition of Prime Cost to be used prior to entering into a contract or sub-contract.

SECTION 1: APPLICATION

1.1 This Definition provides a basis for the valuation of daywork executed under such electrical contracts as provide for its use.
1.2 It is not applicable in any other circumstances, such as jobbing, or other work carried out as a separate or main contract, nor in the case of daywork executed after the date of practical completion.
1.3 The terms 'contract' and 'contractor' herein shall be read as 'sub-contract' and 'sub-contractor' as the context may require.

SECTION 2: COMPOSITION OF TOTAL CHARGES

2.1 The Prime Cost of daywork comprises the sum of the following costs:
 (a) Labour as defined in Section 3.
 (b) Materials and goods as defined in Section 4.
 (c) Plant as defined in Section 5.
2.2 Incidental costs, overheads and profit as defined in Section 6, as provided in the contract and expressed therein as percentage adjustments, are applicable to each of 2.1 (a)-(c).

SECTION 3: LABOUR

3.1 The standard wage rates, emoluments and expenses referred to below and the standard working hours referred to in 3.2 are those laid down for the time being in the rules and determinations or decisions of the Joint Industry Board or the Scottish Joint Industry Board for the Electrical Contracting Industry (or those of such other body as may be appropriate) applicable to the works and relating to the grade of operative concerned at the time when and in the area where daywork is executed.
3.2 Hourly base rates for labour are computed by dividing the annual prime cost of labour, based upon the standard working hours and as defined in 3.4 by the number of standard working hours per annum. See examples.
3.3 The hourly rates computed in accordance with 3.2 shall be applied in respect of the time spent by operatives directly engaged on daywork, including those operating mechanical plant and transport and erecting and dismantling other plant (unless otherwise expressly provided in the contract) and handling and distributing the materials and goods used in the daywork.
3.4 The annual prime cost of labour comprises the following:
 (a) Standard weekly earnings (ie the standard working week as determined at the appropriate rate for the operative concerned).
 (b) Payments in respect of public holidays.
 (c) Any amounts which may become payable by the Contractor to or in respect of operatives arising from operation of the rules etc. referred to in 3.1 which are not provided for in 3.4(a) and (b) nor in Section 6.
 (d) Employer's National Insurance Contributions as applicable to 3.4 (a)-(c).
 (e) Employer's contributions to the Joint Industry Board Combined Benefits Scheme or Scottish Joint Industry Board Holiday and Welfare Stamp Scheme, and holiday payments made to apprentices in compliance with the Joint Industry Board National Working Rules and Industrial Determinations.
 (f) Any contribution, levy or tax imposed by Statute, payable by the Contractor in his capacity as an employer.

## ELECTRICAL INDUSTRY

3.5 Differentials or extra payments in respect of supervisory responsibility are excluded from the annual prime cost (see Section 6). The time of principals and similar categories, when working manually, is admissible under this Section at the rates for the appropriate grades.

SECTION 4: MATERIALS AND GOODS

4.1 The prime cost of materials and goods obtained specifically for the daywork is the invoice cost after deducting all trade discounts and any portion of cash discounts in excess of 5%.
4.2 The prime cost of all other materials and goods used in the daywork is based upon the current market prices plus any appropriate handling charges.
4.3 The prime cost referred to in 4.1 and 4.2 includes the cost of delivery to site.
4.4 Any Value Added Tax which is treated, or is capable of being treated, as input tax (as defined by the Finance Act 1972, or any re-enactment or amendment thereof or substitution therefore) by the Contractor is excluded.

SECTION 5: PLANT

5.1 Unless otherwise stated in the contract, the prime cost of plant comprises the cost of the following:
   (a) Use or hire of mechanically-operated plant and transport for the time employed on and/or provided or retained for the daywork;
   (b) Use of non-mechanical plant (excluding non-mechanical hand tools) for the time employed on and/or provided or retained for the daywork;
   (c) Transport to and from the site and erection and dismantling where applicable.
5.2 The use of non-mechanical hand tools and of erected scaffolding, staging, trestles or the likes is excluded (see Section 6), unless specifically retained for daywork.
5.3 Note: Where hired or other plant is operated by the Electrical Contractor's operatives, such time is to be included under Section 3 unless otherwise provided in the contract.

SECTION 6: INCIDENTAL COSTS, OVERHEADS AND PROFIT

6.1 The percentage adjustments provided in the contract which are applicable to each of the totals of Sections 3, 4 and 5, compromise the following:
   (a) Head Office charges.
   (b) Site staff including site supervision.
   (c) The additional cost of overtime (other than that referred to in 6.2).
   (d) Time lost due to inclement weather.
   (e) The additional cost of bonuses and other incentive payments.
   (f) Apprentices' study time.
   (g) Travelling time and fares.
   (h) Country and lodging allowances.
   (i) Sick pay or insurance in lieu thereof, in respect of apprentices.
   (j) Third party and employers' liability insurance.
   (k) Liability in respect of redundancy payments to employees.
   (l) Employers' National Insurance Contributions not included in 3.4.
   (m) Use and maintenance of non-mechanical hand tools.
   (n) Use of erected scaffolding, staging, trestles or the like (but see 5.2.).
   (o) Use of tarpaulins, protective clothing, artificial lighting, safety and welfare facilities, storage and the like that may be available on site.
   (p) Any variation to basic rates required by the Contractor in cases where the contract provides for the use of a specified schedule of basic plant charges (to the extent that no other provision is made for such variation - see 5.1).
   (q) All other liabilities and obligations whatsoever not specifically referred to in this Section nor chargeable under any other Section.
   (r) Profit.
   (s) In the case of a sub-contract which provides that the sub-contractor shall allow a cash discount, such provision as is necessary for the allowance of the prescribed rate of discount.
6.2 The additional cost of overtime where specifically ordered by the Architect/Supervising Officer shall only be chargeable in the terms of a prior written agreement between the parties.

*Daywork*

**ELECTRICAL INDUSTRY**

Example of calculation of typical standard hourly base rate (as defined in Section 3) for Approved Electrician and Senior Apprentice Stage 2 employed in the Electrical Contracting Industry under JIB Rules applicable to England, Wales, Northern Ireland, Isle of Man and the Channel Islands based upon rates ruling at 4 January 1997 (London Weighing).

|  | *Approved Elect.* | | *Senior Apprentice Stage 2* | |
|---|---|---|---|---|
|  | Rate/hr £ | Rate/annum £ | Rate/hr £ | Rate/annum £ |
| Standard Weekly Earnings | | | | |
| Approved Electrician | | | | |
| 46.2 weeks x 37.5 hrs = 1792.50 hrs | 7.06 | 12655.05 | | |
| Apprentice | | | | |
| 46.2 weeks x 37.5 hrs = 1950 hrs | | | | |
| Less Study Time | | | | |
| 40 weeks @ 75 hrs = 300 hrs | | | | |
| 1650 hrs | | | 4.00 | 6600.00 |
| [See note (c)] | | | | |
| Employer's National Insurance | | | | |
| Contributions @ | 10% | 1265.51 | 5% | 330.00 |
| Employer's Contributions to: | | | | |
| JIB Benefits Scheme 52 wks | 26.39 | 1372.28 | | |
| Annual labour costs as defined | | | | |
| in Section 3 | | £15292.84 | | £6930.00 |
| Hourly base rate defined | £15292.84 = **£8.82** | | £6930.00 = **£4.84** | |
| in Section 3, Clause 3.2 | 1732.50 | | 1432.50 | |

Standard working hours per annum calculated as follows:

| | | | | |
|---|---|---|---|---|
| 52 weeks @ 37½ hours | | | 1960.0 | 1950.0 |
| Less | | | | |
| 4.2 weeks holiday @ 37½ hrs = | 157.5 | | | |
| 8 days public holiday @ 7½ hrs = | 60.0 | 217.50 | 217.50 | |
| | | | 1732.50 | 1732.50 |
| Study time 40 weeks @ 7½ hrs = | | | 300.00 | |
| | | | | 1432.50 |

Notes
(a) It should be noted that all labour costs incurred by the Contractor in his capacity as an Employer, other than those contained in the hourly rate above, must be taken into account under Section 6.
(b) The above example is for the convenience of users only and does not form part of the rules laid down in the Definition; all the basic costs are subject to re-examination according to the time when and in the area where the daywork is executed.
(c) Public Holidays are included in annual earnings.

## Daywork

### ELECTRICAL INDUSTRY

*Example of calculation of typical standard hourly base rate (as defined in Section 3) for Approved Electrician and Electrician employed in the Electrical Contracting Industry under SJIB Rules applicable to Scotland based upon rates ruling at as at 4 January 1997.*

|  | Approved Elect. Rate/hr £ | Approved Elect. Rate/annum £ | Electrician Rate/hr £ | Electrician Rate/annum £ |
|---|---|---|---|---|
| Standard Weekly Earnings 46.2 weeks x 37.5 hrs = 1732.5 hrs | 6.64 | 11503.80 | 6.13 | 10620.25 |
| Employer's National Insurance Contribution @ 10% / 7% | 10% | 1150.38 | 7% | 743.42 |
| Employer's Contribution to: SJIB Holiday and Welfare Scheme, 52 weeks | 30.46 | 1583.92 | 28.51 | 1482.52 |
| Annual labour cost as defined in Section 3 |  | £14238.10 |  | £12846.19 |
| Hourly base rate defined in Section 3, Clause 3.2 | £14238.10 / 1732.50 = | **£8.22** | £12846.19 / 1732.50 = | **£7.42** |

Standard working hours per annum calculated as follows:

| | | |
|---|---|---|
| 52 weeks @ 37.5 hours | | 1950.0 |
| Less | | |
| 4.2 weeks holiday @ 37.5 hours | = 157.5 | |
| 8 days public holiday @ 7.5 hours | = 60.0 | 217.5 |
| | | 1732.50 |

May 1997.

Notes

(a) It should be noted that all labour costs incurred by the Contractor in his capacity as an Employer, other than those contained in the hourly rate above, must be taken into account under Section 6.

(b) The above example is for the convenience of users only and does not form part of the rules laid down in the Definition; all the basic costs are subject to re-examination according to the time when and in the area where the daywork is executed.

# Daywork

## BUILDING INDUSTRY PLANT HIRE COSTS

### SCHEDULE OF BASIC PLANT CHARGES (JANUARY 1990)

This Schedule is published by the Royal Institution of Chartered Surveyors and is for use in connection with Dayworks under a Building Contract.

EXPLANATORY NOTES

1. The rates in the Schedule are intended to apply solely to daywork carried out under and incidental to a Building Contract. They are NOT intended to apply to:
   (i) jobbing or any other work carried out as a main or separate contract; or
   (ii) work carried out after the date of commencement of the Defects Liability Period.

2. The rates in the Schedule are basic and may be subject to an overall adjustment to be quoted by the Contractor prior to the placing of the Contract.

3. The rates apply to plant and machinery already on site, whether hired or owned by the Contractor.

4. The rates, unless otherwise stated, include the cost of fuel and power of every description, lubricating oils, grease, maintenance, sharpening of tools, replacement of spare parts, all consumable stores and for licences and insurances applicable to items of plant. They do not include the costs of drivers and attendants (unless otherwise stated).

5. The rates should be applied to the time during which the plant is actually engaged in daywork.

6. Whether or not plant is chargeable on daywork depends on the daywork agreement in use and the inclusion of an item of plant in this schedule does not necessarily indicate that item is chargeable.

7. Rates for plant not included in the Schedule or which is not on site and is specifically hired for daywork shall be settled at prices which are reasonably related to the rates in the Schedule having regard to any overall adjustment quoted by the Contractor in the Conditions of Contract.

*NOTE: All rates in the schedule are expressed per hour and were calculated during the first quarter of 1989.*

## BUILDING INDUSTRY PLANT HIRE COSTS

### MECHANICAL PLANT AND TOOLS

| Item of plant | Description | | Unit | Rate per hr £ |
|---|---|---|---|---|
| **Bar-bending and Shearing Machines** | Power Driven | | | |
| Bar bending machine | Up to 2 in (51 mm) dia. rods | | each | 2.01 |
| Bar cropper machine | Up to 2 in (51 mm) dia. rods | | each | 1.51 |
| Bar shearing machine | Up to 1½ in (38 mm) dia. rods | | each | 2.00 |
| | Up to 2 in (51 mm) dia. rods | | each | 2.00 |
| **Block & stone splitter** | Hydraulic | | each | 0.83 |
| **Brick Saws** | | | | |
| Brick saw (use of abrasive disc to be charged net & credited) | Power driven (bench type clipper or similar or portable) | | each | 1.83 |
| **Compressors** | | | | |
| Portable compressors (machine only) | *Nominal delivery of free air per min at 100 lb/sq. in. (7 kg/sq.m.) pressure* | | | |
| | (cfm.) | (m3/min) | | |
| | 80/85 | 2.41 | each | 1.30 |
| | 125-140 | 3.50- 3.92 | each | 1.58 |
| | 160-175 | 4.50- 4.95 | each | 2.37 |
| | 250 | 7.08 | each | 3.25 |
| | 380 | 10.75 | each | 4.80 |
| | 600-630 | 16.98-17.84 | each | 6.00 |
| **Lorry Mounted Compressors** (Machine plus lorry only) | 101-150 | 2.86-4.24 | each | 4.12 |
| **Tractor Mounted Compressor** (Machine plus rubber tyred tractor only) | 101-120 | 2.86-3.40 | each | 3.63 |
| **Compressed Air Equipment** (with & including up to 50 ft (15.24 m) of air hose) | | | | |
| Breakers with six steels (Light/Medium/Heavy) | | | each | 0.47 |
| Light pneumatic pick with six steels | | | each | 0.37 |
| Pneumatic clay spade and one blade | | | each | 0.22 |
| Chipping hammer plus six steels | | | each | 0.65 |
| Hand-held rock drill and rod | 15 lb ( 7.0 kg) Class | | each | 0.33 |
| (bits to be paid for at a net | 35 lb (16.0 kg) Class | | each | 0.37 |
| cost and credited) | 45 lb (20.0 kg) Class | | each | 0.37 |
| | 55 lb (25.0 kg) Class | | each | 0.39 |
| Drill, rotary | Up to ¾ in (19 mm) | | each | 0.37 |
| (bits to be paid for at net cost and credited) | Up to 1¼ in (32mm) | | each | 0.73 |
| Sander/Grinder | | | each | 0.39 |
| Scrabbler (heads extra) | Single head | | each | 0.48 |
| | Triple head | | each | 0.65 |

## Daywork

### BUILDING INDUSTRY PLANT HIRE COSTS

**MECHANICAL PLANT AND TOOLS** - *continued*

| Item of plant | Description | | Unit | Rate per hr £ |
|---|---|---|---|---|
| **Compressor Equipment Accessories** | | | | |
| Additional hoses | Per 50 ft (15.0 m) | | each | 0.09 |
| Muffler, tool silencer | | | each | 0.08 |
| **Concrete Breaker** | | | | |
| Concrete breaker, portable hydraulic complete with power pack | | | each | 1.33 |
| **Concrete/Mortar Mixers** | | | | |
| Concrete mixer | Diesel, electric or petrol | | | |
| | cu ft | (cu m) | | |
| Open drum without hopper | 3/2 | (0.09/0.06) | each | 0.40 |
| | 4/3 | (0.12/0.09) | each | 0.44 |
| | 5/3.5 | (0.15/0.10) | each | 0.54 |
| Open drum with hopper | 7/5 | (0.20/0.15) | each | 0.67 |
| Closed drum | 8/6.5 | (0.25/0.18) | each | 0.90 |
| Reversing drum with hopper weigher & feed | 10/7 | (0.28/0.20) | each | 2.44 |
| shovel | 21/14 | (0.60/0.40) | each | 4.17 |
| **Concrete Pump** | | | | |
| Lorry mounted concrete pump (meterage charge to be added net) | | | each | 13.70 |
| **Concrete Equipment** | | | | |
| Vibrator, poker type | Petrol, diesel or electric | | | |
| | Up to 3" diameter | | each | 0.87 |
| | Air, excluding compressor and hose | | | |
| | Up to 3" diameter | | each | 0.65 |
| | Extra heads | | each | 0.50 |
| Vibrator-tamper | With tamping board | | each | 0.87 |
| | Double beam screeder | | each | 1.19 |
| Power float | 29"-36" | | each | 1.13 |
| **Conveyor Belts** | | | | |
| | Power operated up to 25 ft (7.62 m) long, 16 in (400 mm) wide | | each | 3.15 |
| **Cranes** | *Maximum capacity* | | | |
| Mobile, rubber tyred | Up to 15 cwt (762 kg) | | each | 5.15 |
| Lorry mounted, telescopic jib, 2 wheel (All rates inclusive of driver) | 6 tons (tonnes) | | each | 10.47 |
| | 7 tons (tonnes) | | each | 11.22 |
| | 8 tons (tonnes) | | each | 13.67 |
| | 10 tons (tonnes) | | each | 14.99 |
| | 12 tons (tonnes) | | each | 16.91 |
| | 15 tons (tonnes) | | each | 19.21 |
| | 18 tons (tonnes) | | each | 20.34 |
| | 20 tons (tonnes) | | each | 22.04 |
| | 25 tons (tonnes) | | each | 24.86 |

# SPON'S PRICE BOOKS 1998

## SPON'S ARCHITECTS' AND BUILDERS' PRICE BOOK 1998
**123rd edition**
*Davis Langdon & Everest*

The only price book geared to market conditions that affect building tender prices. Spon's A & B provides comprehensive coverage of construction costs from small scale alterations and repairs to the largest residential or commercial developments.

*944 pages
October 1997
Hardback
0-419-23060-2
£72.50*

## SPON'S MECHANICAL AND ELECTRICAL SERVICES PRICE BOOK 1998
**29th edition**
*Davis Langdon & Everest*

"An essential reference for everybody concerned with the calculation of costs of mechanical and electrical works" - Cost Engineer

Outline costs for a wide variety of engineering services are followed by more detailed elemental and quantified analysis.

*560 pages
October 1997
Hardback
0-419-23080-7
£75.00*

## SPON'S LANDSCAPE AND EXTERNAL WORKS PRICE BOOK 1998 [NEW LAYOUT]
**17th edition**
*Derek Lovejoy Partnership and Davis Langdon & Everest*

Now completely revised and expanded. Every rate has been reworked and recalculated. Now includes for the first time a realistic labour rate as well as plant and material costs.

"Surely there can be no office without this publication" - **Landscape Design**

*288 pages
October 1997
Hardback
0-419-23070-X
£62.50*

## SPON'S CIVIL ENGINEERING AND HIGHWAY WORKS PRICE BOOK 1998
**12th edition**
*Davis Langdon & Everest*

"Unquestionably, this will be required reading by all estimators involved in civil engineering works. Quantity surveyors will also find it essential for their shelves." - Civil Engineering Surveyor

*640 pages
October 1997
Hardback
0-419-23050-5
£95.00*

E & FN SPON an imprint of Thomson Professional

## RESPONSE
*SPON's price data now available in an electronic estimating system
Call 0171-522-9966 for details*

# ENERGY MANAGEMENT AND OPERATING COSTS IN BUILDINGS

## Keith J. Moss

Managing the consumption and conservation of energy in buildings must now become the concern of both building managers and occupants. The provision of lighting, hot water supply, communications, cooking, space heating and cooling accounts for 45 per cent of UK energy consumption.

**Energy Management and Operating Costs in Buildings** introduces the reader to the principles of managing and conserving energy consumption in buildings people use for work or leisure. Energy consumption is considered for the provision of space heating, hot water supply ventilation and air conditioning. The author introduces the use of standard performance indicators and energy consumption yardsticks, and discusses the use and application of Degree Days. Following an introduction to the preparation of the energy audit, monitoring and targeting techniques are investigated and analysed.

Readers are not expected to have prior knowledge of the design of building services. Each chapter of the book is set out with:

- nomenclature:
- introduction;
- worked examples and case studies;
- data, text and illustrations appropriate to each topic.

This is a key text for undergraduates on building services engineering, energy management, environmental engineering and related construction courses. It will also be an invaluable work for professional energy managers and building services engineers.

Preface. Acknowledgements. Introduction. The economics of space heating plants. Estimating energy consumption - space heating. Intermittent space heating. Estimating the annual cost for the provision of a hot water supply. Energy consumption for cooling loads. Performance indicators. Energy conservation strategies. Cost benefit analysis. Energy audits. Monitoring and targeting. Appendices. Bibliography. Index

---

**Keith Moss** is a consultant engineer and a visiting lecturer at the University of Bath and City of Bath College.

246x189mm, 200 pages
25 line illustrations
Paperback ISBN 0-419-21770-3
£24.99 June 1997

## New From

### E & FN SPON
An Imprint of Thomson Professional
2-6 Boundary Row
London SE1 8HN

## Daywork

## BUILDING INDUSTRY PLANT HIRE COSTS

### MECHANICAL PLANT AND TOOLS - *continued*

| Item of plant | Description | | | Unit | Rate Per hr £ |
|---|---|---|---|---|---|
| **Cranes** *continued* | | | | | |
| Lorry mounted, telescopic jib, 4 wheel (All rates inclusive of driver) | 10 tons (tonnes) | | | each | 15.30 |
| | 12 tons (tonnes) | | | each | 17.26 |
| | 15 tons (tonnes) | | | each | 19.59 |
| | 20 tons (tonnes) | | | each | 22.48 |
| | 25 tons (tonnes) | | | each | 25.36 |
| | 30 tons (tonnes) | | | each | 28.82 |
| | 45 tons (tonnes) | | | each | 31.12 |
| | 50 tons (tonnes) | | | each | 33.64 |
| | Capacity metre/tonnes | Height under hook above ground (m) | | | |
| Track/mounted tower crane (electric)(Capacity = max lift in tons x max rad at which can be lifted) (All rates inclusive of driver) | 10 | 17 | | each | 7.40 |
| | 15 | 17 | | each | 7.95 |
| | 20 | 18 | | each | 8.50 |
| | 25 | 20 | | each | 10.70 |
| | 30 | 22 | | each | 12.76 |
| | 40 | 22 | | each | 16.75 |
| | 50 | 22 | | each | 21.70 |
| | 60 | 22 | | each | 22.52 |
| | 70 | 22 | | each | 23.07 |
| | 80 | 22 | | each | 23.99 |
| | 110 | 22 | | each | 24.49 |
| | 125 | 30 | | each | 27.20 |
| | 150 | 30 | | each | 29.95 |
| Static tower cranes | To be charged at a percentage of the above rates | | | 90% of the above | |
| **Crane Equipment** | cu yd | (cu m) | | | |
| Tipping bucket (circular) | Up to ¼ | (0.19) | | each | 0.25 |
| | ¾ | (0.57) | | each | 0.28 |
| Skip, muck | Up to ½ | (0.38) | | each | 0.25 |
| | ¾ | (0.57) | | each | 0.25 |
| | 1½ | (0.76) | | each | 0.30 |
| Skip, concrete | Up to ½ | (0.38) | | each | 0.44 |
| | ¾ | (0.57) | | each | 0.53 |
| | 1 | (0.76) | | each | 0.76 |
| | 1½ | (1.15) | | each | 0.61 |
| Skip, concrete lay down or roll over | ½ | (0.38) | | each | 0.44 |
| **Dehumidifiers** | | | | | |
| 110/240V, Water extraction per 24 hrs | 15 galls (68 litres) | | | each | 0.84 |
| | 20 galls (90 litres) | | | each | 0.97 |

## Daywork

### BUILDING INDUSTRY PLANT HIRE COSTS

**MECHANICAL PLANT AND TOOLS** - *continued*

| Item of plant | Description | | Unit | Rate per hr £ |
|---|---|---|---|---|
| **Diamond Drilling and Chasing** | | | | |
| Chasing machine drilling | 6" (152mm) | | each | 1.35 |
| (Diamond core) | Mini rig up to 35mm | | each | 3.16 |
| | 3"-8" (76mm-203mm) electric 110v | | each | 3.33 |
| | 3"-8" (76mm-203mm) Air | | each | 4.23 |
| **Dumpers** | | | | |
| Dumper (site use only excl. Tax, Insurance & extra cost of DERV, etc., when operating on highway) | *Makers capacity* | | | |
| 2 wheel drive | | | | |
| Gravity tip | 15 cwt | (762 kg) | each | 1.04 |
| Hydraulic tip | 20 cwt | (1016 kg) | each | 1.25 |
| Hydraulic tip | 23 cwt | (1168 kg) | each | 1.31 |
| 4 wheel drive | | | | |
| Gravity tip | 23 cwt | (1168 kg) | each | 1.42 |
| Hydraulic tip | 25 cwt | (1270 kg) | each | 1.61 |
| Hydraulic tip | 30 cwt | (1524 kg) | each | 1.83 |
| Hydraulic tip | 35 cwt | (1778 kg) | each | 2.08 |
| Hydraulic tip | 40 cwt | (2032 kg) | each | 2.33 |
| Hydraulic tip | 50 cwt | (2540 kg) | each | 2.96 |
| Hydraulic tip | 60 cwt | (3048 kg) | each | 3.68 |
| Hydraulic tip | 80 cwt | (4064 kg) | each | 4.72 |
| Hydraulic tip | 100 cwt | (5080 kg) | each | 6.38 |
| Hydraulic tip | 120 cwt | (6096 kg) | each | 7.54 |
| **Electric Hand Tools** | | | | |
| Breakers | Heavy: | Kango 1800 | each | 0.87 |
| | | Kango 2500 | each | 0.90 |
| | Medium: | Hilti TP800 | each | 0.73 |
| Rotary Hammer | Heavy: | Kango 950 | each | 0.55 |
| | Medium: | Kango 627/637 | each | 0.47 |
| | Light: | Hilti TE12/TE17 | each | 0.39 |
| Pipe drilling tackle | Ordinary type | | set | 0.51 |
| | Under pressure type | | set | 0.83 |
| **Excavators** | | | | |
| Hydraulic full circle slew | | | | |
| Crawler mounted, backactor | cu yd | (cu m) | | |
| | ½ | (0.50) | each | 5.01 |
| | ¾ | (0.60) | each | 5.92 |
| | ⅞ | (0.70) | each | 8.24 |
| | 1 | (0.75) | each | 13.25 |
| Wheeled tractor type | hydraulic excavator, JCB type 3C or similar | | each | 9.23 |

## BUILDING INDUSTRY PLANT HIRE COSTS

**MECHANICAL PLANT AND TOOLS** - *continued*

| Item of plant | Description | | Unit | Rate per hr £ |
|---|---|---|---|---|
| **Mini Excavators** | | | | |
| Kubota mini excavators | 360 tracked - 1 tonne | | each | 5.38 |
| | 360 tracked - 3.5 tonnes | | each | 7.23 |
| **Forklifts** | *Payload* | *Max lift* | | |
| 2 wheel drive | 20 cwt (1016 kg) | 21 ft 4 in (6.50 m) | each | 4.38 |
| "Rough terrain" | 20 cwt (1016 kg) | 26 ft 0 in (7.92 m) | each | 4.38 |
| | 30 cwt (1524 kg) | 20 ft 0 in (6.09 m) | each | 4.50 |
| | 36 cwt (1829 kg) | 18 ft 0 in (5.48 m) | each | 4.50 |
| | 50 cwt (2540 kg) | 12 ft 0 in (3.66 m) | each | 4.73 |
| 4 wheel drive | 30 cwt (1524 kg) | 20 ft 0 in (6.09 m) | each | 5.60 |
| "Rough terrain" | 40 cwt (2032 kg) | 20 ft 0 in (6.09 m) | each | 5.78 |
| | 50 cwt (2540 kg) | 12 ft 0 in (3.66 m) | each | 7.79 |
| **Hammers** | | | | |
| Cartridge (excluding | Hammer DX450 | | each | 0.40 |
| cartridges and studs) | Hammer DX600 | | each | 0.48 |
| | Hammer spit TS | | each | 0.48 |
| **Heaters, Space** | | | | |
| | Paraffin/electric Btu/hr | | | |
| | 50 000- 75 000 | | each | 0.73 |
| | 80 000-100 000 | | each | 0.77 |
| | 150 000 | | each | 0.88 |
| | 320 000 | | each | 1.20 |
| **Hoists** | | | | |
| Scaffold | Up to 5 cwt (254 kg) | | each | 0.90 |
| Mobile (goods only) | Up to 10 cwt (508 kg) | | each | 1.43 |
| Static (goods only) | 10 to 15 cwt (508 kg - 762 kg) | | each | 1.67 |
| Rack and pinion (goods only) | | | | |
| | 16 cwt (813 kg) | | each | 2.75 |
| | 20 cwt (1016 kg) | | each | 3.00 |
| | 25 cwt (1270 kg) | | each | 3.30 |
| Rack and pinion (goods & passenger) | | | | |
| | 8 person, 1433 lbs (650 kg) | | each | 4.50 |
| | 12 person, 2205 lbs (1000 kg) | | each | 4.95 |
| **Lorries** | *Plated gross vehicle Weight, ton/tonne* | | | |
| Fixed body | Up to 5.50 | | each | 9.60 |
| | Up to 7.50 | | each | 10.20 |
| Tipper | Up to 16.00 | | each | 11.73 |
| | Up to 24.00 | | each | 14.98 |
| | Up to 30.00 | | each | 17.73 |

## BUILDING INDUSTRY PLANT HIRE COSTS

### MECHANICAL PLANT AND TOOLS - *continued*

| Item of plant | Description | | | Unit | Rate per hr £ |
|---|---|---|---|---|---|
| **Pipe Work Equipment** | | | | | |
| Pipe bender | Power driven, 50-150 mm dia | | | each | 0.75 |
| | Hydraulic, 2" capacity | | | each | 0.96 |
| Pipe cutter | Hydraulic | | | each | 1.02 |
| Pipe defrosting equipment | Electrical | | | set | 1.83 |
| Pipe testing equipment | Compressed air | | | set | 1.04 |
| | Hydraulic | | | set | 0.68 |
| Pipe threading equipment | Up to 4" | | | set | 1.65 |
| **Pumps** | | | | | |
| Including 20 ft (6 m) length of suction and/or delivery hose, couplings, valves and strainers | | | | | |
| Diaphragm: | | | | | |
| 'Simplite' 2 in (50 mm) | Petrol or electric | | | each | 0.79 |
| 'Wickham' 3 in (76 mm) | Diesel | | | each | 1.22 |
| Submersible 2 in (50 mm) | Electric | | | each | 0.85 |
| Induced flow: | | | | | |
| 'Spate' 3 in (76 mm) | Diesel | | | each | 1.42 |
| 'Spate' 4 in (102 mm) | Diesel | | | each | 1.88 |
| Centrifugal, self-priming: | | | | | |
| 'Univac' 2 in (50 mm) | Diesel | | | each | 1.87 |
| 'Univac' 4 in (120 mm) | Diesel | | | each | 2.40 |
| 'Univac' 6 in (152 mm) | Diesel | | | each | 3.69 |
| **Pumping Equipment** | in | (mm) | | | |
| Pump hoses, per 20 ft (6 m) | 2 | (51) | | each | 0.15 |
| flexible, suction or delivery, | 3 | (76) | | each | 0.17 |
| inc coupling valve & strainer | 4 | (102) | | each | 0.19 |
| | 6 | (152) | | each | 0.32 |
| **Rammers and Compactors** | | | | | |
| Power rammer, Pegson or similar | | | | each | 0.88 |
| Soil compactor, plate type | | | | | |
| Plate size | | | | | |
| 12 x 13 in (305 x 330 mm) | 172 lb (78 kg) | | | each | 0.78 |
| 20 x 18 in (508 x 457 mm) | 264 lb (120 kg) | | | each | 0.93 |
| **Rollers** | | | | | |
| | cwt | (kg) | | | |
| Vibrating rollers | 7½ - 8¼ | (368-420) | | each | 1.00 |
| Single roller | 10½ | (533) | | each | 1.53 |
| Twin roller | 13¾ | (698) | | each | 1.75 |
| | 16¾ | (851) | | each | 2.01 |
| Twin roller with seat | 21 | (1067) | | each | 2.75 |
| and steering wheel | 27½ | (1397) | | each | 2.88 |
| Pavement rollers | ton | tonne | | | |
| dead weight | 3 | 4 | | each | 2.89 |
| | Over 4 | 6 | | each | 3.75 |
| | Over 6 | 10 | | each | 4.40 |

## Daywork

## BUILDING INDUSTRY PLANT HIRE COSTS

### MECHANICAL PLANT AND TOOLS - continued

| Item of plant | Description | | | Unit | Rate per hr £ |
|---|---|---|---|---|---|
| **Saws, Mechanical** | in | | (m) | | |
| Chain saw | 21 | | (0.53) | each | 0.78 |
| | 30 | | (0.76) | each | 1.12 |
| Bench saw | Up to 20 | | (0.51) blade | each | 0.80 |
| | Up to 24 | | (0.61) blade | each | 1.20 |
| **Screed Pump** | *Maximum delivery* | | | | |
| Wkg vol 7 cu ft | Vertical 300 ft (91 m) | | | | |
| (200 ltrs.) | Horizontal 600 ft (182 m) | | | each | 8.33 |
| Screed pump hose | 50/65 mm dia. | | | | |
| | 13.3 m long | | | each | 1.83 |
| **Screwing Machines** | | | | | |
| | 13-50 mm dia. | | | each | 0.43 |
| | 25-100 mm dia. | | | each | 0.87 |
| **Tractors** | cu yd | | (cu m) | | |
| Shovel, tractor | Up to ¾ | | (0.57) | each | 4.70 |
| (crawler) any type | 1 | | (0.76) | each | 5.70 |
| of bucket | 1¼ | | (0.96) | each | 6.62 |
| | 1½ | | (1.15) | each | 8.15 |
| | 1¾ | | (1.34) | each | 8.98 |
| | 2 | | (1.53) | each | 10.69 |
| | 2¾-3½ | | (2.10-2.70) | each | 17.27 |
| Shovel, tractor | Up to ½ | | (0.38) | each | 3.36 |
| (wheeled) | ¾ | | (0.57) | each | 4.14 |
| | 1 | | (0.76) | each | 5.02 |
| | 2½ | | (1.91) | each | 7.30 |
| Tractor (crawler) with | *Maker's rated flywheel horsepower* | | | | |
| dozer | 75 | | | each | 9.31 |
| | 140 | | | each | 12.52 |
| Tractor-wheeled (rubber tyred) | | Light 48 hp | | each | 4.23 |
| Agricultural type | | Heavy 65 hp | | each | 4.68 |
| **Traffic Lights** | | | | | |
| | Mains/generator 2-way | | | set | 2.23 |
| | Mains/generator 3-way | | | set | 4.40 |
| | Mains/generator 4-way | | | set | 5.45 |
| | Trailer mounted 2-way | | | set | 2.21 |
| **Welding and Cutting and Burning Sets** | | | | | |
| Welding and cutting set (inc. oxygen and acetylene, excl. underwater equip. and thermic boring) | | | | each | 4.33 |
| Welding set, diesel | 300 amp, single operator | | | each | 1.84 |
| (excluding electrodes) | 600 amp, double operator | | | each | 1.89 |
| | 600 amp, double operator | | | each | 2.35 |

*Daywork*

## BUILDING INDUSTRY PLANT HIRE COSTS

### NON-MECHANICAL PLANT

| Item of plant | Description | | Unit | Rate per hr £ |
|---|---|---|---|---|
| **Bar-bending and Shearing Machines** | | | | |
| Bar bending machine manual | Up to 1 in (25 mm) dia. rods | | each | 0.26 |
| Shearing machine, manual | Up to ₑ in (16 mm) dia. rods | | each | 0.15 |
| **Brother or Sling Chains** | | | | |
| | Not exceeding 2 ton/tonne | | set | 0.28 |
| | Exceeding 2 ton/tonne, not exceeding 5 ton/tonne | | set | 0.40 |
| | Exceeding 5 ton/tonne, not exceeding 10 ton/tonne | | set | 0.48 |
| **Drain Testing Equipment** | | | set | 0.42 |
| **Lifting and Jacking Gear** | ton/tonne | | | |
| Pipe winch - inc. shear legs | ½ | | set | 0.35 |
| Pipe winch - inc. gantry | 2 | | set | 0.72 |
| | 3 | | set | 0.84 |
| Chain blocks up to 20 ft (6.10 m) lift | ton/tonne | | | |
| | 1 | | each | 0.35 |
| | 2 | | each | 0.44 |
| | 3 | | each | 0.51 |
| | 4 | | each | 0.66 |
| Pull lift (Tirfor type) | ¾ tons/tonnes | | each | 0.35 |
| | 1½ tons/tonnes | | each | 0.47 |
| | 3 tons/tonnes | | each | 0.61 |
| **Pipe Benders** | | | | |
| | 13-75 mm dia. | | each | 0.32 |
| | 50-100 mm dia. | | each | 0.52 |
| **Plumber's Furnace** | Calor gas or similar | | each | 1.20 |
| **Road Works, Equipment** | | | | |
| Barrier trestles or similar | | | ten | 0.50 |
| Crossing plates (steel sheets) | | | each | 0.32 |
| Danger lamp, inc. oil | | | each | 0.03 |
| Warning signs | | | each | 0.07 |
| Road cones | | | ten | 0.11 |
| Flasher unit (battery to be charged at cost) | | | each | 0.06 |
| Flashing bollard (battery to be charged at cost) | | | each | 0.10 |

*Daywork*

## BUILDING INDUSTRY PLANT HIRE COSTS

### NON-MECHANICAL PLANT - *continued*

| Item of plant | Description | Unit | Rate per hr £ |
|---|---|---|---|
| **Rubbish Chutes** | | | |
| Length of sections 154cm | Standard section | each | 0.15 |
| giving working length of | Brand section | each | 0.20 |
| 130cm | Hopper | each | 0.16 |
| | Fixing frame | each | 0.20 |
| Winch Type A | 20m for erection | each | 0.29 |
| Winch Type B | 40m for erection | each | 0.67 |
| **Scaffolding** | | | |
| Boards | | 100ft (30.48m) | 0.06 |
| Castor wheels | Steel or rubber tyred | 100 | 0.87 |
| Fall ropes | Up to 200 ft (61 m) | each | 0.44 |
| Fittings (inc. couplers, and base plates) | Steel or alloy | 100 | 0.07 |
| Ladders, pole | 20 rung | each | 0.11 |
| | 30 rung | each | 0.15 |
| | 40 rung | each | 0.26 |
| Ladder, extension | Extended length ft (m) | | |
| | 20 (6.10) | each | 0.31 |
| | 26 (7.92) | each | 0.43 |
| | 35 (10.67) | each | 0.64 |
| Putlogs | Steel or alloy | 100 | 0.06 |
| Splithead | Small | 10 | 0.12 |
| | Medium | 10 | 0.15 |
| | Large | 10 | 0.16 |
| Staging, lightweight | | 100ft (30.48 m) | 1.00 |
| Tube | Steel | 100ft | 0.02 |
| | Alloy | (30.48 m) | 0.04 |
| Wheeled tower | Working platform up to 20 ft (6m) | | |
| 7 x 7 ft (2.13 x 2.13 m) | high including castors and boards | each | 0.58 |
| 10 x 10 ft (3.05 x 3.05 m) | | each | 0.59 |
| **Tarpaulins** | | | |
| Trench Struts and Sheets | | 10m2 | 0.04 |
| Adjustable steel trench strut | All sizes from 1 ft (305 mm) closed to 5 ft 6 in (1.68m) extended | 10 | 0.06 |
| Steel trench sheet | 5-14 ft (1.52-4.27 m) lengths | 100ft (30.48 m) | 0.16 |

# Site Management of Building Services Contractors

**Jim Wild, Building Services Management Consultant, Berkshire, UK**

There is an increasingly wide range of building engineering services. When installed, these must sustain and protect a specified internal environment to the satisfaction of the client, designers, insurers and the relevant authorities.

Managing building services contractors can prove to be a minefield. The most successful jobs will always be those where building site managers have first built teams focused on tackling issues that might cause adversarial attitudes later on and jeopardize the project.

The author shows how a simple common management approach can improve site managers' competency in overseeing building services contractors, sub traders and specialists, and maximize the effectiveness of time spent on building services. By providing an account of building services from the site management rather than the corporate viewpoint, this book breaks new ground.

**Site Management of Building Services Contractors:**
- provides step by step guidance from pre award to post handover;
- emphasises system based nature of building services contracts;
- covers risk management throughout the process.

**Contents:**
Overview of building services. Basic planning strategy. Project plans for quality, safety and the environment. Planning and programming. Schedules. Supervision and inspection. Assessing construction progress. Commissioning and its management. The management of defects. Handover. Getting help. Summary. Index

January 1997: 246x189mm
384 pp: 32 line illustrations
Hardback: 0 419 20450 4: £35.00

# E & FN SPON

An Imprint of
Thomson Professional

PART FIVE
# Fees for Professional Services

Quantity Surveyors' Fees, *page 489*

# Heating and Water Services Design in Buildings

**K. Moss**, Visiting Lecturer to City of Bath College and the University of Bath, UK

This book addresses practical design procedures and solutions in heating and water services in buildings. The reader is encouraged to participate in the worked solutions to the numerous examples and case studies given. This book has been written following the authors twenty seven years experience in the industry as a teacher and consultant. Moss has worked with college students, university undergraduates and open learning candidates of all ages.

- **does not include sections of mathematical theory which do not have a practical relevance**

### Contents

Heat requirements of heated buildings in temperate climates. Low-temperature hot water heating systems. Pump and system. High temperature hot water systems. Steam systems. Plant connections and controls. The application of probability and demand units in design. Hot and cold water supply systems utilising the static head. Hot and cold water supply systems using booster pumps. Loose ends.

*E & F N Spon*
*246x189: 264pp :98 line illus: May 1996*
*Paperback: 0-419-20110-6: £19.99*

## QUANTITY SURVEYORS' FEES

In recent years there has been an increasing tendency for Bills of Quantities to be prepared for mechanical and electrical services. Where this is done by Chartered Quantity Surveyors the fees will be in accordance with the appropriate scale of fees issued by The Royal Institution of Chartered Surveyors, 12 Great George Street, London SW1P 3AD.

Reproduced below, by kind permission of the R.I.C.S., are the additional fees where the Chartered Quantity Surveyor executes these services in association with building and civil engineering projects. Services and fees referred to in paragraphs outside the range of those reproduced are for the main building project.

### SCALE 36. INCLUSIVE PROFESSIONAL CHARGES FOR QUANTITY SURVEYING SERVICES FOR BUILDING WORKS

PRE AND POST CONTRACT SERVICES BASED ON BILLS OF QUANTITIES

#### 2.2 Air Conditioning, heating, ventilating and electrical services

(a) When the service outlined in para. 1.3. are provided by the quantity surveyor for the air conditioning, heating, ventilating and electrical services there shall be a fee for these services in addition to the fee calculated in accordance with para. 2.1. as follows:

| Value of Work £ | | Additional fee £ | | £ |
|---|---|---|---|---|
| Up to | 120,000 | | 5.0% | |
| 120,000 – | 240,000 | 6,000 + | 4.7% on balance over | 120,000 |
| 240,000 – | 480,000 | 11,640 + | 4.0% on balance over | 240,000 |
| 480,000 – | 750,000 | 21,240 + | 3.6% on balance over | 480,000 |
| 750,000 – | 1,000,000 | 30,960 + | 3.0% on balance over | 750,000 |
| 1,000,000 – | 4,000,000 | 38,460 + | 2.7% on balance over | 1,000,000 |
| over | 4,000,000 | 119,460 + | 2.4% on balance over | 4,000,000 |

(b) The value of such services, whether the subject of separate tenders or not, shall be aggregated and the total value of the work so obtained used for the purpose of calculating the additional fee chargeable in accordance with para. (a). (Except that when, more than one firm of consulting engineers is engaged on the design of these services, the separate values for which each such firm is responsible shall be aggregated and the additional fees charged shall be calculated independently on each such total value so obtained).

(c) Fees shall be calculated upon the basis of the account for the whole of the air conditioning, heating, ventilating and electrical services for which bills of quantities and final accounts have been prepared by the quantity surveyor.

## Fees for Professional Services

### QUANTITY SURVEYORS' FEES

### SCALE 37. ITEMISED PROFESSIONAL CHARGES FOR QUANTITY SURVEYING SERVICES FOR BUILDING WORK

PRE-CONTRACT SERVICES BASED ON BILLS OF QUANTITIES

2.2 **Air Conditioning, heating, ventilating and electrical services**

(a) Where bills of quantities are prepared by the quantity surveyor for the air conditioning, heating, ventilating and electrical services there shall be a fee for these services (which shall include examining tenders received and reported thereon), in addition to the fee calculated in accordance with para. 2.1., as follows:

| Value of Work £ | | Additional fee | £ |
|---|---|---|---|
| Up to 120,000 | | 2.5% | |
| 120,000 – 240,000 | 3,000 + | 2.25% on balance over | 120,000 |
| 240,000 – 480,000 | 5,700 + | 2.00% on balance over | 240,000 |
| 480,000 – 750,000 | 10,500 + | 1.75% on balance over | 480,000 |
| 750,000 – 1,000,000 | 15,225 + | 1.25% on balance over | 750,000 |
| over 1,000,000 | 18,350 + | 1.15% on balance over | 1,000,000 |

(b) The value of such services, whether the subject of separate tenders or not, shall be aggregated and the total value of the work so obtained used for the purpose of calculating the additional fee chargeable in accordance with para. (a).
(Except that when, more than one firm of consulting engineers is engaged on the design of these services, the separate values for which each such firm is responsible shall be aggregated and the additional fees charged shall be calculated independently on each such total value so obtained).

(c) Fees shall be calculated upon the accepted tender for the whole of the air conditioning, heating, ventilating and electrical services for which bills of quantities have been prepared by the quantity surveyor. In the event of no tender being accepted, fees shall be calculated upon the basis of the lowest original bona fide tender received. In the event of no such tender being received, the fees shall be calculated upon a reasonable valuation of the services based upon the original bills of quantities.

NOTE: In the foregoing context `bona fide tender' shall be deemed to mean a tender submitted in good faith without major errors of computation and not subsequently withdrawn by the tenderer.

(d) When cost planning services are provided by the quantity surveyor for air conditioning, heating, ventilating and electrical services (or for any part of such services) there shall be an additional fee based on the time involved (see paras. 19.1. and 19.2.) Alternatively the fee may be on a lump sum or percentage basis agreed between the employer and the quantity surveyor.

NOTE: The incorporation of figures for air conditioning, heating, ventilating and electrical services provided by the consulting engineer is deemed to be included in the quantity surveyor's services under para. 2.1.

*Fees for Professional Services*

## QUANTITY SURVEYORS' FEES

POST-CONTRACT SERVICES BASED ON BILLS OF QUANTITIES

(Where the Quantity Surveyor also provided the pre-contract services)

5.3 **Air conditioning, heating, ventilating and electrical services**

(a) Where final accounts are prepared by the quantity surveyor for the air conditioning, heating, ventilating and electrical services there shall be a fee for these services, in addition to the fee calculated in accordance with para. 5.2., as follows:

| Value of Work £ | | | Additional fee £ | | | £ |
|---|---|---|---|---|---|---|
| Up to | | 120,000 | | 2.00% | | |
| 120,000 | – | 240,000 | 2,400 + | 1.60% | on balance over | 120,000 |
| 240,000 | – | 1,000,000 | 4,320 + | 1.25% | on balance over | 240,000 |
| 1,000,000 | – | 4,000,000 | 13,820 + | 1.00% | on balance over | 1,000,000 |
| over | | 4,000,000 | 43,820 + | 0.90% | on balance over | 4,000,000 |

(b) The value of such services, whether the subject of separate tenders or not, shall be aggregated and the total value of the work so obtained used for the purpose of calculating the additional fee chargeable in accordance with para. (a).
(Except that when, more than one firm of consulting engineers is engaged on the design of these services, the separate values for which each such firm is responsible shall be aggregated and the additional fees charged shall be calculated independently on each such total value so obtained).

(c) The scope of the scale of the services to be provided by the quantity surveyor under para. (a) above shall be deemed to be equivalent to those described for the basic scale for post-contract services.

(d) When the quantity surveyor is required to prepare periodic valuations of materials or goods off site, an additional fee shall be charged based on the time involved (see paras. 19.1 and 19.2).

(e) The basic scale for post-contract services included for a simple routine of periodically estimating final costs. When the employer specifically requests a cost monitoring service which involves the quantity surveyor in additional or abortive measurement an additional fee shall be based on the time involved (see paras. 19.1 and 19.2), or alternatively on a lump sum or percentage basis agreed between the employer and the quantity surveyor.

(f) Fees shall be calculated upon the basis of the account for the whole of the air conditioning, heating, ventilating and electrical services for which final accounts have been prepared by the quantity surveyor.

## Fees for Professional Services
## QUANTITY SURVEYORS' FEES

BILLS OF APPROXIMATE QUANTITIES, INTERIM CERTIFICATES AND FINAL ACCOUNTS

### 7.2 Air conditioning, heating, ventilating and electrical services

(a) Where bills of approximate quantities and final accounts are prepared by the quantity surveyor for the air conditioning, heating, ventilating and electrical services there shall be a fee for these services, in addition to the fee calculated in accordance with para. 7.1., as follows:

| Value of Work £ | | | Additional fee | | £ |
|---|---|---|---|---|---|
| Up to | 120,000 | | 4.50% | | |
| 120,000 – | 240,000 | 5,400 + | 3.85% | on balance over | 120,000 |
| 240,000 – | 480,000 | 10,020 + | 3.25% | on balance over | 240,000 |
| 480,000 – | 750,000 | 17,820 + | 3.00% | on balance over | 480,000 |
| 750,000 – | 1,000,000 | 25,920 + | 2.50% | on balance over | 750,000 |
| 1,000,000 – | 4,000,000 | 32,170 + | 2.15% | on balance over | 1,000,000 |
| over | 4,000,000 | 96,670 + | 2.05% | on balance over | 4,000,000 |

(b) The value of such services, whether the subject of separate tenders or not, shall be aggregated and the value of the work so obtained used for the purpose of calculating the additional fee chargeable in accordance with para. (a). (Except that when, more than one firm of consulting engineers is engaged on the design of these services, the separate values for which each such firm is responsible shall be aggregated and the additional fees charged shall be calculated independently on each such total value so obtained).

(c) The scope of the services to be provided by the quantity surveyor under para. (a) above shall be deemed to be equivalent to those described for the basic scale for pre-contract and post-contract services.

(d) When the quantity surveyor is required to prepare valuations of materials or goods off site, an additional fee shall be charged based on the time involved (see paras. 19.1 and 19.2).

(e) The basic scale for post-contract services includes for a simple routine of periodically estimating final costs. When the employer specifically requests a cost monitoring services which involves the quantity surveyor in additional or abortive measurement, an additional fee shall be based on the time involved (see paras. 19.1 and 19.2), or alternatively on a lump sum or percentage basis agreed between the employer and the quantity surveyor.

(f) Fees shall be calculated upon the basis of the account for the whole of the air conditioning, heating, ventilating and electrical services for which final accounts have been prepared by the quantity surveyor.

(g) When cost planning services are provided by the quantity surveyor for air conditioning, heating, ventilating and electrical services (or for any part of such services) there shall be an additional fee based on the time involved (see paras. 19.1 and 19.2), or alternatively on a lump sum or percentage basis agreed between the employer and quantity surveyor.

NOTE: The incorporation of figures for air conditioning, heating, ventilating and electrical services provided by the consulting engineer is deemed to be included in the quantity surveyor's services under para 7.1.

## QUANTITY SURVEYORS' FEES

### SCALE 38. ITEMISED CHARGES FOR QUANTITY SURVEYING SERVICES FOR CIVIL ENGINEERING WORKS

PRE-CONTRACT SERVICES BASED ON BILLS OF QUANTITIES

2.2  **Mechanical and electrical installations**

(a) When bills of quantities giving estimated quantities of the work required are prepared by the quantity surveyor for the mechanical and electrical services installed in structures for domestic or environmental purposes common to buildings generally (i.e. purposes other than industrial processes) there shall be a fee for these services (which shall include examining tenders received and reporting thereon) in addition to the fee calculated in accordance with paragraph 2.1 as follows:

| Value of Work £ | | | Additional fee | | | £ |
|---|---|---|---|---|---|---|
| Up to | | 120,000 | | 1.80% | | |
| 120,000 | - | 240,000 | 2,160 + | 1.55% | on balance over | 120,000 |
| 240,000 | - | 480,000 | 4,020 + | 1.30% | on balance over | 240,000 |
| 480,000 | - | 750,000 | 7,140 + | 1.20% | on balance over | 480,000 |
| 750,000 | - | 1,000,000 | 10,380 + | 1.00% | on balance over | 750,000 |
| 1,000,000 | - | 4,000,000 | 12,880 + | 0.85% | on balance over | 1,000,000 |
| over | | 4,000,000 | 38,380 + | 0.82% | on balance over | 4,000,000 |

(b) When bills of quantities giving estimated quantities of the work required are prepared for machinery and other permanent mechanical and electrical installation or plant is connection with industrial processes other than those included in paragraph 2.1 then a fee in addition to that envisaged under paragraph 2.1 shall be negotiated.

(c) Fees shall be calculated upon the accepted tender for each individual contract.

POST CONTRACT SERVICES BASED ON BILLS OF QUANTITIES

4.2  **Mechanical and electrical installations**

(a) When final accounts are prepared by the quantity surveyor for mechanical and electrical services installed in structures for domestic or environmental purposes common to buildings generally (i.e. purpose other than industrial processes) there shall be a fee for these services in addition to the fee calculated in accordance with paragraph 4.1 as follows:

| Value of Work £ | | | Category II fee | | | £ |
|---|---|---|---|---|---|---|
| Up to | | 120,000 | | 2.70% | | |
| 120,000 | - | 240,000 | 3,240 + | 2.30% | on balance over | 120,000 |
| 240,000 | - | 480,000 | 6,000 + | 1.95% | on balance over | 240,000 |
| 480,000 | - | 750,000 | 10,680 + | 1.80% | on balance over | 480,000 |
| 750,000 | - | 1,000,000 | 15,540 + | 1.50% | on balance over | 750,000 |
| 1,000,000 | - | 4,000,000 | 19,290 + | 1.30% | on balance over | 1,000,000 |
| over | | 4,000,000 | 58,290 + | 1.23% | on balance over | 4,000,000 |

(b) When final accounts are required to be prepared by the quantity surveyor for machinery and other permanent mechanical and electrical installations or plant in connection with industrial processes other than those included in paragraph 4.1 then a fee in addition to that envisaged under paragraph 4.1. shall be negotiated.

(c) Fees shall be calculated upon the basis of the accounts for each individual contract for which the final account has been prepared by the quantity surveyor.

# Building Services Engineering Spreadsheets

**D.V. Chadderton**, Chartered building services engineer, Victoria, Australia, formerly at the Southampton Institute of Higher Education, Southampton, UK

Building Services Engineering Spreadsheets is a versatile, user friendly tool for design calculations. Spreadsheet application software is readily understandable since each formula is readable in the location where it is used. Each step in the development of these engineering solutions is fully explained.

- **fills the gap between manual calculation methods using a calculator and specifically engineered software costing thousands of pounds**

### Contents
Contents. Preface. Acknowledgements. Introduction. Units and constants. Symbols. Chapter 1. Computer and spreadsheet use. Chapter 2. Thermal transmittance. Chapter 3. Heat gain. Chapter 4. Combustion of a fuel. Chapter 5. Building heat loss. Chapter 6. Fan and system selection. Chapter 7. Air duct network. Chapter 8. Water pipe sizing. Chapter 9. Lighting. Chapter 10. Electrical cable sizing. References. Index. Answers.

*E & F N Spon*
*most major IBM compatible spreadsheet applications*
*246x189: approx. 300pp: 63 line illus, 3 halftone illus: September 1997*
*Paperback: 0-419-22620-6: c. £29.95*

# Tables and Memoranda

Conversion Tables, *page 497*
Formulae, *page 500*
Fractions, Decimals and Millimetre Equivalents, *page 501*
Imperial Standard Wire Gauge, *page 502*
Water Pressure Due to Height, *page 503*
Table of Weights for Steelwork, *page 504*
Dimension and Weights of Copper and Stainless Steel Pipes, *page 509*
Dimensions of Steel Pipes, *page 510*
Approximate Metres per Tonne of Tubes, *page 511*
Flange Dimension Chart, *page 512*
Minimum Distances Between Supports/Fixings, *page 513*
Litres of Water Storage Required per Person per Building Type, *page 513*
Cold Water Plumbing - Insulation Thickness Required Against Frost, *page 514*
Capacity and Dimensions of Galvanised Mild Steel Cisterns, *page 514*
Capacity of Cold Water Polypropylene Storage Cisterns, *page 514*
Storage Capacity and Recommended Power of Hot Water Storage Boilers, *page 515*
Thickness of Thermal Insulation for Heating Installations, *page 516*
Capacities and Dimensions of :
Galvanised Mild Steel Indirect Cylinders, *page 518*
Copper Indirect Cylinders, *page 518*
Recommended Air Conditioning Design Loads, *page 519*

# The Technology of Building Defects

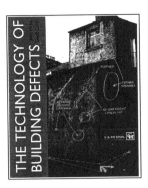

**J. Hinks**, Reader in Facilities Management, Department of Building Engineering and Surveying, Heriot-Watt University, Edinburgh, UK
**G. Cook**, Senior Lecturer, Department of Construction Management and Engineering, University of Reading, UK

The Technology of Building Defects has been developed to provide a unique stand alone review of building defects. It gives the reader a comprehensive understanding of how and why building defects occur. Defects are considered as part of the whole building rather than in isolation. General education objectives are set out which offer the reader the opportunity of self-assessment and build up an understanding of a range of technical topics concerned with building defects. This is a one stop resource which dispenses with the need to consult a mass of different information sources.

*E & F N Spon*
*246x189: c. 304pp: 95 line illus, 46 halftone illus: October 1997*
*Paperback: 0-419-19780-X: £24.99*

*Tables and Memoranda*

## CONVERSION TABLES

Example : 1mm = 0.039in - 1in = 25.4mm

1 LINEAR

| | | | | |
|---|---|---|---|---|
| 0.039 | in | » 1 « | mm | 25.4 |
| 0.281 | ft | » 1 « | metre | 0.305 |
| 1.094 | yd | » 1 « | metre | 0.914 |

2 WEIGHT

| | | | | |
|---|---|---|---|---|
| 0.020 | cwt | » 1 « | kg | 50.802 |
| 0.984 | ton | » 1 « | tonne | 1.016 |
| 2.205 | lb | » 1 « | kg | 0.454 |

3 CAPACITY

| | | | | |
|---|---|---|---|---|
| 1.76 | pt | » 1 « | litre | 0.568 |
| 0.220 | gal | » 1 « | litre | 4.546 |

4 AREA

| | | | | |
|---|---|---|---|---|
| 0.002 | in² | » 1 « | mm² | 645.16 |
| 10.764 | ft² | » 1 « | m² | 0.093 |
| 1.196 | yd² | » 1 « | m² | 0.836 |
| 2.471 | acre | » 1 « | ha | 0.405 |
| 0.386 | mile² | » 1 « | km² | 2.59 |

5 VOLUME

| | | | | |
|---|---|---|---|---|
| 0.061 | in³ | » 1 « | cm³ | 16.387 |
| 35.315 | ft³ | » 1 « | m³ | 0.028 |
| 1.308 | yd³ | » 1 « | m³ | 0.765 |

6 POWER

| | | | | |
|---|---|---|---|---|
| 1.310 | HP | » 1 « | kW | 0.746 |

## CONVERSION TABLES

### LENGTH    Conversion Factors

| | | | | | | | | |
|---|---|---|---|---|---|---|---|---|
| Centimetre | (cm) | 1 in | = | 2.54 cm | : 1 cm | = | 0.3937 | in |
| Metre | (m) | 1 ft | = | 0.3048 m | : 1 m | = | 3.2808 | ft |
| Kilometre | (km) | 1 yd | = | 0.9144 m | : 1 m | = | 1.0936 | yd |
| Kilometre | (km) | 1 mile | = | 1.6093 km | : 1 km | = | 0.6214 | mile |

NOTE :

| | | | | | |
|---|---|---|---|---|---|
| 1 cm | = | 10 mm | 1 ft | = | 12 in. |
| 1 m | = | 100 cm | 1 yd | = | 3 ft |
| 1 km | = | 1000 m | 1 mile | = | 1760 yd |

### AREA

| | | | | | | | |
|---|---|---|---|---|---|---|---|
| Square Millimetre | (mm²) | 1 in² | = | 645.2 mm² | : 1 mm² | = | 0.0016 in² |
| Square Centimetre | (cm²) | 1 in² | = | 6.4516 cm² | : 1 cm² | = | 0.1550 in² |
| Square Metre | (m²) | 1 ft² | = | 0.0929 m² | : 1 m² | = | 10.764 ft² |
| Square Metre | (m²) | 1 yd² | = | 0.8361 m² | : 1 m² | = | 1.1960 yd² |
| Square Kilometre | (km²) | 1 mile² | = | 2.590 km² | : 1 km² | = | 0.3861 mile² |

NOTE :

| | | | | | |
|---|---|---|---|---|---|
| 1 cm² | = | 100 mm² | 1 ft² | = | 144 in² |
| 1 m² | = | 10000 cm² | 1 yd² | = | 9 ft² |
| 1 km² | = | 100 ha | 1 mile² | = | 640 acre |
| | | | 1 acre | = | 4840 yd² |

### VOLUME

| | | | | | | | |
|---|---|---|---|---|---|---|---|
| Cubic Centimetre | (cm³) | 1 cm³ | = | 0.0610 in³ | : 1 in³ | = | 16.387 cm³ |
| Cubic Decimetre | (dm³) | 1 dm³ | = | 0.0353 ft³ | : 1 ft³ | = | 28.329 dm³ |
| Cubic Metre | (m³) | 1 m³ | = | 35.315 ft³ | : 1 ft³ | = | 0.0283 m³ |
| Cubic Metre | (m³) | 1 m³ | = | 1.3080 yd³ | : 1 yd³ | = | 0.7646 m³ |
| Litre | (L) | 1 L | = | 1.76 pint | : 1 pint | = | 0.5683 L |
| Litre | (L) | 1 L | = | 2.113 US pt | : 1 pint | = | 0.4733 US L |

NOTE :

| | | | | | |
|---|---|---|---|---|---|
| 1 dm³ | = | 1000 cm³ | 1 ft³ | = | 1728 in³ |
| 1 m³ | = | 1000 dm³ | 1 yd³ | = | 27 ft³ |
| 1 L | = | 1 dm³ | 1 pint | = | 20 fl oz |
| 1 HL | = | 100 L | 1 gal | = | 8 pints |

### MASS

| | | | | | | | |
|---|---|---|---|---|---|---|---|
| Milligram | (mg) | 1 mg | = | 0.0154 grain | : 1 grain | = | 64.935 mg |
| Gram | (g) | 1 g | = | 0.0353 oz | : 1 oz | = | 28.35 g |
| Kilogram | (kg) | 1 kg | = | 2.2046 lb | : 1 lb | = | 0.4536 kg |
| Tonne | (t) | 1 t | = | 0.9842 ton | : 1 ton | = | 1.016 t |

NOTE :

| | | | | | |
|---|---|---|---|---|---|
| 1 g | = | 1000 mg | 1 oz | = | 437.5 grains |
| 1 kg | = | 1000 g | 1 lb | = | 16 oz |
| 1 t | = | 1000 kg | 1 stone | = | 14 lb |
| | | | 1 cwt | = | 112 lb |
| | | | 1 ton | = | 20 cwt |

*Tables and Memoranda*

**CONVERSION TABLES**

FORCE

Conversion Factors

| | | | | | | | | | |
|---|---|---|---|---|---|---|---|---|---|
| Newton | (N) | 1 lb f | = | 4.448 | N | : 1 kg f | = | 9.807 | N |
| Kilonewton | (kN) | 1 lb f | = | 0.004448 | kN | : 1 ton f | = | 9.964 | kN |
| Meganewton | (mN) | 100 ton f | = | 0.9964 | mN | | | | |

PRESSURE AND STRESS

| | | | | | |
|---|---|---|---|---|---|
| Kilonewton per square metre (kN/m²) | 1 lb f/in² | = | 6.895 | kN/m² | : |
| | 1 bar | = | 100 | kN/m² | : |
| Meganewton per square metre (mN/m²) | 1 ton f/ft² | = | 107.3 | kN/m² | : = 0.1073 mN/m² |
| | 1 kg f/cm² | = | 98.07 | kN/m² | : |
| | 1 lb f/ft² | = | 0.0479 | kN/m² | : |

TEMPERATURE

Degree Celcius (ᴱC)    ᴱC = 5 ÷ 9 (ᴱF - 32ᴱ)    ᴱF = 9 ÷ 5 (ᴱC + 32ᴱ)

---

# Keep your figures up to date, free of charge

This section, and most of the other information in this Price Book, is brought up to date every three months, until the next annual edition, in the *Price Book Update*.

The *Update* is available free to all Price Book purchasers.

To ensure you receive your copy, simply complete the reply card from the centre of the book and return it to us.

## Tables and Memoranda

## FORMULAE

Two dimensional figures

| Figure | Area |
|---|---|
| Triangle | 0.5 x base x height, <br> or $\sqrt{(s(s-a)(s-b)(s-c))}$ where s = 0.5 x the sum of the three sides and a,b and c are the lengths of the three sides, <br><br> or $a^2 = b^2 + c^2 - 2 \times bc \times COS\ A$ where A is the angle opposite side a |
| Hexagon | 2.6 x (side)$^2$ |
| Octagon | 4.83 x (side)$^2$ |
| Trapezoid | height x 0.5 (base + top) |
| Circle | 3.142 x radius$^2$ or 0.7854 x diameter$^2$ <br> (circumference = 2 x 3.142 x radius or 3.142 x diameter) |
| Sector of a circle | 0.5 x length of arc x radius |
| Segment of a circle | area of sector - area of triangle |
| Ellipse of a circle | 3.142 x AB <br> (where A = 0.5 x height and B = 0.5 x length) |
| Spandrel | 3/14 x radius$^2$ |

Three dimensional figure

| Figure | Volume | Surface Area |
|---|---|---|
| Prism | Area of base x height | circumference of base x height |
| Cube | (length of side) cubed | 6 x (length of side)$^2$ |
| Cylinder | 3.142 x radius$^2$ x height | 2 x 3.142 x radius x (height - radius) |
| Sphere | 4/3 x 3.142 x radius$^3$ | 4 x 3.142 x radius$^2$ |
| Segment of a sphere | [(3.142 x h)/6] x (3 x r$^2$ + h$^2$) | (2 x 3.142 x r x h) |
| Pyramid | 1/3 x (area of base x height) | 0.5 x circumference of base x slant height |

## FRACTIONS, DECIMALS AND MILLIMETRE EQUIVALENTS

| Fractions | Decimals | mm | Fractions | Decimals | mm |
|---|---|---|---|---|---|
| 1/64 | 0.015625 | 0.396875 | 33/64 | 0.515625 | 13.096875 |
| 1/32 | 0.03125 | 0.79375 | 17/32 | 0.53125 | 13.49375 |
| 3/64 | 0.046875 | 1.190625 | 35/64 | 0.546875 | 13.890625 |
| 1/16 | 0.0625 | 1.5875 | 9/16 | 0.5625 | 14.2875 |
| 5/64 | 0.078125 | 1.984375 | 37/64 | 0.578125 | 14.684375 |
| 3/32 | 0.09375 | 2.38125 | 19/32 | 0.59375 | 15.08125 |
| 7/64 | 0.109375 | 2.778125 | 39/64 | 0.609375 | 15.478125 |
| 1/8 | 0.125 | 3.175 | 5/8 | 0.625 | 15.875 |
| 9/64 | 0.140625 | 3.571875 | 41/64 | 0.640625 | 16.271875 |
| 5/32 | 0.15625 | 3.96875 | 21/32 | 0.65625 | 16.66875 |
| 11/64 | 0.171875 | 4.365625 | 43/64 | 0.671875 | 17.065625 |
| 3/16 | 0.1875 | 4.7625 | 11/16 | 0.6875 | 17.4625 |
| 13/64 | 0.203125 | 5.159375 | 45/64 | 0.703125 | 17.859375 |
| 7/32 | 0.21875 | 5.55625 | 23/32 | 0.71875 | 18.25625 |
| 15/64 | 0.234375 | 5.953125 | 47/64 | 0.734375 | 18.653125 |
| 1/4 | 0.25 | 6.35 | 3/4 | 0.75 | 19.05 |
| 17/64 | 0.265625 | 6.746875 | 49/64 | 0.765625 | 19.446875 |
| 9/32 | 0.28125 | 7.14375 | 25/32 | 0.78125 | 19.84375 |
| 19/64 | 0.296875 | 7.540625 | 51/64 | 0.796875 | 20.240625 |
| 5/16 | 0.3125 | 7.9375 | 13/16 | 0.8125 | 20.6375 |
| 21/64 | 0.328125 | 8.334375 | 53/64 | 0.828125 | 21.034375 |
| 11/32 | 0.34375 | 8.73125 | 27/32 | 0.84375 | 21.43125 |
| 23/64 | 0.359375 | 9.128125 | 55/64 | 0.859375 | 21.828125 |
| 3/8 | 0.375 | 9.525 | 7/8 | 0.875 | 22.225 |
| 25/64 | 0.390625 | 9.921875 | 57/64 | 0.890625 | 22.621875 |
| 13/32 | 0.40625 | 10.31875 | 29/32 | 0.90625 | 23.01875 |
| 27/64 | 0.421875 | 10.71563 | 59/64 | 0.921875 | 23.415625 |
| 7/16 | 0.4375 | 11.1125 | 15/16 | 0.9375 | 23.8125 |
| 29/64 | 0.453125 | 11.50938 | 61/64 | 0.953125 | 24.209375 |
| 15/32 | 0.46875 | 11.90625 | 31/32 | 0.96875 | 24.60625 |
| 31/64 | 0.484375 | 12.30313 | 63/64 | 0.984375 | 25.003125 |
| 1/2 | 0.5 | 12.7 | 1.0 | 1 | 25.4 |

## IMPERIAL STANDARD WIRE GAUGE (SWG)

| SWG No | Diameter in | Diameter mm | SWG No | Diameter in | Diameter mm |
|---|---|---|---|---|---|
| 7/0 | 0.5 | 12.7 | 23 | 0.024 | 0.61 |
| 6/0 | 0.464 | 11.79 | 24 | 0.022 | 0.559 |
| 5/0 | 0.432 | 10.97 | 25 | 0.02 | 0.508 |
| 4/0 | 0.4 | 10.16 | 26 | 0.018 | 0.457 |
| 3/0 | 0.372 | 9.45 | 27 | 0.0164 | 0.417 |
| 2/0 | 0.348 | 8.84 | 28 | 0.0148 | 0.376 |
| 1/0 | 0.324 | 8.23 | 29 | 0.0136 | 0.345 |
| 1 | 0.3 | 7.62 | 30 | 0.0124 | 0.315 |
| 2 | 0.276 | 7.01 | 31 | 0.0116 | 0.295 |
| 3 | 0.252 | 6.4 | 32 | 0.0108 | 0.274 |
| 4 | 0.232 | 5.89 | 33 | 0.01 | 0.254 |
| 5 | 0.212 | 5.38 | 34 | 0.009 | 0.234 |
| 6 | 0.192 | 4.88 | 35 | 0.008 | 0.213 |
| 7 | 0.176 | 4.47 | 36 | 0.008 | 0.193 |
| 8 | 0.16 | 4.06 | 37 | 0.007 | 0.173 |
| 9 | 0.144 | 3.66 | 38 | 0.006 | 0.152 |
| 10 | 0.128 | 3.25 | 39 | 0.005 | 0.132 |
| 11 | 0.116 | 2.95 | 40 | 0.005 | 0.122 |
| 12 | 0.104 | 2.64 | 41 | 0.004 | 0.112 |
| 13 | 0.092 | 2.34 | 42 | 0.004 | 0.102 |
| 14 | 0.08 | 2.03 | 43 | 0.004 | 0.091 |
| 15 | 0.072 | 1.83 | 44 | 0.003 | 0.081 |
| 16 | 0.064 | 1.63 | 45 | 0.003 | 0.071 |
| 17 | 0.056 | 1.42 | 46 | 0.002 | 0.061 |
| 18 | 0.048 | 1.22 | 47 | 0.002 | 0.051 |
| 19 | 0.04 | 1.016 | 48 | 0.002 | 0.041 |
| 20 | 0.036 | 0.914 | 49 | 0.001 | 0.031 |
| 21 | 0.032 | 0.813 | 50 | 0.001 | 0.025 |
| 22 | 0.028 | 0.711 | | | |

## WATER PRESSURE DUE TO HEIGHT

### Imperial

| Head Feet | Pressure lb/in² | Head Feet | Pressure lb/in² |
|---|---|---|---|
| 1 | 0.43 | 70 | 30.35 |
| 5 | 2.17 | 75 | 32.51 |
| 10 | 4.34 | 80 | 34.68 |
| 15 | 6.5 | 85 | 36.85 |
| 20 | 8.67 | 90 | 39.02 |
| 25 | 10.84 | 95 | 41.18 |
| 30 | 13.01 | 100 | 43.35 |
| 35 | 15.17 | 105 | 45.52 |
| 40 | 17.34 | 110 | 47.69 |
| 45 | 19.51 | 120 | 52.02 |
| 50 | 21.68 | 130 | 56.36 |
| 55 | 23.84 | 140 | 60.69 |
| 60 | 26.01 | 150 | 65.03 |
| 65 | 28.18 | | |

### Metric

| Head m | Pressure bar | Head m | Pressure bar |
|---|---|---|---|
| 0.5 | 0.049 | 18.0 | 1.766 |
| 1.0 | 0.098 | 19.0 | 1.864 |
| 1.5 | 0.147 | 20.0 | 1.962 |
| 2.0 | 0.196 | 21.0 | 2.06 |
| 3.0 | 0.294 | 22.0 | 2.158 |
| 4.0 | 0.392 | 23.0 | 2.256 |
| 5.0 | 0.491 | 24.0 | 2.354 |
| 6.0 | 0.589 | 25.0 | 2.453 |
| 7.0 | 0.687 | 26.0 | 2.551 |
| 8.0 | 0.785 | 27.0 | 2.649 |
| 9.0 | 0.883 | 28.0 | 2.747 |
| 10.0 | 0.981 | 29.0 | 2.845 |
| 11.0 | 1.079 | 30.0 | 2.943 |
| 12.0 | 1.177 | 32.5 | 3.188 |
| 13.0 | 1.275 | 35.0 | 3.434 |
| 14.0 | 1.373 | 37.5 | 3.679 |
| 15.0 | 1.472 | 40.0 | 3.924 |
| 16.0 | 1.57 | 42.5 | 4.169 |
| 17.0 | 1.668 | 45.0 | 4.415 |

| | | |
|---|---|---|
| 1 bar | = | 14.5038 lbf/in² |
| 1 metre | = | 3.2808 ft  or  39.3701 in |
| 1 foot | = | 0.3048 metres |
| 1 lbf/in² | = | 0.06895 bar |
| 1 in wg | = | 2.5 mbar (249.1 N/m²) |

## TABLE OF WEIGHTS FOR STEELWORK

### Mild Steel Bar

| Diameter (mm) | Weight (kg/m) | Diameter (mm) | Weight (kg/m) |
|---|---|---|---|
| 6 | 0.22 | 20 | 2.47 |
| 10 | 0.62 | 25 | 3.85 |
| 12 | 0.89 | 30 | 5.55 |
| 16 | 1.58 | 32 | 6.31 |

### Mild Steel Flat

| Size (mm) | Weight (kg/m) | Size (mm) | Weight (kg/m) |
|---|---|---|---|
| 15 x 3 | 0.36 | 15 x 5 | 0.59 |
| 20 x 3 | 0.47 | 20 x 5 | 0.79 |
| 25 x 3 | 0.59 | 25 x 5 | 0.98 |
| 30 x 3 | 0.71 | 30 x 5 | 1.18 |
| 40 x 3 | 0.94 | 40 x 5 | 1.57 |
| 45 x 3 | 1.06 | 45 x 5 | 1.77 |
| 50 x 3 | 1.18 | 50 x 5 | 1.96 |
| 20 x 6 | 0.94 | 20 x 8 | 1.26 |
| 25 x 6 | 1.18 | 25 x 8 | 1.57 |
| 30 x 6 | 1.41 | 30 x 8 | 1.88 |
| 40 x 6 | 1.88 | 40 x 8 | 2.51 |
| 45 x 6 | 2.12 | 45 x 8 | 2.83 |
| 50 x 6 | 2.36 | 50 x 8 | 3.14 |
| 55 x 6 | 2.60 | 55 x 8 | 3.45 |
| 60 x 6 | 2.83 | 60 x 8 | 3.77 |
| 65 x 6 | 3.06 | 65 x 8 | 4.08 |
| 70 x 6 | 3.30 | 70 x 8 | 4.40 |
| 75 x 6 | 3.53 | 75 x 8 | 4.71 |
| 100 x 6 | 4.71 | 100 x 8 | 6.28 |
| 20 x 10 | 1.57 | 20 x 12 | 1.88 |
| 25 x 10 | 1.96 | 25 x 12 | 2.36 |
| 30 x 10 | 2.36 | 30 x 12 | 2.83 |
| 40 x 10 | 3.14 | 40 x 12 | 3.77 |
| 45 x 10 | 3.53 | 45 x 12 | 4.24 |
| 50 x 10 | 3.93 | 50 x 12 | 4.71 |
| 55 x 10 | 4.32 | 55 x 12 | 5.12 |
| 60 x 10 | 4.71 | 60 x 12 | 5.65 |
| 65 x 10 | 5.10 | 65 x 12 | 6.12 |
| 70 x 10 | 5.50 | 70 x 12 | 6.59 |
| 75 x 10 | 5.89 | 75 x 12 | 7.07 |
| 100 x 10 | 7.85 | 100 x 12 | 9.42 |

## TABLE OF WEIGHTS FOR STEELWORK

### Mild Steel Equal Angle

| Size (mm) | Weight (kg/m) | Size (mm) | Weight (kg/m) |
|---|---|---|---|
| 13 x 13 x 3 | 0.56 | 60 x 60 x 10 | 8.69 |
| 20 x 20 x 3 | 0.88 | 70 x 70 x 10 | 10.30 |
| 25 x 25 x 3 | 1.11 | 75 x 75 x 10 | 11.05 |
| 30 x 30 x 3 | 1.36 | 80 x 80 x 10 | 11.90 |
| 40 x 40 x 3 | 1.82 | 90 x 90 x 10 | 13.40 |
| 45 x 45 x 3 | 2.06 | 100 x 100 x 10 | 15.00 |
| 50 x 50 x 3 | 2.30 | 120 x 120 x 10 | 18.20 |
|  |  | 150 x 156 x 10 | 23.00 |
| 30 x 30 x 6 | 2.56 |  |  |
| 40 x 40 x 6 | 3.52 | 75 x 75 x 12 | 13.07 |
| 45 x 45 x 6 | 4.00 | 80 x 80 x 12 | 14.00 |
| 50 x 50 x 6 | 4.47 | 90 x 90 x 12 | 15.90 |
| 60 x 60 x 6 | 5.42 | 100 x 120 x 12 | 21.90 |
| 70 x 70 x 6 | 6.38 | 120 x 120 x 12 | 21.60 |
| 75 x 75 x 6 | 6.82 | 150 x 150 x 12 | 27.30 |
| 80 x 80 x 6 | 7.34 | 200 x 200 x 12 | 36.74 |
| 90 x 90 x 6 | 8.30 |  |  |
| 40 x 40 x 8 | 4.55 |  |  |
| 50 x 50 x 8 | 5.82 |  |  |
| 60 x 60 x 8 | 7.09 |  |  |
| 70 x 70 x 8 | 8.36 |  |  |
| 75 x 75 x 8 | 8.96 |  |  |
| 80 x 80 x 8 | 9.63 |  |  |
| 90 x 90 x 8 | 10.90 |  |  |
| 100 x 100 x 8 | 12.20 |  |  |
| 120 x 120 x 8 | 14.70 |  |  |

### Mild Steel Unequal Angle

| Size (mm) | Weight (kg/m) | Size (mm) | Weight (kg/m) |
|---|---|---|---|
| 40 x 25 x 6 | 2.79 | 100 x 65 x 10 | 12.30 |
| 50 x 40 x 6 | 4.24 | 100 x 75 x 10 | 13.00 |
| 60 x 30 x 6 | 3.99 | 125 x 75 x 10 | 15.00 |
| 65 x 50 x 6 | 5.16 | 150 x 75 x 10 | 17.00 |
| 75 x 50 x 6 | 5.65 | 150 x 90 x 10 | 18.20 |
| 80 x 60 x 6 | 6.37 | 200 x 100 x 10 | 23.00 |
| 125 x 75 x 6 | 9.18 |  |  |
| 75 x 50 x 8 | 7.39 | 100 x 75 x 12 | 15.40 |
| 80 x 60 x 8 | 8.34 | 125 x 75 x 12 | 17.80 |
| 100 x 65 x 8 | 9.94 | 150 x 75 x 12 | 20.20 |
| 100 x 75 x 8 | 10.60 | 150 x 90 x 12 | 21.60 |
| 125 x 75 x 8 | 12.20 | 200 x 100 x 12 | 27.30 |
| 137 x 102 x 8 | 14.88 | 200 x 150 x 12 | 32.00 |

## TABLE OF WEIGHTS FOR STEELWORK

### Rolled Steel Channels

| Size (mm) | Weight (kg/m) | Size (mm) | Weight (kg/m) |
|---|---|---|---|
| 32 x 27 | 2.80 | 178 x 76 | 20.84 |
| 38 x 19 | 2.49 | 178 x 79 | 26.81 |
| 51 x 25 | 4.46 | 203 x 76 | 23.82 |
| 51 x 38 | 5.80 | 203 x 89 | 29.78 |
| 64 x 25 | 6.70 | 229 x 76 | 26.06 |
| 76 x 38 | 7.46 | 229 x 89 | 32.76 |
| 76 x 51 | 9.45 | 254 x 76 | 28.29 |
| 102 x 51 | 10.42 | 254 x 89 | 35.74 |
| 127 x 64 | 14.90 | 305 x 89 | 41.67 |
| 152 x 76 | 17.88 | 305 x 102 | 46.18 |
| 152 x 89 | 23.84 | 381 x 102 | 55.10 |

### Rolled Steel Joists

| Size (mm) | Weight (kg/m) | Size (mm) | Weight (kg/m) |
|---|---|---|---|
| 76 x 38 | 6.25 | 152 x 76 | 17.86 |
| 76 x 76 | 12.65 | 152 x 89 | 17.09 |
| 102 x 44 | 7.44 | 152 x 127 | 37.20 |
| 102 x 64 | 9.65 | 178 x 102 | 21.54 |
| 102 x 102 | 23.06 | 203 x 102 | 25.33 |
| 127 x 76 | 13.36 | 203 x 152 | 52.03 |
| 127 x 114 | 26.78 | 254 x 114 | 37.20 |
| 127 x 114 | 29.76 | 254 x 203 | 81.84 |
| | | 305 x 203 | 96.72 |

### Universal Columns

| Size (mm) | Weight (kg/m) | Size (mm) | Weight (kg/m) |
|---|---|---|---|
| 152 x 152 | 23.00 | 254 x 254 | 89.00 |
| 152 x 152 | 30.00 | 254 x 254 | 107.00 |
| 152 x 152 | 37.00 | 254 x 254 | 132.00 |
| 203 x 203 | 46.00 | 254 x 254 | 167.00 |
| 203 x 203 | 52.00 | 305 x 305 | 97.00 |
| 203 x 203 | 60.00 | 305 x 305 | 118.00 |
| 203 x 203 | 71.00 | 305 x 305 | 137.00 |
| 203 x 203 | 86.00 | 305 x 305 | 158.00 |
| 254 x 254 | 73.00 | 305 x 305 | 198.00 |

## TABLE OF WEIGHTS FOR STEELWORK

### Universal Beams

| Size (mm) | Weight (kg/m) | Size (mm) | Weight (kg/m) |
|---|---|---|---|
| 203 x 133 | 25.00 | 305 x 127 | 48.00 |
| 203 x 133 | 30.00 | 305 x 165 | 40.00 |
| 254 x 102 | 22.00 | 305 x 165 | 46.00 |
| 254 x 102 | 25.00 | 305 x 165 | 54.00 |
| 254 x 102 | 28.00 | 356 x 127 | 33.00 |
| 254 x 146 | 31.00 | 356 x 127 | 39.00 |
| 254 x 146 | 37.00 | 356 x 171 | 45.00 |
| 254 x 146 | 43.00 | 356 x 171 | 51.00 |
| 305 x 102 | 25.00 | 356 x 171 | 57.00 |
| 305 x 102 | 28.00 | 356 x 171 | 67.00 |
| 305 x 102 | 33.00 | 381 x 152 | 52.00 |
| 305 x 127 | 37.00 | 381 x 152 | 60.00 |
| 305 x 127 | 42.00 | 381 x 152 | 67.00 |

### Circular Hollow Sections

| Size (mm) | Weight (kg/m) | Size (mm) | Weight (kg/m) |
|---|---|---|---|
| 21.3 x 3.2 | 1.43 | 76.1 x 3.2 | 5.75 |
| 26.9 x 3.2 | 1.87 | 76.1 x 4.0 | 7.11 |
| 33.7 x 2.6 | 1.99 | 76.1 x 5.0 | 8.77 |
| 33.7 x 3.2 | 2.41 | 88.9 x 3.2 | 6.76 |
| 33.7 x 4.0 | 2.93 | 88.9 x 4.0 | 8.36 |
| 42.4 x 2.6 | 2.55 | 88.9 x 5.0 | 10.30 |
| 42.4 x 3.2 | 3.09 | 114.3 x 3.6 | 9.83 |
| 42.4 x 4.0 | 3.79 | 114.3 x 5.0 | 13.50 |
| 48.3 x 3.2 | 3.56 | 114.3 x 6.3 | 16.80 |
| 48.3 x 4.0 | 4.37 | 139.7 x 5.0 | 16.60 |
| 48.3 x 5.0 | 5.34 | 139.7 x 6.3 | 20.70 |
| 60.3 x 3.2 | 4.51 | 139.7 x 8.0 | 26.00 |
| 60.3 x 4.0 | 5.55 | 139.7 x 10.0 | 32.00 |
| 60.3 x 5.0 | 6.82 | 168.3 x 5.0 | 20.10 |

## Tables and Memoranda

## TABLE OF WEIGHTS FOR STEELWORK

### Square Hollow Sections

| Size (mm) | Weight (kg/m) | Size (mm) | Weight (kg/m) |
|---|---|---|---|
| 20 x 20 x 2.0 | 1.12 | 90 x 90 x 3.6 | 9.72 |
| 20 x 20 x 2.6 | 1.39 | 90 x 90 x 5.0 | 13.30 |
| 30 x 30 x 2.6 | 2.21 | 90 x 90 x 6.3 | 16.40 |
| 30 x 30 x 3.2 | 2.65 | 100 x 100 x 4.0 | 12.00 |
| 40 x 40 x 2.6 | 3.03 | 100 x 100 x 5.0 | 14.80 |
| 40 x 40 x 3.2 | 3.66 | 100 x 100 x 6.3 | 18.40 |
| 40 x 40 x 4.0 | 4.46 | 100 x 100 x 8.0 | 22.90 |
| 50 x 50 x 3.2 | 4.66 | 100 x 100 x 10.0 | 27.90 |
| 50 x 50 x 4.0 | 5.72 | 120 x 120 x 5.0 | 18.00 |
| 50 x 50 x 5.0 | 6.97 | 120 x 120 x 6.3 | 22.30 |
| 60 x 60 x 3.2 | 5.67 | 120 x 120 x 8.0 | 27.90 |
| 60 x 60 x 4.0 | 6.97 | 120 x 120 x 10.0 | 34.20 |
| 60 x 60 x 5.0 | 8.54 | 150 x 150 x 5.0 | 22.70 |
| 70 x 70 x 3.2 | 7.46 | 150 x 150 x 6.3 | 28.30 |
| 70 x 70 x 5.0 | 10.10 | 150 x 150 x 8.0 | 35.40 |
| 80 x 80 x 3.6 | 8.59 | 150 x 150 x 10.0 | 43.60 |
| 80 x 80 x 5.0 | 11.70 | | |
| 80 x 80 x 6.3 | 14.40 | | |

### Rectangular Hollow Sections

| Size (mm) | Weight (kg/m) | Size (mm) | Weight (kg/m) |
|---|---|---|---|
| 50 x 30 x 2.6 | 3.03 | 120 x 80 x 5.0 | 14.80 |
| 50 x 30 x 3.2 | 3.66 | 120 x 80 x 6.3 | 18.40 |
| 60 x 40 x 3.2 | 4.66 | 120 x 80 x 8.0 | 22.90 |
| 60 x 40 x 4.0 | 5.72 | 120 x 80 x 10.0 | 27.90 |
| 80 x 40 x 3.2 | 5.67 | 150 x 100 x 5.0 | 18.70 |
| 80 x 40 x 4.0 | 6.97 | 150 x 100 x 6.3 | 23.30 |
| 90 x 50 x 3.6 | 7.46 | 150 x 100 x 8.0 | 29.10 |
| 90 x 50 x 5.0 | 10.10 | 150 x 100 x 10.0 | 35.70 |
| 100 x 50 x 3.2 | 7.18 | 160 x 80 x 5.0 | 18.00 |
| 100 x 50 x 4.0 | 8.86 | 160 x 80 x 6.3 | 22.30 |
| 100 x 50 x 5.0 | 10.90 | 160 x 80 x 8.0 | 27.90 |
| 100 x 60 x 3.6 | 8.59 | 160 x 80 x 10.0 | 34.20 |
| 100 x 60 x 5.0 | 11.70 | 200 x 100 x 5.0 | 22.70 |
| 100 x 60 x 6.3 | 14.40 | 200 x 100 x 6.3 | 28.30 |
| 120 x 60 x 3.6 | 9.72 | 200 x 100 x 8.0 | 35.40 |
| 120 x 60 x 5.0 | 13.30 | 200 x 100 x 10.0 | 43.60 |
| 120 x 60 x 6.3 | 16.40 | | |

## DIMENSIONS AND WEIGHTS OF COPPER PIPES TO BS 2871 PART 1

| Outside Diameter (mm) | Internal Diameter (mm) | Weight per Metre (kg) | Internal Diameter (mm) | Weight per Metre (kg) | Internal Diameter (mm) | Weight per Metre (kg) |
|---|---|---|---|---|---|---|
| | Table X | | Table Y | | Table Z | |
| 6 | 4.80 | 0.0911 | 4.40 | 0.1170 | 5.00 | 0.0774 |
| 8 | 6.80 | 0.1246 | 6.40 | 0.1617 | 7.00 | 0.1054 |
| 10 | 8.80 | 0.1580 | 8.40 | 0.2064 | 9.00 | 0.1334 |
| 12 | 10.80 | 0.1914 | 10.40 | 0.2511 | 11.00 | 0.1612 |
| 15 | 13.60 | 0.2796 | 13.00 | 0.3923 | 14.00 | 0.2031 |
| 18 | 16.40 | 0.3852 | 16.00 | 0.4760 | 16.80 | 0.2918 |
| 22 | 20.22 | 0.5308 | 19.62 | 0.6974 | 20.82 | 0.3589 |
| 28 | 26.22 | 0.6814 | 25.62 | 0.8985 | 26.82 | 0.4594 |
| 35 | 32.63 | 1.1334 | 32.03 | 1.4085 | 33.63 | 0.6701 |
| 42 | 39.63 | 1.3675 | 39.03 | 1.6996 | 40.43 | 0.9216 |
| 54 | 51.63 | 1.7691 | 50.03 | 2.9052 | 52.23 | 1.3343 |
| 76.1 | 73.22 | 3.1287 | 72.22 | 4.1437 | 73.82 | 2.5131 |
| 108 | 105.12 | 4.4666 | 103.12 | 7.3745 | 105.72 | 3.5834 |
| 133 | 130.38 | 5.5151 | -- | -- | 130.38 | 5.5151 |
| 159 | 155.38 | 8.7795 | -- | -- | 156.38 | 6.6056 |

## DIMENSIONS OF STAINLESS STEEL PIPES TO BS 4127.

| Outside Diameter (mm) | Maximun Outside Diameter (mm) | Minimum Outside Diameter (mm) | Wall Thickness (mm) | Working Pressure (bar) |
|---|---|---|---|---|
| 6 | 6.045 | 5.94 | 0.6 | 330 |
| 8 | 8.045 | 7.94 | 0.6 | 260 |
| 10 | 10.045 | 9.94 | 0.6 | 210 |
| 12 | 12.045 | 11.94 | 0.6 | 170 |
| 15 | 15.045 | 14.94 | 0.6 | 140 |
| 18 | 18.045 | 17.94 | 0.7 | 135 |
| 22 | 22.055 | 21.95 | 0.7 | 110 |
| 28 | 28.055 | 27.95 | 0.8 | 121 |
| 35 | 35.070 | 34.96 | 1.0 | 100 |
| 42 | 42.070 | 41.96 | 1.1 | 91 |
| 54 | 54.090 | 53.94 | 1.2 | 77 |

## DIMENSIONS OF STEEL PIPES TO BS 1387

| Nominal Size | Approx. Outside Diameter | Outside Diameter | | | | Thickness | | |
|---|---|---|---|---|---|---|---|---|
| | | Light | | Medium and Heavy | | Light | Medium | Heavy |
| | | Maximum | Minimum | Maximum | Minimum | | | |
| mm | mm | mm | mm | mm | mm | mm | mm | mm |
| 6 | 10.20 | 10.10 | 9.70 | 10.40 | 9.80 | 1.80 | 2.00 | 2.65 |
| 8 | 13.50 | 13.60 | 13.20 | 13.90 | 13.30 | 1.80 | 2.35 | 2.90 |
| 10 | 17.20 | 17.10 | 16.70 | 17.40 | 16.80 | 1.80 | 2.35 | 2.90 |
| 15 | 21.30 | 21.40 | 21.00 | 21.70 | 21.10 | 2.00 | 2.65 | 3.25 |
| 20 | 26.90 | 26.90 | 26.40 | 27.20 | 26.60 | 2.35 | 2.65 | 3.25 |
| 25 | 33.70 | 33.80 | 33.20 | 34.20 | 33.40 | 2.65 | 3.25 | 4.05 |
| 32 | 42.40 | 42.50 | 41.90 | 42.90 | 42.10 | 2.65 | 3.25 | 4.05 |
| 40 | 48.30 | 48.40 | 47.80 | 48.80 | 48.00 | 2.90 | 3.25 | 4.05 |
| 50 | 60.30 | 60.20 | 59.60 | 60.80 | 59.80 | 2.90 | 3.65 | 4.50 |
| 65 | 76.10 | 76.00 | 75.20 | 76.60 | 75.40 | 3.25 | 3.65 | 4.50 |
| 80 | 88.90 | 88.70 | 87.90 | 89.50 | 88.10 | 3.25 | 4.05 | 4.85 |
| 100 | 114.30 | 113.90 | 113.00 | 114.90 | 113.30 | 3.65 | 4.50 | 5.40 |
| 125 | 139.70 | -- | -- | 140.60 | 138.70 | -- | 4.85 | 5.40 |
| 150 | 165.1* | -- | -- | 166.10 | 164.10 | -- | 4.85 | 5.40 |

* 165.1mm (6.5in) outside diameter is not generally recommended except where screwing to BS 21 is necessary.

All dimensions are in accordance with ISO R65 except approximate outside diameters which are in accordance with ISO R64.

Light quality is equivalent to ISO R65 Light Series II.

*Tables and Memoranda*

## APPROXIMATE METRES PER TONNE OF TUBES TO BS 1387

| Nom. Size mm | BLACK | | | | | | GALVANISED | | | | | |
|---|---|---|---|---|---|---|---|---|---|---|---|---|
| | Plain/screwed ends | | | Screwed & socketed | | | Plain/screwed ends | | | Screwed & socketed | | |
| | Light m | Medium m | Heavy m | Light m | Medium m | Heavy m | Light m | Medium m | Heavy m | Light m | Medium m | Heavy m |
| 6 | 2765 | 2461 | 2030 | 2743 | 2443 | 2018 | 2604 | 2333 | 1948 | 2584 | 2317 | 1937 |
| 8 | 1936 | 1538 | 1300 | 1920 | 1527 | 1292 | 1826 | 1467 | 1254 | 1811 | 1458 | 1247 |
| 10 | 1483 | 1173 | 979 | 1471 | 1165 | 974 | 1400 | 1120 | 944 | 1386 | 1113 | 939 |
| 15 | 1050 | 817 | 688 | 1040 | 811 | 684 | 996 | 785 | 665 | 987 | 779 | 661 |
| 20 | 712 | 634 | 529 | 704 | 628 | 525 | 679 | 609 | 512 | 673 | 603 | 508 |
| 25 | 498 | 410 | 336 | 494 | 407 | 334 | 478 | 396 | 327 | 474 | 394 | 325 |
| 32 | 388 | 319 | 260 | 384 | 316 | 259 | 373 | 308 | 254 | 369 | 305 | 252 |
| 40 | 307 | 277 | 226 | 303 | 273 | 223 | 296 | 268 | 220 | 292 | 264 | 217 |
| 50 | 244 | 196 | 162 | 239 | 194 | 160 | 235 | 191 | 158 | 231 | 188 | 157 |
| 65 | 172 | 153 | 127 | 169 | 151 | 125 | 167 | 149 | 124 | 163 | 146 | 122 |
| 80 | 147 | 118 | 99 | 143 | 116 | 98 | 142 | 115 | 97 | 139 | 113 | 96 |
| 100 | 101 | 82 | 69 | 98 | 81 | 68 | 98 | 81 | 68 | 95 | 79 | 67 |
| 125 | -- | 62 | 56 | -- | 60 | 55 | -- | 60 | 55 | -- | 59 | 54 |
| 150 | -- | 52 | 47 | -- | 50 | 46 | -- | 51 | 46 | -- | 49 | 45 |

The figures for `plain or screwed ends' apply also to tubes to BS 1775 HFW of equivalent size and thickness.

## FLANGE DIMENSION CHART TO BS 4504 & BS 10

### Normal Pressure Rating (N.P.6) 6 Bar

| Nom. Size | Flange Outside Diam. | Table 6/2 Forged Welding Neck | Table 6/3 Plate Slip | Table 6/4 Forged Bossed Screwed | Table 6/5 Forged Bossed Slip on | Table 6/8 Plate Blank | Raised Face Diam. | Raised Face T'ness | Pitch Circle Diam. | Nr. Bolt Holes | Size of Bolt |
|---|---|---|---|---|---|---|---|---|---|---|---|
| 15  | 80  | 12 | 12 | 12 | 12 | 12 | 40  | 2 | 55  | 4  | M10 x 40 |
| 20  | 90  | 14 | 14 | 14 | 14 | 14 | 50  | 2 | 65  | 4  | M10 x 45 |
| 25  | 100 | 14 | 14 | 14 | 14 | 14 | 60  | 2 | 75  | 4  | M10 x 45 |
| 32  | 120 | 14 | 16 | 14 | 14 | 14 | 70  | 2 | 90  | 4  | M12 x 45 |
| 40  | 130 | 14 | 16 | 14 | 14 | 14 | 80  | 3 | 100 | 4  | M12 x 45 |
| 50  | 140 | 14 | 16 | 14 | 14 | 14 | 90  | 3 | 110 | 4  | M12 x 45 |
| 65  | 160 | 14 | 16 | 14 | 14 | 14 | 110 | 3 | 130 | 4  | M12 x 45 |
| 80  | 190 | 16 | 18 | 16 | 16 | 16 | 128 | 3 | 150 | 4  | M16 x 55 |
| 100 | 210 | 16 | 18 | 16 | 16 | 16 | 148 | 3 | 170 | 4  | M16 x 55 |
| 125 | 240 | 18 | 20 | 18 | 18 | 18 | 178 | 3 | 200 | 8  | M16 x 60 |
| 150 | 265 | 18 | 20 | 18 | 18 | 18 | 202 | 3 | 225 | 8  | M16 x 60 |
| 200 | 320 | 20 | 22 | -- | 20 | 20 | 258 | 3 | 280 | 8  | M16 x 60 |
| 250 | 375 | 22 | 24 | -- | 22 | 22 | 312 | 3 | 335 | 12 | M16 x 65 |
| 300 | 440 | 22 | 24 | -- | 22 | 22 | 365 | 4 | 395 | 12 | M20 x 70 |

### Normal Pressure Rating (N.P.16) 16 Bar

| Nom. Size | Flange Outside Diam. | Table 6/2 Forged Welding Neck | Table 6/3 Plate Slip | Table 6/4 Forged Bossed Screwed | Table 6/5 Forged Bossed Slip on | Table 6/8 Plate Blank | Raised Face Diam. | Raised Face T'ness | Pitch Circle Diam. | Nr. Bolt Holes | Size of Bolt |
|---|---|---|---|---|---|---|---|---|---|---|---|
| 15  | 95  | 14 | 14 | 14 | 14 | 14 | 45  | 2 | 55  | 4  | M12 x 45 |
| 20  | 105 | 16 | 16 | 16 | 16 | 16 | 58  | 2 | 65  | 4  | M12 x 50 |
| 25  | 115 | 16 | 16 | 16 | 16 | 16 | 68  | 2 | 75  | 4  | M12 x 50 |
| 32  | 140 | 16 | 16 | 16 | 16 | 16 | 78  | 2 | 90  | 4  | M16 x 55 |
| 40  | 150 | 16 | 16 | 16 | 16 | 16 | 88  | 3 | 100 | 4  | M16 x 55 |
| 50  | 165 | 18 | 18 | 18 | 18 | 18 | 102 | 3 | 110 | 4  | M16 x 60 |
| 65  | 185 | 18 | 18 | 18 | 18 | 18 | 122 | 3 | 130 | 4  | M16 x 60 |
| 80  | 200 | 20 | 20 | 20 | 20 | 20 | 138 | 3 | 150 | 8  | M16 x 60 |
| 100 | 220 | 20 | 20 | 20 | 20 | 20 | 158 | 3 | 170 | 8  | M16 x 65 |
| 125 | 250 | 22 | 22 | 22 | 22 | 22 | 188 | 3 | 200 | 8  | M16 x 70 |
| 150 | 285 | 22 | 22 | 22 | 22 | 22 | 212 | 3 | 225 | 8  | M20 x 70 |
| 200 | 340 | 24 | 24 | -- | 24 | 24 | 268 | 3 | 280 | 12 | M20 x 75 |
| 250 | 405 | 26 | 26 | -- | 26 | 26 | 320 | 3 | 335 | 12 | M24 x 90 |
| 300 | 460 | 28 | 28 | -- | 28 | 28 | 378 | 4 | 395 | 12 | M24 x 90 |

## MINIMUM DISTANCES BETWEEN SUPPORTS/FIXINGS

| Material | BS Nominal Pipe Size | | Pipes - Vertical | Pipes - Horizontal on to low gradients |
|---|---|---|---|---|
| | inch | mm | support distance in metres | support distance in metres |
| Copper | 0.50 | 15.00 | 1.90 | 1.30 |
| | 0.75 | 22.00 | 2.50 | 1.90 |
| | 1.00 | 28.00 | 2.50 | 1.90 |
| | 1.25 | 35.00 | 2.80 | 2.50 |
| | 1.50 | 42.00 | 2.80 | 2.50 |
| | 2.00 | 54.00 | 3.90 | 2.50 |
| | 2.50 | 67.00 | 3.90 | 2.80 |
| | 3.00 | 76.10 | 3.90 | 2.80 |
| | 4.00 | 108.00 | 3.90 | 2.80 |
| | 5.00 | 133.00 | 3.90 | 2.80 |
| | 6.00 | 159.00 | 3.90 | 2.80 |
| muPVC | 1.25 | 32.00 | 1.20 | 0.50 |
| | 1.50 | 40.00 | 1.20 | 0.50 |
| | 2.00 | 50.00 | 1.20 | 0.60 |
| Polypropylene | 1.25 | 32.00 | 1.20 | 0.50 |
| | 1.50 | 40.00 | 1.20 | 0.50 |
| uPVC | -- | 82.40 | 1.20 | 0.50 |
| | -- | 110.00 | 1.80 | 0.90 |
| | -- | 160.00 | 1.80 | 1.20 |

## LITRES OF WATER STORAGE REQUIRED PER PERSON PER BUILDING TYPE

| Type of Building | Storage per person litres |
|---|---|
| Houses and flats | 90 |
| Hostels | 90 |
| Hotels | 135 |
| Nurses home and medical quarters | 115 |
| Offices with canteen | 45 |
| Offices without canteen | 35 |
| Restaurants, per meal served | 7 |
| Boarding schools | 90 |
| Day schools | 30 |

## COLD WATER PLUMBING - INSULATION THICKNESS REQUIRED AGAINST FROST

| Bore of Tube | | Pipework within Buildings - Declared Thermal Conductivity (W/m degrees C) | | |
|---|---|---|---|---|
| inch | mm | Up to 0.040 | 0.041 to 0.055 | 0.056 to 0.070 |
| 0.50 | 15 | 32 | 50 | 75 |
| 0.75 | 20 | 32 | 50 | 75 |
| 1.00 | 25 | 32 | 50 | 75 |
| 1.25 | 32 | 32 | 50 | 75 |
| 1.50 | 40 | 32 | 50 | 75 |
| 2.00 | 50 | 25 | 32 | 50 |
| 2.50 | 65 | 25 | 32 | 50 |
| 3.00 | 80 | 25 | 32 | 50 |
| 4.00 | 100 | 19 | 25 | 38 |

## CAPACITY AND DIMENSIONS OF GALVANISED MILD STEEL CISTERNS - BS 417

| Capacity (litres) | BS type (SCM) | Dimensions | | |
|---|---|---|---|---|
| | | Length (mm) | Width (mm) | Depth (mm) |
| 18 | 45 | 457 | 305 | 305 |
| 36 | 70 | 610 | 305 | 371 |
| 54 | 90 | 610 | 406 | 371 |
| 68 | 110 | 610 | 432 | 432 |
| 86 | 135 | 610 | 457 | 482 |
| 114 | 180 | 686 | 508 | 508 |
| 159 | 230 | 736 | 559 | 559 |
| 191 | 270 | 762 | 584 | 610 |
| 227 | 320 | 914 | 610 | 584 |
| 264 | 360 | 914 | 660 | 610 |
| 327 | 450/1 | 1220 | 610 | 610 |
| 336 | 450/2 | 965 | 686 | 686 |
| 423 | 570 | 965 | 762 | 787 |
| 491 | 680 | 1090 | 864 | 736 |
| 709 | 910 | 1070 | 889 | 889 |

## CAPACITY OF COLD WATER POLYPROPYLENE STORAGE CISTERNS - BS 4213

| Capacity (litres) | BS type (PC) | Maximum Height mm |
|---|---|---|
| 18 | 4 | 310 |
| 36 | 8 | 380 |
| 68 | 15 | 430 |
| 91 | 20 | 510 |
| 114 | 25 | 530 |
| 182 | 40 | 610 |
| 227 | 50 | 660 |
| 273 | 60 | 660 |
| 318 | 70 | 660 |
| 455 | 100 | 760 |

## STORAGE CAPACITY AND RECOMMENDED POWER OF HOT WATER STORAGE BOILERS

| Type of Building | | Storage at 65°C (litres / person) | Boiler power to 65°C (kW / person) |
|---|---|---|---|
| Flats and Dwellings | | | |
| (a) | Low Rent Properties | 25 | 0.5 |
| (b) | Medium Rent Properties | 30 | 0.7 |
| (c) | High Rent Properties | 45 | 1.2 |
| Nurses Homes | | 45 | 0.9 |
| Hostels | | 0.7 | 0.7 |
| Hotels | | | |
| (a) | Top Quality - Up Market | 1.2 | 1.2 |
| (b) | Average Quality - Low Market | 0.9 | 0.9 |
| Colleges and Schools | | | |
| (a) | Live-in Accomodation | 0.7 | 0.7 |
| (b) | Public Comprehensive | 0.1 | 0.1 |
| Factories | | 0.1 | 0.1 |
| Hospitals | | | |
| (a) | General | 1.5 | 1.5 |
| (b) | Infectious | 1.5 | 1.5 |
| (c) | Infirmaries | 0.6 | 0.6 |
| (d) | Infirmaries with Laundary | 0.9 | 0.9 |
| (e) | Maternity | 2.1 | 2.1 |
| (f) | Mental | 0.7 | 0.7 |
| Offices | | 0.1 | 0.1 |
| Sports Pavilions | | 0.3 | 0.3 |

## THICKNESS OF THERMAL INSULATION FOR HEATING INSTALLATIONS

| Size of Tube mm | Declared Thermal Conductivity | | | |
|---|---|---|---|---|
| | Up to 0.025 | 0.026 to 0.040 | 0.041 to 0.055 | 0.056 to 0.070 |
| LTHW Systems | Minimum thickness of insulation | | | |
| 15 | 25 | 25 | 38 | 38 |
| 20 | 25 | 32 | 38 | 38 |
| 25 | 25 | 38 | 38 | 38 |
| 32 | 32 | 38 | 38 | 50 |
| 40 | 32 | 38 | 38 | 50 |
| 50 | 38 | 38 | 50 | 50 |
| 65 | 38 | 50 | 50 | 50 |
| 80 | 38 | 50 | 50 | 50 |
| 100 | 38 | 50 | 50 | 63 |
| 125 | 38 | 50 | 50 | 63 |
| 150 | 50 | 50 | 63 | 63 |
| 200 | 50 | 50 | 63 | 63 |
| 250 | 50 | 63 | 63 | 63 |
| 300 | 50 | 63 | 63 | 63 |
| Flat Surfaces | 50 | 63 | 63 | 63 |

| MTHW Systems and Condensate | Minimum thickness of insulation | | | |
|---|---|---|---|---|
| 15 | 25 | 38 | 38 | 38 |
| 20 | 32 | 38 | 38 | 50 |
| 25 | 38 | 38 | 38 | 50 |
| 32 | 38 | 50 | 50 | 50 |
| 40 | 38 | 50 | 50 | 50 |
| 50 | 38 | 50 | 50 | 50 |
| 65 | 38 | 50 | 50 | 50 |
| 80 | 50 | 50 | 50 | 63 |
| 100 | 50 | 63 | 63 | 63 |
| 125 | 50 | 63 | 63 | 63 |
| 150 | 50 | 63 | 63 | 63 |
| 200 | 50 | 63 | 63 | 63 |
| 250 | 50 | 63 | 63 | 75 |
| 300 | 63 | 63 | 63 | 75 |
| Flat Surfaces | 63 | 63 | 63 | 75 |

## THICKNESS OF THERMAL INSULATION FOR HEATING INSTALLATIONS

| Size of Tube mm | Declared Thermal Conductivity | | | |
|---|---|---|---|---|
| | Up to 0.025 | 0.026 to 0.040 | 0.041 to 0.055 | 0.056 to 0.070 |
| HTHW Systems and Steam | Minimum thickness of insulation | | | |
| 15 | 38 | 50 | 50 | 50 |
| 20 | 38 | 50 | 50 | 50 |
| 25 | 38 | 50 | 50 | 50 |
| 32 | 50 | 50 | 50 | 63 |
| 40 | 50 | 50 | 50 | 63 |
| 50 | 50 | 50 | 75 | 75 |
| 65 | 50 | 63 | 75 | 75 |
| 80 | 50 | 63 | 75 | 75 |
| 100 | 63 | 63 | 75 | 100 |
| 125 | 63 | 63 | 100 | 100 |
| 150 | 63 | 63 | 100 | 100 |
| 200 | 63 | 63 | 100 | 100 |
| 250 | 63 | 75 | 100 | 100 |
| 300 | 63 | 75 | 100 | 100 |
| Flat Surfaces | 63 | 75 | 100 | 100 |

## CAPACITIES AND DIMENSIONS OF GALVANISED MILD STEEL INDIRECT CYLINDERS TO BS 1565

| Approximate Capacity (litres) | BS Size No. | Internal Diameter (mm) | External height over dome (mm) |
|---|---|---|---|
| 109 | BSG.1M | 457 | 762 |
| 136 | BSG.2M | 457 | 914 |
| 159 | BSG.3M | 457 | 1067 |
| 227 | BSG.4M | 508 | 1270 |
| 273 | BSG.5M | 508 | 1473 |
| 364 | BSG.6M | 610 | 1372 |
| 455 | BSG.7M | 610 | 1753 |
| 123 | BSG.8M | 457 | 838 |

## CAPACITIES AND DIMENSIONS OF COPPER INDIRECT CYLINDERS (COIL TYPE) TO BS 1566

| Approximate Capacity (litres) | BS Type | External Diameter (mm) | External height over dome (mm) |
|---|---|---|---|
| 96 | 0 | 300 | 1600 |
| 72 | 1 | 350 | 900 |
| 96 | 2 | 400 | 900 |
| 114 | 3 | 400 | 1050 |
| 84 | 4 | 450 | 675 |
| 95 | 5 | 450 | 750 |
| 106 | 6 | 450 | 825 |
| 117 | 7 | 450 | 900 |
| 140 | 8 | 450 | 1050 |
| 162 | 9 | 450 | 1200 |
| 190 | 10 | 500 | 1200 |
| 245 | 11 | 500 | 1500 |
| 280 | 12 | 600 | 1200 |
| 360 | 13 | 600 | 1500 |
| 440 | 14 | 600 | 1800 |

## RECOMMENDED AIR CONDITIONING DESIGN LOADS

| Building Type | Design Loading |
|---|---|
| Computer rooms | 10 m² of floor area per ton |
| Restaurants | 16 m² of floor area per ton |
| Banks (main area) | 22 m² of floor area per ton |
| Large Office Buildings (exterior zone) | 25 m² of floor area per ton |
| Supermarkets | 30 m² of floor area per ton |
| Large Office Block (interior zone) | 32 m² of floor area per ton |
| Small Office Block (interior zone) | 35 m² of floor area per ton |

# Index

| | | | | |
|---|---|---|---|---|
| A.B.S. Pipe | | | Ball Float Valves | 229 |
| waste pipes | 141 | | Batten Lampholders | 461 |
| waste pipe fittings | 141 | | Battery Pack | 378 |
| water/gas pipes | 149 | | Bench Trunking | 399 |
| water/gas pipe fittings | 149 | | Black Steel Pipe; Screwed | |
| flanges | 156 | | pipes | 163 |
| Accelerator Pumps | 259 | | flanges | 164 |
| Accessories for Ethernet | 366 | | tubular fittings | 165 |
| Acoustic Louvres | 315 | | steel fittings | 166 |
| Acknowledgements | xi | | malleable iron fittings | 168 |
| Adaptable Boxes | | | Black Steel Pipe; Welded | |
| cast iron, black | 386 | | pipes | 174 |
| cast iron, galvanised | 387 | | flanges | 175 |
| PVC | 388 | | welded fittings | 177 |
| sheet steel black | 387 | | labours | 179 |
| sheet steel galvanised | 388 | | Blank Flanges | |
| Addressable Call Point | 379 | | metric | 175 |
| Air Circuit Breakers | 439 | | imperial | 176 |
| Air Conditioning | | | Boilers | |
| approximate estimating | 10 | | cast iron section | 113 |
| recommended loads | 519 | | domestic water, solid fuel fired | 116 |
| Air Curtains | | | domestic water, gas fired | 112 |
| industrial grade | 136 | | Chimney and Flues | 117 |
| commercial grade | 137 | | Bolted Connections | |
| Air Distribution Equipment | 306 | | black steel; screwed | 185 |
| All-in Manhour Costs | | | black steel; welded | 176 |
| mechanical | 103 | | galvanised steel | 182 |
| electrical | 341 | | Branch Weld | 180 |
| All-in rates | 9 | | Break Glass Unit | 378 |
| Aluminium Diffusers | 322 | | Bucholz Relay | 346 |
| Anodised Aluminium Trunking | 394 | | Bus-Bar; prewired | 424 |
| Anti-vibration Mountings | 337 | | Bus-Bar Systems | 424 |
| Approximate Estimating | | | Butt Weld | 179 |
| cost indices | 4 | | Butterfly Type Damper | 337 |
| directions | 3 | | | |
| elemental and all-in costs | 7 | | Cable | |
| elemental costs | | | bell and telephone | 351 |
| data processing centre | 37 | | coaxial, television, radio, frequency, video | 352 |
| hotel | 35 | | data transmission | 362 |
| hospital (Example 1) | 39 | | drop | 374 |
| hospital (Example 2) | 44 | | HOFR | 418 |
| office | 34 | | lead covered | 404 |
| outline costs | 6 | | mineral insulated copper | 412 |
| regional variations | 5 | | outlet drop data cable | 363 |
| supplement all-in rates | | | patch | 375 |
| catering equipment | 31 | | PVC insulated non-sheathed | 417 |
| escalators | 33 | | PVC insulated, parallel twin | 422 |
| lifts | 32 | | VR insulated, tough rubber sheathed | 421 |
| medical gases | 29 | | XLPE, SWA | 405 |
| Approximate Metres per tonne of tube | 511 | | CableTerminations | 404 |
| Architrave Plate Switch | 456 | | Cable Tiles | 463 |
| Arts and Drama Centres | 6 | | Cable Tray | |
| Attenuators | 304 | | galvanised steel | 428 |
| Automatic Air Vents | 246 | | rigid PVC | 430 |
| Automatic Sump Pump | 257 | | GRP | 432 |
| Axial Flow Fans | 308 | | | |

# INDEX

| | | | | |
|---|---|---|---|---|
| Cable Trunking | | | Conduit Boxes | |
|   anodised aluminium | 394 | |   cast iron | 386 |
|   fittings | 390 | |   sheet steel | 387 |
|   multi-compartment | 390 | |   PVC | 388 |
|   PVC | 396 | |   heavy duty | 383 |
|   single compartment | 389 | |   medium duty | 385 |
|   steel | 389 | | Connectors | 366 |
|   underfloor | 395 | | Connection Units | 460 |
| Calorifiers | 269 | | Conservator | 346 |
| Capillary Type Fittings | | | Consumer Units | 445 |
|   copper | 202 | | Contactor Relays | 452 |
|   stainless steel | 197 | | Contents | vii |
| Carbon Steel Pipes | 194 | | Control Dampers | 299 |
| Cast Iron Adaptable Boxes | 386 | | Control Valves | 234 |
| Cast Iron Pipes | | | Control Components | 338 |
|   pipes | 161 | | Convector Units | 130 |
|   fittings | 161 | | Conversion Tables | 497 |
| Catering Equipment | 31 | | Cooker Control Units | 462 |
| Ceiling Mounted Diffusers | 322 | | Copper Bus-Bar Systems | 424 |
| Ceiling Roses | 461 | | Copper Cylinders | 265 |
| Ceiling Switches | 456 | | Copper Pipe | |
| Centrifugal Heating Pumps | 255 | |   table W, microbore | 200 |
| Check Valves | 231 | |   table X | 201 |
| Chilled Water Installations | 19 | |   table X, plastic coated | 201 |
| Chimneys | | |   table Y | 201 |
|   domestic gas boilers | 124 | |   table Y, plastic coated | 202 |
|   industrial and commercial | | |   dimensions, weights | 509 |
|   oil and gas appliances | 117 | | Copper Pipe Fittings | |
| Circuit Breakers | | |   capillary | 202 |
|   hand operated | 439 | |   compression | 210 |
|   high voltage | 436 | |   DZR compression | 211 |
|   low voltage | 439 | |   flanges | 213 |
|   minature | 441 | |   high duty capillary | 208 |
|   moulded case | 441 | |   labours | 215 |
| Circulators | 258 | | Cost Indices | 4 |
| Cisterns | | | Cost Summary | 34 |
|   capacity | 514 | | Cover Plates | 223 |
|   fibreglass | 262 | | Cylinders | |
|   polypropylene | 263 | |   copper; direct | 265 |
| Clock Systems | 27 | |   copper; indirect | 267 |
| Clocks | | |   mild steel; indirect | 266 |
|   battery driven | 354 | |   storage | 265 |
|   digital | 355 | | | |
|   elapsed time | 355 | | | |
|   master | 354 | | Dampers | |
|   matching | 355 | |   control | 299 |
|   wall | 354 | |   fire | 299 |
|   weather resistant | 355 | | Data Outlet Face Plate | 355 |
| Cocks | 242 | | Data Processing Centre | 37 |
| Cold Water Services | 7 | | Data Transmission | 369 |
| Cold Water Storage Tanks | 261 | | Daywork | 467 |
| Cold Water Supply Set | 260 | | Detectors | 379 |
| Combination Hot & Cold Water Tanks | 261 | | Dial Thermometers | 251 |
| Commissioning Valves | 233 | | Diffusers | |
| Compression Type Fittings | | |   circular | 322 |
|   copper | 210 | |   perforated | 323 |
|   stainless steel | 199 | |   plastic | 324 |
| Conduit | | |   rectangular | 322 |
|   steel, black enamelled | 383 | |   slot | 323 |
|   steel, galvanised | 383 | | Digital Slave Clocks | 355 |
|   PVC | 386 | | Dimension of Copper and Stainless | |
|   flexible | 386 | |   Steel Pipes | 509 |

# INDEX

| | | | | |
|---|---|---|---|---|
| Dimension of Steel Pipes | 509 | Fibreglass Insulation | 327 | |
| Dimmer Switch | 457 | Fibreglass Slabs | 336 | |
| Distribution Boards | 443 | Filters | 313 | |
| District General Hospital | 39 | Fire Alarm Systems | 28 | |
| Domestic Central Heating | 18 | Fire Blankets | 110 | |
| Domestic Heating Pumps | 257 | Fire Dampers | 299 | |
| Domestic Heating Pumps | 233 | Fire Extinguishers | 110 | |
| Dry Rising Main | 107 | Fire Hose Reels | 106 | |
| Driving Stud | 381 | Fire Hydrants | 111 | |
| Ductwork | | Fire Protection | 8 | |
|   access doors | 303 | Fittings | | |
|   acoustic louvres | 315 |   cable tray | 431 | |
|   air distribution equipment | 306 |   cable trunking | 389 | |
|   approximate estimating | 19 |   ductwork | | |
|   circular section, class A | 288 |     rectangular; class A | 274 | |
|   circular section, class C | 292 |     rectangular; class C | 281 | |
|   dampers | 299 |     circular; class A | 289 | |
|   diffusers | 322 |     flat oval | 295 | |
|   external louvres | 326 |   flush floor trunking | 395 | |
|   fittings | |   ladder rack | 428 | |
|     rectangular ductwork, class A | 274 |   PVC trunking | 396 | |
|     rectangular ductwork, class C | 281 |   underfloor trunking | 395 | |
|     circular ductwork, class A | 289 | Flange dimension chart | 512 | |
|     circular ductwork, class C | 292 | Flanges | | |
|     flat oval ductwork | 295 |   ABS | 156 | |
|   flat oval section | 295 |   blank | 175 | |
|   rectangular section, Class A | 273 |   bolted connections | 176 | |
|   rectangular section, Class C | 281 |   bronze | 213 | |
|   grilles | 317 |   dimensions | 512 | |
| | |   screwed, black steel | 164 | |
| | |   screwed, galvanised steel | 182 | |
| | |   welded | 175 | |
| Earthing Device | 440 | Flats (L.A.) | 6 | |
| Earth Electrodes | 381 | Flexible Conduit | 386 | |
| Earth Plate | 381 | Flexible Connectors | 250 | |
| Electrical Heating | 27 | Flexible Cord | 421 | |
| Electrical Installations | 34 | Fluorescent Luminaries | | |
|   rates of wages | 83 |   batten type | 453 | |
|   working rules | 89 |   corrosion resistant | 453 | |
| Electronic Thermostats | 338 |   flameproof | 454 | |
| Elemental Rates | 7 |   modular | 453 | |
| Elevation Rod | 380 | Formulae | 500 | |
| Emergency Lighting | 27 | Fractions, Decimals & Millimetric equivalents | 501 | |
| Energy Management Systems | 38 | Fuel storage tanks | 105 | |
| Escalators | 33 | Fuse Links | | |
| Ethernet Systems | 362 |   HRC | 451 | |
| Expansion Joints | 249 |   motor protection | 451 | |
| External Lighting | 28 | | | |
| External Services | 43 | | | |
| Extinguishers | 110 | | | |
| Factories | 6 | Galvanised Flush Floor Trunking | 395 | |
| Fans | | Galvanised Cisterns | 266 | |
|   axial flow fans | 308 | Galvanised Conduit | 383 | |
|   roof fans | 308 | Galvanised Steel Pipe | | |
|   toilet fans | 310 |   pipes | 181 | |
|   kitchen fans | 311 |   flanges | 182 | |
|   multivent fans | 312 |   tubular fittings | 183 | |
| Fan Convectors | 130 |   malleable iron fittings | 186 | |
| Fees for Professional Services | 489 | Galvanised Steel Trunking | 389 | |
| Fibreglass Cisterns | 262 | Galvanised Calorifiers | 271 | |

## INDEX

| | | | | |
|---|---|---|---|---|
| Gas Circuit Breakers | 436 | Labours | | |
| Gas Installation | 42 |   black steel | 179 | |
| Gate Valves | 237 |   copper | 215 | |
| Generating Sets | 345 |   hangers and brackets | 222 | |
| Glandless Domestic Heating Pump | 257 | Labour Constant | 101 | |
| Glandless Accelerator Pumps | 259 | Labour Cost | 101 | |
| Globe Valves | 240 | Labour Rates | | |
| Gridswitch | 455 |   ductwork | 104 | |
| Grilles | |   electrical | 341 | |
|   aluminium | 317 |   mechanical | 103 | |
|   plastic | 324 |   plumbing | 103 | |
| Guidelines for Network Design | 356 | Ladder Rack | | |
| | |   fittings | 434 | |
| | |   galvanised | 434 | |
| | | Lattice Earth Plate | 381 | |
| | | Lift and Conveyor Installations | 32 | |
| | | Lighting | | |
| Heaters | |   controllers | 455 | |
|   air curtains | 136 |   switches | 455 | |
|   fan convectors | 130 | Lightning Protection | | |
|   perimeter heaters | 128 |   installations | 7 | |
|   radiant panels | 132 |   conductors | 380 | |
|   radiant strip heaters | 134 | Line Conditioners | 449 | |
|   unit heaters | 135 | Line Tap | 366 | |
| Heat Exchangers | 264 | Line Terminators | 366 | |
| Heating and Heat Source | 7 | List Price | 101 | |
| HOFR Cables | 418 | Local Area Networking | 362 | |
| Hose Reels | 106 | Local Bridge Filters | 368 | |
| Hospitals | 6 | Local Heating Units | 135 | |
| Hotel | 35 | Local Switches | 455 | |
| Hot Water Source | 7 | Louvres | | |
| Houses (L.A.) | 6 |   acoustic | 315 | |
| HV Circuit Breakers | |   external | 326 | |
|   3 Phase | 436 | L.V. Circuit Breakers | | |
| | |   non automatic operated | 439 | |
| Immersion Thermostats | 339 | | | |
| Imperial Standard Wire Gauge | 502 | Mains Switchgear | 436 | |
| Indirect Cylinders | | Malleable Iron Pipe Fittings | | |
|   steel | 266 |   for black steel pipes | 168 | |
|   copper | 267 |   for galvanised steel pipes | 186 | |
| Induction System | 10 |   for MDPE pipes | 153 | |
| Insulation | | Manual Call Point | 379 | |
|   ductwork | 336 | Master Clock | 354 | |
|   valve boxes | 333 | Material Cost | 101 | |
|   pipe; plant rooms | 327 | MDPE Pipe | | |
|   pipe; non-plant rooms | 329 |   blue pipes | 149 | |
| | |   blue fittings | 149 | |
| | |   plastic fittings | 150 | |
| | |   yellow pipes | 151 | |
| Junction Boxes | |   yellow fittings | 152 | |
|   fex glassfibre | 464 |   malleable iron fittings | 153 | |
|   underfloor | 395 | Mechanical Installations | | |
|   sheet steel | 464 |   rates of wages | 53 | |
| | |   working rules | 60 | |
| | | Mechanical Plant and Tools | 477 | |
| Kitchen Fans | 311 | Medical Gases | 29 | |

## INDEX

| | | | |
|---|---|---|---|
| Medium & Low Temperature H.W. Heating | | stainless steel | 197 |
| | | Pipeline Ancillaries | 225 |
| fan convectors | 130 | Pipe Fittings | |
| perimeter heating | 128 | ABS | |
| radiant panels | 132 | waste | 143 |
| radiant strip heaters | 134 | water/gas | 154 |
| Microbore Copper Pipe | | black steel | |
| fittings | 202 | screwed | 165 |
| manifolds | 200 | tubular | 161 |
| pipe | 200 | welded | 177 |
| Mineral Fibre Insulation | 327 | capillary | |
| Mineral Insulated Cables | 412 | copper | 202 |
| Minature Circuit Breakers | 441 | stainless steel | 197 |
| Mini Trunking | 398 | cast iron | 161 |
| Minimum Distance between Supports | 513 | compression | |
| Moulded Case Circuit Breakers | 441 | copper | 210 |
| Motorised Control Valves | 234 | stainless steel | 199 |
| Multiport Transceivers | 365 | copper | |
| Mutiport Repeater/Hubs | 366 | capillary | 202 |
| Multiple/Muti-Station Access Units | 376 | compression | 210 |
| Multiple Point Air Terminal | 380 | DZR compression | 211 |
| Multi-vent fans | 312 | high duty capillary | 208 |
| Museums | 6 | galvanised steel | 183 |
| | | malleable iron | |
| | | for black steel pipes | 172 |
| | | for galvanised steel pipes | 186 |
| Net Price | 101 | for MDPE pipes | 153 |
| Non-mechanical Plant | 484 | MDPE | |
| Non-storage Calorifiers | 269 | blue | 149 |
| | | yellow | 151 |
| | | polypropylene | |
| | | waste | 143 |
| Offices | 6 | overflow | 144 |
| Outline Costs | 3 | PVC-ABS | 145 |
| | | PVC-U | |
| | | waste, solvent joints | 138 |
| | | waste, ring seal joints | 139 |
| Packaged Units | | water/gas | 157 |
| air handling | 19 | stainless steel | |
| steam boiler | 127 | capillary | 197 |
| Patch Panels | 369 | compression | 199 |
| Pipes | | Perforated Diffusers | 323 |
| ABS | | Pipe Fixings | 217 |
| waste | 141 | Pipe Hangers | 222 |
| water/gas | 153 | Pipe Rollers/Chairs | 224 |
| black steel | | Pipe Supports | 217 |
| screwed | 163 | Plastic Pipe Fittings | 157 |
| welded | 174 | Plate Switches | 455 |
| cast iron | 161 | Polypropylene Pipe | |
| copper | 200 | overflow pipes | 144 |
| galvanised steel | 181 | overflow fittings | 145 |
| MDPE | | traps | 146 |
| blue | 149 | waste pipe | 143 |
| yellow | 151 | waste pipe fittings | 143 |
| polypropylene | | Power Charger | 378 |
| waste | 143 | Power Supplies | 24 |
| overflow | 144 | Preface | v |
| PVC-ABS | 145 | Pressed Steel Radiators | 132 |
| PVC-U | | Pressure Gauges | 251 |
| waste, solvent joints | 138 | Pressure Relief Valves | 241 |
| waste, ring-seal joints | 139 | Pre-Wired Busbars | 427 |
| water/gas | 156 | Protective Installations | 34 |

# INDEX

| | | | |
|---|---|---|---|
| Pumps | 255 | Special Services | 36 |
| Push Button Stations | 452 | Splash proof Switches | 456 |
| PVC Adaptable Boxes | 374 | Sprinkler Heads | 108 |
| PVC Bench Trunking | 399 | Sprinkler Valves | 108 |
| PVC Mini Trunking | 398 | Square Metre Rates | 6 |
| PVC Trunking | 396 | Stainless Steel Pipe | |
| PVC Underfloor Trunking | 400 | dimensions, weights | 509 |
| PVC-ABS Pipe | | grade 304 | 197 |
| pipes | 141 | grade 316 | 197 |
| fittings | 141 | capillary fittings | 197 |
| PVC-U Pipes | | compression fittings | 199 |
| waste pipes, ring seal joints | 139 | Standard Wire Gauge | 502 |
| waste fittings; ring seal joints | 140 | Standby Diesel Generating Sets | 345 |
| waste pipes; solvent joints | 138 | Steam Strainers | 247 |
| waste fittings; solvent joints | 138 | Steam Traps | |
| water/gas pipes | 156 | cast iron | 226 |
| water/gas fittings | 157 | stainless steel | 247 |
| | | Steel Cable Tray | 428 |
| | | Steelwork, weights of | 504 |
| | | Step down transformer | 346 |
| Quantity Surveyors Fees | 489 | Stopcocks | 242 |
| | | Strainers | 247 |
| | | Sump Pumps | 257 |
| | | Support & Fixings, minimum distance | 513 |
| | | Supply Grilles | 317 |
| Radiant Strip Heaters | 134 | Surface Thermostats | 338 |
| Radiators/Radiant Panels | 132 | Surface Trunking | 389 |
| Radiator Thermostats | 339 | Surge Protectors | 365 |
| Radiator Valves | 227 | Suspended Lightweight Trunking | 394 |
| Rates of Wages | 101 | Switches | 455 |
| Regional Variations | 5 | Switch Fuses | 437 |
| Regulators | 225 | Switchgear & Distribution | 437 |
| Relief Valves | 241 | Synchronisation Panel | 345 |
| Repeaters/Retimers | 367 | Sysmatic; 10Base5 Ethernet | 358 |
| Repeater Panel | 378 | Sysmatic; 10Base2 Ethernet | 359 |
| Residual Current Circuit Breaker | 441 | Sysmatic; 10Base VG Ethernet | 360 |
| Residential Silencer | 345 | Sysmatic; Token Ring | 361 |
| Rising Main Bus-Bar System | 426 | | |
| Room Thermostats | 338 | | |
| Roof Fans | 308 | | |
| | | Tables and Memoranda | 495 |
| | | Table of Weights | 504 |
| | | Tanks | |
| Safety and Relief Valves | 241 | fuel storage tanks | 105 |
| Schedule of Plant Charges | 476 | sectional fibre tanks | 261 |
| Sectional Tanks | 263 | Tapered Rod | 380 |
| Sensors | 325 | Tap-Off Rod | 424 |
| Service Outlet Boxes | 402 | Telephone | |
| Shaver Unit | 461 | cable | 351 |
| Sheet Steel Adaptable Boxes | 387 | installations | 34 |
| Sight Glasses | 248 | outlet | 461 |
| Siren | 379 | Temperature Regulators | 225 |
| Slave Clocks | 354 | Terminal Strips | 389 |
| Slot Diffusers | 323 | Test Clamps | 380 |
| Socket Outlet | | Theatres | 6 |
| metalclad | 458 | Thermal Insulation | 327 |
| moulded cover plate | 458 | Thermal Wheels | 264 |
| RCD protected | 459 | Thermometers | 251 |
| weatherproof | 459 | Thermostatic Regulators | 225 |
| Soil and Waste Pipes | 138 | Thermostats | |
| Sound Attenuators | 304 | electrical | 338 |
| Space Heating | 18 | immersion | 339 |

# INDEX

| | |
|---|---|
| Thermostats (contd) | |
|   radiator | 339 |
|   room | 338 |
| Toilet Fans | 310 |
| Token Ring System | 371 |
| Trace Heating System | 382 |
| Transceiver Body | 367 |
| Transformers | 346 |
| Traps | 146 |
| Trunking Fittings | 390 |
| Tubes, weights | 504 |
| T.V. Socket Outlet | 461 |
| | |
| Underfloor Ducting | 395 |
| Underfloor Ducting Fittings | 395 |
| Underfloor Trunking PVC | 385 |
| Uninterruptable Power Supplies | |
|   single phase | 349 |
|   rotary | 350 |
|   static | 349 |
| Unit Heaters | 135 |
| Universities | 6 |
| Update | 101 |
| UPVC | See PVC-U |
| | |
| Valves | |
|   ball float valves | 229 |
|   check valves | 231 |
|   commissioning valves | 233 |

| | |
|---|---|
| Valves (contd) | |
|   control valves | 234 |
|   gate valves | 237 |
|   globe valves | 240 |
|   plug valves | 236 |
|   pressure reducing valves | 241 |
|   radiator valves | 227 |
|   regulating valves | 225 |
|   relief valves | 241 |
|   safety valves | 241 |
|   sprinkler | 108 |
|   thermostatic | 225 |
| Variable Air Volume System | 10 |
| Ventilating Systems | 36 |
| Voltage Stabilisers | 449 |
| | |
| Wage Rates | |
|   mechanical | 103 |
|   electrical | 341 |
| Warehouses | 6 |
| Water Boilers | 112 |
| Water and Gas Pipes | 149 |
| Water and Dust Tight Switches | 456 |
| Water Installations | 34 |
| Water Pressure Due to Height | 503 |
| Water Storage Capacities | 513 |
| Water Supply Sets | 260 |
| Weatherproof Junction Boxes | 464 |
| Weather Louvres | 326 |
| Weights for Steelwork | 504 |
| | |
| Zone Control Panels | 378 |